Biogeochemistry

SECOND EDITION

In memory of my father, William L. Schlesinger, M.D.
13 November 1911–18 September 1996
He knew the world as it was,
and he could see where we are headed

Biogeochemistry

An Analysis of Global Change

SECOND EDITION

William H. Schlesinger
Division of Earth and Ocean Sciences
Nicholas School of the Environment and Earth Sciences
Duke University
Durham, North Carolina

ACADEMIC PRESS

An Imprint of Elsevier

Amsterdam Boston Heidelberg London New York Oxford
Paris San Diego San Francisco Singapore Sydney Tokyo

This book is printed on acid-free paper. 

Academic Press
An Imprint of Elsevier
525 B Street, Suite 1900, San Diego, California 92101-4495, USA
http://www.academicpress.com

Academic Press
84 Theobald's Road, London WC1X 8RR, UK
http://www.academicpress.com

Library of Congress Cataloging-in-Publication Data

Schlesinger, William H.
Biogeochemistry : an analysis of global change / by William H. Schlesinger. – 2nd ed.
p. cm.
Includes bibliographical references (p.) and index.
ISBN-13: 978-0-12-625155-5 ISBN-10: 0-12-625155-X (alk. paper)
1. Biogeochemistry. I. Title.
QH343.7.S35 1997
574.5'222—dc20

96-43747
CIP

ISBN-13: 978-0-12-625155-5
ISBN-10: 0-12-625155-X

PRINTED IN THE UNITED STATES OF AMERICA
07 9

Contents

Preface

This is a textbook about the chemistry of the surface of the Earth. Outside of a few infalling meteors and outflying spaceships, the Earth is a closed chemical system in which the reactions that maintain life are fueled by sunlight. During the last 4 billion years, there have been remarkable changes in the chemical milieu at the surface of the Earth, where all life is found. With the origin of photosynthetic organisms and the first appearance of oxygen in Earth's atmosphere, the environment and the evolutionary potential for all subsequent life on the planet changed profoundly. Throughout most of Earth's history, however, changes in its chemical characteristics have occurred relatively slowly, with ample time for evolutionary change to keep pace.

Times are different now. With the advent of industrialization and an exponentially increasing population, a single species—the human—is usurping an extraordinary fraction of the resources that support all life on this planet. Satellite views of the Earth show broadscale destruction of tropical rainforests, expansion of deserts, and smoggy cities. The planet is changing rapidly under our inattentive stewardship.

Somehow, in the current, heated political arena, global *change* has become equated with global *warming*. Certainly, climate change is a real possibility, but we should not let the equivocal debate over global warming divert our attention from the fact that the Earth's basic chemistry is changing rapidly and all about us. Ice core records show that many gases had stable concentrations in the atmosphere for 10,000 years. Now, the concentrations of CO_2, CH_4, and N_2O are increasing at rates of up to 1% per year. We know that the production and consumption of these gases are controlled by the biosphere. When we see their concentrations increasing rapidly, we should recognize that something has affected life on this planet—not just in our neighborhood swamp, but globally. The biosphere is unhealthy.

I wrote the first edition of this textbook for college-level and graduate students interested in global change. The rapid pace of progress in our understanding of earth system science demands an update of that edition. Nearly every page of this book has been revised as a result of recent advances in our scientific understanding of the Earth's function. As I revised the text, I gave special attention to expanded treatments of stratospheric chemistry (Chapter 3), nitrogen cycling in the soil (Chapter 6), primary production in the sea (Chapter 9), the global water cycle (Chapter 10), the response of land plants to rising CO_2 (Chapters 5 and 11), and human perturbations of the global nitrogen cycle (Chapter 12).

Like that of its predecessor, the organization of this book follows the structure of a class in biogeochemistry that I have taught for many years at Duke University. The book covers the basics about the effects of life on the chemistry of Earth and the impact of humans in altering the chemistry of the global environment. To understand the Earth as a closed chemical system, we must recognize the interactions between chemical reactions at the Earth's surface. Following my class syllabus, I have organized the book into two sections: Part I covers the microbial and chemical reactions that occur on land, in the sea, and in the atmosphere. Part II is a set of shorter chapters that link the mechanistic understanding of the earlier chapters to a large-scale, synthetic view of global biogeochemical cycles.

Throughout this book, I give special emphasis to the chemical reactions that link the elements that are important to life. In several locations, I show how computer models can be used to help in understanding elemental cycling and ecosystem function. Many of these models are based on biochemistry and interactions among the biochemical elements. The models are useful in extrapolating small-scale observations to the global level. Therefore, the intent of this book is to link disparate fields ranging from geomicrobiology to global change ecology—all of which are now part of the science of biogeochemistry.

With a look to the future, I demonstrate how satellite technology is useful in understanding global biogeochemistry, and I show the important role that NASA's Earth Observing System (EOS) will play in studies of global change. Although I discuss how the Earth's chemical system affects and is affected by climate, this is not a book about global warming. Similarly, while I show the effects of humans on global chemistry, there is little emphasis on the traditional, local problems of water and air pollution.

This text will provide the framework for a class in biogeochemistry. It is meant to be supplemented by readings from the current literature, so that areas of specific interest or recent progress can be explored in more detail. Although not encyclopedic, the text includes a large number of references to aid students and others who wish to enter the current literature. Reflecting the book's interdisciplinary nature, I have made a special effort to provide abundant cross-referencing of chapters, figures, and tables throughout.

As with the first edition, I hope that this book will stimulate a new generation of students to address the science and policy of global change.

WHS
18 March 1996
Durham, North Carolina

Acknowledgments

My interest in ecology has been stimulated by a large number of teachers who were influential at critical stages of my scientific career. Among them are Jim Eicher, Joe Chadbourne, John Baker, Russ Hansen, Bill Reiners, Noye Johnson, Bob Reynolds, and Peter Marks. Over the years, workshops and informal conversations with Dan Botkin, Gene Likens, Dan Livingstone, Jerry Melillo, Peter Vitousek, and George Woodwell have made me recognize that the Earth is a single, interactive chemical system that is changing rapidly without proper stewardship. A number of colleagues have provided me with references, photographs, special insights, helpful reviews, and critical comments—large and small—on all or parts of this book, especially on the first edition. Among them are Jeff Andrews, Bob Bell, Bill Bowman, Nina Buchmann, Indy Burke, Bruce Corliss, Randy Dahlgren, Evan DeLucia, Steve Faulkner, Peter Haff, Kevin Harrison, John Houghton, Dan Jacobs, Eric Kasischke, Emily Klein, Bill Labiosa, Dan Livingstone, Susan Lozier, Pam Matson, Pat Megonigal, James J. Morgan, Cheryl Palm, Milan Pavich, Bill Peterjohn, Roger Pocklington, Jane Raikes, Joan Riera, Jim Siedow, Sarah Townsend, Cornelis H. van der Weijden, and Mark Walbridge. Dawn Cardascia of NASA's Earth Science Support Office kindly provided the color plates. Various graduate students at Duke who have enrolled in "272" have been among my most careful critics, offering especially useful advice on how to present difficult concepts more clearly. Lisa Dellwo Schlesinger also helped make the text more understandable to the general reader. I thank them all.

PART I

Processes and Reactions

1

Introduction

Introduction

At the present time, there is little scientific doubt that the composition of the atmosphere and reactions among atmospheric constituents are changing as a result of human activities. For example, the reduction of stratospheric ozone over the South Pole appears directly related to the release of chlorofluorocarbons, which are used in a variety of products in the industrialized world. Whenever we discover that humans have changed their global environment, we build upon our recognition that living organisms, including humans, can affect the conditions of an entire planet, the Earth. Beyond human effects, the influence of all life on Earth is so pervasive that scientists have come to realize that there are few chemical reactions on the surface of the Earth that are not affected by biota. Many of the conditions on Earth that we regard as "normal" are the product of at least 3.5 billion years of life on Earth (Reiners 1986). Today, living systems exert a major control on the composition of the oceans and the atmosphere and on the rate of weathering of the Earth's crust. When we study the geochemistry of the Earth's surface, it is unavoidable that we are students of *bio*geochemistry.

Encompassing chemical reactions in the atmosphere, the oceans, crustal minerals, and living organisms, biogeochemistry is a unique, interdisciplinary science. Traditional approaches of experimentation and strong inference cannot be used to study global biogeochemistry; there is only one Earth! Working on different facets, teams of biogeochemists must assemble a model of the whole, from a reductionist study of its parts. Modeling is an essential tool of the biogeochemist. Models often help to extend the results of small-scale measurements and field experiments to regional and global estimates. Often, we can test the validity of models by observations of the Earth at the global level, using satellite technology. In many cases, historical records of the Earth's past also provide a test for models that attempt to describe its present or future condition. The ultimate goal of biogeochemistry is to understand the processes controlling the chemical environment in which we live.

A Model for the Earth as a Biogeochemical System

Garrels and Lerman (1981) developed a simple model for the biogeochemistry of the Earth's surface which includes interactions between the atmospheric, oceanic, and crustal compartments and the biosphere (Fig. 1.1). The model assumes that the atmosphere and the oceans have not shown large changes in their composition during geologic time. Of course, we know that this has not always been true, but for the last 60 million years or so, there is good geologic evidence that this assumption is reasonable (Holland et al. 1986). With these constraints, the model couples reactions in the atmosphere and oceans to seven compartments that represent major crustal minerals, such as gypsum ($CaSO_4 \cdot 2H_2O$), pyrite (FeS_2), and calcium carbonate ($CaCO_3$). For instance, if the weathering of limestone transfers

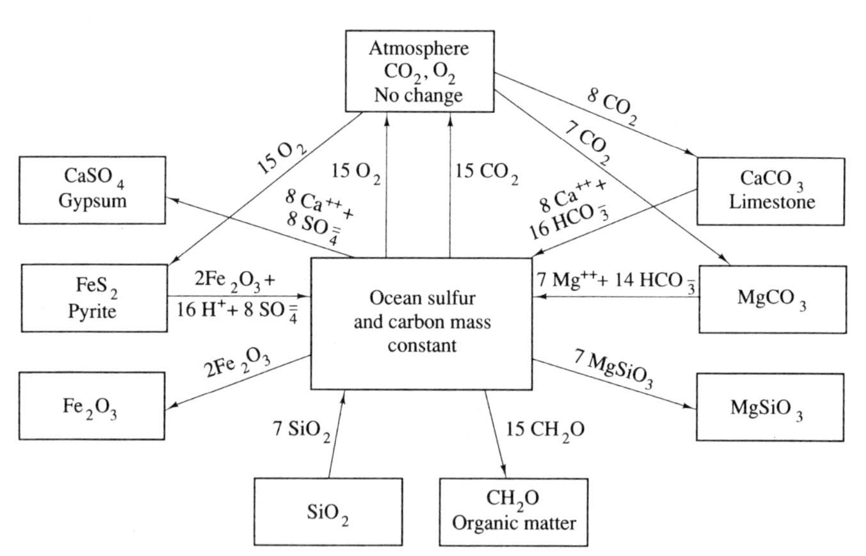

Figure 1.1 A model of major sedimentary reservoirs and the transfers between them that are associated with an increase in the mass of the biosphere by 15 moles. From Garrels and Lerman (1981).

8 units of Ca^{2+} to the world's oceans and the Ca content of seawater does not change, then the same amount of Ca must be deposited in a sedimentary mineral on the seafloor.

Organic materials, living and dead, constitute the biosphere, which appears in the model compartment labeled CH_2O, representing the approximate stoichiometric composition of living tissues. Changes in the mass of the biosphere through geologic time are indicated by net transfers of material into and out of that compartment.

Consider the increase in the total mass of organic matter that must have occurred during the Carboniferous Period, when large areas of land were covered by swamps. Here, dead vegetation accumulated as peat that was later transformed into coal. Storage of carbon in dead materials, detritus, represents an increase in the mass of the biosphere. With no change in the CO_2 content of the atmosphere or the CO_2 dissolved in the oceans as HCO_3^-, the carbon added to the biosphere must have been derived from the weathering of carbonate minerals. Weathering of carbonate minerals, however, would also transfer Ca and Mg to the oceans, and to maintain constant seawater chemistry the model predicts that Ca must be deposited as $CaSO_4$ and that Mg must be deposited in silicate minerals by various reactions that occur in ocean sediments (Chapter 9). To deposit $CaSO_4$ with no change in the SO_4^{2-} content of the world's oceans, sulfur must be derived from another pool. Oxidative weathering of pyrite (FeS_2) would supply SO_4 to the oceans, also consuming some of the oxygen that would have been added to the atmosphere by photosynthesis. The remaining oxygen would be consumed in the deposition of Fe_2O_3. Note that the total O_2 consumed by these reactions is in molar stoichiometric balance with the carbon stored in organic matter by photosynthesis, so the atmospheric content of O_2 does not change.

This model illustrates how minerals such as magnesium silicates, not traditionally the focus of biological studies, are linked to the activities of the biosphere. Certainly, it is legitimate to ask whether this is a reasonable model for the linkage of chemical reactions on Earth. Support for the model would be found if large geologic deposits of gypsum are associated with periods in which there were large net stores of organic carbon, since the model predicts a coupled balance,

$$\text{pyrite} + \text{carbonates} \rightleftarrows \text{gypsum} + \text{organic carbon}, \quad (1.1)$$

through geologic time. Indeed, Garrels and Lerman (1981) show that the molar ratio of organic carbon to gypsum has remained fairly constant through geologic time, with large deposits of gypsum associated with the Carboniferous Period, when large amounts of organic carbon were stored in coal (Fig. 1.2).

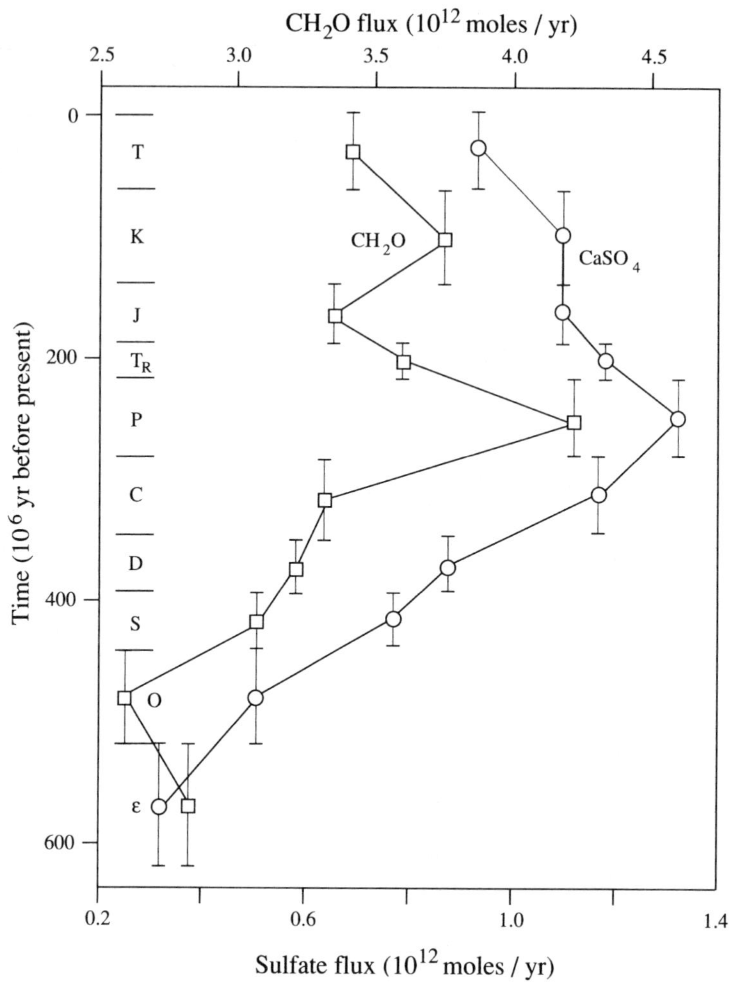

Figure 1.2 Garrels and Lerman (1981) use the isotopic ratio in the sedimentary record of organic carbon to calculate the rate of the accumulation of organic carbon through geologic time. Independently, they use their model (Fig. 1.1) to calculate the sedimentary accumulation of $CaSO_4 \cdot 2H_2O$ (gypsum) over the same interval. Deposition of organic carbon and $CaSO_4$ appear to have varied in parallel for the last 500 million years.

The model developed by Garrels and Lerman (1981) reminds us that the size of the biosphere waxes and wanes as a result of the balance between photosynthesis and respiration. During the history of life on Earth, the mass of the biosphere has increased at times when high rates of photosynthesis have resulted in a net storage of organic carbon and the release of free O_2 as a by-product:

$$CO_2 + H_2O \rightarrow CH_2O + O_2\uparrow. \tag{1.2}$$

The metabolic activities of microbes and higher animals—heterotrophic respiration—convert organic carbon back to CO_2 and H_2O. Fires can also perform this reaction abiotically and very quickly. When we compare the conditions on Earth to those on other planets (Chapter 2), we will see that the storage of organic carbon and the release of free O_2 are the essence of life; evidence for a significant production of either material on another planet would be strongly suggestive of life there as well (Horowitz 1977, Klein 1979, Sagan et al. 1993).

Currently, we are reducing the amount of organic carbon stored in the biospheric compartment by burning coal and oil deposited during earlier geologic times. As a result, the concentration of atmospheric carbon dioxide is increasing (Fig. 1.3). Specifically, atmospheric CO_2 is increasing because the rate of fossil fuel combustion exceeds the rate at which other compartments of the Earth's biogeochemical system (Fig. 1.1) can take up carbon so as to moderate or buffer changes in the atmosphere. However, if we were to stop burning all fossil fuels, the CO_2 concentration in the atmosphere would be expected to decline slowly to a stable level—only somewhat higher than the preindustrial level—as CO_2 is taken up by the oceans (Laurmann 1979).

Cycles in Biogeochemistry

Since its organization as a planet, the Earth has been exposed to cyclic phenomena (Degens et al. 1981, Harrington 1987). Some cycles, such as

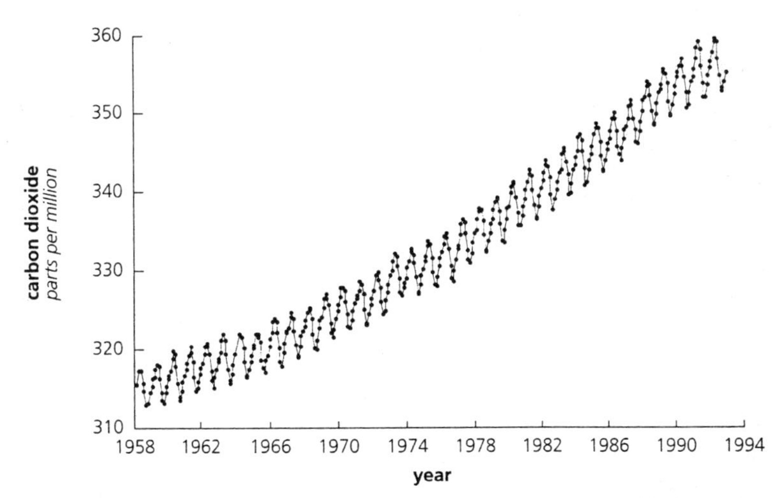

Figure 1.3 The concentration of atmospheric CO_2 measured at monthly intervals at the Mauna Loa Observatory in Hawaii, expressed in parts per million (ppm) of dry air. The annual oscillation reflects the seasonal cycles of photosynthesis and respiration by land biota in the northern hemisphere, while the overall increase is largely due to the burning of fossil fuels. From Keeling and Whorf (1994).

the daily rotation of the Earth on its axis and its annual rotation about the Sun, are now so obvious that it seems surprising that they were mysterious to philosophers and scientists throughout much of human history. Considering longer periods of time, we now believe that glacial cycles are linked to small variations in the Earth's orbit that alter the receipt of radiation from the Sun (Berger 1978, Harrington 1987).

Another long-term cycle on Earth involves the interaction of carbon dioxide with the Earth's crust. CO_2 in the atmosphere dissolves in rainwater to form carbonic acid (H_2CO_3), which reacts with the minerals exposed on land in the process known as rock weathering (Chapter 4). The dissolved products of rock weathering are carried by rivers to the sea (Fig. 1.4). In the oceans, calcium carbonate is deposited in marine sediments, which in time are subducted into the lower crust. Here, the sediments are metamorphosed; calcium and silicon are converted back into the primary minerals of silicate rock, and the carbon is returned to the atmosphere as CO_2 in volcanic emissions. On Earth, the entire oceanic crust appears to circulate through this pathway in about 100–200 million years (Li 1972, Howell and Murray 1986). The presence of life on Earth does not speed the turning of this cycle, but it may increase the amount of material moving in the various pathways by increasing the rate of rock weathering on land and the rate of carbonate precipitation in the oceans.

Models such as this are known as steady-state models, in the sense that they normally show equal transfers of material along the flowpaths and no change in the mass of various compartments over time. In fact, such a model potentially offers a degree of self-regulation to the system, because any period of high CO_2 emissions from volcanoes should lead to greater rates of rock weathering, removing CO_2 from the atmosphere and restoring balance to the system. Biogeochemists often use steady-state assumptions to simplify models of biogeochemical cycles, but in many cases the assumption of a steady state is not valid during transient periods of rapid change. For example, high rates of volcanic and tectonic activity may have resulted in a temporary increase in atmospheric CO_2 and a period of global warming during the Eocene (Owen and Rea 1985).

The biosphere is always changing in response to cycles. In plants, photosynthesis dominates over respiration in the daytime; the reverse is true at night. During the summer, total photosynthesis in the northern hemisphere exceeds respiration by decomposers. This results in a temporary storage of carbon in plant tissues and a seasonal decrease in atmospheric CO_2, which shows a minimum value during August of each year (Fig. 1.3). The annual cycle is completed during the winter months, when atmospheric CO_2 returns to higher levels as decomposition continues during the time that many plants are dormant or leafless. Certainly, it would be a mistake to model the annual activity of the biosphere by considering only the summertime conditions.

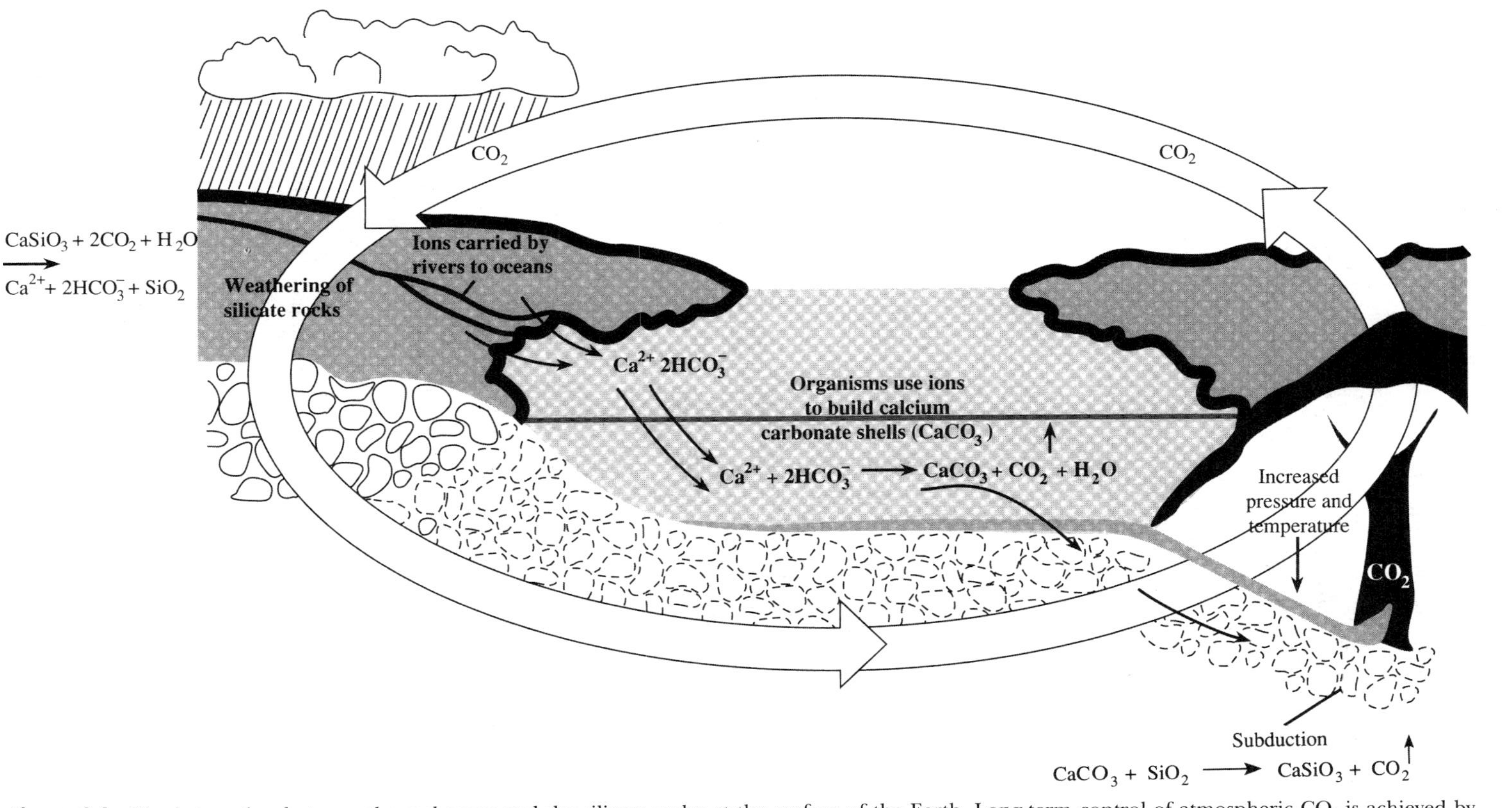

Figure 1.4 The interaction between the carbonate and the silicate cycles at the surface of the Earth. Long-term control of atmospheric CO_2 is achieved by dissolution of CO_2 in surface waters and its participation in the weathering of rocks. This carbon is carried to the sea as bicarbonate (HCO_3^-), and it is eventually buried as part of carbonate sediments in the oceanic crust. CO_2 is released back to the atmosphere when these rocks undergo metamorphism at high temperature and pressures deep in the Earth. Modified from Kasting et al. (1988).

In a longer time frame, the biosphere has increased and decreased in size during glacial cycles (Faure 1990), and the storage of organic carbon increased strongly during the Carboniferous Period, when most of the major economic deposits of coal were laid down. The unique conditions of the Carboniferous Period are poorly understood, but it is certainly possible that such conditions are part of a long-term cycle that might return again. Significantly, unless we recognize the existence and periodicity of cycles—and adjust our models accordingly—we may err in our assumptions regarding a steady state in the Earth's biogeochemistry.

All current observations of global change must be evaluated in the context of underlying cycles and potentially non-steady-state conditions in the Earth system. The current changes in atmospheric CO_2 are best viewed in the context of cyclic changes seen during the last 220,000 years—the historical record that is obtained from bubbles of air trapped in the polar ice caps. These bubbles have been collected from various layers of ice in a core taken near Vostok, Antarctica, and analyzed for CO_2 and other trace gases (Fig. 1.5). During the entire 220,000-year period, the concentration of atmospheric CO_2 appears to have oscillated between high values during warm periods and lower values during glacial intervals. During the last glacial epoch (20,000–50,000 years ago), CO_2 ranged from 180 to 200 ppm in the atmosphere. CO_2 rose dramatically at the end of the last glacial epoch (10,000 years ago) and was relatively stable at 280 ppm until the Industrial Revolution. The increase in CO_2 at the end of the last glacial epoch may have contributed to the global warming that melted the continental ice sheets (Sowers and Bender 1995).

When viewed in the context of this cycle, we can see that the recent increase in atmospheric CO_2, to today's value of about 360 ppm, has occurred at an exceedingly rapid rate, and it carries the planet into a range of concentrations never before experienced during the evolution of modern human social and economic systems. If the past is an accurate predictor of the future, higher atmospheric CO_2 will lead to global warming, but any observed changes in global climate must also be evaluated in the context of long-term cycles in climate with many possible causes (Hansen et al. 1981, Schneider 1994, Stouffer et al. 1994).

Because the atmosphere is well mixed, changes in its composition are perhaps our best evidence of the ability of humans to alter the environment globally. Concern about global change is greatest when we see changes in atmospheric constituents such as carbon dioxide (0.4%/yr), methane (CH_4; about 1%/yr), and nitrous oxide (N_2O; about 0.3%/yr), for which we see little or no precedent in the geologic record. However, humans have also changed other aspects of the Earth's natural biogeochemistry. For example, when human activities increase the erosion of soil, we alter the natural rate of sediment delivery to the oceans and the deposition of sediments on the seafloor. As in the case of atmospheric CO_2, evidence for global changes

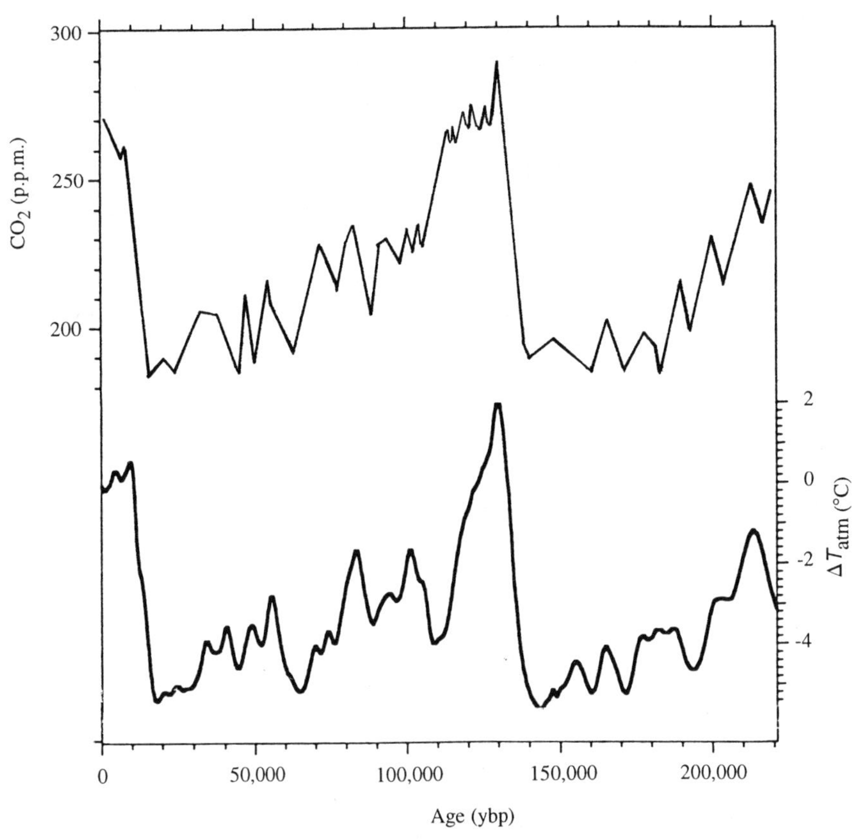

Figure 1.5 Variations in atmospheric CO_2 in bubbles of gas collected from the Vostok ice core, and variations in mean air temperature at the South Pole as calculated from isotope ratios in the ice, over the past 220,000 years. Modified from Jouzel et al. (1993).

in erosion induced by humans must be considered in the context of long-term oscillations in the rate of crustal exposure, weathering, and sedimentation due to changes in climate and sea level (Worsley and Davies 1979, Clemens et al. 1993).

Extraction of fossil fuels and the mining of metal ores cause an increase in the natural rate at which these materials in the Earth's crust would be uplifted and exposed to weathering at the surface (Bertine and Goldberg 1971). For example, the industrial use of lead (Pb) has increased the transport of Pb in world rivers by about a factor of 10 (Martin and Meybeck 1979). Recent changes in the content of lead in coastal sediments appear directly related to fluctuations in the use of Pb by humans, especially in leaded gasoline (Trefry et al. 1985)—trends superimposed on underlying natural variations in the transport of Pb in world rivers.

Recent estimates suggest that the global cycles of many metals have been significantly increased by human activities (Table 1.1). Some of these metals are released to the atmosphere and deposited in remote locations (Boutron et al. 1994). For example, the concentration of mercury (Hg) in the Greenland ice cap is significantly greater in ice layers deposited in the last 100 years (Weiss et al. 1971). Recognizing that the deposition of Hg in the Antarctic ice cap shows large variations over the past 34,000 years (Vandal et al. 1993), we must evaluate any recent increase in Hg deposition in the context of past cyclic changes in Hg transport through the atmosphere. Again, human-induced changes in the movement of materials through the atmosphere must be placed in the context of natural cycles in Earth system function (Nriagu 1989).

Thermodynamics

Two basic laws of physical chemistry, the laws of thermodynamics, tell us that energy can be converted from one form to another and that chemical reactions should proceed spontaneously in the direction of lower free energy, G. The lowest free energy of a given reaction represents its equilibrium, and it is found in the mix of chemical species that show maximum bond strength and maximum disorder among the components. In the face of these basic laws, living systems and the conditions on the surface of the Earth exist in a nonequilibrium condition.

Even the simplest cell is an ordered system; a membrane separates an inside from an outside, and the inside contains a mix of very specialized molecules. Biological molecules are collections of compounds with relatively weak bonds. For instance, to break the covalent bonds between two carbon atoms requires 83 kcal/mole, versus 192 kcal/mole for each of the double bonds between carbon and oxygen in CO_2 (Davies 1972, Morowitz 1968).

Table 1.1 Movement of Selected Elements through the Atmosphere[a]

	Natural			Anthropogenic		
		Volcanic				
Element	Continental dust	Dust	Gas	Industrial particles	Fossil fuel	Ratio anthropogenic: natural
Al	356,500	132,750	8.4	40,000	32,000	0.15
Fe	190,000	87,750	3.7	75,000	32,000	0.38
Cu	100	93	0.012	2,200	430	13.63
Zn	250	108	0.14	7,000	1,400	23.46
Pb	50	8.7	0.012	16,000	4,300	345.83

[a] All data are expressed in 10^8 g/yr. From Lantzy and MacKenzie (1979).

In living tissue most of the bonds between C, H, N, O, P, and S, the major biochemical elements, are reduced, or "electron-rich" bonds (Chapter 7). It is an apparent violation of the laws of thermodynamics that the reduced bonds in the molecules of living organisms exist in the presence of a strong oxidizing agent in the form of O_2 in the atmosphere. Thermodynamics would predict a spontaneous reaction between these components to produce CO_2, H_2O, and NO_3^-—molecules with much stronger bonds. In fact, upon the death of an organism, this is exactly what happens! Living organisms must continuously process energy to counteract the basic laws of thermodynamics that would otherwise produce disordered systems with oxidized molecules.

Photosynthetic organisms capture the energy in sunlight and convert the bonds between carbon and oxygen in CO_2 to the weak, reduced biochemical bonds that characterize life (Eq. 1.2). Heterotrophic organisms obtain energy by capitalizing on the natural tendency for electrons to flow from reduced bonds to oxidizing substances, such as O_2. Thus, these organisms oxidize the bonds in organic matter and convert the carbon back to CO_2. A variety of other metabolic pathways have evolved using transformations among other compounds (Chapter 2), but in every case metabolic energy is obtained from the flow of electrons between compounds in oxidized or reduced states. Metabolism is possible because living systems can sequester high concentrations of oxidized and reduced substances from their environment. Without membranes to compartmentalize living cells, thermodynamics would predict a uniform mix; energy transformations, such as respiration, would be impossible.

Free oxygen appeared in the Earth's surface environments sometime after the appearance of autotrophic, photosynthetic organisms. Free O_2 is one of the most oxidizing substances known, and the movement of electrons from reduced substances to O_2 releases large amounts of free energy. Thus, large releases of free energy are found in aerobic metabolism, including the efficient metabolism of eukaryotic cells. The appearance of eukaryotic cells was not immediate; the fossil record suggests that they evolved nearly 2 billion years after the appearance of the simplest living cells (Knoll 1992). Presumably the evolution of eukaryotic cells was possible only after the accumulation of sufficient O_2 in the environment to sustain aerobic metabolic systems. In turn, aerobic metabolism offered large amounts of energy that could allow the elaborate structure and activity of higher organisms.

Gaia

In a provocative book, *Gaia,* published in 1979, James Lovelock focused scientific attention on the fact that the chemical conditions of the present-day Earth, especially in the atmosphere, are extremely unusual and in disequilibrium with respect to thermodynamics. The 21% atmospheric content of O_2 is the most obvious result of living organisms, but the content of other gases, including NH_3 and CH_4, is higher than one would expect in an O_2-rich atmosphere. This level of O_2 in our atmosphere is maintained despite known reactions that should consume O_2 in reaction with crustal minerals and organic carbon. Further, Lovelock suggested that the *albedo* (reflectivity) of Earth must be regulated by the biosphere, because the planet has shown relatively small changes in surface temperature despite large fluctuations in the Sun's radiation during the history of life on Earth (Watson and Lovelock 1983).

Gaia suggests that the conditions of our planet are so unusual that they could only be expected to result from activities of the biosphere. Indeed, *Gaia* suggests that the biosphere evolved to regulate conditions within a range favorable for the continued persistence of life on Earth. The planet functions as a kind of "superorganism." Reflecting the vigor and excitement of a new scientific field, other workers have strongly disagreed—not denying that biotic factors have strongly influenced the conditions on Earth, but not accepting the hypothesis of purposeful self-regulation of the planet (Kirchner 1989, Schneider and Boston 1991). Like all models, *Gaia* remains as a provocative hypothesis, and the rapid pace at which humans are changing the function of the biosphere should alarm us all.

2

Origins

Introduction

Six elements, C, H, O, N, P, and S, are the major constituents of living tissue and account for 95% of the biosphere. At least 20 other elements (see inside front cover) are known to be essential to life, and it is possible that this list may grow slightly as we come to recognize the role of additional trace elements in biochemistry (McDowell 1992). In the periodic table all of the biologically relevant elements are found at atomic numbers less than that of iodine at 53. Even though living organisms affect the distribution and abundance of some of the heavier elements, we can speak of the chemistry of life as the chemistry of "light" elements (Deevey 1970a). One initial constraint on the composition of life must have been the relative abundance of chemical elements in our galaxy. Later as the Sun and its planets formed and differentiated, the chemical composition of the Earth's surface determined the environment in which life arose.

In this chapter we will examine models that astrophysicists suggest for the origin of the elements. Then, we will examine models for the formation of the solar system and the planets. There is good evidence that the conditions on the surface of the Earth changed greatly during the first billion years or so after its formation—well before life arose. Early differentiation of the Earth, cooling of its surface, and changes caused by the evolution and proliferation of life have strongly determined the conditions on our planet today. In this chapter, we will consider the origin of the major metabolic pathways that characterize life and affect Earth's biogeochemistry. The chapter ends with a discussion of the planetary evolution that has occurred on the Earth compared to its near neighbors—Mars and Venus.

Origins of the Elements

Any theory for the origin of the chemical elements must account for their relative abundance in the Universe. Estimates of the cosmic abundance of elements are made by examining the spectral emission from the stars of distant galaxies, as well as the emission from our Sun (Ross and Aller 1976). Analyses of meteorites also provide important information on the composition of the solar system (Fig. 2.1). Two points are obvious: (1) with three exceptions (Li, Be, and B), the light elements, i.e., those with an atomic number <30, are far more abundant than the heavy elements, and (2) especially among the light elements, the even-numbered elements are more abundant than the odd-numbered elements of similar atomic weight.

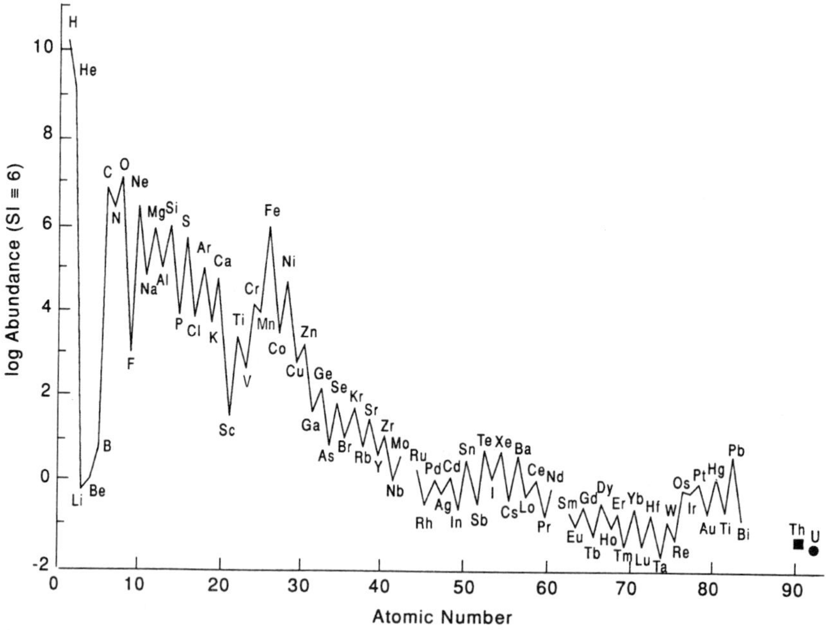

Figure 2.1 The relative abundance of elements in the solar system as a function of atomic number. Abundances are plotted logarithmically and scaled so that silicon (Si) = 1,000,000. From a drawing by Brownlee (1992) based on the data of Anders and Grevesse (1989).

A central theory of astrophysics is that the origin of the Universe began with a gigantic explosion, "The Big Bang," about 13 to 18 billion years ago (van den Bergh 1992). The Big Bang initiated the fusion of hypothetical fundamental particles, known as quarks, to form protons (^{1}H) and neutrons, and it allowed the fusion of protons and neutrons to form some simple atomic nuclei (^{2}H, ^{3}He, ^{4}He, and ^{7}Li; Malaney and Fowler 1988, Pagel 1993, Copi et al. 1995). After the Big Bang, the Universe began to expand outward, and there was a rapid decline in the temperatures and pressures that would be needed to produce heavier elements by fusion in interstellar space.[1] Moreover, the elements with atomic masses of 5 and 8 are unstable, so no fusion of the abundant initial products of the Big Bang (i.e., ^{1}H and ^{4}He) could yield an appreciable, persistent amount of a heavier element. Thus, the theory of the Big Bang can explain the origin of elements up to ^{7}Li, but the origin of heavier elements had to await the formation of stars in the Universe—about 1 billion years later.

A model for the synthesis of heavier elements was first proposed by Burbidge et al. (1957), who outlined a series of pathways that could occur in the interior of stars during their evolution (Fowler 1984, Wallerstein 1988). As a star ages, the abundance of hydrogen in the core declines as it is converted to helium by fusion. As the heat from nuclear fusion decreases, the star begins to collapse inward under its own gravity. This collapse increases the internal temperature and pressure until He begins to be converted, or "burn," to form carbon (C) in a two-step reaction. First,

$$^{4}He + {}^{4}He \rightleftarrows {}^{8}Be. \tag{2.1}$$

Then, while most ^{8}Be decays spontaneously back to helium, the momentary existence of small amounts of ^{8}Be under these conditions allows reaction with another helium to produce carbon:

$$^{8}Be + {}^{4}He \rightarrow {}^{12}C. \tag{2.2}$$

The main product of helium burning is ^{12}C; however, a small amount of ^{16}O is also formed by the addition of ^{4}He to ^{12}C. As the supply of helium begins to decline, a second phase of stellar collapse is followed by the initiation of a sequence of further "burning" reactions (Fowler 1984). First, fusion of two ^{12}C forms ^{24}Mg, some of which decays to ^{20}Ne (neon) by loss of an alpha (^{4}He) particle. Subsequently, oxygen burning produces

[1] Some widely publicized estimates made using the Hubble Space Telescope suggested that the Universe might be expanding outward more rapidly than previously thought, and thus, might be only 10 billion years old (Freedman et al. 1994). This would be younger than independent estimates of the age of stars in our galaxy (Bolte and Hogan 1995). More recent work now suggests that the expansion of the Universe is not quite so rapid, so that the "Big Bang" model for the nucleosynthesis of light elements appears robust.

^{32}S, which forms an appreciable amount of ^{28}Si (silicon) by loss of an alpha particle (Woosley 1986).

A variety of fusion reactions in stars are thought to be responsible for the synthesis of the even-numbered elements up to iron (Fe) (Fowler 1984). These fusion reactions release energy and produce increasingly stable nuclei (Fig. 2.2). To make a nucleus heavier than Fe requires energy, so when a star's core is dominated by Fe, it can no longer burn. This leads to the catastrophic collapse and explosion of a star, which we recognize as a supernova. Heavier elements are apparently formed by the successive capture of neutrons by Fe, either deep in the interior of stable stars or during the explosion of a supernova (Rank et al. 1988, Woosley and Phillips 1988). A supernova casts all portions of the star into space as hot gases (Chevalier and Sarazin 1987).

This model explains a number of observations about the abundance of the chemical elements in the Universe. First, the abundance of elements declines logarithmically with increasing mass from hydrogen and helium, the original building blocks of the Universe. However, as the Universe ages, more and more of the hydrogen will be converted to heavier elements during the evolution of stars. Astrophysicists can recognize younger, second-generation stars, such as our Sun, that have formed from the remnants of previous supernovas because they contain a higher abundance of iron and heavier elements than older, first-generation stars, in which the initial hydrogen-burning reactions are still predominant (Penzias 1979).

Second, because the first step in the formation of all the elements beyond lithium is the fusion of nuclei with an even number of atomic mass (i.e., ^{4}He, ^{12}C), the even-numbered light elements are relatively abundant in the cosmos. The odd-numbered light elements are formed by the fission of heavier even-numbered nuclei. Figure 2.2 shows that in most cases an odd-numbered nucleus is slightly less stable than its even-numbered "neighbors," so we should expect odd-numbered nuclei to be less abundant. For example, phosphorus is formed in the reaction

$$^{16}O + ^{16}O \rightarrow ^{32}S \rightarrow ^{31}P + ^{1}H. \quad (2.3)$$

Thus, phosphorus is much less abundant than the adjacent elements in the periodic table, Si and S (Fig. 2.1). It is interesting to speculate that the low cosmic abundance of P may account for the fact that P is often in short supply for the biosphere on Earth today.

Finally, the low cosmic abundance of Li, Be, and B is due to the fact that the initial fusion reactions pass over nuclei with atomic masses of 5 to 8, forming ^{12}C as shown in Eqs. 2.1 and 2.2. Apparently, most Li, Be, and B are formed by spallation—the fission of heavier elements that are hit by cosmic rays in interstellar space (Olive and Schramm 1992, Reeves 1994, Chaussidon and Robert 1995).

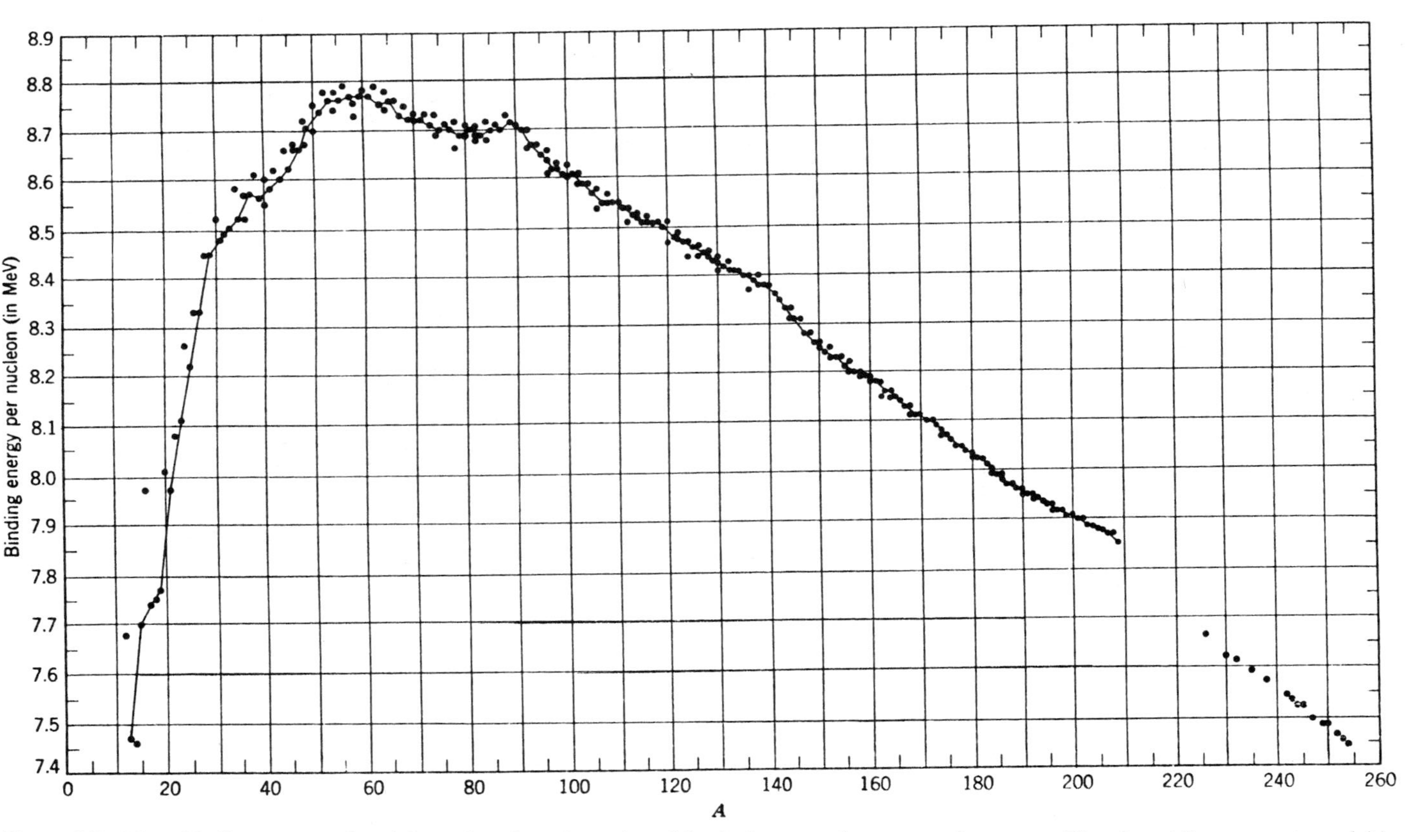

Figure 2.2 Mean binding energy of nuclei as a function of atomic weight A, the sum of protons and neutrons. The plotted line connects nuclei in which A is odd. Note that for the relatively light elements, when A is even, the nuclei are generally more stable than when A is odd. From Friedlander et al. (1964).

This model for the origin and the cosmic abundance of the elements offers some initial constraints for biogeochemistry. All things being equal, we might expect that the chemical environment in which life arose would approximate the cosmic abundance of elements. Thus, the evolution of biochemical molecules might be expected to capitalize on the light elements that were abundant in the primordial environment. It is then of no great surprise that no element heavier than Fe is more than a trace constituent in living tissue and that among the light elements, no Li or Be, and only traces of B, are essential components of biochemistry. The composition of life is remarkably similar to the composition of the Universe; as put by William Fowler in 1984, we are all "a little bit of stardust."

Origin of the Solar System and the Earth

Our Galaxy is probably about 10 billion years old (Oswalt et al. 1996), but the solar system appears to be only half that age, about 4.6 billion years old (Allègre et al. 1995). Current models for the origin of the solar system suggest that the Sun and its planets formed from the remnants of a supernova which left a collection of hot gases and dust particles in space (Chevalier and Sarazin 1987, Snow and Witt 1995). As the Sun and the planets began to condense, each developed a gravitational field that helped capture materials that added to its initial mass. The mass concentrated in the Sun apparently allowed condensation to pressures that reinitiated the fusion of hydrogen to helium (Christensen-Dalsgaard et al. 1996) Some astronomers have pointed out that the hydrogen-rich atmosphere on Jupiter is similar to that on "Brown Dwarfs"—stars that never "ignited."

The planets of our solar system appear to have formed from the coalescing of dust and small bodies, known as planetesimals, that formed within the primitive solar cloud (Press and Siever 1986). Recent observations suggest that a similar process is now occurring around another star in our galaxy, β Pictoris (Lagage and Pantin 1994), and that planets surround other stars (Mayor and Queloz 1995, Beckwith and Sargent 1996). Overall, the original nebula is likely to have been composed of about 98% gaseous elements (H, He, and noble gases), 1.5% ice (H_2O, NH_3, and CH_4), and 0.5% solid materials, but the composition of each planet was determined by its position relative to the Sun and the rate at which the planet grew (McSween 1989, Harper and Jacobsen 1996).

The "inner" planets (Mercury, Venus, Earth, and Mars) seemed to have formed in an area where the solar nebula was very hot, perhaps at a temperature close to 1200 K (Boss 1988). Mercury formed closest to the Sun, and it is dominated by metallic Fe and other materials that condense at high temperatures. The high density of Mercury (5.4 g/cm^3) contrasts with the lower average density of the larger, outer planets that captured a greater fraction of lighter constituents from the initial solar cloud. Jupiter contains much hydrogen and helium. Its average density is 1.25 g/cm^3, and its overall composition does not appear too different from the cosmic abundance of elements (Lunine 1989, but see also Niemann et al. 1996).

Apparently the inner portion of the solar nebula was too hot to contain a large complement of light, volatile elements that could be captured during planetary growth (Humayun and Clayton 1995). Venus, Earth, and Mars, are all depleted in light elements compared to the cosmic abundances, and they are dominated by silicate minerals that condense at intermediate temperatures and contain large amounts of FeO (McSween 1989). The mean density of Earth is about 5.5 g/cm^3. Thus, from the initial solar cloud of elements, the initial chemical composition of the Earth was a selective mix, peculiar to the orbit of the incipient planet.

The bulk of the Earth's mass appears to have accumulated within about 100 million years of its initial formation (Allègre et al. 1995). There are several theories that account for the origin and differentiation of Earth. One theory suggests that the Earth may have grown by homogeneous accretion; that is, throughout its early history, the Earth may have captured planetesimals that were relatively similar in composition (Stevenson 1983). Kinetic energy generated during the collision of planetesimals (Wetherill 1985), as well as the heat generated from radioactive decay in its interior (Hanks and Anderson 1969), would heat the primitive Earth to the melting point of iron, nickel, and other metals, forming a magma ocean. These heavy elements were "smelted" from the materials arriving from space and sank to the interior of the Earth to form the core (Agee 1990, Newsom and Sims 1991, Minarik et al. 1996). As the Earth cooled, lighter minerals progressively solidified to form a mantle with the approximate composition of perovskite ($MgSiO_3$) and olivine ($FeMgSiO_4$) and a crust dominated by aluminosilicate minerals of lower density and the approximate composition of feldspar (Chapter 4). Thus, despite the abundance of iron in the cosmos and in the Earth as a whole, the crust of the Earth is largely composed of Si, Al, and O (Fig. 2.3). Even today, the aluminosilicate rocks of the crust "float" on the heavier semifluid rocks of the mantle, resulting in the drift of continents on the Earth's surface (Fig. 2.4; Bowring and Housh 1995).

A second theory for the origin of Earth suggests that the characteristics of planetesimals and other materials contributing to the growth of the planet were not uniform through time. Theories of heterogeneous accretion suggest that materials in the Earth's mantle arrived later than those of the core (Harper and Jacobsen 1996), and that a late veneer delivered by a class of meteors known as carbonaceous chondrites was responsible for the light elements in the Earth's crust (Anders and Owen 1977, Wetherill 1994). Both theories recognize that our planet was very hot during its early history, but theories of heterogeneous accretion do not require a complete melting and reorganization of the Earth to differentiate the core, the mantle, and the crust.

Consistent with either theory are several lines of evidence that the primitive Earth was devoid of an atmosphere. During its early history, the gravita-

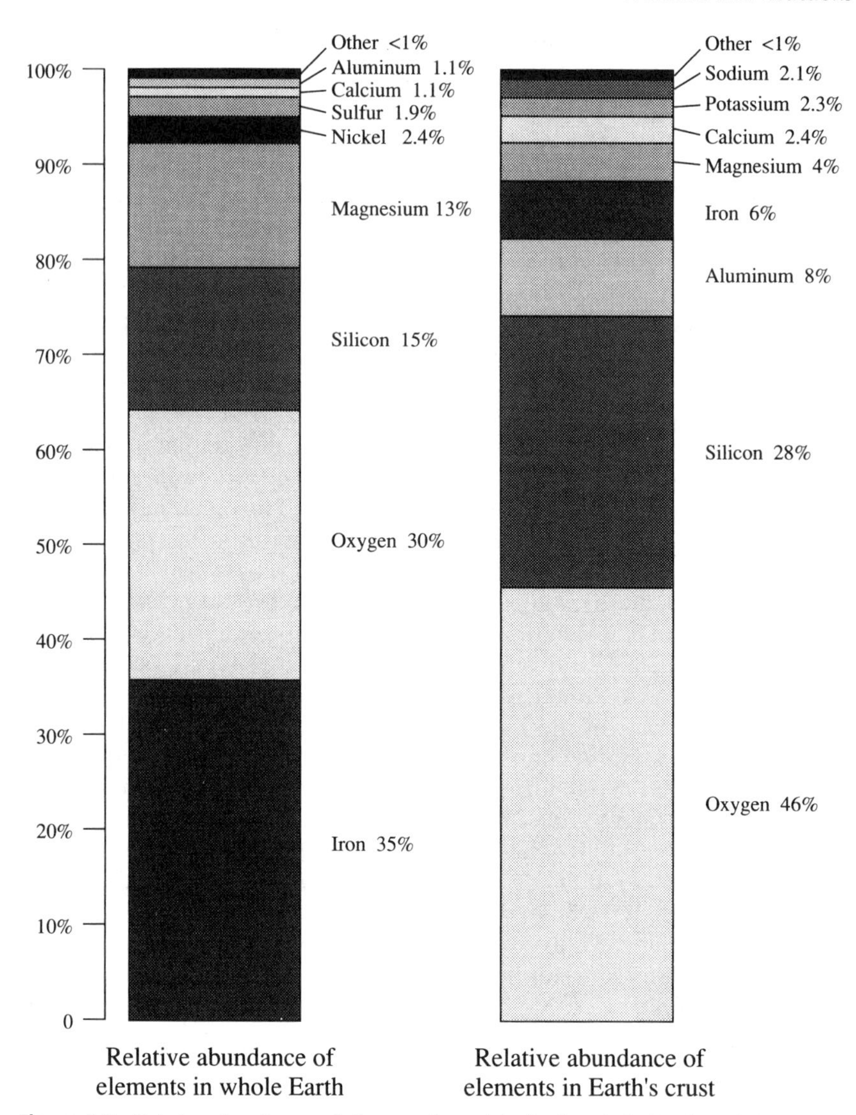

Figure 2.3 Relative abundance of elements by weight in the whole Earth and the Earth's crust. From Earth 4/E. By Frank Press and Raymond Siever. Copyright 1986 by W. H. Freeman and Company. Reprinted by permission.

tional field on Earth would have been too weak to retain gaseous elements, and the incoming planetesimals are likely to have been too small and too hot to carry an envelope of volatile elements. If a significant fraction of today's atmosphere were derived from the original solar cloud, we might expect that its gases would exist in proportion to their solar abundances (Fig. 2.1). Here, ^{20}Ne is of particular interest, because it is not produced

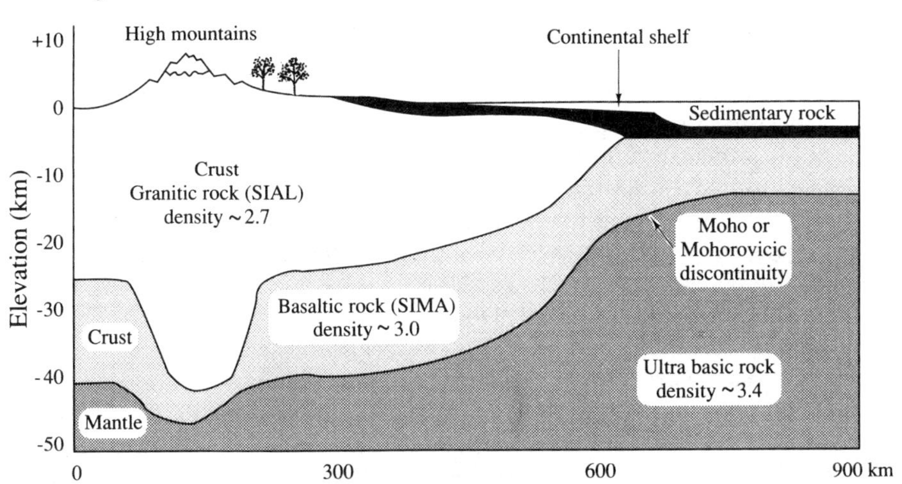

Figure 2.4 A geologic profile of the Earth's surface. On land the crust is dominated by granitic rocks, largely composed of silicon and aluminum (SIAL). The oceanic crust is dominated by basaltic rocks, with a large proportion of silicon and magnesium (SIMA). Both granite and basalt have a lower density than the upper mantle, which contains ultrabasic rocks with the approximate composition of olivine ($FeMgSiO_4$). From Howard and Mitchell (1985).

by any known radioactive decay, it is too heavy to escape from the Earth's gravity, and as an inert gas, it is not likely to have been consumed in any reaction with crustal minerals (Walker 1977). Thus, the present-day abundance of ^{20}Ne in the atmosphere is likely to represent its primary abundance—that derived from the solar cloud. Assuming that other solar gases were delivered to the Earth in a similar manner, we can calculate the total mass of the primary atmosphere by multiplying the mass of ^{20}Ne in today's atmosphere by the ratio of each of the other gases to ^{20}Ne in the solar abundance. For example, the solar ratio of nitrogen to neon is 0.91 (Fig. 2.1). If the present-day atmospheric mass of neon, 6.5×10^{16} g, is all from primary sources, then $(0.91) \times (6.5 \times 10^{16}$ g) should be the mass of nitrogen that is also of primary origin. The product, 5.9×10^{16} g, is much less than the observed atmospheric mass of nitrogen, 39×10^{20} g. Thus, most of the nitrogen in today's atmosphere must have been derived from other sources.

Origin of the Atmosphere and Oceans

The Earth's inventory of "light" elements is likely to have been delivered to the planet as constituents of the silicate minerals in carbonaceous chondrites. Even today, many silicate minerals in the Earth's mantle carry elements such as oxygen and hydrogen as part of their crystalline structure (Bell and Rossman 1992, Meade et al. 1994). Of particular interest to biogeochemistry, carbonaceous chondrites typically contain from 0.5 to

3.6% carbon and 0.01 to 0.28% N (Anders and Owen 1977), which may represent the major source of these elements in the present-day biosphere. It is likely that the bulk of the Earth's gaseous compounds were first released from rocks to the atmosphere as the growing mass of the Earth was melted and transformed during the early differentiation of the crust and mantle (Fanale 1971, Walker 1977, Stevenson 1983, Kuramoto and Matsui 1996). Although it appears that more than 80% of the primordial volatiles have been degassed from the mantle (Ozima and Podosek 1983), volcanic emissions of inert gases such as ^{3}He and ^{20}Ne indicate that some degassing still occurs (Lupton and Craig 1981, Honda et al. 1991).

Today, a variety of gases are released during volcanic eruptions at the Earth's surface. For example, Table 2.1 gives the composition of volcanic gases emitted from the Kudryavy volcano in the western Pacific Ocean (Taran et al. 1995). Characteristically, water vapor dominates the emissions, but small quantities of C, N, and S gases are also present. Volcanic emissions, representing degassing of the Earth's interior, are consistent with the observation that the Earth's atmosphere is of secondary origin.

The atmospheric content of ^{40}Ar is suggestive of the extent of crustal degassing. Like ^{20}Ne, this noble element is too heavy to escape the gravity of the Earth. The cosmic abundance of ^{40}Ar is small compared to that of ^{36}Ar; on Earth ^{40}Ar appears to be wholly the result of the decay of ^{40}K in crustal minerals. Thus, the atmospheric content of ^{36}Ar should represent the proportion that is due to the residual primary atmosphere, whereas the content of ^{40}Ar is indicative of the proportion due to crustal degassing.[2] The ratio of ^{40}Ar/^{36}Ar on Earth is nearly 300, suggesting that 99.7% of the Ar in our present atmosphere is derived from the interior of the Earth. The ratio of N_2 to ^{40}Ar in the Earth's mantle is close to the atmospheric value—implying a common source for both elements in the atmosphere (Marty 1995). The Viking spacecraft measured a ratio of 2750 for ^{40}Ar versus ^{36}Ar in the atmosphere on Mars (Owen and Biemann 1976), which is consistent with the emerging belief that Mars also underwent crustal degassing but has now lost a large portion of its atmosphere (Carr 1987).

Comets may have also been a source of volatile elements for Earth (Delsemme 1992, Owen et al. 1992, Fomenkova et al. 1994). Chyba (1987) estimated the rate of cometary bombardment of Earth from the craters on the Moon and calculated that nearly all of the water on Earth could have been delivered by comets. Of course, the high-speed impact of comets would be likely to drive some volatiles back into space, but overall, the net effect would lead to accumulations of water and volatile elements on Earth (Chyba 1990a).

It appears that most of the Earth's mass is likely to have been received during heavy meteor bombardment in the first billion years or so of its

[2] A small amount of ^{40}Ar has been destroyed by cosmic rays, producing ^{35}S in the atmosphere (Tanaka and Turekian 1991).

Table 2.1 Composition of Volcanic Gases Released from the Kudryavy and Other Volcanoes

Volcano	Units	H_2O	H_2	CO_2	SO_2	H_2S	HCl	HF	N_2	NH_3	O_2	Ar	CH_4	Reference
Kudryavy, Russia	mole %	95.00	0.56	2.00	1.32	0.41	0.3700	0.030	0.21	—	0.03	0.002	0.002	Taran et al. (1995)
Nevado del Ruiz, Colombia	wgt. %	94.90		2.91	2.74	0.80	0.0052							Williams et al. (1986)
Kamchatka, Russia	vol. %	78.60	3.01	4.87	0.03	0.16	0.5700	0.056	11.87	0.11	0.01	0.060	0.440	Dobrovolsky (1994)

history. The present-day receipt of extraterrestrial materials (approximately 5×10^{10} g/yr; Esser and Turekian 1988, Love and Brownlee 1993) is much too low to account for the Earth's mass (6×10^{27} g), even if it has continued for all of Earth's history. Based on the age distribution of craters on the Moon, there is good evidence that meteor bombardment slowed about 3.8 billion years ago (Neukum 1977).

As long as the Earth was very hot, all volatiles remained in the atmosphere, but when the surface temperature cooled to 100°C, water could condense out of the primitive atmosphere to form the oceans. This must have been a rainstorm of true global proportion! Although heat from large, late-arriving meteors may have caused temporary revaporization of the earliest oceans (Sleep et al. 1989), the rock record suggests that liquid water was present on the Earth's surface 3.8 billion years ago, and has been ever since. Consistent with the belief that most of the crustal outgassing was completed early in the Earth's history, the oceans are likely to have achieved close to their present volume in a relatively short period of time (Holland 1984).

Various other gases would quickly enter the primitive ocean as a result of their high solubility in water; e.g.,

$$CO_2 + H_2O \rightleftarrows H_2CO_3 \rightleftarrows H^+ + HCO_3^- \tag{2.4}$$

$$HCl + H_2O \rightleftarrows H_3O^+ + Cl^- \tag{2.5}$$

$$SO_2 + H_2O \rightleftarrows H_2SO_3. \tag{2.6}$$

These reactions removed a large proportion of reactive water-soluble gases from the atmosphere, as predicted by Henry's Law for the partitioning of gases between gaseous and dissolved phases,

$$S = kP, \tag{2.7}$$

where S is the solubility of a gas in a liquid, k is the solubility constant, and P is the overlying pressure in the atmosphere. Under 1 atm of partial pressure, the solubilities of CO_2, HCl, and SO_2 in water are 1.4, 700, and 94.1 g/liter, respectively, at 25°C. When dissolved in water, all of these gases form acids, which would be neutralized by immediate reaction with the surface minerals on Earth. Thus, the Earth's earliest atmosphere is likely to have been dominated by N_2, which has relatively low solubility in water (0.018 g/liter at 25°C).

Because many gases dissolve so readily in water, an estimate of the total extent of crustal degassing through geologic time must consider the mass of the atmosphere, the mass of oceans, and the mass of volatile elements that are now contained in sedimentary minerals, such as $CaCO_3$, which have been deposited from seawater (Li 1972). By this account, the mass of the present-day atmosphere (5.14×10^{21} g; Trenberth and Guillemot 1994) represents less than 1% of the total degassing of the Earth's crust (Table 2.2). The oceans and various marine sediments contain nearly all of the remainder.

As defined in Chapter 1, substances that are electron-rich are known as reduced substances, while those that are electron-poor are known as oxidizing substances. Oxidizing substances tend to draw electrons from reduced substances. Despite uncertainty about the exact composition of the Earth's earliest atmosphere, several lines of evidence suggest that when life arose the atmosphere was a moderately reducing mix of gases—N_2, CO_2, and H_2O (Holland 1984). The oxidation state of iron found as olivine ($FeMgSiO_4$) in the mantle is not likely to have allowed an abundance of more reduced gases, such as H_2, CH_4, and NH_3, in volcanic emissions (Kasting 1993). Moreover, any H_2 that was released would have been likely to escape from the Earth's gravitational field, leaving little to react with CO_2 to produce methane (Warneck 1988, p. 607):

$$4H_2 + CO_2 \rightarrow CH_4 + 2H_2O. \tag{2.8}$$

On the primitive Earth, ammonia (NH_3) would have been subject to photolysis in the upper atmosphere, producing N_2 with the loss of H_2 to space. Thus, it is likely that the atmosphere in which life arose was dominated by N_2, with lesser, equilibrium proportions of H_2O and CO_2, and trace quantities of other gases from volcanic emissions that were continuing at that time (Hunten 1993). There was certainly no O_2; the small concentrations produced by the photolysis of water in the upper atmosphere would rapidly be consumed in the oxidation of reduced gases and crustal minerals (Walker 1977, Kasting and Walker 1981).

During its early evolution as a star, the Sun's luminosity was as much as 30% lower than at present. We might expect that the Earth was colder, but the fossil record indicates a continuous presence of liquid water on the Earth's surface since 3.8 billion years ago. One explanation is that the primitive atmosphere contained much higher concentrations of CO_2 and other greenhouse gases than today (Walker 1985). This CO_2 would trap outgoing infrared radiation and produce global warming through the "greenhouse" effect (Fig. 3.2). In fact, even today the presence of water vapor and CO_2 in the atmosphere creates a significant greenhouse effect on Earth. Without these gases, Earth's temperature would be about 33°C cooler, and the planet would be covered with ice (Ramanathan 1988).

There are few direct indications of the composition of the earliest seawater. Like seawater today, the Precambrian ocean is likely to have contained a substantial amount of chloride. HCl and Cl_2 emitted by volcanoes would dissolve in water, forming Cl^- (Eq. 2.5; see also Eq. 7.11). The dissolution

Table 2.2 Total Inventory of Volatiles at the Surface of the Earth[a]

Reservoir	H_2O	CO_2	C	O_2	N	S	Cl	Ar	Total (rounded)
Atmosphere (cf. Table 3.1)	1.3	0.28	—	119	387	—	—	6.6	514
Oceans	135,000	19.3[b]	0.07	256[c]	2[d]	128[e]	2610	—	138,000
Land plants	0.1	—	0.06	—	0.0004	—	—	—	0.16
Soils	12	0.40[f,g]	0.15	—	0.0095	—	—	—	12.6
Freshwater (including ice and groundwater)	4,850	—	—	—	—	—	—	—	4,850
Sedimentary rocks	15,000[h]	32,400[g]	1560	4745[i]	100[h]	744[j]	500[h]	—	55,000
Total (rounded)	155,000	32,400	1560	5120	490	872	3100	7	198,000
See also	Fig. 10.1	Fig. 11.1	Fig. 11.1	Fig. 2.7		Table 13.1			

[a] All data are expressed as 10^{19} g, with values derived from this text unless noted otherwise.
[b] Assumes the pool of inorganic C is in the form of HCO_3^-.
[c] Oxygen content of dissolved SO_4^{2-}.
[d] Dissolved N_2.
[e] S content of SO_4^{2-}.
[f] Desert soil carbonates.
[g] Assumes 60% of $CaCO_3$ is carbon and oxygen.
[h] Walker (1977).
[i] O_2 held in sedimentary Fe_2O_3 + evaporitic $CaSO_4$.
[j] S content of $CaSO_4$ and FeS_2.

of these and other gases in water (Eqs. 2.4–2.6) produces acids, which should have reacted with minerals of the Earth's crust, releasing Na^+, Mg^{2+}, and other cations by chemical weathering (Chapter 4). Carried by rivers, these cations would accumulate in seawater, until their concentrations increased to levels that would precipitate secondary minerals. For instance, sedimentary accumulations of $CaCO_3$ of Precambrian age indicate that the primitive oceans had substantial concentrations of Ca^{2+} (Walker 1983). Thus, it is likely that the dominant cations (Na, Mg, and Ca) and the dominant anion (Cl) in Precambrian seawater were similar to those in seawater today (Holland 1984). Only SO_4^{2-} seems to have been less concentrated in the Precambrian ocean (Grotzinger and Kasting 1993, Walker and Brimblecombe 1985, Habicht and Canfield 1996).

Origin of Life

Working with Harold Urey in the early 1950s, Stanley Miller added the probable constituents of the primitive atmosphere and oceans to a laboratory flask and subjected the mix to an electric discharge to represent the effects of lightning. After several days, Miller found that simple, reduced organic molecules had been produced (Miller 1953, 1957). This experiment, simulating the conditions on the early Earth, suggested that the organic constituents of living organisms could be produced abiotically.

The experiment has been repeated in many laboratories, and with many combinations of conditions (Chang et al. 1983). Ultraviolet light can substitute for electrical discharges as an energy source; a high flux of ultraviolet light would be expected on the primitive Earth in the absence of an ozone (O_3) shield in the stratosphere (Chapter 3). Additional energy for abiotic synthesis may have been derived from the impact of late-arriving meteors and comets passing through the atmosphere (Chyba and Sagan 1992). The mix of atmospheric constituents taken to best represent the primitive atmosphere is controversial, although an acceptable yield of simple organic molecules has been produced in experiments using mildly reducing atmospheres (Pinto et al. 1980). These experiments are never successful when free O_2 is included; O_2 rapidly oxidizes the simple organic products before they can accumulate.

Interplanetary dust particles and cometary ices also contain a wide variety of simple organic molecules (Cox 1989), and various amino acids are found in carbonaceous chondrites, suggesting that these abiotic synthesis reactions may have occurred repeatedly in the galaxy (Kvenvolden et al. 1970, Orgel 1994). Significantly, it is possible that some of the organic molecules in chondrites and comets survived passage through the Earth's atmosphere, contributing to the initial inventory of organic molecules on its surface (Anders 1989, Chyba and Sagan 1992). Even if the total mass received was small, exogenous sources of organic molecules are important, for they may have served as catalysts, speeding the rate of abiotic synthesis on Earth.

A wide variety of simple organic molecules have now been produced under abiotic conditions in the laboratory (Dickerson 1978). In many cases hydrogen cyanide and formaldehyde are important initial products that polymerize to produce simple sugars such as ribose and more complex molecules such as amino acids and nucleotides. Methionine, a sulfur-containing amino acid, has been synthesized abiotically (Van Trump and Miller 1972), and short chains of amino acids have been linked by condensation reactions involving phosphates (Rabinowitz et al. 1969, Lohrmann and Orgel 1973). An early abiotic role for organic polyphosphates in synthesis speaks strongly for the origin of adenosine triphosphate (ATP) as the energizing reactant in virtually all biochemical reactions that we know of today (Dickerson 1978).

Clay minerals, with their surface charge and repeating crystalline structure, may have acted to concentrate simple, polar organic molecules from the primitive ocean, making assembly into more complicated forms, such as RNA and protein, more likely (Cairns-Smith 1985, Pitsch et al. 1995, Ferris et al. 1996). Metal ions such as zinc and copper can enhance the binding of nucleotides and amino acids to clays (Lawless and Levi 1979). It is interesting to speculate why, however, nature incorporates only the "left-handed" forms of amino acids in proteins, when equal forms of L- and D-enantiomers are produced by abiotic synthesis (Fig. 2.5).

Just as droplets of cooking oil form "beads" on the surface of water, it has long been known that organic polymers will spontaneously form coacervates, which are colloidal droplets small enough to remain suspended in water. Coacervates are perhaps the simplest systems that might be said to be "bound," as if by a membrane, providing an inside and an outside. Yanagawa et al. (1988) describe several experiments in which protocellular structures with lipoprotein envelopes were constructed in the laboratory. In such structures, the concentration of substances will differ between the inside (hydrophobic) and the outside (hydrophilic) as a result of the

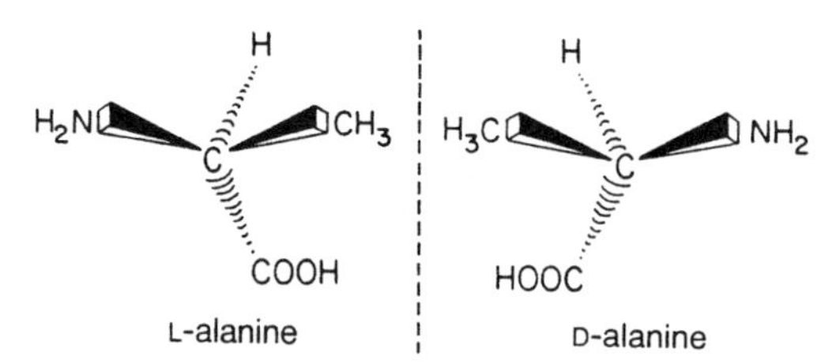

Figure 2.5 The left-handed (L) and right-handed (D) forms, known as enantiomers, of the amino acid alanine. No rotation of these molecules can allow them to be superimposed. Although both forms are found in the extraterrestrial organic matter of carbonaceous chondrites, all life on Earth incorporates only the L-form in proteins. From Chyba (1990b).

differing solubility of substances in an organic medium and water, respectively.

Some organic molecules produced in the laboratory will self-replicate, suggesting potential mechanisms that may have increased the initial yield of organic molecules from abiotic synthesis (Hong et al. 1992, Orgel 1992, Lee et al. 1996). Other laboratories have produced simple organic structures, known as micelles, that will self-replicate their external framework (Bachmann et al. 1992). There is good reason to believe that the earliest genetic material controlling self-replication may not have been DNA, but a related molecule, RNA, which can also perform catalytic activities (de Duve 1995, Robertson and Miller 1995). Although the assembly of simple organic molecules into a self-replicating, metabolizing, and membrane-bound form that we might call life has so far eluded experimental approaches, it is possible that researchers will produce an organic form with many of these characteristics within the next few years.

A traditional view holds that life arose in the sea, and that biochemistry preferentially incorporated constituents that were abundant in seawater. For example, Banin and Navrot (1975) point out the striking correlation between the abundance of elements in today's biota and the solubility of elements in ocean water. Elements with low ionic potential (i.e., ionic charge/ionic radius) are found as soluble cations (Na^+, K^+, Mg^{2+}, and Ca^{2+}) in seawater and as important components of biochemistry. Other elements, including C, N, and S, that form soluble oxyanions in seawater (HCO_3^-, NO_3^-, SO_4^{2-}) are also abundant biochemical constituents. Molybdenum is much more abundant in biota than one might expect based on its crustal abundance; molybdenum forms the soluble molybdate ion (MoO_4^{2-}) in ocean water. In contrast, aluminum (Al) and silicon (Si) form insoluble hydroxides in seawater. They are found at low concentrations in living tissue, despite relatively high concentrations in the Earth's crust (Hutchinson 1943). Indeed, most elements that are rare in seawater are familiar poisons to living systems (e.g., Be, As, Hg, Pb, and Cd).

Although phosphorus forms a oxyanion, PO_4^{3-}, it may never have been particularly abundant in seawater, owing to its tendency to bind to other minerals (Griffith et al. 1977). Unique properties of phosphorus may account for its major role in biochemistry, despite its limited solubility in seawater and its relatively low geochemical abundance on Earth. With three ionized groups, phosphoric acid can link two nucleotides in DNA, with the third negative site acting to prevent hydrolysis and maintain the molecule within a cell membrane (Westheimer 1987). These ionic properties also allow phosphorus to serve in intermediary metabolism and energy transfer in ATP.

In sum, if one begins with the cosmic abundance of elements as an initial constraint, and the partitioning of elements during the formation of the Earth as subsequent constraints, then solubility in water appears to be a final constraint in determining the relative abundance of elements in the geochemical arena in which life arose. Those elements that were abundant in seawater are also important biochemical constituents. Phosphorus appears as an important exception—an important biochemical constituent that has been in short supply for much of the Earth's biosphere through geologic time.

Evolution of Metabolic Pathways

Awramik et al. (1983) report that 3.5-billion-year-old rocks collected in western Australia contain microfossils that may be our earliest direct evidence of life. Their specimens resemble some modern cyanobacteria (blue-green algae). The first living organisms on Earth may have resembled the Archaebacteria that survive today in anaerobic hydrothermal (volcanic) environments at temperatures above 90°C (Brock 1985). Archaebacteria are distinct from eubacteria due to a lack of a muramic acid component in the cell wall and to a distinct r-RNA sequence (Fox et al. 1980). Both halophilic (salt-tolerant) and thermophilic (heat-tolerant) forms of Archaebacteria are known. Huber et al. (1989) describe methanogenic Archaebacteria growing at 110°C near a deep sea hydrothermal vent in Italy. This environment may resemble that of some of the earliest habitats for life on Earth.

The most primitive metabolic pathway probably involved the production of methane by splitting simple organic molecules, such as acetate, that would have been present in the oceans from abiotic synthesis:

$$CH_3COOH \rightarrow CO_2 + CH_4. \tag{2.9}$$

Organisms using this metabolism were scavengers of the products of abiotic synthesis and obligate heterotrophs, sometimes classified as chemoheterotrophs. The modern fermenting bacteria in the order Methanobacteriales may be our best present-day analogues.

Longer pathways of anaerobic metabolism, such as glycolysis, probably followed with increasing elaboration and specificity of enzyme systems. Oxidation of simple organic molecules in anaerobic respiration was coupled to the reduction of inorganic substrates from the environment. For example, sometime after the appearance of methanogenesis from acetate splitting, methanogenesis by CO_2 reduction,

$$CO_2 + 4H_2 \rightarrow CH_4 + 2H_2O, \tag{2.10}$$

probably arose among early heterotrophic microorganisms. Generally this reaction occurs in two steps: eubacteria convert organic matter to acetate, H_2, and CO_2, and then Archaebacteria transform these to methane, following Eq. 2.10 (Wolin and Miller 1987). Note that methanogenesis by CO_2 reduction is more complicated than that from acetate splitting and would require a more complex enzymatic catalysis. Both pathways of methanogenesis are found among the fermenting bacteria that inhabit wetlands and coastal ocean sediments today (see Chapters 7 and 9). Recently, microbial communities performing methanogenesis by CO_2 reduction were found in basaltic rocks at a depth of 1000 m, where H_2 is available from geologic sources (Stevens and McKinley 1995). This microbial population is functionally isolated from the rest of the biosphere; it is perhaps analogous to one of the first microbial communities on Earth.

The sulfate-reducing pathway,

$$2CH_2O + 2H^+ + SO_4^{2-} \rightarrow H_2S + 2CO_2 + 2H_20, \quad (2.11)$$

is also found in Archaebacteria, and on the basis of the S-isotope ratios in preserved sediments, it appears to have arisen at least 2.4 billion years ago (bya) (Cameron 1982). Its later appearance may be related to the time needed to accumulate sufficient SO_4^{2-} in ocean waters to make this an efficient means of metabolism (Habicht and Canfield 1996). This biochemical pathway has recently been found in a group of thermophilic Archaebacteria isolated from the sediments of hydrothermal vent systems in the Mediterranean Sea, where a hot, anaerobic and acidic microenvironment may resemble the conditions of the primitive Earth (Stetter et al. 1987, Jørgensen et al. 1992, Elsgaard et al. 1994).

Before the advent of atmospheric O_2, the primitive oceans are likely to have contained low concentrations of available nitrogen—largely in the form of nitrate (NO_3^-; Kasting and Walker 1981, but see also Yung and McElroy 1979). Thus, the earliest organisms had limited supplies of nitrogen available for protein synthesis. There is little firm evidence that dates the origin of nitrogen fixation, in which certain bacteria break the inert, triple bond in N_2 and reduce the nitrogen to NH_3, but today this reaction is performed by bacteria that require strict local anaerobic conditions. The reaction,

$$N_2 + 8H^+ + 8e^- + 16ATP \rightarrow 2NH_3 + H_2 + 16ADP + 16P_i, \quad (2.12)$$

requires the enzyme complex known as nitrogenase, which consists of two proteins incorporating iron and molybdenum in their molecular structure (Georgiadis et al. 1992, Kim and Rees 1992). A cofactor, vitamin B_{12} that contains cobalt, is also essential (Jurgensen 1973, Palit et al. 1994). Nitrogen fixation requires the expenditure of large amounts of energy; breaking the N_2 bond requires 226 kcal/mole (Davies 1972). Modern nitrogen-fixing cyanobacteria couple nitrogen fixation to their photosynthetic reaction;

other nitrogen-fixing organisms are frequently symbiotic with higher plants (Chapter 6).

Despite various pathways of anaerobic metabolism, the opportunities for heterotrophic metabolism must have been limited in a world where organic molecules were only available as a result of abiotic synthesis. Natural selection would strongly favor autotrophic systems that could supply their own reduced, organic molecules for metabolism.

We might expect that the earliest photosynthetic reaction was based on sulfur, since the free energy of reaction, *G*, with hydrogen sulfide is less positive than that with water (Schidlowski 1983). This reaction,

$$CO_2 + 2H_2S \xrightarrow{\text{Sunlight}} CH_2O + 2S + H_2O, \tag{2.13}$$

was probably performed by sulfur bacteria, not unlike the anaerobic forms of green and purple sulfur bacteria of today. These bacteria would have been particularly abundant around shallow submarine volcanic emissions of reduced gases, including H_2S.

Several lines of evidence suggest that photosynthesis by sulfur bacteria and oxygen-producing photosynthesis by cyanobacteria were both found in ancient seas (Schopf 1993). Both forms of photosynthesis produce organic carbon in which ^{13}C is depleted relative to its abundance in dissolved bicarbonate (HCO_3^-), and there are no other processes known to produce such strong fractionations between the stable isotopes of carbon (Schidlowski 1988). Fossil organic matter with such depletion is found in rocks dating back to 3.8 bya (Mojzsis et al. 1996). This discrimination, which is about −2.8% (−28‰) in the dominant form of present-day photosynthesis, is based on the slower diffusion of $^{13}CO_2$ relative to $^{12}CO_2$ and the greater affinity of the carbon-fixation enzyme, ribulose bisphosphate carboxylase, for the more abundant $^{12}CO_2$ (Fig. 2.6, Chapter 5).

Evidence for oxygen-producing photosynthesis, in particular, is found in metamorphosed sedimentary rocks at least 3.5 billion years old, in which layered deposits of Fe_2O_3 are found in siliceous rocks known as chert. Under the anoxic conditions of the primitive Earth, Fe^{2+} released during rock weathering and from submarine hydrothermal emissions would be soluble and accumulate in seawater. With the advent of oxygen-producing photosynthesis, O_2 would be available to oxidize Fe^{2+} and deposit Fe_2O_3 in the sediments of the primitive ocean. These deposits are known as Banded Iron Formations (BIF), reaching a peak occurrence in rocks of 2.5 to 3.0 bya (Walker et al. 1983). Most of the major economic deposits of iron ore in the United States (Minnesota), Australia, and South Africa date to formations of this age (Meyer 1985). Other biotic mechanisms for the deposition of Fe_2O_3 are possible (e.g., Widdel et al. 1993), but massive, worldwide deposits of the Banded Iron Formation are generally taken as

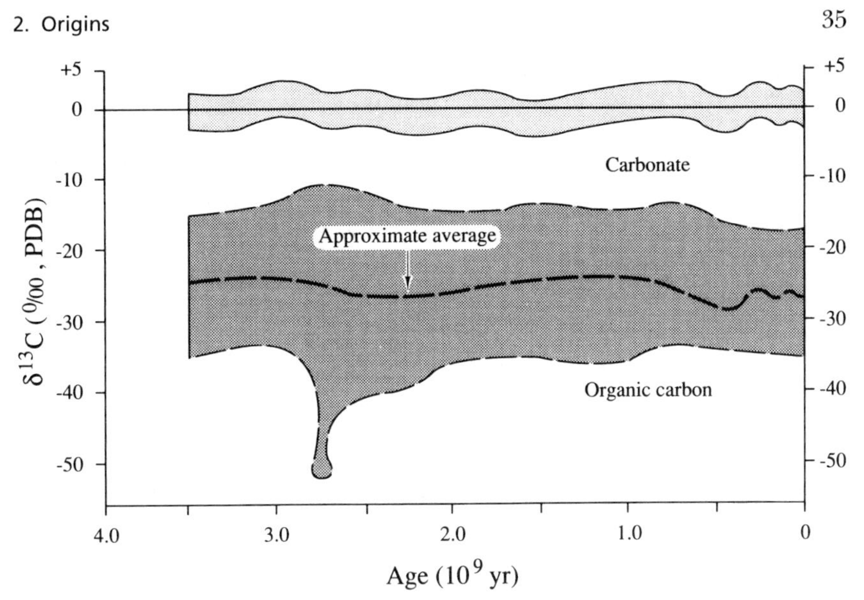

Figure 2.6 The isotopic composition of carbon in fossil organic matter and marine carbonates through geologic time, showing the range (shaded) among specimens of each age. The isotopic composition is shown as the ratio of ^{13}C to ^{12}C, relative to the ratio in an arbitrary standard (PDB belemite), which is assigned a ratio of 0. Carbon in organic matter is 2.8% less rich in ^{13}C than the standard, and this depletion is expressed as $-28‰$ $\delta^{13}C$ (see Chapter 5). From Schidlowski (1983).

evidence for the presence of oxygen-producing photosynthesis based on the photochemical splitting of water:

$$H_2O + CO_2 \xrightarrow{\text{Sunlight}} CH_2O + O_2\uparrow . \tag{2.14}$$

Despite the relatively large energy barrier inherent in the reaction, there must have been strong selection for photosynthesis based on the splitting of water, particularly as the limited supplies of H_2S in the primitive ocean were removed by sulfur bacteria (Schidlowski 1983). Water offered an inexhaustible supply of substrate for photosynthesis. Evidence of its occurrence in rocks 3.5 billion years old suggests that oxygen-producing photosynthesis was among the first metabolic pathways to evolve.

Despite the early evolution of oxygen-producing photosynthesis, the Earth's atmosphere seems to have remained anoxic until about 2.0 billion years ago. Until recently, most researchers attributed the lack of oxygen solely to its reaction with reduced iron (Fe^{2+}) in seawater and the deposition of Fe_2O_3 in Banded Iron Formations (Cloud 1973, François and Gérard 1986). Now it seems that the quantity of Fe^{2+} in the sea may have been insufficient to remove all the O_2 (Towe 1990, Kump and Holland 1992).

Oxidation of other reduced species, perhaps sulfide (S^{2-}), may have also played a role, accounting for the slow buildup of SO_4^{2-} in Precambrian seawater (Walker and Brimblecombe 1985). Several recent papers also postulate an early evolution of aerobic respiration, closely coupled to local sites of O_2 production, which may have held the concentration of O_2 at low levels (Towe 1990, Castresana and Saraste 1995). However, only when the oceans were swept clear of reduced substances could excess O_2 accumulate in seawater and diffuse to the atmosphere.

The first O_2 that reached the atmosphere was probably immediately involved in oxidation reactions with reduced atmospheric gases and with exposed crustal minerals of the barren land (Holland et al. 1989). Oxidation of reduced minerals, such as pyrite (FeS_2), would transfer SO_4^{2-} and Fe_2O_3 to the oceans in riverflow. Deposits of Fe_2O_3 that are found in alternating layers with other sediments of terrestrial origin constitute Red Beds, that are found beginning at 2.0 bya and indicate aerobic terrestrial weathering. It is noteworthy that the earliest occurrence of Red Beds roughly coincides—with little overlap—with the latest deposition of Banded Iron Formation, further evidence that the oceans were swept clear of reduced Fe before O_2 began to diffuse to the atmosphere.

Oxygen began to accumulate to its present-day atmospheric level of 21% when the rate of O_2 production by photosynthesis exceeded its rate of consumption by the oxidation of reduced substances. Atmospheric oxygen may have reached 21% as early as the Silurian—about 430 million years ago (see inside back cover), and it is not likely to have fluctuated outside the range of 15 to 35% ever since (Berner and Canfield 1989). What maintains the concentration at such stable levels? Walker (1980) examined all the oxidation/reduction reactions affecting atmospheric O_2, and suggested that the balance is due to the negative feedback between O_2 and the long-term net burial of organic matter in sedimentary rocks. If O_2 rises, less organic matter escapes decomposition, stemming a further rise in O_2. We will examine these processes in more detail in Chapters 3 and 11, but here it is interesting to note the significance of an atmosphere with 21% O_2. Lovelock (1979) points out that with $<15\%$ O_2 fires would not burn and at $>25\%$ O_2 even wet organic matter would burn freely (Watson et al. 1978). Either scenario would result in a profoundly different world than that of today.

The release of O_2 by photosynthesis is perhaps the single most significant effect of life on the geochemistry of the Earth's surface. The accumulation of free O_2 in the atmosphere has established the oxidation state for most of the Earth's surface for the last 2 billion years. However, of all the oxygen ever evolved from photosynthesis, only about 2% resides in the atmosphere today; the remainder is buried in various oxidized sediments, including Banded Iron formations and Red Beds (Fig. 2.7). The total inventory of free oxygen that has ever been released on the Earth's surface is, of course,

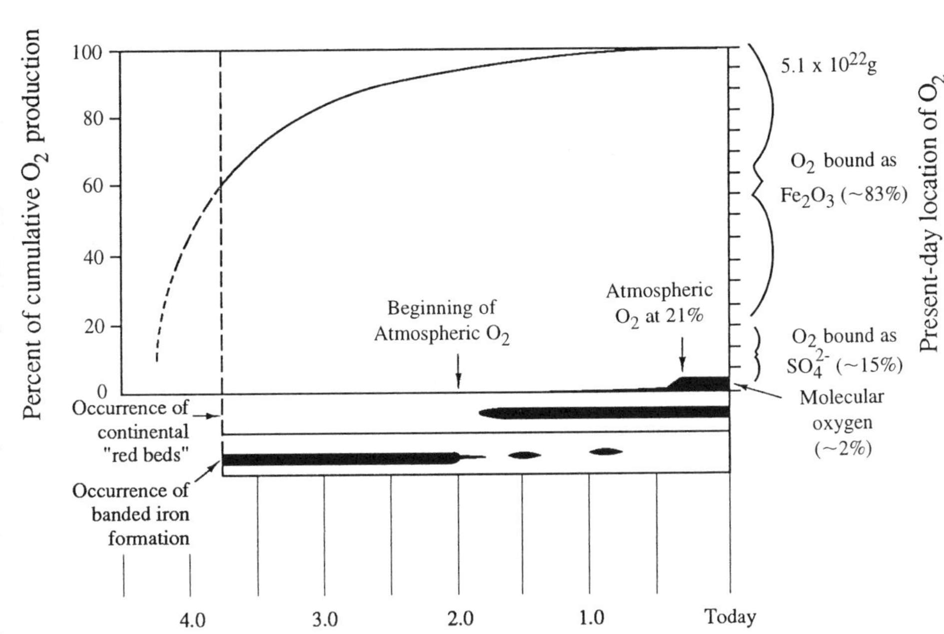

Figure 2.7 Cumulative history of O_2 released by photosynthesis through geologic time. Of more than 5.1×10^{22} g of O_2 released, about 98% is contained in seawater and sedimentary rocks, beginning with the occurrence of Banded Iron Formations at least 3.5 billion years ago (bya). Although O_2 was released to the atmosphere beginning about 2.0 bya, it was consumed in terrestrial weathering processes to form Red Beds, so that the accumulation of O_2 to present levels in the atmosphere was delayed to 400 mya. Modified from Schidlowski (1980).

balanced stoichiometrically by the storage of reduced carbon in the Earth's crust, including coal, oil, and other reduced compounds of biogenic origin (e.g., sedimentary pyrite). The sedimentary storage of organic carbon is now estimated at 1.56×10^{22} g (Des Marais et al. 1992; see also Table 2.2), representing the cumulative net production of *bio*geochemistry since the origin of life.

The release of free oxygen as a by-product of photosynthesis also dramatically altered the evolution of life on Earth. The pathways of anaerobic respiration by methanogenic bacteria and photosynthesis by sulfur bacteria are poisoned by O_2. These organisms generally lack catalase and have only low levels of superoxide dismutase—two enzymes that protect cellular structures from damage by highly oxidizing compounds such as O_2 (Fridovich 1975). Today these organisms are confined to local anoxic environments. Alternatively, eukaryotic metabolism is possible at O_2 levels that are about 1% of present levels (Berkner and Marshall 1965, Chapman and Schopf 1983). Fossil evidence of eukaryotic organisms is found in rocks of

1.7 to 1.9 billion years ago (Knoll 1992), and perhaps even as old as 2.1 billion years ago (Han and Runnegar 1992). The rate of evolution of amino acid sequences among major groups of organisms also suggests that prokaryotes and eukaryotes diverged 2 billion years ago (Doolittle et al. 1996). All these dates are generally consistent with the end of deposition of the Banded Iron Formation and the presence of O_2 in the atmosphere as indicated by Red Beds.

O_2 in the environment allowed eukaryotes to localize their heterotrophic respiration in mitochondria, providing an efficient means of metabolism and allowing a rapid proliferation of higher forms of life. Similarly, more efficient photosynthesis in the chloroplasts of eukaryotic plant cells presumably enhanced the production and further accumulation of atmospheric oxygen.

O_2 in the stratosphere is subject to photochemical reactions leading to the formation of ozone (Chapter 3). Today, stratospheric ozone provides an effective shield for much of the ultraviolet radiation from the Sun that would otherwise reach the Earth's surface and destroy most life. Before the O_3 layer developed, the earliest colonists on land may have resembled the microbes and algae that inhabit desert rocks of today (e.g., Friedmann 1982, Bell 1993). Although there is some fossil evidence for the occurrence of extensive microbial communities on land during the Precambrian (Horodyski and Knauth 1994), it is unlikely that higher organisms were able to colonize land abundantly until the ozone shield developed. Multicellular organisms are found in ocean sediments dating to about 680 million years ago (mya), but the colonization of land by higher plants was apparently delayed until the Silurian (Gensel and Andrews 1987). A proliferation of plants on land followed the development of lignified, woody tissues (Lowry et al. 1980) and the origin of effective symbioses with mycorrhizal fungi that allow plants to obtain phosphorus from unavailable forms in the soil (Pirozynski and Malloch 1975, Simon et al. 1993; Chapter 6).

Oxygen also allowed the evolution of several new biochemical pathways of critical significance to the global cycles of biogeochemistry. Two forms of aerobic biochemistry constitute chemoautotrophy. One based on sulfur or H_2S,

$$2S + 2H_2O + 3O_2 \rightarrow 2SO_4^{2-} + 4H^+, \quad (2.15)$$

is performed by various species of *Thiobacilli* (Ralph 1979). The protons generated are coupled to energy-producing reactions, including the fixation of CO_2 to organic matter. On the primitive Earth, these organisms could capitalize on elemental sulfur deposited from anaerobic photosynthesis (Eq. 2.13), and today they are found in local environments where elemental sulfur or H_2S is present, including some deep-sea hydrothermal vents (Chapter 9).

Also important are the reactions involving nitrogen transformations by *Nitrosomonas* and *Nitrobacter* bacteria:

$$2NH_4^+ + 3O_2 \rightarrow 2NO_2^- + 2H_2O + 4H^+ \quad (2.16)$$

and

$$2NO_2^- + O_2 \rightarrow 2NO_3^-. \quad (2.17)$$

These reactions constitute *nitrification,* and the energy released is coupled to the fixation of carbon by these chemoautotrophic bacteria. The nitrate produced by these reactions is highly soluble in water, and it is the dominant form of inorganic nitrogen delivered in riverflow to the oceans (Chapters 8 and 12).

Today, an anaerobic, heterotrophic reaction called *denitrification* is performed by bacteria, commonly of the genus *Pseudomonas,* found in soils and wet sediments (Knowles 1982). Although the denitrification reaction,

$$5CH_2O + 4H^+ + 4NO_3^- \rightarrow 2N_2 + 5CO_2 + 7H_2O, \quad (2.18)$$

requires anoxic environments, denitrifiers are only facultatively anaerobic. Several lines of evidence suggest that denitrification may have appeared later than the strictly anaerobic pathways of methanogenesis and sulfate reduction (Betlach 1982). Most denitrifiers such as *Pseudomonas* are found among the eubacteria, which appear more advanced than Archaebacteria. Moreover, denitrification would have been efficient only after relatively high concentrations of NO_3 had accumulated in the primitive ocean, which is likely to have contained low NO_3 at the start (Kasting and Walker 1981). Thus, the evolution of denitrification may have been delayed until sufficient O_2 was present in the environment to drive the nitrification reactions (Eqs. 2.16 and 2.17). It is interesting to note that having evolved in a world dominated by O_2, the enzymes of today's denitrifying organisms are not destroyed, but merely inactivated, by O_2 (Bonin et al. 1989, McKenney et al. 1994). Indeed, most denitrifying organisms switch to aerobic respiration when O_2 is present.

All metabolic activity is based on the transfer of electrons between oxidized and reduced substances that organisms isolate from their environment (Fig. 2.8). The diversity of metabolic reactions outlined in this chapter shows how evolution has capitalized on the wide variety of the electron transfers that are possible in different environments—largely determined by the presence or absence of O_2. However, the reactions are linked—oxidized products of one pathway become substrates for reduction in another. For example, denitrification, the anaerobic metabolism that pro-

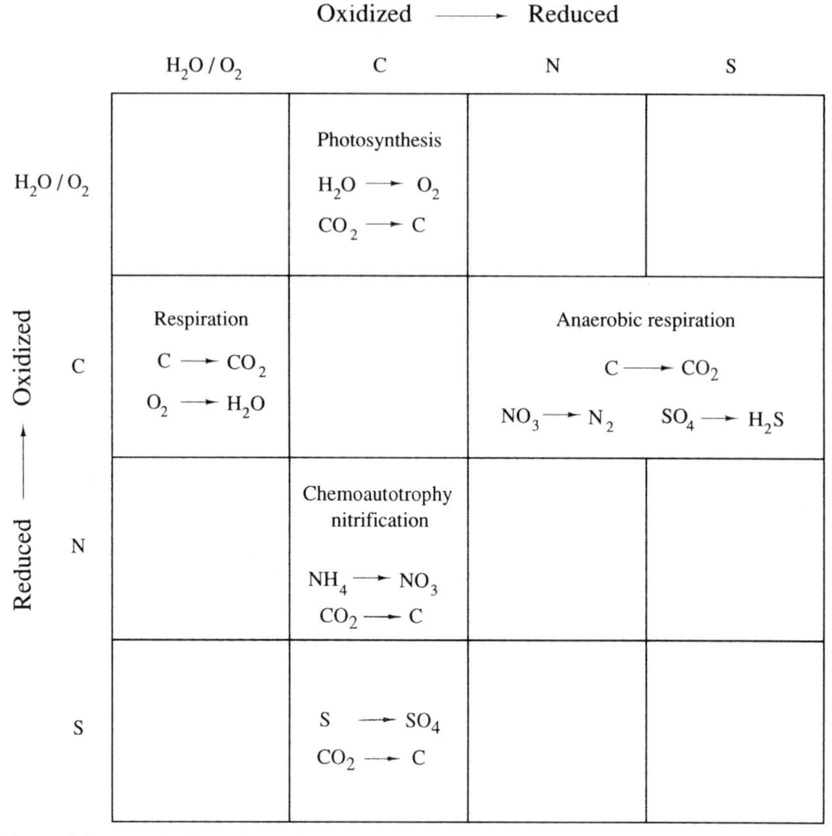

Figure 2.8 Metabolic pathways that couple oxidations of C, N, and S on the Earth's surface. For each pathway, the constituent at the top is transformed from an oxidized form obtained from the environment to a reduced form, released to the environment. At the same time, the constituent at left is transformed from a reduced form to an oxidized form. Modified from Schlesinger (1989).

duces N_2 in wetland soils, depends on the availability of NO_3^- by microbial nitrification, an aerobic reaction occurring in adjacent, well-drained soils. Throughout this text, we will see that global biogeochemical cycles are driven by transformations of substances between oxidized and reduced forms in different environments.

Comparative Planetary History: Earth, Mars, and Venus

In the release of free O_2 to the atmosphere, life has profoundly affected the conditions on the surface of the Earth. But what might have been the conditions on Earth in the absence of life? Some indication is given by our neighboring planets, Mars and Venus, which are the best replicates of the underlying geochemical arena on Earth. We are fairly confident that there has never been life on these planets, so their surface composition represents the cumulative effect of 4.6 billion years of abiotic processes.

Table 2.3 compares a number of properties and conditions on Earth, Mars and Venus. Two properties characterize the atmosphere of these planets: the total mass (or pressure) and the proportional abundance of the constituents. The present atmosphere on Mars is only about 0.76% as massive as that on Earth (Hess et al. 1976). We might expect a less massive atmosphere on Mars than on Earth, because the gravitational field is weaker on a smaller planet. Mars probably began with a smaller allocation of primary gases during planetary formation, and we might expect that a small planet would retain less internal heat to drive tectonic activity and crustal outgassing after its formation (Anders and Owen 1977, Owen and Biemann 1976). Indeed, Mars shows only limited evidence of volcanic activity during the past two billion years, and estimates of the cumulative generation of magma are substantially lower for Mars (0.17 km^3/yr) than for Earth (26–34 km^3/yr) or Venus (<19 km^3/yr) (Greeley and Schneid 1991).

We might also expect that the surface temperature on Mars would be colder than that on Earth, because the planet is much farther from the Sun. The average temperature on Mars, −53°C at the site of the 1976 Viking landing (Kieffer 1976), ensures that water is frozen over most of the Martian surface at all seasons. In the absence of liquid water, one might expect that the atmosphere on Mars would be mostly dominated by CO_2, which readily dissolves in seawater on Earth (Eq. 2.4). Indeed, CO_2 constitutes a major proportion of the thin atmosphere of Mars, and the observed fluctuations of the Martian ice cap appear due to seasonal variations in the amount of CO_2 that is frozen out of its atmosphere (James et al. 1992).

Table 2.3 Some Characteristics of the Inner Planets

	Mars[a]	Earth	Venus[b]
Distance to the sun (10^6 km)	228	150	108
Surface temperature (°C)	−53	16	474
Radius (km)	3390	6371	6049
Atmospheric pressure (bars)	0.007	1	92
Atmospheric mass (g)	2.4×10^{19}	5.1×10^{21}	5.3×10^{23}
Atmospheric composition (% wt.)			
CO_2	95	0.036	98
N_2	2.5	78	2
O_2	0.25	21	0
H_2O	0.10	<1	0.05

[a] From Owen and Biemann (1976).
[b] From Nozette and Lewis (1982).

Several attributes of Mars are anomalous. First, with most of the water and CO_2 now trapped on the surface, why is N_2 such a minor component of the atmosphere on Mars? Second, why do the surface conditions on Mars indicate a period when liquid water was most certainly present on its surface (Carr 1987, V.R. Baker et al. 1991, Zent 1996)? Could it be that a more massive early atmosphere may once have allowed a significant "greenhouse effect" on Mars and warmer surface temperatures than today (Pollack et al. 1987)? Such a scenario could explain an early abundance of liquid water, but if it is correct, why did Mars lose its atmosphere and cool to its present surface temperature of $-53°C$?

Losses of atmospheric gases from Mars may have resulted from several processes. A thick atmosphere on Mars may have been lost to space as a result of catastrophic impacts during its early history (Carr 1987, Melosh and Vickery 1989) or by a process known as "sputtering" driven by solar wind (Kass and Yung 1995, Hutchins and Jakosky 1996). Impacts are consistent with the low abundance of noble gases on Mars relative to the concentrations in carbonaceous chondrites and the Sun (Hunten 1993). Catastrophic loss of an early atmosphere is also consistent with the observation that nearly all of the present-day atmosphere on Mars is secondary, as evidenced by the $^{40}Ar/^{36}Ar$ ratio of 2750 (Owen and Biemann 1976, Owen et al. 1977).

Loss of water from Mars may have also occurred as the water vapor in its atmosphere underwent photolysis by ultraviolet light. Observations of analogous processes on Earth are instructive. In the upper atmosphere on Earth, small amounts of water vapor are subject to photodisassociation, with the loss of H_2 to space. However, because the upper atmosphere is cold, little water vapor is present, and the process has been minor throughout the Earth's history. If this process were significant on Mars, we would expect that the loss of ^1H would be more rapid than that of ^2H, leaving a greater proportion of ^2H_2O in the planetary inventory. Owen et al. (1988) found that the ratio of ^2H (deuterium) to ^1H on Mars is about 6× that on Earth, suggesting that Mars may have once possessed a large inventory of water that has been lost to space. Although a small amount of O_2 is found in the Martian atmosphere (Table 2.3), most of the oxygen produced from the photolysis of water has probably oxidized minerals of the crust, viz.,

$$4FeO + O_2 \rightarrow 2Fe_2O_3, \tag{2.19}$$

giving Mars its reddish color. A small amount of oxygen has also been detected in the atmosphere of Jupiter's moon Europa, where it appears to originate from a similar photolytic process (Hall et al. 1995).

Nitrogen may have also been lost from Mars as N_2 underwent photodisassociation in the upper atmosphere, forming monomeric N. This process occurs on Earth as well, but even N is too heavy to escape the Earth's gravitational field and quickly recombines to form N_2. With its smaller size, Mars allows the loss of N. Relative to the Earth, a higher proportion of $^{15}N_2$ in the Martian atmosphere is suggestive of this process, since the escape of ^{15}N would be slower than that of ^{14}N, which has a lower atomic weight (McElroy et al. 1976).

With losses of H_2O and N_2 to space, it is not surprising that the Martian atmosphere is dominated by CO_2. What is surprising is that the atmospheric mass is so low. As much as 3 bars of CO_2 may have been degassed from the interior of Mars, but only about 10% of that amount appears to be frozen in the polar ice caps and the soil (Kahn 1985). Some CO_2 may have been lost to space (Kass and Yung 1995), but during an earlier period of moist conditions, CO_2 may have also reacted with the crust of Mars, weathering rocks and forming carbonate minerals on its surface (Pollack et al. 1990, Romanek et al. 1994). With the loss of tectonic activity on Mars, there was no mechanism to release this CO_2 back to the atmosphere, as we have on Earth (cf. Fig. 1.4).

In sum, various lines of evidence suggest that Mars had a higher inventory of volatiles early in its history, but most of the atmosphere has been lost to space or in reactions with its crust. The presence of water on Mars may have once offered an environment conducive to the evolution of life (Davis et al. 1996), although evidence for life on Mars is rather scant (McKay et al. 1996). The loss of water from Mars would remove a large component of greenhouse warming from the planet. The thin atmosphere that remains is dominated by CO_2, but it offers little greenhouse warming—raising the temperature of Mars only about 10°C over what might be seen if Mars had no atmosphere at all (Houghton 1986). Our best estimates suggest that the volume of CO_2 now frozen in the polar ice caps (100 mbar) and soils (300 mbar) on Mars is insufficient to supply the 2 bars of atmospheric CO_2 that would be necessary for the greenhouse effect to raise the temperature of the planet above the freezing point of water (Pollack et al. 1987). Thus, it would be difficult to use planetary-level engineering to establish a large, self-sustained greenhouse effect on Mars, allowing humans to colonize the planet (McKay et al. 1991).

On Venus, the ratio of the mass of the atmosphere to the mass of the planet (1.09×10^{-4}) is slightly less than the ratio of the total mass of volatiles on Earth (Table 2.2) to the mass of the Earth (3.3×10^{-4}). These values suggest a similar degree of crustal degassing on these planets. Unlike the Earth, the high surface temperature of 474°C on Venus ensures that its present inventory of volatiles resides entirely in its atmosphere. Indeed, the atmospheric pressure on Venus is nearly 100X that of Earth (Table 2.3).

The massive atmosphere on Venus is dominated by CO_2, conferring a large greenhouse warming and surface temperatures well in excess of that predicted for a nonreflective body at the same distance from the Sun (54°C; Houghton 1986).[3] The relative abundance of CO_2 and N_2 in the atmosphere of Venus is similar to that in the total inventory of volatiles on Earth (Oyama et al. 1979, Pollack and Black 1982). What is unusual about Venus is the low abundance of water in its atmosphere. Was Venus wet in the past?

The ratio of ^2H (deuterium) to ^1H on Venus is >100× higher than that on Earth (Donahue et al. 1982, McElroy et al. 1982, de Bergh et al. 1991), suggesting that Venus, like Mars, may have possessed a large inventory of water in the past, but lost water through a process that differentiates between the isotopes of hydrogen. With the warm initial conditions on Venus, a large amount of the water vapor in the atmosphere may have been subject to photodisassociation, causing the planet to dry out through its history (Kasting et al. 1988). The oxygen released during the photodisassociation of water has probably reacted with crustal minerals (Donahue et al. 1982).

At the surface temperatures found on Venus, little CO_2 can react with its crust (cf. Fig. 1.4), so high concentrations of CO_2 remain in the atmosphere (Nozette and Lewis 1982). Various other gases, such as SO_2, that are found dissolved in seawater on Earth also reside as gases in the atmosphere on Venus (Oyama et al. 1979). Continuing volcanic releases of CO_2 have accumulated in the atmosphere to produce a runaway greenhouse effect in which increasing temperatures allow an increasing potential for the atmosphere to hold CO_2 and other gases (Walker 1977). Thus, the current temperature on Venus, 474°C, is much greater than we would predict if Venus had no atmosphere.

Certainly the most unusual characteristic of the Earth's atmosphere is the presence of large amounts of O_2, which is an unequivocal indication of life on this planet (Sagan et al. 1993). Having examined the conditions on Mars and Venus, we can now offer some speculation on the conditions that might exist on a lifeless Earth. At a distance of 150×10^6 km from the Sun, the surface temperature on Earth, assuming no reflectivity to incoming solar radiation, would be close to the freezing point of water (Houghton 1986). Such cold conditions would seem to ensure that the atmosphere on Earth has never contained much water vapor, so little water has been lost to space as a result of photolysis in the upper atmosphere. Despite the small amount of H_2O in Earth's atmosphere, the atmosphere confers enough greenhouse warming to the planet to have maintained liquid oceans for most of its history. Thus, even on a lifeless Earth, most

[3] The widely cited report of the Intergovernmental Panel on Climate Change (IPCC) indicates that the surface temperature on Venus in the absence of a greenhouse effect would be −47°C (Houghton et al. 1990). This is lower than the value given here because the IPCC report accounts for the reflectivity of the thick cloud layer on Venus, whereas my value considers the equilibrium temperature for a black body absorber in the orbit of Venus.

of the inventory of volatiles would reside in the oceans. The atmosphere on a lifeless Earth would be dominated by N_2, which is only slightly soluble in water. The size of Earth and its gravitational field ensure that photolysis of N_2 does not result in the loss of N from the planet. Moreover, the rate of fixation of nitrogen by lightning in an atmosphere without O_2 appears too low to remove a significant portion of N_2 from the atmosphere (Kasting and Walker 1981, Chapter 12). Thus, the main effect of life has been to dilute the initial nitrogen-rich atmosphere on Earth with a large quantity of O_2 (Walker 1984).

How long will the Earth be hospitable to life? Barring some unforeseen catastrophe wrought by humans, we can speculate that the biosphere will persist as long as our planet harbors liquid water on its surface. Eventually, however, a gradual increase in the Sun's luminosity will warm the Earth, causing a photolytic loss of water from the upper atmosphere and the demise of life—perhaps after another 2.5 billion years (Lovelock and Whitfield 1982, Caldeira and Kasting 1992).

Summary

In this chapter we have reviewed theories for the formation and differentiation of the early Earth. In the process of planetary formation, certain elements were concentrated near its surface and only some elements were readily soluble in seawater. Thus, the environment in which life arose is a special mix taken from the geochemical abundance of elements that was available on Earth. Simple organic molecules can be produced by physical processes in the laboratory; presumably similar reactions occurred in the oceans on the primitive Earth. Life may have arisen by the abiotic assembly of these constituents into simple forms, resembling the most primitive bacteria that we know of today. Essential to living systems is the processing of energy, which is likely to have begun with the heterotrophic consumption of molecules found in the environment. A persistent scarcity of such molecules is likely to have led to selection for the autotrophic production of energy by various pathways, including photosynthesis. Autotrophic photosynthesis appears to be responsible for nearly all the production of O_2, which has accumulated in the Earth's atmosphere over the last 2 billion years. The major biogeochemical cycles on Earth are mediated by organisms, whose metabolic activities couple the oxidation and reduction of substances isolated from the environment.

Recommended Readings

Bengtson, S. (ed.). 1994. *Early Life on Earth.* Columbia University Press, New York.

Brown, G.C. and A.E. Mussett. 1981. *The Inaccessible Earth.* Unwin Hyman Publishers, London.

Cox, P.A. 1989. *The Elements.* Oxford University Press, Oxford.

Faure, G. 1991. *Principles and Applications of Inorganic Geochemistry.* Macmillan, New York.

Frausto de Silva, J.J.R. and R.J.P. Williams. 1991. *The Biological Chemistry of Life.* Oxford University Press, Oxford.

Holland, H.D. 1984. *The Chemical Evolution of the Atmosphere and Oceans.* Princeton University Press, Princeton, New Jersey.

Lewis, J.S. and R.G. Prinn. 1984. *Planets and Their Atmospheres.* Academic Press, Orlando.

Schopf, J.W. and C. Klein. (eds.). 1992. *The Proterozoic Biosphere: A Multidisciplinary Study.* Cambridge University Press, Cambridge.

Space Studies Board. 1990. *The Search for Life's Origins.* National Academy Press, Washington, D.C.

3

The Atmosphere

Introduction

There are several reasons to begin our treatment of biogeochemistry with a consideration of the atmosphere. The atmosphere has evolved as a result of the history of life on Earth (Chapter 2), and there is good evidence that it is now changing rapidly as a result of human activities. The atmosphere controls Earth's climate and ultimately determines the environment in which we live. Further, the atmosphere is relatively well mixed, so changes in its composition can be taken as a first index of changes in biogeochemical processes at the global level. The circulation of the atmosphere transports biogeochemical constituents between land and sea, resulting in a global circulation of elements.

We will begin our discussion with a brief consideration of the structure, circulation, and composition of the atmosphere. Then, we will examine reactions that occur among various gases, especially in the lower atmosphere. Many of these reactions remove constituents from the atmosphere, depositing them on the surface of the land and sea. In the face of constant losses, the composition of the atmosphere is maintained by biotic processes that supply gases to the atmosphere. We will mention the sources of atmospheric gases here briefly, but they will be treated in more detail in later chapters of this book, especially as we examine the microbial reactions that occur in soils, wetlands, and ocean sediments.

Structure and Circulation

The atmosphere is held on the Earth's surface by the gravitational attraction of the Earth. At any altitude, the downward force (F) is related to the mass (M) of the atmosphere above that point,

$$F = M(g), \tag{3.1}$$

where g is the acceleration due to gravity (980 cm/sec^2 at sea level). Pressure (force per unit area) decreases with increasing altitude because the mass of the overlying atmosphere is smaller (Walker 1977). The decline in atmospheric pressure (P in bars) with altitude (A in km) is approximated by the logarithmic relation

$$\log P = -0.06(A), \tag{3.2}$$

over the whole atmosphere (Fig. 3.1).

Although the chemical composition of the atmosphere is relatively uniform, when we visit high mountains, we often say that the atmosphere seems "thinner" than at sea level. The abundance of molecules in each volume of the atmosphere is greater at sea level, because it is compressed by the pressure of the overlying atmosphere. Thus, the lower atmosphere, the *troposphere,* contains about 80% of the atmospheric mass (Warneck 1988), and jet aircraft flying at high altitudes require cabin pressurization for their passengers.

Certain atmospheric constituents, such as ozone, absorb portions of the radiation that the Earth receives from the Sun, so only about half of the Sun's radiation penetrates the atmosphere to be absorbed at the Earth's surface (Fig. 3.2). The land and ocean surfaces reradiate long-wave (heat) radiation to the atmosphere, so the atmosphere is heated from the bottom and is warmest at the Earth's surface (Fig. 3.1). Because warm air is less

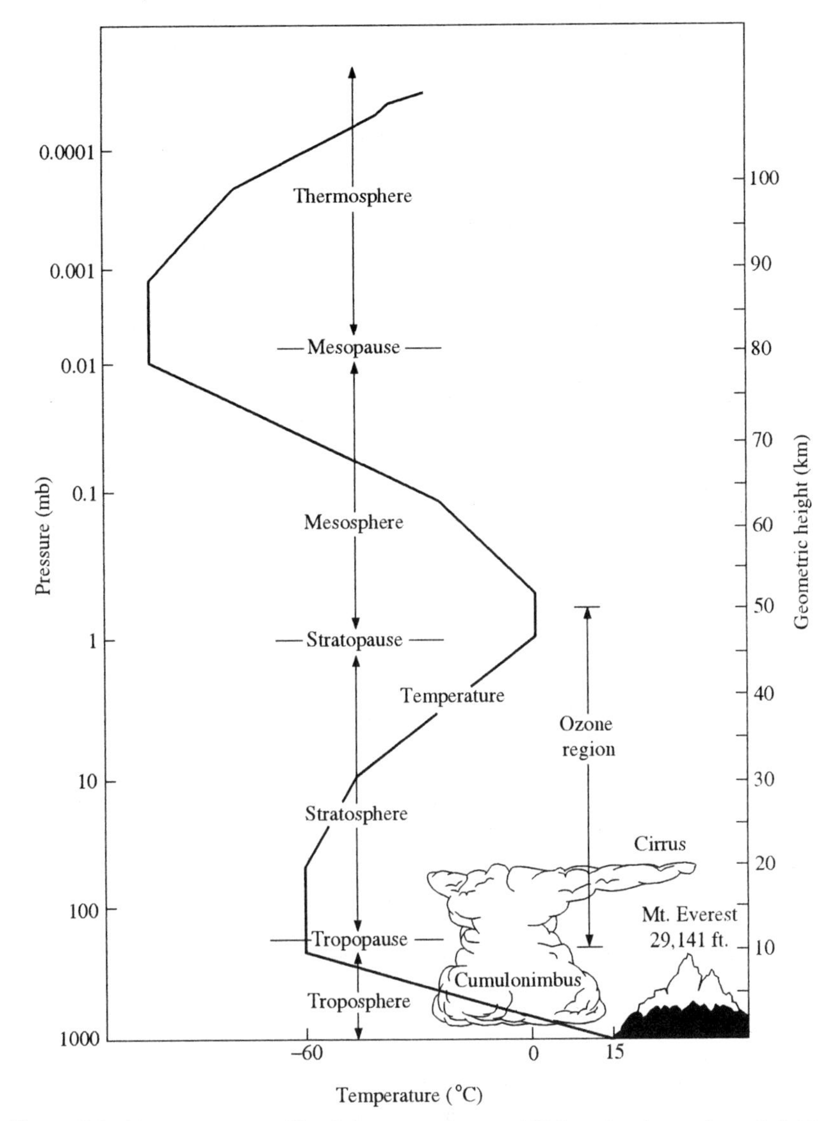

Figure 3.1 Temperature profile of the atmosphere to 100 km, showing major subdivisions of the atmosphere. Note the logarithmic decline in pressure (left-hand axis) as a function of altitude.

dense and rises, the troposphere is well mixed. The top of the troposphere extends to 10–15 km, varying seasonally and with latitude. The temperature of the upper troposphere is about −60°C, which ensures that the atmosphere above 10 km contains only small amounts of water vapor.

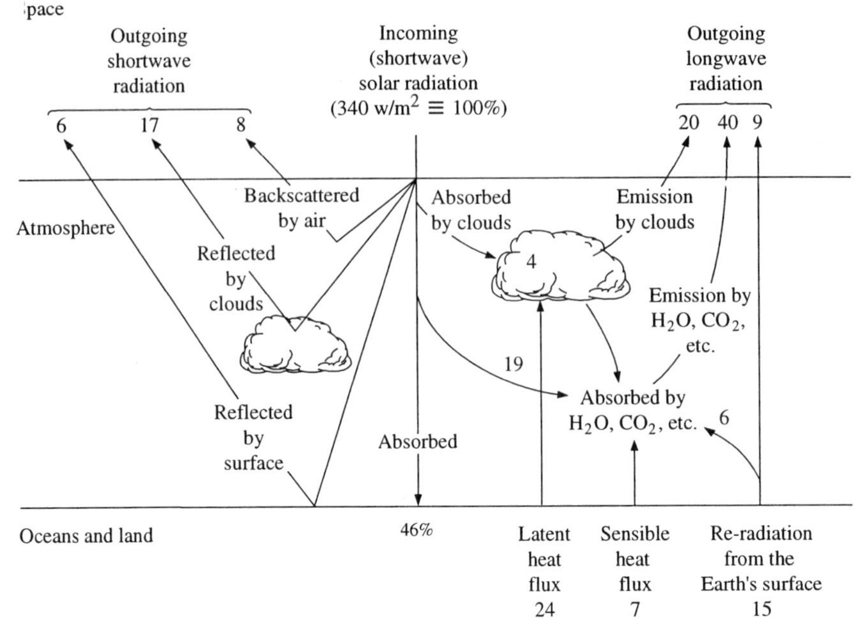

Figure 3.2 The radiation budget for Earth, showing the proportional fate of the energy that the Earth receives from the Sun. Each year, the Earth receives about 340 W/m^2 of radiation, mostly in short wavelengths. About a third of this radiation is reflected back to space, and the remainder is absorbed by the atmosphere (23%) or the surface (46%). Long-wave (infrared) radiation is emitted from the Earth's surface, some of which is absorbed by atmospheric gases, warming the atmosphere (the greenhouse effect). The atmosphere emits long-wave radiation, so that the total energy received is balanced by the total energy emitted from the planet. Modified from MacCracken (1985).

Above the troposphere, the stratosphere is defined by the zone in which temperatures increase with altitude, extending to about 50 km (Fig. 3.1). The increase is largely due to the absorption of ultraviolet light by ozone. Vertical mixing in the stratosphere is limited, as is exchange across the boundary between the troposphere and the stratosphere, the *tropopause*. Thus, materials that enter the stratosphere remain there for long periods, allowing transport around the globe.

The thermal mixing of the troposphere is largely responsible for the global circulation of the atmosphere, as well as local weather patterns (Figs. 3.3a and 3.3b). The large annual receipt of solar energy at the equator causes warming of the atmosphere (sensible heat) and the evaporation of large amounts of water, carrying latent heat, from tropical oceans and rainforests. As this warm, moist air rises, it cools, producing a large amount of precipitation in equatorial regions. Having lost its moisture, the rising air mass moves both north and south, away from the equator. In a belt centered on approximately 30° N or S latitude, these dry air masses sink to the Earth's surface, undergoing compressional heating. Most of the world's major deserts are associated with the downward movement of hot, dry air at this latitude. A similar, but much weaker, circulation pattern is found at the poles, where cold air sinks and moves north or south along the Earth's surface to lower latitudes. Known as Direct Hadley cells, these tropical and polar circulation patterns drive an indirect circulation in each hemisphere between 40° and 60° latitude, producing the regional storm systems and the prevailing west winds that we experience in the temperate zone (Fig. 3.3c).

The tropospheric air in each hemisphere mixes on a time scale of a few months (Warneck 1988). Each year, there is also complete mixing of tropospheric air between the northern and the southern hemispheres across the intertropical convergence zone (ITCZ). If a gas shows a higher concentration in one hemisphere, we can infer that a large natural or human source must exist in that hemisphere, overwhelming the tendency for atmospheric mixing to equalize the concentrations (Fig. 3.4).

Exchange between the troposphere and the stratosphere is driven by several processes (Warneck 1988). In the tropical Hadley cells, rising air masses carry some tropospheric air to the stratosphere (Holton et al. 1995). The strength of the updraft varies seasonally, as a result of variations in the radiation received from the Sun. When the height of the tropopause drops, tropospheric air is trapped in the stratosphere, or vice versa. There is also exchange across the tropopause due to large-scale wind movements (Appenzeller and Davies 1992), thunderstorms (Dickerson et al. 1987), and eddy diffusion (Warneck 1988).

Exchange between the troposphere and the stratosphere has been examined by following the fate of industrial pollutants released to the troposphere and radioactive contaminants released to the stratosphere by tests of atomic weapons during the 1950s and early 1960s (Warneck 1988). In these considerations, the concept of mean residence time is useful. For any biogeochemical reservoir, mean residence time (MRT) is defined as

$$\text{MRT} = \text{Mass}/\text{flux}, \tag{3.3}$$

where flux may be either the input or the loss from the reservoir.[1] The input of tropospheric air to the stratosphere amounts to about 75% of the stratospheric mass each year, leading to a mean residence time of 1.3 years for stratospheric air (Warneck 1988). Thus, if a large volcano injects sulfur

[1] Assuming exponential decay of a tracer from a reservoir that is in steady state, the fractional loss per year ($-k$) is equal to the reciprocal of the mean residence time in years, i.e., 1/MRT. The amount remaining in the reservoir at any time t (in years) as a fraction of the original content is equal to e^{-kt}, the half-life of the reservoir in years is $0.693/k$, and 95% will have disappeared from the reservoir after $3/k$ years.

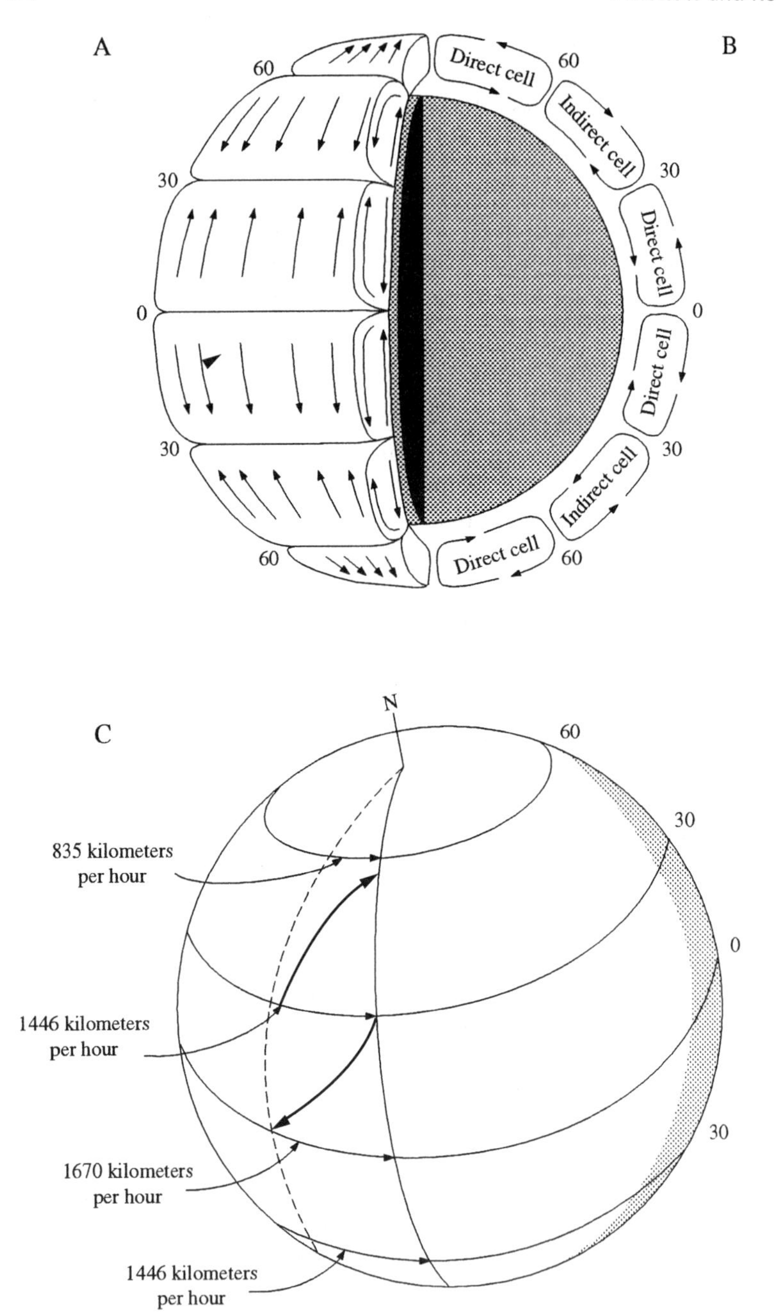

Figure 3.3 Generalized pattern of global circulation showing (a) surface patterns, (b) vertical patterns, and (c) origin of the Coriolis force. As air masses move across different latitudes, they are deflected by the Coriolis force, which arises because of the different speeds of the Earth's rotation at different latitudes. For instance, if you were riding on an air mass moving

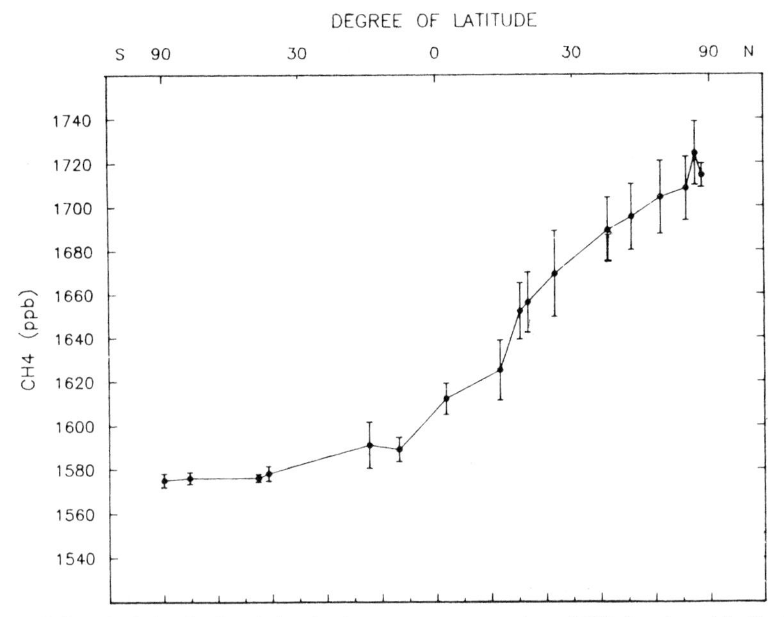

Figure 3.4 The latitudinal variation in the mean concentration of CH_4 (methane) in Earth's atmosphere. From Steele et al. (1987).

dioxide into the stratosphere, about half will remain after 1 year and about 5% will remain after 4 years.

Atmospheric Composition

Gases

Table 3.1 gives the globally averaged concentration of some important gases in the atmosphere. Three gases—nitrogen, oxygen, and argon—make up 99% of the atmospheric mass of 5.14×10^{21} g (Trenberth and Guillemot 1994). These gases are relatively unreactive; their mean residence times in the atmosphere are much longer than the rate of atmospheric mixing. Thus, the concentrations of N_2, O_2, and all noble gases (He, Ne, Ar, Kr, and Xe) are nearly uniform globally.

---

at a constant speed south from 30° N, you would begin your journey seeing 1446 km of the Earth's surface pass to the east every hour. By the time your air mass reached the equator, 1670 km would be passing to the east each hour. While moving south at a constant velocity, you would find that you had traveled 214 km west of your expected trajectory. The Coriolis force means that all movements of air in the northern hemisphere are deflected to the right; those in the southern hemisphere are deflected to the left. Modified from Oort (1970) and Gross (1977).

Table 3.1 Global Average Concentration of Well-Mixed Atmospheric Constituents[a]

Compounds	Formula	Concentration	Total mass (g)
Major constituents (%)			
Nitrogen	N_2	78.084	3.87×10^{21}
Oxygen	O_2	20.946	1.19×10^{21}
Argon	Ar	0.934	6.59×10^{19}
Parts-per-million constituents (ppm = 10^{-6})			
Carbon dioxide	CO_2	360	2.80×10^{18}
Neon	Ne	18.2	6.49×10^{16}
Helium	He	5.24	3.70×10^{15}
Methane	CH_4	1.75	4.96×10^{15}
Krypton	Kr	1.14	1.69×10^{16}
Parts-per-billion constituents (ppb = 10^{-9})			
Hydrogen	H_2	510	1.82×10^{14}
Nitrous oxide	N_2O	311	2.42×10^{15}
Xenon	Xe	87	2.02×10^{15}
Parts-per-trillion constituents (ppt = 10^{-12})			
Carbonyl sulfide	COS	500	5.30×10^{12}
Chlorofluorocarbons			
CFC 11	CCl_3F	280	6.79×10^{12}
CFC 12	CCl_2F_2	550	3.12×10^{13}
Methylchloride	CH_3Cl	620	5.53×10^{12}
Methylbromide	CH_3Br	11	1.84×10^{11}

[a] Those with a mean residence time >1 year. Assuming a dry atmosphere with a molecular weight of 28.97. The overall mass of the atmosphere sums to 514×10^{19} g (Trenberth and Guillemot 1994).

Several hundred trace gases, including a wide variety of volatile hydrocarbons (Greenberg and Zimmerman 1984, Chameides et al. 1992), have also been identified in the Earth's atmosphere. Most of these gases have short mean residence times, so it is not surprising that they are minor constituents in the atmosphere. The concentration of such gases varies in space and time. For instance, we expect high concentrations of certain pollutants (ozone, carbon monoxide, etc.) over cities, and high concentrations of some reduced gases (methane and hydrogen sulfide) over swamps and other areas of anaerobic decomposition (e.g., Harriss et al. 1982, Steudler and Peterson 1985). Winds mix the concentrations of these gases to their lower, average tropospheric background concentration within a short distance downwind of local sources. Thus, we can best perceive global changes in atmospheric composition, such as the current increase in CH_4, by making long-term measurements in remote locations.

Junge (1974) related geographic variations in the atmospheric concentration of various gases to their estimated mean residence time in the atmosphere (Fig. 3.5). Gases that have short mean residence times are highly variable from place to place, whereas those that have long mean residence times, relative to atmospheric mixing, show relatively little variation. For example, the average volume of water in the atmosphere is equivalent to about 13,000 km^3 at any time, or 25 mm above any point on the Earth's surface (Speidel and Agnew 1982). The average daily precipitation would be about 2.7 mm, if it were deposited evenly around the globe. Thus, the mean residence time for water vapor in the atmosphere is

$$25 \text{ mm}/2.7 \text{ mm day}^{-1} = 9.3 \text{ days}. \quad (3.4)$$

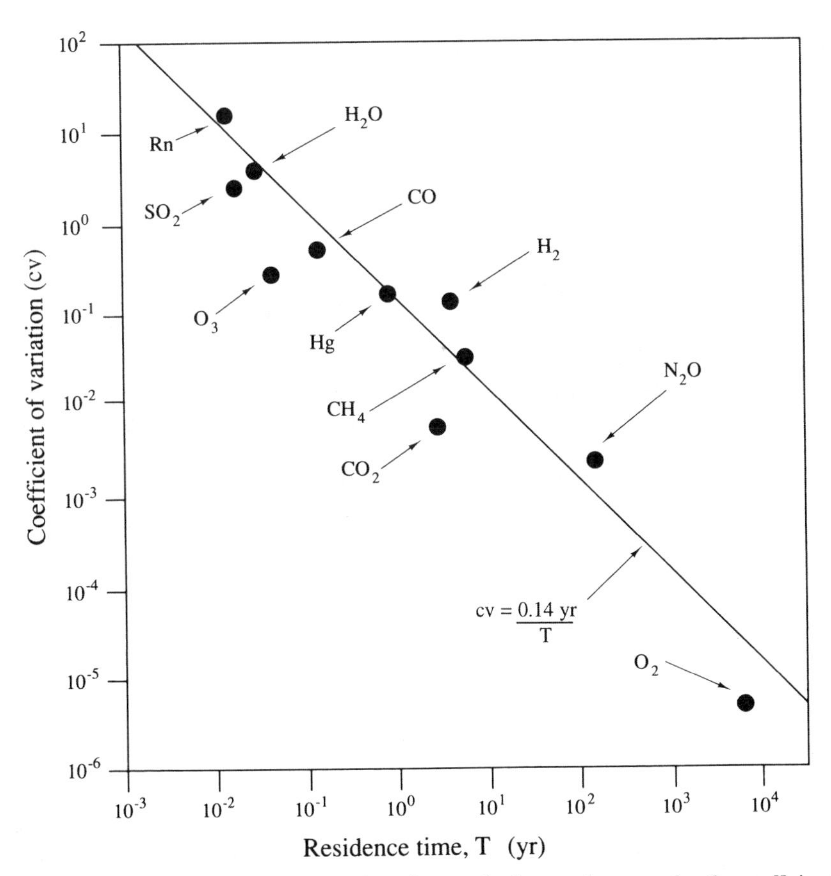

Figure 3.5 Variability in the concentration of atmospheric gases (expressed as the coefficient of variation among measurements) as a function of their estimated mean residence times in the atmosphere. Modified from Junge (1974), as updated by Slinn (1988).

This is a short time compared to the circulation of the troposphere, so we should expect water vapor to show highly variable concentrations in space and time (Fig. 3.5).

The mean residence time for carbon dioxide is about 5 years—only slightly longer than the mixing time for the atmosphere. Owing to the seasonal uptake of CO_2 by plants, CO_2 shows a minor seasonal and latitudinal variation ($\pm$ about 1%) in its global concentration of 360 ppm (Fig. 3.6). In contrast, painstaking analyses are required to show *any* variation in the concentration of O_2, because the amount in the atmosphere is so large and its mean residence time, 4000 years, is so much longer than the mixing time of the atmosphere (Keeling and Shertz 1992).

Gases with mean residence times $\ll$1 year in the troposphere do not persist long enough for appreciable mixing into the stratosphere. Indeed, one of the most valuable, but dangerous, industrial properties of the chlorofluorocarbons is that they are chemically inert in the troposphere (Rowland 1989). This allows chlorofluorocarbons to mix into the stratosphere, where they destroy ozone in reaction with ultraviolet light.

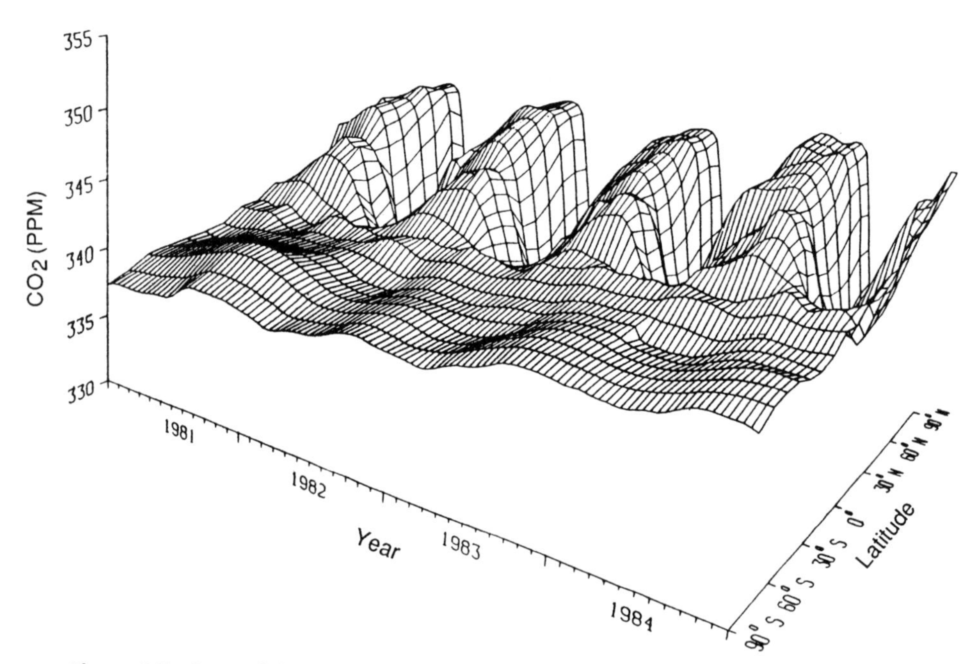

Figure 3.6 Seasonal fluctuations in the concentration of atmospheric CO_2 (1981–1984), shown as a function of 10° latitudinal belts (Conway et al. 1988). Note the smaller amplitude of the fluctuations in the southern hemisphere, reaching peak concentrations during northern hemisphere minima.

Aerosols

In addition to gaseous components, the atmosphere contains particles, known as aerosols, that arise from a variety of sources (Table 3.2). Soil minerals are dispersed by wind erosion (deflation weathering) from arid and semi-arid regions (Pye 1987, Tegen and Fung 1995). Particles with a diameter $<1.0\ \mu m$ are held aloft by turbulent motion and subject to long-range transport. Schütz (1980) estimates that 1×10^{15} g/yr of soil particles enter the atmosphere from arid regions, and about 20% of these particles are involved in long-range transport. Dust from the deserts of central Asia falls in the Pacific ocean (Duce et al. 1980), where it contributes much of the iron needed by oceanic phytoplankton (Chapter 9). Similarly, dust from the Sahara supplies nutrients to phytoplankton in the Atlantic ocean

Table 3.2 Global Emissions of Aerosols[a]

Source	Global flux (10^{12}g/yr)
Natural sources	
Primary aerosols	
Soil dust	1500
Seasalt	1300
Volcanic dust	33
Organic particles	50
Secondary aerosols	
Sulfates from volatile organic sulfides (e.g., $(CH_3)_2S$)	90
Sulfates from SO_2	12
Organic condensates	55
Nitrates from NO_x	22
Sum of natural sources	3070
Anthropogenic sources	
Primary aerosols	
Industrial particles	100
Soot	20
Particles from forest fires	80
Secondary aerosols	
Sulfates from SO_2	140
Nitrates from NO_x	36
Organic condensates	10
Sum of anthropogenic sources	390
Total	3460

[a] From Jonas et al. (1995).

(Talbot et al. 1986) and phosphorus to Amazon rainforests (Swap et al. 1992). Typically, while it is in transit, soil dust warms the atmosphere over land and cools the atmosphere over the oceans, which have lower surface albedo (reflectivity) (Ackerman and Chung 1992, Kellogg 1992).

An enormous quantity of particles enter the atmosphere from the ocean as a result of tiny droplets that become airborne with the bursting of bubbles at the surface (MacIntyre 1974, Wu 1981). As the water evaporates from these bubbles, the salts crystallize to form seasalt aerosols, which carry the approximate chemical composition of seawater (Glass and Matteson 1973, Möller 1990). As in the case of soil dust, most seasalt aerosols are relatively large and settle from the atmosphere quickly, but a significant proportion remain in the atmosphere for global transport. Möller (1990) estimates a total seasalt production of 10×10^{15} g/yr, which carries about 200×10^{12} g of chloride from sea to land. Other global estimates of seasalt production are somewhat lower (Table 3.2; see also Fig. 3.12).

Forest fires produce particles of charcoal that are carried throughout the troposphere, and small organic particles (soot) are produced by the condensation of volatile hydrocarbons from the smoke of forest fires (Hahn 1980, Cachier et al. 1989). Forest fires in the Amazon are thought to release as much as 1×10^{13} g of particulate matter to the atmosphere each year (Kaufman et al. 1990). It is likely that the global production of aerosols from forest fires has increased markedly in this century as a result of higher rates of biomass burning in the tropics (Andreae 1991, Cahoon et al. 1992). Aerosols from these fires may affect regional patterns of rainfall (Cachier and Ducret 1991) and global climate (Penner et al. 1992). At the same time, in the temperate zone, control of forest fires has reduced the aerosol loading to the atmosphere over the last century (Clark and Royall 1994).

Volcanoes disperse finely divided rock materials over large areas (Table 3.3), contributing to soil development in regions that are downwind from major eruptions (Watkins et al. 1978, Dahlgren and Ugolini 1989, Zobel and Antos 1991). Volcanic gases and ash that are transported to the stratosphere by violent eruptions undergo global transport, potentially affecting climate for several years (Minnis et al. 1993, Langway et al. 1995, McCormick et al. 1995).

Small particles are also produced by reactions between gases in the atmosphere. For instance, when SO_2 is oxidized to sulfuric acid (H_2SO_4) in the atmosphere, particles rich in $(NH_4)_2SO_4$ may be produced by a subsequent reaction with atmospheric ammonia (NH_3):

$$2NH_3 + H_2SO_4 \rightarrow (NH_4)_2SO_4. \tag{3.5}$$

Sulfate aerosols are also produced during the oxidation of dimethylsulfide released from the ocean (Chapter 9). These aerosols increase the albedo of the Earth's atmosphere, so estimates of the abundance of sulfate aerosols

Table 3.3 Composition of an Airborne Particulate Sample Collected during the Eruption of Mt. St. Helens on May 19, 1980[a]

Constituent	Particulate sample	Average ash
	Major elements (%)	
SiO_2	≡65.0	65.0
Fe_2O_3	6.7	4.81
CaO	3.0	4.94
K_2O	2.0	1.47
TiO_2	0.42	0.69
MnO	0.054	0.077
P_2O_5[b]	—	0.17
	Trace elements (ppm)	
S	3220	940
Cl	1190	660
Cu	61	36
Zn	34	53
Br	<8	~1
Rb	<17	32
Sr	285	460
Zr	142	170
Pb	36	8.7

[a] Average ash is shown for comparison. From Fruchter et al. (1980).

[b] From Hooper et al. (1980). Copyright 1980 by the AAAS.

are an important component of global climate models (Kiehl and Briegleb 1993, Mitchell et al. 1995).

Finally, a wide variety of particles are produced from human industrial processes, especially the burning of coal (Hulett et al. 1980, Shaw 1987). Globally, the release of particles during the combustion of fossil fuels rivals the mobilization of elements by rock weathering at the Earth's surface (Bertine and Goldberg 1971). Recently, the mass of industrial aerosols has declined in many developed countries where pollution controls have been instituted (Fig. 3.7). One of the most widespread anthropogenic aerosols, particles of lead from automobile exhaust, has declined in global abundance over the past 20 years due to a decline in the use of leaded gasoline (Boutron et al. 1991). Overall, human activities probably account for 10–20% of the burden of aerosols in today's atmosphere (Table 3.2).

Small particles are much more numerous in the atmosphere than large particles, but it is the large particles that contribute the most to the total airborne mass (Warneck 1988). The mass of aerosols declines with increas-

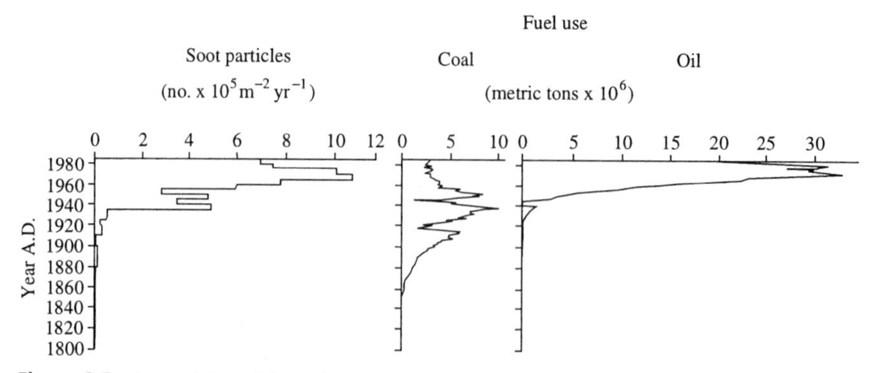

Figure 3.7 Annual deposition of soot in Lake Koltjarn, as recorded in sediment layers, and the annual consumption of coal and oil in Sweden since 1800. The use of fossil fuel and the deposition of soot have both declined in recent years. From Renberg and Wik (1984).

ing altitude from values ranging between 1 and 50 $\mu g/m^3$ near unpolluted regions of the Earth's surface. Although there is an inverse relation between the size of particles and their persistence in the atmosphere, the overall mean residence time for tropospheric aerosols is about 5 days (Warneck 1988). Thus, aerosols are not uniform in their distribution in the atmosphere. As a result of their longer mean residence time, small particles have the greatest influence on Earth's climate and global biogeochemical transport through the atmosphere.

The composition of the tropospheric aerosol varies greatly depending upon the proximity of continental, maritime, or anthropogenic sources (Heintzenberg 1989). Over land, aerosols are often dominated by soil minerals and human pollutants (Shaw 1987, Gillette et al. 1992). Over the ocean, the composition of aerosols is a mixture of contributions from silicate minerals of continental origin and seasalt from the ocean (Andreae et al. 1986). Various workers have used ratios among the elemental constituents of aerosols to deduce the relative contribution of different sources (e.g., Moyers et al. 1977, Rahn and Lowenthal 1984).

Aerosols are important in reactions with atmospheric gases and as nuclei for the condensation of raindrops. The latter are known as cloud condensation nuclei, often abbreviated CCN. Raindrops are formed when water vapor begins to condense on aerosols >0.1 μm in diameter. As raindrops enlarge and fall to the ground, they collide with other particles and absorb atmospheric gases. Soil dusts often contain a large portion of insoluble material (Reheis and Kihl 1995), but seasalt aerosols and those derived from pollution sources are readily soluble and contribute to the dissolved chemical content of rainwater. Reactions of atmospheric gases with aerosols or raindrops are known as *heterogeneous gas reactions.* Such reactions are responsible for the ultimate removal of many reactive gases from the atmosphere.

Biogeochemical Reactions in the Troposphere

Major Constituents—N_2, O_2, and Ar

It is perhaps not surprising that the major constituents of the atmosphere, N_2, O_2, and Ar, are all relatively unreactive, showing nearly uniform concentrations and long mean residence times in the atmosphere. From a biogeochemical perspective, N_2 is practically inert; reactive N is found only in molecules such as NH_3 and NO. Collectively these compounds are known as "odd" nitrogen, because the molecules have an odd number of N atoms (versus N_2 or N_2O). In fact, despite the abundance of nitrogen in the atmosphere, N_2 is so inert that the availability of odd nitrogen is one of the primary factors that limits the growth of plants on land and in the oceans (Delwiche 1970, Vitousek and Howarth 1991). Among atmospheric gases, only argon and the other noble gases are less reactive.

Conversion of N_2 to reactive compounds, *N-fixation,* occurs in lightning bolts, but the estimated global production of NO by lightning ($<3 \times 10^{12}$ g N/yr; Chapter 12) is too low to account for a significant turnover of N_2 in today's atmosphere. By far the most important source of fixed nitrogen for the biosphere derives from the bacteria that convert N_2 to NH_3 in the process of biological nitrogen fixation (Eq. 2.12). The global rate of biological N-fixation is poorly known, because it must be extrapolated from small-scale measurements to the entire surface of the Earth. Including human activities, global nitrogen fixation is not likely to exceed 300×10^{12} g N/yr, with the production of synthetic nitrogen fertilizer now accounting for about $\frac{1}{3}$ of the total (Chapter 12).

Denitrification (Eq. 2.18) returns N_2 from the biosphere to the atmosphere. In the absence of denitrification, the global rate of nitrogen fixation would remove the pool of N_2 from the atmosphere in about 20 million years. At present, we have little evidence that the rate of either N-fixation or denitrification changes significantly in response to changes in the concentration of N_2 in the atmosphere, so it would seem that the biosphere plays a minor role in maintaining a stable concentration of atmospheric N_2 (Walker 1984).

In Chapter 2 we discussed the accumulation of O_2 in the atmosphere during the evolution of life on Earth. The atmosphere now contains only a small portion of the total O_2 released by photosynthesis through geologic time (Fig. 2.7). However, the atmosphere contains much more O_2 than can be explained by the storage of carbon in land plants today. The instantaneous combustion of all the organic matter now stored on land would reduce the atmospheric oxygen content by only 0.03% (Chapter 5).

The accumulation of O_2 is the result of the long-term burial of reduced carbon in ocean sediments (Berner 1982), which contain nearly all of the reduced, organic carbon on Earth (Table 2.2). This organic matter is largely derived from photosynthesis in the sea, because the transport of organic carbon from land to sea in the world's rivers is very small (Schlesinger and Melack 1981). The rate of burial is determined by the area of the ocean floor that is subject to anoxic conditions (Walker 1977, 1980). Because that area varies inversely with the concentration of atmospheric O_2, the balance between the burial of organic matter and its oxidation maintains O_2 at a steady-state concentration of about 21% (see also Chapters 9 and 11).

A large amount of O_2 has been consumed in weathering of reduced crustal minerals, especially Fe and S, through geologic time (Fig. 2.7); the current rate of exposure of these minerals would consume all atmospheric oxygen in about 2 million years (see Fig. 11.7). However, the rate of exposure is not likely to vary greatly in response to changes in atmospheric O_2, so weathering is not the major factor controlling O_2 in the atmosphere. In sum, despite the potential reactivity of O_2, its rate of reaction with reduced compounds is rather slow, and O_2 is a stable component of the atmosphere. The mean residence time of O_2 in the atmosphere is on the order of 4000 years (cf. Fig. 3.5).

Carbon Dioxide

Carbon dioxide is not reactive with other gases in the atmosphere. The concentration of CO_2 is affected by interactions with the Earth's surface, including the reactions of the carbonate–silicate cycle (Fig. 1.4), gas exchange with seawater following Henry's Law (Eq. 2.7), and annual cycles of photosynthesis and respiration by land plants (Fig. 3.6). For the Earth's land surface, our best estimates of plant uptake (60×10^{15} g C/yr; Chapter 5) suggest a mean residence time of about 12.5 years before a hypothetical molecule of CO_2 in the atmosphere is captured by photosynthesis. The annual exchange of CO_2 with seawater, particularly in areas of cold, downwelling water and high productivity (Chapter 9), is about 1.5× as large as the annual uptake of CO_2 by land plants. Both plant and ocean uptake are likely to increase with increasing concentrations of atmospheric CO_2, potentially buffering fluctuations in its concentration. Following Eq. 3.3, the mean residence time for CO_2, determined by the total flux from the atmosphere (the sum of land and ocean uptake), is about 5 years. The carbonate–silicate cycle (Fig. 1.4) also buffers the concentration of CO_2 in the atmosphere, but does not affect the concentration of atmospheric CO_2 significantly in periods of less than 100,000 years. We will compare the relative importance of these processes in more detail in Chapter 11, which examines the global carbon cycle.

The current increase in atmospheric CO_2 is a non-steady-state condition, caused by the combustion of fossil fuels and destruction of land vegetation (Schimel et al. 1995). CO_2 is released by these processes faster than it can be taken up by land vegetation and the sea. If these activities were to cease, atmospheric CO_2 would return to a steady state, and after several hundred years nearly all of the CO_2 released by humans would reside in the oceans. In the meantime, higher concentrations of CO_2 are likely to cause significant atmospheric warming through the "greenhouse effect" (Fig. 3.2).

Trace Biogenic Gases

Volcanoes are the original source of volatiles in the Earth's atmosphere (Chapter 2) and are a small continuing source of some of the reduced gases (H_2S, H_2, NH_3, CH_4) that are found in the atmosphere today (Table 2.1). These and other trace gases are found at concentrations well in excess of what is predicted from equilibrium geochemistry in an atmosphere with 21% O_2 (Table 3.4). In most cases, the observed atmospheric concentrations are supplied by the biosphere, particularly by microbial activity. Methane is largely produced by anaerobic decomposition in wetlands (Chapters 7 and 11), nitrogen oxides by soil microbial transformations (Chapter 6), carbon monoxide by combustion of biomass and fossil fuels (Chapters 5 and 11), and volatile hydrocarbons, especially isoprene, by vegetation and human industrial activities (Chapter 5). The production of trace gases containing N and S contributes to the global cycling of these elements, which is controlled by the biosphere (Deevey 1970b, Crutzen 1983).

Unlike major atmospheric constituents, many of the trace biogenic gases in the atmosphere are highly reactive, showing short mean residence times

Table 3.4 Some Trace Biogenic Gases in the Atmosphere

Compound	Formula	Concentration (ppb)		Mean residence time	Percentage of sink due to OH
		Expected[a]	Actual[b]		
Carbon compounds					
Methane	CH_4	10^{-148}	1750	9 yr	90
Carbon monoxide	CO	10^{-51}	45–250	60 days	80
Isoprene	$CH_2{=}C(CH_3){-}CH{=}CH_2$		0.2–10.0	<1 day	100
Nitrogen compounds					
Nitrous oxide	N_2O	10^{-22}	311	120 yr	0
Nitric oxides	NO_x	10^{-13}	0.02–10.0	1 day	100
Ammonia	NH_3	10^{-63}	0.08–5.0	5 days	<2
Sulfur compounds					
Dimethylsulfide	$(CH_3)_2S$		0.004–0.06	1 day	50
Hydrogen sulfide	H_2S		<0.04	4 days	100
Carbonyl sulfide	COS	0	0.50	5 yr	22
Sulfur dioxide	SO_2	0	0.02–0.10	3 days	50

[a] Approximate values in equilibrium with an atmosphere containing 21% O_2 (Chameides and Davis 1982).

[b] For short-lived gases, the value is the range expected in remote, unpolluted atmospheres.

and variable concentrations in space and time (Fig. 3.5). Concentrations of these gases in the atmosphere are determined by the balance between local sources and chemical reactions—known as *sinks*—that remove these gases from the atmosphere. Losses from the troposphere are largely driven by oxidation reactions and the capture of the reaction products by rainfall. Currently the concentration of nearly all these constituents is increasing as a result of human activities, suggesting that humans are affecting biogeochemistry at the global level (Rasmussen and Khalil 1986, Mooney et al. 1987).

Despite its abundance in the atmosphere, O_2 is too unreactive to oxidize reduced gases by direct reaction in the atmosphere. However, through a variety of reactions driven by sunlight, small amounts of oxygen are converted to ozone (O_3), and further reactions yield hydroxyl radicals (OH) (Logan 1985, Thompson 1992). Ozone and OH are the primary species that oxidize many of the trace gases to CO_2, HNO_3, and H_2SO_4.

It is important to understand the natural production, occurrence, and reactions of ozone in the atmosphere. Nearly daily we read seemingly contradictory reports of the harmful effects of ozone depletion in the stratosphere and harmful effects of ozone pollution in the troposphere. In each case, human activities are upsetting the natural concentrations of ozone that are critical to atmospheric biogeochemistry.

Most ozone is produced by the reaction of sunlight with O_2 in the stratosphere, as described in the next section. Some of this ozone is transported to the Earth's surface by the mixing of stratospheric and tropospheric air. However, observations of ozone in the smog of polluted cities such as Los Angeles alerted atmospheric chemists to reactions by which ozone is produced in the troposphere (Warneck 1988). When NO_2 is present in the atmosphere it is dissociated by sunlight ($h\nu$),

$$NO_2 + h\nu \rightleftarrows NO + O, \tag{3.6}$$

followed by a reaction producing ozone:

$$O + O_2 \rightleftarrows O_3. \tag{3.7}$$

This reaction sequence is an example of a *homogeneous gas reaction,* i.e., a reaction between atmospheric constituents that are all in the gaseous phase. The net reaction is

$$NO_2 + O_2 \rightleftarrows NO + O_3, \tag{3.8}$$

which is an equilibrium reaction, so high concentrations of NO tend to drive the reaction backward. Seinfeld (1989) indicates that the concentration of O_3 is determined by

$$O_3\ (\text{ppm}) = 0.021\ [NO_2]/[NO]. \tag{3.9}$$

Both NO_2 and NO, collectively known as NO_x, are found in polluted air, in which they are derived from industrial and automobile emissions.[2] Small concentrations of both of these constituents are also found in the natural atmosphere, where they are derived from forest fires, lightning discharges, and microbial processes in the soil (Chapter 6). Thus, the production of ozone from NO_2 has probably always occurred in the troposphere, and the present-day concentrations of tropospheric ozone have simply increased as industrial emissions have raised the concentration of NO_2 and other precursors to O_3 formation (Volz and Kley 1988, Hough and Derwent 1990, Thompson 1992).

Ozone is subject to further photochemical reaction,

$$O_3 + h\nu \rightleftarrows O_2 + O(^1D), \tag{3.10}$$

where $h\nu$ is ultraviolet light with wavelengths <310 nm and $O(^1D)$ is an excited atom of oxygen. Reaction of $O(^1D)$ with water yields hydroxyl radicals:

$$O(^1D) + H_2O \rightleftarrows 2OH. \tag{3.11}$$

Hydroxyl radicals may further react to produce HO_2 and H_2O_2,

$$2OH + 2O_3 \rightleftarrows 2HO_2 + 2O_2 \tag{3.12}$$

$$2HO_2 \rightleftarrows H_2O_2 + O_2, \tag{3.13}$$

which are other short-lived oxidizing compounds in the atmosphere (Thompson 1992).

Hydroxyl radicals exist with a mean concentration of 9.7×10^5 molecules/cm^3 (Prinn et al. 1995). The highest concentrations occur in daylight (Platt et al. 1988, Mount 1992) and at tropical latitudes, where the concentration of water vapor is greatest (Hewitt and Harrison 1985). The average OH radical persists only for a few seconds in the atmosphere, so concentrations of OH are highly variable. Local concentrations can be measured using beams of laser-derived light, which is absorbed as a function of the number of OH radicals in its path (Dorn et al. 1988, Mount 1992). The global mean concentration of OH radicals must be estimated indirectly. For this purpose, atmospheric chemists rely on methylchloroform (trichloroethane), a gas that is known to result only from human activity. Methylchloroform has a mean residence time of about 4.8 years (Prinn et al.

[2] NO_x (pronounced "knox") refers to the sum of NO + NO_2; NO_y is used to refer to the sum of NO_x plus all other oxidized forms of nitrogen [e.g., HNO_3 and peroxyacetyl nitrate (PAN)] in the atmosphere.

1995), so it is well mixed in the atmosphere. In the laboratory, it reacts with OH,

$$OH + CH_3CCl_3 \rightarrow H_2O + CH_2CCl_3, \tag{3.14}$$

and the rate constant, K, for the reaction can be carefully measured as 0.85×10^{-14} cm^3 $molecule^{-1}$ sec^{-1} at 25°C (Talukdar et al. 1992). Then, knowing K, the rate of industrial production of CH_3CCl_3, and the rate of its accumulation in the atmosphere, one can calculate the concentration of OH that must be present; viz.,

$$OH = \frac{(\text{Production} - \text{Accumulation})}{K} \tag{3.15}$$

Hydroxyl radicals are the major source of oxidizing power in the troposphere. For example, in an unpolluted atmosphere, hydroxyl radicals destroy methane in a series of reactions,

$$CH_4 + OH \rightarrow CH_3 + H_2O \tag{3.16}$$

$$CH_3 + O_2 \rightarrow CH_3O_2 \tag{3.17}$$

$$CH_3O_2 + HO_2 \rightarrow CH_3O_2H + O_2 \tag{3.18}$$

$$CH_3O_2H \rightarrow CH_3O + OH \tag{3.19}$$

$$CH_3O + O_2 \rightarrow CH_2O + HO_2, \tag{3.20}$$

for which the net reaction is

$$CH_4 + O_2 \rightarrow CH_2O + H_2O. \tag{3.21}$$

Note that the hydroxyl radical has acted as a catalyst to initiate the oxidation of CH_4 and its by-products by O_2. Other volatile hydrocarbons, known as nonmethane hydrocarbons (NMHCs), released from vegetation (Lamb et al. 1987a, Guenther et al. 1994) and human activities (Piccot et al. 1992) are also oxidized through this pathway (Altshuller 1991).

The formaldehyde that is produced in these reactions is further oxidized to carbon monoxide,

$$CH_2O + OH + O_2 \rightarrow CO + H_2O + HO_2, \tag{3.22}$$

and CO is oxidized by OH to produce CO_2,

$$CO + OH \rightarrow H + CO_2 \tag{3.23}$$

$$H + O_2 \rightarrow HO_2 \tag{3.24}$$

$$HO_2 + O_3 \rightarrow OH + 2O_2, \tag{3.25}$$

for which the net reaction is,

$$CO + O_3 \rightarrow CO_2 + O_2. \tag{3.26}$$

Thus, OH acts to scrub the atmosphere of a wide variety of reduced carbon gases, ultimately oxidizing their carbon atoms to carbon dioxide.

Hydroxyl radicals also react with NO_2 and SO_2 in homogeneous gas reactions:

$$NO_2 + OH \rightarrow HNO_3 \tag{3.27}$$

$$SO_2 + OH \rightarrow HSO_3. \tag{3.28}$$

The reaction with NO_2 is very fast, and it produces nitric acid that is removed from the atmosphere by a heterogeneous interaction with raindrops. The reaction with SO_2 is much slower, accounting for the long-distance transport of SO_2 as a pollutant in the atmosphere (Rodhe et al. 1981). HSO_3 is eventually converted to SO_4^{2-} which is removed from the atmosphere by rainfall (Warneck 1988). SO_2 is also removed from the atmosphere by reaction with H_2O_2 in raindrops (Eqs. 3.12 and 3.13), forming H_2SO_4 (Chandler et al. 1988). Similarly, hydrogen sulfide (H_2S) and dimethylsulfide $(CH_3)_2S$, released from anaerobic soils (Chapter 7) and the ocean surface (Chapter 9), are removed by reactions with OH and other oxidizing compounds, eventually leading to the deposition of H_2SO_4 (Toon et al. 1987). Thus, OH radicals cleanse the atmosphere of trace N and S gases by converting them to "acid anions" (NO_3^-, SO_4^{2-}) in the atmosphere.

The vast majority of OH radical in the atmosphere is consumed in reactions with CO and CH_4. Although the concentration of methane is much higher than that of carbon monoxide in unpolluted atmospheres, the reaction of OH with CO is much faster. The speed of reaction of CO with OH accounts for the short mean residence time of CO in the atmosphere (Table 3.4). The mean residence time for methane is much longer, accounting for its more uniform distribution in the atmosphere (Fig. 3.5). One explanation for the current increase in methane in the atmosphere is that the anthropogenic release of CO consumes OH radicals previously available for the oxidation of methane (Khalil and Rasmussen 1985), but other measurements suggest that OH concentrations have remained fairly constant over the last decade (Prinn et al. 1995).

In unpolluted atmospheres, all these reactions consume OH. In "dirty" atmospheres, a different set of reactions pertains, in which there can be a net *production* of O_3, and thus OH. When the concentration of NO is >3–8 ppt (= dirty), the oxidation of carbon monoxide begins by reaction with hydroxyl radical and proceeds as follows (Crutzen and Zimmermann 1991):

$$CO + OH \rightarrow CO_2 + H \tag{3.29}$$

$$H + O_2 \rightarrow HO_2 \tag{3.30}$$

$$HO_2 + NO \rightarrow OH + NO_2 \tag{3.31}$$

$$NO_2 + h\nu \rightarrow NO + O \tag{3.32}$$

$$O + O_2 \rightarrow O_3. \tag{3.33}$$

The net reaction is

$$CO + 2O_2 \rightarrow CO_2 + O_3. \tag{3.34}$$

Similarly, the oxidation of methane in the presence of high concentrations of NO proceeds through a large number of steps, yielding a net reaction of

$$CH_4 + 4O_2 \rightarrow CH_2O + H_2O + 2O_3. \tag{3.35}$$

Crutzen (1988) points out that the oxidation of one molecule of CH_4 could consume up to 3.5 molecules of OH and 1.7 molecules of O_3 when the NO concentration is low, whereas it would yield a net gain of 0.5 OH and 3.7 O_3 in polluted environments (see also Wuebbles and Tamaresis 1993). Although they were first discovered in urban areas, the reactions of "dirty" atmospheres are likely to be relatively widespread in nature. NO is produced naturally by soil microbes (Chapter 6) and forest fires. Concentrations of NO >3–8 ppt are present over most of the Earth's land surface (Torres and Buchan 1988, Chameides et al. 1992). In the presence of NO, oxidation of volatile hydrocarbons emitted from vegetation, and CO emitted from both vegetation and forest fires, can account for unexpectedly high concentrations of O_3 over rural areas of the southeastern United States

(Fig. 3.8) (Jacob et al. 1993, Kleinman et al. 1994) and in remote tropical regions (Crutzen et al. 1985, Zimmerman et al. 1988, Jacob and Wofsy 1990, M.O. Andreae et al. 1994). In urban areas, where the concentration of NO_x is especially high due to industrial pollution, effective control of atmospheric O_3 levels may also depend on the regulation of volatile hydrocarbons (Chameides et al. 1988, Seinfeld 1989).

Although these reactions are both numerous and interrelated, further complexity stems from the presence of clouds, which promote heterogeneous interactions with water droplets in the atmosphere. Lelieveld and Crutzen (1991) suggest that when water droplets are present, there is no gain of OH radical in the atmosphere, even in the presence of NO_2 (Eqs. 3.8, 3.10, and 3.11), because O_3 preferentially reacts with H_2O_2 in cloud drops:

$$O_3 + H_2O_2 \rightleftarrows 2O_2 + H_2O. \tag{3.36}$$

Clouds fill only about 15% of the atmosphere, but they may significantly reduce the global oxidation capacity of the troposphere (Lelieveld and Crutzen 1990).

Understanding changes in the concentration of OH and other oxidizing species in the atmosphere is critical to predicting future trends in the concentration of trace gases, such as CH_4, that potentially contribute to greenhouse warming. Some models predict an increase in OH (Prinn et al. 1992), and O_3 (Isaksen and Hov 1987, Hough and Derwent 1990, Thompson 1992) in the atmosphere as a result of increasing human emissions of NO, creating dirty atmospheric conditions over much of the planet. Indeed, measurements near Paris in the late 1800s indicate lower concentrations of tropospheric ozone than today (Volz and Kley 1988). The models are also consistent with indirect observations that the global concentration of OH has remained fairly stable in recent years, despite increasing emissions of reduced gases that should scrub OH from the atmosphere (Prinn et al. 1995).

Some of the O_3 produced over the continents undergoes long-distance transport (Jacob et al. 1993, Parrish et al. 1993, Dickerson et al. 1995),

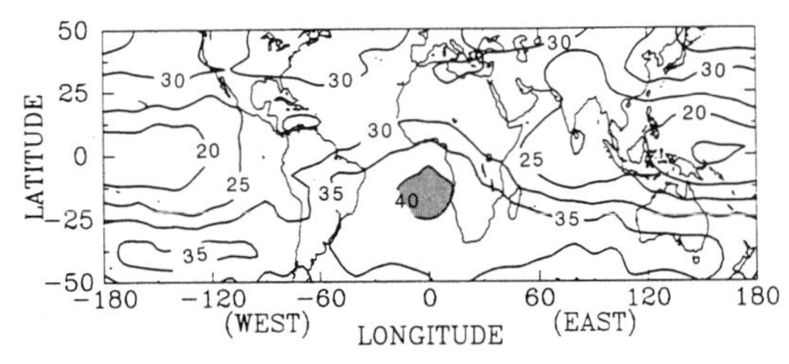

Figure 3.8 Distribution of tropospheric ozone during September through November, for the years 1979 to 1989 as determined by satellite measurements. Note the high concentration downwind (west) of central Africa, where biomass burning is widespread. From Fishman et al. (1990).

resulting in the appearance of O_3 and its by-products at considerable distances from their source. Concentrations of H_2O_2, derived from OH (Eqs. 3.12 and 3.13), have increased in layers of Greenland ice deposited during the last 200 years, suggesting a greater oxidizing capacity in the northern hemisphere as a result of human activities (Fig. 3.9; see also Jiang and Yung 1996). Other workers disagree, finding that local atmospheric conditions, rather than global changes in transport from polluted areas, determine the oxidizing capacity of the atmosphere over much of the planet (Oltmans and Levy 1992, Ayers et al. 1992).

Biogeochemical Reactions in the Stratosphere

Ozone

Ozone is produced in the stratosphere by the disassociation of oxygen atoms that are exposed to short-wave solar radiation. The reaction accounts for most of the absorption of ultraviolet sunlight ($h\nu$) at wavelengths of 180–240 nm and proceeds as follows:

$$O_2 + h\nu \rightarrow O + O \tag{3.37}$$

$$O + O_2 \rightarrow O_3. \tag{3.38}$$

Some ozone from the stratosphere mixes down into the troposphere, where the natural production of O_3 (Eqs. 3.6–3.8) is much slower because less

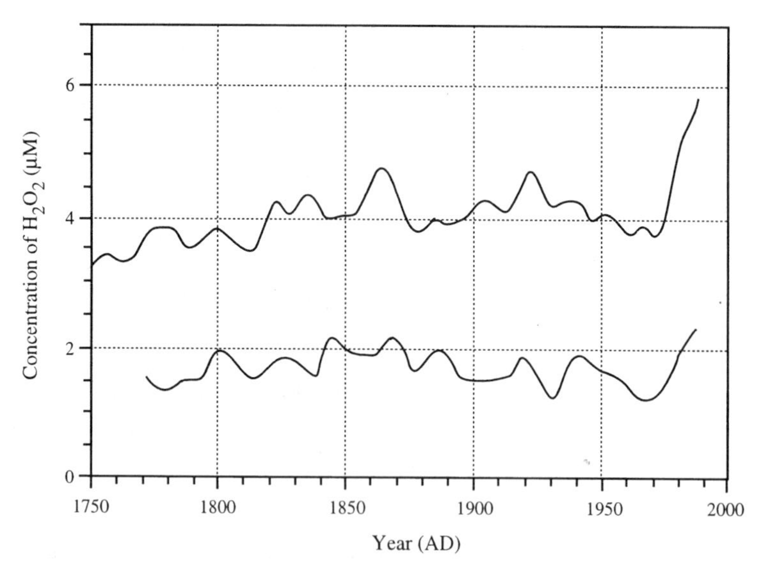

Figure 3.9 Variation in the mean annual H_2O_2 concentration in two cores from the Greenland ice pack over the past 200 years. Modified from Sigg and Neftel (1991).

ultraviolet light is available (Cicerone 1987). The remaining ozone is destroyed by a variety of reactions in the stratosphere. Absorption of ultraviolet light at wavelengths between 200 and 320 nm destroys ozone:

$$O_3 + h\nu \rightarrow O_2 + O \quad (3.39)$$

$$O + O_3 \rightarrow O_2 + O_2. \quad (3.40)$$

This absorption warms the stratosphere (Fig. 3.1) and protects the Earth's surface from the ultraviolet portion of the solar spectrum that is most damaging to living tissue (uvB).

Stratospheric ozone is also destroyed by reaction with OH,

$$O_3 + OH \rightarrow HO_2 + O_2 \quad (3.41)$$

$$HO_2 + O_3 \rightarrow OH + 2O_2, \quad (3.42)$$

and by reactions stemming from the presence of nitrous oxide (N_2O), which mixes up from the troposphere. Tropospheric N_2O is produced in a variety of ways (Chapters 6 and 12), but it is inert in the lower atmosphere. The only known sink for N_2O is photolysis in the stratosphere. About 80% of the N_2O reaching the stratosphere is destroyed in a reaction producing N_2 (Warneck 1988),

$$N_2O \rightarrow N_2 + O(^1D), \quad (3.43)$$

and about 20% in a reaction with the $O(^1D)$ produced in Eq. 3.43:

$$N_2O + O(^1D) \rightarrow 2NO. \quad (3.44)$$

The nitric oxide (NO) produced from N_2O destroys ozone in a series of reactions,

$$NO + O_3 \rightarrow NO_2 + O_2 \quad (3.45)$$

$$O_3 \rightarrow O + O_2 \quad (3.46)$$

$$NO_2 + O \rightarrow NO + O_2, \quad (3.47)$$

for which the net reaction is,

$$2O_3 \rightarrow 3O_2. \quad (3.48)$$

Note that the mean residence time of NO in the troposphere is too short for an appreciable amount to reach the stratosphere, where it might contribute to the destruction of ozone. Nearly all the NO in the stratosphere is produced in the stratosphere from N_2O. Eventually NO_2 is removed from the stratosphere by reacting with OH to produce nitric acid (Eq. 3.27).

Finally, stratospheric ozone is destroyed by chlorine, which acts as a catalyst in the reaction,

$$Cl + O_3 \rightarrow ClO + O_2 \quad (3.49)$$

$$O_3 + h\nu \rightarrow O + O_2 \quad (3.50)$$

$$ClO + O \rightarrow Cl + O_2, \quad (3.51)$$

for a net reaction of,

$$2O_3 + h\nu \rightarrow 3O_2. \quad (3.52)$$

Although each Cl produced may destroy many molecules of O_3, Cl is eventually converted to HCl and removed from the stratosphere by downward mixing and heterogeneous interaction with cloud drops in the troposphere (Rowland 1989, Solomon 1990).

The balance between ozone production (Eqs. 3.37 and 3.38) and the various reactions that destroy ozone maintains a steady-state concentration of O_3 of approximately 7×10^{18} molecules/m^3, peaking at 30 km altitude (Cicerone 1987). Although the photochemical production of O_3 is greatest at the equator, the density of the ozone layer is thickest at the poles (Cicerone 1987).

Recent measurements suggest that the total density of ozone molecules in the atmospheric column has declined significantly over Antarctica (Fig. 3.10) and perhaps globally (Stolarski et al. 1992, Jones and Shanklin 1995, Bojkov and Fioletov 1995). The decline (0.3%/yr) is unprecedented and represents a perturbation of global biogeochemistry. Destruction of ozone is likely to lead to an increased flux of ultraviolet radiation to the Earth's surface (Correll et al. 1992, Kerr and McElroy 1993) and to lower stratospheric temperatures, which may alter global heat balance (Ramanathan 1988). Greater uvB radiation at the Earth's surface may already be reducing marine production in Antarctic waters (R.C. Smith et al. 1992) and perhaps the reproductive capacity of amphibians worldwide (Blaustein et al. 1994). Because previous, steady-state ozone concentrations were maintained in the face of natural photochemical reactions that produce and consume ozone, attention has focused on disruptions of this balance by human activities (McElroy and Salawitch 1989).

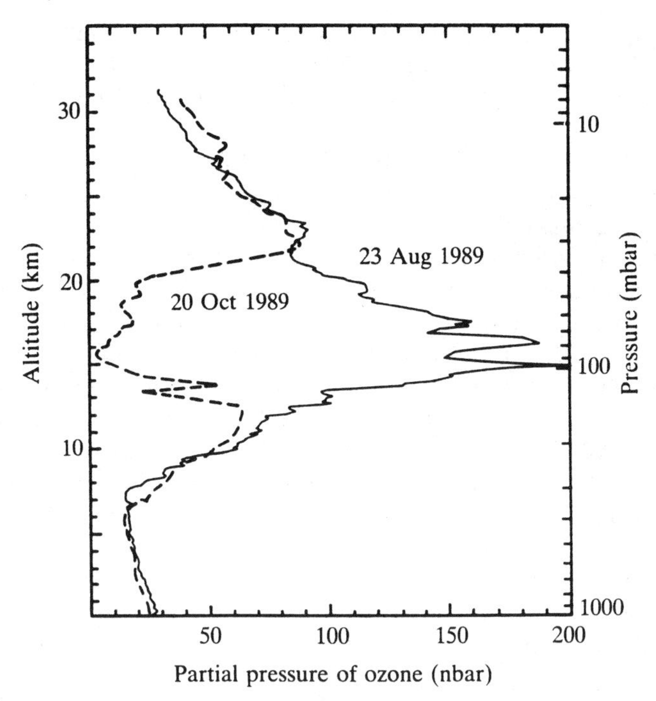

Figure 3.10 Ozone concentrations over McMurdo station, Antarctica in the spring of 1989. Note the near-complete loss of ozone at 15 km altitude (Deshler et al. 1990).

Chlorofluorocarbons (freons), which are produced as aerosol propellants, refrigerants, and solvents, have no known natural source in the atmosphere (Prather 1985). These compounds are chemically inert in the troposphere, so they eventually mix into the stratosphere where they are decomposed by photochemical reactions producing active chlorine (Molina and Rowland 1974, Rowland 1989, 1991),

$$CCl_2F_2 \rightarrow Cl + CClF_2, \tag{3.53}$$

which can destroy ozone by the reactions of Eqs. 3.49 to 3.51. These reactions are greatly enhanced in the presence of ice particles, which accounts for the first observations of the O_3 "hole" in the springtime over Antarctica (Farman et al. 1985, Solomon et al. 1986). In a dry atmosphere, ClO reacts with NO_2 to form $ClONO_2$, an inactive compound that removes both gases from O_3 destruction. In the presence of ice clouds, $ClONO_2$ breaks down,

$$ClONO_2 + HCl \rightarrow Cl_2 + HNO_3 \tag{3.54}$$

$$Cl_2 + h\nu \rightarrow 2Cl, \tag{3.55}$$

producing active chlorine for ozone destruction (Molina et al. 1987, Solomon 1990). Significantly, during the last 40 years, levels of active chlorine have increased in a mirror image to the loss of ozone from the stratosphere (Fig. 3.11).

The relative importance of chlorofluorocarbons versus natural sources of chlorine in the stratosphere is apparent in a global budget for atmospheric chlorine (Fig. 3.12). Seasalt aerosols are the largest natural source of chlorine in the troposphere, but they have such a short mean residence time that they do not contribute Cl to the stratosphere. There is also no good reason to suspect that they have increased in abundance in the last few decades. Similarly, industrial emissions of HCl are rapidly removed from the troposphere by rainfall. Especially violent volcanic eruptions can inject gases directly into the stratosphere, sometimes adding to stratospheric Cl (Johnston 1980, Mankin and Coffey 1984). However, in most cases only a small amount of Cl reaches the stratosphere, because various processes remove HCl from the rising volcanic plume (Tabazadeh and Turco 1993). After the Mt. Pinatubo eruption, which released 4.5×10^{12} g of HCl, stratospheric Cl increased by $<1\%$ (Mankin et al. 1992). The only significant natural source of Cl in the stratosphere stems from the production of methylchloride by marine algae (Wuosmaa and Hager 1990), higher plants

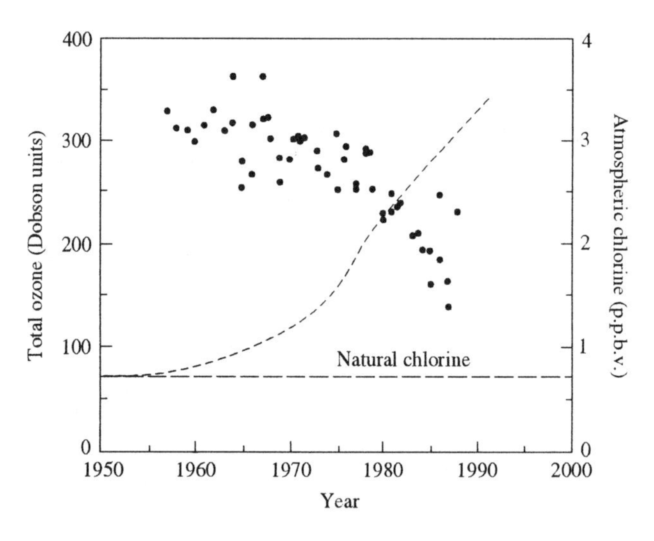

Figure 3.11 The decline in atmospheric O_3 (●) over Antarctica since 1958 corresponds to an increase in chlorine (—) in the stratosphere. The customary unit for the total number of ozone molecules in an atmospheric column, the Dobson, is equivalent to 2.69×10^{16} molecules/cm^2 of the Earth's surface. Modified from Solomon (1990).

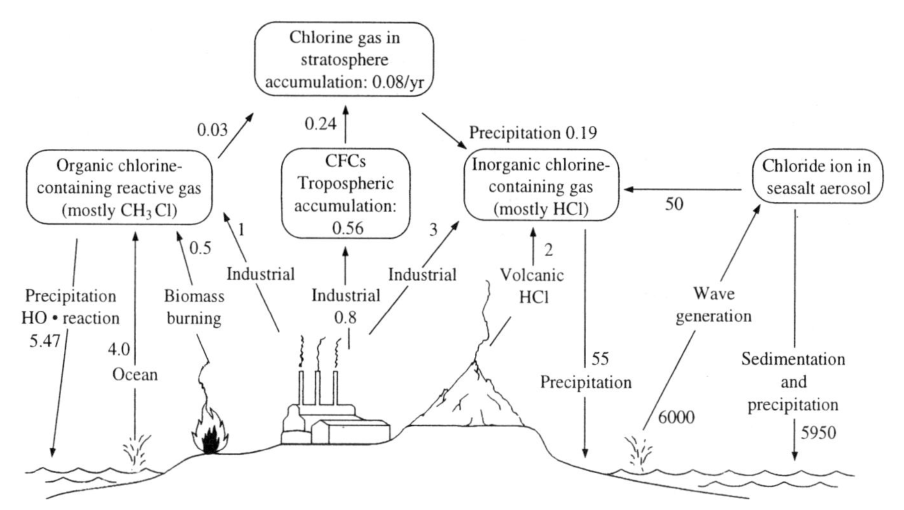

Figure 3.12 A global budget for Cl in the troposphere and the stratosphere. All data are given in 10^{12} g Cl/yr. Modified from Graedel and Crutzen (1993) and Graedel and Keene (1995).

(Saini et al. 1995), and forest fires (Crutzen et al. 1979, Laursen et al. 1992, Rudolph et al. 1995). Methylchloride has a mean residence time of about 1.8 years in the atmosphere, so a small portion of CH_3Cl mixes into the stratosphere.

In the global budget, the relatively small industrial production of chlorofluorocarbons is the dominant source of Cl delivered to the stratosphere (Fig. 3.12; Russell et al. 1996). Increasing concentrations of these compounds have been strongly implicated in ozone destruction (Rowland 1989). Happily, with the advent of the Montreal protocol, which limits the use of these compounds worldwide, there is already some evidence that the growth rate of these compounds in the atmosphere is slowing (Elkins et al. 1993).

Similar reactions are possible with compounds containing bromine; in fact, Br compounds may be even more potent in the destruction of stratospheric O_3 than Cl (Wennberg et al. 1994). Industry is a large source of methylbromide (CH_3Br), which is used as an agricultural fumigant (Yagi et al. 1995). Bromoform ($CHBr_3$) produced by marine algae (Sturges et al. 1992, Manley et al. 1992) and methylbromide by biomass burning (Manö and Andreae 1994) also contribute to the atmospheric Br budget. Sinks of CH_3Br include uptake by the oceans (Lobert et al. 1995, Anbar et al. 1996) and soils (Shorter et al. 1995). Concentrations of methylbromide in the atmosphere are increasing at about 3%/yr (Khalil et al. 1993a). The global budget of CH_3Br and its mean residence time (0.8–2 years) in the atmosphere are poorly known, but some CH_3Br may persist long enough to reach the stratosphere, where it can lead to ozone destruction. Among other halogen-containing gases, the lifetimes of methyliodide produced by marine phytoplankton (Campos et al. 1996) and various inorganic fluoride compounds are too short for appreciable mixing into the stratosphere. Indeed, the observed increase of fluoride in the stratosphere appears solely due to the presence of chlorofluorocarbons, and it is an independent verification of their destruction in the stratosphere by ultraviolet light (Zander et al. 1994).

Satellite observations have greatly aided our understanding of changes in stratospheric ozone. The loss of ozone from the atmosphere has been monitored since 1979 by the Total Ozone Mapping Satellite (TOMS) that records the abundance of O_3 in a column extending from the bottom to the top of the atmosphere (Gleason et al. 1993). TOMS showed that the loss of ozone accelerated in the presence of stratospheric aerosols produced by the Pinatubo volcano (Herman et al. 1993, Brasseur and Granier 1992). Recently, the Upper Atmosphere Research Satellite (UARS) has observed ozone depletions over the Arctic apparently by similar processes as at the South Pole (Waters et al. 1993, Brune et al. 1991, Manney et al. 1994).

Stratospheric Sulfur Compounds

Sulfate aerosols in the stratosphere are important to the albedo of the Earth (Warneck 1988). A layer of sulfate aerosols, known as the Junge layer, is found in the stratosphere at about 20–25 km altitude. Its origin is twofold. Large volcanic eruptions can inject SO_2 into the stratosphere, where it is oxidized to sulfate (Eq. 3.28). Large eruptions have the potential to increase the abundance of stratospheric sulfate by 100-fold (Arnold and Bührke 1983, Hofmann and Rosen 1983), and the sulfate aerosols persist in the stratosphere for several years, cooling the planet (Minnis et al. 1993, McCormick et al. 1995). During periods without volcanic activity, the dominant source of stratospheric sulfate is carbonyl sulfide (COS) that mixes up from the troposphere, where it originates from a variety of sources (Chapter 13). Most sulfur gases are so reactive that they do not reach the stratosphere, but COS has a mean residence time of about 5 years in the atmosphere (Table 3.4).

Carbonyl sulfide that reaches the stratosphere is oxidized by photolysis, forming sulfate aerosols that contribute to the Junge layer (Chin and Davis 1993). Eventually, these aerosols are removed from the stratosphere by downward mixing of stratospheric air. There is some evidence that the column density of sulfate aerosols in the upper atmosphere has increased during the last few decades (Hofmann 1990). The increase in stratospheric SO_4 may result from an increasing use of high-altitude aircraft (Hofmann 1991) and other human perturbations of the global sulfur budget. Estimated sources of COS exceed known sinks (Chin and Davis 1993), but there is no observable increase in COS globally (Rinsland et al. 1992).

Models of the Atmosphere and Global Climate

A large number of models have been developed to explain the physical properties and chemical reactions in the atmosphere. When these models attempt to predict the characteristics in a single column of the atmosphere, they are known as one-dimensional (1D) and radiative-convective models. For example, Figure 3.2 is a simple 1D model for the greenhouse effect, which assumes that the behavior of the Earth's atmosphere can be approximated by average values applied to the entire surface. Two-dimensional models (2D) can be developed using the vertical dimension and a single horizontal dimension (e.g., latitude) to examine the change in atmospheric characteristics across a known distance of the Earth's surface (e.g., Brasseur and Hitchman 1988, Hough and Derwent 1990). On a regional scale these are particularly useful in following the fate of pollution emissions (e.g., Rodhe et al. 1981). Three-dimensional models (3D) attempt to follow the fate of particular parcels of air as they move both horizontally and vertically in the atmosphere. Dynamic 3D models are known as *general circulation models* (GCMs) for the globe (Fig. 3.13).

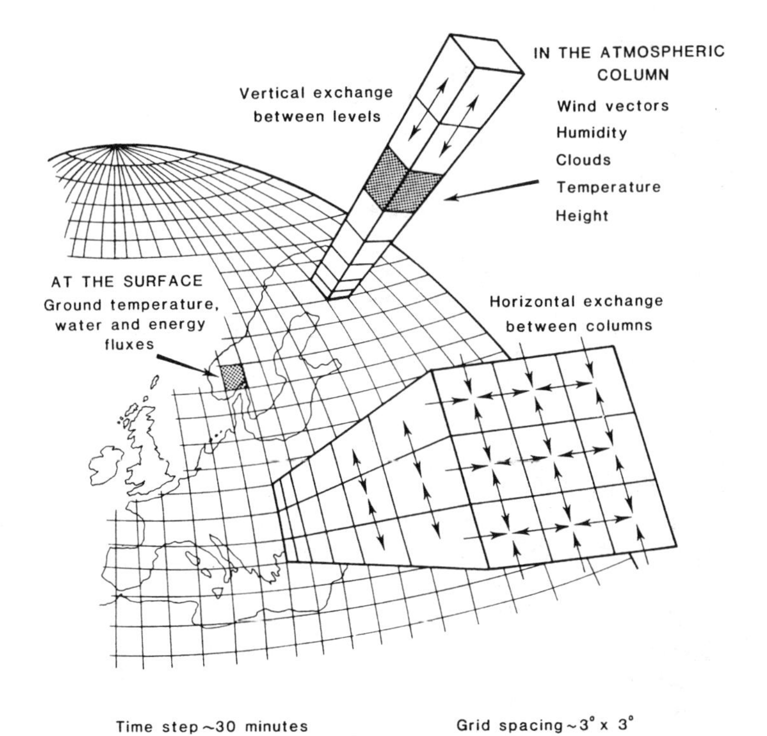

Figure 3.13 Conceptual structure of a dynamic, three-dimensional general circulation model for the Earth's atmosphere, indicating the variables that must be included for a global model to function properly. From Henderson-Sellers and McGuffie (1987).

Many models are constructed to include both chemical reactions and physical phenomena, such as the circulation of the atmosphere due to temperature differences. Chemical transformations are parameterized using the rate and equilibrium coefficients for the reactions that we have examined in this chapter. Because there are a large number of reactions, most of these models are quite complex (e.g., Logan et al. 1981, Isaksen and Hov 1987, Lelieveld and Crutzen 1990), but they give useful predictions of future atmospheric composition when the input of several constituents is changing simultaneously.

Nearly all models suggest that substantial warming of the atmosphere (1.5 to 5.5°C) will accompany increasing concentrations of CO_2, N_2O, CH_4, and chlorofluorocarbons (Houghton et al. 1990). The warming results from the absorption of infrared (heat) radiation emitted from the surface of the Earth (Fig. 3.2). Warming will be greatest near the poles, where there is normally the greatest net loss of infrared radiation relative to incident sunlight (Manabe and Wetherald 1980). Presumably the oceans will warm more slowly than the atmosphere, but eventually warmer ocean waters will allow greater rates of evaporation, increasing the circulation of water in the global hydrologic cycle (Graham 1995; see also Chapter 10). Water vapor also absorbs infrared radiation, so it is likely to further accelerate the potential greenhouse effect (Raval and Ramanathan 1989, Rind et al. 1991). Thus, most models predict that higher concentrations of CO_2 and other trace gases in the atmosphere will make the Earth a warmer and more humid planet (Chapter 10).

Differential warming of the atmosphere will change global patterns of precipitation and evapotranspiration (Manabe and Wetherald 1986, Rind et al. 1990), causing substantial changes in the climate of most areas outside the tropics. How rapidly these changes in climate occur will be moderated by the thermal buffer capacity of the world's oceans, which can absorb enormous quantities of heat. However, the magnitude of the potential changes in climate is much larger than most changes in global climate during the last 2 million years. Although it is difficult to demonstrate any trend toward global warming from satellite observations of the planet (Spencer and Christy 1990), the global record from local weather stations suggests that warming is already in progress (Fig. 3.14). Many climatologists believe that we will see an unambiguous validation of this global experiment, in excess of normal climatic oscillations, before the end of this century (Hansen et al. 1981, Ramanathan 1988, Schneider 1994, Santer et al. 1996).

Of course, such predictions are not made without disagreement (Luther and Cess 1985). One of the largest uncertainties in climate models is the effect of tropospheric aerosols, particularly sulfate aerosols, which are reflective to incoming solar radiation. Through soil disturbance, humans have increased the abundance of soil dusts that are distributed globally by

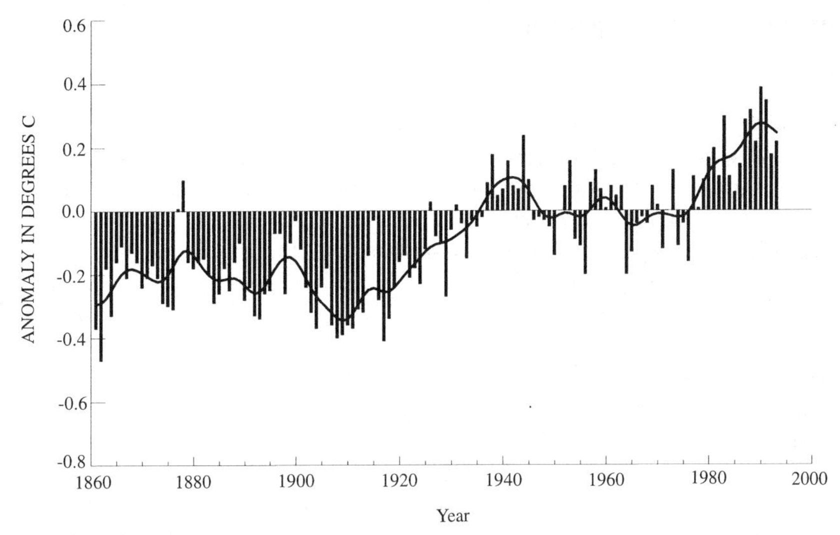

Figure 3.14 Mean global temperature, derived from measurements made on land and at sea, for the period 1861–1993, shown as deviations from the mean in the interval from 1951 to 1980. From Hadley Center for Climate Prediction and Research, United Kingdom.

wind (Tegen et al. 1996, Sokolik and Toon 1996). Despite pollution controls, sulfate aerosols have also increased in many areas of the troposphere due to human industrial activities (Langner et al. 1992). In addition to their direct effect on albedo, these aerosols increase the abundance of clouds, which amplify the reflectivity of the planet (Charlson et al. 1992). Global climate models that include the effects of aerosols produce the closest agreement with the observed trends in global temperature over the last several decades (Mitchell et al. 1995). In these models, the reflectivity of low clouds slows the rate of warming of the Earth that is expected as a result of increasing concentrations of greenhouse gases (Ramanathan et al. 1989).

Incoming solar radiation delivers about 340 W/m^2 to the Earth (Fig. 3.2). The natural greenhouse effect warms the planet about 33°C by trapping 153 W/m^2 of outgoing radiation (Ramanathan 1988). The greenhouse effect due to increasing concentrations of atmospheric trace gases currently adds about 2.1 W/m^2 to the natural greenhouse, while increasing clouds may reflect 0.3 W/m^2 to outer space (Kiehl and Briegleb 1993). It is interesting to note that aerosol concentrations were higher (De Angelis et al. 1987, Petit et al. 1990) and CO_2 concentrations were lower (Fig. 1.4) during the last glacial period, but the cause and effect relation of these observations is unclear.

Atmospheric Deposition

Processes

Elements of biogeochemical interest are deposited on the Earth's surface as a result of rainfall, dry deposition (sedimentation), and the direct absorption of gases. The importance of each of these processes differs for different regions and for different elements (Gorham 1961). Over time, cumulative deposition from the atmosphere accounts for a large fraction of the nitrogen and sulfur that is contained in terrestrial ecosystems (Chapter 6).

The chemical composition of rainfall has received great attention, as a result of widespread concern about dissolved constituents that lead to "acid rain." The dissolved constituents in rainfall are often separated into two fractions. The *rainout* component consists of constituents derived from cloud processes, such as the nucleation of raindrops. The *washout* component is derived from below cloud level, by scavenging of aerosol particles and the dissolution of gases in raindrops as they fall (Brimblecombe and Dawson 1984, Shimshock and de Pena 1989). The relative contribution of these fractions varies depending upon the length of the rainstorm. As washout cleanses the lower atmosphere, the content of dissolved materials in rainfall declines. Thus, the concentration of dissolved constituents in precipitation is inversely related to the rate of precipitation (Gatz and Dingle 1971) and to the total volume collected (Likens et al. 1984, Lesack and Melack 1991, Minoura and Iwasaka 1996). The concentration of dissolved constituents also varies inversely as a function of mean raindrop size (Georgii and Wötzel 1970). This inverse relation explains why extremely high concentrations of dissolved constituents are found in fog waters (Weathers et al. 1986, Waldman et al. 1982). Capture of fog and cloud water by vegetation dominates the deposition of nutrient elements from the atmosphere in some high-elevation and coastal ecosystems (Lovett et al. 1982, Azevedo and Morgan 1974, Waldman et al. 1985, Miller et al. 1993a).

The relative efficiency of scavenging by rainwater is often expressed as the washout ratio:

$$\text{Washout} = \frac{\text{Ionic concentration in rain (mg/liter)}}{\text{Ionic concentration in air (mg/m}^3\text{)}}. \qquad (3.56)$$

With units of m^3/liter, this ratio gives an indication of the volume of atmosphere cleansed by each liter of rainfall. Large ratios are generally found for ions that are derived from relatively large aerosols or from highly water-soluble gases in the atmosphere. Snowfall is generally less efficient at scavenging than rainfall.

The deposition of nutrients by precipitation is often called wetfall; dryfall is the result of gravitational sedimentation of particles during periods without rain (Hidy 1970). Dryfall of dusts in areas downwind of arid lands is often spectacular; Liu et al. (1981) reported 100 g m^{-2} hr^{-1} of dustfall in Beijing, China, as a result of a single dust storm on 18 April 1980. Enormous deposits of wind-deposited soil, known as loess, were laid down during glacial periods, when large areas of semiarid land were subject to wind erosion (Pye 1987, Simonson 1995). Today, various elements necessary for plant growth are released by chemical weathering of soil minerals in these deposits (Chapter 4).

The dryfall received in many areas contains a significant fraction that is easily dissolved by soil waters and immediately available for plant uptake. Despite the high rainfall found in the southeastern United States, Swank and Henderson (1976) reported that 19–64% of the total annual atmospheric deposition of ions such as Ca, Na, K, and Mg, and up to 89% of the deposition of P, was derived from dryfall. In many forests, the majority of the annual uptake and circulation of these elements in vegetation may be derived from the atmosphere, even though rock weathering dominates overall inputs to the ecosystem and the content of runoff waters (Miller et al. 1993). Dryfall inputs of P may assume special significance to plant growth in areas where the release of P from rock weathering is very small (Newman 1995). Dry deposition contributes about 30 to 60% of the deposition of sulfur in New Hampshire (Likens et al. 1990, cf. Tanaka and Turekian 1995) and 30% of the total input of acidic substances in southern Canada (Sirois and Barrie 1988).

Dryfall is often measured in collectors that are designed to close during rainstorms. When open to the atmosphere, these instruments capture particles that are deposited vertically, known as sedimentation. In natural ecosystems, dryfall is also derived by the capture of particles on vegetation surfaces. When vegetation captures particles that are moving horizontally in the airstream, the process is known as impaction (Hidy 1970). Impaction is a particularly important process in the capture of seasalt aerosols near the ocean (Art et al. 1974, Potts 1978).

In addition to the uptake of CO_2 in photosynthesis, vegetation also absorbs N- and S-containing gases directly from the atmosphere (Whelpdale and Shaw 1974, Hosker and Lindberg 1982, Lindberg et al. 1986). Uptake of pollutant SO_2 and NO_2 by vegetation is particularly important in humid regions (McLaughlin and Taylor 1981, Rondón and Granat 1994), where plant stomata remain open for long periods. Lovett and Lindberg (1986, 1993) found that uptake of HNO_3 vapor accounted for 75% of the annual dry deposition of nitrogen (4.8 kg/ha) in a deciduous forest in Tennessee, and dry deposition was nearly half of the total annual deposition of nitrogen from the atmosphere. Vegetation also can be a source (Farquhar et al. 1979, Heckathorn and DeLucia 1995) or a sink for atmospheric NH_3 (Denmead et al. 1976), depending on the ambient concentration in the atmosphere (Langford and Fehsenfeld 1992, Sutton et al. 1993).

Total capture of dry particles and gases by land plants is difficult to measure. When rainfall is collected inside a forest, it contains materials that have been deposited on the plant surfaces, but also large quantities of elements that are derived from the plants themselves (Parker 1983, Chapter 6). Artificial collectors (surrogate surfaces) are often used to approximate the capture by vegetation (White and Turner 1970, Vandenberg and Knoerr 1985, Lindberg and Lovett 1985). The capture on known surfaces can be compared to the airborne concentrations to calculate a deposition velocity (Sehmel 1980):

$$\text{Deposition Velocity} = \frac{\text{Rate of dryfall (mg/cm}^2\text{/sec)}}{\text{Concentration in air (mg/cm}^3\text{)}}. \quad (3.57)$$

In units of cm/sec, these velocities can be multiplied by the estimated surface area of vegetation (cm^2) and the concentration in the air to calculate total deposition for an ecosystem. For example, Lovett and Lindberg (1986) used a deposition velocity of 2.0 cm/sec to calculate a nitrogen deposition of 3.0 kg/ha/yr in a forest with a leaf area index of 5.8 m^2/m^2 and an ambient concentration of 0.82 μg N m^{-3} in the form of nitric acid vapor. It is often unclear if deposition velocities measured using artificial surfaces apply to natural surfaces (e.g., bark), and accurate estimates of the surface area of vegetation are difficult (Whittaker and Woodwell 1968). Clearly, further work on dry deposition is needed (Lovett 1994).

Atmospheric deposition on the surface of the sea is often estimated from collections of wetfall and dryfall on remote islands (Duce et al. 1991). The surface of the sea can also exchange gases with the atmosphere (Liss and Slater 1974), often acting as a sink for atmospheric SO_2 (Beilke and Lamb 1974) and a source of NH_3 (Quinn et al. 1987, 1988).

Regional Patterns

Regional patterns of rainfall chemistry in the United States reflect the relative importance of different constituent sources and deposition processes in different areas (Munger and Eisenreich 1983; Fig. 3.15). Coastal areas are dominated by atmospheric inputs from the sea, with large depositions of Na, Mg, Cl, and SO_4 that are the major constituents in the seasalt aerosol (Junge and Werby 1958, Hedin et al. 1995). Areas of arid and semiarid land show high concentrations of soil-derived constituents, such as Ca (Young et al. 1988, Sequeira 1993, Gillette et al. 1992). Areas downwind of regional pollution show exceedingly low pH and high concentrations of SO_4^{2-} and NO_3^- (Schwartz 1989, Ollinger et al. 1993).

The ratio among ionic constituents in rainfall can be used to trace their origin. Except in unusual circumstances nearly all the sodium (Na) in rainfall is derived from the ocean. When magnesium is found in a ratio of 0.12 with respect to Na—the ratio in seawater (Table 9.1)—one may presume that the Mg is also of marine origin. In the southeastern United States, however, Mg/Na ratios in wetfall range from 0.29 to 0.76 (Swank and Henderson 1976). Here the Mg content has increased relative to Na, presumably because the airflow that brings precipitation to this region has crossed the United States, picking up Mg from soil dust and other sources. Schlesinger et al. (1982) used this approach to deduce nonmarine sources of Ca and SO_4 in the rainfall in coastal California (Fig. 3.16).

Iron (Fe) and Al are largely derived from the soil, and ratios of various ions to these elements in soil can be used to predict their expected concentrations in rainfall when soil dust is a major source (Lawson and Winchester 1979, Warneck 1988). High concentrations of Al in dryfall on Hawaii were traced to springtime dust storms on the central plains of China (Parrington et al. 1983). Windborne particles of soil and vegetation contribute significantly to the global transport of trace metals in the atmosphere (Nriagu 1989).

In many areas downwind of pollution, a strong correlation between H^+ and SO_4^{2-} is the result of the production of H_2SO_4 during the oxidation of SO_2 and its dissolution in rainfall (Cogbill and Likens 1974, Gorham et al. 1984, Irwin and Williams 1988). Nitrate (NO_3^-) also contributes to the strong acid content in rainfall (HNO_3). These constituents depress the pH of rainfall below 5.6, which would be expected for water in equilibrium with atmospheric CO_2 (Galloway et al. 1976). Ammonia is a net source of alkalinity in rainwater, since its dissolution produces OH^-:

$$NH_3 + H_2O \rightleftarrows NH_4^+ + OH^-. \tag{3.58}$$

Thus, the pH of rainfall is determined by the concentration of strong acid anions that are not balanced by NH_4^+ and Ca^{2+} (from $CaCO_3$), viz. (from Gorham et al. 1984),

$$H^+ = [NO_3^- + 2SO_4^{2-}] - [NH_4^+ + 2Ca^{2+}]. \tag{3.59}$$

Globally, about 22% of the atmosphere's acidity is neutralized by NH_3 (Chapter 13), with a higher proportion in the southern hemisphere where there is less industrial pollution (Savoie et al. 1993). In the eastern United States, the acidity of rainfall is often directly correlated to the concentration of SO_4^{2-}. In the western United States, when high concentrations of SO_4^{2-} are derived from industrial pollution (Epstein and Oppenheimer 1986, Oppenheimer et al. 1985), the rainfall is usually less acidic, because the acid-forming anions have reacted with soil aerosols containing $CaCO_3$ (Young et al. 1988, Reheis and Kihl 1995).

The concentrations of constituents in the Greenland snowpack reflect the changes in the abundance of anthropogenic pollutants due to industrialization in the northern hemisphere (Herron et al. 1977, Mayewski et al.

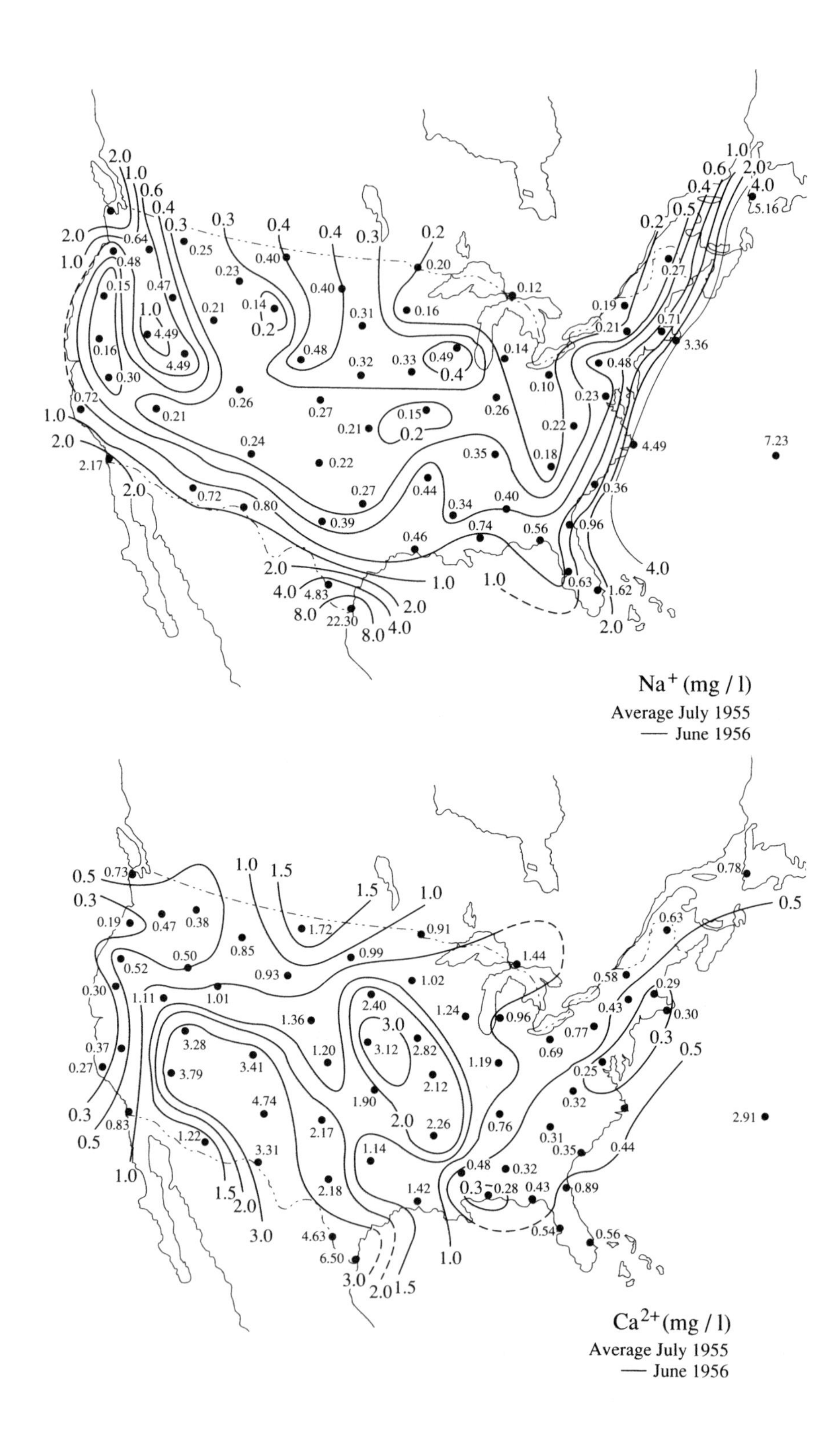
Na+ (mg / l)
Average July 1955
— June 1956
Ca2+ (mg / l)
Average July 1955
— June 1956

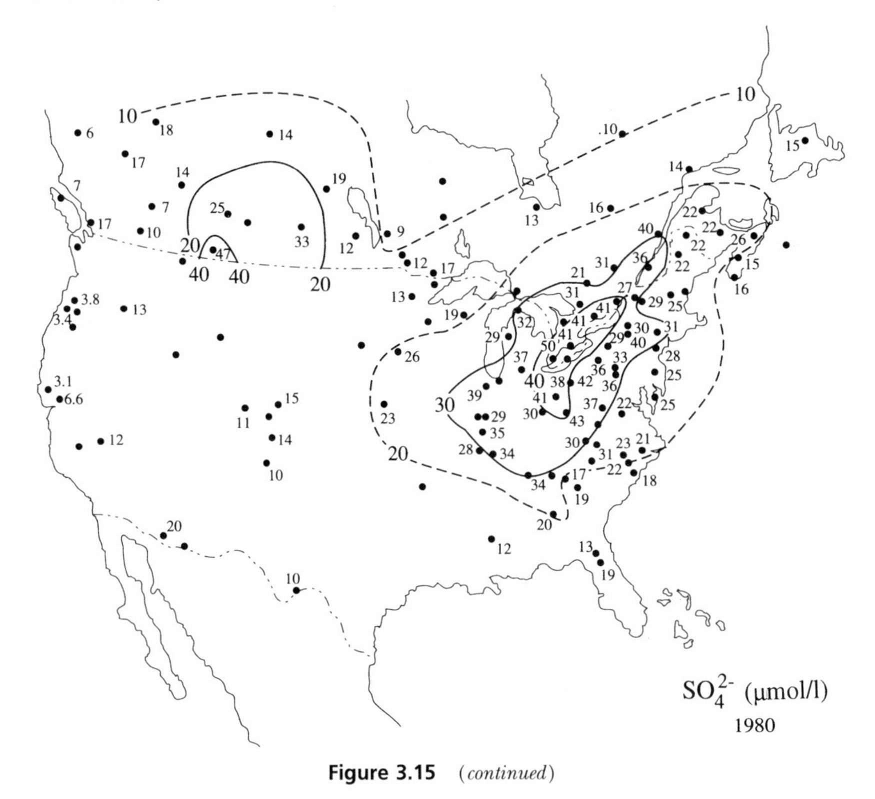

Figure 3.15 *(continued)*

1986, Laj et al. 1992). In recent ice on Greenland, SO_4 and NO_3 are enriched 3 to 4x over 18th century values. There are no apparent changes in the deposition of these ions in the southern hemisphere as recorded by Antarctic ice (Langway et al. 1994). Similarly, the uppermost sediments in lakes of the northern hemisphere contain higher concentrations of many trace metals, presumably from industrial sources (Galloway and Likens 1979, Swain et al. 1992). Long-term records of precipitation chemistry are rare, but the collections at the Hubbard Brook Ecosystem in central New Hampshire (USA) suggest a recent decline in the concentrations of Pb and SO_4 that may reflect improved control of emissions (Likens et al. 1984). Over most of the same period, however, concentrations of cations have also decreased, so the acidity of rainfall shows little change (Hedin et al. 1994).

Figure 3.15 Geographic pattern in the concentration of some major constituents in U.S. precipitation. Na and Cl are from Junge and Werby (1958) and SO_4 is modified from Barrie and Hales (1984).

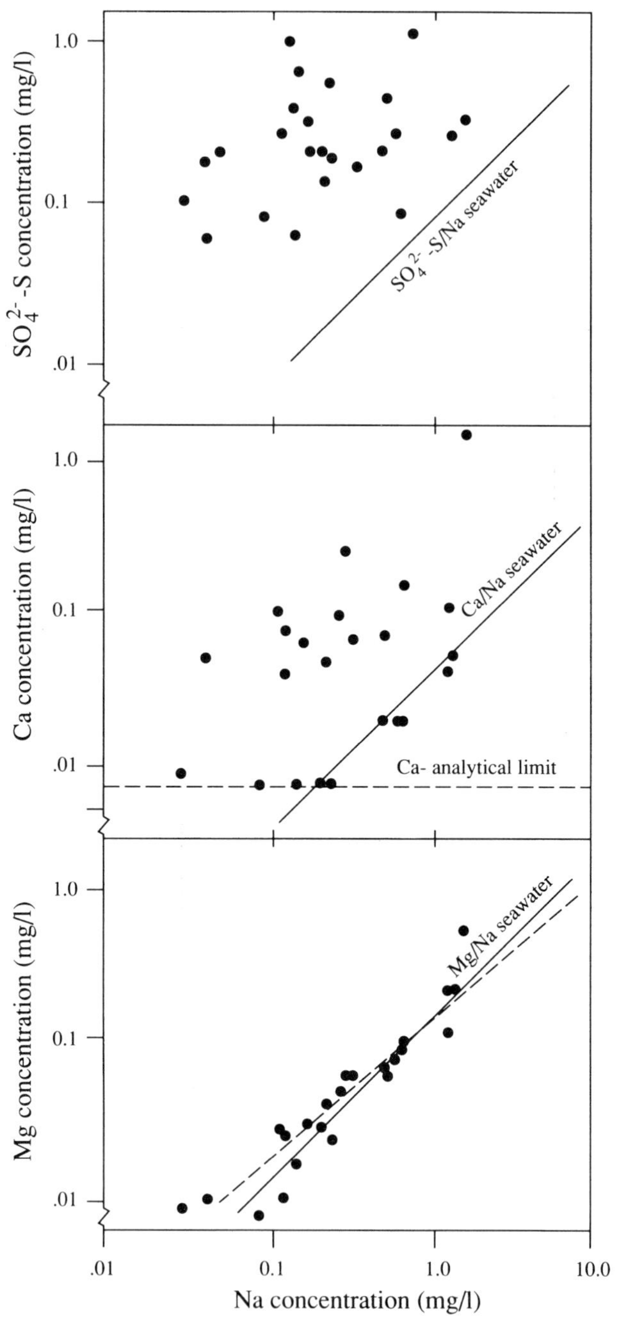

Figure 3.16 Concentrations of SO_4, Ca, and Mg in wetfall precipitation near Santa Barbara, California, plotted as a logarithmic function of Na concentrations in the same collections (Schlesinger et al. 1982). The solid line represents the ratio of these ions to Na in seawater. Ca and SO_4 are enriched in wetfall relative to seawater, whereas Mg shows a correlation (dashed) that is not significantly different from the ratio expected in seawater.

The long-term records suggest that many natural ecosystems, land and water, currently receive a greater input of N, S, and other elements of biogeochemical importance than before widespread emissions from human activities. Pollutant emissions have more than doubled the annual input of S-containing gases to the atmosphere globally (Chapter 13). The deposition of these compounds is localized (Barrie and Hales 1984). Galloway et al. (1984) calculate that the deposition of SO_4^{2-} in the eastern United States is enriched by 2 to 16 times over background conditions. The western North Atlantic ocean receives about 20 to 40% of the sulfur and nitrogen oxides emitted in eastern North America (Galloway and Whelpdale 1987, Shannon and Sisterson 1992). Deposition of nitrogen in fixed compounds might be expected to enhance the growth of forests, but in combination with acidity, this fertilization effect may lead to deficiencies of P, Mg, and other plant nutrients (Waring and Schlesinger 1985). These interactions are discussed in more detail in Chapters 6 and 9.

Summary

In this chapter we have examined the physical structure, circulation and composition of the atmosphere. Major constituents, such as N_2, are rather unreactive and have long mean residence times in the atmosphere. CO_2 is largely controlled by plant photosynthetic uptake and by its dissolution in waters on the surface of the Earth. The atmosphere contains a variety of minor constituents, many of which are reduced gases. These gases are highly reactive in homogeneous reactions with hydroxyl (OH) radicals and heterogeneous reactions with aerosols and cloud droplets, which scrub them from the atmosphere. Changes in the concentration of many trace gases are indicative of global change, perhaps leading to future climatic warming and higher surface flux of ultraviolet light. The oxidized products of trace gases are deposited in land and ocean ecosystems, resulting in the input of N, S, and other elements of biogeochemical significance. Pollution of the atmosphere by the release of oxidized gases containing N and S as a result of human activities results in acid deposition in downwind ecosystems. The enhanced deposition of N and S represents altered biogeochemical cycling on a regional and global basis.

Recommended Readings

Graedel, T.E. and P.J. Crutzen. 1993. *Atmospheric Change.* Freeman, New York.

Henderson-Sellers, A. and K. McGuffie. 1987. *A Climate Modelling Primer.* Wiley, New York.

Houghton, J.T., L.G. Meira Filho, J. Bruce, H. Lee, B.A. Callander, E. Haites, N. Harris and K. Maskell. (eds.). 1995. *Climate Change 1994.* Cambridge University Press, Cambridge.

Turco, R.P. 1996. *Earth under Siege.* Oxford University Press, Oxford.

Walker, J.C.G. 1977. *Evolution of the Atmosphere.* Macmillan, New York.

Warneck, P. 1988. *Chemistry of the Natural Atmosphere.* Academic Press, London.

Wayne, R.P. 1991. *Chemistry of Atmospheres.* Clarendon Press, Oxford.

4

The Lithosphere

Introduction

Since early geologic time, the atmosphere has interacted with the exposed crust of the Earth, causing rock weathering. Many of the volcanic gases in the Earth's earliest atmosphere dissolved in water to form acids that could react with surface minerals (Chapter 2). Later, as oxygen accumulated in the atmosphere, rock weathering occurred as a result of the oxidation of reduced minerals, such as pyrite, that were exposed at the surface of the Earth. At least since the advent of land plants, rock minerals have also been exposed to high concentrations of carbon dioxide in the soil as a result of the metabolic activities of soil microbes and plant roots. Today, carbonic acid (H_2CO_3), derived by reaction of CO_2 with soil water, is a major determinant of rock weathering in most ecosystems. In recent years, humans have added large quantities of NO_x and SO_2 to the atmosphere, causing acid rain (Chapter 3) and increasing the rate of rock weathering in many areas (Cronan 1980).

Siever (1974) proposed a basic equation to summarize the close linkage between the Earth's atmosphere and its crust:

$$\text{Igneous rocks} + \text{acid volatiles} = \text{sedimentary rocks} + \text{salty oceans}. \quad (4.1)$$

This formula recognizes that through geologic time the primary minerals of the Earth's crust have been exposed to reactive, acid-forming C, N, and S gases of the atmosphere. The products of the reaction are carried to the oceans, where they accumulate as dissolved salts or in ocean sediments (Li 1972). Large amounts of sedimentary rock have formed through geologic time; indeed, about 75% of the rocks now exposed on land are sedimentary rocks that have been uplifted by tectonic activity (Blatt and Jones 1975). These sedimentary rocks are subject to further weathering reactions with acid volatiles, in accord with Siever's basic equation. Eventually, geologic processes carry sedimentary rocks to the deep Earth, where CO_2 is released and the solid constituents are converted back to primary minerals under great heat and pressure (Fig. 1.4; see also Siever 1974).

This chapter reviews the basic types of rock weathering on land and the processes that drive these weathering reactions. Rock weathering is important to the bioavailability of elements that have no gaseous forms (e.g., Ca, K, Fe, and P; Table 4.1). Thus, weathering is basic to soil fertility,

Table 4.1 Approximate Mean Composition of the Earth's Continental Crust[a]

Constituent	Percentage composition
Si	28.8
Al	7.96
Fe	4.32
Ca	3.85
Na	2.36
Mg	2.20
K	2.14
Ti	0.40
P	0.076
Mn	0.072
S	0.070

[a] Data from Wedepohl (1995).

biological diversity, and agricultural productivity (Huston 1993). Reactions between soil waters and the solid materials in soil determine the availability of essential elements to biota and the losses of these elements in runoff.

Conversely, land plants and soil microbes affect rock weathering and soil development, potentially regulating global biogeochemistry and the Earth's climate (Fig. 1.4). In this chapter, we examine soil development in the major ecosystems on Earth. Finally, we will estimate the global rate of rock weathering in an attempt to determine the annual new supply of biochemical elements on land and the total delivery of weathering products to rivers and the sea.

Rock Weathering

Upon geologic uplift and exposure at the Earth's surface, all rocks undergo weathering, a general term that encompasses a wide variety of processes that decompose rocks. Mechanical weathering is the fragmentation or loss of materials without a chemical reaction; in laboratory terminology it is equivalent to a physical change. Mechanical weathering is important in extreme and highly seasonal climates and in areas with much exposed rock. Wind abrasion is a form of mechanical weathering in arid environments, whereas the expansion of frozen water in rock crevices—often resulting in fractured rocks—is an important form of mechanical weathering in cold climates. Plant roots also fragment rock when they grow in crevices.

Where the products of mechanical weathering are not rapidly removed by erosion, thick soils develop, and the landscape is said to be "transport-limited." In other areas, finely divided rock and soil are removed by erosion—the loss of particulate solids from the ecosystem. The products of mechanical weathering may also be lost in catastrophic events such as landslides (Swanson et al. 1982). Indicative of the high rates of mechanical weathering and erosion at high elevations, small rivers draining mountainous regions often carry exceptionally large sediment loads to the sea (Milliman and Syvitski 1992).

Chemical Weathering

Chemical weathering occurs when the minerals in rocks and soils react with acidic and oxidizing substances. Usually chemical weathering involves water, and various constituents are released as dissolved ions that are available for uptake by biota or loss in streamwaters. In many cases, mechanical weathering is important in exposing the minerals in rocks to chemical attack (e.g., Miller and Drever 1977, Anbeek 1993). If the rate of mechanical weathering is slow, it may limit the rate of chemical weathering and lower the fertility of soils.

The rate of chemical weathering depends on the mineral composition of rocks. Igneous and metamorphic rocks contain primary minerals (e.g., olivine and plagioclase) that were formed under conditions of high temperature and pressure deep in the Earth. These primary silicate minerals are crystalline in structure, and they are found in two classes—the ferromagnesian or *mafic* series and the plagioclase or *felsic* series—depending on the crystal structure and the presence of magnesium versus aluminum in the crystal lattice (Fig. 4.1). Among these minerals, the rate of weathering tends to follow the reverse order of the sequence of mineral formation during the original cooling and crystallization of rock; i.e., minerals that condensed first are the most susceptible to weathering reactions (Goldich

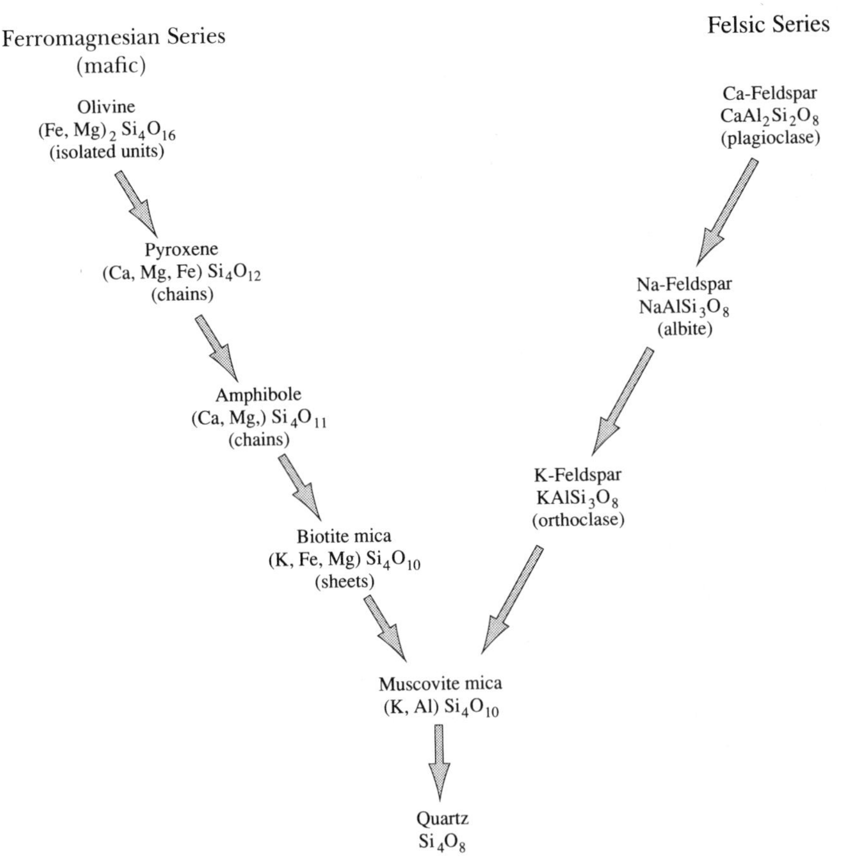

Figure 4.1 Silicate minerals are divided into two classes, the ferromagnesian series and the felsic series, based on the presence of Mg or Al in the crystal structure. Among the ferromagnesian series, minerals that exist as isolated crystal units (e.g., olivine) are most susceptible to weathering, while those showing linkage of crystal units and a lower ratio of oxygen to silicon are more resistant. Among the felsic series, Ca-feldspar (plagioclase) is more susceptible to weathering than Na-feldspar (albite) and K-feldspar (orthoclase). Quartz is the most resistant of all. This weathering series is the reverse of the order in which these minerals are precipitated during the cooling of magma.

1938). Minerals formed during rapid, early crystallization of magma at high temperatures contain few bonds that link the units of their crystalline structure. They also have frequent substitutions of various cations [e.g., Ca, Na, K, Mg, and trace metals (Fe and Mn)] in their crystal lattice, distorting its shape and increasing its susceptibility to weathering. Thus, for example, olivine, which is formed under conditions of great heat and pressure deep within the Earth, is most likely to weather rapidly when exposed at the Earth's surface.

In rocks or soils of mixed composition, chemical weathering is concentrated on the relatively labile minerals, while other minerals may be unaffected (April et al. 1986, White et al. 1996). The dissolution and loss of some constituents may result in a reduction of the density of bedrock, with little collapse or loss of the initial rock volume. The product of such isovolumetric weathering, *saprolite,* composes the lower soil profile of many regions, especially in the southeastern United States (Gardner et al. 1978, Velbel 1990, Stolt et al. 1992). In other cases, the removal of some constituents is accompanied by the collapse of the soil profile and an apparent increase in the concentration of the elements that remain (e.g., Zr, Ti, and Fe; Brimhall et al. 1991).

Quartz is very resistant to chemical weathering and often remains when other minerals are lost (Fig. 4.1). Quartz is a relatively simple silicate mineral consisting only of silicon and oxygen in tetrahedral crystals that are linked in three dimensions. In many cases, the sand fraction of soils is largely composed of quartz crystals that remain following the chemical weathering and loss of other constituents during soil development (Brimhall et al. 1991).

In addition to mineralogy, rock weathering also depends on climate (Peltier 1950, Drever and Zobrist 1992). Chemical weathering involves chemical reactions, so it is not surprising that it occurs most rapidly under conditions of higher temperature and precipitation. Chemical weathering is more rapid in tropical forests than in temperate forests, and more rapid in most forests than in grasslands or deserts. White and Blum (1995) show that the loss of Si in streamwater, which is a good index of chemical weathering, is directly related to precipitation over much of the Earth's surface (Fig. 4.2).

The dominant form of chemical weathering is the carbonation reaction, driven by the formation of carbonic acid, H_2CO_3, in the soil solution:

$$H_2O + CO_2 \rightleftarrows H^+ + HCO_3^- \rightleftarrows H_2CO_3. \qquad (4.2)$$

Because plant roots and soil microbes release CO_2 to the soil, the concentration of H_2CO_3 in soil waters is often much greater than that in equilibrium with atmospheric CO_2 at 360 ppm (or 0.036%) (Castelle and Galloway 1990, Amundson and Davidson 1990, Piñol et al. 1995). Buyanovsky and

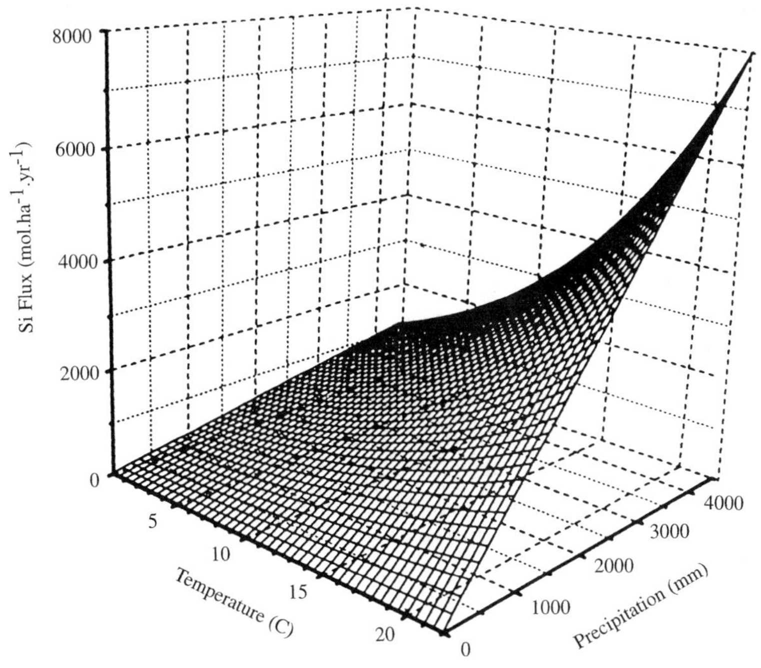

Figure 4.2 Loss of silicon (SiO_2) in runoff as a function of mean annual temperature and precipitation in various areas of the world. Modified from White and Blum (1995).

Wagner (1983) report seasonal CO_2 concentrations of greater than 7.0% in the soil beneath wheat fields in Missouri. Such high concentrations of CO_2 can extend to considerable depths in the soil profile, affecting the weathering of underlying rock (Sears and Langmuir 1982, Richter and Markewitz 1995). Wood and Petraitis (1984) found CO_2 concentrations of 1.0% at 36 m, which they link to the downward transport of organic materials that subsequently decompose at depth. Solomon and Cerling (1987) found that high concentrations of CO_2 accumulated in the soil under a mountain snowpack, potentially leading to significant weathering during the winter.

Examining data from a variety of ecosystems, Brook et al. (1983) suggest that the average concentration of soil CO_2 varies as a function of actual evapotranspiration of the site (Fig. 4.3). Plant growth is greatest in warm and wet climates. These areas maintain the highest levels of soil CO_2 and the greatest rates of carbonation weathering (Johnson et al. 1977). However, even in arid regions, rock weathering appears to be controlled by carbonation weathering (Routson et al. 1975). By maintaining high concentrations of CO_2 in the soil, living organisms exert biotic control over the geochemical process of rock weathering on land—a good example of the

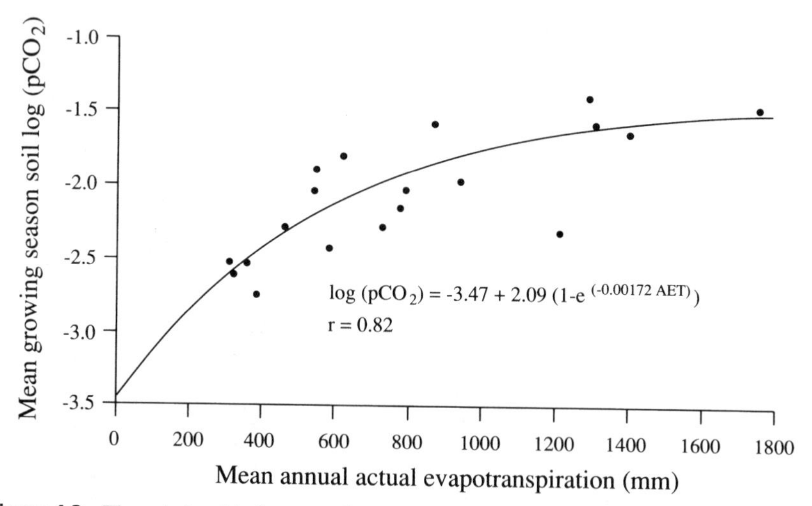

Figure 4.3 The relationship between the mean concentration of CO_2 in the soil pore space and the actual evapotranspiration of the site for various ecosystems of the world. From Brook et al. (1983).

importance of understanding biogeochemistry (Schwartzman and Volk 1989).

Carbonic acid attacks silicate rocks. For example, weathering of the Na-feldspar, albite, proceeds as

$$2NaAlSi_3O_8 + 2H_2CO_3 + 9H_2O \rightarrow$$
$$2Na^+ + 2HCO_3^- + 4H_4SiO_4 + Al_2Si_2O_5(OH)_4. \quad (4.3)$$

During this process a primary mineral is converted to a secondary mineral, kaolinite, by the removal of Na^+ and soluble silica. A sign that carbonation weathering has occurred is the observation that HCO_3^- is the dominant anion in runoff waters. Formation of the secondary mineral, kaolinite, involves hydration with H^+ and water. The secondary mineral has a lower ratio of Si to Al, as a result of the loss of some Si to streamwaters. Because only some of the constituents of the primary mineral are released, this type of weathering reaction is known as an *incongruent dissolution*. Under conditions of high rainfall, as in the humid tropics, kaolinite may undergo a second incongruent dissolution to form another secondary mineral, gibbsite:

$$Al_2Si_2O_5(OH)_4 + 5H_2O \rightarrow 2H_4SiO_4 + Al_2O_3 \cdot 3H_2O. \quad (4.4)$$

Some weathering reactions involve congruent dissolutions. In moist climates, limestone undergoes a relatively rapid congruent dissolution during carbonation weathering:

$$CaCO_3 + H_2CO_3 \rightarrow Ca^{2+} + 2HCO_3^-. \quad (4.5)$$

Olivine ($FeMgSiO_4$) undergoes congruent dissolution in water, releasing Fe, Mg, and Si (Grandstaff 1986). The magnesium and silicon are lost in runoff waters, but usually the Fe reacts with oxygen, resulting in the precipitation of Fe_2O_3 in the soil profile. Often this reaction is mediated by the chemoautotrophic bacterium *Thiobacillus ferrooxidans* (Eq. 2.15; Temple and Colmer 1951, Ralph 1979). Similarly, pyrite (FeS_2) undergoes a congruent reaction during its oxidation:

$$4FeS_2 + 8H_2O + 15O_2 \rightarrow 2Fe_2O_3 + 16H^+ + 8SO_4^{2-}. \qquad (4.6)$$

The H^+ produced in this reaction accounts for the acidity of runoff from many mining operations. As with the weathering of olivine, the Fe from weathered pyrite is subsequently precipitated as Fe_2O_3 in the soil profile or streambed (Bloomfield 1972, Garrels and MacKenzie 1971, Johnson et al. 1992).

In addition to carbonic acid, living organisms release a variety of organic acids to the soil solution that can be involved in the weathering of silicate minerals (Ugolini and Sletten 1991). Many simple organic compounds, including acetic and citric acids, are released from plant roots (Smith 1976, Tyler and Ström 1995). Organic acids from plant roots and microbes can weather biotite mica, releasing K (Boyle and Voigt 1973, April and Keller 1990). Soil microbes produce fulvic and humic acids during decomposition of plant remains (Chapter 5). Phenolic acids (i.e., tannins) are released during the decomposition of plant litter (Schlesinger 1985a), and many fungi release oxalic acid that results in chemical weathering (Cromack et al. 1979, Lapeyrie et al. 1987, Welch and Ullman 1993).

In addition to their contributions to total acidity, organic acids speed weathering reactions in the soil by combining with some weathering products, in a process called *chelation*. When Fe and Al combine with fulvic acid, they are mobile and move to the lower soil profile in percolating water (Dahlgren and Walker 1993, Lundström 1993). When these elements are involved in chelation, their inorganic concentration in the soil solution remains low, an equilibrium between dissolved products and mineral forms is not achieved, and chemical weathering may continue unabated (Berggren and Mulder 1995). Grandstaff (1986) found that additions of small concentrations of EDTA (an organic chelation agent) to weathering solutions increased the dissolution of olivine by 110 times over inorganic conditions. Fulvic and humic acids increase the weathering of a variety of silicate minerals, including quartz, particularly when the soil solution is neutral or slightly acid (Tan 1980, Bennett et al. 1988, Wogelius and Walther 1991, Welch and Ullman 1993).

Organic acids often dominate the acidity of the upper soil profile, while carbonic acid is important below (Ugolini et al. 1977). In general, organic acids dominate the weathering processes in cool temperate forests where decomposition processes are slow and incomplete, whereas carbonic acid drives the chemical weathering in tropical forests where lower concentrations of fulvic acids remain after the decomposition of plant debris (Johnson et al. 1977).

There is much debate about changes in the rate of chemical weathering through geologic time, especially as a result of the evolution of vascular land plants (Berner 1992, Drever 1994). Greater rock weathering, beginning with the initial colonization of land by plants, may have been responsible for a decline in atmospheric CO_2 300–400 million years ago, owing to the consumption of CO_2 in carbonation weathering (Knoll and James 1987, Berner 1992, 1994, Mora et al. 1996). Various workers have found evidence for higher rates of rock weathering under areas of vegetation (Crawley et al. 1969, Cochran and Berner 1992), suggesting that higher plants add large amounts of CO_2 to the soil from root metabolism and from the decomposition of plant debris by microbes. This view holds that the process of photosynthesis acts to speed the transfer of an acid volatile, CO_2, from the atmosphere to the soil profile. However, even before the advent of vascular land plants, the Earth's surface may have been covered with algae and lichens, producing relatively high levels of CO_2 in the soil (Schwartzman and Volk 1989, Keller and Wood 1993, Horodyski and Knauth 1994).

Other workers suggest that temperature and precipitation are the dominant factors controlling the rate of chemical weathering, with plants playing a lesser role (Brady and Carroll 1994, Drever 1994). The main effect of high CO_2 in the Earth's atmosphere may be to speed the rate of rock weathering by raising Earth's temperature (cf. Figs. 1.4 and 4.2). In either scenario, carbonation weathering consumes CO_2, so the rate of rock weathering on Earth may have a major long-term effect on global climate (Brady 1991).

Secondary Minerals

Secondary minerals are formed as by-products of weathering at the Earth's surface. Usually the formation of secondary minerals begins near the site where primary minerals are being attacked, perhaps even originating as coatings on the crystal surfaces (Casey et al. 1993). Many types of secondary minerals can form in soils during chemical weathering. Although weathering leads to the loss of Si as a dissolved constituent in streamwater (Fig. 4.2), some Si is often retained in the formation of secondary minerals (Eq. 4.3).

The secondary minerals in temperate forest soils are often dominated by layered silicate or "clay" minerals. These exist as small (<0.002 mm) particles that control the structural and chemical properties of soils. In general, two types of layers characterize the crystalline structure of secondary, aluminosilicate clay minerals—Si layers and layers dominated by Al, Fe, and Mg. These layers are held together by shared oxygen atoms. Clay minerals and the size of their crystal units are recognized by the number, order, and ratio of these layers (Birkeland 1984). Moderately weathered soils are often dominated by secondary minerals such as montmorillonite and illite, which have a 2 : 1 ratio of Si- to Al-dominated layers. More strongly weathered soils, such as in the southeastern United States, are dominated by kaolinite clays with a 1 : 1 ratio of layers, reflecting a greater loss of Si.

When secondary minerals incorporate elements of biochemical interest, one cannot assume that the release of those elements from primary minerals is reflected by an immediate increase in the pool of ions available for uptake by plants. Magnesium is often fixed in the crystal lattice of montmorillonite, whereas illite contains K (Martin and Sparks 1985, Harris et al. 1988). These are common secondary minerals in temperate soils. Similarly, although little nitrogen is contained in primary minerals, some 2 : 1 clay minerals incorporate N as fixed ammonium (NH_4) in their crystal lattice. Fixed ammonium can represent more than 10% of the total N in some soils (Stevenson 1982, Smith et al. 1994). The release of fixed ammonium from clay minerals is slow, but recognizing the widespread nitrogen limitation on land (Chapter 6), the dynamics of fixed ammonium may play an important role in determining the availability of N for plant growth (Mengel and Scherer 1981, Baethgen and Alley 1987, Green et al. 1994).

In contrast to the loss of Si and other cations (e.g., Ca and Na) to runoff waters, Al and Fe are relatively insoluble in soils unless they are involved in chelation relations with organic matter (Huang 1988, Alvarez et al. 1992, Allan and Roulet 1994, Ross and Bartlett 1996). In the absence of chelation reactions, these elements tend to accumulate in the soil as oxides. Initially, free Fe accumulates in amorphous and poorly crystallized forms, known as ferrihydrite, which are often quantified by extraction in a weak oxalate solution (Shoji et al. 1993, Birkeland 1984). With increasing time, most Fe is found in crystalline oxides and hydroxides, which are traditionally extracted using a reducing solution of citrate–dithionate. Some of these mineral transformations involve bacteria, and thus are biogeochemical in nature (Fassbinder et al. 1990).

Crystalline oxides and hydrous oxides of Fe (e.g., goethite and hematite) and Al (e.g., gibbsite and boehmite) are common in many tropical soils, where high temperatures and rainfall cause relatively rapid decomposition of plant debris and few organic acids remain to chelate and mobilize Fe and Al. Under these climatic conditions, the secondary clay minerals typical of temperate zone soils are subject to weathering, with the near-complete removal of Si, Ca, K, and other basic cations in streamwater (Eq. 4.4). However, in an interesting example of the importance of biota in soil development, Lucas et al. (1993) show that kaolinite may persist in the upper horizons of some rainforest soils due to the plant uptake of Si from the lower soil profile and the return of Si to the soil surface in plant debris.

Phosphorus Minerals

Phosphorus deserves special attention, since it is often in limited supply for plant growth. The only primary mineral with significant phosphorus content is apatite, which can undergo carbonation weathering in a congruent reaction, releasing P:

$$Ca_5(PO_4)_3OH + 4H_2CO_3 \rightarrow 5Ca^{2+} + 3HPO_4^{2-} + 4HCO_3^- + H_2O. \quad (4.7)$$

Although this phosphorus may be accumulated by biota, a large proportion is involved in reactions with other soil minerals, leading to its precipitation in unavailable forms. Phosphorus may be bound by iron and aluminum oxides, accounting for the low availability of phosphorus in many tropical soils (Sanchez et al. 1982a, Smeck 1985). This *occluded* phosphorus is held in the interior of crystalline Fe and Al oxides and is essentially unavailable to biota. Nonoccluded phosphorus includes forms that are held on the surface of soil minerals by a variety of reactions, including anion adsorption (see below). As seen in Fig. 4.4, the maximum level of available phosphorus in the soil solution is found at a pH of about 7.0. In acid soils, P availability is controlled by direct precipitation with iron and aluminum (Lindsay and Moreno 1960), whereas in alkaline soils phosphorus is often precipitated with calcium minerals (Cole and Olsen 1959, Lajtha and Bloomer 1988), and phosphorus may be deficient for optimal plant growth (Tyler 1994).

Walker and Syers (1976) diagram the general evolution of phosphorus availability during the weathering of rocks containing apatite (Fig. 4.5). Apatite weathers rapidly, giving rise to phosphorus contained in various other forms and to a decline of total phosphorus in the system due to losses in runoff. P released from apatite is initially held in nonoccluded forms or taken up by biota (organic P). With time, oxide minerals accumulate, and phosphorus is precipitated in occluded forms. At the later stages of weathering and soil development, occluded and organic P dominate the forms of P remaining in the system (Cross and Schlesinger 1995, Crews et al. 1995). At this stage, almost all available phosphorus may be found in organic forms in the upper soil profile, while phosphorus found at lower depths is largely bound with secondary minerals (Yanai 1992). Plant growth may depend almost entirely on the release of phosphorus from dead organic matter, defining a biogeochemical cycle of phosphorus in the upper soil horizons (Wood et al. 1984).

As seen for the weathering of silicate minerals, organic acids can influence the release of phosphorus during rock weathering. Jurinak et al. (1986) show how the production of oxalic acid by plant roots can lead to the weathering of P from apatite. Oxalate production is directly related to soil phosphorus availability in sandy soils of Florida (Fox and Comerford 1992a). Organic acids can inhibit the crystallization of Al and Fe oxides,

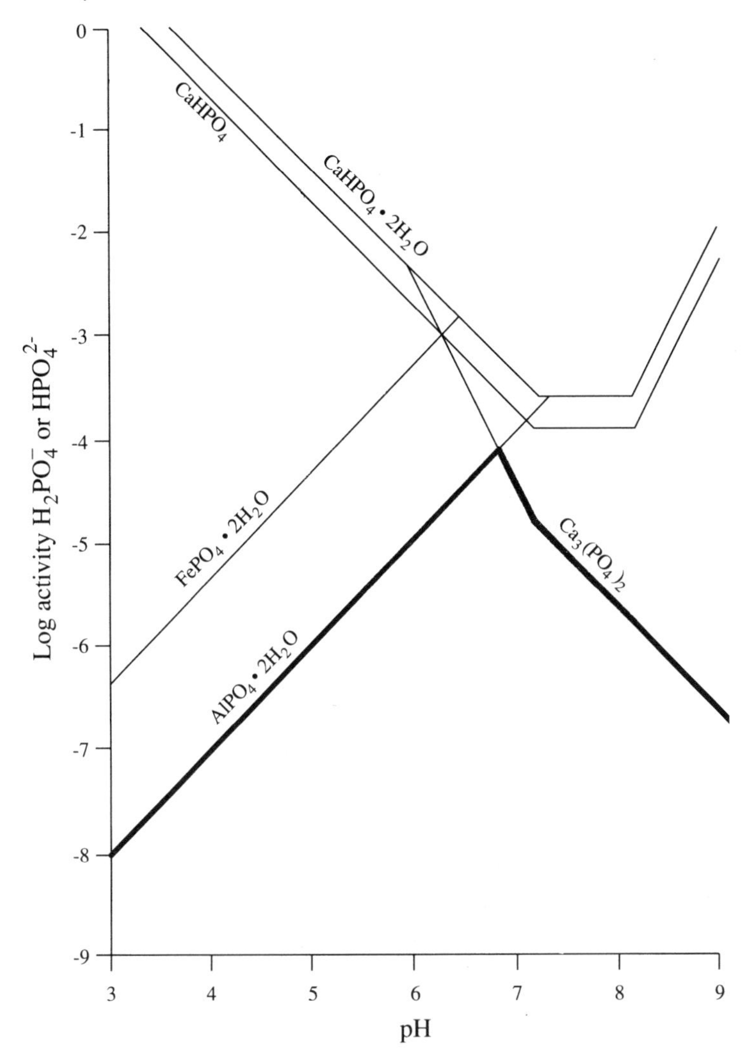

Figure 4.4 The solubility of phosphorus in the soil solution as a function of pH. Precipitation with Al sets the upper limit on dissolved phosphate at low pH (bold line); precipitation with Ca sets a limit at high pH. Phosphorus is most available at a pH of about 7.0. Modified from Lindsay and Vlek (1977).

reducing the rate of phosphorus occlusion (Schwertmann 1966, Kodama and Schnitzer 1977, 1980), and allowing noncrystalline (amorphous) forms to dominate the P-adsorption capacity of the soil (Walbridge et al. 1991, Yuan and Lavkulich 1994). Also, phosphorus may be more available in the presence of organic acids, such as oxalate, which remove Fe and Ca from the soil solution by chelation and precipitation (Graustein et al. 1977).

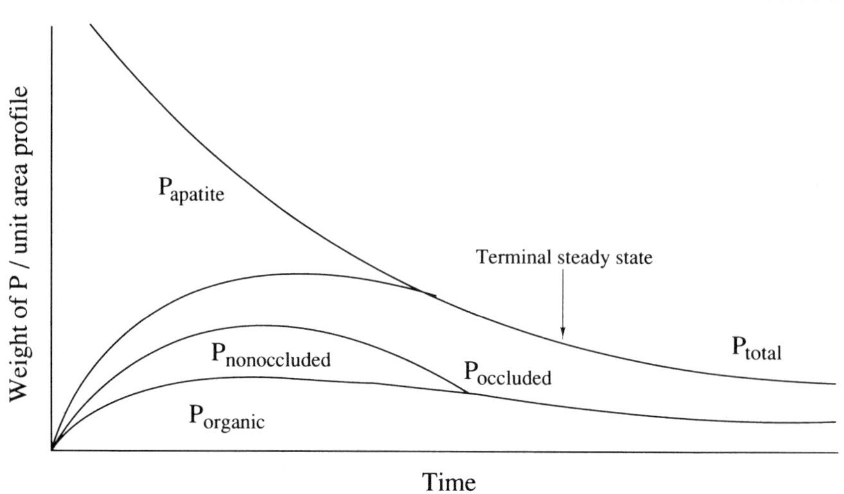

Figure 4.5 Changes in the forms of phosphorus found during soil development on sand dunes in New Zealand. Modified from Walker and Syers (1976).

The production and release of oxalic acid by mycorrhizal fungi (Chapter 6) explains their importance to the phosphorus nutrition of higher plants (Bolan et al. 1984, Cromack et al. 1979) and the greater availability of phosphorus under fungal mats (R.F. Fisher 1972, 1977). Some workers believe that the mobilization of phosphorus by symbiotic fungi was a precursor to the successful establishment of plants on land (Chapter 2).

Soil Chemical Reactions

Following their release by weathering, the availability of essential biochemical elements to biota is controlled by a number of reactions that determine the equilibrium between concentrations in the soil solution and contents that are associated with the soil mineral or organic materials. In contrast to the kinetics of weathering reactions, soil exchange reactions occur relatively rapidly (Furrer et al. 1989). The specific soil reactions differ depending on how soil development has progressed under the influence of climate, time, biota, topographic position, and the parent material of the soil (Jenny 1980).

Cation Exchange Capacity

The layered silicate clay minerals that dominate temperate zone soils possess a net negative charge that attracts and holds cations dissolved in the soil solution. This negative charge has several origins; most of it arises from ionic substitutions within silicate clays, especially in 2 : 1 clays. For example, when Mg^{2+} substitutes for Al^{3+} in montmorillonite, there is an unsatisfied negative charge in the internal crystal lattice. This negative charge is permanent in the sense that it arises inside the crystal structure and cannot be neutralized by covalent bonding of cations from the soil solution. Permanent charge is expressed as a zone or "halo" of negative charge surrounding the surface of clay particles in the soil.

A second source of negative charge is found at the edges of clay particles, where hydroxide (-OH) groups are often exposed to the soil solution. Depending on the pH of the solution, the H^+ ion may be more or less strongly bound to this group. In most cases, a considerable number of the H^+ are dissociated, leaving negative charges ($-O^-$) that can attract and bind cations (e.g., Ca^{2+}, K^+, and NH_4^+). This cation exchange capacity is known as pH-dependent charge. The binding is reversible and exists in equilibrium with ionic concentrations in the soil solution.

In many temperate soils, a large amount of cation exchange capacity is also contributed by soil organic matter. These are pH-dependent charges originating from the phenolic (-OH) and organic acid (-COOH) groups of soil humic materials. In some sandy soils, as in central Florida, and in many highly weathered soils, nearly all cation exchange is the result of soil organic matter (e.g., Daniels et al. 1987, Richter et al. 1994). Organic matter is also the major source of cation exchange in desert soils that contain a relatively small amount of secondary clay minerals as a result of limited chemical weathering.

The total negative charge in a soil is expressed as mEq/100 g or cmol(+)/kg of soil, which constitutes the cation exchange capacity (CEC). Exchange of cations occurs as a function of chemical mass balance with the soil solution. Elaborate models of ion exchange have been developed by soil chemists (Sposito 1984). In general, cations are held and displace one another in the sequence

$$Al^{3+} > H^+ > Ca^{2+} > Mg^{2+} > K^+ > NH_4^+ > Na^+ \qquad (4.8)$$

on cation exchange sites. This sequence assumes equal molar concentrations in the initial soil solution and can be altered by the presence of large quantities of the more weakly held ions. Agricultural liming, for example, is an attempt to displace Al^{3+} ions from the exchange sites by "swamping" the soil solution with excess Ca^{2+}. In most cases, few cation exchange sites are actually occupied by H^+, which reacts with soil minerals, releasing Al and other cations.

Cations other than Al and H are informally known as base cations, since they tend to form bases [e.g., $Ca(OH)_2$] when they are released to the soil solution (Birkeland 1984, p. 23). The percentage of the total cation exchange capacity occupied by base cations is termed *base saturation.* Both cation exchange capacity and base saturation increase during initial soil development on newly exposed parent materials. However, as the weathering of soil minerals continues, cation exchange capacity and base saturation decline (Bockheim 1980). Temperate forest soils dominated by 2:1 clay minerals have greater cation exchange capacity than those dominated by 1:1 clay minerals such as kaolinite. Highly weathered soils in the humic tropics are dominated by aluminum hydroxide minerals, which offer essentially no cation exchange capacity in the mineral fraction at their natural soil pH.

Soil Buffering

Cation exchange capacity acts to buffer the acidity of many temperate soils. When H^+ is added to the soil solution, it exchanges for cations, especially Ca, on clay minerals and organic matter (Bache 1984, James and Riha 1986). Over a wide range of pH, temperate soils maintain a constant value (k) for the expression

$$pH - 1/2(pCa) = k, \tag{4.9}$$

which is known as the lime potential. This expression suggests that when H^+ is added to the soil solution (lower pH), the concentration of Ca^{2+} increases in the soil solution (lower pCa), so that k remains constant. The 1/2 reflects the valence of Ca^{2+} versus H^+. As long as there is sufficient base saturation (e.g., >15%), buffering by CEC explains why the pH of many temperate soils shows relatively little change when they are exposed to acid rain (Federer and Hornbeck 1985, David et al. 1991, A.H. Johnson et al. 1994, Likens et al. 1996).

In strongly acid soils, as in the humid tropics, there is little CEC to buffer the soil solution. These soils are buffered by various geochemical reactions involving aluminum (Fig. 4.6). Aluminum is not a base cation inasmuch as its release to the soil solution leads to the formation of H^+, as Al^{3+} is precipitated as aluminum hydroxide:

$$Al^{3+} + H_2O \rightleftarrows Al(OH)^{2+} + H^+ \tag{4.10}$$

$$Al(OH)^{2+} + H_2O \rightleftarrows Al(OH)_2^+ + H^+ \tag{4.11}$$

$$Al(OH)_2^+ + H_2O \rightleftarrows Al(OH)_3 + H^+. \tag{4.12}$$

These reactions account for the acidity of many soils in the humid tropics (Sanchez et al. 1982a), but note that the reactions are reversible, so that the soil solution is also buffered against additions of H^+ by the dissolution of aluminum hydroxide. The acid rain received by the northeastern United States appears to dissolve gibbsite [$Al(OH)_3$] from many forest soils, leading to high concentrations of Al^{3+} that are toxic to fish in lakes and streams

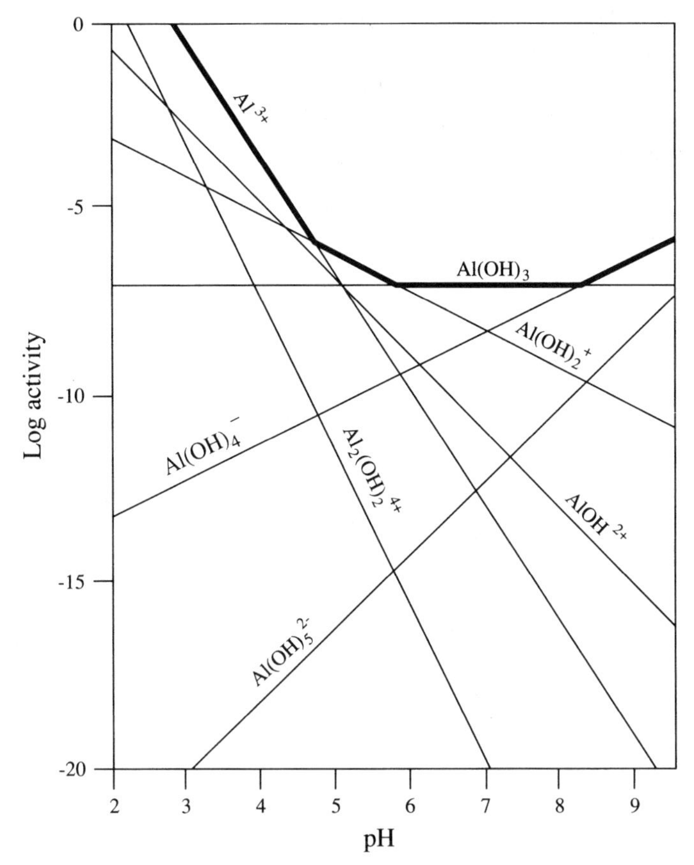

Figure 4.6 The solubility of aluminum as a function of pH. For pH in the neutral range, gibbsite [$Al(OH)_3$] controls aluminum solubility, and there is little Al^{3+} in solution. Al^{3+} become more soluble at pH < 4.7. From Lindsay (1979).

at high elevations. As streamwaters flow to lower elevations, H^+ is consumed in weathering reactions with various silicate minerals, streamwater pH increases, and aluminum hydroxides are precipitated (N.M. Johnson et al. 1981). In other regions, dissolution of Al–organic complexes (rather than gibbsite) appears to control the concentration of dissolved Al^{3+} in the soil solution (Mulder and Stein 1994, Allan and Roulet 1994).

Anion Adsorption Capacity

In contrast to the permanent, negative charge in soils of the temperate zone, tropical soils dominated by oxides and hydrous oxides of iron and aluminum show variable charge, depending on soil pH (Uehara and Gillman 1981, Sollins et al. 1988). Under acid conditions these soils possess positive charge, as a result of the association of H^+ with the surface hydroxide groups (Fig. 4.7). With an experimental increase in pH, a soil sample is observed to pass through a zero point of charge (ZPC), where the number of cation and anion exchange sites is equal. These soils develop cation exchange capacity at high pH. For gibbsite, the ZPC occurs around pH 9.0, so significant anion adsorption capacity (AAC) is present in acid tropical soils in most field situations. It is important to recognize that these reactions occur on soil constituents everywhere, but the ZPC of layered silicate minerals or soil organic matter occurs at $pH < 2.0$, so they offer little anion adsorption capacity in most temperate soils (Sposito 1984, Polubesova et al. 1995).

The ZPC of a bulk soil sample will depend on the relative mix of various minerals and organic matter. Tropical soils in Costa Rica show a ZPC at a pH of about 4.0, as a result of their mixture of soil organic matter and gibbsite (Sollins et al. 1988). Some anion adsorption capacity is found in temperate soils when iron and aluminum oxides and hydroxides occur in the lower soil profile (D.W. Johnson et al. 1981, 1986). Anion adsorption capacity is typically greater on poorly crystalline forms of Fe and Al (oxalate-extractable), which have greater surface area than crystalline forms (dithionate-extractable) (Parfitt and Smart 1978, D. W. Johnson et al. 1986). Potential adsorption of various anions, including sulfate from acid rain, is positively correlated to the oxalate-extractable Al in a variety of soils (Harrison et al. 1989, Courchesne and Hendershot 1989, MacDonald and Hart 1990, Walbridge et al. 1991).

Anion adsorption follows the sequence

$$PO_4^{-3} > SO_4^{2-} > Cl^- > NO_3^-, \tag{4.13}$$

which accounts for the low availability of phosphorus in many tropical soils. Frequently anion exchange is described using a Langmuir model (Fig. 4.8), in which the content of anions held on exchange sites is expressed as a function of the concentration in the solution (Travis and Etnier 1981, Reuss and Johnson 1986, Autry and Fitzgerald 1993). Phosphorus, sulfate, and selenate (SeO_4^{2-}) are so strongly held that the binding is known as *specific adsorption* or ligand exchange and is thought to replace -OH groups on the surface of the minerals (Fig. 4.9; Hingston et al. 1967, Guadalix and Pardo 1991). Thus, the adsorption of SO_4^{2-} from acid rain is associated with an increase in soil pH, a decline in apparent ZPC, and higher cation exchange capacity (e.g., Marcano-Martinez and McBride 1989, David et al. 1991, see also Melamed et al. 1994). All these anions are also involved in nonspecific adsorption, which is more readily reversible with changes in their concentration in the soil solution. Phosphorus held on anion adsorption sites by either mechanism is known as nonoccluded phosphorus (see above).

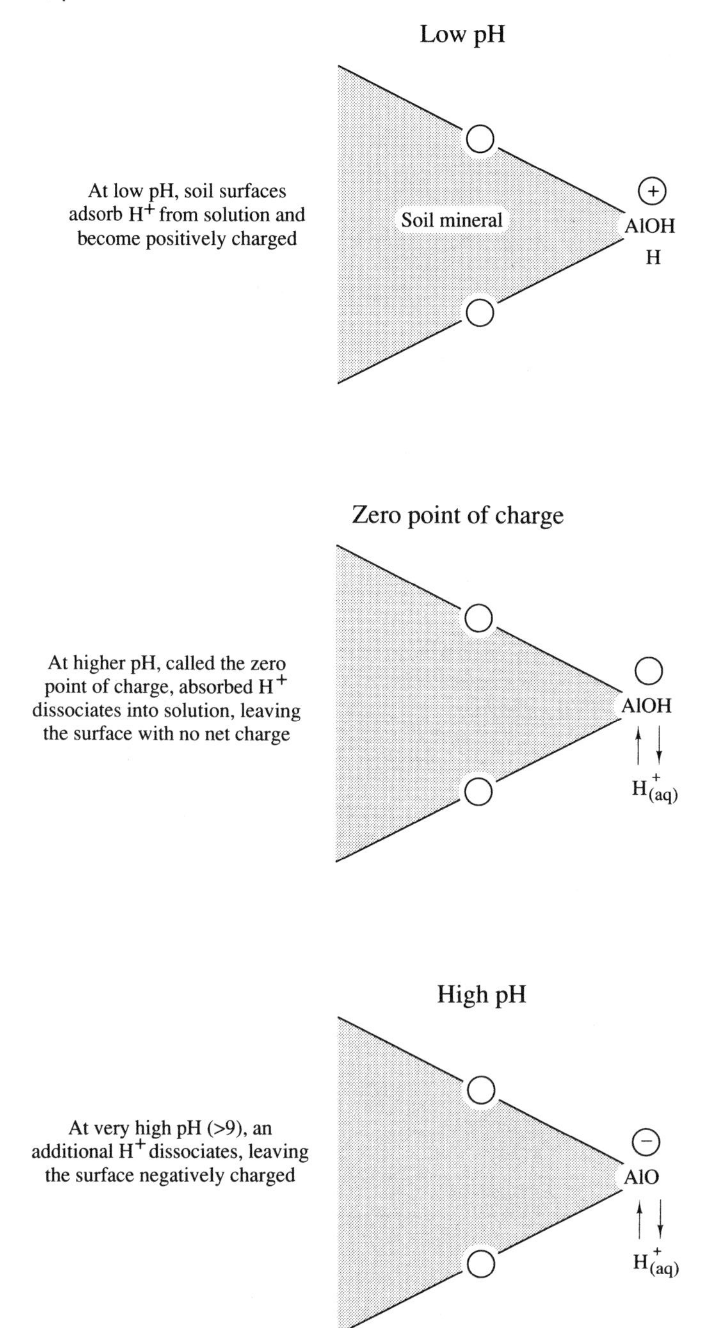

Figure 4.7 Variation in surface charge on iron and aluminum hydroxides as a function of the pH of the soil solution. From Johnson and Cole (1980).

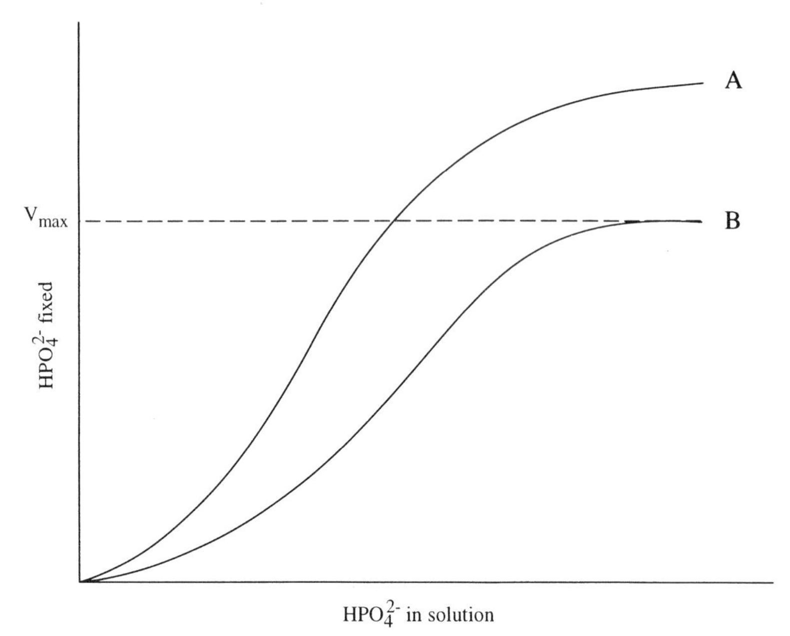

Figure 4.8 The Langmuir adsorption isotherm is used to compare the affinity of soils for anions as a function of the concentration of the anion in solution. In this diagram, soil B has a lower affinity for phosphorus than soil A; at equal concentrations of phosphorus in solution, more P will be available in soil B. Conversely, if these soils are exposed to long-term additions of solutions with a given phosphorus concentration, it will take longer for soil A to equilibrate with that solution (see Johnson and Cole 1980, Reuss and Johnson 1986). In each case, the rapid soil accumulation of P at low concentrations is presumably related to specific-adsorption, while nonspecific processes dominate the exchange at higher concentrations (Dobrovolsky 1994, p. 147).

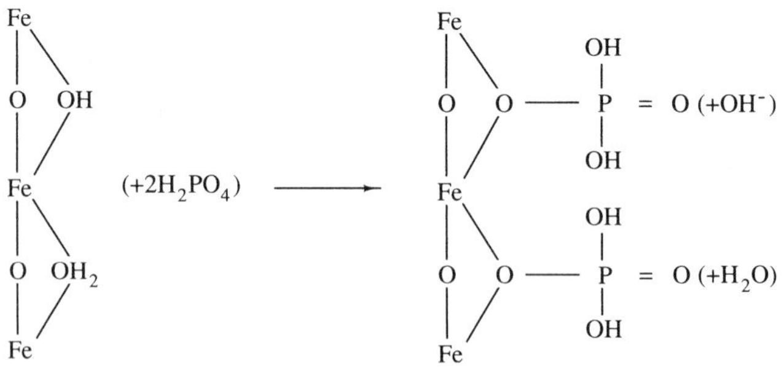

Fe O OH Fe O OH_2 Fe $(+2H_2PO_4)$ ⟶ Fe O O — P(OH)(OH) = O $(+OH^-)$ Fe O O — P(OH)(OH) = O $(+H_2O)$ Fe

Figure 4.9 The specific adsorption of phosphate by iron sesquioxides may release OH^- or H_2O to the soil solution. From Binkley (1986).

Anion adsorption capacity is inhibited by soil organic matter, especially organic anions, which tend to bind to the reactive surfaces of Fe and Al minerals (Johnson and Todd 1983, Hue 1991, Karltun and Gustafsson 1993, Gu et al. 1995). Soils or soil layers that are rich in organic matter are less efficient in anion adsorption than those dominated solely by Fe and Al oxide and hydroxide minerals. Percolating waters often carry SO_4^{2-} from the upper, organic layers of the soil to lower depths, where it is captured by Fe and Al minerals (Dethier et al. 1988, Vance and David 1991). Thus, anion adsorption capacity is determined by the effects of soil organic matter on a variety of soil properties: increasing AAC by inhibiting the crystallization of Fe and Al minerals, but reducing AAC by binding to the anion exchange sites (D.W. Johnson et al. 1986).

Soil Development

Soils usually consist of a number of layers, or horizons, that collectively constitute the complete soil profile, or pedon. Rock weathering, water movement, and organic decomposition all influence the development of a soil profile under varying climatic conditions (Jenny 1980). Of course, today, humans also have dramatic effects on soil development in many regions (Amundson and Jenny 1991). Recognition of the processes that occur in the different layers of soil is an essential part of understanding biogeochemical cycles on land. In this section, we consider soil development in forests, grasslands, and deserts.

Forests

In forests it is often easy to separate an organic layer, the forest floor, from the underlying layers of mineral soil, but these two major categories can be further subdivided. In the *forest floor,* the L- or O_i-layer consists of fresh, undecomposed plant debris, easily recognized by species.[1] The F- or O_e-layer lies immediately below the L-layer and consists of fragmented organic matter in a stage of partial decomposition. This layer is dominated by organic materials in cellular form, and fungi and bacteria are common. The designation of F-layer is derived from "fermentation," but this does not imply that the environment for microbial processes is anaerobic. Beneath the F-layer lies the H-, O_a-, or humus, layer, primarily consisting of amorphous, resistant products of decomposition, with a lower proportion of organic matter in cellular form. The lower portion of the H-horizon often

[1] The taxonomy of soils and soil horizons has undergone substantial revisions in recent years. In this text, I give both old (e.g., L) and new (O_i) equivalents, so that the student can equate early literature to more recent studies. A convenient table of equivalent terms is given by Guthrie and Witty (1982). The *Glossary of Soil Science Terms* (Soil Science Society of America, 1984) and *Keys to Soil Taxonomy* (U.S. Department of Agriculture, 1994) are also useful sources of information on the taxonomy of soils and their horizons.

contains an increasing proportion of inorganic mineral soil constituents. Differentiation of the H-layer from the uppermost layer of mineral soil is sometimes difficult, but a greater predominance of organic content versus mineral content is a useful criterion.

Not all forest soils show the differentiation of all layers of the forest floor. The thickness and presence of the layers also vary throughout the year, especially in regions where plant litterfall is strongly seasonal. In some tropical forests decomposition of fresh litter is so rapid that there is little surface litter (Olson 1963, Vogt et al. 1986). On the other hand, slow decomposition in coniferous forests, especially in the boreal zone, results in the accumulation of a thick forest floor, known as a *mor,* that is sharply differentiated from the underlying soil (Romell 1935). Much of the arctic zone is characterized by waterlogged soils, in which the entire rooting zone is composed of organic materials. Such peatland soils are known as *Histosols.* We will treat the special properties of waterlogged organic soils in Chapter 7.

The upper mineral soil is designated as the A-horizon, which often contains a significant organic fraction. It may vary in thickness from several centimeters to 1 m. The A-horizon is recognized as a zone of removal or *eluvial* processes. In most temperate regions soil water percolating through the forest floor contains organic acids derived from the microbial decomposition of litter (Vance and David 1991). These organic acids dominate the weathering of soil minerals in the A-horizon. Solutions collected beneath the A-horizon contain cations and silicate, derived from weathering reactions (Table 4.2). Iron and Al may also be removed from the A-horizon by chelation with fulvic acids that percolate downward from the forest floor

Table 4.2 Chemical Composition of Precipitation, Soil Solutions, and Groundwater in a 175-yr-old *Abies amabilis* Stand in Northern Washington[a]

Solution	pH	Total cations (mEq/liter)	Soluble ions (mg/liter)			Total (mg/liter)	
			Fe	Si	Al	N	P
Precipitation							
Above canopy	5.8	0.03	<0.01	0.09	0.03	0.60	0.01
Below canopy	5.0	0.10	0.02	0.09	0.06	0.40	0.05
Forest floor	4.7	0.14	0.04	3.50	0.79	0.54	0.04
Soil							
15 cm E	4.6	0.12	0.04	3.55	0.50	0.41	0.02
30 cm B_s	5.0	0.08	0.01	3.87	0.27	0.20	0.02
60 cm B3	5.6	0.25	0.02	2.90	0.58	0.37	0.03
Groundwater	6.2	0.26	0.01	4.29	0.02	0.14	0.01

[a] Data from Ugolini et al. (1977), *Soil Sci.* **124,** 291–302. Copyright (1977) Williams & Wilkins.

(Ugolini et al. 1977, Antweiler and Drever 1983, Driscoll et al. 1985). Downward transport of Fe and Al in conjunction with organic matter is known as *podzolization* (Chesworth and Macias-Vasquez 1985, Ugolini and Dahlgren 1987, Jersak et al. 1995).

Although it is found throughout the world, podzolization is particularly intense in subarctic (boreal) and cool temperate forests (e.g., Ugolini et al. 1987, Evans 1980, De Kimpe and Martel 1976). Many of these areas are characterized by coniferous forests which produce litterfall that is rich in phenolic compounds and organic acids (Cronan and Aiken 1985). In these ecosystems, decomposition is slow and incomplete, and large quantities of fulvic acid percolate from the forest floor into the underlying A-horizon. The pH of the soil solution is often as low as 4.0 (Dethier et al. 1988, Vance and David 1991). When the removal of Fe, Al, and organic matter is very strong, a whitish layer is easily recognized at the base of the A-horizon. This horizon is sometimes designated as an A_e- or E- (eluvial) horizon, which may consist entirely of quartz grains that are resistant to weathering and relatively insoluble under acid conditions (Pedro et al. 1978). These eluvial horizons reflect the importance of biota in soil development; in the absence of organic chelation, one would expect Fe and Al to accumulate as weathering products and for Si to dominant the losses from the upper soil profile.

During soil development, substances leached from the A- and E-horizons are deposited in the underlying B-horizons (Jersak et al. 1995). These are defined as the zone of deposition or the *illuvial* horizons, where secondary clay minerals accumulate. Clay minerals arrest the downward movement of dissolved organic compounds that are carrying Fe and Al (Greenland 1971, Chesworth and Macias-Vasquez 1985, Cronan and Aiken 1985, Schulthess and Huang 1991). Typically Fe minerals precipitate first, and Al moves lower in the profile (Adams et al. 1980, McDowell and Wood 1984, Olsson and Melkerud 1989, K.R. Law et al. 1991), but the mechanism underlying this difference is not entirely clear. Strongly podzolized soils, *Spodosols*, are characterized by a dark spodic horizon designated Bhs that is rich in Fe and organic matter. The development of a spodic horizon requires from 350 (Singleton and Lavkulich 1987) to several thousand years (Protz et al. 1984, 1988, Barrett and Schaetzl 1992). In New England forests, the accumulation of organic matter in the spodic horizon appears to limit the loss of dissolved organic carbon in streams (McDowell and Wood 1984, Chapter 8).

In warmer climates, decomposition is more rapid, smaller quantities of fulvic acids remain to percolate through the A-horizon, podzolization is less intense, and there is no sharply defined E-horizon (Pedro et al. 1978). Podzolization is found in only a few tropical soils, usually those that develop on sandy parent materials (Bravard and Righi 1989). In most areas of the tropics, decomposition is so complete that there is almost no soluble organic acid percolating through the soil profile (Johnson et al. 1977). In the absence of podzolization, Fe and Al are not removed from the upper soil profile by chelation; rather, they are precipitated as oxides and hydroxides that accumulate in the zone of active weathering. Cations and Si are lost to runoff.

In the absence of a spodic horizon, soils in warm temperate forests are often classified as *Alfisols*. These soils have high base saturation (>35%) in the subsoil and accumulations of clay in one or more layers of the B-horizon, which are known as argillic horizons and designated Bt. In areas of particularly intense or long-term weathering, such as in the southeastern United States, *Ultisols* develop. These are recognized by a high content of iron oxide in the lower profile, which gives a yellowish to deep-reddish coloration to the B-horizon. These soils have low base saturation and are transitional to the highly weathered soils of the tropics (Markewich and Pavich 1991).

Soils in tropical forests may be many meters in depth, because in many areas they have developed over millions of years, without disturbances such as glaciation (Birkeland 1984). In the absence of clear zones of eluviation and illuviation, the distinction of A- and B-horizons is difficult. Long periods of intense weathering have removed cations and silicon from the entire soil profile. The climate of most tropical regions includes high precipitation, and the solubility of Si increases with increasing temperature (Meybeck 1980, Fig. 4.2). Many tropical soils are classified as *Oxisols,* on the basis of high contents of Fe and Al oxides throughout the soil profile (Richter and Babbar 1991). Over large portions of lowland tropical rainforest, soils are acid (with Al buffering), low in base cations, and infertile with P deficiency (Sanchez et al. 1982a). Under extreme conditions, these soils are known informally as laterite.

A comparative index of soil formation and the degree of podzolization is seen in the ratio of Si to sesquioxides (Fe and Al) in the soil profile (Table 4.3). In boreal forest soils, Si is relatively immobile and Fe and Al are removed, which results in high values for this ratio in the A-horizon. The accumulation of secondary minerals such as montmorillonite in the moderately weathered soils of the glaciated portion of the United States yields Si/sesquioxide ratios of two to four, as a result of the ratio of Si to Al in the crystal lattice. Silicon/sesquioxide ratios are lower in more highly weathered soils. During soil development in the southeastern United States, low ratios characterize older soils in which kaolinite (a 1 : 1 mineral) has accumulated as a secondary mineral (Fig. 4.10). Tropical Oxisols and Ultisols have very low values for this ratio in all horizons, because they are dominated by kaolinite and iron and aluminum minerals.

Below the B-horizons, the C-horizon consists of coarsely fragmented soil material with little organic content. When the soil has developed from local materials, the C-horizon shows mineralogical similarity to the underlying parent rock. In contrast, when the parent materials have been deposited

Table 4.3 Silicon/Sesquioxide ($Al_2O_3 + Fe_2O_3$) Ratios for the A- and B-Horizons of Some Soils in Different Climatic Regions[a]

Region	Number of sites	Mean Si/sesquioxide ratio		Reference
		A-horizon	B-horizon	
Boreal	1	12.0	8.1	Wright et al. (1959)
Boreal	1	9.3	6.7	Leahey (1947)
Cool-temperate	4	4.07	2.28	Mackney (1961)
Warm-temperate	6	3.77	3.15	Tan and Troth (1982)
Tropical	5	1.47	1.61	Tan and Troth (1982)

[a] Note that the removal of Al and Fe results in high values in boreal and cool temperate soils, especially in the A-horizon. Lower values characterize tropical soils and there is little differentiation between horizons as a result of the removal of Si from the entire profile in long periods of weathering. Modified from Waring and Schlesinger (1985).

by transport (e.g., glacial till), there may be little correspondence between the C-horizon and the underlying bedrock. In either case, carbonation weathering tends to predominate in the C-horizon, which may consist of a thick saprolite of "rotten" rock (Ugolini et al. 1977). Weathering can extend to a depth of hundreds of meters in the soil profile (Richter and Markewitz 1995), and it controls the chemistry of groundwater (Chapter 7).

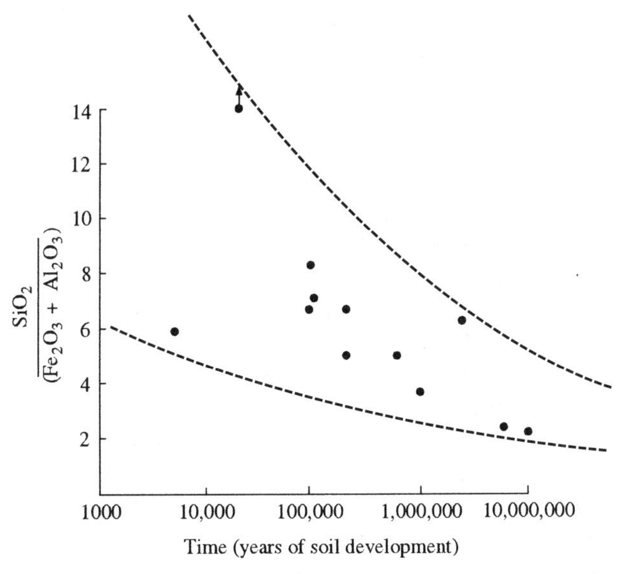

Figure 4.10 Changes in the Si-to-sesquioxide ratio during the long-term development of soils in the southeastern United States (Markewich and Pavich 1991). The data are recalculated from Markewich et al. (1989).

The distribution of soils forms a continuous gradient over broad geographic regions, in response to parent materials, topography, climate, vegetation, and time (Jenny 1980). For example, the degree of podzolization varies beneath deciduous and coniferous forests in the same geographic region (Stanley and Ciolkosz 1981, De Kimpe and Martel 1976). Soil profile development on steep slopes is often incomplete as a result of landslides and other mechanical weathering events. Soils in floodplain areas and those that have received deposits of volcanic ash may have "buried" horizons (Dahlgren and Ugolini 1989). In areas of recent disturbance, soils with little or no profile development are known as *Inceptisols* and *Entisols,* respectively. In all cases one must remember that soil profile development is slow compared to changes in vegetation.

Soil profile development is also affected by human activities. In the Piedmont region of the southeastern United States, the forest floor often resides directly on top of the B-horizon. Here, the A-horizon was lost by erosion during past agricultural use. In the northeastern United States, forest Spodosols are now exposed to acid rain. At high elevations, the acidity of the solution percolating through the forest floor and A-horizon is not dominated by organic or carbonic acids, but by "strong" acids such as H_2SO_4 (see Chapter 13). At normal levels of acidity, aluminum is mobile only as an organic chelate, and it is precipitated in the B-horizon. Under the present conditions of higher acidity, Al is mobile as Al^{3+}, which is carried through the lower profile to streamwaters with SO_4^{2-} as a balancing anion (Fig. 4.6; Johnson et al. 1972, Cronan 1980, Reuss et al. 1987). The overall rate of chemical weathering has increased (Cronan 1980, April et al. 1986).

Grasslands

In contrast to soil development in forests, where precipitation greatly exceeds evapotranspiration and excess water is available for soil leaching and runoff, soil development in grasslands proceeds under conditions of relative drought. The products of chemical weathering are not rapidly leached from the soil profile, so soils remain near neutral pH with high base saturation. High contents of Ca and other cations tend to flocculate clay minerals and fulvic acids in the upper profile (Oades 1988), limiting podzolization and the downward movement of weathering products. Often there is little development of a Bt horizon in grassland soils until the upper profile has been leached free of Ca. Overall, the intensity of chemical weathering and podzolization in grassland soils is much less than in forests (Madsen and Nørnberg 1995).

Trends in the development of grassland soils are best seen by examining a transect across the midportion of the United States. Mean annual precipitation decreases westward from the tallgrass prairie to the shortgrass prairie at the base of the Rocky Mountains. Honeycutt et al. (1990) show that the thickness of the soil profile—measured by the depth to the peak content of clay—decreases from east to west along this gradient (Fig. 4.11). Along the same gradient, base saturation, pH, and the content of calcium increase (Ruhe 1984). Similar trends are seen along a gradient of decreasing precipitation in tropical grasslands of eastern Africa (Scott 1962). In the western Great Plains, the leaching of the soil profile is so limited that Ca precipitates as $CaCO_3$, which accumulates in the lower profile in calcic horizons, designated Bk, and informally known as caliche.

The climatic regime of grasslands results in lower levels of primary production than in forests, and smaller quantities of plant residues are added to the soil each year. Nevertheless, grassland soils contain large stores of organic matter, because the limited availability of water also results in slower rates of decomposition (Chapter 5). Most grassland soils in the temperate zone are classified as *Mollisols,* on the basis of a high content of organic carbon and high base saturation in the surface layers.

As seen for forest ecosystems, soil properties in grasslands vary locally as a result of differences in underlying parent materials and hillslope positions. Schimel et al. (1985) show how the downslope movement of materials results in thin soils on hillslopes and accumulations of organic matter and nitrogen in local depressions. Similarly, Aguilar and Heil (1988) found that grassland soils derived from sandstone had lower contents of organic carbon, nitrogen, and total P than soils derived from fine textured materials,

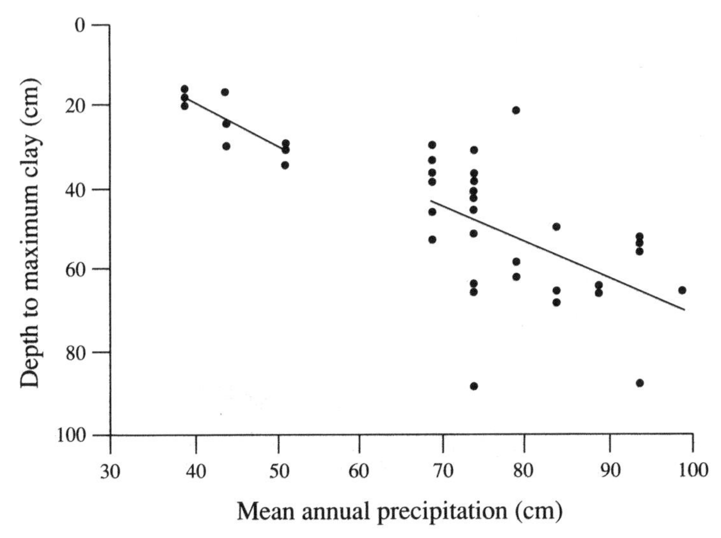

Figure 4.11 The depth to the peak content of clay in the soil profile, an index of weathering and soil development, decreases from east to west across the Great Plains of the United States as a function of the decrease in mean annual precipitation. From Honeycutt et al. (1990).

such as shale, which have a higher content of clay minerals. The proportion of P contained in organic forms was greater for soils derived from sandstone. These differences in soil properties can strongly affect the local productivity of grasslands.

Deserts

The trends in soil development that are seen with increasing aridity in grasslands reach their extreme expression in deserts. There is little runoff from most desert regions, so the soils retain a record of the vertical leaching and horizontal redistributions of materials across the landscape, much like a chromatographic column used in a laboratory. Over much of the southwestern United States, soils develop from materials that have been deposited by alluvial transport from adjacent mountain ranges (Cooke and Warren 1973). Soil development can be studied by examining an age or chronosequence of soils that are derived from similar parent materials (Fig. 4.12). Recently deposited soils show little profile development and are classified as Entisols; older soils show several distinct horizons and are classified as *Aridisols* (Dregne 1976).

Chemical weathering proceeds slowly in deserts, but the small amount of water percolating through these soils transports substances both vertically in the profile and horizontally across the landscape. As water is removed by plant uptake, soluble substances precipitate. Typically, well-developed soils contain $CaCO_3$ horizons that show progressive development and cementation through time (Gile et al. 1966). Depth to the $CaCO_3$ horizon shows a direct relation to mean annual rainfall and wetting of the soil profile (Arkley 1963, 1967, Schlesinger 1982). Beneath the $CaCO_3$, one

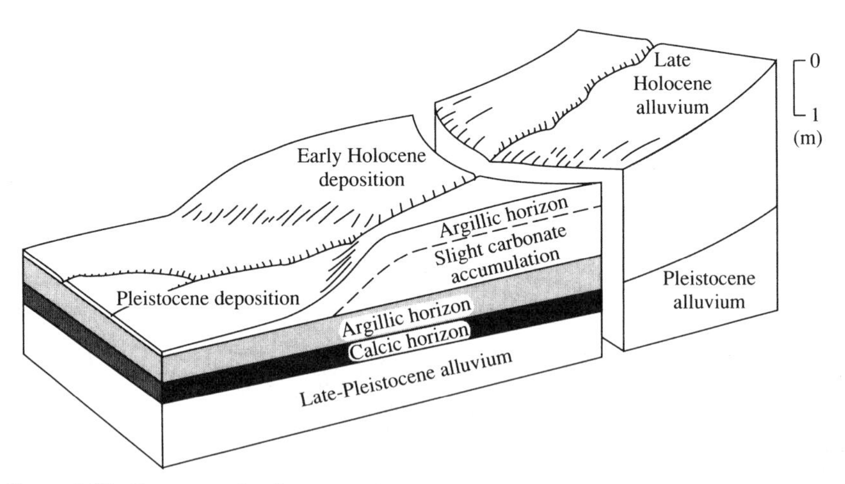

Figure 4.12 Sequence of soil age and formation on alluvial material in the Chihuahuan desert of New Mexico. From Lajtha and Schlesinger (1988).

may find horizons in which $CaSO_4 \cdot 2H_2O$ (gypsum) or NaCl are dominant, reflecting the greater solubility and downward movement of these salts (Yaalon 1965, Marion and Schlesinger 1994). Similar patterns are seen across the landscape, where Na, Cl, and SO_4 are carried to intermittent lakes in basin lows, while Ca remains in the upland soils of the adjacent piedmont (Drever and Smith 1978, Eghbal et al. 1989).

Despite sparse plant cover, much of the nutrient cycling in desert ecosystems is controlled by biota. With widespread root systems, desert shrubs accumulate nutrients from a large area and concentrate dead organic matter in the local area beneath their canopy. Most of the annual turnover of N, P, and other elements is controlled by biogeochemical processes in these "islands of fertility" (Klopatek 1987, Burke 1989, Schlesinger et al. 1996). In the soil solution beneath shrubs, the deposition of $CaCO_3$ is inhibited by the presence of dissolved organic materials in the soil solution (Inskeep and Bloom 1986, Reddy et al. 1990, Suarez et al. 1992), and calcic horizons are found deeper in the soil profile. Because $CaCO_3$ precipitates in desert soils in equilibrium with CO_2 derived from plant root respiration,

$$2CO_2 + 2H_2O \rightleftarrows 2H^+ + 2HCO_3^- \tag{4.14}$$

$$Ca^{2+} + 2HCO_3^- \rightleftarrows CaCO_3 + H_2O + CO_2, \tag{4.15}$$

this carbonate carries a carbon isotopic signature that can be traced to photosynthesis (Chapter 2) and used as an index of past climate and vegetation (Quade et al. 1989). Seasonal variations in the activity of plant roots (Schlesinger 1985b) and soil microbes (Monger et al. 1991) affect $CaCO_3$ deposition and other aspects of soil development in deserts, despite the outward appearance that abiotic processes should be dominant.

Soil development in desert ecosystems occurs slowly, due to limited weathering and leaching of the soil profile. However, desert soils frequently contain clay minerals deposited in eolian dust, giving the appearance that substantial weathering has occurred (Singer 1989). Most calcic horizons are $>$10,000 years old, and the $CaCO_3$ has accumulated at rates of 1.0 to 5.0 g m^{-2} yr^{-1} from the downward transport of Ca-rich minerals deposited from the atmosphere (Schlesinger 1985b, Chadwick et al. 1994). In the southwestern United States, the horizons of many desert soils are thought to have been formed under conditions of greater rainfall during the latest Pleistocene glaciation (Chadwick et al. 1995). For instance, many desert soils contain horizons of illuvial clay and iron oxide minerals, indicating greater rates of weathering and illuviation than occur in the modern climate (Nettleton et al. 1975, McFadden and Hendricks 1985, Harden et al. 1991, Graham and Franco-Vizcaíno 1992).

Models of Soil Development

The processes underlying soil profile development are conducive to simulation modeling. Models of soil chemistry include the weathering reactions described earlier in this chapter and equilibrium constants for the exchange of cations and anions between the soil solution and the mineral phases (Furrer et al. 1989). Depending on the time scale of the simulation, a model of soil chemistry usually routes daily or annual precipitation sequentially through the soil profile, where the solution achieves an equilibrium with the soil minerals in each horizon. Removal of water from the soil profile is calculated from estimates of evaporation from the soil surface, plant uptake, and runoff to streams. Simulation models of soil processes have been constructed to predict soil profile development and to calculate losses of dissolved constituents in forested regions subject to acid rain (Reuss 1980, Cosby et al. 1985, 1986, David et al. 1988).

The long-term development of soils in arid regions is simulated in a model, CALDEP, developed by Marion et al. (1985), in which daily precipitation achieves equilibrium with carbonate biogeochemistry as it percolates through the soil profile. Plant root respiration is explicitly included in the calculation, and it varies seasonally and with depth in the profile. Plants also control the loss of water from the soil surface by evapotranspiration. The model suggests that the $CaCO_3$ horizon will be deeper in a soil profile developed from coarse-textured parent materials, which allow greater percolation of water, and when plant root respiration varies seasonally, showing high values of soil CO_2 during the growing season.

Using current climatic conditions to parameterize precipitation and evaporation, the model was run to simulate 500 years of soil profile development (Fig. 4.13). It predicted mean depths to the $CaCO_3$ that were much shallower than observed in a sample of 16 desert soils from Arizona. When the model was reparameterized using the cool, wet conditions that are thought to have been widespread in the southwestern U.S. during the latest Pleistocene glaciation, the predicted depth of $CaCO_3$ closely matched that found in the field. These conditions produced greater percolation of soil moisture and lower rates of evaporation from the soil surface. Such models are only as good as the data used in the simulations, and rarely can models establish the importance of processes unequivocally. Nevertheless, models are useful for hypothesis development and for organizing research priorities. CALDEP suggests that most $CaCO_3$ horizons were formed during the Pleistocene, when deserts received more precipitation than today. This suggestion is consistent with the ^{14}C age of many desert soil carbonates (Schlesinger 1985b).

Weathering Rates

Rock weathering and soil formation are difficult to study because the processes occur slowly and the soil profile is nearly impossible to sample without

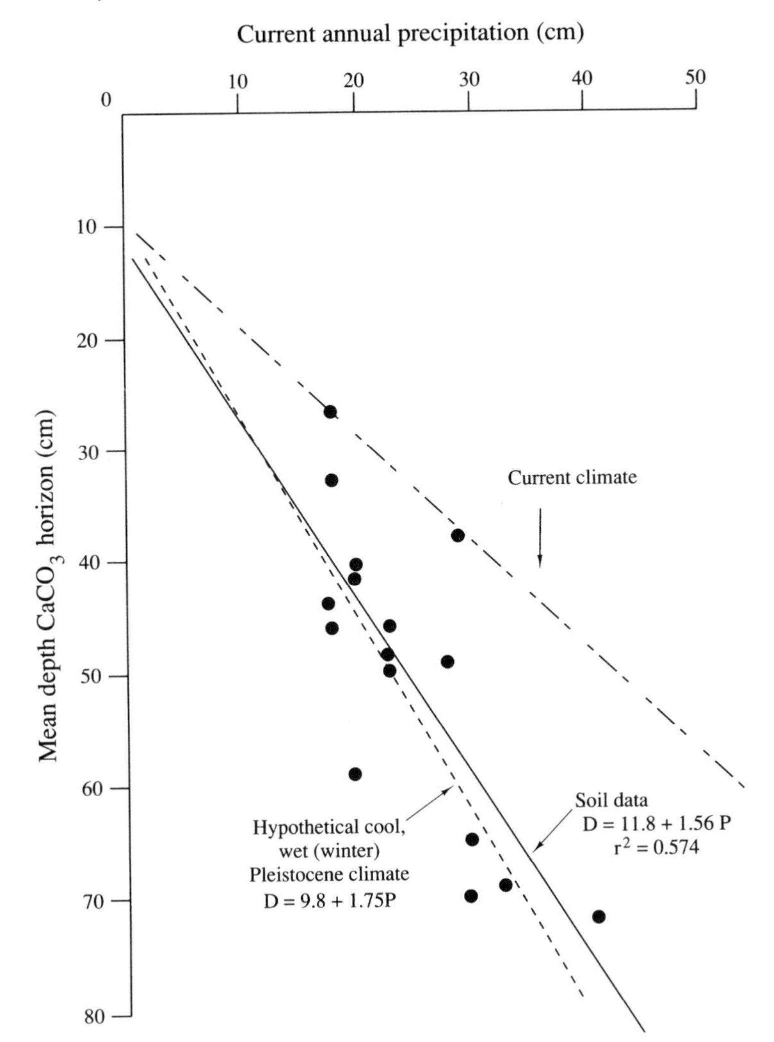

Figure 4.13 Depth to $CaCO_3$ in desert soils of Arizona (USA) as a function of mean annual precipitation. The dashed line (---) shows the prediction from the CALDEP model using current precipitation regimes. The solid line (—) shows the best fit to actual data reported from the field. The dotted line (---) shows the predictions when the model is run with hypothetical climatic data from the latest Pleistocene pluvial period. Modified from Marion et al. (1985).

disturbing many of the chemical reactions of interest. Nevertheless, estimates of weathering are essential to understanding the biogeochemistry of local watersheds, where essential elements for biota are derived from the underlying rock. Often we must infer the probable weathering reactions and estimate weathering rates from what remains in the soil profile and what is lost to streamwater. Estimates of the dissolved and suspended load of rivers allow us to calculate a global estimate of weathering, which supplies nutrient elements to land biota (Chapter 6), rivers (Chapter 8), and the sea (Chapter 9).

Chemical Weathering Rates

One of the best-known attempts to calculate the rate of rock weathering began in 1963, when Gene Likens, Herbert Bormann, and Noye Johnson quantified the chemical budgets for the Hubbard Brook forest in New Hampshire (Likens and Bormann 1995). Here, a number of comparable watersheds are underlain by an impermeable bedrock with no flow to groundwater. These workers reasoned that if the atmospheric inputs of chemical elements were subtracted from the streamwater losses, the difference should reflect the annual release from rock weathering. They estimated the rate of rock weathering using the equation

$$\text{Weathering} = \frac{(\text{Ca lost in streamwater}) - (\text{Ca received in precipitation})}{(\text{Ca in parent material}) - (\text{Ca in residual material in soil})}. \tag{4.16}$$

The solution of this equation, however, shows rather different amounts of weathering when the calculations are performed using different rock-forming elements (Table 4.4). The observed losses of calcium and sodium in streamwater imply higher rates of weathering than what is calculated using potassium and magnesium in the same equation. Johnson et al. (1968) suggest that the latter elements are accumulating in secondary minerals (illite and vermiculite) in the soil. In addition, trees may take up and store essential elements in long-lived tissues (e.g., wood growth), temporarily reducing the loss of some elements in streamwater (Taylor and Velbel 1991; see also Chapter 6).

Table 4.4 Calculation of the Rate of Primary Mineral Weathering, Using the Streamwater Losses and Mineral Concentrations of Cationic Elements[a]

Element	Annual net loss ($kg\ ha^{-1}\ yr^{-1}$)	Concentration in rock (kg/kg of rock)	Concentration in soil (kg/kg of soil)	Calculated rock weathering ($kg\ ha^{-1}\ yr^{-1}$)
Ca	8.0	0.014	0.004	800
Na	4.6	0.016	0.010	770
K	0.1	0.029	0.024	20
Mg	1.8	0.011	0.001	180

[a] Data from Johnson et al. (1968).

Release from rock weathering is the dominant source of Ca, Mg, K, Fe, and P in streamwaters draining the Hubbard Brook Experimental Forest, whereas deposition from the atmosphere is the dominant input for Cl, S, and N, which have a small content in rocks (Table 4.5). In forests not subject to regional inputs of acid rain, the proportion of sulfur that is derived from the atmosphere may be somewhat lower than that at Hubbard Brook (e.g., Mitchell et al. 1986). A large number of watershed studies have been conducted, allowing similar calculations of weathering rates for a variety of ecosystems (Table 4.6; Likens and Bormann 1995, Henderson et al. 1978, Feller and Kimmins 1979, Velbel 1992). In most areas underlain by silicate rocks, the loss of elements in streamwater relative to their concentration in bedrock follows the order

$$\mathrm{Ca} > \mathrm{Na} > \mathrm{Mg} > \mathrm{K} > \mathrm{Si} > \mathrm{Fe} > \mathrm{Al}, \tag{4.17}$$

but the order is affected by the specific composition of bedrock and the secondary minerals that are formed in the soil profile (Holland 1978, Harden 1988, Hudson 1995). This general order reflects the tendency for Ca- and Na-silicates to weather easily and for the limited incorporation of Ca and Na in secondary minerals. In most cases, Fe and Al are retained in the lower soil profile as oxides and hydroxides, which are essentially immobile (Chesworth et al. 1981).

In calculating weathering rates, it is often useful to examine the watershed budgets for Cl and Si. In nearly all cases, the atmospheric inputs of Si are trivial, so Si in streamwater is an index of chemical weathering, which is its only source (e.g., Fig. 4.2). In contrast, the Cl content of rocks is normally very low. Nearly all Cl^- in streamwater is derived from the atmosphere, passing through the system unimpeded by biotic activity or geochemical

Table 4.5 Inputs and Outputs of Elements from the Hubbard Brook Experimental Forest, New Hampshire[a]

	Inputs (%)		
	Atmosphere	Weathering	Output as a percent of input
Ca	9	91	59
Mg	15	85	78
K	11	89	24
Fe	0	100	25
P	1	99	1
S	96	4	90
N	100	0	19
Na	22	78	98
Cl	100	0	74

[a] Data from Likens et al. (1981).

Table 4.6 Net Transport (Export Minus Atmospheric Deposition) of Major Ions, Soluble Silica, and Suspended Solids from Various Watersheds of Forested Ecosystems[a]

Watershed characteristics	Caura River, Venezuela	Gambia River, W. Africa	Catoctin Mtns., Maryland	Hubbard Brook, New Hampshire
Size (km^2)	47,500	42,000	5.5	2
Precipitation (cm)	450	94	112	130
Vegetation	Tropical forest	Savanna forest	Temperate forest	Temperate forest
Net dissolved transport ($kg\ ha^{-1}\ yr^{-1}$)				
Na	19.4	3.9	7.3	5.9
K	13.6	1.4	14.1	1.5
Ca	14.2	4.0	11.9	11.7
Mg	5.7	2.0	15.6	2.7
HCO_3^-	124.0	20.3	78.1	7.7
Cl^-	− 1.4	0.6	16.6	−1.6
SO_4^{2-}	1.5	0.4	21.2	14.8
SiO_2	195.7	15.0	56.1	37.7
Total transport ($kg\ ha^{-1}\ yr^{-1}$)	372.7	47.6	220.9	80.4

[a] Modified from Lewis et al. (1987).

reactions (Lockwood et al. 1995). Thus, a balanced budget for Cl is indicative of an accurate hydrologic budget for the watershed (Juang and Johnson 1967).

Because temperate forest soils are dominated by clay minerals with permanent negative charge, the loss of cations is determined by the availability of anions that "carry" cations through the soil profile to streamwaters (Gorham et al. 1979, Johnson and Cole 1980, Terman 1977). In most cases, the dominant anion in soil water is bicarbonate (HCO_3^-); thus, the activities of plant roots and soil biota control chemical weathering and the composition of streamwater. The high streamwater concentration of HCO_3^- in the rainforest of Venezuela (Table 4.6) reflects the importance of carbonation weathering in tropical ecosystems (Fig. 4.3). The total mobilization of cations and silicon in the Venezuelan study is also high, consistent with our expectations of rapid chemical weathering in tropical climates (Fig. 4.2). In many cases, the adsorption of SO_4^{2-} in tropical soils reduces its importance as a mobile anion in the streamwaters and groundwaters draining these regions (Szikszay et al. 1990; Table 4.6). In Chapter 6, we will see how an increase in the availability of NO_3^-—a mobile anion—increases the loss of cations following disturbances, such as forest cutting.

In both the northeastern United States and much of Europe, losses of base cations appear to have increased in recent years due to the large amounts of H^+ and SO_4^{2-} delivered to the soil by acid rain (Tamm and Hällbäcken 1986, Sjöström and Qvarfort 1992, Wright et al. 1994). Some of the cations are lost from the cation exchange capacity, while others are derived from increased rates of rock weathering under the influence of strong acids (Miller et al. 1993b, Likens et al. 1996). The cations move through the soil to streamwater, with SO_4^{2-} acting as a balancing anion. In contrast, in the more highly weathered soils of the southeastern United States, increased losses of cations from acid rain are less dramatic because SO_4^{2-} is held on the anion adsorption sites of soil minerals, so it cannot act as a mobile anion (Reuss and Johnson 1986, Harrison et al. 1989, Cronan et al. 1990). As a historical index of anion adsorption in Tennessee, D.W. Johnson et al. (1981) found a lower content of SO_4^{2-} in the soils beneath a house built in 1890 (67 mg/kg) compared to that in adjacent soils (195 mg/kg) that have been exposed to acid rain throughout this century. Presumably, when this anion adsorption capacity is saturated, increasing losses of cations and SO_4^{2-} can also be expected in this region (Ryan et al. 1989).

In soils of the humid tropics, dominated by variable-charge minerals, one might expect that the abundance of mobile *cations* might determine the loss of anions from adsorption sites. Indeed, the loss of nitrate is retarded as water passes through experimental columns of tropical soils (Wong et al. 1990, Bellini et al. 1996), and Matson et al. (1987) suggested that the loss of NO_3^- following forest cutting in Costa Rica was reduced by a high soil anion adsorption capacity.

Plant uptake of essential elements and the retention of cations and anions on soil minerals complicate the use of streamwater concentrations to calculate the rate of chemical weathering over short periods of time (Taylor and Velbel 1991, Gardner 1990). The eventual loss of cations to riverflow, however, explains the decline in base saturation and pH during soil development (Bockheim 1980). Losses of dissolved constituents from terrestrial ecosystems represent the products of chemical weathering and constitute *chemical denudation* of the landscape.

A relative index of chemical weathering is calculated by summing the annual losses of various elements in streamwater. In comparisons of ecosystems of the world, chemical denudation is found to increase with increasing runoff (Fig. 4.2; see also Chapter 8). Total dissolved transport ranges from 47.6 to 372.7 kg ha^{-1} yr^{-1} among the watersheds of Table 4.6, and Alexander (1988) found that chemical denudation ranged from 20 to 200 kg ha^{-1} yr^{-1} among 18 undisturbed ecosystems in a variety of climatic regimes. Globally, rivers transport about 4×10^{15} g of dissolved substances to the oceans each year—an average of about 270 kg ha^{-1} yr^{-1} for the Earth's land surface (Table 4.7).

The chemical weathering of primary minerals in igneous rocks accounts for 27% of the dissolved constituents delivered to the ocean, while the

Table 4.7 Chemical and Mechanical Denudation of the Continents

Continent	Chemical denudation[a]		Mechanical denudation[b]		Ratio mechanical/ chemical
	Total (10^{14}g/yr)	Per unit area (kg ha^{-1} yr^{-1})	Total (10^{14}g/yr)	Per unit area (kg ha^{-1} yr^{-1})	
North America	7.0	330	14.6	840	2.1
South America	5.5	280	17.9	1000	3.3
Asia	14.9	320	94.3	3040	6.3
Africa	7.1	240	5.3	350	0.7
Europe	4.6	420	2.3	500	0.5
Australia	0.2	20	0.6	280	3.0
Total	39.3	267	135.0	918	3.4

[a] From Garrels and MacKenzie (1971).
[b] From Milliman and Meade (1983).

chemical weathering of sedimentary rocks accounts for the remainder (Li 1972, Meybeck 1987), roughly in proportion to their exposure on land (Blatt and Jones 1975, Bluth and Kump 1991). Because chemical weathering involves the reaction between atmospheric constituents and rock minerals, weathering of 100 kg of igneous rock results in 113 kg of sediments that are deposited in the ocean and about 2.5 kg of salts that are added to seawater (Li 1972). Thus, a significant fraction of the transport of total dissolved substances in rivers (Table 4.7) is derived from the atmosphere and does not represent true chemical denudation of the continents (Berner and Berner 1987).

Total Denudation Rates

In addition to chemical denudation, a large amount of material derived from mechanical weathering is eroded from land and carried in rivers as the particulate or suspended load. These materials have received less attention from biogeochemists, because their elemental contents are not immediately available to biota. However, the *total* denudation of land is dominated by the products of mechanical weathering, which exceeds chemical weathering by three to four times, worldwide (Table 4.7). The mean rate of total continental denudation is about 1000 kg ha^{-1} yr^{-1}, with approximately 75% carried in the suspended sediments in rivers (Alexander 1988, Wakatsuki and Rasyidin 1992; Table 4.7). In many areas the current rate of erosion is substantially higher than a few centuries ago as a result of clearing of vegetation for agriculture (Lal 1995, Pimentel et al. 1995, Milliman and Syvitski 1992).

The importance of mechanical weathering increases with elevation; differences in mean elevation among the continents explain much of the variation in mechanical weathering in Table 4.7. Tamrazyan (1989) suggests that the total transport of suspended materials in all rivers of the world is 16.2×10^{15} g/yr. Milliman and Meade (1983) suggest a slightly lower value—13.5×10^{15} g/yr—with 70% carried by the rivers of southeast Asia (Fig. 4.14). Assuming that the specific gravity of suspended sediment is 2.5 g/cm^3, these estimates are 4 to 5 times higher than estimates (1.27 km^3/yr) of the volume of deep ocean sediments derived from land (Howell and Murray 1986). Presumably, most sediment is deposited near the shore, in continental shelf deposits (Chapter 9). Gregor (1970) suggests that the global rate of sediment transport, an overall measure of mechanical weathering, may have been about four times greater before the land surface was colonized by plants. Indeed, today, especially high concentrations of suspended sediment are seen in rivers draining arid and semi-arid regions where vegetation is sparse (Milliman and Meade 1983).

Because Fe, Al, and Si are only slightly soluble in water, particulate and suspended sediments account for most of the removal of these elements from terrestrial ecosystems (Table 4.8). Suspended sediments are also enriched in phosphorus, owing to chemical reactions between dissolved P and various soil minerals (Avnimelech and McHenry 1984, Sharpley 1985). As a result of various human activities, the river transport of many metals (e.g., Cu, Pb, and Zn) is now greater than the transport under preindustrial conditions (Table 4.9), but it is interesting to note that the concentration of Pb in rivers has declined recently, presumably as a result of the decreased use of leaded gasoline in automobiles (Smith et al. 1987, Trefrey et al. 1985).

Summary

In this chapter we have seen that the rate of weathering and soil development is strongly affected by biota, particularly through carbonation weathering and the production of organic acids in the soil profile. It is tempting to speculate that the rate of carbonation weathering was lower before the advent of land plants, when it depended solely on the downward diffusion of atmospheric CO_2 through the soil profile. However, at periods in the Earth's early history, the concentration of atmospheric CO_2 was most certainly higher than today, yielding high rates of carbonation weathering. Weathering is also driven by the availability of water. The high concentration of CO_2 on Venus (Table 2.3) is ineffective in weathering because the surface of the planet is dry (Nozette and Lewis 1982).

By mining its surface and extracting buried fossil fuels from the Earth, humans have increased the global rates of chemical and mechanical weathering and added significant quantities of dissolved materials to global riverflow (Bertine and Goldberg 1971, Martin and Meybeck 1979). Human exposure and erosion of soils in agricultural use have increased the global denudation due to mechanical weathering by a factor of about 2 (Gregor 1970, Milliman and Syvitski 1992), leading to increases in the rate of sediment accumulation in estuaries and river deltas (Chapter 8).

Chemical weathering is a source of essential elements for the biochemistry of life, but streamwater runoff removes these elements from the land surface. Chemical reactions among soil constituents and uptake by biota determine the rate of loss,

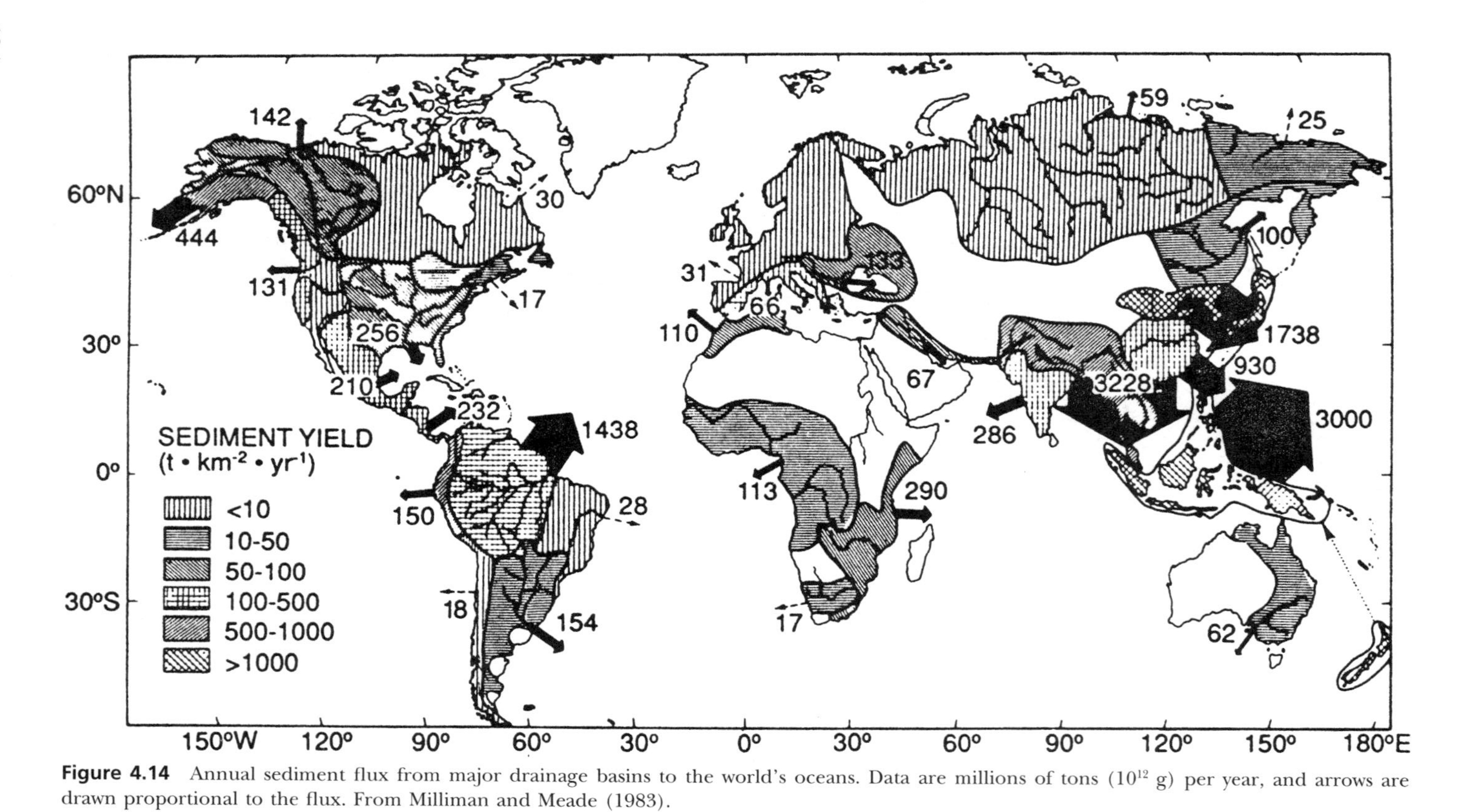

Figure 4.14 Annual sediment flux from major drainage basins to the world's oceans. Data are millions of tons (10^{12} g) per year, and arrows are drawn proportional to the flux. From Milliman and Meade (1983).

Table 4.8 Concentrations of Major Elements in Continental Rocks and Soils and in River Dissolved and Particulate Matter[a]

	Continents		Rivers				Ratios	
Element	Surficial rock concentration (mg/g)	Soil concentration (mg/g)	Particulate concentration (mg/g)	Dissolved concentration (mg/liter)	Particulate load (10^6 tons/yr)	Dissolved load (10^6 tons/yr)	River particulate/rock	Particulate/(particulate + dissolved)
Al	69.3	71.0	94.0	0.05	1457	2	1.35	.999
Ca	45.0	15.0	21.5	13.40	333	501	0.48	.40
Fe	35.9	40.0	48.0	0.04	744	1.5	1.33	.998
K	24.4	14.0	20.0	1.30	310	49	0.82	.86
Mg	16.4	5.0	11.8	3.35	183	125	0.72	.59
Na	14.2	5.0	7.1	5.15	110	193	0.50	.36
Si	275.0	330.0	285.0	4.85	4418	181	1.04	.96
P	0.61	0.8	1.15	0.025	18	1.0	1.89	.82

[a] From Berner and Berner (1987).

Table 4.9 Estimates of Some Elemental Fluxes to the Ocean in Rivers (10^{12} g yr^{-1})[a]

	Ca	Na	Mg	Si	Fe	Cu	Pb	Zn
River particulate load	345	110	209	4430	733	1.55	2.3	5.4
River dissolved load	495	131	129	203	1.5	0.37	0.04	1.1
Total river load	840	241	338	4630	734	1.9	2.3	6.5
Theoretical load[b]	946	298	345	5780	754	0.67	0.33	2.6
Discrepancy	N.S.	N.S.	N.S.	N.S.	N.S.	+1.2	+2.0	+3.9
World mining production	—	—	—	—	—	4.4	3.0	3.9

[a] From Martin and Meybeck (1979). N.S., not significant.
[b] Based on weathering of average rock.

but the inevitable removal of cations results in lower soil pH and base saturation through time (Bockheim 1980). Phosphorus is particularly critical as a soil nutrient, because it is not abundant in crustal rocks and is easily precipitated in unavailable forms in the soil. Old soils in highly weathered landscapes are composed of resistant, residual Fe and Al oxide minerals. In these soils, P is often deficient for plant growth.

Recommended Readings

Birkeland, P.W. 1984. *Soils and Geomorphology.* Oxford University Press, Oxford.
Garrels, R.M. and F.T. MacKenzie. 1971. *Evolution of Sedimentary Rocks.* W.W. Norton, New York.
Jenny, H. 1980. *The Soil Resource.* Springer-Verlag, New York
Likens, G.E. and F.H. Bormann. 1995. *Biogeochemistry of a Forested Ecosystem.* 2nd ed. Springer-Verlag, New York.
Lindsay, W.L. 1979. *Chemical Equilibria in Soils.* Wiley, New York.
Reuss, J.O. and D.W. Johnson. 1986. *Acid Deposition and the Acidification of Soils and Waters.* Springer-Verlag, New York.
Sposito, G. 1989. *The Chemistry of Soils.* Oxford University Press, Oxford.
Sverdrup. H.U. 1990. *The Kinetics of Base Cation Release Due to Chemical Weathering.* Lund University Press, Lund, Sweden.

5

The Biosphere: The Carbon Cycle of Terrestrial Ecosystems

Introduction

Photosynthesis is the biogeochemical process that acts to transfer carbon from its oxidized form, CO_2, in the atmosphere to the reduced (organic) forms that result in plant growth. Directly or indirectly, photosynthesis provides the energy for all other forms of life in the biosphere, and the use of plant products for food, fuel, and shelter brings photosynthesis into our daily lives. The growth of plants affects the composition of the atmosphere (Chapter 3) and the development of soils (Chapter 4), linking photosynthesis to other aspects of global biogeochemistry. Indeed, the presence of organic carbon in soils and O_2 in our atmosphere provides the striking contrast between the *bio*geochemistry on Earth and the simple geochemistry that characterizes our neighboring planets.

In this chapter we will consider the measurement of net primary production—the rate of accumulation of organic carbon in the tissues of land plants. A similar treatment of photosynthesis in the world's oceans is given in Chapter 9. The rate of plant growth varies widely over the land surface. Deserts and continental ice masses may have little or no net primary production, while tropical rainforests can show annual production of >1000 g C/m^2.

Various environmental factors affect the global rate of net primary productivity on land and the total storage of organic carbon in plant tissues (biomass), dead plant parts (detritus), and soil organic matter. As any home gardener knows, light and water are important, but plant growth is also determined by the stock of available nutrients in the soil. These nutrients are ultimately derived from the atmosphere or from the underlying bedrock (Table 4.5). The overall storage of carbon on land is determined by the balance between primary production and decomposition, which returns carbon to the atmosphere as CO_2 (Schlesinger 1977).

Photosynthesis

Containing a central atom of magnesium, the chlorophyll molecule is a prime example of how plants have incorporated an abundant product of rock weathering as an essential element in biochemistry. When photosynthetic pigments absorb sunlight, a few of the chlorophyll molecules are oxidized—passing an electron to a sequence of electron transfer proteins that ultimately lead to the reduction of a high-energy molecule, known as nicotinamide adenine dinucleotide phosphate (NADP), to NADPH. These chlorophyll molecules regain an electron from a water molecule, which is split by an enzyme containing manganese, calcium, and chlorine in a recently postulated asymmetrical three-dimensional structure (Pecoraro 1988, Yachandra et al. 1993). This reaction is the origin of O_2 in the Earth's atmosphere:

$$2H_2O \rightarrow 4H^+ + 4e^- + O_2 \uparrow . \quad (5.1)$$

In all cases, the photosynthetic pigments and proteins are embedded in a cell membrane, which allows protons (e.g., H^+ of Eq. 5.1) to build up to high concentrations on one side of the membrane and for this potential energy to be used to synthesize another high-energy compound, adenosine triphosphate or ATP. In higher plants, the accumulation of protons occurs within the chloroplasts of leaf cells, whereas in photosynthetic bacteria, the reaction is conducted across the external cell membrane.

The high-energy compounds NADPH and ATP are then used by a suite of enzymes to reduce CO_2 and build carbohydrate molecules. The reaction begins with the enzyme ribulose bisphosphate carboxylase, also known as *Rubisco,* which adds CO_2 to the basic carbohydrate unit.[1] The overall reaction for photosynthesis is

$$CO_2 + H_2O \rightarrow CH_2O + O_2, \tag{5.2}$$

but we should remember that the process occurs in two stages. First, the capture of light energy allows water molecules to be split and high-energy molecules to form. This reaction is followed by carbon reduction, in which CO_2 is converted to carbohydrate.

The carbon dioxide used in photosynthesis diffuses into plant leaves through pores, stomates, that are generally found on the lower surface of broad-leaf plants (Fig. 5.1). One factor that determines the rate of photosynthesis is the stomatal aperture, which plant physiologists express as stomatal *conductance* in units of cm/sec. Stomatal conductance is controlled primarily by the availability of water to the plant and the concentration of CO_2 inside the leaf, where it is consumed by photosynthesis. When well-watered plants are actively photosynthesizing, internal CO_2 is relatively low and stomates show maximum conductance. Under such conditions, the amount and activity of ribulose bisphosphate carboxylase may determine the rate of photosynthesis (Sharkey 1985).

Water-Use Efficiency in Photosynthesis

There is a trade-off in photosynthesis; when plant stomates are open, allowing CO_2 to diffuse inward, O_2 and H_2O diffuse outward to the atmosphere. The loss of water through stomates, transpiration, is a major mechanism by which soil moisture is returned to the atmosphere (Table 8.1). In the Hubbard Brook Experimental Forest in New Hampshire (see Chapter 4), about 25% of the annual precipitation is lost by plant uptake and transpiration; streamflow increased by 26–40% when the forest was cut (Pierce et al. 1970). Because water is often in short supply for plant growth (Kramer

[1] For understanding global biogeochemistry, we focus on the photosynthesis of C3 plants, which account for the overwhelming proportion of plant biomass and net primary productivity on Earth. C3 plants are so named because the first product of the photosynthetic reaction is a carbohydrate containing three carbon atoms. However, some plant species, largely warm-climate grasses, conduct photosynthesis by another biochemical pathway, known as C4 photosynthesis (Ehleringer and Monson 1993). C4 plants may account for 21% of global net primary production (Lloyd and Farquhar 1994), but their contribution to global biomass is small because most species are not woody. The overall photosynthetic reaction is identical to Eq. 5.2, but C4 plants have different water-use efficiency and a different isotopic fractionation in their tissues (average −12‰). In fact, various workers have shown that the isotopic ratio of plant debris preserved in soils can be used to trace changes in the distribution of C3 and C4 plants due to past shifts in climate (e.g., Quade et al. 1989, Ambrose and Sikes 1991).

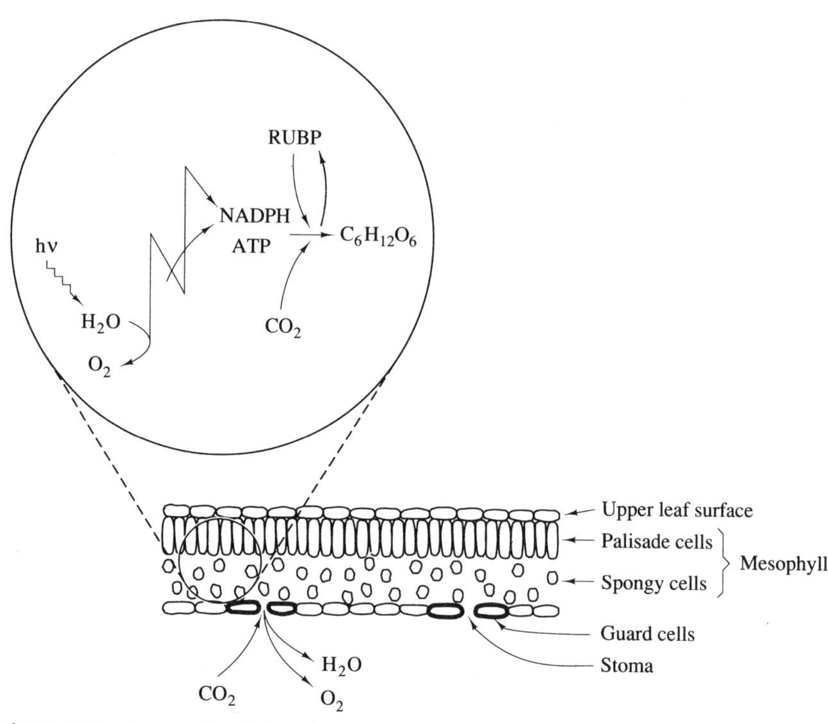

Figure 5.1 Cross-sectional view of a typical plant leaf, showing the upper (palisade) layer of cells, in which photosynthesis occurs, and the guard cells, which control the diffusion of CO_2 (in) and H_2O and O_2 (out) through stomates on the lower surface. A summary of the photosynthetic reaction occurring in the chloroplasts of the palisade cells is shown in the insert.

1982), these large losses of water by plants are somewhat surprising. One might expect natural selection for more efficient use of water by plants, especially in dry environments.

Plant physiologists express the loss of water relative to photosynthesis as water-use efficiency (WUE), viz.,

$$\text{WUE} = \text{mmoles of } CO_2 \text{ fixed/moles of } H_2O \text{ lost.} \tag{5.3}$$

For most plants, water-use efficiency typically ranges from 0.86 to 1.50 mmol/mol, depending upon environmental conditions (Osmond et al. 1982). Water-use efficiency is higher at lower stomatal conductance. Rising concentrations of CO_2 in the atmosphere allow the same rate of photosynthesis to occur at lower stomatal conductance, thus increasing WUE (Bazzaz 1990, Ceulemans and Mousseau 1994). There is also some evidence that the number of stomates per unit of leaf surface has declined as atmospheric CO_2 has risen during the Industrial Revolution (Woodward 1987, 1993, Peñuelas and Matamala 1990). The olive leaves preserved in King Tut's tomb (1327 B.C.) have a higher density of stomates than the leaves of the same species growing in Egypt today (Beerling and Chaloner 1993).

Equation 5.3 largely applies to short-term experiments in the laboratory. For the biogeochemist, long-term average water-use efficiency may be estimated from the carbon isotope composition of plant tissues. This method is based on the observation that the diffusion of $^{12}CO_2$, a lighter molecule, is more rapid than that of $^{13}CO_2$, which composes about 1.1% of atmospheric CO_2. Thus, in a given period of time more $^{12}CO_2$ enters the leaf than $^{13}CO_2$. Inside the leaf, ribulose bisphosphate carboxylase also has a higher affinity for $^{12}CO_2$. As a result of these factors, plant tissue contains a lower proportion of $^{13}CO_2$ than the atmosphere by about 2% (= 20‰) (O'Leary 1988). The discrimination (fractionation) between carbon isotopes is expressed relative to an accepted standard as

$$\delta^{13}C = \left[\frac{^{13}C/^{12}C_{\text{sample}} - {}^{13}C/^{12}C_{\text{standard}}}{^{13}C/^{12}C_{\text{standard}}} \right] \times 1000 \tag{5.4}$$

using the units of parts per thousand parts (‰). Because atmospheric CO_2 shows an isotopic ratio of −8.0‰ versus the standard, most plant tissues show a $\delta^{13}C$ of ca. −28‰ [i.e., (−8‰) + (−20‰)]. Sedimentary organic carbon with this isotopic signature is useful in determining the antiquity of photosynthesis as a biochemical process (Fig. 2.6).

The discrimination between $^{12}CO_2$ and $^{13}CO_2$ during photosynthesis is greatest when stomatal conductance is high (Fig. 5.2). When stomates are partially or completely closed, nearly all of the CO_2 inside the leaf reacts with ribulose bisphosphate carboxylase, and there is less fractionation of the isotopes. Thus, the isotopic ratio of plant tissue is directly related to the average stomatal conductance during its growth, providing a long-term index of water-use efficiency (Farquhar et al. 1989). Significantly, $\delta^{13}C$ values of preserved plant materials indicate that the water-use efficiency of plants has increased as the concentration of atmospheric CO_2 rose at the end of the last glacial period (Van de Water et al. 1994) and during the last several hundred years (Peñuelas and Azcón-Bieto 1992).

Nutrient-Use Efficiency

Over a broad range of plant species, the rate of photosynthesis is directly correlated to leaf nitrogen content when both are expressed on a mass basis (Reich et al. 1992, 1995; Fig. 5.3). Most leaf nitrogen is contained in enzymes; by itself, ribulose bisphosphate carboxylase usually accounts for 20 to 30% of leaf nitrogen (Evans 1989). Seemann et al. (1987) found that photosynthetic potential is directly related to the content of ribulose bisphosphate carboxylase and leaf nitrogen in several species, suggesting

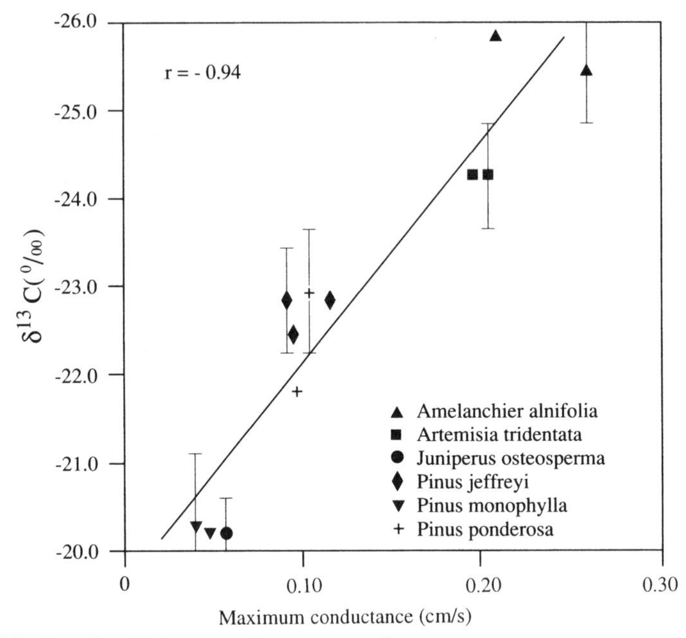

Figure 5.2 Relationship between the content of ^{13}C in plant tissues (expressed as $\delta^{13}C$) and stomatal conductance for a variety of plant species in western Nevada. Modified from DeLucia et al. (1988).

that the availability of nitrogen determines leaf enzyme contents and, thus, the rate of photosynthesis in land plants. In addition to nitrogen, leaf phosphorus content may be an important determinant of photosynthetic capacity in some species (Reich and Schoettle 1988, DeLucia and Schlesinger 1995, Raaimakers et al. 1995). Despite their central role in the molecules of photosynthesis, magnesium and manganese are seldom in short supply for plant growth.

Because most land plants grow under conditions of nitrogen deficiency, we might expect adjustments in nutrient use to maximize photosynthesis. The rate of photosynthesis per unit of leaf nitrogen—the slope of the line in Fig. 5.3—is one measure of nutrient-use efficiency (NUE) (Evans 1989). Overall, the data of Figure 5.3 would seem to indicate that most species have similar photosynthetic NUE, but subtle variations in NUE are seen among different types of plants (Reich et al. 1995) and among plants grown at different levels of fertility (Reich et al. 1994). For many plant species, when leaf nutrient content increases (by fertilization), NUE declines (Ingestad 1979a, Lajtha and Whitford 1989). Nutrient-use efficiency also appears inversely correlated to WUE across many species (Field et al. 1983, DeLucia and Schlesinger 1991).

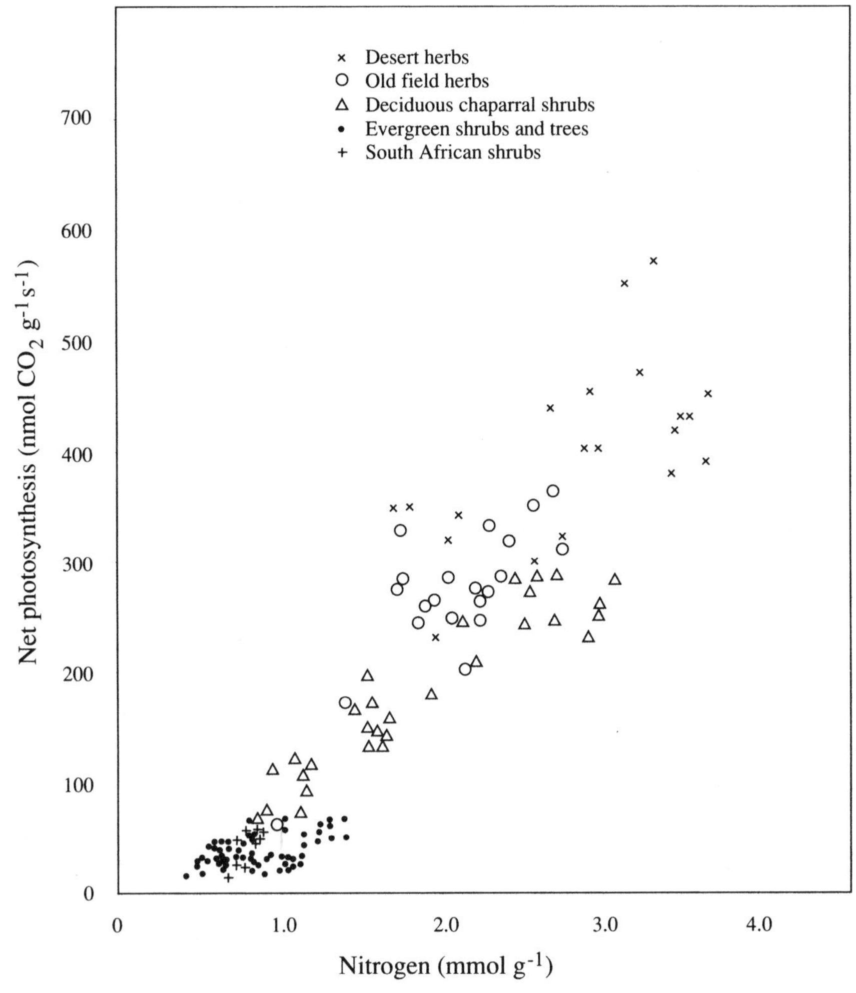

Figure 5.3 Relationship between net photosynthesis and leaf nitrogen content among 21 species from different environments. From Field and Mooney (1986).

Respiration

Photosynthesis is usually measured by placing leaves or whole plants in closed chambers and measuring the uptake of CO_2 or release of O_2. The rates are a measure of *net* photosynthesis by the plant, i.e., the fixation of carbon in excess of the simultaneous release of CO_2 by plant metabolism. Plant metabolism, known as respiration, is largely the result of mitochondrial activity in plant cells, and it is correlated to the nitrogen content, which is a good index of metabolic activity in most plant tissues (Fig. 5.4; see also Ryan 1995). In woody plants, a large fraction of the respiration is contributed by stems and roots owing to their large contribution to total

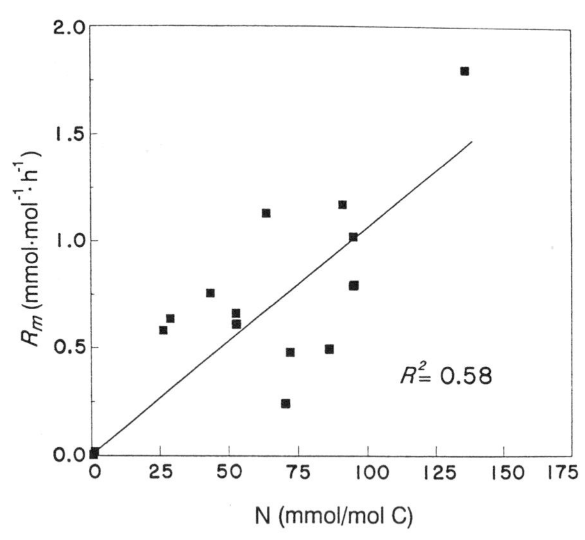

Figure 5.4 Respiration of plant tissues as a function of their nitrogen content. Modified from Ryan (1991).

plant biomass (Amthor 1984, Waring and Schlesinger 1985). For leaf tissues, rates of respiration are higher in the daytime than during the night as a result of the additional process of photorespiration (Sharkey 1988).

Independent measures of respiration suggest that about one-half of the gross carbon fixation by photosynthesis is used by plants, so the actual rate of photosynthesis is often twice that which is measured as plant growth (Farrar 1985, Amthor 1989). For long-lived woody plants, maintenance respiration increases with stand age, consuming an increasing fraction of the gross photosynthesis and contributing to the reduction in the rate of plant growth with age (Kira and Shidei 1967, Waring and Schlesinger 1985). Plant respiration generally increases with increasing temperature, accounting for high rates of respiration in tropical forests and potentially higher rates of plant respiration with global warming (Ryan 1991, Ryan et al. 1994, 1995). Respiration may also increase as plants allocate more tissue to sapwood in hotter, drier environments (Callaway et al. 1994).

Net Primary Production

The rate of photosynthesis measured by physiologists in the laboratory is analogous to the rate of *net primary production* (NPP) measured by ecologists in the field. For plants in nature, we say that

$$\underset{\text{(GPP)}}{\text{Gross primary production}} - \underset{(\text{R}_\text{p})}{\text{plant respiration}} = \underset{\text{(NPP)}}{\text{net primary production.}} \tag{5.5}$$

Net primary production is, however, not directly equivalent to plant growth as measured by foresters, ranchers, and farmers. Some fraction of NPP is lost to herbivores and in the death and loss of tissues, known collectively as litterfall. Foresters frequently call the NPP that remains the *true increment,* which may add to the accumulation of biomass over many years. When mortality occurs during forest development, the true increment is the net increase in the mass of woody tissue in living plants, after subtracting the mass of individuals that die over the same interval.

The annual accumulation of organic matter per unit of land is a measure of NPP, often expressed in units of g m^{-2} yr^{-1}. Plant tissue typically contains about 45 to 50% carbon, so division by two is a convenient way to convert units of organic matter to carbon fixation (Reichle et al. 1973a). Net primary production can also be expressed in units of energy, by measurements of the caloric content of various plant tissues (Paine 1971, Darling 1976). Calories are particularly useful for expressing the efficiency of photosynthesis relative to the receipt of sunlight energy. Net primary production typically increases as a function of intercepted radiation (e.g., Runyon et al. 1994), but even in forests, photosynthesis usually captures only about 1% of the total energy received in sunlight (Botkin and Malone 1968, Reiners 1972).

The measurement of NPP in the field is not easy, but the methods for estimating aboveground NPP are well developed and reviewed extensively elsewhere for forests (Whittaker and Marks 1975) and grasslands (Singh et al. 1975). Traditional methods for forests and shrublands involve the harvest of vegetation and calculation of the annual growth of wood and the mass of foliage at the peak of annual leaf display. Independent estimates of the seasonal loss of plant parts can be obtained from collections of plant litterfall through the year. In grasslands, there is little or no true increment, and estimates of net primary production generally involve the difference between the mass of tissue harvested from small plots at the beginning and the end of the growing season (e.g., Wiegert and Evans 1964, Lauenroth and Whitman 1977). These estimates must be corrected for the consumption and loss of tissues during the same period.

Allocation of net primary production varies with vegetation type and age. In forests, 25 to 35% of aboveground production is found in leaves (Whittaker et al. 1974), with this percentage tending to decrease with stand age. Allocation to foliage in shrublands is generally greater, ranging from 35 to 60% in desert and chaparral shrubs (Whittaker and Niering 1975, Gray 1982). In grassland communities, essentially all aboveground net primary production is found in photosynthetic tissue.

Comparing plant communities in different regions, Jordan (1971) found that the proportional allocation of NPP to wood growth was greater in boreal forests than in the tropics—i.e., there is greater wood production per unit of foliage in boreal forests. As a result of their massive structure and high environmental temperatures, tropical forests may expend a greater percentage of their gross primary production in respiration (Whittaker and Marks 1975, Ryan et al. 1994), leaving less for wood growth. Webb et al. (1983) found a logarithmic relationship between total aboveground NPP and foliage biomass for a variety of plant communities in North America, with some deserts showing exceptionally high values of this ratio (Fig. 5.5). Compared to communities with abundant precipitation, however, desert shrublands show relatively low allocation of NPP to wood production (Jordan 1971), perhaps as a result of a large allocation to roots (Wallace et al. 1974, Caldwell and Camp 1974, Runyon et al. 1994).

Root growth is especially difficult to study, and many estimates of NPP include data only for the aboveground tissues. Nevertheless, when roots have been examined carefully, the annual growth and turnover of root tissues account for a significant fraction of the NPP in most communities. In forests the proportional allocation of photosynthate to root growth varies as an inverse function of site fertility (Axelsson 1981, Gower et al. 1992), although the absolute amount of root growth is greatest on sites with

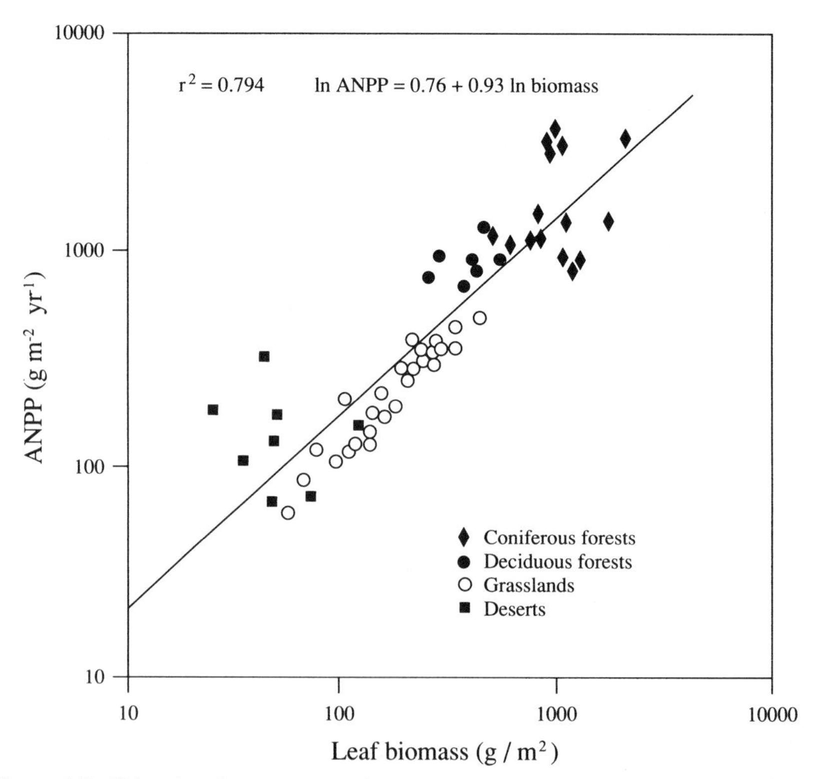

Figure 5.5 Using data from a variety of ecosystems in North America, Webb et al. (1983) found a strong relation between the annual aboveground net primary production and the biomass of foliage.

high NPP (Raich and Nadelhoffer 1989). Edwards and Harris (1977) reported that the growth and death of roots delivered 733 g C m^{-2} yr^{-1} to the soil in a forest in Tennessee, where the aboveground production was 685 g C m^{-2} yr^{-1} (Reichle et al. 1973a). Similarly, roots composed more than half of the NPP in coniferous forests in Washington (Table 5.1) and in the deciduous forest at Hubbard Brook (Fahey and Hughes 1994). An even larger proportion of NPP is allocated to root growth in many grassland ecosystems (Lauenroth and Whitman 1977, Warembourg and Paul 1977). Unfortunately, for the purpose of global estimates, there are no obvious general correlations that allow us to predict the allocation of NPP to aboveground and belowground tissues worldwide (Nadelhoffer and Raich 1992, Gower et al. 1996).

In recent years, ecologists have developed a new approach to estimating the net carbon balance of whole ecosystems—plants as well as soil. The approach is based on the observation that during the day, atmospheric

Table 5.1 Net Primary Production in 23- and 180-yr-old *Abies amabilis* forests in the Cascade Mountains, Washington[a]

	23-yr-old		180-yr-old	
	g m^{-2} yr^{-1}	% of total	g m^{-2} yr^{-1}	% of total
Aboveground				
Biomass increment				
Tree total	426		232	
Shrub stems	6		<1	
Total	432	18.37	232	9.33
Detritus production				
Litterfall	151		218	
Mortality	30			
Herb layer turnover	32		5	
Total	213	9.06	223	8.97
Total aboveground	645	27.42	455	18.30
Belowground				
Roots				
Fine (≤2 mm)	650	27.64	1290	51.87
Fibrous-textured	571		1196	
Mycorrhizal (host tissue)	79		94	
Coarse (>2 mm)	358		324	
Angiosperm fine root turnover	373		44	
Total root turnover	1381	58.72	1658	66.67
Mycorrhizal fungal component	326	13.86	374	15.04
Total belowground	1707	72.58	2032	81.70
Ecosystem total	2352		2487	

[a] From Vogt et al. (1982).

CO_2 typically shows a concentration gradient from approximately 360 ppm, the tropospheric background (Table 3.1), to lower values in the plant canopy as a result of CO_2 uptake by vegetation. This gradient develops despite atmospheric mixing that should otherwise cause the concentrations of CO_2 to be uniform with height. At any rate of mixing, the strength of the gradient is related to the net carbon uptake by vegetation, allowing an estimate of NPP. During the night, the gradient is often reversed, as plant and soil respiration continue in the absence of vegetation uptake (Woodwell and Dykeman 1966, Reiners and Anderson 1968). Known as the eddy-correlation technique, this approach has been applied to a variety of forest and grassland ecosystems (Baldocchi et al. 1987, Wofsy et al. 1988, Kim et al. 1992, Hollinger et al. 1994, Grace et al. 1995).

Using the eddy-correlation approach, Goulden et al. (1996) summed hourly and daily carbon exchange to provide annual estimates of net carbon accumulation in a deciduous forest in Massachusetts. Over five years the forest took up between 1070 and 1210 g C m^{-2} yr^{-1} (GPP), but plant and soil respiration returned 810 to 1140 g C m^{-2} yr^{-1} to the atmosphere, so the net accumulation of carbon in the ecosystem—the true increment—was only 140 to 280 g C m^{-2} yr^{-1}. This value is similar to independent estimates of wood growth in the forest; but, remember that this is less than total NPP, which also includes the production of leaves and other short-lived tissues (Valentini et al. 1996).

Remote Sensing of Primary Production and Biomass

Harvest measurements of NPP are labor intensive and necessarily applied only to small areas. Productivity of vegetation may vary greatly over the landscape, so regional estimates of productivity by harvest become prohibitively expensive. Nevertheless, for understanding global change, regional and global estimates of NPP are essential, and various methods using remote sensing to provide integrated estimates of NPP over large areas are currently under development.

The basis of satellite measurements of NPP is the differential absorption of light by chlorophyll and other leaf pigments. Green plants look green because chlorophyll preferentially absorbs light in the blue and red portions of the solar spectrum, reflecting a large portion of the green light to our eyes. Despite its strong absorption of red light (760 nm), chlorophyll shows little absorption of infrared light with wavelengths of 800 to 1200 nm. Thus, to provide an index of the underlying greenness of the Earth's surface, the LANDSAT (Thematic Mapper) satellites have measured surface reflectance in discrete portions of the visible and infrared spectrum, labeled TM1-7 in Fig. 5.6. Bare soil should show similar reflectance in the TM4 and TM3 wavebands, whereas vegetation shows a TM4/TM3 ratio $\gg$ 1.0 as a result of the absorption of red light by chlorophyll.

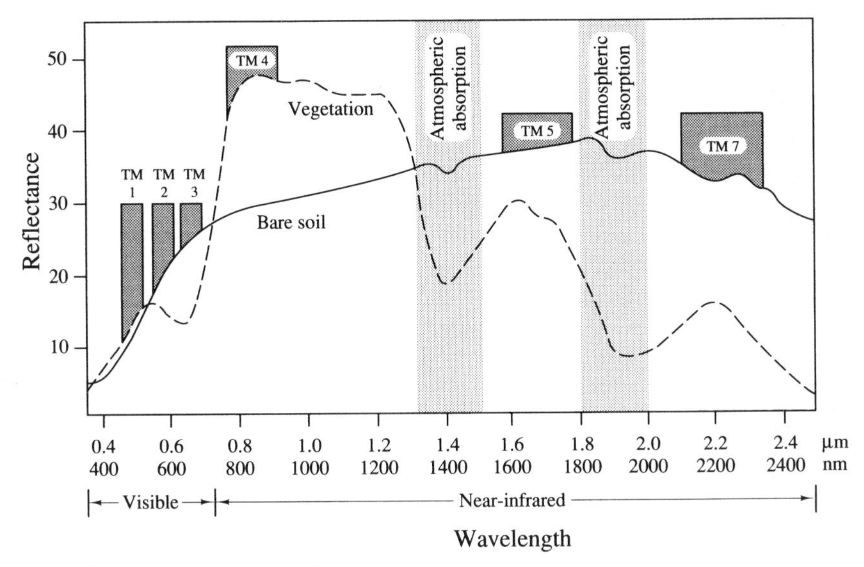

Figure 5.6 A portion of the solar spectrum showing the typical reflectance from soil (—) and leaf (– – –) surfaces and the portions of the spectrum that are measured by the LANDSAT satellite.

The LANDSAT instrument measures the reflectance in each waveband for a 30 × 30-m plot or *pixel* of land. In the northwestern United States, the TM4/TM3 ratio was directly correlated to leaf area in 17 coniferous forests studied by harvest measurements (Fig. 5.7). In each forest, leaf area was expressed as the leaf-area index (LAI) in units of m^2/m^2—the area of leaves above a square meter of ground surface. Previous studies had shown a direct relation between LAI and NPP in these forests (Fig. 5.8), so the potential extrapolation from satellite measurements of LAI to regional estimates of NPP is obvious. Indeed, Cook et al. (1989) found a good relationship between thematic mapper data and regional estimates of NPP across a range of ecosystem types in North America.

Goward et al. (1985) followed a similar approach using data from an advanced very-high-resolution radiometer (AVHRR) carried on the NOAA-7 satellite. They calculate a normalized difference vegetation index (NDVI),

$$\text{NDVI} = (\text{NIR} - \text{VIS})/(\text{NIR} + \text{VIS}), \tag{5.6}$$

where NIR is reflectance in the near-infrared and VIS is reflectance in the visible wavebands, respectively. This index minimizes the effects of variations in background reflectance and emphasizes variations in the data that occur because of the density of green vegetation. Their data allow global mapping of a "greenness" index for the Earth's land surface (Plate 1).

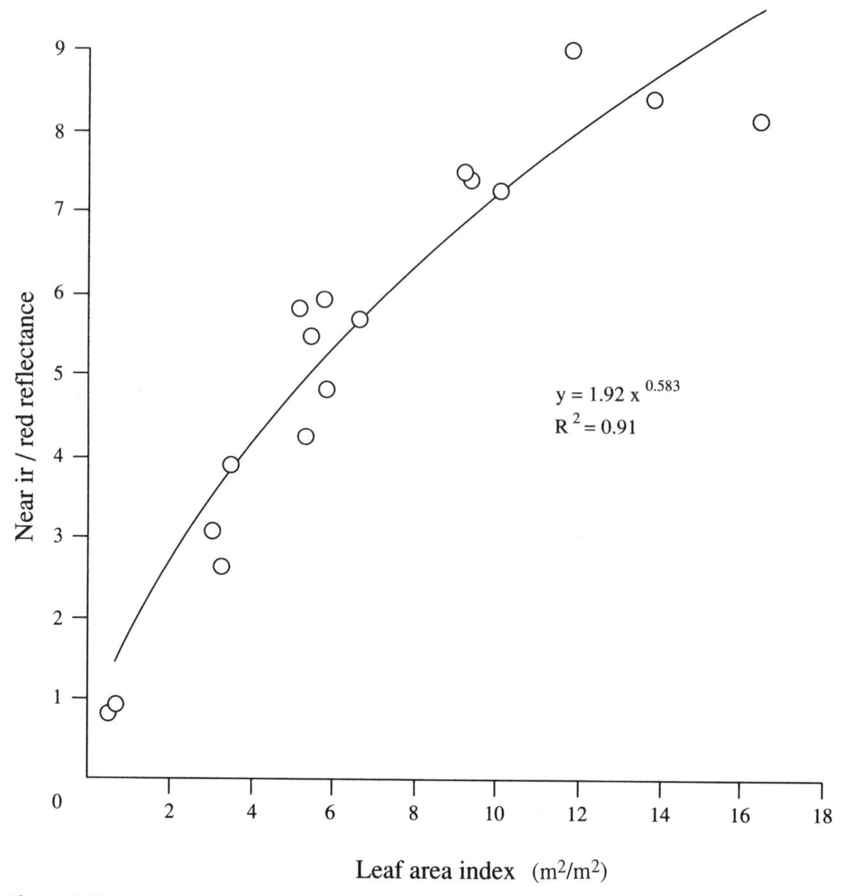

Figure 5.7 The ratio of light reflected in the near infrared and red spectral bands (wavebands TM4 and TM3 of the LANDSAT satellite; see Fig. 5.6) is related to LAI for forest stands in the northwestern United States. From Peterson et al. (1987).

Satellite measurements of "greenness" may lead to global estimates of NPP, assuming that "greenness" is directly related to leaf area and that LAI is good predictor of NPP (Figs. 5.5 and 5.8). Integrations of NDVI measured at frequent intervals over the growing season show a direct correlation to regional average values of NPP measured by harvest methods (Goward et al. 1985, Box et al. 1989). Recently, Fung et al. (1987) have shown that the seasonal patterns of NDVI for the latitudinal bands of the globe are consistent with the magnitude of the seasonal oscillation of atmospheric CO_2 measured at various latitudinal stations (Fig. 3.6). Although the LANDSAT data have finer resolution than those gathered by AVHRR (1.1 km^2), the AVHRR data are often more useful in regional and global estimates because the number of pixels covering the land surface remains manageable during data processing. Running et al. (1989) used

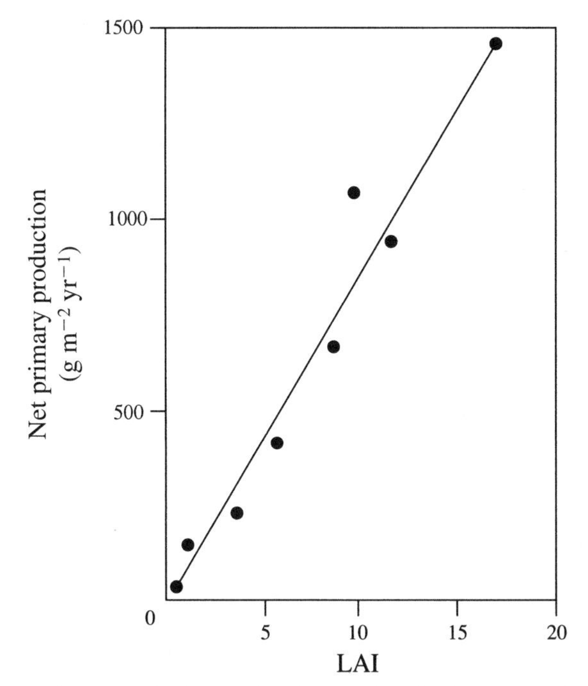

Figure 5.8 Net primary production is directly related to leaf-area index for forests in the northwestern United States. Modified from Gholz (1982).

AVHRR data to estimate the leaf-area index of forests in western Montana and used the LAI data in a model for forest growth to calculate regional evapotranspiration, NPP, and water-use efficiency.

Remote sensing of leaf area and NPP has proven easier than similar measurements of biomass. However, live woody vegetation absorbs and reflects microwave energy as a function of its height and the volume of water-filled tissue, and this observation has been used as the basis for remote sensing of forest biomass. In most cases a microwave emitter is mounted on an aircraft, which also carries a sensor to measure the proportion of the emitted radiation that is reflected back to the source. Known as synthetic aperture radar (SAR), this technique has been used successfully to measure vegetation biomass, especially the regrowth of forests after clearing (Fig. 5.9).

Global Estimates of Net Primary Production and Biomass

Although remote sensing techniques will undoubtedly offer future refinements, most current estimates of global net primary production and biomass are based on compilations of data from harvest measurements. One of the first compilations, that by Whittaker and Likens (1973), is widely cited, but it has been modified repeatedly to incorporate new data from around the

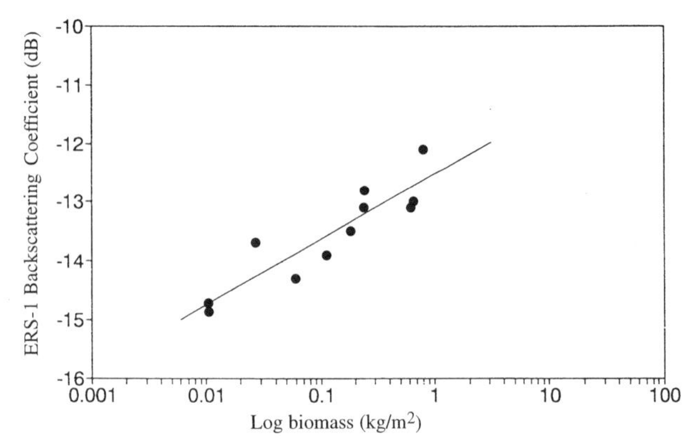

Figure 5.9 The reflected microwave radiation (backscattering coefficient) measured by an airborne synthetic aperture radar (SAR) for stands of young loblolly pine (*Pinus taeda*) in central North Carolina. Modified from Kasischke et al. (1994).

world (Table 5.2). Most of the estimates of global NPP are in a range of 45 to 65 $\times$ 10^{15} g C/yr. Olson et al. (1983) offer the most comprehensive estimate of the total biomass of plants on land; their value is 560 $\times$ 10^{15} g C. The ratio of biomass/NPP is an estimate of the mean residence time for an atom of carbon in plant tissues (cf. Eq. 3.3). The global values yield an overall mean residence time of about 9 years, but this value varies from about 4 in deserts to >20 in some forests (Table 5.2). Of course, we

Table 5.2 Primary Production and Biomass Estimates for the World[a]

Ecosystem	Area (10^{12} m^2)	Mean plant biomass (kg C/m^2)	Carbon in vegetation (10^{15} g)	Mean net primary production (g C/m^2/yr)	Net primary productivity (10^{15} g C/yr)
Tropical wet and moist forest	10.4	15	156.0	800	8.3
Tropical dry forest	7.7	6.5	49.7	620	4.8
Temperate forest	9.2	8	73.3	650	6.0
Boreal forest	15.0	9.5	143.0	430	6.4
Tropical woodland and savanna	24.6	2	48.8	450	11.1
Temperate steppe	15.1	3	43.8	320	4.9
Desert	18.2	0.3	5.9	80	1.4
Tundra	11.0	0.8	9.0	130	1.4
Wetland	2.9	2.7	7.8	1300	3.8
Cultivated land	15.9	1.4	21.5	760	12.1
Rock and ice	15.2	0	0.0	0	0.0
Global total	145.2		558.8		60.2

[a] From Houghton and Skole (1990).

must remember that these are weighted averages. In forests some tissues, such as leaves, may last only a few months, while wood may last for centuries.

Estimates such as those in Table 5.2 are calculated by classifying the land vegetation into a small number of categories and by assigning a mean value to the NPP and biomass of each category based on data from the widest possible number of field studies. The classification of vegetation is arbitrary, and estimates of the land area in each category often vary considerably (Golley 1972). Moreover, the NPP data often do not reflect the full range of variation in the field, because ecologists tend to select mature, well-developed stands for study.

Brown and Lugo (1984) have considered the effect of differences in classification and stand selection on estimates of the biomass of tropical forests. Their data are considerably lower than those reported in Table 5.2. Revised, lower estimates of the biomass in boreal forests (Botkin and Simpson 1990) and deciduous forests (Botkin et al. 1993) of North America are also available. We can expect that global estimates of NPP and biomass by remote sensing will help resolve some of these differences. Similarly, improved classifications of the Earth's vegetation, based on its physiological response to climate, will allow for more consistent compilation of global values for net primary production and biomass (Prentice 1990, Prentice et al. 1992, Neilson 1995, DeFries et al. 1995).

The data in Table 5.2 suggest that the primary productivity of forests is greatest in the tropics and declines with increasing latitude to low values in boreal forests and shrub tundra. Along a gradient of decreasing precipitation, NPP declines from forests to grasslands, showing very low values in most deserts. Wetland vegetation has high NPP; we will examine wetlands in more detail in Chapter 7.

Evidence for the importance of temperature and moisture is seen in regional comparisons of productivity, especially patterns along gradients of elevation. Whittaker (1975) found that NPP declined with increasing elevation in the forested mountains of the eastern United States, presumably reflecting the influence of declining temperatures (i.e., a shorter growing season). In the southwestern United States, where precipitation is more limited, NPP tends to increase with elevation in communities ranging from desert shrublands to montane forests (Whittaker and Niering 1975). Sala et al. (1988) show a direct relation between net primary production and precipitation within the grasslands of the central United States. In forests of the northwestern United States, NPP and LAI are directly related to site water balance, which is the difference between precipitation inputs and losses of soil moisture during the growing season (Grier and Running 1977, Gholz 1982).

Rosenzweig (1968) combined temperature and precipitation to calculate actual evapotranspiration, which shows a positive correlation to NPP in temperate zone ecosystems (Webb et al. 1978). The overall strength of the relation may partially derive from the influence of these variables on microbial processes that speed nutrient turnover in the soil (Chapter 6). Nutrient availability often determines local differences in net primary productivity among sites within the temperate zone (e.g., Pastor et al. 1984). In tropical rainforests, where both light and moisture are abundant, the relationship of NPP to these variables is weak, and local soil conditions determining fertility are potentially most important (Brown and Lugo 1982).

One of the earliest systematic attempts to estimate global NPP and biomass focused on climatic variables. Lieth (1975) related NPP at 52 field sites to the mean annual temperature and precipitation recorded in nearby weather stations (Figs. 5.10 and 5.11). He considered temperature to be an index of solar irradiance and to determine the length of the growing season (cf. Bonan 1993). These data provided Lieth with equations that he could use with local weather data to predict productivity in other areas of the world. A map of global productivity was developed using the lower of the two predictions of NPP at each site, to reflect a temperature or moisture limitation on NPP. The global map of NPP (Fig. 5.12) is surprisingly similar to the satellite picture of "greenness" (Plate 1). Lieth's (1975) approach suggests that light and moisture are the main factors determining NPP, with soil nutrients playing a lesser role. His global map suggests a total terrestrial NPP of about 63×10^{15} g C yr^{-1}, assuming that the world's land vegetation is undisturbed by humans (Esser et al. 1982).

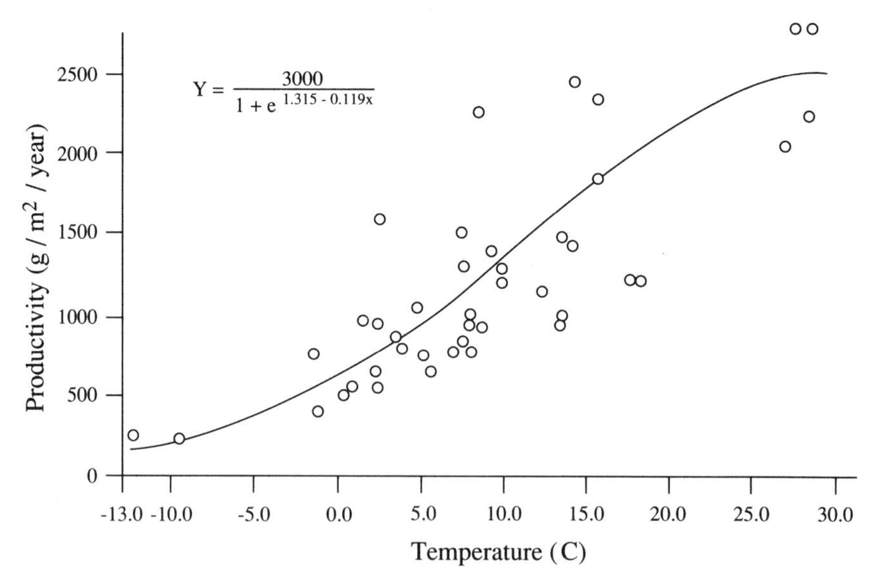

Figure 5.10 Relationship between NPP determined by harvest and mean annual temperature for 52 studies on various continents. From Lieth (1975).

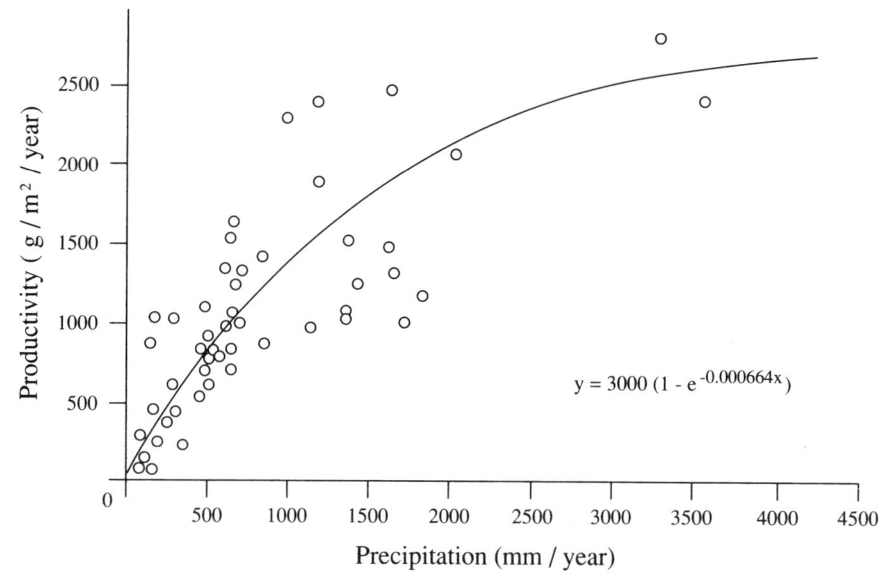

Figure 5.11 Relationship between NPP and mean annual precipitation for 52 locations around the world. From Lieth (1975).

Recently, several attempts have been made to estimate global NPP by incorporating a large number of factors in global simulation models. Melillo et al. (1993) suggest that for any site,

$$NPP = NPP_{(max)} \times PAR \times LAI \times T \times CO_2 \times H_2O \times NA, \quad (5.7)$$

where PAR is photosynthetically active radiation, T is temperature, CO_2 is the atmospheric concentration of CO_2, H_2O is soil moisture, and NA is an index of nutrient availability. Assuming that the Earth's vegetation is undisturbed by humans, they compiled or estimated these data for >56,000 pixels of the Earth's land surface to calculate a total productivity of 53.2 $\times 10^{15}$ g C/yr. A similar approach, using AVHRR to estimate LAI globally, arrives at 48 $\times 10^{15}$ g C as a global estimate of NPP—with 70% between 30° N and 30° S latitude (Potter et al. 1993, Field et al. 1995). The difference between these values may well represent the effect of human disturbance of the land. These models can be modified to incorporate future changes in the Earth's condition, such as the increase in atmospheric CO_2 and potential changes in the Earth's temperature and precipitation. For instance, with changes in climate and rising CO_2, Melillo et al. (1993) predict an increase in global NPP to more than 60 $\times 10^{15}$ g C/yr, assuming no shifts in the distribution of vegetation and no disturbance of the land by humans.

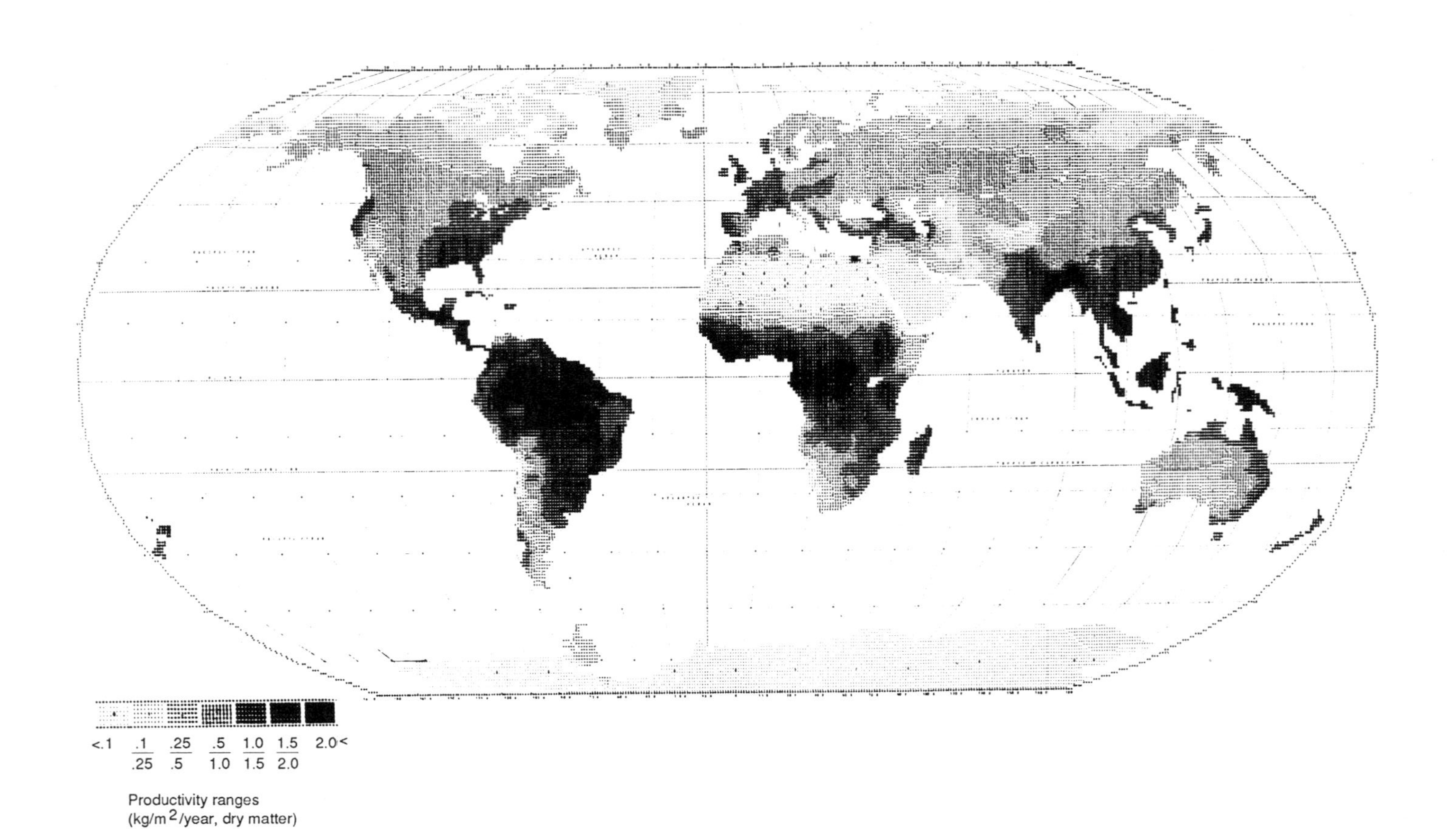

Figure 5.12 A map of terrestrial net primary production, predicted using the relationships in Figs. 5.10 and 5.11 and local weather data. From Lieth (1975).

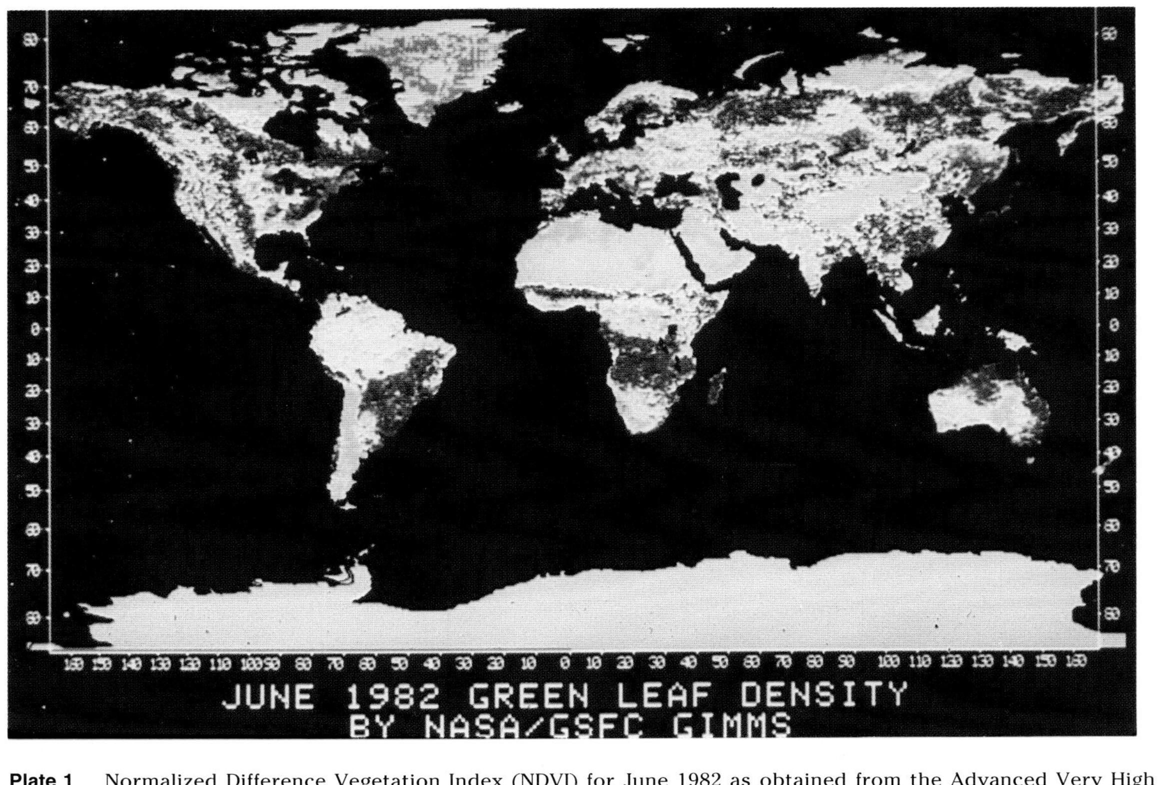

Plate 1 Normalized Difference Vegetation Index (NDVI) for June 1982 as obtained from the Advanced Very High Resolution Radiometer on the NOAA satellite. Note that the greatest vegetation density is colored white and blue, whereas green and yellow indicate lower leaf area. The Northern Hemisphere is in mid-summer. From NASA, 1987, Moderate-Resolution Imaging Spectrometer, Instrument Report, Washington D.C.

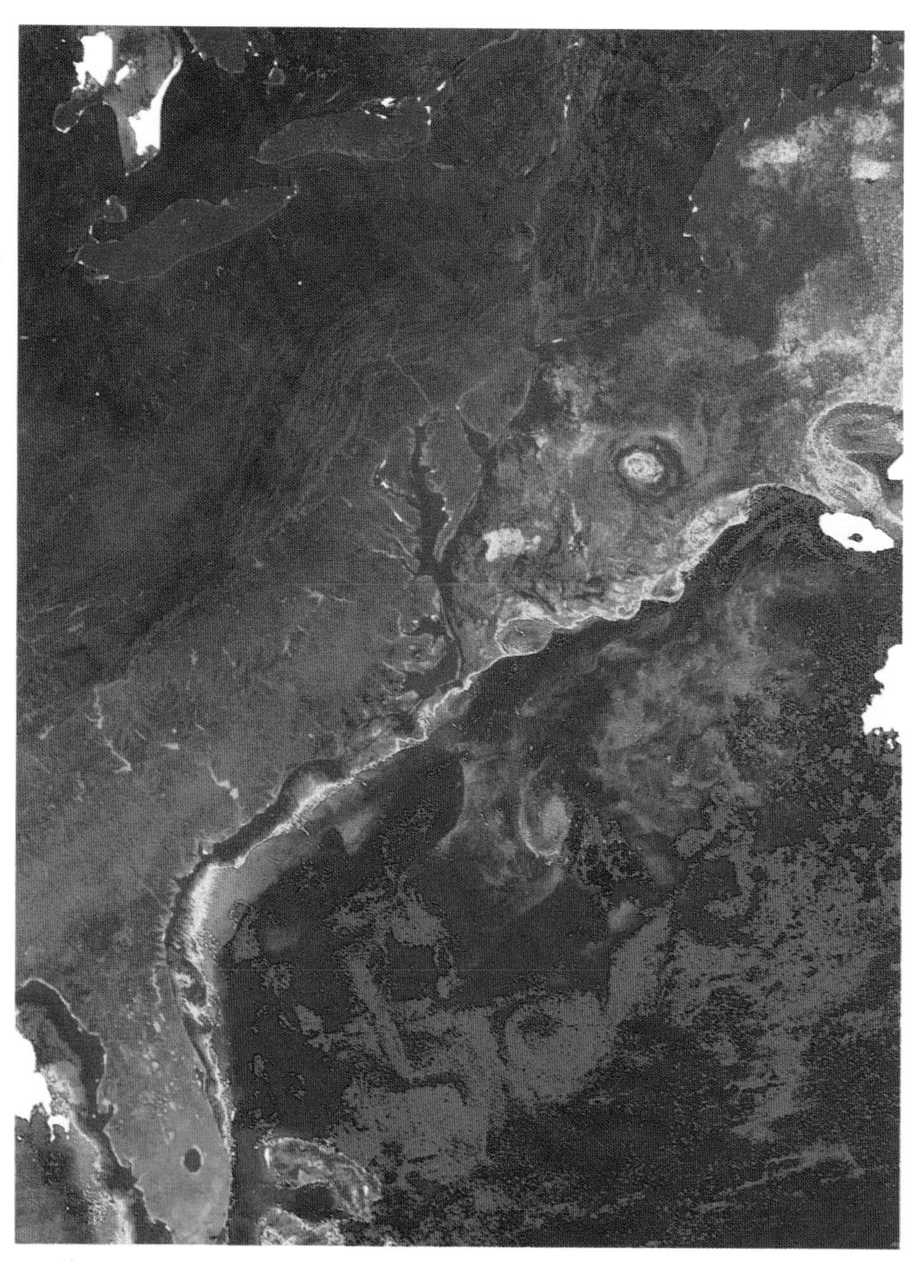

Plate 2 Distribution of chlorophyll in the western North Atlantic Ocean during May 1981, as recorded by the Coastal Zone Color Scanner (CZCS) on the Nimbus-7 satellite. Areas rich in phytoplankton are shown as red (> 1mg chlorophyll/m^3); blue and purple areas have lower chlorophyll concentrations (< 0.01 mg/m^3). Note the high productivity of coastal areas, especially from North Carolina to Maine. From NASA, 1987, High-Resolution Imaging Spectrometer, Instrument Panel Report, Washington, D.C.

Net Primary Production and Global Change

The direct harvest of plants for food, fuel, and shelter accounts for about 3.3×10^{15} g C/yr or about 6% of the terrestrial productivity worldwide (Vitousek et al. 1986). As a result of inadvertent activities, such as wildfires and pollution, humans may have reduced total net primary production by up to 25 to 40% (Vitousek et al. 1986, Dobrovolsky 1994, p. 67). Vitousek et al. (1986) suggest that this is probably the largest diversion of primary production to support a single species in the history of life on Earth, certainly a thought-provoking percentage given the current continuing rate of human population growth.

The effect of humans on biomass and net primary productivity is reflected by changes in the global carbon cycle and in the composition of the atmosphere (Chapters 3 and 11). Most of the increase in atmospheric carbon dioxide is due to the burning of fossil fuels, but a significant portion is also due to the destruction of plant biomass, especially in tropical forests. Although fast-growing successional vegetation is found on most areas that are harvested, the rate of carbon accumulation in these areas does not equal the rate of carbon loss during harvest, so there is a net transfer of carbon from biomass to atmospheric CO_2 (Houghton et al. 1983, Harmon et al. 1990). Simply put, the carbon storage in agricultural crops or forest regrowth is less than the carbon contained in the original forest biomass (Table 5.2).

Houghton et al. (1983) attempt to account for the effect of humans on changes in world biomass between 1860 and 1980, compiling land-use statistics to calculate the rate of agricultural expansion and forest harvest. They suggest that world biomass has been reduced by about 110×10^{15} g C since 1860, equivalent to 13% of their estimate of preindustrial biomass (827×10^{15} g C). The release of carbon from land, including the release from soils, was estimated to be in the range of 1.8 to 4.7×10^{15} g C/yr, compared to a release from fossil fuels of 5×10^{15} g C/yr in 1980 (Rotty and Masters 1985). More recent estimates of the net release of carbon from vegetation and soils are somewhat lower, reflecting better estimates of deforestation (Houghton et al. 1987, Houghton 1993). For the decade of the 1980s, the net release was about 1.6×10^{15} g C/yr, reflecting a 3% imbalance between NPP and decomposition on the Earth's land surface (Houghton 1995). Much of the current destruction occurs in tropical forests, which is why an accurate estimate of tropical forest biomass and its rate of deforestation is critical to our understanding of changes in the global carbon cycle (Houghton 1993, Skole and Tucker 1993).

The net destruction of terrestrial vegetation over the last century is reflected in changes in the isotopic composition of atmospheric CO_2. Remembering that photosynthesis discriminates against $^{13}CO_2$ in favor of $^{12}CO_2$, we can expect that plant tissues and fossil fuels are depleted in ^{13}C

and dilute the atmospheric content of $^{13}CO_2$ when they are burned. In addition, fossil fuels contain no ^{14}C; that radioactive isotope decays away with a half-life of 5700 years. Thus, the burning of fossil fuels also dilutes the atmospheric content of $^{14}CO_2$. Conveniently, tree rings, gas bubbles trapped in polar ice cores, and direct measurements of atmospheric CO_2 since the mid-1950s provide a record of the atmospheric content of $^{13}CO_2$ and $^{14}CO_2$ during the recent past (Leavitt and Long 1988, Siegenthaler and Oeschger 1987). The ^{13}C content of the atmosphere (i.e., $\delta^{13}C$, Eq. 5.4) has declined about -0.034%/yr during the last several decades (Keeling et al. 1989). This dilution is greater than what is expected from fossil fuels alone, suggesting a net release from terrestrial vegetation.

Release of carbon from forest destruction in the tropics could, of course, be balanced by the abandonment of farmland and the permanent regrowth of vegetation elsewhere. Forest regrowth in the southeastern United States has apparently been a sink for atmospheric carbon of about 0.07×10^{15} g/yr during this century (Delcourt and Harris 1980). Globally, afforestation may sequester about 0.7×10^{15} g C/yr (Dixon et al. 1994; see also Chapter 11). We can expect the strength of this regional carbon sink to diminish as reforestation is complete and most forests become mature (Schiffman and Johnson 1989). Of course, the effect of the net sink is lost if these forests are harvested and the wood is converted to short-lived products, such as paper, that are burned.

Some workers have suggested that the primary production of land vegetation will increase as the concentration of atmospheric CO_2 rises, stimulating photosynthesis by a greater delivery of CO_2 to the enzyme ribulose bisphosphate carboxylase (Amthor 1995). When plants are maintained at optimal conditions, the theoretical growth increase should be about 40% when atmospheric CO_2 is double the ambient value (Woodward et al. 1991). With irrigation and fertilization, many crop plants show a response of this magnitude in the field (Strain and Cure 1985, Idso and Idso 1994, Wullschleger et al. 1995). There are few long-term experiments with trees, but several indicate a similar response (Curtis 1996, Norby 1996). If this effect is significant globally, then increased productivity by undisturbed vegetation could sequester CO_2 released by fossil fuels and forest destruction (Idso and Kimball 1993). Evidence for a CO_2 stimulation of forest growth, as recorded by tree-ring width, is mixed, with some studies showing increases (Graybill and Idso 1993), while others show little or no change over the last 100 years (Graumlich 1991, Jacoby and D'Arrigo 1995). In many areas, the growth of vegetation is limited by other factors, and CO_2 may have little long-term direct effect (Kramer 1981, Körner 1993, Thomas et al. 1994, Brown 1991). Experimental studies of plant growth at high CO_2 often find an increased allocation of NPP to root tissues (Norby et al. 1992, Rogers et al. 1994). These roots may elevate the concentrations of CO_2 in the soil pore space and add to the organic matter stored in soil (D.W. Johnson et al. 1994, Wood et al. 1994).

Elevated CO_2 concentrations should increase water-use efficiency of vegetation, since stomata show partial closure at high CO_2 concentrations. Higher water-use efficiency by terrestrial vegetation could leave greater amounts of moisture in the soil, contributing to an increase in the volume of runoff and global riverflow (Idso and Brazel 1984, Probst and Tardy 1987; Chapter 10). Unfortunately, as for the CO_2 response, most of our work has focused on agricultural plants grown under laboratory conditions. There are few studies that document the response of whole ecosystems to increasing CO_2. Soil moisture and streamflow may not increase if plants maintain more leaves at high CO_2, compensating for lower transpiration losses per unit of leaf area (Allen 1990, Ellsworth et al. 1995).

Potentially more significant are changes in the distribution of terrestrial vegetation that may occur as a result of a CO_2-induced global warming (Overpeck et al. 1991, VEMAP 1995). T.M. Smith et al. (1992) examined the current distribution of world vegetation types and the expected changes in their distribution with changes in global precipitation and temperature. In the northern hemisphere, a northward shift in the distribution of productive forests may increase the rate of carbon storage in some areas, potentially sequestering 180×10^{15} g C when the vegetation has fully adjusted to the global warming expected with a doubling of atmospheric CO_2. Significantly, however, land vegetation may be an important source of CO_2 during the adjustment period (Smith and Shugart 1993, Pastor and Post 1993), as warming of the land surface proceeds warming of the oceans and drought becomes widespread (Rind et al. 1990; see Chapter 10). This scenario has dramatic implications for the distribution and productivity of agricultural lands worldwide (R.M. Adams et al. 1990, Rosenzweig and Parry 1994). Changes in the position of well-defined vegetation boundaries, such as continental treeline and the borders of deserts, may be our first indications of global climate change (Tucker et al. 1991, MacDonald et al. 1993, Lescop-Sinclair and Payette 1995, Schlesinger and Gramenopoulos 1996).

The Fate of Net Primary Production

As communities of long-lived plants develop on land, a certain fraction of net primary production is allocated to perennial, woody tissues that accumulate as biomass through time. Plant communities achieve a steady state in living biomass when the allocation to woody tissue is balanced by the death and loss of older parts (Fig. 5.13). At that point, there is no true increment in biomass, although dead organic matter may still be accumulating in the soil. Odum (1969) summarized these trends in community development, suggesting that increasing fractions of gross primary

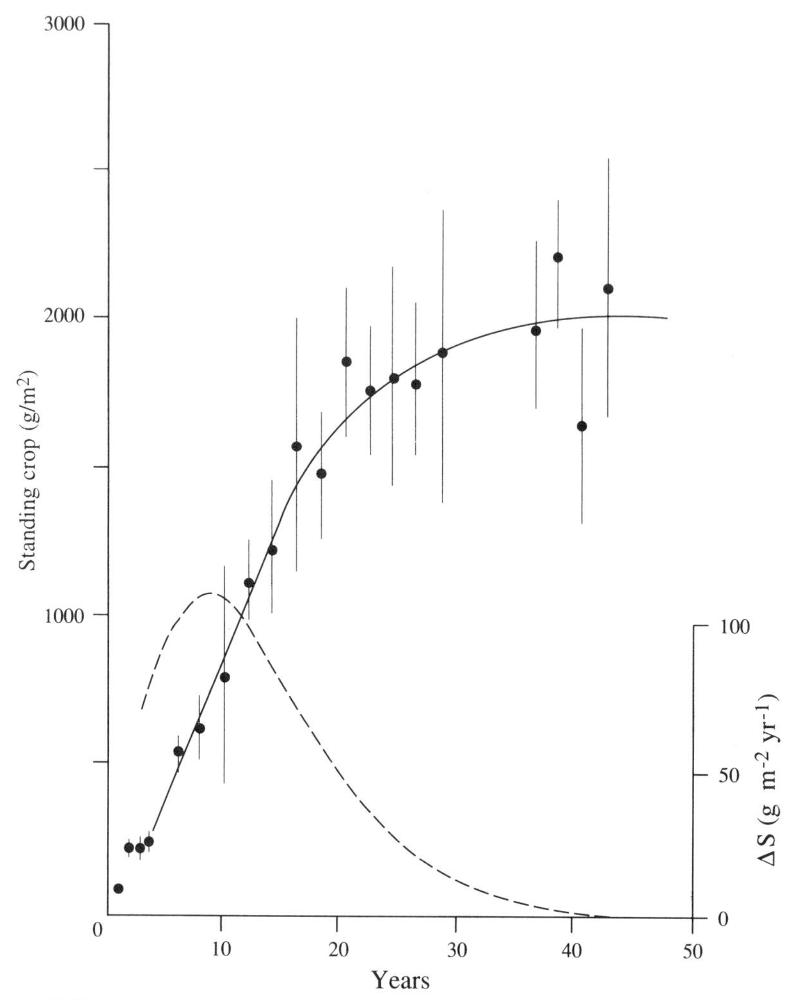

Figure 5.13 Biomass (solid line) and true increment (dashed line) of the aboveground components of a *Calluna* shrubland during 50 years of recovery after fire. From Chapman et al. (1975).

production are lost to plant respiration and decomposition through time (Fig. 5.14). His work defines *net ecosystem production* (NEP) as

$$\text{NEP} = \text{NPP} - (R_{\text{h}} + R_{\text{d}}), \tag{5.8}$$

where R_{h} is respiration of herbivores and R_{d} is respiration of decomposers. Remembering that

$$\text{NPP} = \text{GPP} - R_{\text{p}}, \tag{5.9}$$

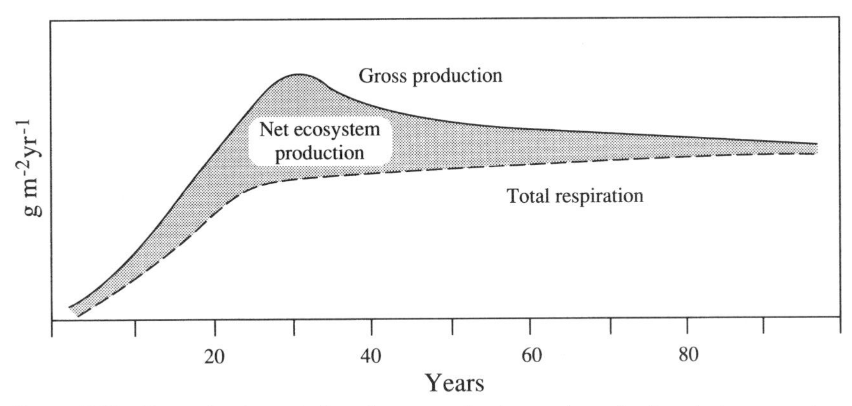

Figure 5.14 Generalized trends in primary production and respiration during ecosystem development. Modified from Odum (1969).

we can say that

$$\text{NEP} = \text{GPP} - R_t, \tag{5.10}$$

where R_t is the total respiratory loss of CO_2 from the ecosystem.

These relationships suggest that increments in organic matter are possible only during the early stages of plant community development. In older communities, there is no true increment to live biomass, and nearly all the NPP is delivered to the soil, where it is decomposed. The role of animals is relatively minor. Although herbivory may play a role in controlling forest productivity and nutrient cycling (Chapter 6), the consumption of plant tissues by herbivores is nearly always <20% of NPP (e.g., Mispagel 1978, McNaughton et al. 1989, Cyr and Pace 1993). By consuming leaf area and root tissues, however, herbivores may have an indirect effect on NPP that is larger than estimates of their direct consumption (Reichle et al. 1973b, Llewellyn 1975, Ingham and Detling 1990). Globally, herbivores consume about 3×10^{15} g C/yr (Whittaker and Likens 1973).

Fires are analogous to respiration by herbivores and decomposers, rapidly converting long-term accumulations of NEP to CO_2 and restarting the process of ecosystem development. Globally, the consumption of NPP by fire is estimated to lie between 2 and 5×10^{15} g C/yr (Crutzen and Andreae 1990, Hao and Liu 1994), and in recent years humans appear to have increased the frequency and area of land that is subjected to fire, especially in the tropics (Andreae 1991, Prins and Menzel 1994). Whereas boreal forests may have once acted to store atmospheric CO_2 in biomass, a recent increase in the rate of burning at northern latitudes now makes these ecosystems a potential net source of CO_2 to the atmosphere (Kasischke et al. 1995).

It is also worth noting, briefly, that vegetation is a source of a wide variety of volatile organic compounds in the atmosphere, including nonmethane hydrocarbons (Greenberg and Zimmerman 1984, Guenther et al. 1994) and carbon monoxide (Jacob and Wofsy 1990). Isoprene is especially abundant in the southeastern United States, where pine forests are widespread (Lamb et al. 1987a, Guenther et al. 1994). Emissions of isoprene, which seem associated with plant response to high temperatures (Sharkey and Singsaas 1995), probably account for about half of the total emission of volatile organic compounds from live vegetation worldwide (Miyoshi et al. 1994, Guenther et al. 1995). Emissions of reduced carbon gases represent NPP that escapes from the terrestrial biosphere to be oxidized in the atmosphere by hydroxyl radicals (Chapter 3). Globally, the emission of reduced carbon compounds from natural vegetation may be as large as 1.15×10^{15} g C/yr, or about 2% of NPP (Crutzen et al. 1985, Guenther et al. 1995).

Production of Detritus

The largest fraction of NPP is delivered to the soil as dead organic matter. Global patterns in the deposition of plant litterfall are similar to global patterns in net primary production (Esser et al. 1982). The deposition of litterfall declines with increasing latitude from tropical to boreal forests (Bray and Gorham 1964, Van Cleve et al. 1983, Lonsdale 1988). Leaf tissues account for about 70% of aboveground litterfall in forests (O'Neill and De Angelis 1981, Meentemeyer et al. 1982), but the deposition of woody litter tends to increase with forest age, and fallen logs may be a conspicuous component of the forest floor in old-growth forests (Lang and Forman 1978, Harmon et al. 1986). In grassland ecosystems, where little of the aboveground production is contained in perennial tissues, the annual litterfall is nearly equal to annual net primary production. In most areas, the annual growth and death of fine roots contributes a large amount of detritus to the soil, which has been overlooked by studies that only consider aboveground litterfall (Vogt et al. 1986, Nadelhoffer and Raich 1992). Following the approach of models for net primary production, Meentemeyer et al. (1982) used actual evapotranspiration to predict global patterns of plant litterfall and to estimate 54.8×10^{15} g for the annual production of aboveground litterfall worldwide.

The Decomposition Process

Most detritus, whether from litterfall or root turnover, is delivered to the upper layers of the soil where it is subject to the decomposition by microfauna, bacteria, and fungi (Swift et al. 1979, Schaefer 1990). Decomposition leads to the release of CO_2, H_2O, and nutrient elements, and to the microbial production of highly resistant organic compounds known as *humus.*

Humus compounds accumulate in the lower soil profile (Chapter 4) and compose the bulk of soil organic matter (Schlesinger 1977). The dynamics of the pool of carbon in soils is best viewed in two stages—processes leading to rapid turnover of the majority of litter at the surface and processes leading to the slower production, accumulation, and turnover of humus at depth.

The litterbag approach is widely used to study decomposition at the surface of the soil. Fresh litter is confined in mesh bags that are placed on the ground and collected for measurements at periodic intervals (Singh and Gupta 1977). Simple models of decay are based on an exponential pattern of loss, where the fraction remaining after 1 year is given by

$$X/X_o = e^{-k}. \tag{5.11}$$

An alternative, the mass-balance approach, suggests that the annual decomposition should equal the annual input of fresh debris, so that the mass of detritus stays constant. Under these assumptions, a constant fraction, k, of the detrital mass decomposes, so that

$$\text{litterfall} = k(\text{detrital mass}), \tag{5.12}$$

or

$$\frac{\text{litterfall}}{\text{detrital mass}} = k. \tag{5.13}$$

When the detritus is in steady state, the values for k calculated from the litterbag and mass-balance approaches should be equivalent, and mean residence time for plant debris is $1/k$ (Olson 1963; see also footnote, p. 51). Vogt et al. (1983) shows the importance of fine roots in the calculation of mean residence times by the mass-balance approach. When root turnover was estimated in a montane fir forest, the mean residence time for organic matter in the forest floor was 8.2 to 15.6 years, compared to 31.7 to 68.6 years calculated from aboveground litter alone.

With either approach, when decomposition rates are rapid, there is little surface accumulation and values for k are greater than 1.0 (e.g., in tropical rain forests; Cuevas and Medina 1988). In such systems, decomposition has the potential to respire more than the annual input of carbon in litterfall. In contrast, in some peatlands values for k are very small (e.g., 0.001; Olson 1963). Decomposition in grasslands shows a range of 0.20 to 0.60 in values for k (Vossbrinck et al. 1979, Seastedt 1988), but values for deserts may be as high as 1.00 due to the action of termites and photooxidation of litter by ultraviolet light (Schaefer et al. 1985). Esser et al. (1982) suggest a global mean residence time of 3 years (i.e., $k = 0.33$) for carbon on the surface of the soil.

Decomposition rates vary as a function of temperature, moisture, and the chemical composition of the litter material. Microbial activity increases exponentially with increasing temperature (e.g., Edwards 1975). This relation often shows a Q_{10} of 2.0, i.e., a doubling in activity per 10°C increase in temperature (Singh and Gupta 1977, Raich and Schlesinger 1992, Kirschbaum 1995). Van Cleve et al. (1981) found that the thickness of the forest floor in black spruce forests in Alaska was inversely related to the cumulative degree days favorable to decomposition each year. In contrast, soil moisture often limits the rate of decomposition in arid and semiarid regions (Wildung et al. 1975, Santos et al. 1984, Amundson et al. 1989). Field experiments suggest that moisture assumes increasing importance when temperate forest soils are warmed (Peterjohn et al. 1994).

Meentemeyer (1978a) compiled data from various decomposition studies to relate surface decomposition to actual evapotranspiration, and used the resulting equation to predict regional patterns of decomposition (Fig. 5.15). His predictions are consistent with observations of surface litter in much of the United States (e.g., Lang and Forman 1978). Actual evapotranspiration is also a good predictor of decomposition in Europe (Berg et al. 1993). Improvements in these predictions are found when chemical parameters such as lignin and nitrogen are also considered (Meentemeyer 1978b, Melillo et al. 1982), but we will defer a discussion of the dynamics of nutrient elements during decomposition until Chapter 6.

Humus Formation and Soil Organic Matter

Plant litter and soil microbes constitute the cellular fraction of soil organic matter. As decomposition proceeds, there is an increasing content of noncellular organic matter—humus—that appears to result from microbial activity. The structure of humus is poorly known, but it contains numerous aromatic rings with phenolic (—OH) and organic acid (—COOH) groups (Flaig et al. 1975, Stevenson 1986). As we saw in Chapter 4, these radicals offer a major source of cation exchange capacity in many soils. Some tentative models have been proposed for the complete molecular structure of humus (Schulten and Schnitzer 1993), but many scientists believe that a large portion of soil humus is amorphous, with no consistent molecular weight or repeating units of structure. Recent progress in elucidating the chemical structure of humus has been made using ^{13}C nuclear magnetic resonance (NMR) spectroscopy and pyrolysis-field ionization mass spectrometry (Py-FIMS) (Schnitzer and Schulten 1992). The most recalcitrant fractions of soil humus appear to have a large component of polymethylene (C=C=C) groups that are synthesized by microorganisms (Baldock et al. 1992).

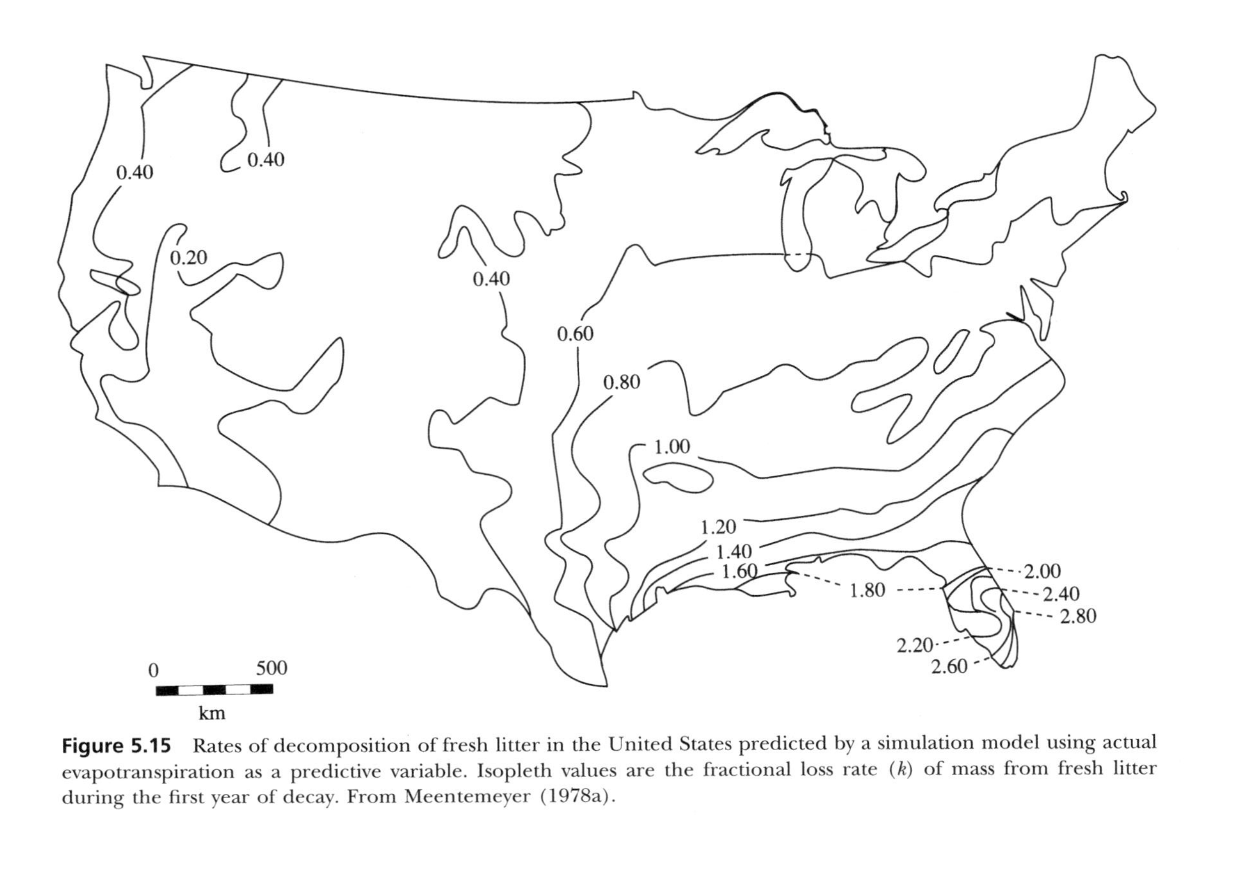

Figure 5.15 Rates of decomposition of fresh litter in the United States predicted by a simulation model using actual evapotranspiration as a predictive variable. Isopleth values are the fractional loss rate (k) of mass from fresh litter during the first year of decay. From Meentemeyer (1978a).

Traditional chemical characterizations of humus have been based on the solubility of humic and fulvic acid components in alkaline and acid solutions, respectively (Fig. 5.16). The acid-soluble component of humus, primarily fulvic acid, controls the downward movement of Fe and Al in soils (Chapter 4). Fulvic acids often account for a large fraction of the soil organic matter in the lower soil profile (Beyer et al. 1993), where they are complexed with clay minerals and calcium (Oades 1988). This humus is very resistant to microbial attack. Campbell et al. (1967) extracted humic materials from a forest soil in Saskatchewan and measured a mean ^{14}C age of 250 to 940 years.

Under most vegetation, the mass of humus in the soil profile exceeds the combined content of organic matter in the forest floor and aboveground biomass (Schlesinger 1977). Table 5.3 provides a global inventory of plant detritus and soil organic matter, totaling 1456×10^{15} g C. Alternative estimates based on soil groups or climatic regions are similar (Post et al. 1982, Eswaran et al. 1993, Batjes 1996). The global estimate of soil organic matter, divided by the estimate of global litterfall, suggests a mean residence time of about 30 years for the total pool of organic carbon in soils, but the mean residence time varies over several orders of magnitude between the surface litter and the various humus fractions (Fig. 5.17).

The incorporation of nuclear-bomb-derived radiocarbon (^{14}C) into different fractions of soil organic matter shows promise as a means of estimating

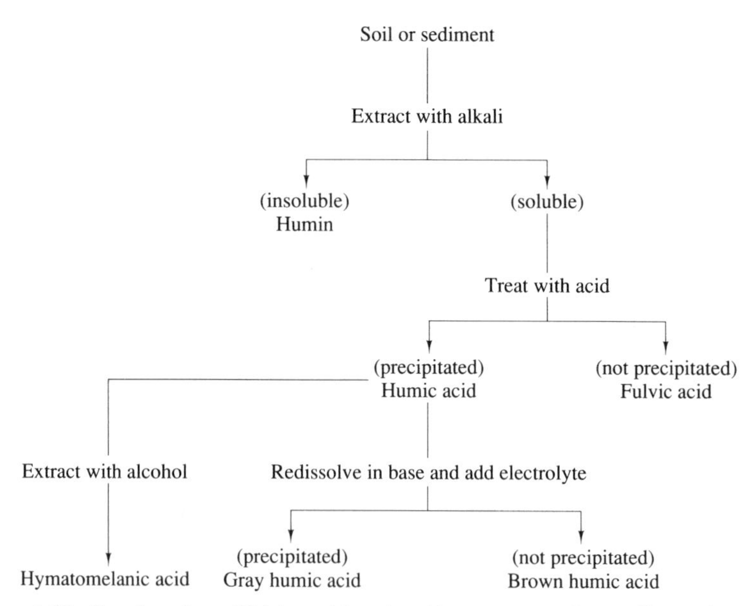

Figure 5.16 Fractionation of fulvic and humic acid components from soil organic matter. From Stevenson (1986).

Table 5.3 Distribution of Soil Organic Matter by Ecosystem Types[a]

Ecosystem type	Mean soil organic matter (kg C m^{-2})	World area (ha × 10^8)	Total world soil organic carbon (mt C × 10^9)	Amount in surface litter (mt C × 10^9)
Tropical forest	10.4	24.5	255	3.6
Temperate forest	11.8	12	142	14.5
Boreal forest	14.9	12	179	24.0
Woodland and shrubland	6.9	8.5	59	2.4
Tropical savanna	3.7	15	56	1.5
Temperate grassland	19.2	9	173	1.8
Tundra and alpine	21.6	8	173	4.0
Desert scrub	5.6	18	101	0.2
Extreme desert, rock, and ice	0.1	24	3	0.02
Cultivated	12.7	14	178	0.7
Swamp and marsh	68.6	2	137	2.5
Totals		147	1456	55.2

[a] From Schlesinger (1977).

their turnover (Trumbore 1993, Harrison et al. 1993). O'Brien and Stout (1978) used radiocarbon dating to find that 16% of the organic matter in a pasture soil had a minimum age of 5700 years, while the rest was of recent

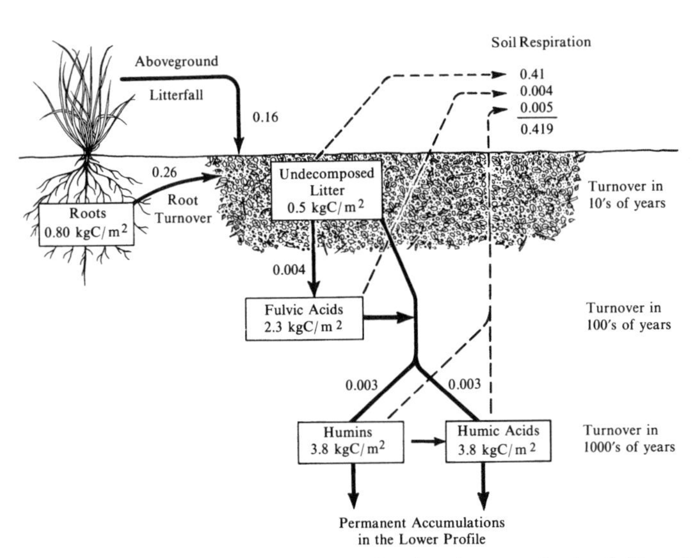

Figure 5.17 Turnover of litter and soil organic fractions in a grassland soil. Note that mean residence time can be calculated for each fraction from measurements of the quantity in the soil and the annual production or loss (respiration) from that fraction. Flux estimates are in kg C m^{-2} yr^{-1}. From Schlesinger (1977).

origin and concentrated near the surface. Because of different turnover times, decomposition constants, *k*, for surface litter cannot be applied to the entire mass of organic matter in the soil profile.

Field measurements of the flux of CO_2 from the soil surface provide an estimate of the total respiration in the soil, and a potential alternative approach to estimating the turnover of the humus pool (Raich and Schlesinger 1992). Most of the production of CO_2 occurs in the surface litter where decomposition is rapid and a large proportion of the fine root biomass is found (Bowden et al. 1993). Edwards and Sollins (1973) found that only 17% of the annual production of CO_2 in a temperate forest soil was contributed by soil layers below 15 cm. Flux of CO_2 from the deeper soil layers is presumably due to the decomposition of humus substances. Production of CO_2 in the soil leads to the accumulation of CO_2 in the soil pore space, which drives carbonation weathering in the lower profile (Chapter 4).

Unfortunately, the respiration of living roots makes it difficult to use estimates of CO_2 flux in calculations of turnover of the soil organic pool. In a compilation of values, Schlesinger (1977) found that CO_2 evolution exceeded the deposition of aboveground litter by a factor of about 2.5 (Fig. 5.18). The additional CO_2 is presumably derived from root metabolism

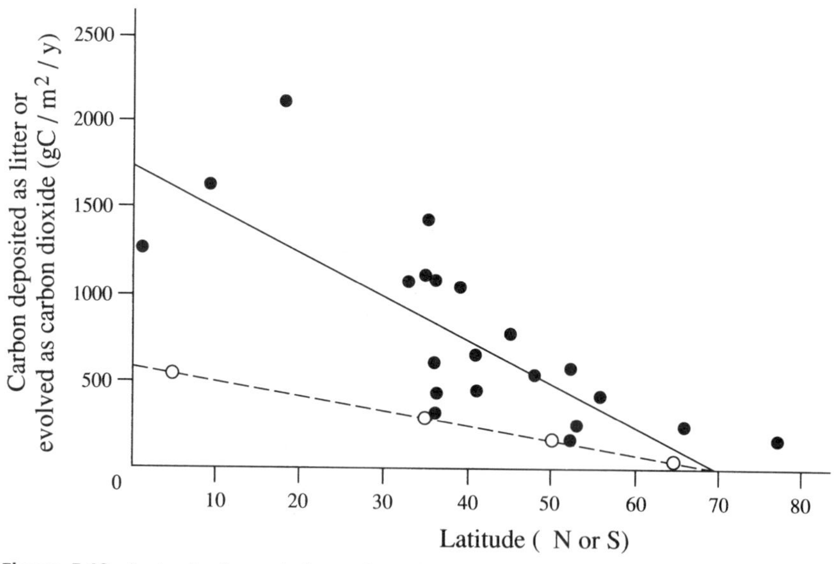

Figure 5.18 Latitudinal trends for carbon dynamics in forest and woodland soils of the world. The dashed line shows the mean annual input of organic carbon to the soil by litterfall. The solid line shows the loss of carbon, measured as the flux of CO_2 from the surface. The difference between these lines represents the loss of CO_2 from root respiration and from the respiration of root detritus and mycorrhizae. From Schlesinger (1977).

and the decomposition of root detritus (Raich and Nadelhoffer 1989). Global soil respiration is 68 to 77 $\times 10^{15}$ g C/yr (Raich and Schlesinger 1992, Raich and Potter 1995), with about $\frac{1}{3}$ derived from the respiration of live roots and the remainder from decomposition. Soil respiration shows a strong correlation with temperature and NPP in world ecosystems (cf. Fig. 4.3).

The global distribution of soil organic matter shows how moisture and temperature control the balance between primary production and decomposition in surface and lower soil layers. Accumulations of soil organic matter are greatest in wetland ecosystems and least in deserts (Table 5.3). Among forests, accumulations increase from tropical to boreal climates. Net primary productivity shows the opposite trend, so the accumulation of soil organic matter is largely due to differences in decomposition. Thus, compared to the process of primary production, soil microbes are more sensitive to regional differences in temperature and moisture (Fig. 5.18). Worldwide, the accumulation of soil organic matter seems more related to factors controlling decomposition than to the NPP of terrestrial ecosystems (Cebrián and Duarte 1995).

Parton et al. (1987) developed a model based on the differential turnover of soil organic fractions to predict the accumulation of soil organic matter in grassland ecosystems. Accurate predictions were achieved when temperature, moisture, soil texture, and plant lignin content were included as variables. Despite relatively low NPP, soils of temperate grasslands contain large amounts of soil organic matter (Sanchez et al. 1982b), due to relatively low rates of decomposition and a larger fraction of plant debris that is derived from root turnover (Oades 1988). In contrast, tropical grasslands and savannas have relatively low soil organic content, perhaps due to frequent fire (Kadeba 1978, Jones 1973).

Storage of soil organic matter represents the net ecosystem production (NEP) in terrestrial ecosystems. Although many wetland ecosystems may show long-term net accumulations (Chapter 7), the mass of soil organic matter in most upland ecosystems is likely to have been fairly constant before widespread human disturbance of soils. Studies of soil chronosequences suggest that humus accumulates at a rates of about 1–12 g C m^{-2} yr^{-1} during soil development (Table 5.4; Chadwick et al. 1994), with the highest rates under cool, wet conditions. Globally, the annual net production of humus substances is $<0.4 \times 10^{15}$ g C/yr (Schlesinger 1990). When soils show a steady state in soil organic content, the production of humic compounds must be equal to their removal from soils by erosion. Thus, an estimate of the transport of organic carbon in rivers is an alternative upper limit for terrestrial NEP (Lugo and Brown 1986). Recent estimates of the global transport of organic carbon in rivers are also about 0.4×10^{15} g C/yr (Schlesinger and Melack 1981, Meybeck 1982), so either approach suggests that terrestrial NEP for the globe is not likely to be more

Table 5.4 Long-Term Rates of Accumulation of Organic Carbon in Holocene-Age Soils

Ecosystem type	Vegetation in terminal state	Soil origin	Accumulation interval (yr)	Long-term rate of accumulation ($g\ C\ m^{-2}\ yr^{-1}$)
Tundra	Polar desert	Glacial retreat	8,000	0.2
	Polar desert	Glacial retreat	9,000	0.2
	Polar desert	Glacial retreat	2,600	2.4
	Sedge moss	Glacial retreat	1,000	2.4
	Sedge moss	Glacial retreat	9,000	1.1
	Sedge moss	Glacial retreat	8,700	0.7*
Boreal forest	Spruce	Glacial retreat	3,500	11.7
	Spruce-fir	Glacial retreat	5,435	0.8
	Spruce-fir	Glacial retreat	2,740	2.2
Temperate forest	Broadleaf evergreen	Volcanic ash	1,277	12.0
	Coniferous	Volcanic mudflow	1,200	10.0
	Deciduous	Alluvium	1,955	5.1
	Deciduous	Dunes	10,000	0.7
	Podocarpus	Dunes	10,000	2.1
	Angophora	Dunes	4,200	1.7
	Eucalyptus	Dunes	6,500	1.4
	Eucalyptus	Dunes	5,500	2.1
	Low forest	Glacial deposits	9,000	2.5
Tropical forest	*Metrosideros*	Volcanic ash	3,500	2.5
	Rain forest	Volcanic ash	8,620	2.3
Temperate grassland	*Chionochloa*	Glacial deposits	9,000	2.2
Temperate desert	Grassland	Alluvium	3,040	0.8

From Schlesinger (1990); citations to original literature are given therein.
* Corrected from value given in original publication.

than 0.7% of NPP (Schlesinger 1990, Dobrovolsky 1994, p. 114). Despite the stability of humus substances in the soil profile, the low rate of accumulation of soil organic matter in upland soils speaks strongly for the efficiency of decomposers using aerobic metabolic pathways of degradation (Gale and Gilmour 1988).

For areas covered by the last continental glaciation, the total accumulation of soil organic matter represents NEP for the last 10,000 years. The maximum extent of the last glacial, covering $29.5 \times 10^6\ km^2$ of the present land area (Flint 1971), now contains roughly 400×10^{15} g C, or about 25% of the carbon contained in all soils of the world (Table 5.3). In these areas, soil organic matter has accumulated at rates of about 1.35 g C m^{-2} yr^{-1} during the Holocene period. The current rate of storage in northern ecosystems (0.04×10^{15} g C/yr) is too small to be a significant sink for human releases of CO_2 to the atmosphere from fossil fuels, nor is it likely to have increased significantly during the last century (Gorham 1991, Harden et al. 1992).

Total storage of carbon in soils, 1456×10^{15} g or 121×10^{15} moles, can account for only 0.3% of the O_2 content of the atmosphere, given that the storage of organic carbon and the release of O_2 occur on a mole-for-mole basis during photosynthesis (Eq. 5.2). Thus, accumulations of atmospheric O_2 cannot be the result of the storage of organic carbon on land. Long-term storage of organic carbon appears to be dominated by accumulations in anoxic marine sediments (Chapter 9).

Soil Organic Matter and Global Change

When soils are brought under cultivation, their content of soil organic matter declines (Fig. 5.19). Losses from many soils are typically 20 to 30% within the first few decades of cultivation (Schlesinger 1986, Mann 1986, Detwiler 1986, Davidson and Ackerman 1993, Scholes and Scholes 1995). The loss is greatest during the first few years of cultivation. Eventually a new, lower level of soil organic matter is achieved that is in equilibrium with the lower production of plant detritus and the greater rates of decomposition under cropland (Jenkinson and Rayner 1977). Some of the soil organic matter is lost in erosion, but most is probably oxidized to CO_2 and released to the atmosphere. With about 10% of the world's soils under cultivation (Table 5.3), losses of organic matter from agricultural soils have been a major component of the past increase in atmospheric CO_2 (Schlesinger 1984). The current rate of release from soils, as much as 0.8×10^{15} g C/yr, is largely dependent upon the current rate at which natural ecosystems, especially in the tropics, are being converted to agriculture. Especially large losses of soil carbon are seen when organic soils in wetlands and peatlands are drained (Armentano and Menges 1986, Hutchinson 1980).

The dynamics of soil organic matter are illustrated by the pattern of loss after land is converted to agriculture. Recall that soil organic matter consists of a labile and a resistant fraction. The labile fraction is composed of fresh plant materials that are subject to rapid decomposition, whereas the resistant fraction is composed of humic materials that are often complexed with clay minerals. Rather than using biochemical fractionations (Fig. 5.16), some workers have used size or density fractionation to quantify the labile and resistant organic matter. Density fractionations are performed by adding soil samples to solutions of increasing specific gravity and collecting the material that floats to the surface (Spycher et al. 1983). In size fraction-

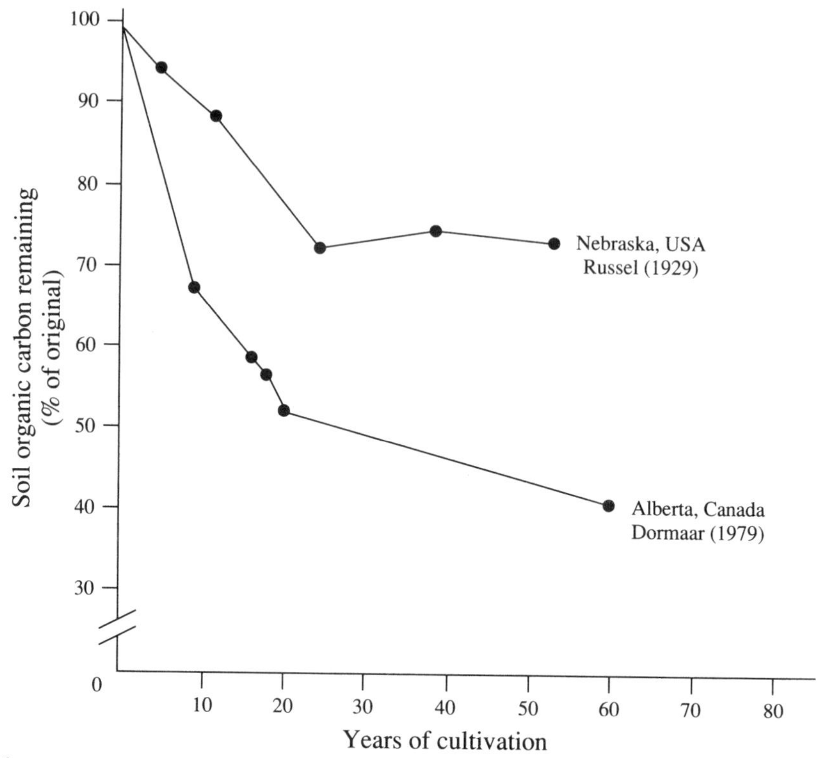

Figure 5.19 Decline in soil organic matter following conversion of native soil to agriculture in two grassland soils. From Schlesinger (1986).

ation, soils are passed through screens of varying mesh (Tisdall and Oades 1982, Elliott 1986). Most of the turnover of soil organic matter is in the "light" or large fractions that represent fresh plant materials (Foster 1981, Tiessen and Stewart 1983, Dalal and Mayer 1986a, 1986b). The "heavy" fraction is composed of polysaccarides (sugars) and humic materials that are complexed with clay minerals to form microaggregates of relatively high specific gravity (Tisdall and Oades 1982, Tiessen and Stewart 1988). The radiocarbon age of the different size or weight fractions indicates their rate of turnover. Anderson and Paul (1984) reported a ^{14}C age of 1255 years for organic matter in the clay fraction of a soil for which the overall age was 795 years. The decline in soil organic matter in agricultural soils is largely due to losses from the light fraction (Buyanovsky et al. 1994, Cambardella and Elliott 1994). Significantly, soil organic matter can accumulate fairly rapidly when agricultural soils are abandoned (Table 5.5).

The formation of charcoal during forest fires is a source of recalcitrant organic matter which accumulates in soils. Fearnside et al. (1993) calculate

Table 5.5 Accumulation of Soil Organic Matter in Abandoned Agricultural Soils and in Other Disturbed Sites, Which are Allowed to Return to Native Vegetation

Ecosystem type	Previous land use	Period of abandonment (yr)	Rate of accumulation (g C m^{-2} yr^{-1})	Reference
Subtropical forest	Cultivation	40	30–50	Lugo et al. (1986)
Temperate deciduous forest	Cultivation	100	45	Jenkinson (1990)
Temperate coniferous forest	Cultivation	50	21–26	Schiffman and Johnson (1989)
Temperate coniferous forest	Diked soils	100	26	Beke (1990)
Temperate deciduous forest	Mine spoils	50	55	Leisman (1957)
Temperate grassland	Mine spoils	28–40	28	Anderson (1977)
Temperate grassland	Cultivation	53	1.55	Burke et al. (1995)
Temperate grassland	Cultivation	5	110.0	Gebhart et al. (1994)

that about 2.7% of the aboveground biomass in tropical rainforests is converted to charcoal when they burn. Pine forests on the Virginia Piedmont contain 0.46 kg C/m^2 as charcoal in the forest floor (Schiffman and Johnson 1989), and Sanford et al. (1985) found 0.23 to 0.70 kg C/m^2 of charcoal in three soils of Brazilian rainforest. Unfortunately, we know relatively little about the long-term dynamics of charcoal in soils (Shneour 1966). Although the charcoal content of soils is typically only a small portion of the total pool of organic matter (Table 5.3), the apparent rate of accretion of charcoal (0.4 to 2.8 g C m^{-1} yr^{-1}) in the Brazilian rainforest soils is similar to the overall rate of humus formation in many ecosystems (Table 5.4). However, despite increasing rates of biomass burning by humans, it is not likely that an additional storage of carbon in charcoal is a significant flux in models of the global carbon cycle (Suman et al. 1997; Chapter 11).

In addition to changes in the pool of organic carbon as a result of cultivation and fire, soil carbon will change with climatic warming, which should stimulate rates of decomposition and the loss of soil organic matter in many ecosystems (Schleser 1982, Jenkinson et al. 1991, Peterjohn et al. 1994, Kirschbaum 1995, Trumbore et al. 1996). The effect of temperature will interact with other factors. For example, several experiments show that warming of organic soils in the tundra will increase the flux of CO_2 from the soil, but the loss is especially large if the water table is also lowered due to melting of permafrost (Billings et al. 1982, Moore and Dalva 1993, Funk et al. 1994). Oechel et al. (1993, 1995) suggest that a loss of carbon from some tundra soils is associated with unusually warm conditions over the last few decades. However, when tundra ecosystems are exposed to both high temperature and elevated CO_2, at levels chosen to resemble presumed atmospheric conditions of the next century, they show a net storage of carbon (Oechel et al. 1994). Inasmuch as many tundra ecosystems are nutrient limited (Billings et al. 1984), an increased storage of carbon in vegetation may be allowed by a greater rate of decomposition and nutrient release in warmer soils. The most accurate predictions of future changes in net ecosystem production may be derived from simulation models that include these interactive factors (e.g., Pastor and Post 1986, Seastedt et al. 1994, Schimel et al. 1994). Changes in soil carbon storage will be closely associated with changes in the distribution and productivity of vegetation discussed earlier.

Summary

Photosynthesis provides the energy that powers the biochemical reactions of life. That energy is captured from sunlight. Globally, net primary production of about 60×10^{15} g C yr^{-1} is available in the terrestrial biosphere. Although that is a large value, NPP typically captures less than 1% of the available sunlight energy. Most of the remaining energy evaporates water and heats the air, resulting in the global circulation of the atmosphere (Chapters 3 and 10). Thus, the terrestrial biosphere is fueled by a relatively inefficient initial process.

During photosynthesis, plants take up moisture from the soil and lose it to the atmosphere in the process of transpiration. Available moisture appears to be a primary factor determining the display of leaf area and NPP (Figs. 5.5 and 5.8). Among communities with adequate soil moisture, net primary production is determined by the length of the growing season and mean annual temperature; both are an index of the receipt of solar energy. Soil nutrients appear to be of secondary importance to NPP on land, perhaps because plants have various adaptations for obtaining and recycling nutrients efficiently when they are in short supply (Chapter 6).

Most net primary production is delivered to the soil where it is decomposed by a variety of organisms. The decomposition process is remarkably efficient, so only small amounts of NPP are added to the long-term storage of soil organic matter or humus. Soil organic matter consists of a dynamic pool near the surface, in which there is rapid turnover of fresh plant detritus and little long-term accumulation, and a large refractory pool of humic substances that are dispersed throughout the soil profile. Thus, the turnover time of organic carbon in the soil ranges from about 3 years for the litter to thousands of years for humus.

Humans have altered the processes of net primary production and decomposition on land, resulting in the transfer of organic carbon to the atmosphere, and perhaps a permanent reduction in the global rate of NPP. This disruption has produced global changes in the biogeochemical cycle of carbon, but little change in the atmospheric concentration of O_2.

Recommended Readings

Reichle, D.E. (ed.). 1981. *Dynamic Properties of Forest Ecosystems.* Cambridge University Press, Cambridge.

Solomon, A.H. and H. Shugart (eds.). 1993. *Vegetation Dynamics and Global Change.* Chapman and Hall, New York.

Swift, M.J., O.W. Heal, and J.M. Anderson. 1979. *Decomposition in Terrestrial Ecosystems.* University of California Press, Berkeley.

Waring, R.H. and W.H. Schlesinger. 1985. *Forest Ecosystems.* Academic Press, Orlando, Florida.

Whittaker, R.H. 1975. *Communities and Ecosystems.* 2nd ed. Macmillan, New York.

6

The Biosphere: Biogeochemical Cycling on Land

Introduction

Living tissue is primarily composed of carbon, hydrogen, and oxygen in the approximate proportion of CH_2O, but more than 20 other elements are necessary for biochemical reactions and for the growth of structural biomass. For example, phosphorus (P) is required for adenosine triphosphate (ATP), the universal molecule for energy transformations in organisms, and calcium (Ca) is a major structural component in both plants and animals. The proteins found in plants and animals contain about 16% nitrogen (N) by weight. Earlier we saw that the protein ribulose bisphosphate carboxylase (RuBP) controls the rate of carbon fixation during photosynthesis in many plant species (Chapter 5). The link between C and N that begins in cellular biochemistry extends to the global biogeochemical cycles of these elements.

The various elements essential to biochemical structure and function are often found in predictable proportions in living tissues (e.g., wood, leaf, bone, and muscle; Garten 1976, Reiners 1986). For instance, the ratio of C to N in leaf tissue is about 50. At the global level, our estimate of net primary production (NPP), 60×10^{15} g C/yr, implies that about 1200×10^{12} g of nitrogen must be supplied to plants each year through biogeochemical cycling to achieve the level of NPP that we observe. As we shall see, the availability of some elements, such as N and P, is often limited, and the supply of these elements may control the rate of net primary production in terrestrial ecosystems. Conversely, for elements that are typically available in greater quantities, e.g., Ca and S, the rate of net primary production often determines the rate of cycling in the ecosystem and losses to streamwaters. In every case, the biosphere exerts a strong control on the geochemical behavior of the major elements of life. Much less biological control is seen in the cycling of elements such as sodium (Na) and chloride (Cl), which are less important constituents of biomass (Gorham et al. 1979).

In earlier chapters we saw that the atmosphere is the dominant source of C, N, and S in terrestrial ecosystems, and that rock weathering is the major source for most of the remaining biochemical elements (e.g., Ca, Mg, K, Fe, P). In any terrestrial ecosystem, the receipt of elements from the atmosphere and the lithosphere represents an input of new quantities of nutrients for plant growth.[1] However, as a result of internal cycling and retention of past inputs, plant growth is not solely dependent upon new inputs to the system. In fact, the annual circulation of important elements such as N within an ecosystem is often 10 to 20× greater than the amount received from outside the system (Table 6.1). This large internal, or *intrasystem cycle,* is achieved by the long-term retention of elements received from the atmosphere and the lithosphere. Important biochemical elements are accumulated in terrestrial ecosystems by biotic uptake, whereas nonessential elements pass through these systems under simple geochemical control (Johnson 1971, Vitousek and Reiners 1975).

[1] Unlike the well-known models developed for the Hubbard Brook Ecosystem (e.g., Likens and Bormann 1995), the nutrient budgets in this book consider rock weathering as an *external* source of nutrients that enter a terrestrial ecosystem each year (Gorham et al. 1979).

Table 6.1 Percentage of the Annual Requirement of Nutrients for Growth in the Northern Hardwoods Forest at Hubbard Brook, New Hampshire, That Could Be Supplied by Various Sources of Available Nutrients[a]

Process	N	P	K	Ca	Mg
Growth requirement (Kg ha^{-1} yr^{-1})	115.4	12.3	66.9	62.2	9.5
Percentage of the requirement that could be supplied by:					
Intersystem inputs					
Atmospheric	18	0	1	4	6
Rock weathering	0	1	11	34	37
Intrasystem transfers					
Reabsorptions	31	28	4	0	2
Detritus turnover (includes return in throughfall and stemflow)	69	67	87	85	87

[a] Calculated using Eqs. 6.2 and 6.3. Reabsorption data are from Ryan and Bormann (1982). Data for N, K, Ca, and Mg are from Likens and Bormann (1995) and for P from Yanai (1992).

In this chapter we will examine the cycle of biochemical elements in terrestrial ecosystems. We will begin by examining aspects of plant uptake, allocations during growth, and losses due to the death of plants and plant tissues. Then, we will examine how elements such as N, P, and S in dead organic matter are transformed in the soil, leading to their release for plant uptake or for loss from the ecosystem. We will stress interactions between carbon and other biochemical elements and examine how land plants have adapted to the widespread limitations of N and P in most ecosystems. A brief examination will be given to how biogeochemical processes may control the distribution of plants and animals on land.

Biogeochemical Cycling in Land Plants

Nutrient Uptake

It is easy to forget the essential, initial role played by plants in all of biochemistry. Plants obtain essential elements from the soil (e.g., N from NO_3^-) and incorporate them into biochemical molecules (e.g., amino acids) (Oaks 1994). Animals may eat plants, and each other, and synthesize new amino acids, but the building blocks of the amino acids in animal protein are those originally synthesized in plants. Only in isolated instances, for example, in animals at natural salt licks, do we find a direct transfer of elements from inorganic form to animal biochemistry (Jones and Hanson 1985). There are no vitamin pills in the natural biosphere!

Soil chemical reactions, such as ion exchange and mineral solubility, set the initial constraints on the availability of essential elements for plant uptake. However, when plant uptake of an element such as phosphorus is rapid, additional phosphorus may enter the soil solution from the dissolution of minerals, and plants can release organic compounds that enhance the solubility of various nutrient elements from soil minerals (Chapter 4).

Delivery of ions to plant roots can occur by several pathways (Barber 1962). The concentration of some elements in the soil solution is such that their passive uptake with water is adequate for plant nutrition (Turner 1982). In some cases, the delivery is excessive, and the ions are actively excluded at the root surface. For example, it is not unusual to see accumulations of Ca, as $CaCO_3$, surrounding the roots of desert shrubs growing in calcareous soils (Klappa 1980, Wullstein and Pratt 1981). In contrast, for N, P, and K the concentration in the soil solution is often much too low for adequate delivery in the transpiration stream, and plant uptake is enhanced by enzymes that carry ions across the root membrane using active transport (Ingestad 1982, Robinson 1986, Chapin 1988).

The enzymes embedded in root membranes achieve increasing rates of nutrient uptake as a function of increasing concentrations in the soil solution until the activity of the enzyme system is saturated. Chapin and Oechel (1983) found that populations of the arctic sedge, *Carex aquatilis,* from colder habitats had higher rates of uptake than those from warmer habitats, presumably reflecting adaptation to the lower availability of phosphorus in cold soils (Fig. 6.1).

The uptake of N and P is so rapid and the concentrations in the soil solution are typically so low that these elements are effectively absent in the soil solution surrounding roots, and the rate of uptake is determined by diffusion to the root from other areas (Nye 1977). Phosphate is particularly immobile in most soils, and the rate of diffusion strongly limits its supply to plant roots (Robinson 1986). Although adaptations for more efficient root enzymes are seen in some species (Pennell et al. 1990), the most apparent response of plants to low nutrient concentrations is an increase in the root/shoot ratio, which increases the volume of soil exploited and decreases diffusion distances (Chapin 1980, Clarkson and Hanson 1980, Robinson 1994). In many species the relative growth rate of roots determines the uptake of nitrogen and phosphorus (Newman and Andrews 1973; Fig. 6.2), and roots show rapid proliferation in nutrient-rich patches (Jackson et al. 1990, Pregitzer et al. 1993, Bates and Lynch 1996).

Higher plants and soil microbes exude enzymes into the soil that can release inorganic phosphorus from organic matter. These extracellular enzymes are known as *phosphatases,* which have different forms in acid and alkaline soils (Malcolm 1983, Tarafdar and Claassen 1988, Dinkelaker and Marschner 1992, Duff et al. 1994). In many cases, root phosphatase activity is inversely proportional to available soil P (McGill and Cole 1981, Fox and Comerford 1992b). For example, phosphatase activity rises with the accumulation of organic matter in soils during the development of *Eucalyptus* plantations after fire (Polglase et al. 1992). Phosphatase activity associ-

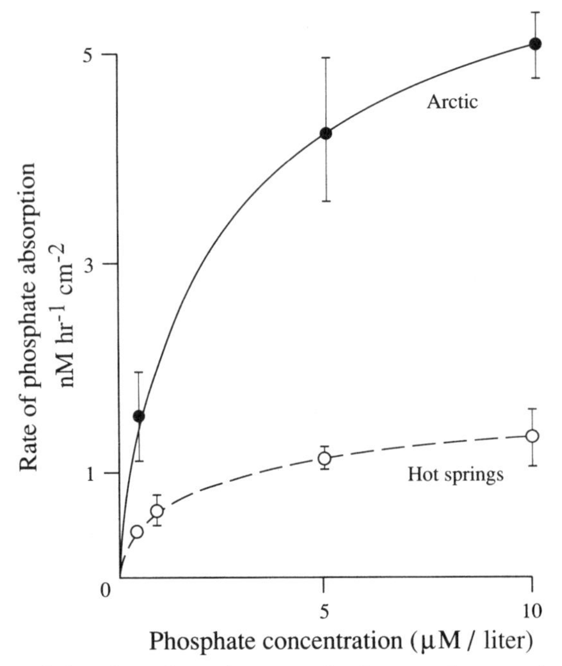

Figure 6.1 Rate of phosphate absorption per unit of root surface area in populations of *Carex aquatilis* from cold (Arctic) and warm (Hot Springs) habitats measured at 5°C. From Chapin (1974).

ated with root surfaces is particularly significant to plants in phosphorus-poor habitats, and it may supply up to 69% of the annual phosphorus demand of some tundra plants (Kroehler and Linkins 1991).

Nutrient Balance

In addition to an adequate supply of nutrient elements, plant growth is affected by the balance of nutrients in the soil (Shear et al. 1946). For seedlings of several tree species, Ingestad (1979b) found that a solution containing 100 parts N, 15 parts P, 50 parts K, 5 parts Ca and Mg, and 10 parts S was ideal for maximum growth. However, unless the supply of a nutrient reaches very low levels, plants usually do not show deficiency symptoms; they simply grow more slowly (Clarkson and Hanson 1980). Inherent slow growth is a characteristic of plants adapted to infertile habitats, and it often persists even when nutrients are added experimentally (Chapin et al. 1986a).

Because more nutrients occur as positively charged ions than as negatively charged ions in the soil solution, one might expect that plant roots would develop a charge imbalance as a result of nutrient uptake. When ions such as K^+ are removed from the soil solution in excess of the uptake of negatively charged ions, the plant releases H^+ to maintain an internal balance of

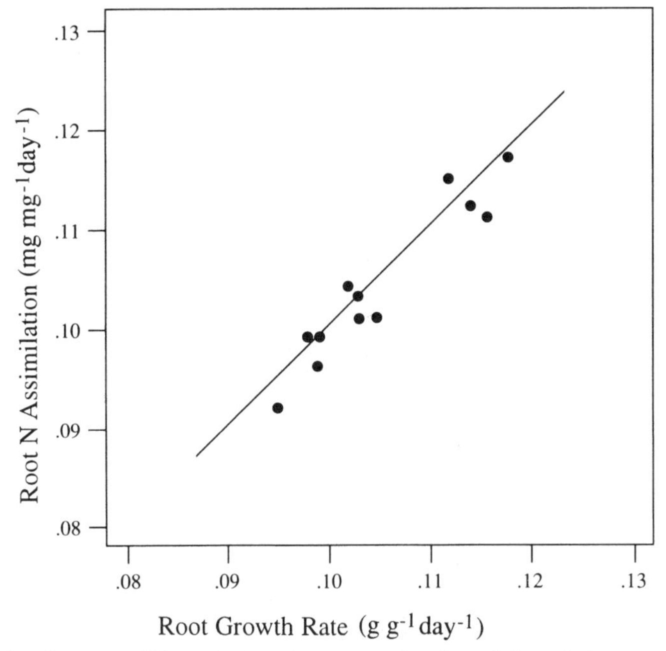

Figure 6.2 The rate of N uptake in tobacco as a function of the relative growth rate of roots. From Raper et al. (1978).

charge (Maathuis and Sanders 1994). This H^+ may, in turn, replace K^+ on a cation exchange site, driving another K^+ into the soil solution. The high concentration of N in plant tissues causes the form in which N is taken up to dominate this process (Table 6.2). Oaks (1992) has shown how plants that use NH_4^+ as a N source tend to acidify the immediate zone around their roots. The uptake of NO_3^- has the opposite effect as a result of plant releases of HCO_3^- and organic acids to balance the negative charge (Nye 1981, Hedley et al. 1982a).

Table 6.2 Chemical Composition and Ionic Imbalance for Perennial Ryegrass[a]

	N	P	S	Cl	K	Na	Mg	Ca
Percent in leaf tissue	4.00	0.40	0.30	0.20	2.50	0.20	0.25	1.00
Equivalent weight (g)	14.00	30.98	16.03	35.46	39.10	22.99	12.16	20.04
mEq present	285.7	12.9	18.7	5.6	63.9	8.8	20.6	49.9
Sum of mEq	±285.7		−37.2			+143.1		

Imbalance in mEq %
(a) where ammonium nitrogen is taken up: 285.7 + 143.1 − 37.2 = +391.6
(b) where nitrate nitrogen is taken up: 143.1 − 285.7 − 37.2 = −179.8

[a] From Middleton and Smith (1979).

Nitrogen Assimilation

Among various habitats, the availability of soil nitrogen as NH_4^+ or NO_3^- differs largely depending upon the environmental conditions that affect the conversion of NH_4^+ to NO_3^- in the microbial process known as nitrification (Eqs. 2.16–17). For example, in the waterlogged soils of the tundra, almost all nitrogen is found as NH_4^+ (Barsdate and Alexander 1975), whereas in some deserts and forests, only NO_3^- is important (Virginia and Jarrell 1983, Nadelhoffer et al. 1984). Many species show a preference for NO_3^-, although species occurring in sites where nitrification is slow or inhibited often tend to show superior growth with ammonium (Haynes and Goh 1978, Adams and Attiwill 1982, Falkengren-Grerup 1995). A few unusual, insectivorous plants obtain their N by digesting captured organisms. Dixon et al. (1980) found that 11 to 17% of the annual uptake of N in *Drosera erythrorhiza* (sundew) could be obtained from captured insects. Direct uptake of amino acids from the soil solution is also known for some arctic plants (Chapin et al. 1993, Kielland 1994, Atkin 1996).

Inside the plant, both forms of inorganic N are converted to amino groups ($-NH_2$) that are attached to soluble organic compounds. In many woody species these conversions occur in the roots, and N is transported to the shoot as amides, amino acids, and ureide compounds in the xylem stream (Andrews 1986). However, in some species N in the xylem is found as NO_3^-, and the reduction of NO_3^- to $-NH_2$ occurs in leaf tissues (Smirnoff et al. 1984). Eventually, most plant N is incorporated into protein.

The conversion of NO_3^- to $-NH_2$ is a biochemical reduction reaction that requires metabolic energy and is catalyzed by an enzyme, *nitrate reductase*. One might puzzle why most plants do not show a clear preference for NH_4^+, which is assimilated more easily. Several explanations have been offered. Recall that NH_4^+ interacts with soil cation exchange sites, whereas NO_3^- is highly mobile in most soils. The rate of delivery of NO_3^- to the root by diffusion or mass flow is much higher than that of NH_4^+ under otherwise equivalent conditions (Raven et al. 1992). Plants that utilize NH_4^+ may have to compensate for the differences in diffusion by investing more energy in root growth (Gijsman 1990, Oaks 1992, Bloom et al. 1993). Uptake of NO_3^- also avoids the competition that occurs in root enzyme carriers between NH_4^+ and other positively charged nutrient ions. For example, the presence of large amounts of K^+ in the soil solution can reduce the uptake of NH_4^+ (Haynes and Goh 1978). Finally, relatively low concentrations of NH_4^+ are potentially toxic to plant tissues. These potential disadvantages in the uptake of NH_4^+ may explain why many plants take up NO_3^- when thermodynamic calculations suggest that metabolic costs of reducing NO_3^- are significantly greater than for plants that assimilate NH_4^+ directly (Middleton and Smith 1979, Gutschick 1981, Bloom et al. 1992).

It is unclear why so many species concentrate nitrate reductase in their roots, when the same reaction performed in leaf tissues, where it can be coupled to the photosynthetic reaction, is energetically much less costly (Gutschick 1981, Andrews 1986). Addition of NO_3^- to the soil often induces the production of root enzymes for NO_3^- uptake and the synthesis of more nitrate reductase in plant tissues (Lee and Stewart 1978, Hoff et al. 1992, Oaks 1994). There is some evidence that the proportion of nitrate reductase in the shoot increases at high levels of available NO_3^- (Andrews 1986).

Nitrogen Fixation

Several types of bacteria and blue-green algae possess the enzyme *nitrogenase,* which converts atmospheric N_2 to NH_3 (see Eq. 2.12). Some of these exist as free-living (asymbiotic) forms in soils, but others, such as *Rhizobium* and *Frankia,* form symbiotic associations with the roots of higher plants. The symbiotic bacteria reside in root nodules that are easily recognized in the field. Nitrogen fixation is especially well known among species of legumes (Leguminosae).

Nitrogen that enters terrestrial ecosystems by fixation is a "new" input in the sense that it is derived from outside the boundaries of the ecosystem—i.e., from the atmosphere. The reduction of N_2 to NH_3 has large metabolic costs that require the respiration of organic carbon. Nevertheless, Gutschick (1981) suggests that symbiotic fixation in higher plants is not greatly less efficient than the uptake of NO_3^- for those species in which the nitrate reductase is located in plant roots. Only a few land plants support symbiotic nitrogen fixation, and it is interesting to speculate why nitrogen fixation is not more widespread, when nitrogen limitations of net primary production are so frequent (Vitousek and Howarth 1991). Globally, plants "spend" only about 2.5% of NPP on nitrogen fixation (Gutschick 1981).

The energy cost of nitrogen fixation links this biogeochemical process to the availability of organic carbon, provided by net primary production. In plants with symbiotic nitrogen fixation, the rate of N-fixation is often directly related to the rate of photosynthesis and the efficiency of plant growth (Bormann and Gordon 1984). Free-living heterotrophic bacteria that conduct asymbiotic nitrogen fixation are usually found in organic soils or local areas with high levels of organic matter that provide a ready source of energy (Granhall 1981). For instance, nitrogen fixation is frequently observed in rotten logs (Roskoski 1980, Silvester et al. 1982, Griffiths et al. 1993), where it is probably associated with anaerobic cellulolytic bacteria (Leschine et al. 1988).

In both symbiotic and asymbiotic forms, nitrogen fixation is generally inhibited at high levels of available nitrogen (Cejudo et al. 1984). In many cases the rate of fixation appears to be controlled by the N:P ratio in the soil (Chapin et al. 1991, V.H. Smith 1992); for instance, added phosphorus stimulates asymbiotic N-fixation in prairie soils (Fig. 6.3). In bacteria, phosphorus appears to activate the gene for the synthesis of nitrogenase (Stock

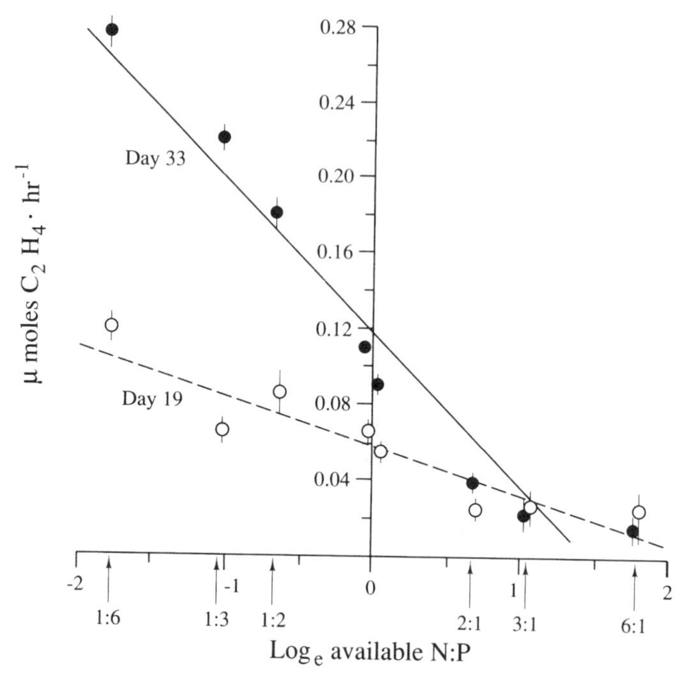

Figure 6.3 Acetylene reduction, an index of nitrogen fixation by asymbiotic N-fixing bacteria, as a function of the N : P ratio in soil. From Eisele et al. (1989).

et al. 1990), illustrating how the linkage between the global cycles of nitrogen and phosphorus has a basis in molecular biology. Requirements for Mo and Fe as structural components of nitrogenase also link nitrogen fixation to the availability of these elements in natural ecosystems. Some plants with symbiotic N-fixing bacteria appear to acidify their rooting zone to make Fe and P more available (Ae et al. 1990, Raven et al. 1990, Gillespie and Pope 1990). Silvester (1989) suggests that low availability of Mo may limit asymbiotic N-fixation in forests of the northwestern United States.

The isotopic ratio of N in plant tissues is expressed as $\delta^{15}N$, using a calculation analogous to that which we saw for the isotopes of carbon in Chapter 5. In the case of nitrogen, the standard is the atmosphere, which contains 99.63% ^{14}N and 0.37% ^{15}N. Nitrogenase shows only a slight discrimination between the isotopes of N, that is, between $^{15}N_2$ and $^{14}N_2$ (Handley and Raven 1992), but differences in the isotopic ratio of nitrogen among plant species can be used to suggest which species may be involved in nitrogen fixation in the field (Virginia and Delwiche 1982, Yoneyama et al. 1993). Nitrogen-fixing species typically show values of $\delta^{15}N$ that are close to the atmospheric ratio (Shearer and Kohl 1989), whereas nonfixing species show a wide range of values depending on the rate of nitrogen mineralization in the soil (Garten and van Miegroet 1994) (Fig. 6.4, see also p. 204). Shearer et al. (1983) used the difference in isotopic ratio between *Prosopis* grown in the laboratory without added N (i.e., all nitrogen was derived from fixation) and the same species in the field to estimate that the field plants derived 43 to 61% of their nitrogen from fixation. Of course, when nitrogen-fixing plants die, their nitrogen content is available for other species in the ecosystem (Huss-Danell 1986, van Kessel et al., 1994). Lajtha and Schlesinger (1986) found that the desert shrub, *Larrea tridentata,* growing adjacent to nitrogen-fixing *Prosopis* had lower $\delta^{15}N$ than when *Larrea* were growing alone.

Nitrogenase activity can be measured using the acetylene-reduction technique, which is based on the observation that this enzyme also converts acetylene to ethylene under experimental conditions. Plants or nodules are placed in small chambers or small chambers are placed over field plots, and the conversion of injected acetylene to ethylene over a known time period is measured using gas chromatography. The conversion of acetylene (in moles) is not exactly equivalent to the potential rate of fixation of N_2 because the enzyme has different affinities for these substrates. However, appropriate conversion ratios can be determined using other techniques. For instance, investigators have added $^{15}N_2$, the heavy stable isotope of N, to chambers and used the increase in organic compounds containing ^{15}N

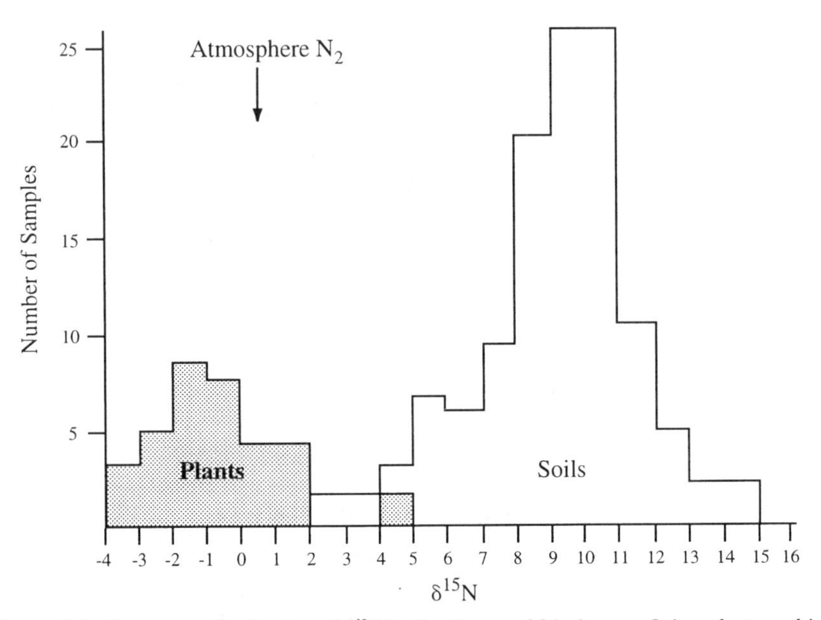

Figure 6.4 Frequency distribution of $\delta^{15}N$ in the tissues of 34 nitrogen-fixing plants and in the organic matter of 124 soils from throughout the United States. Plotted using data from Shearer and Kohl (1988, 1989).

in test plants or soil as a measure of nitrogen fixation (e.g., Silvester et al. 1982, Zechmeister-Boltenstern and Kinzel 1990).

Asymbiotic bacteria and blue-green algae are widespread, and their nitrogen fixation can be an important source of N for some terrestrial ecosystems. Exceptionally high rates of fixation have been recorded in blue-green algae crusts that cover the soil surface in some desert ecosystems (Rychert et al. 1978); however, in most cases the total input from asymbiotic fixation is in the range of 1 to 5 kg N ha^{-1} yr^{-1} (Boring et al. 1988, Cushon and Feller 1989, Jeffries et al. 1992). In most regions, this input is about equivalent to the annual deposition of nitrogen in wetfall and dryfall from the atmosphere.

The importance of nitrogen fixation in terrestrial ecosystems varies widely depending upon the presence of species that harbor symbiotic bacteria. Some of the greatest rates of N-fixation are seen in species that invade after disturbance, where high light levels allow maximum photosynthesis (Vitousek and Howarth 1991). For example, in the recovery of Douglas fir forests after fire, Youngberg and Wollum (1976) found that the nodulated shrub, *Ceanothus velutinus,* contributed up to 100 kg N ha^{-1} yr^{-1} on some sites. Invasion of the exotic, nitrogen-fixing tree, *Myrica faya,* provides important inputs of nitrogen (18 kg ha^{-1} yr^{-1}) on fresh volcanic ashflows on Hawaii (Vitousek et al. 1987). In most cases the importance of plants with symbiotic nitrogen fixation declines with the recovery of mature vegetation, and their occurrence in undisturbed communities is limited. In this regard, the widespread occurrence of leguminous species in mature tropical forests is deserving of further study. The sporadic occurrence of symbiotic nitrogen fixation in terrestrial ecosystems makes it difficult to extrapolate from local studies to provide a global estimate of its importance. Burns and Hardy (1975) suggest that the global rate of N-fixation (asymbiotic + symbiotic) in natural ecosystems may supply 100×10^{12} g N/yr, or about 10% of the annual plant demand for nitrogen on land (Chapter 12).

Mycorrhizal Fungi

Symbiotic associations between fungi and higher plants are found in most ecosystems (Harley and Smith 1983, Allen 1992). This symbiosis is important for the nutrition of plants, and may have even determined the origin of land plants (Pirozynski and Malloch 1975). There are several forms of symbiosis. In temperate regions, many trees harbor ectotrophic mycorrhizal fungi. These fungi form a sheath around the active fine roots and extend additional hyphae into the surrounding soil. In many areas, especially the tropics, plants possess endotrophic mycorrhizal fungi in which the hyphae actually penetrate cells of the root. By virtue of their large surface area and efficient absorption capacity, mycorrhizal fungi are able to obtain soil nutrients and transfer these to the higher plant root. Harrison and van

Buuren (1995) describe the molecular structure of a transporter protein in mycorrhizae that moves phosphorus from the soil into roots. Recent work suggests that mycorrhizal fungi are also directly involved in the decomposition of soil organic materials through the release of extracellular enzymes such as cellulases and phosphatases (Antibus et al. 1981, Dodd et al. 1987) and in the weathering of soil minerals through the release of organic acids (Bolan et al. 1984, Illmer et al. 1995; see also Chapter 4). It is important to remember that most of these reactions are also associated with plant roots; mycorrhizae simply enhance their occurrence in the rhizosphere, increasing the overall rate of plant nutrient uptake (Bolan 1991). In return, mycorrhizal fungi depend upon the host plant for supplies of carbohydrate.

The importance of mycorrhizae in infertile sites is well known. Many species of pine require ectotrophic mycorrhizae, which perhaps accounts for the success of pines in nutrient-poor soils. Most tropical trees appear to require endotrophic mycorrhizal associations for proper growth (Janos 1980), and mycorrhizal fungi are widespread among the *Eucalyptus* species growing in the low-phosphorus soils of Australia. Berliner et al. (1986) report the complete exclusion of *Cistus incanus* from basaltic soils in Israel due to a failure of mycorrhizal development. The same species grows well on adjacent calcareous soils, or in basaltic soils supplied with fertilizer.

Mycorrhizal fungi are especially important in the transfer of those soil nutrients with low diffusion rates in the soil. A large number of studies document the importance of mycorrhizae in P nutrition (Koide 1991), but absorption of N and other nutrients is also known (Bowen and Smith 1981, Ames et al. 1983). Some plants with mycorrhizal fungi show higher levels of various nutrients in foliage, but frequently the enhanced uptake of nutrients results in higher rates of growth (Schultz et al. 1979). Rose and Youngberg (1981) provide an insightful experiment with *Ceanothus velutinus* growing in nitrogen-deficient soils with and without mycorrhizae and symbiotic nitrogen-fixing bacteria (Table 6.3). The highest rates of growth were seen when both of these symbiotic associations were present, which also allowed a decrease in the root/shoot ratio. Nitrogen fixation enhanced the uptake of phosphorus by mycorrhizal fungi. These results illustrate the strong interactions between N, P, and C in the nutrition of higher plants.

Under conditions of nutrient deficiency, plant growth usually slows, whereas photosynthesis continues at relatively high rates (Chapin 1980), and the content of soluble carbohydrate in the plant increases. Marx et al. (1977) found that high concentrations of carbohydrate in root tissues of loblolly pine (*Pinus taeda*) stimulated mycorrhizal infection (Fig. 6.5). Thus, internal plant allocation of carbohydrates to roots may result in increased nutrient uptake by mycorrhizae and an alleviation of nutrient deficiencies.

Mycorrhizal fungi use a fraction of the carbon fixed by the host plant, representing a drain on net primary production that might otherwise be

Table 6.3 Effects of Mycorrhizae and N-Fixing Nodules on Growth and Nitrogen Fixation in *Ceanothus velutinus* Seedlings[a]

	Control	+Mycorrhizae	+Nodules	+Mycorrhizae and nodules
Mean shoot dry weight (mg)	72.8	84.4	392.9	1028.8
Mean root dry weight (mg)	166.4	183.4	285.0	904.4
Root/shoot	2.29	2.17	0.73	0.88
Nodules per plant	0	0	3	5
Mean nodule weight (mg)	0	0	10.5	44.6
Acetylene reduction (mg/nodule/hr)	0	0	27.85	40.46
Percent mycorrhizal colonization	0	45	0	80
Nutrient concentration (in shoot, %)				
N	0.32	0.30	1.24	1.31
P	0.08	0.07	0.25	0.25
Ca			1.07	1.15

[a] From Rose and Youngberg (1981).

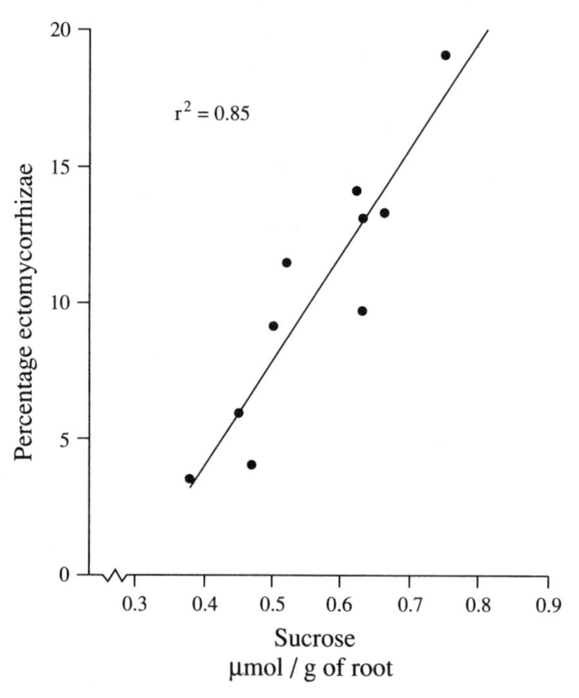

Figure 6.5 Relationship between infection of the roots of loblolly pine by ectomycorrhizal fungi and the sucrose concentration in the root. From Marx et al. (1977).

allocated to growth (Rygiewicz and Andersen 1994). That the cost of symbiotic fungi is significant is underscored by experiments in which the degree of colonization has declined and plant growth has increased when plants were fertilized (e.g., Blaise and Garbaye 1983). Vogt et al. (1982) found that mycorrhizal biomass was only 1% of the ecosystem total in a fir forest, but the growth of mycorrhizal fungi utilized about 15% of the net primary production (see Table 5.1). Unfortunately, we have few data from which a global estimate of the effect of mycorrhizae on net primary production might be calculated.

Nutrient Allocations and Cycling in Land Vegetation

The Annual Intrasystem Cycle

The uptake of nutrients from the soil is allocated to the growth of new plant tissues. Although short-lived tissues (leaves and fine roots) compose a small fraction of total plant biomass, they receive the largest proportion of the annual nutrient uptake. Growth of leaves and roots received 87% of the N and 79% of the P allocated to new tissues in a deciduous forest in England (Cole and Rapp 1981, p. 404). In a perennial grassland dominated by *Bouteloua gracilis* in Colorado, new growth of aboveground tissues accounted for 67% of the annual uptake of N (Woodmansee et al. 1978).

When leaf buds break and new foliage begins to grow, the leaf tissues often have high concentrations of N, P, and K. As the foliage matures, these concentrations often decrease (van den Driessche 1974). Some of these changes are due to the increasing accumulation of photosynthetic products with time and to leaf thickening during development. Leaf mass per unit area (mg/cm^2) may increase as much as 50% during the growing season and then decline as the leaf senesces (Smith et al. 1981). The initial concentrations of N and P are diluted as the leaf tissues accumulate carbohydrates and cellulose. In contrast, the concentrations of some nutrients, e.g., Ca, Mg, and Fe, often increase with leaf age (van den Driessche 1974). Increases in calcium concentration with leaf age result from secondary thickening, including calcium pectate deposition in cell walls, and from increasing storage of calcium oxalate in cell vacuoles.

Although there are variations among species, nutrient concentrations in mature foliage are related to the rate of photosynthesis (Chapter 5) and plant growth (e.g., Tilton 1978), and analysis of foliage is often used as an index of site fertility (Van den Driessche 1974). Leaf concentrations of trace metals often reflect the content of the underlying soil, such that leaf tissues are useful for mineral prospecting in some areas (Cannon 1960, Brooks 1973). Among tropical forests, concentrations of major nutrients in leaves are significantly higher on more fertile soils (Vitousek and Sanford 1986). Yin (1993) found that C/N and C/P ratios in the foliage of deciduous trees varied systematically with lower values among species in colder habitats than in the tropics. The higher leaf nutrient contents in colder climates may allow for higher photosynthetic rates and rapid growth of these species in response to a short growing season (Mooney and Billings 1961, Reich et al. 1995, Körner 1989).

Upon fertilization with a specific nutrient, the concentrations of other leaf nutrients can show surprising changes. For example, leaf N increased when Miller et al. (1976) fertilized Corsican pine (*Pinus nigra*) with nitrogen, but in the same samples, concentrations of P, Ca, and Mg declined. When nitrogen fertilization of N-deficient stands stimulates photosynthesis, the concentrations of other nutrients in foliage may be diluted by new accumulations of carbohydrate (Fowells and Krauss 1959, Timmer and Stone 1978, Jarrell and Beverly 1981). In some cases, uptake of P from the soil may fall behind the rates needed for maximum growth at the newly established levels of N availability. In other cases, improvements in plant nitrogen status enhance the uptake of other elements as well (e.g., Table 6.3). Plant response to single element fertilizations are illustrative of the importance of considering balanced nutrition for maximum plant growth.

Once leaves are fully expanded, changes in the nutrient content per unit of leaf area indicate movements of nutrients between the foliage and the stem. Woodwell (1974) found that oak leaves rapidly accumulated N during the early summer, presumably as a component of photosynthetic enzymes. The leaf content of N, P, and K remained relatively constant at high levels throughout the growing season, but was strongly removed from leaves in autumn. Such losses often represent active withdrawal of nutrients from foliage for reuse during the next year. Some trace micronutrients are also withdrawn before leaf-fall (Killingbeck 1985), but usually reabsorption of foliar Ca and Mg is limited. Fife and Nambiar (1984) observed that reabsorption of N, P, and K was not just related to leaf senescence in *Pinus radiata;* these nutrients could also move from the early to the later tissues produced during the same growing season.

Leaf nutrient contents are also affected by rainfall that leaches nutrients from the leaf surface (Tukey 1970, Parker 1983). In particular, seasonal changes in the content of K, which is highly soluble and especially concentrated in cells near the leaf surface, may represent leaching. The losses of nutrients in leaching often follow the order

$$K \gg P > N > Ca. \tag{6.1}$$

Leaching rates generally increase as foliage senesces before abscission; thus, care must be taken to recognize changes due to leaching versus changes due to active nutrient withdrawals (Ostman and Weaver 1982).

Nutrient losses by leaching differ among leaf types. Luxmoore et al. (1981) calculated lower rates of leaching loss from pines than from broadleaf deciduous species in a forest in Tennessee. Such differences may be due to differences in leaf nutrient concentration, surface-area-to-volume ratio, surface texture, and leaf age. Among the trees of the humid tropics, the smooth surface of broad sclerophylls may be an adaptive response to reducing leaching by minimizing the length of time that rainwater is in contact with the leaf surface (Dean and Smith 1978). Species-specific differences in rates of leaching may explain differences in the epiphyte loads on different trees (Benzing and Renfrow 1974, Schlesinger and Marks 1977, Awasthi et al. 1995).

Rainwater that passes through a vegetation canopy is called *throughfall,* which is usually collected in funnels or troughs placed on the ground. Throughfall contains nutrients leached from leaf surfaces and is most important in the cycling of nutrients such as K (Parker 1983, Schaefer and Reiners 1989). In forests, rainwater that travels down the surface of stems is called *stemflow.* The concentrations of nutrients in stemflow waters are high, but usually much more water reaches the ground as throughfall. Stemflow is significant to the extent that it returns highly concentrated nutrient solutions to the soil at the base of plants (Gersper and Holowaychuk 1971).

Leaching varies seasonally depending on forest type and climate. Not surprisingly, in temperate deciduous forests, the greatest losses are during the summer months (Lindberg et al. 1986). Some of the nutrient content in throughfall is derived from aerosols that are deposited on leaf surfaces (Chapter 3). Indeed, Lindberg and Garten (1988) found that about 85% of the apparent loss of sulfate from a forest canopy was due to dry deposition on leaf surfaces, and some studies have used the SO_4^{2-} content of throughfall to estimate dry deposition in the canopy (Garten et al. 1978, Ivens et al. 1990, Neary and Gizyn 1994). For most elements, however, leaching of nutrients from vegetation makes it difficult to use nutrient concentrations in the rainfall collected under a canopy to calculate dry deposition on leaf surfaces (Chapter 3). In some cases leaves appear to take up nutrients from rainfall, particularly soluble forms of N (Carlisle et al. 1966, Miller et al. 1976, Olson et al. 1981, Lovett and Lindberg 1993).

Litterfall

When the biomass of vegetation is not changing, the annual production of new tissues is balanced by the senescence and loss of plant parts (Chapter 5). In the intrasystem cycle, plant litterfall is the dominant pathway for nutrient return to the soil, especially for N and P (Fig. 6.6). Below ground, root death also makes a major contribution of nutrients to the soil each year (Cox et al. 1978, Vogt et al. 1983, Burke and Raynal 1994).

The nutrient concentrations in litterfall differ from the nutrient concentrations in mature foliage by the reabsorption of constituents during leaf

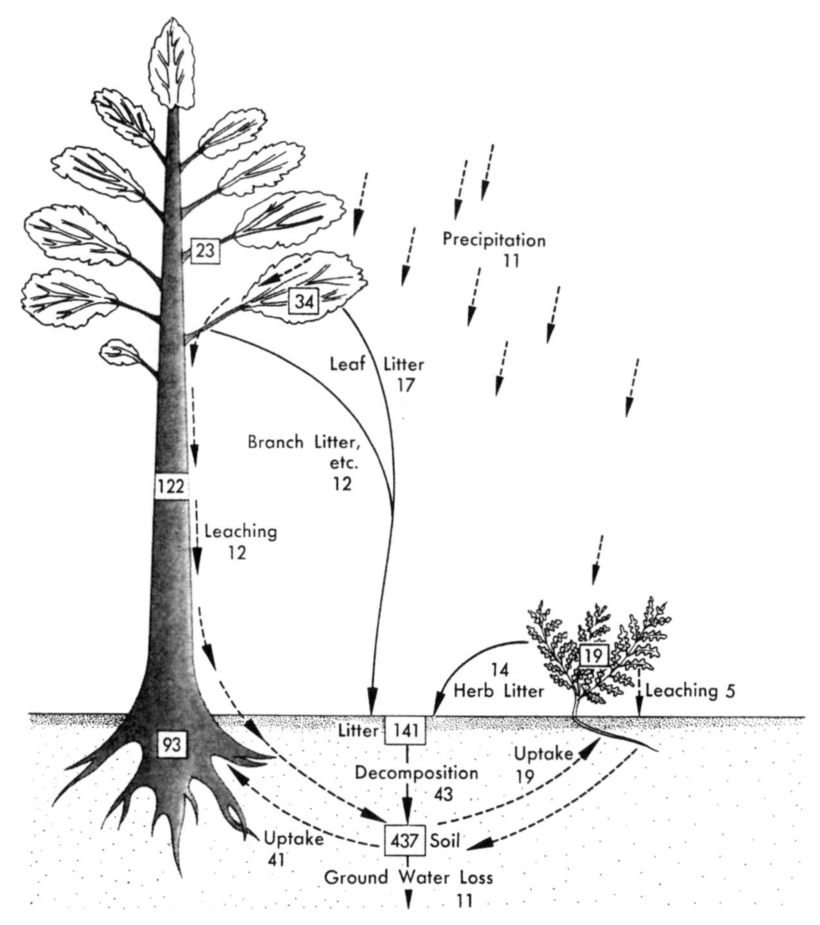

Figure 6.6 The intrasystem cycle for Ca in a forest ecosystem in Great Britain. Pools are shown in kg/ha and annual flux in kg ha^{-1} yr^{-1}. From Whittaker (1970).

senescence (Killingbeck 1996). In the tundra shrub *Eriophorum vaginatum,* Chapin et al. (1986b) found that all organic N and P compounds decreased to a similar extent during leaf senescence, suggesting that reabsorption is not simply limited to certain biochemical compounds that are particularly susceptible to hydrolysis. Nutrient reabsorption potentially confers a second type of nutrient-use efficiency on vegetation (see Chapter 5 for nutrient-use efficiency in photosynthesis). Nutrients that are reabsorbed can be used in net primary production in future years, increasing the carbon fixed per unit of nutrient uptake.

In a wide range of species, Aerts (1996) found a mean fractional reabsorption of 50% N and 52% P during leaf senescence. Somewhat lower values are seen in a California shrubland (Table 6.4), in the Hubbard Brook forest

(Table 6.1), and in grassland ecosystems (Woodmansee et al. 1978). Lajtha (1987) found exceptionally high values for P reabsorption (72–86%) in the desert shrub *Larrea tridentata,* growing in calcareous soils in which P availability is limited due to the precipitation of calcium phosphate minerals (see Fig. 4.4). DeLucia and Schlesinger (1995) report 94% reabsorption of leaf P in *Cyrilla racemiflora* in a P-limited bog in the southeastern United States.

Plants grown with low nutrient availability or occurring on infertile sites tend to have low nutrient concentrations in mature leaves and litter; they generally reabsorb a smaller *amount* but a larger *proportion* of the nutrient pool in senescent leaves, compared to individuals of the same species under conditions of greater nutrient availability (Chapin 1988, Boerner 1984, Pugnaire and Chapin 1993, Killingbeck 1996). In most cases, however, species appear to have only a limited ability to adjust the efficiency of reabsorption of leaf nutrients as a function of site fertility (Chapin and Kedrowski 1983, Birk and Vitousek 1986, Chapin and Moilanen 1991).

Differences in nutrient-use efficiency in reabsorption between nutrient-rich and nutrient-poor sites are not as likely to be due to a direct response of plants as to the tendency for species with higher inherent capabilities for nutrient reabsorption to dominate nutrient-poor sites (Pastor et al. 1984, Chapin et al. 1986a, Schlesinger et al. 1989). In a compilation of data from various forest ecosystems of the world, Vitousek (1982) found that the C/N ratio of leaf litterfall varied by a factor of 4, declining as an inverse function of the apparent nutrient availability of the site. Because the nutrient concentrations in mature foliage seldom vary by more than a factor of 2, his correlation suggests that species under nutrient-poor conditions reabsorb a greater proportion of leaf N before leaf-fall. Nutrient-rich sites are associated with high productivity and abundant nutrient circulation, but low nutrient-use efficiency. Later studies report a similar pattern for phosphorus in tropical forests (Vitousek 1984; Fig. 6.7). As a result of mycorrhizal associations and internal conservation of P, it appears that tropical trees are well adapted to P-deficient soils, which are widespread in these regions (Cuevas and Medina 1986).

Mass Balance of the Intrasystem Cycle

The annual circulation of nutrients in land vegetation, the intrasystem cycle (Fig. 6.6), can be modeled using the mass-balance approach. Nutrient requirement is equal to the peak nutrient content in newly produced tissues during the growing season (Table 6.1). Nutrient uptake cannot be measured directly, but uptake must equal the annual storage in perennial tissues, such as wood, plus the replacement of losses in litterfall and leaching; thus,

$$\text{uptake} = \text{retained} + \text{returned}. \qquad (6.2)$$

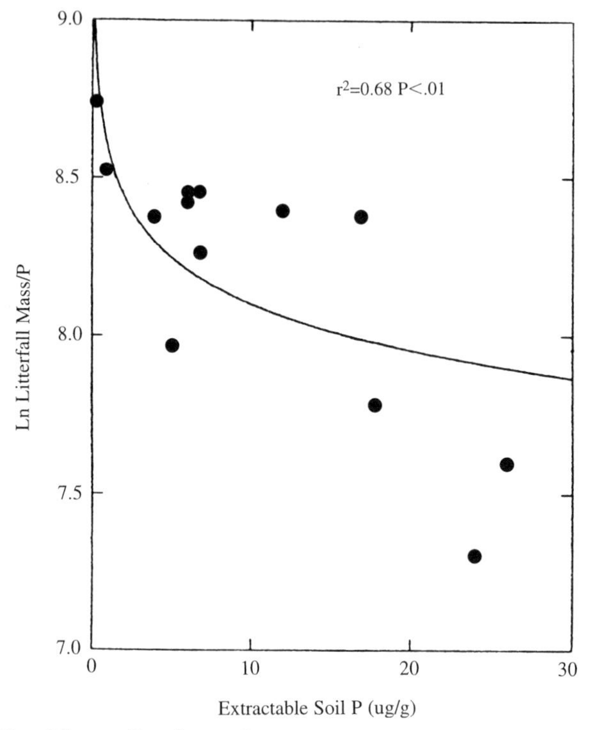

Figure 6.7 Litterfall mass/P ratio as a function of soil phosphorus availability in 13 humid tropical forests. From Silver (1994).

Uptake is less than the annual requirement by the amount reabsorbed from leaf tissues before abscission, viz.,

$$\text{requirement} = \text{uptake} + \text{retranslocation}. \quad (6.3)$$

The requirement is the nutrient flux needed to complete a mass balance; it should not be taken as indicative of biological requirements. In fact, this equation can be solved for nonessential elements such as Na. For a forest in Tennessee, the mass-balance approach was used to show that net accumulations of Ca and Mg in vegetation were directly related to decreases in the content of exchangeable Ca and Mg in the soil during 11 years of growth (Johnson et al. 1988). Similarly, Richter et al. (1994) used the mass-balance approach to estimate the weathering of soil minerals that was necessary to supply K to forest vegetation in South Carolina.

The mass-balance approach was used to analyze the internal storage and the annual transfers of nutrients in a California shrubland (Table 6.4). These data serve to summarize many aspects of the intrasystem cycle. Note that 71% of the annual requirement of N is allocated to foliage, whereas

Table 6.4 Nutrient Cycling in a 22-yr-old Stand of the Chaparral Shrub *Ceanothus megacarpus* near Santa Barbara, California[a]

	Biomass	N	P	K	Ca	Mg
Atmospheric input (g m^{-2} yr^{-1})						
Deposition		0.15		0.06	0.19	0.10
N-fixation		0.11				
Total input		0.26		0.06	0.19	0.10
Compartment pools (g/m^2)						
Foliage	553	8.20	0.38	2.07	4.50	0.98
Live wood	5929	32.60	2.43	13.93	28.99	3.20
Reproductive tissues	81	0.92	0.08	0.47	0.32	0.06
Total live	6563	41.72	2.89	16.47	33.81	4.24
Dead wood	1142	6.28	0.46	2.68	5.58	0.61
Surface litter	2027	20.5	0.6	4.7	26.1	6.7
Annual flux (g m^{-2} yr^{-1})						
Requirement for production						
Foliage	553	9.35	0.48	2.81	4.89	1.04
New twigs	120	1.18	0.06	0.62	0.71	0.11
Wood increment	302	1.66	0.12	0.71	1.47	0.16
Reproductive tissues	81	0.92	0.08	0.47	0.32	0.07
Total in production	1056	13.11	0.74	4.61	7.39	1.38
Reabsorption before abscission		4.15	0.29	0	0	0
Return to soil						
Litter fall	727	6.65	0.32	2.10	8.01	1.41
Branch mortality	74	0.22	0.01	0.15	0.44	0.02
Throughfall		0.19	0	0.94	0.31	0.09
Stemflow		0.24	0	0.87	0.78	0.25
Total return	801	7.30	0.33	4.06	9.54	1.77
Uptake (=increment + return)		8.96	0.45	4.77	11.01	1.93
Streamwater loss (g m^{-2} yr^{-1})		0.03	0.01	0.06	0.09	0.06
Comparisons of turnover and flux						
Foliage requirement/total requirement (%)		71.3	64.9	61.0	66.2	75.4
Litter fall/total return (%)		91.1	97.0	51.7	84.0	79.7
Uptake/total live pool (%)		21.4	15.6	29.0	32.6	45.5
Return/uptake (%)		81.4	73.3	85.1	86.6	91.7
Reabsorption/requirement (%)		31.7	39.0	0	0	0
Surface litter/litter fall (yr)	2.8	3.1	1.9	1.2	3.3	4.8

[a] Modified from Gray (1983) and Schlesinger et al. (1982).

much less is allocated to stem wood. Nevertheless, total nutrient storage in short-lived tissues is small compared to storage in wood, which has lower nutrient concentrations than leaf tissue but has accumulated during 22 years of growth. For most nutrients in this ecosystem, the storage in wood increases by about 5% each year. In this community the nutrient flux in stemflow is unusually large, but the total annual return in leaching is relatively small, except for K. Despite substantial reabsorption of N and P before leaf abscission, litterfall is the dominant pathway of return of these elements to the soil from the aboveground vegetation. It appears that Ca is actively exported to the leaves before abscission (i.e., requirement $<$ uptake). In this shrubland, annual uptake is 16 to 46% of the total storage in vegetation, but 73 to 92% of the uptake is returned each year. As in most studies, some of these calculations would be revised if belowground transfers were better understood.

Nutrient cycling changes during the development of vegetation, as the allocation of net primary production changes. During forest regrowth after disturbance, the leaf area develops rapidly, and the nutrient movements dependent upon leaf area (i.e., litterfall and leaching) are quickly reestablished (Marks and Bormann 1972, Boring et al. 1981). Gholz et al. (1985) found that the proportion of the annual requirement met by internal cycling (i.e., nutrient reabsorption from leaves) increased with time during the development of pine forests in Florida. Nutrients are accumulated most rapidly during the early development of forests, and more slowly as the aboveground biomass reaches a steady state (Gholz et al. 1985, Pearson et al. 1987, Reiners 1992). Percentage turnover in vegetation declines as the mass and nutrient storage in vegetation increase. In mature forests, leaf biomass is $<$5% of the total, and leaves contain only 5 to 20% of the total nutrient pool in vegetation (Waring and Schlesinger 1985).

Vitousek et al. (1988) have compiled data showing the proportions of biomass (i.e., carbon) and major nutrient elements in various types of mature forest (Table 6.5). The nutrient ratios vary over a surprisingly small range, so the global pattern of element stocks in vegetation is similar to that for biomass: i.e., tropical $>$ temperate $>$ boreal forests (Table 5.2). It is important to remember that these ratios are calculated for the total plant biomass; the concentrations of nutrients in leaf tissues are higher and the C/N and C/P ratios in leaves are correspondingly smaller. Thus, nutrient ratios for whole-plant biomass increase with time as the vegetation becomes increasingly dominated by structural tissues with lower nutrient content (Vitousek et al. 1988, Reiners 1992).

Nutrient-Use Efficiency

A mass balance for the intrasystem cycle of vegetation allows us to calculate an integrated measure of nutrient-use efficiency by vegetation—net primary production per unit nutrient uptake. This measure is affected by various factors that we have examined individually, including the rate of photosynthesis per unit leaf nutrient (Chapter 5), uptake per unit of root growth (Fig. 6.2), and leaching and nutrient retranslocations from leaves. As a result of changes in these factors, net primary production per unit of nitrogen or phosphorus taken from the soil increased by factors of 5 and 10, respectively, during the growth of pine forests in central Florida (Gholz et al. 1985).

Table 6.5 Biomass and Element Accumulation in Biomass of Mature Forests[a]

Forest biome	Number of stands	Total biomass (t/ha)	Percent of total biomass				Mass ratio		
			Leaf	Branch	Bole	Roots	C/N	C/P	N/P
Northern/subalpine conifer	12	233	4.5	10.2	62.8	22.6	143	1246	8.71
Temperate broadleaf deciduous	13	286	1.1	16.2	63.1	19.5	165	1384	8.40
Giant temperate conifer	5	624	2.5	10.2	66.4	20.8	158	1345	8.53
Temperate broadleaf evergreen	15	315	2.7	14.7	66.2	16.5	159	1383	8.73
Tropical/subtropical closed forest	13	494	1.9	21.8	59.8	16.4	161	1394	8.65
Tropical/subtropical woodland and savanna	13	107	3.6	19.1	60.4	16.9	147	1290	8.80

[a] From Vitousek et al. (1988).

Among temperate forests, the annual circulation of nutrients in coniferous forests is much lower than the circulation in deciduous forests, largely as a result of lower leaf turnover in coniferous forest species (Cole and Rapp 1981). Leaching losses are also lower in coniferous forests (Parker 1983). Photosynthesis per unit of leaf nitrogen also tends to be greater in coniferous species (Reich et al. 1995). Together these mechanisms result in greater nutrient-use efficiency in coniferous forests compared to deciduous forests of the world (Table 6.6). Higher nutrient-use efficiency in coniferous species may explain their frequent occurrence on nutrient-poor sites and in boreal climates where soil nutrient turnover is slow. Significantly, larch (*Larix* sp.), one of the few deciduous species in the boreal forest, has exceptionally high fractional reabsorption of foliar nutrients (Carlyle and Malcolm 1986).

The high nutrient-use efficiency of most conifers may also extend to the occurrence of broad-leaf evergreen vegetation on nutrient-poor soils in other climates (Monk 1966, Beadle 1966, Goldberg 1982, 1985, DeLucia and Schlesinger 1995). Escudero et al. (1992) suggest that leaf longevity was the most important factor increasing nutrient-use efficiency among various trees and shrubs in central Spain (cf. Reich et al. 1992), since deciduous and evergreen species had roughly similar amounts of nutrient reabsorption during leaf senescence (del Arco et al. 1991).

For biogeochemical cycling in vegetation, we have seen that the leaves and fine roots contain only a small portion of the nutrient content in biomass, but the growth, death, and replacement of these tissues largely determine the annual intrasystem cycle of nutrients. Net primary production is positively correlated to soil N availability in both coniferous and deciduous forests (Cole and Rapp 1981, Zak et al. 1989), but differences in nutrient reabsorption tend to weaken the correlation, so that light and moisture are the primary determinants of net primary production on a global basis (Figs. 5.10 and 5.11). When nutrient concentrations in litter are low, as might be expected after reabsorption of nutrients, decomposition is

Table 6.6 Net Primary Production (kg ha^{-1} yr^{-1}) per Unit of Nutrient Uptake Used as an Index of Nutrient-Use Efficiency to Compare Deciduous and Coniferous Forests[a]

	Production per unit nutrient uptake				
Forest type	N	P	K	Ca	Mg
Deciduous	143	1859	216	130	915
Coniferous	194	1519	354	217	1559

[a] From Cole and Rapp (1981).

slower. Thus, intrasystem cycling contains a positive feedback to the extent that an increase in nutrient-use efficiency by vegetation may reduce the future availability of soil nutrients for plant uptake (Shaver and Melillo 1984).

Biogeochemical Cycling in the Soil

Soil Microbial Biomass and Decomposition Processes

Despite new inputs from the atmosphere and from rock weathering, and plant adaptations to minimize the loss of nutrients, most of the annual nutrient requirement of land plants is supplied from the decomposition of dead materials in the soil (Table 6.1). Decomposition of dead organic matter completes the intrasystem cycle by releasing nutrient elements for plant uptake. *Decomposition* is a general term that refers to the breakdown of organic matter. *Mineralization* is a more specific term[2] that refers to processes that release carbon as CO_2 and nutrients in inorganic form, e.g., P as PO_4^{3-}. A variety of soil animals, including earthworms, fragment and mix fresh litterfall (Swift et al. 1979, Hole 1981); however, the main biogeochemical transformations are performed by fungi and bacteria in the soil. Most of the mineralization reactions are the result of the activity of extracellular degradative enzymes, released by soil microbes (Burns 1982, Linkins et al. 1990, Sinsabaugh et al. 1993, Deng and Tabatabai 1994). During the course of decomposition, humus compounds are synthesized by microbial activity (Chapter 5).

Total microbial biomass (bacteria + fungi) typically composes <3% of the organic carbon found in soils (Wardle 1992, Zak et al. 1994). High levels of microbial biomass are found in most forest soils and lower values in deserts (Insam 1990, Gallardo and Schlesinger 1992). Fungi dominate over bacteria in most well-drained, upland soils (Anderson and Domsch 1980). Reuss and Seagle (1994) found a direct correlation between soil microbial biomass and soil respiration in the grasslands of the African Serengeti, and microbial biomass is often measured as an index of its activity.

Determination of microbial biomass is usually performed by one of several techniques involving fumigation with chloroform (Jenkinson and Powlson 1976, Martens 1995). For instance, in a subdivided soil sample, total soluble nitrogen (NH_4 + NO_3 + dissolved organic N) is measured before and after fumigation with chloroform. A higher content in the fumigated sample is assumed to result from the lysis of microbes that were killed by chloroform (Brooks et al. 1985, Joergensen 1996). Microbial biomass is then calculated by assuming a standard nitrogen content in microbial

[2] This use of the term mineralization differs from its common usage in the literature of geology, in which mineralization refers to various processes (e.g., precipitation from hydrothermal fluids) that result in the deposition of metals in an ore deposit of economic significance.

tissue and a correction factor, K_n, to account for microbial N that is not immediately released by fumigation (Voroney and Paul 1984, Shen et al. 1984, Joergensen and Mueller 1996). The technique is justified by the observation of relatively constant C/N and C/P ratios in soil microbial biomass from many different environments (e.g., Fig. 6.8).

Soil microbes have high nutrient concentrations relative to the organic matter they decompose (Anderson and Domsch 1980, Diáz-Raviña et al. 1993). Microbial biomass contained 2.5 to 5.6% of the organic carbon, but up to 19.2% of the organic phosphorus in tropical soils of central India (Srivastava and Singh 1988). During the decomposition of plant material,

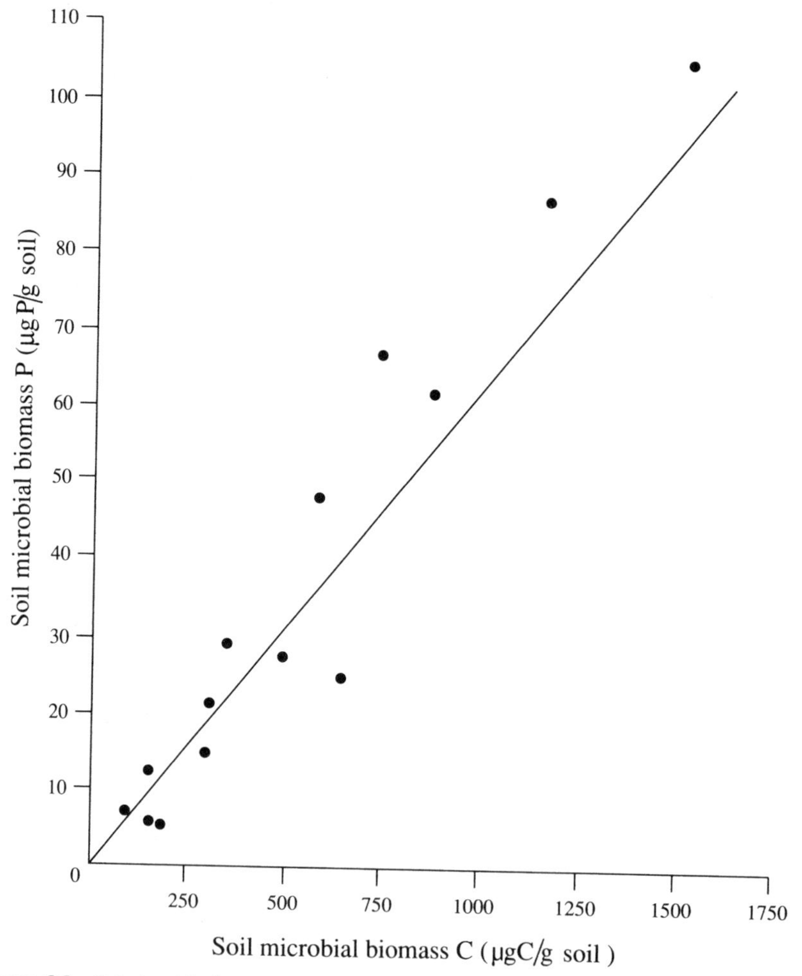

Figure 6.8 Relationship between the phosphorus and carbon contained in the microbial biomass of 14 soils. From Brooks et al. (1984).

respiration of soil microbes converts organic carbon to CO_2, while the N and P contents are retained in microbial biomass. When the decomposition of fresh litter is observed in litterbags (Chapter 5), the C/N and C/P ratios decline as decomposition proceeds and as the remaining materials are progressively dominated by microbial biomass that has colonized and grown on the substrate (Table 6.7; Sinsabaugh et al. 1993).

The accumulation of N, P, and other nutrients in soil microbes is known as *immobilization*. Immobilization is most significant for N and P, which are limiting to microbial growth, and usually less obvious for Mg and K, which are available in greater quantities (Jorgensen et al. 1980, Staaf and Berg 1982, O'Connell 1988). In the process of immobilization, soil microbes not only retain the nutrients released from their substrate, but also accumulate nutrients from the soil solution (Fig. 6.9). Microbial uptake of NH_4^+ is rapid, sequestering available NH_4^+ that might otherwise be available for plant uptake or nitrifying bacteria (Jackson et al. 1989, Schimel and Firestone 1989, Qualls et al. 1991). In cases of net accumulation by microbes, the nutrient content of the substrate appears to increase during the initial phases of decomposition (e.g., Aber and Melillo 1980, Schlesinger 1985a, Berg 1988).

During decomposition, a fraction of the substrate is converted to fulvic and humic compounds (Chapter 5) that have high N content (Schulten and Schnitzer 1993) and long-term stability in the soil profile (Fig. 6.9). Decaying plant litter appears to adsorb Al and Fe (Rustad 1994, Laskowski

Table 6.7 Ratios of Nutrient Elements to Carbon in the Litter of Scots Pine (*Pinus sylvestris*) at Sequential Stages of Decomposition[a]

	C/N	C/P	C/K	C/S	C/Ca	C/Mg	C/Mn
			Needle litter				
Initial	134	2630	705	1210	79	1350	330
After incubation of:							
1 yr	85	1330	735	864	101	1870	576
2 yr	66	912	867	ND	107	2360	800
3 yr	53	948	1970	ND	132	1710	1110
4 yr	46	869	1360	496	104	704	988
5 yr	41	656	591	497	231	1600	1120
			Fungal biomass				
Scots pine forest	12	64	41	ND	ND	ND	ND

[a] Some values for fungal tissues are also given. Note that C/N and C/P ratios decline, which indicates retention of these nutrients as C is lost, whereas C/Ca and C/K ratios increase, which indicates that these nutrients are lost more rapidly than carbon. From Staaf and Berg (1982).

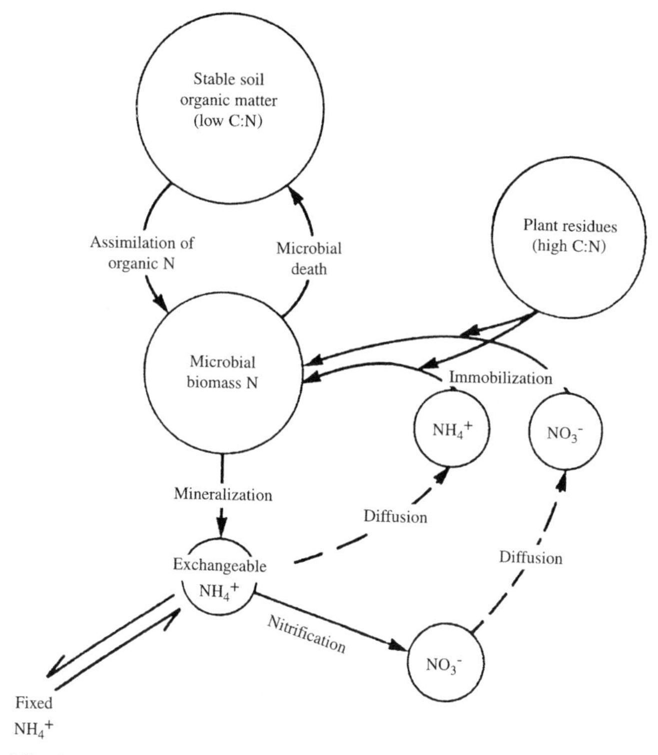

Figure 6.9 A conceptual model of the soil nitrogen cycle. From Drury et al. (1991).

et al. 1995), perhaps in compounds that are precursors to the fulvic acids that carry Al and Fe to the lower soil profile in the process of podzolization (Chapter 4).

When microbial growth slows, there is little further nutrient immobilization. When the substrate is exhausted and the microbial population dies, N is released as NH_4^+ from dead microbial tissue (Ladd et al. 1981, van Veen et al. 1987). The net mineralization of N often begins with C/N ratios near 30 : 1, but this can vary depending on the substrate and the assimilation efficiency of the decomposer (Rosswall 1982). Using ^{15}N as a tracer, Marumoto et al. (1982) have shown that much of the N mineralized in the soil is released from dead microbes and not directly from soil organic matter. The presence of soil animals that feed on bacteria and fungi can increase the rates of release of N and P from microbial tissues (Cole et al. 1978, Anderson et al. 1983).

Plant litter with higher concentrations of nutrients decomposes more rapidly, and net mineralization is likely to begin earlier. Fallen logs, on the other hand, have low N contents and long-term immobilization of N is especially evident during log decay (Lambert et al. 1980, Fahey 1983,

Schimel and Firestone 1989). Ecologists have long used the C/N ratio of litterfall as an index of its potential rate of decomposition (Taylor et al. 1989, Enriquez et al. 1993). More recently, Melillo et al. (1982) have used the lignin/nitrogen ratio in litterfall as a predictor of the rate of decomposition in various ecosystems (Fig. 6.10). These relationships may allow us to predict the rate of decomposition over large regions, where the canopy lignin and nitrogen contents have been measured by remote sensing (Wessman et al. 1988a).

Immobilization of nutrients predominates in the layer of fresh litter on the soil surface, while mineralization of N, P, and S is usually greatest in the lower forest floor (Federer 1983, Qualls et al. 1991). Fulvic acids and other dissolved organic compounds transport nutrients to the lower soil horizons (Schoenau and Bettany 1987, Qualls et al. 1991), where they add to the nutrient pool in humus. Sollins et al. (1984) found that the "light" fraction of soil organic matter, representing fresh plant residues, had a higher C/N ratio and lower mineralization than the "heavy" fraction, composed of humic substances (Chapter 5).

Release of N, P, and S from soil organic matter is likely to occur at different rates (McGill and Cole 1981). Nitrogen is largely bound directly to C in amino groups ($-C-NH_2$). Thus, N is mineralized as a result of the balance between the degradation of organic substances for energy and the synthesis of protein by microbes. Although much of the S in organic matter

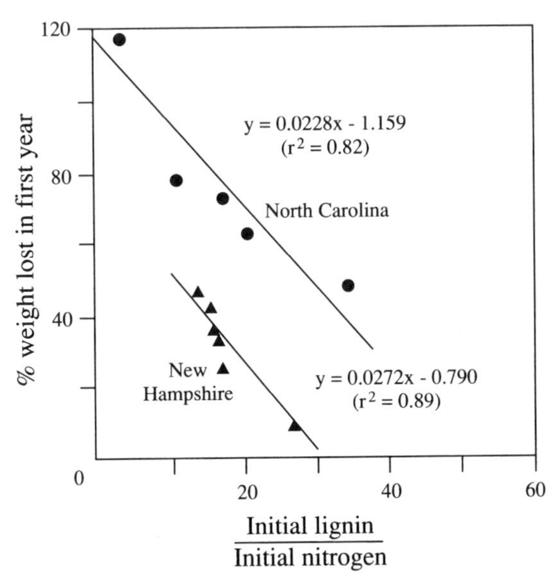

Figure 6.10 Decomposition of leaf litter as a function of the lignin/nitrogen ratio in fresh litterfall of various forest species in New Hampshire and North Carolina. From Melillo et al. (1982).

is also bound directly to C, much of the P is found in ester linkages (i.e., –C–O–P). These groups may be mineralized by the release of extracellular enzymes (e.g., phosphatases) in response to specific microbial demand for nutrients. Release of acid phosphatases by soil microbes is directly related to levels of soil organic matter (Tabatabai and Dick 1979, Polglase et al. 1992). For P, organic transformations are increasingly important as soils age and inorganic P is complexed into secondary minerals (Cross and Schlesinger 1995; see also Fig. 4.5).

Differential losses of nutrients and nutrient immobilizations mean that the loss of mass from litterbags cannot be directly equated with the proportional release of its original nutrient contents (Jorgensen et al. 1980, Rustad 1994). Table 6.8 shows the mean residence time for organic matter and its nutrient content in the surface litter of various ecosystems. Some nutrients, such as K, that are easily leached from litter may show mineralization rates in excess of the loss of litter mass. Others such as N turn over more slowly due to immobilization in microbial tissues. Vogt et al. (1986) suggest that immobilization of N is greatest in temperate and boreal forests, whereas immobilization of P is more important in tropical forests.

In Chapter 5 we saw that the pool of soil organic matter greatly exceeds the mass of live biomass in most ecosystems. As a result of its high nutrient content, humus also dominates the storage of biogeochemical elements in most ecosystems. Generally, the ratio of C, N, P, and S in humus is close to 140:10:1.3:1.3 (Stevenson 1986, Schulten and Schnitzer 1993). The global pool of nitrogen in soil, 95×10^{15} g (Post et al. 1985), dwarfs the pool of nitrogen in vegetation, 3.5×10^{15} g [calculated using the global biomass of 560×10^{15} g C (Table 5.2) and a C/N ratio in vegetation of 160 (Table 6.5)]. Aboveground biomass contains only 4 to 8% of the total

Table 6.8 Mean Residence Time (yr) for Organic Matter and Nutrients in the Surface Litter of Forest and Woodland Ecosystems[a]

	Mean residence time (yr)					
Region	Organic matter	N	P	K	Ca	Mg
Boreal forest	353	230	324	94	149	455
Temperate forest						
Coniferous	17	17.9	15.3	2.2	5.9	12.9
Deciduous	4	5.5	5.8	1.3	3.0	3.4
Mediterranean	3.8	4.2	3.6	1.4	5.0	2.8
Tropical rainforest	0.4	2.0	1.6	0.7	1.5	1.1

[a] Values are calculated by dividing the forest floor mass by the mean annual litterfall. Boreal and temperate values are from Cole and Rapp (1981), tropical values are from Edwards and Grubb (1982) and Edwards (1977, 1982), and Mediterranean values are from Gray and Schlesinger (1981).

quantity of N in temperate forests (Cole and Rapp 1981) and up to 32% in tropical forests (Edwards and Grubb 1982). Owing to the stability of humus substances in the soil, the large nutrient pool in humus turns over very slowly. Typically soil microbes mineralize about 1–3% of the pool of nitrogen in the soil each year (Connell et al. 1995).

Simple measurements of extractable nutrients, such as NH_4^+ or PO_4^{3-}, are unlikely to give a good index of nutrient availability in terrestrial ecosystems. These nutrients are subject to active uptake by plant roots, immobilization by soil microbes, and a variety of other processes that rapidly remove available forms from the soil solution. At any moment, the quantity extractable from a soil sample may be only a small fraction of that which is made available by mineralization during the course of a growing season (Davidson et al. 1990). Thus, studies of biogeochemical cycling in the soil need to be based on measurements that record the dynamic nature of nutrient turnover.

Nitrogen Cycling

The mineralization of N from decomposing materials begins with the release of NH_4^+ by heterotrophic microbes. This process is known as *ammonification.* Subsequently, a variety of processes affect the concentration NH_4^+ in the soil solution, including uptake by plants, immobilization by microbes, and fixation in clay minerals (Chapter 4). Some of the remaining NH_4^+ may undergo nitrification, in which the oxidation of NH_4^+ to NO_3^- is coupled to the fixation of carbon by chemoautotrophic bacteria in the genera *Nitrosomonas* and *Nitrobacter* (Meyer 1994; see Eqs. 2.16 and 2.17). In some cases, organic N is also oxidized by heterotrophic nitrification, producing NO_3^- (Schimel et al. 1984, Duggin et al. 1991). Nitrate may be taken up by plants and microbes or lost from the ecosystem in runoff waters or in emissions of N-containing gases. Nitrate taken up by soil microbes (immobilization) is reduced to NH_4^+ by nitrate reductase and used in microbial growth (Davidson et al. 1990, DeLuca and Keeney 1993, Downs et al. 1996). This process is known as *assimilatory reduction.* At any time, the extractable quantities of NH_4^+ and NO_3^- in the soil represent the net result of all of these processes. A low concentration of NH_4^+ is not necessarily an indication of low mineralization rates, because it can also indicate rapid nitrification or plant uptake (Rosswall 1982, Davidson et al. 1990).

A variety of techniques are available to study the transformations of nitrogen in the soil (Binkley and Hart 1989). Many workers have used the "buried-bag" approach to examine net mineralization. A soil sample is subdivided and part is extracted immediately, usually with KCl, to measure the available NH_4^+ and NO_3^-. The remaining soil is replaced in the field in a polyethylene bag, which is permeable to O_2 but not to H_2O. After a short period, usually 30 days, the second bag is retrieved and analyzed for the forms of available N. An increase in the quantity of available N in the buried bag is taken to represent *net* mineralization, i.e., the mineralization in excess of microbial immobilization, in the absence of plant uptake. Repeated samples taken through an annual cycle allow an estimate of annual net mineralization, which can be correlated with plant uptake and cycling (Pastor et al. 1984).

Although this technique has proven useful in a variety of studies, it is not without problems. During the course of incubation, soil moisture content in the buried bag does not fluctuate as it does in the natural ecosystem, and the original soil sample inevitably contains some fine root material that is severed during collection. Field measurements can also be performed in tubes (Raison et al. 1987) or trenched plots (Vitousek et al. 1982). In the latter, a block of soil, often 1 m^2, is isolated on all sides by trenching and the trenches are lined with plastic to prevent the invasion of roots. Plants rooted in this plot are removed, but the area is not otherwise disturbed. Periodic measurements of NH_4^+ and NO_3^- indicate rates of mineralization and nitrification in the absence of plant uptake. Trenching also eliminates the plant uptake of water, so this approach measures microbial activity at artificially high soil moisture content, with potential losses from the ecosystem due to leaching and denitrification.

An expensive, but promising, approach involves the use of $^{15}NH_4^+$ to label the initial pool of available NH_4^+ in the soil (Van Cleve and White 1980, Davidson et al. 1991a). After a period of time, the pool is remeasured for the ratio of $^{15}NH_4^+$ to $^{14}NH_4^+$. A decline in proportion of $^{15}NH_4^+$ is assumed to result from the microbial mineralization of NH_4^+ from the pool of N in soil organic matter, which is dominated by ^{14}N. This gives a measure of total (gross) mineralization under natural field conditions. Using this approach, Davidson et al. (1992) found that net mineralization was only 14% of the total in a coniferous forest; the remainder was immobilized by the microbial community. Similarly, ^{15}N can be used to label the pool of soil NO_3^-, and net nitrification is measured as the label is diluted by $^{14}NO_3^-$. Net nitrification can also be studied by measuring changes in the concentration of NH_4^+ and NO_3^- after application of compounds that specifically inhibit nitrification, including nitrapyrin (Bundy and Bremner 1973), acetylene (Berg et al. 1982), or chlorate (Belser and Mays 1980).

Mineralization and nitrification have been studied in a wide variety of ecosystems (Vitousek and Melillo 1979, Robertson 1982b, Vitousek and Matson 1988, Davidson et al. 1992). Generally, net mineralization is directly related to the total content of organic nitrogen in the soil (e.g., Marion and Black 1988, McCarty et al. 1995), but mineralization is also closely linked to the availability of carbon. Vegetation with a high C/N ratio in litterfall often shows low rates of mineralization in the soil (Gosz 1981, Vitousek et al. 1982). When field plots are fertilized with sugar, net mineralization and nitrification slow, due to increased immobilization of NH_4^+ by soil microbes (Schlesinger and Peterjohn 1991, DeLuca and Keeney 1993, Zagal and Persson 1994). Fertilization of a Douglas fir forest with sugar resulted in lower N content in leaves and greater nutrient reabsorption before leaf-fall (Turner and Olson 1976), showing a direct link between microbial processes in the soil and the nutrient-use efficiency of vegetation.

Although soil microbial populations may adapt to a wide variety of field conditions, nitrification is generally lower at low pH, low O_2, low soil moisture content, and high litter C/N ratios (Wetselaar 1968, Rosswall 1982, Robertson 1982a, Bramley and White 1990). Vitousek and Matson (1988) found high rates of mineralization and nitrification in most tropical forests, but Marrs et al. (1988) reported that net mineralization and nitrification were inhibited by exceptionally high soil water contents in montane tropical forests of Costa Rica. Nitrification rates are high when NH_4^+ is readily available (Robertson and Vitousek 1981), but the concentrations of other nutrients generally have little effect (Robertson 1982b, 1984, Christensen and MacAller 1985).

A large amount of effort has been directed toward understanding the control of nitrification following disturbances, such as forest harvest or fire (Vitousek and Melillo 1979, Vitousek et al. 1982). When vegetation is removed, soil temperature and moisture contents are generally higher, and rapid ammonification increases the availability of NH_4^+. Subsequently, nitrification may be so rapid that uptake by regrowing vegetation and immobilization by soil microbes are insufficient to prevent large losses of NO_3^- in streamwater following disturbance. However, not all disturbed sites show large losses of NO_3^-. In pine forests in the southeastern United States, microbial immobilization in harvest debris accounted for 83% of the uptake of ^{15}N that was applied as an experimental tracer following forest harvest (Vitousek and Matson 1984). Microbial immobilization also retards the loss of nitrate following burning of tallgrass prairie (Seastedt and Hayes 1988).

In general, nitrification and losses of NO_3^- in streamwater are greatest in forests with high nitrogen availability prior to disturbance (Krause 1982, Vitousek et al. 1982). Rates of nitrification decline during the early recovery of vegetation, and only minor differences are seen between middle- and old-age forests (Robertson and Vitousek 1981, Christensen and MacAller 1985, Davidson et al. 1992). There is some evidence that nitrification is inhibited by terpenoid and tannin compounds released by some types of vegetation (Olson and Reiners 1983, White 1986, 1988, Northup et al. 1995), but little evidence for a direct inhibition of nitrification by mature vegetation, as predicted by Rice and Pancholy (1972).

Increases in nitrification following disturbance affect other aspects of ecosystem function. Nitrification generates acidity (Eq. 2.16), so losses of NO_3^- in streamwater are often accompanied by increased losses of cations, which are removed from cation exchange sites in favor of H^+ (Likens et al. 1970). Streamwater losses of nearly all biogeochemical elements increased following harvest of the Hubbard Brook forest in New Hampshire. Sulfate was a curious exception (Fig. 6.11). Nodvin et al. (1988) showed that the decline in streamwater SO_4^{2-} concentrations after forest harvest was a result of the acidity generated from nitrification, which increased soil anion adsorption capacity (Mitchell et al. 1989, Chapter 4). These observations are a good example of a link between the biogeochemical cycles of N and S in terrestrial ecosystems.

Emission of Nitrogen Gases from Soils

During transformations of nitrogen in the soil, a variety of nitrogen gases, including NH_3, NO, N_2O, and N_2, are produced as products and by-products

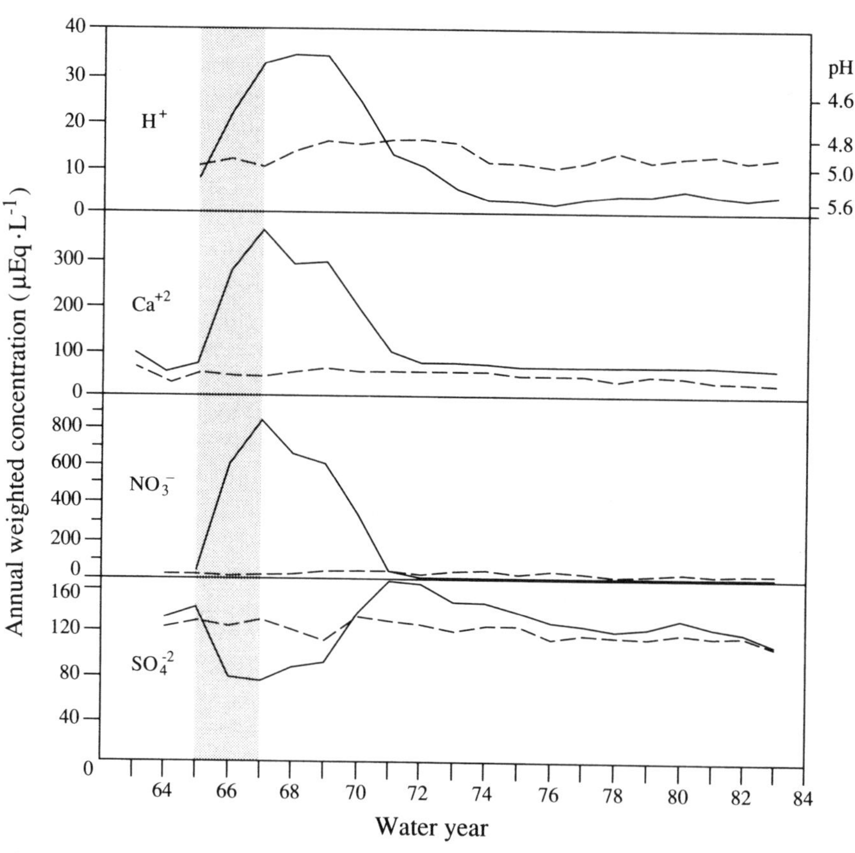

Figure 6.11 Concentrations of H^+, Ca^{2+}, NO_3^-, and SO_4^{2-} in the Hubbard Brook experimental forest for the years 1964-1984. Streams draining undisturbed forest are shown with the dashed line. The solid line depicts the concentrations in a stream draining a single watershed that was disturbed between 1965 and 1967 (shaded). Losses of Ca and NO_3^- increased strongly during the period of disturbance, and then recovered to normal values as the vegetation regenerated. The budget for SO_4^{2-} shows greater retention during and after the period of disturbance, presumably as a result of increased acidity and anion adsorption capacity in the soil. Modified from Nodvin et al. (1988).

of microbial activity (Fig. 6.12). Some of these may escape from the ecosystem, contributing to a loss of local soil fertility. More significantly, terrestrial ecosystems are a significant source of these gases in the atmosphere (Chapters 3 and 12).

In soils, ammonium may be converted to ammonia gas (NH_3), which is lost to the atmosphere. The reaction,

$$NH_4^+ + OH^- \rightleftarrows NH_3 \uparrow + H_2O, \quad (6.4)$$

is favored in dry soils and deserts, where accumulations of $CaCO_3$ maintain alkaline pH (Schlesinger and Peterjohn 1991). Low cation exchange capacity and low rates of nitrification maximize the production and loss of NH_3 (Nelson 1982, Fleisher et al. 1987). Small losses of NH_3 have been measured in a variety of natural forest and grassland soils worldwide (Schlesinger and Hartley 1992). Losses of NH_3 are greatest in fertilized soils and during the decomposition of urea excreted by wild and domestic animals (Terman 1979). Vegetation can also be a source of NH_3 during leaf senescence (Whitehead et al. 1988, Heckathorn and DeLucia 1995), but some of the NH_3 emitted by soils may be taken up by plants, so in many cases terrestrial ecosystems are only a small net source of NH_3 to the atmosphere (Langford et al. 1992, Sutton et al. 1993). During the loss of NH_3 from soils, isotopic fraction occurs, leaving soils with high $\delta^{15}N$ (Mizutani et al. 1986, Mizutani and Wada 1988).

The loss of ammonia is typically <1 kg ha^{-1} yr^{-1} in soils that are not impacted by fertilizers or domestic animals. The global flux from natural soils is about 10×10^{12} g N/yr (Schlesinger and Hartley 1992). This flux to the atmosphere is significant inasmuch as NH_3 is the only substance that is a net source of alkalinity in the atmosphere, where it can reduce the acidity of rain (Eqs. 3.5 and 3.59). Extremely high NH_3 volatilization from barns and animal feedlots may result in enhanced atmospheric deposition of NH_4^+ in areas immediately downwind (Draaijers et al. 1989). Somewhat counterintuitively, these large inputs of NH_4^+ may acidify soils, as the NH_4^+

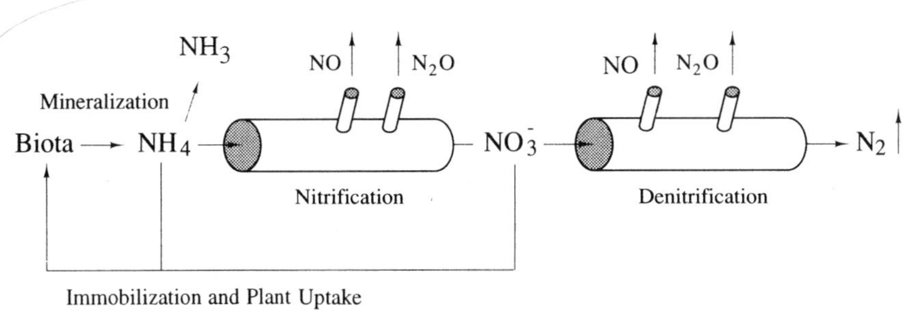

Figure 6.12 Microbial processes that yield nitrogen gases during nitrification and denitrification in the soil. Modified from Firestone and Davidson (1989).

is nitrified and the nitrate is taken up by vegetation (van Breemen et al. 1982, Verstraten et al. 1990).

Both nitric oxide (NO) and nitrous oxide (N_2O) are generated as microbial by-products of nitrification, with NO generally being the more abundant (E.J. Williams et al. 1992). Typically, about 1–3% of the nitrogen passing through the nitrification pathway is volatilized as NO each year (Baumgärtner and Conrad 1992, Hutchinson et al. 1993), and the global flux from natural soils is estimated to be about $10–12 \times 10^{12}$ g N/yr (Prather et al. 1995, Potter et al. 1996). The flux of NO from soils is highest under conditions that stimulate nitrification, including fertilization with NH_4^+ (Skiba et al. 1993). Among various ecosystems, the flux appears to increase as a function of soil temperature (Slemr and Seiler 1991, E.J. Williams et al. 1992) and immediately upon the wetting of dry soils (Davidson et al. 1991b, 1993). When the atmospheric concentration of nitric oxide is high, some NO is taken up by plants and soils (Rondón and Granat 1994, Slemr and Seiler 1991, Yienger and Levy 1995). The atmospheric concentration that produces no net uptake or loss is known as the *compensation point,* but in most cases the background concentration in the atmosphere, about 10 ppbv (Table 3.4), is below the compensation point, so terrestrial ecosystems are a net source of NO to the atmosphere (Kaplan et al. 1988, Duyzer and Fowler 1994). In remote areas, the concentration of NO is greatest in the lower atmosphere, and it declines with altitude (Luke et al. 1992). [Recall that NO plays a major role in the chemistry of ozone in the troposphere (Chapter 3)].

Nitrate is also converted to NO, N_2O, and N_2 in the process of denitrification (Knowles 1982, Firestone 1982, Ye 1994). This reaction (Eq. 2.18) is performed by soil bacteria that are aerobic heterotrophs in the presence of O_2, but facultative anaerobes at low concentrations of O_2. During anoxia, heterotrophic activity continues, with nitrate serving as the terminal electron acceptor in metabolism. The structure of the Cu-containing denitrification enzyme, nitrite reductase, has recently been published (Godden et al. 1991). Because NO_3^- is reduced, but not incorporated into microbial tissue, denitrification is also known as *dissimilatory nitrate reduction.* Bacteria in the genus *Pseudomonas* are the best known denitrifiers, but many others are reported (Knowles 1982, Tiedje et al. 1989).

For a long time, denitrification was thought to occur only in flooded, anoxic soils (Chapter 7), and its importance in upland ecosystems was overlooked. Now, soil scientists have shown that oxygen diffusion to the center of soil aggregates is so slow that anoxic microsites are common, even in well-drained soils (Fig. 6.13; Tiedje et al. 1984, Sexstone et al. 1985a). Thus, denitrification is widespread in terrestrial ecosystems, especially those in which organic carbon and nitrate are readily available in the soil (Burford and Bremner 1975, Carter et al. 1995, Wagner et al. 1996). Davidson and Swank (1987) found that additions of NO_3^- stimulated denitri-

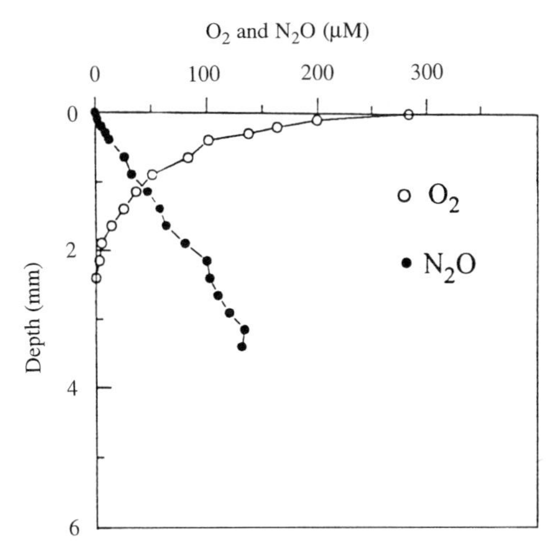

Figure 6.13 Concentrations of O_2 and N_2O determined in a soil aggregate as a function of the depth of penetration of a microelectrode. From Højberg et al. (1994).

fication in forest soils of western North Carolina, and the addition of organic carbon stimulated denitrification in the mineral soil. Rainfall generally increases the rate of denitrification, because the diffusion of oxygen is slower in wet soils (Sexstone et al. 1985b, Smith and Tiedje 1979, Rudaz et al. 1991, Peterjohn and Schlesinger 1991).

The relative importance of denitrification as a source of NO, N_2O, and N_2 varies depending upon environmental conditions (Firestone and Davidson 1989, Bonin et al. 1989). In Germany, well-drained soils with near-neutral pH produced NO only from nitrification, whereas in acid, anoxic soils, NO was produced from denitrification (Remde and Conrad 1991). In studies of a semidesert ecosystem, Mummey et al. (1994) found that nitrification accounted for 61–98% of the N_2O produced in moist soils, but denitrification was the predominant reaction in saturated conditions (cf. Skiba et al. 1993). Typically, in denitrification, the production of N_2O dwarfs the production of NO, so the total (nitrification + denitrification) and proportional loss of NO from soils declines with increasing moisture content, while the flux of N_2O increases (Fig. 6.14; Potter et al. 1996). Matson and Vitousek (1987) found a direct relation of N_2O production and nitrogen mineralization in comparisons of various tropical forests, implying that nitrification was the source (Fig. 6.15), but in the wet soils of Amazon rainforests, N_2O appeared to be mostly from denitrification (Livingston et al. 1988, Keller et al. 1988).

Factors affecting the relative loss of N_2O and N_2 by denitrification are poorly understood, but they include soil pH and the relative abundance of NO_3^- and O_2 as oxidants and organic carbon as a reductant (Firestone

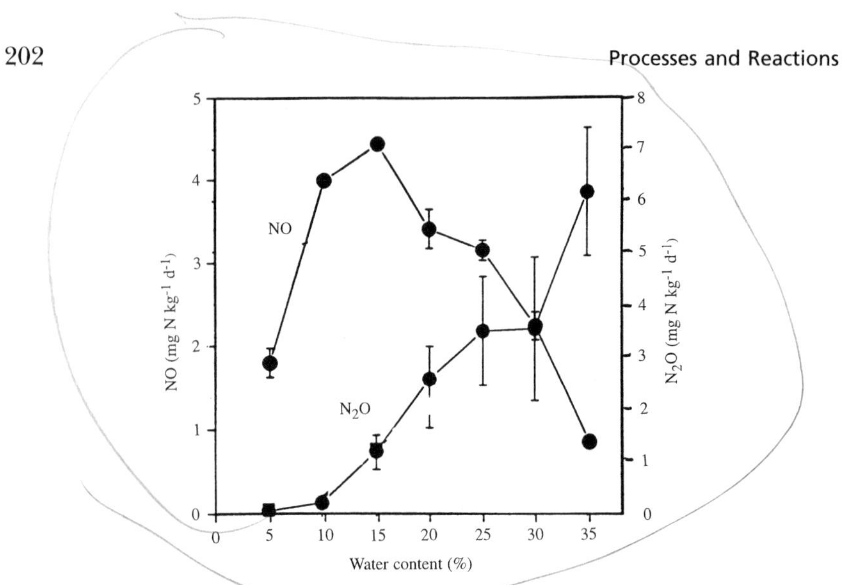

Figure 6.14 Flux of NO and N_2O from denitrification in a clay-loam soil as a function of soil water content under anoxic conditions. Modified from Drury et al. (1992).

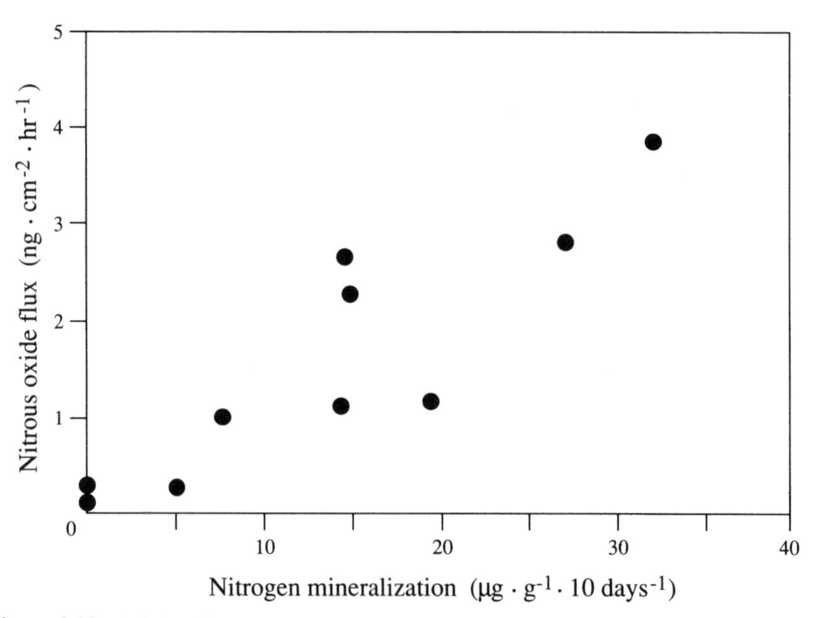

Figure 6.15 Relationship between nitrogen mineralization measured in laboratory incubations and the loss of N_2O from 10 tropical forest soils. From Matson and Vitousek (1987).

et al. 1980, McKenney et al. 1994, Conrad 1994, Chen et al. 1995). When NO_3^- is abundant relative to the supply of organic carbon, N_2O can be an important product (Firestone and Davidson 1989). In studies of forest soils in New Hampshire (Melillo et al. 1983) and Michigan (Merrill and Zak 1992), N_2O was the only significant product of denitrification. Overall, the ratio of N_2/N_2O produced in denitrification varies widely (Weier et al. 1993), but the median value appears to be about 22 (Chapter 12). The total loss of N from soils by denitrification is typically <2 kg ha^{-1} yr^{-1} in forests and grasslands (Robertson and Tiedje 1984, Vermes and Myrold 1992, Bowden 1986). Globally, the loss of N_2 from soils (>100 × 10^{12} g N/yr) dwarfs the loss of N_2O (6.1-9.5 × 10^{12} g N/yr) or NO (10-12 × 10^{12} g N/yr). Denitrification is the major process that returns N_2 to the atmosphere, completing the global biogeochemical cycle of nitrogen (Bowden 1986, Chapter 12). The flux of N_2O is significant inasmuch as its concentration in the atmosphere is increasing, and N_2O plays an important role as a greenhouse gas and as a component of ozone chemistry in the stratosphere (Eqs. 3.43–3.44).

Losses of NO and N_2O increase with factors that increase the rate of nitrification in soils, including the clearing, cultivation, and fertilization of agricultural soils (Bremner and Blackmer 1978, Conrad et al. 1983, Mosier et al. 1991, Bronson and Mosier 1993, Clayton et al. 1994). Shepherd et al. (1991) report that 11% of fertilizer N was lost as NO and 5% as N_2O in some cultivated fields in Ontario. When tropical forests are cleared the losses of NO and N_2O increase dramatically (Robertson and Tiedje 1988, Luizão et al. 1989, Sanhueza et al. 1994, Keller and Reiners 1994). Thus, fertilized and cleared soils may be responsible for the rising concentration of N_2O in the atmosphere (Chapter 12).

Field measurements of denitrification are usually based on the observation that acetylene blocks the conversion of the intermediate denitrification product, N_2O, to the final product, N_2 (Fig. 6.12; Yoshinari and Knowles 1976, Burton and Beauchamp 1984, Tiedje et al. 1989). Thus, following application of acetylene to laboratory soils or field plots, the sole product of denitrification is N_2O, which is easy to measure with gas chromatography against its background concentration of about 311 ppb in the atmosphere. The incubations must be short, because acetylene also blocks nitrification and the rate measured during a long-term incubation would be affected by a decline in the pool of NO_3^- that is available for denitrification (Davidson et al. 1986). Denitrification can also be estimated by the application of $^{15}NO_3^-$ to field plots and by measurements of the release of ^{15}N gases or the decline in $^{15}NO_3^-$ remaining in the soil (Parkin et al. 1985, Mosier et al. 1986, Remde and Conrad 1991). Short-term incubation studies using the radioactive isotope ^{13}N also show promise for understanding the role of nitrification and denitrification in the production of N_2O and N_2 (Speir et al. 1995).

Field estimates of denitrification are complicated by high spatial variability. At the local scale, a large portion of the total variability is found at distances of <10 cm, which Parkin et al. (1987) link to the local distribution of soil aggregates that provide anaerobic microsites. In one case, Parkin (1987) found that 85% of the total denitrification in a 15-cm-diameter soil core was located under a 1-cm^2 section of decaying pigweed (*Amaranthus*) leaf! In desert ecosystems, soil nitrogen content and nitrification are localized under shrubs, and denitrification is largely confined to those areas (Virginia et al. 1982).

Robertson et al. (1988) documented the pattern of mineralization, nitrification, and denitrification in a field in Michigan. All these processes showed large spatial variation, but the coefficient-of-variation for denitrification, 275%, was the largest measured. Significant correlations were found among these processes. Soil respiration and potential nitrification explained 37% of the variation in denitrification, presumably due to the dependence of this process on organic carbon and NO_3^- as substrates. The high variability in these processes makes it difficult to use measurements from a few sample chambers to calculate a mean or total flux from an ecosystem (Ambus and Christensen 1994). Groffman and Tiedje (1989) suggest that correlations of denitrification to soil texture may allow the most accurate extrapolations of laboratory measurements to regional estimates of gaseous N loss.

At a larger scale, high rates of denitrification are often confined to particular landscape positions, where conditions are favorable. For example, Peterjohn and Correll (1984) suggest that the runoff of nitrate from agricultural fields was largely denitrified in streamside forests, minimizing the losses in rivers (Pinay et al. 1993, Schipper et al. 1993, Jordan et al. 1993). In calculating regional averages for denitrification, investigators must evaluate the relative contributions from local areas of high and low activity (e.g., Matson et al. 1991, Yavitt and Fahey 1993).

As in the case of NH_3 volatilization, the losses of N gases as products and by-products of nitrification and denitrification leave soils enriched in ^{15}N (Högberg 1991). During decomposition, soil microbes mineralize ^{14}N in favor of ^{15}N, which increases in the undecomposed residue (Nadelhoffer and Fry 1988). Feeding on the products of microbial mineralization, plants are typically less concentrated in ^{15}N than the soils in which they grow (Garten and Van Miegroet 1994). Denitrifying bacteria further fractionate among the isotopes of available nitrogen—that is, between $^{14}NO_3^-$ and $^{15}NO_3^-$ (Handley and Raven 1992). Preference for $^{14}NO_3^-$ leads to positive $\delta^{15}N$ in most soils (Fig. 6.4), as $^{14}N_2$ is lost from the soil by denitrification (Shearer and Kohl 1988). Evans and Ehleringer (1993) show a strong inverse relation between $\delta^{15}N$ and the nitrogen content in soils (Fig. 6.16), suggesting that soils with low nitrogen are enriched in ^{15}N as a result of the loss of N gases (Garten 1993). Strong enrichments in ^{15}N are also seen in saturated soils

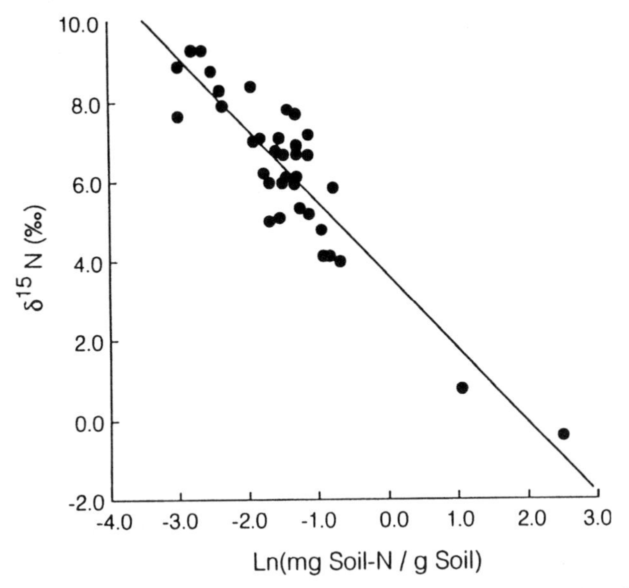

Figure 6.16 $\delta^{15}N$ of soil organic matter as a function of the total N content in soils of juniper woodlands of Utah (USA). From Evans and Ehleringer (1993).

with low redox potential (Chapter 7) compared to adjacent well-drained soils in the same field (Sutherland et al. 1993).

Phosphorus Cycling

Transformations of organic phosphorus in the soil are difficult to study because of the reaction of inorganic phosphorus with various soil minerals (Figs. 4.4 and 6.17). A few workers have examined phosphorus mineralization in soils using the buried-bag approach (e.g., Pastor et al. 1984), but in many cases there is no apparent mineralization because of the immediate complexation of P with soil minerals. Thus, most studies of phosphorus cycling have followed the decay of radioactively labeled plant materials (Harrison 1982) or measured the dilution of radioactive ^{32}P that is applied to the soil pool as a tracer (Walbridge and Vitousek 1987). With the isotope-dilution technique, one must assume that ^{32}P equilibrates with all the chemical pools in the soil, and that the only dilution of its concentration is by the mineralization of organic phosphorus. Unfortunately, these assumptions are not always valid, making the technique difficult to apply in many instances (Walbridge and Vitousek 1987).

Despite the shortcomings of using simple extractions to measure soil nutrients, many workers have used a sequential extraction scheme to quantify phosphorus availability in the soil (Hedley et al. 1982b, Stevenson 1986). Extraction with 0.5 *M* $NaHCO_3$ is a convenient index of labile inorganic

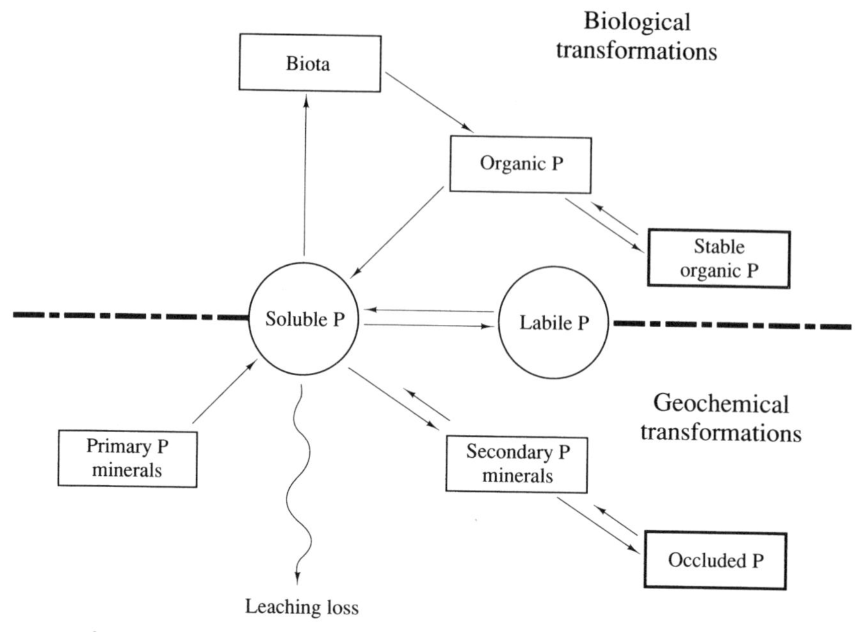

Figure 6.17 Phosphorus transformations in the soil. From Smeck (1985).

and soluble organic phosphorus in many soils (Olsen et al. 1954, Sharpley et al. 1987). Organic P is often determined as the difference between PO_4 in a sample that has been combusted at high temperatures and that in an untreated sample (Stevenson 1986), and microbial P by the change in extractable phosphorus after fumigation with chloroform (Brooks et al. 1982, 1984). Extraction with NaOH (to raise pH and lower anion adsorption capacity) indicates the amount of P that is held on Fe and Al minerals, while extraction with HCl releases P from many Ca-bound forms, including $CaCO_3$ (Tiessen et al. 1984, Lajtha and Schlesinger 1988, Cross and Schlesinger 1995). Acid-extractable phosphorus also includes P derived from apatite (Chapter 4), including secondary hydroxyapatite [$Ca_5OH(PO_4)_3$] from bones and fluoroapatite [$Ca_5F(PO_4)_3$] in teeth. These secondary biominerals in soils are sometimes used by archeologists to determine the location of past human settlements (Sjöberg 1976).

In most ecosystems the phosphorus available for biogeochemical cycling is held in organic forms (Chapin et al. 1978, Wood et al. 1984, Yanai 1992, Gressel et al. 1996). Walbridge et al. (1991) found that up to 35% of the organic P in the undecomposed litter in a warm-temperate forest was held in microbial biomass. Fulvic acids carry organic phosphorus compounds from the forest floor to the lower soil profile, where they accumulate in humus (Schoenau and Bettany 1987, Qualls et al. 1991). Gallardo and

Schlesinger (1994) found that additions of inorganic P increased the microbial biomass in the lower horizons of a forest soil in North Carolina, where the mineralogy is dominated by Fe and Al-oxide minerals with strong phosphorus adsorption capacity. Earlier we discussed the ability of various organisms to release phosphatase enzymes and organic acids that mineralize P from organic and inorganic forms (see also Chapter 4). Net mineralization of P from organic matter usually begins at C/P ratios that are <200.

Despite folklore to the contrary, the production of phosphine gas (PH_3) is impossible in upland soils, requiring extremely low redox potential (Bartlett 1986, Chapter 7). There are scattered reports that such conditions exist in sewage treatment ponds (Dévai et al. 1988) and salt marshes (Dévai and DeLaune 1995), but the movement of phosphorus as a gas is of negligible importance in its regional or global cycle (Chapter 12).

Sulfur Cycling

Like the phosphorus cycle, the cycle of sulfur in the soil is affected by both chemical and biological reactions. Sulfur is derived from atmospheric deposition (Chapter 3) and from the weathering of sulfur-bearing minerals in rocks (Chapter 4). The concentration of SO_4^{2-} in the soil solution exists in equilibrium with sulfate adsorbed on soil minerals (Chapter 4). Plant uptake of SO_4^{2-} is followed by assimilatory reduction and incorporation of carbon-bonded sulfur into the amino acids cysteine and methionine, which are constituents of protein (Johnson 1984). The molecular structure of the S-reducing protein contains Fe as a cofactor (Crane et al. 1995). A small quantity of sulfur in plants is also found in ester-bonded sulfates ($-C-O-SO_4$), and when soil sulfate concentrations are high, plants may accumulate SO_4 in leaf tissues (Turner et al. 1980).

In most soils the majority of the pool of soil S is held in organic forms (Watwood et al. 1988, Bartel-Ortiz and David 1988, Mitchell et al. 1992, Houle and Carignan 1992). Decomposition of plant tissues is accompanied by microbial immobilizations (Saggar et al. 1981, Staaf and Berg 1982, Fitzgerald et al. 1984). Wu et al. (1995) used ^{35}S as an isotopic tracer to show high rates of microbial immobilization of $^{35}SO_4^{2-}$ when glucose was added to soils. Typically, mineralization of SO_4^{2-} begins at C/S ratios <200 (Stevenson 1986). Sulfur in soil humus shows higher $\delta^{34}S$ than soil sulfate, suggesting that soil microbes discriminate against this rare, heavy isotope of S in favor of ^{32}S during decomposition of organic matter (Mayer et al. 1995). Downward movement of fulvic acids appears to transport ester sulfates to the lower soil profile (Schoenau and Bettany 1987).

In forest soils, the microbial immobilization of added $^{35}SO_4^{2-}$ is greatest in the upper soil profile and anion adsorption of inorganic SO_4^{2-} dominates the B horizons, where sesquioxide minerals are present (S.C Schindler et al. 1986, Randlett et al. 1992). In most cases, the majority of microbial S is found in carbon-bonded form (David et al. 1982, Watwood et al. 1988, S.C. Schindler et al. 1986, Dhamala and Mitchell 1995). However, at the Cowetta Experimental Forest in North Carolina, a large portion of the immobilization of sulfur by soil microbes formed ester sulfates (Fitzgerald et al. 1985, Watwood and Fitzgerald 1988), conferring a significant sink for SO_4^{2-} deposited from the atmosphere (Swank et al. 1984). Despite the predominance of organic forms, the pool of SO_4^{2-} in most soils is not insignificant. In the study of a forest in Tennessee, Johnson et al. (1982) found that the pool of adsorbed SO_4^{2-} was larger than the total pool of S in vegetation by a factor of 15.

To maintain a charge balance, plant uptake and reduction of SO_4^{2-} consume H^+ from the soil, whereas the mineralization of organic sulfur returns H^+ to the soil solution, producing no net increase in acidity (Binkley and Richter 1987). In contrast, reduced inorganic sulfur is found in association with some rock minerals (e.g., pyrite), and the oxidative weathering of reduced sulfide minerals accounts for highly acidic solutions draining mine tailings (Eqs. 2.15 and 4.6). This oxidation is performed by chemoautotrophic bacteria, generally in the genus *Thiobacillus*.

Production of reduced sulfur gases such as H_2S, COS (carbonyl sulfide), and $(CH_3)_2S$ (dimethylsulfide) is largely confined to wetland soils, since highly reducing, anaerobic conditions are required (Chapter 7). Globally upland soils are only a small source of sulfur gases in the atmosphere (Lamb et al. 1987b, Goldan et al. 1987, Staubes et al. 1989). However, many plants (e.g., garlic) produce a variety of volatile organic sulfur compounds. The smell of CS_2 (carbon disulfide) is often found when excavating the roots of the tropical tree *Stryphnodendron excelsum* (Haines et al. 1989), and many plant leaves are known to release sulfur gases during photosynthesis (Winner et al. 1981, Garten 1990, Kesselmeier et al. 1993). Often the total net flux of sulfur gases from an ecosystem (soil + plant) is estimated by examining the vertical profile of gas concentrations in the atmosphere (e.g., Andreae and Andreae 1988). Hydrogen sulfide appears to dominate the release of sulfur gases from plants (Delmas and Servant 1983, Andreae et al. 1990, Rennenberg 1991). Terrestrial ecosystems also appear to be a source of $(CH_3)_2S$ during the day (Andreae et al. 1990, Berresheim and Vulcan 1992, Kesselmeier et al. 1993), but vegetation is a major sink for COS globally (Chin and Davis 1993, Chapter 13).

Transformations in Fire

During fires, nutrients are lost in gases and in the particles of smoke, and soil nutrient availability increases with the addition of ash to soil (Raison 1979, Woodmansee and Wallach 1981). Following fire, there is often increased runoff and erosion from bare, ash-covered soils. High rates of nitrification in nutrient-rich soils can stimulate the loss of NO and N_2O from burned soils (Anderson et al. 1988, Levine et al. 1988).

Widespread increases in the rate of forest burning worldwide have the potential to deplete the nutrient content of soils and add trace gases to the atmosphere. However, before human intervention, fires were a natural part of the environment in many regions; thus, nutrient losses as a result of fire occurred at infrequent but somewhat regular intervals (Clark 1990). Using a mass-balance approach we can estimate the length of time that it takes to replace the nutrients that are lost in a single fire. For instance, 11 to 40 kg/ha of N are lost in small ground fires in southeastern pine forests (Richter et al. 1982), equivalent to 3 to 12 times the annual deposition of N from the atmosphere in this region (Swank and Henderson 1976).

When leaves and twigs are burned under laboratory conditions, up to 90% of their N content can be lost, presumably as N_2 or as one or more forms of nitrogen oxide gases (DeBell and Ralston 1970, Lobert et al. 1990). Sulfur is also released as SO_2 (Sanborn and Ballard 1991, Andreae 1991). Forest fires volatilize nitrogen in proportion to the heat generated and the organic matter consumed (DeBano and Conrad 1978, Raison et al. 1985, Kauffman et al. 1993); the rate of loss changes dramatically as fires pass from flaming to smoldering phases (Crutzen and Andreae 1990). Typically N losses in forest fires range from 100 to 300 kg/ha, or 10 to 40% of the amount in aboveground vegetation and surface litter. Especially large losses have recently been reported from slash fires in the Amazon rainforest (Kauffman et al. 1993).

Studies of the gaseous products of fires are often conducted by flying aircraft through the smoke plume to gather gas samples (e.g., Cofer et al. 1990, Nance et al. 1993, Hurst et al. 1994). The enrichment of CO_2 and CO over the atmospheric background is measured, as well as the ratio of other gases to CO_2 in the smoke (e.g., NH_3/CO_2). Assuming that the carbon in the fuel is all converted to CO_2 and CO, the loss of other fuel constituents as gases and particles can be calculated from estimates of the biomass carbon consumed and the ratio of the constituent in question to CO_2+CO in smoke (Laursen et al. 1992, Delmas et al. 1995). Thus, global estimates of the volatilization of nitrogen from forest fires can be calculated from global estimates of the amount of carbon lost in forest fires each year (Crutzen and Andreae 1990). N_2 dominates the gaseous loss of nitrogen (Kuhlbusch et al. 1991), constituting a form of "pyrodenitrification" that removes fixed nitrogen from the biosphere (Chapters 3 and 12). The losses of other nitrogen gases in forest fires account for 3% (N_2O) to 15% (NO_x) of the annual atmospheric input of these species to the atmosphere (Chapter 12). Forest fires are also a major global source of CH_4 (Delmas et al. 1991, Quay et al. 1991), CH_3Br (Manö and Andreae 1994), and CO (Seiler and Conrad 1987), but a relatively small source of SO_2 (Crutzen and Andreae 1990).

Air currents and updrafts during fire carry particles of ash that remove other nutrients from the site. These losses are usually much smaller than

N losses (Arianoutsou and Margaris 1981, Gaudichet et al. 1995). Expressed as a percentage of the amount present in aboveground vegetation and litter before fire, the total loss in gases and particulates often follows the order $N \gg K > Mg > Ca > P > 0\%$. Differential rates of loss change the balance of nutrients available in the soil after fire (Raison et al. 1985), and nutrient losses to the atmosphere in fire may enhance the atmospheric deposition in adjacent locations (Clayton 1976, Lewis 1981).

Depending on intensity, fire kills aboveground vegetation and transfers varying proportions of its mass and nutrient content to the soil as ash. There are a large number of changes in chemical and biological properties of soil as a result of the addition of ash (Raison 1979). Cations and P may be readily available in ash, which usually increases soil pH (Woodmansee and Wallach 1981). Burning increases extractable P, but reduces the levels of organic P and phosphatase activity in soil (DeBano and Klopatek 1988, Saá et al. 1993, Serrasolsas and Khanna 1995). Nitrogen in the ash is subject to rapid mineralization and nitrification (Christensen 1973, 1977, Dunn et al. 1979, Matson et al. 1987), so available NH_4^+ and NO_3^- usually increase after fire, even though total soil N may be lower (Covington and Sackett 1992). The increase in available nutrients as a result of ash-fall is usually short-lived, as nutrients are taken up by vegetation or lost to leaching and erosion (Lewis 1974, Christensen 1977, Uhl and Jordan 1984).

Streamwater runoff is often greater after fire because of reduced water losses in transpiration. High nutrient availability in the soil coupled with greater runoff can lead to large nutrient losses from the ecosystem. The loss of nutrients in runoff depends on many factors, including the season, rainfall pattern, and the growth of postfire vegetation (Dyrness et al. 1989). Wright (1976) noted significant increases in the loss of K and P from burned forest watersheds in Minnesota. These losses were greatest in the first 2 years after fire; by the third year there was actually less P lost from burned watersheds than from adjacent mature forests, presumably due to uptake by regrowing vegetation (McColl and Grigal 1975, cf. Saá et al. 1994). Although there are exceptions, the relative increase in the loss of Ca, Mg, Na, and K in runoff waters after fire often exceeds that of N and P (Chorover et al. 1994).

Ice cores and sediments contain a historical record of biomass burning. Layers of ice with buried ash often contain especially high concentrations of NH_4, indicating substantial NH_3 volatilization during fires (Legrand et al. 1992, Whitlow et al. 1994). Cores taken from the Greenland ice cap show episodes of increased biomass burning that appear to be related to the European colonization of North America (Whitlow et al. 1994). Lake and ocean sediments contain layers of buried ash that indicate the frequency of fires (Bryne et al. 1977, Clark and Royall 1994). It is likely that humans have significantly increased the rate of biomass burning worldwide, especially as a result of tropical deforestation (Crutzen and Andreae 1990,

Cahoon et al. 1992). Improved estimates of the frequency and area of fires are likely to derive from satellite remote sensing (Prins and Menzel 1994).

The Role of Animals

Discussions of terrestrial biogeochemistry center on the role of plants and soil microbes. Having seen that animals harvest 10–20% of net primary production (Chapter 5), it is legitimate to ask if they might play a significant role in nutrient cycling. Certainly an impressive nutrient influx is observed in the soils below roosting birds (Gilmore et al. 1984, Mizutani and Wada 1988, Lindeboom 1984). In Yellowstone National Park, elk appear to redistribute plant materials among habitats in the landscape, increasing the nitrogen content and nitrogen mineralization in soils where they congregate (Frank et al. 1994).

Various workers have suggested that the grazing of vegetation, especially by insects, stimulates the intrasystem cycle of nutrients and might even be advantageous for terrestrial vegetation (Owen and Wiegert 1976). Trees that are susceptible to herbivory are often those that are deficient in minerals or otherwise stressed (Waring and Schlesinger 1985). Periodic herbivory may stimulate nutrient return to the soil and alleviate nutrient deficiencies (Mattson and Addy 1975). Risley and Crossley (1988) noted significant premature leaf-fall in a forest that was subject to insect grazing. These leaves delivered large quantities of nutrients to the soil, since nutrient reabsorption had not yet occurred. In the same forest, Swank et al. (1981) noted an increase in streamwater nitrate when the trees were defoliated by grazing insects.

In extreme cases, defoliations may be the dominant form of nutrient turnover in the ecosystem (Hollinger 1986). Usually, however, the role of grazing animals in terrestrial ecosystems is rather minor (Gosz et al. 1978, Woodmansee 1978, Pletscher et al. 1989), and certainly of limited benefit to plants (Lamb 1985). In fact, plants often show large allocations of net primary production to defensive compounds (Coley et al. 1985) and higher net primary production when they are relieved of insect pests (Morrow and LaMarche 1978, Marquis and Whelan 1994).

The role of animals in litter decomposition is much more significant (Swift et al. 1979, Hole 1981, Seastedt and Crossley 1980). Nematodes, earthworms, and termites are particularly widespread and important in the initial breakdown of litter and the turnover of nutrients in the soil. Schaefer and Whitford (1981) found that termites were responsible for the turnover of 8% of litter N in a desert soil (Fig. 6.18). An additional 2% of the pool of nitrogen in surface litter was transported belowground by their burrowing activities. When termites were excluded by the application of pesticides, decomposition slowed and surface litter accumulated. Because soil animals have short lifetimes, their nutrient contents are rapidly decomposed and returned to the intrasystem cycle (Seastedt and Tate 1981).

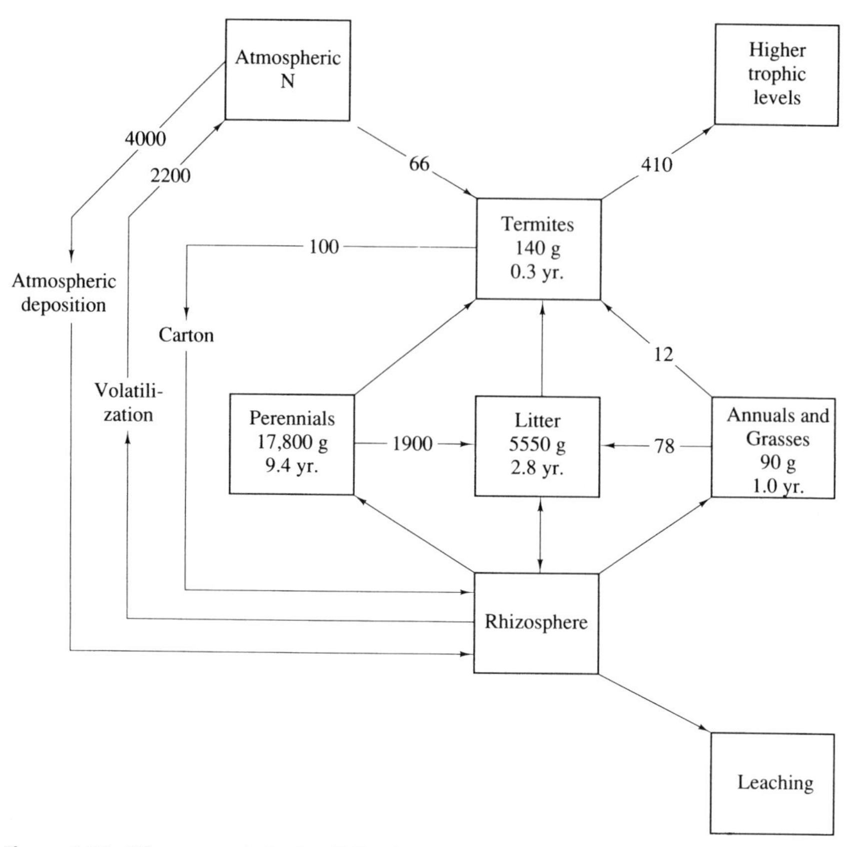

Figure 6.18 Nitrogen cycle in the Chihuahuan Desert of New Mexico, showing the role of termites in nitrogen transformations. Annual flux of nitrogen is shown along arrows in g N/ha; nitrogen pools are shown in boxes with turnover time in years. From Schaefer and Whitford (1981).

It is interesting to view the biogeochemistry of animals from another perspective: what is the role of biogeochemistry in determining the distribution and abundance of animals? The death of ducks and cattle feeding in areas of high soil selenium (Se) suggests that such interactions might be of widespread significance.

Plants have no essential role for sodium in their biochemistry, and naturally have low Na contents due to limited uptake and exclusion at the root surface (Smith 1976). On the other hand, sodium is an important, essential element for all animals. The wide ratio between the Na content of herbivores to that in their foodstuffs suggests that Na might limit mammal populations generally. Observations of Na deficiency are supported by the interest that many animals show in natural salt licks (Jones and Hanson 1985, Freeland et al. 1985, Smedley and Eisner 1995) and Na-rich plants

(Botkin et al. 1973). Weir (1972) suggested that the distribution of elephants in central Africa was at least partially dependent on sodium in seasonal waterholes, and McNaughton (1988) found that the abundance of ungulates in the Serengeti area was linked to Na, P, and Mg in plant tissues available for grazing. Thus, animal populations may be affected by the biogeochemical cycling of Na in natural ecosystems. Aumann (1965) found high rodent populations in areas of Na-rich soils, and speculated that the increased abundance of rodents in the eastern United States during the 1930s might have been due to a large deposition of Na-rich soil dust that was derived from the prairies during the "Dust-Bowl." Such a case would link the abundance of animals to the biogeochemistry of soils and to soil erosion by wind in a distant region.

An enormous literature exists on the characteristics of plant tissues that are selected for food. Seasonal variations in the plants selected as food by large mammals may help them to avoid mineral deficiency (McNaughton 1990, Ben-Shahar and Coe 1992, Grasman and Hellgren 1993). Many studies report that herbivory is centered on plants with high nitrogen contents (Mattson 1980, Lightfoot and Whitford 1987), suggesting that animal populations might be limited by N. However, the preference for such tissues may be related more to the high water (Scriber 1977) and low phenolic contents (Jonasson et al. 1986) of those tissues than to a specific search for leaves with high amino acid content to support the protein requirement of animals. Grazing often reduces plant photosynthesis while nutrient uptake continues, resulting in high nutrient contents in the aboveground tissues that remain (McNaughton and Chapin 1985). Grazing may even enhance nitrogen uptake in some species (Jaramillo and Detling 1988). In both cases, consumers increase the nutritional quality of the forage available for future consumption, although the quantity of defensive compounds may also increase (White 1984, Seastedt 1985).

Calculating Landscape Mass Balance and Responses to Global Change

Elements are retained in terrestrial ecosystems when they play an essential functional role in biochemistry or are incorporated into organic matter. The pool of nutrients held in the soil and vegetation is many times larger than the annual receipt of nutrients from the atmosphere and rock weathering. Turnover times (mass/input) range from 21 yr for Mg to >100 yr for P in the vegetation and forest floor of the Hubbard Brook Experimental Forest in New Hampshire (Likens and Bormann 1995, Yanai 1992). In contrast, for nonessential elements, such as Na, the turnover time is rapid (1.2 yr), because these elements are not retained by biota or incorporated into humus. Some nonessential—even toxic—elements such as lead (Pb) bind to soil organic matter and may accumulate in an ecosystem (Smith and Siccama 1981, Lindberg and Turner 1988, Friedland and Johnson 1985, Dörr and Münnich 1989). Even though these elements are not involved in biochemistry, their retention in the ecosystem is the result of biotic activity. Thus, a study of their movement on the surface of the Earth is also in the realm of biogeochemistry.

Annual mineralization, plant uptake, and plant death result in a large internal cycle of elements in most ecosystems. Annual nitrogen inputs are typically 1 to 5 kg ha^{-1} yr^{-1}, while mineralization of soil nitrogen is 50 to 100 kg ha^{-1} yr^{-1} (Bowden 1986). Despite such large movements of available nutrients within the ecosystem, there are usually only small losses of N, P, and K in streamwater (Table 6.9). The minor losses of plant nutrients in streamwater speak strongly for the efficiency of biological processes that retain the elements essential to biochemistry. Only a few studies have included measurements of gaseous flux in ecosystem nutrient budgets. Losses in denitrification may explain why the retention of N applied in fertilizer is often somewhat lower than that of other elements (e.g., P and K), which have no gaseous phase (Stone and Kszystyniak 1977, Keeney 1980, Melin et al. 1983, Neilsen et al. 1992). Globally, denitrification may explain the tendency for the growth of most vegetation to be N limited, despite efficient plant uptake of N from the soil and only minor losses in streamwater (Chapter 12).

Allan et al. (1993) compared the mass balance of elements in small patches of forest occupying rock outcrops in Ontario. Areas of bare rock showed net losses of various elements, whereas adjacent patches of forest showed accumulations of N, P, and Ca in vegetation. We should not, however, expect that the essential biological nutrients will accumulate indefinitely in all ecosystems. The incorporation of N and P in biomass should be greatest when structural biomass and soil organic matter are accumulating rapidly—that is, in young ecosystems where there is positive net ecosystem production (Chapter 5). Losses of N should be higher in mature, steady-

Table 6.9 Annual Chemical Budgets for Undisturbed Forests (Total Streamwater Losses Minus Atmospheric Deposition)[a]

Location and reference	Precipitation (cm)	Chemical (kg ha^{-1} yr^{-1})			
		Ca	Cl	N	P
British Columbia (Feller and Kimmins 1979)	240	15.8	2.9	−2.6	0
Oregon (Martin and Harr 1988)	219	41.2	—	−1.2	0.3
New Hampshire (Likens et al. 1977)	130	11.7	−1.6	−16.7	0
North Carolina (Swank and Douglas 1977)	185	3.9	1.7	−5.5	−0.1
Venezuela (Lewis et al. 1987, Lewis 1988)	450	14.2	−1.4	8.5	0.32

[a] Negative values indicate net accumulations in the ecosystem.

state ecosystems where the total biomass is stable (Vitousek and Reiners 1975). Using the mass-balance approach, where

$$\text{input} - \text{output} = \Delta \text{ storage}, \tag{6.5}$$

Vitousek (1977) found greater losses of available N from old-growth forests than from younger sites in New Hampshire. Hedin et al. (1995) confirm higher nutrient losses in old-growth forests of Chile, and Lewis (1986) suggested that relatively high losses of N and P from the Caura River in Venezuela were due to the mature vegetation covering most of the watershed (Table 6.9). In seasonal climates, losses of N, P, and K are usually minor during the growing season and greater during the winter period of plant dormancy (Likens and Bormann 1995, Likens et al. 1994). Often there is little seasonal variation in the loss of Na and Cl, which pass through the system under simple geochemical control (Juang and Johnson 1967, Gardner 1981, Johnson et al. 1969, Belillas and Rodà 1991). Streamwater losses of nutrients give old-growth ecosystems the appearance of being "leaky," but it is important to recognize that outputs represent the excess of inputs over the seasonal demand for nutrients by vegetation and soil microbes (Gorham et al. 1979).

Among elements in short supply to biota, nitrogen is unique in that it is only derived from the atmosphere (Table 4.5). Net primary production in some temperate forests appears to show a correlation to N inputs in precipitation (Cole and Rapp 1981). Comparing forests from Oregon, Tennessee, and North Carolina, Henderson et al. (1978) noted strong N retention in each, despite a 10-fold difference in N input from the atmosphere. The data suggest that plant growth is limited by N in each region. In contrast, losses of Ca were always a large percentage of the annual amount cycling in these forests. Especially on limestone soils, ample supplies of Ca were derived from rock weathering and Ca was not in short supply. Thus, abundant (e.g., Ca) and non-essential (e.g., Na) elements are most useful in estimating the rate of rock weathering (Chapter 4), whereas biogeochemistry controls the loss of scarce elements that are essential to life.

While most studies of ecosystem mass balance have considered single watersheds, Robertson and Rosswall (1986) have compiled an input-output budget for nitrogen in all of west Africa, south of the Sahara Desert (Fig. 6.19). They found that N-fixation and precipitation dominated the sources of nitrogen in this region, while fire (8.3×10^{12} g N/yr) rivers (1.5×10^{12} g N/yr), and denitrification (1.1×10^{12} g N/yr) dominated the losses. By including volatile losses, the nitrogen budget was balanced within 1% for the entire region. Similarly, Riggan et al. (1985) developed a nitrogen budget for the Los Angeles Basin, in which volatilization of NO_2 in automobile exhaust and its subsequent deposition in chaparral ecosystems were major cycling processes.

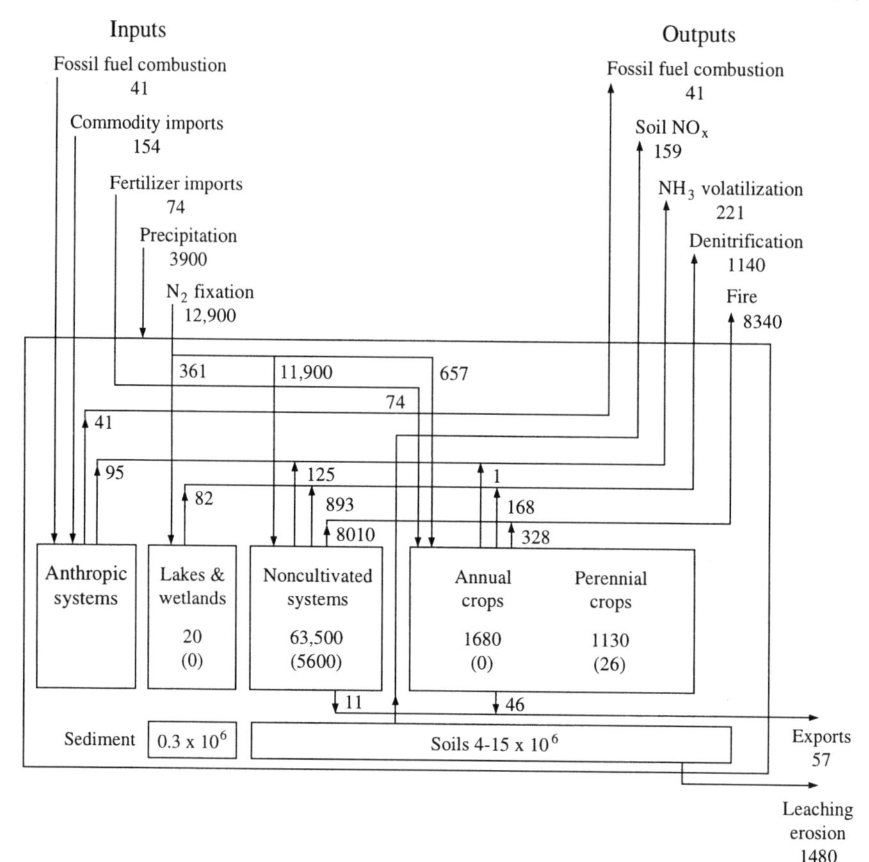

Figure 6.19 The nitrogen budget for west Africa in 1978. All flux estimates are in units of 10^6 kg/yr. Pool values are in 10^6 kg and increments to the pools are in parentheses. Modified from Robertson and Rosswall (1986).

Many of the transformations in biochemistry involve oxidation and reduction reactions that generate or consume acidity (H^+). For instance, H^+ is produced during nitrification and consumed in the plant uptake and reduction of NO_3^-. Binkley and Richter (1987) review these processes and demonstrate how ecosystem budgets for H^+ may be useful as an index of net change in ecosystem function, particularly as soils acidify during ecosystem development (Chapter 4). H^+-ion budgets are also useful as an index of human impact. For example, a net increase in acidity is expected when excess NH_4^+ deposition is subject to nitrification, with the subsequent loss of NO_3^- in streamwater (van Breemen et al. 1982). H^+ budgets are analogous to measurements of human body temperature. When we see a change, we suspect that the ecosystem is stressed, but we must look carefully within the system for the actual diagnosis.

Acid Rain and Nitrogen Saturation

Recent studies suggest that forest growth has declined in areas that are downwind of air pollution. In addition to the direct effects of ozone, nitric oxide, and other gaseous pollutants on plant growth, plants in these areas are subject to "acid rain." Acid rain is characterized by low pH, as a result of NO_3^- and SO_4^{2-} that are derived from the incorporation of gaseous pollutants in raindrops (Chapter 3). The chemical inputs in acid rain may affect several aspects of the mineral nutrition of plants, leading to changes in their growth rate.

Inputs of H^+ in acid rain increase the rate of weathering of soil minerals, the release of cations from cation exchange sites, and the movement of Al^{3+} in the soil solution (Chapter 4). High concentrations of Al^{3+} may reduce the plant uptake of Ca and other cations (Godbold et al. 1988, Bondietti et al. 1989). In the northeastern United States, forest growth appears to decline as a result of an increased Al/Ca ratio in the soil solution (Shortle and Smith 1988, Cronan and Grigal 1995). Depending upon the underlying parent rocks, the soil exchange capacity may be depleted of various nutrient cations (Miller et al. 1993b, Wright et al. 1994, Likens et al. 1996). Bernier and Brazeau (1988a, 1988b) link dieback of sugar maple to deficiencies of K on areas of low-K rocks and to deficiencies of Mg on low-Mg granites in southeastern Quebec. Magnesium deficiencies are also seen in the forests of central Europe, where forest decline is linked to an imbalance in the supply of Mg and N to plants (Oren et al. 1988, Berger and Glatzel 1994). Graveland et al. (1994) suggest that the effects of acid rain are also seen at higher trophic levels; in the Netherlands birds showed poor reproduction in forests subject to acid rain as a result of a decline in the abundance of snails, which are the main source of Ca for eggshell development.

Inputs of N from the atmosphere in the northeastern United States and central Europe (10–50 kg N ha^{-1} yr^{-1}) are 5 to 20× greater than those under pristine conditions. Many workers have speculated that over the long term the nitrogen deposited in acid rain may act as fertilizer—stimulating the growth of trees in areas where forest growth is limited by nitrogen (Kauppi et al. 1992, Townsend et al. 1996). Fertilizer experiments show that some of the added nitrogen also accumulates in soil organic matter (Proe et al. 1992, Preston and Mead 1994, Nadelhoffer et al. 1995, Mäkipää 1995, Downs et al. 1996, Koopmans et al. 1996). Eventually, however, symptoms of forest decline are observed as the ecosystem becomes saturated with nitrogen (Aber et al. 1989). Nitrification rates increase dramatically, increasing the loss of NO_3^- in streamwaters (Van Vuuren et al. 1992, Aber et al. 1993, Durka et al. 1994, Hedin et al. 1995, Nohrstedt et al. 1996, Peterjohn et al., 1996).

Long-term deposition of nitrogen can lead to the loss of fine-root biomass and deficiencies of other nutrients in plants (Schulze 1989, van Dijk et al.

1990). Excessive N deposition may also lead to the loss of mycorrhizal fungi (Littke et al. 1984), which may exacerbate P deficiency. Along three gradients of increasing air pollution in southern California, Zinke (1980) showed that N content in the foliage of Douglas fir increased from 1% to more than 2%, while the P content decreased abruptly, changing the ratios of N/P from about 7 in relatively pristine areas to 20–30 in polluted areas. Such an imbalance in leaf N/P ratios is also seen in the Netherlands, in areas of excessive inputs of NH_4^+ from the atmosphere (Mohren et al. 1986). These hypotheses should be tested by examining the forest response to experimental additions of P in affected areas.

Summary: Integrative Models of Terrestrial Nutrient Cycling

Interactions between plants, animals, and soil microbes link the internal biogeochemistry of terrestrial ecosystems. Under conditions of low nutrient availability, plants have low nutrient contents and higher nutrient reabsorption before leaf-fall, reflecting a higher nutrient-use efficiency by vegetation (Fig. 6.20). In some cases these characteristics can be induced by experimental treatments that reduce nutrient availability. For instance, when Douglas fir were fertilized with sugar, which increases the C/N ratio in the soil and the immobilization of N by microbes, reabsorption of foliar N increased, implying greater nutrient-use efficiency by the trees (Turner and Olson 1976). Internal cycling by the vegetation may partially alleviate nutrient deficiencies, but decomposition of nutrient-poor litterfall is slow, further exacerbating the low availability of nutrients in the soil. Thus, nutrient-poor sites are likely to be occupied by vegetation that is specially adapted for long-

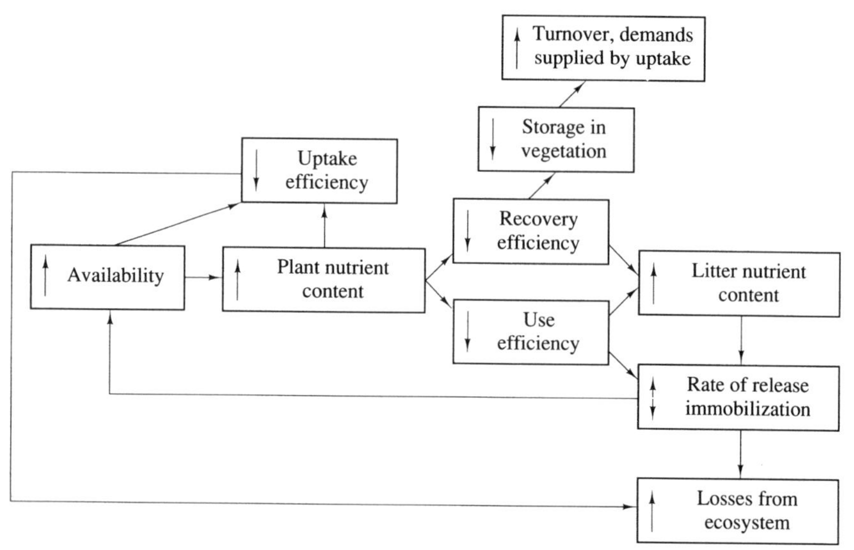

Figure 6.20 Changes in internal nutrient cycling that are expected with changes in nutrient availability. From Shaver and Melillo (1984).

term persistence under such conditions (Chapin et al. 1986a). Higher plant growth at elevated atmospheric CO_2 (Chapter 5) may increase the C/N ratio of plant litter and slow the rate of nutrient turnover in terrestrial ecosystems globally (Coûteaux et al. 1991, Cotrufo et al. 1994).

The role of biogeochemistry in controlling the distribution and characteristics of vegetation is seen at varying scales. Continental distributions of vegetation, such as the widespread dominance of conifers in the boreal regions, are likely to be related to the higher nutrient-use efficiency of evergreen vegetation under conditions of limited nutrient turnover in the soil. The effect of soil properties on the regional distribution of vegetation is seen in the occurrence of evergreen vegetation on nutrient-poor, hydrothermally altered soils in arid and semiarid climates (Billings 1950, Goldberg 1982, 1985, Schlesinger et al. 1989). Fine-scale spatial heterogeneity in soil properties, as recorded by Robertson et al. (1988) for a field in Michigan, has been linked to the maintenance of diversity in land plant communities (Tilman 1982, 1985), and several early studies show the importance of local soil conditions to the distribution and abundance of forest and grassland herbs (Snaydon 1962, Pigott and Taylor 1964, Lechowicz and Bell 1991). Additions of fertilizer tend to reduce the species diversity of plant communities (Tilman 1987, Huenneke et al. 1990, Schlesinger 1994).

Linkages among components of the intrasystem cycle suggest that an integrative index of terrestrial biogeochemistry might be derived from the measure of a single component, such as the chemical characteristics of the leaf canopy (Matson et al. 1994). Changes in canopy characteristics might provide an index of the effects of acid rain or other pollutants on nutrient cycling. Variations of leaf C/N ratio across sites provide a convenient index of the intrasystem cycle of nutrients. Wessman et al. (1988b) have measured the nitrogen and lignin content of foliage by analyzing the spectral reflectance of tissues in the laboratory, as a first step toward developing an index of forest canopies by remote sensing. Their data show a strong correlation between nitrogen and lignin measured by infrared reflectance and by traditional laboratory analyses. These workers then used an aircraft-borne spectrophotometer to obtain the reflectance spectra of forest canopies in Wisconsin. Canopy lignin, calculated by applying the laboratory calibration to the remote sensing images, was highly correlated to soil nitrogen mineralization that had been measured in these stands in earlier studies (Fig. 6.21; Pastor et al. 1984). Recognizing that decomposition is frequently controlled by the lignin and nitrogen content of litter (Fig. 6.10), these data suggest that remote sensing of canopy characteristics has potential for comparative regional studies of nutrient cycling in different plant communities. Myrold et al. (1989) found that a variety of soil properties were related to canopy characteristics that could be measured by remote sensing, and Reiners et al. (1989) used thematic mapper images (Chapter 5) to classify landscape units for regional estimates of nitrogen cycling in Wyoming. Studies such as these reinforce our appreciation of the linkage between vegetation and soil characteristics, as outlined in Fig. 6.20.

Various models demonstrate other linkages between plant and soil processes in terrestrial biogeochemistry. Walker and Adams (1958) suggested that the level of available phosphorus during soil development was the primary determinant of terrestrial net primary production, since nitrogen-fixing bacteria depend on a supply of organic carbon and available phosphorus. They use the level of organic carbon

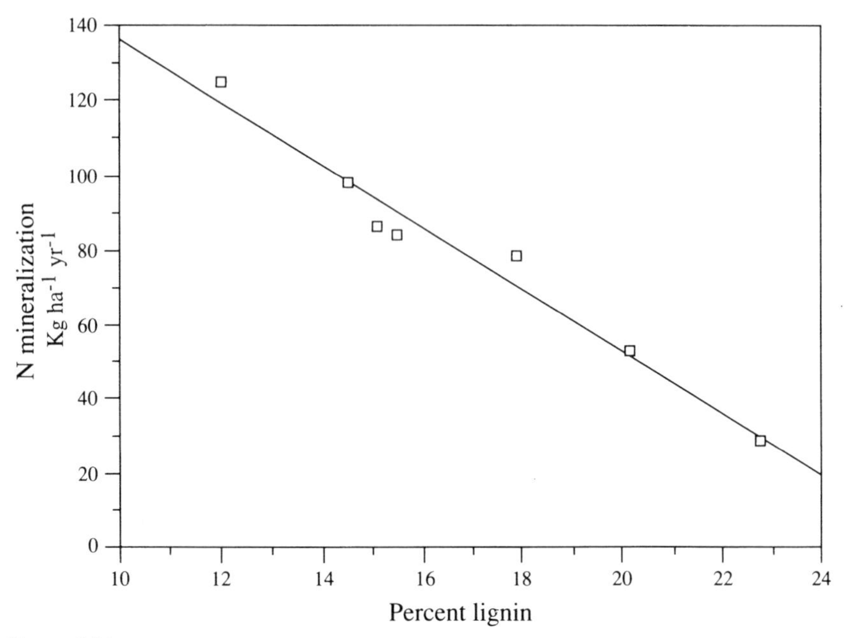

Figure 6.21 Nitrogen mineralization in seven Wisconsin forests, measured by Pastor et al. (1984) using buried bags, as a function of canopy lignin. From Wessman et al. (1988a).

in the soil as an index of terrestrial productivity and suggest that organic carbon will peak midway during soil development and then decline as an increasing fraction of the phosphorus is rendered unavailable by precipitation with secondary minerals (Fig. 4.5).

Numerous workers have examined the Walker and Adams (1958) hypothesis in various ecosystems. Tiessen et al. (1984) found that available phosphorus explained 24% of the variability of organic carbon in a collection of 168 soils from eight different soil orders. Roberts et al. (1985) found a similar relationship between bicarbonate-extractable P and organic carbon in several grassland soils of Saskatchewan. Raghubanshi (1992) found that phosphorus was well correlated to soil organic matter, soil nitrogen, and nitrogen mineralization rates in dry tropical forests of India. Thus, available phosphorus explains some, but not all, of the variation in soil organic carbon, which is ultimately derived from the production of vegetation. The linkage of phosphorus and carbon is likely to be strongest during early soil development, when both organic phosphorus and carbon are accumulating. The importance of organic phosphorus increases during soil development, and through the release of phosphatase enzymes, vegetation interacts with the soil pool to control the mineralization of P. Thus, in mature soils, net primary production is more likely to be limited by other factors.

Parton et al. (1988) present a model linking the cycling of C, N, P, and S in grassland ecosystems. The flow of carbon is shown in Fig. 6.22. The nitrogen cycling submodel has a similar structure, since the model assumes that most nitrogen is bonded directly to carbon in amino groups (McGill and Cole 1981). Lignin controls

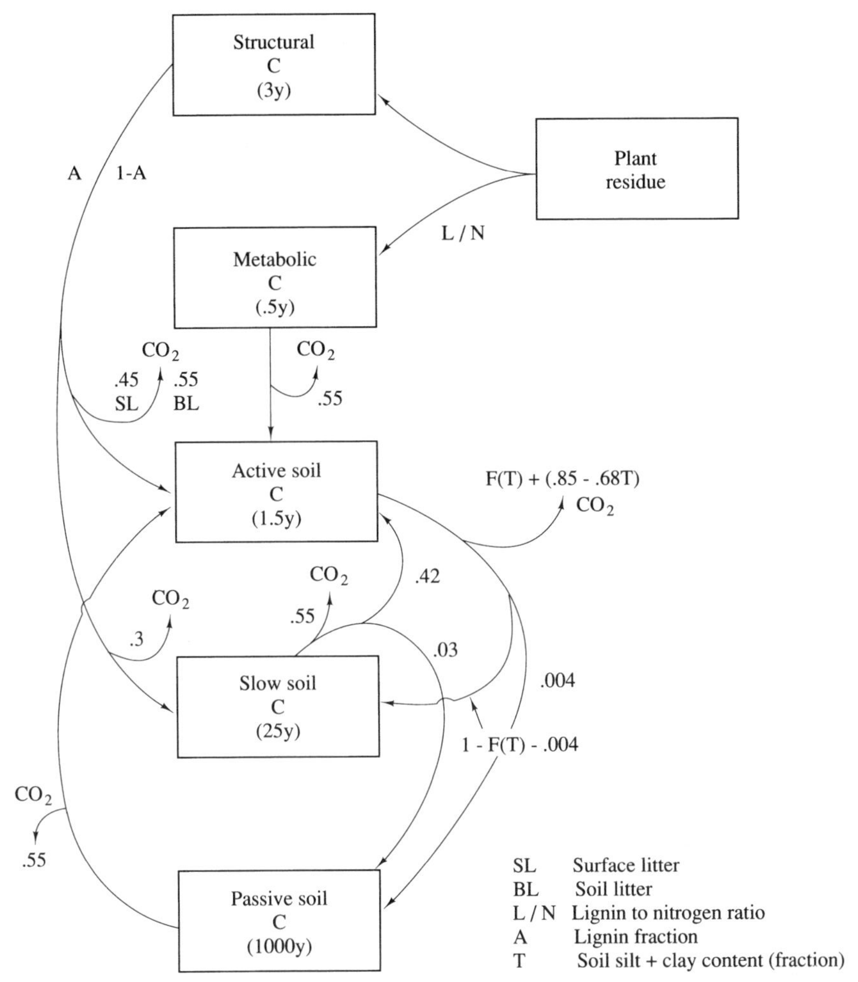

Figure 6.22 Flow diagram for carbon in the CENTURY model. The proportion of carbon moving along each flowpath is shown as a fraction, and turnover times for reservoirs are shown in parentheses. From Parton et al. (1988).

decomposition rates and nitrogen is mineralized from soil pools when critical C/N ratios are achieved during the respiration of carbon. Phosphorus availability is controlled by a modification of a model first presented by Cole et al. (1977), which includes C/P control over mineralization of organic pools and geochemical control over the availability of inorganic forms as in Fig. 6.17. However, unlike N, C/P ratios in plant tissues and soil organic matter are allowed to vary widely as a function of P availability.

The complete model was used to predict patterns of primary production and nutrient mineralizations during 10,000 years of soil development (Fig. 6.23). Net primary production and accumulations of soil organic matter are strongly linked

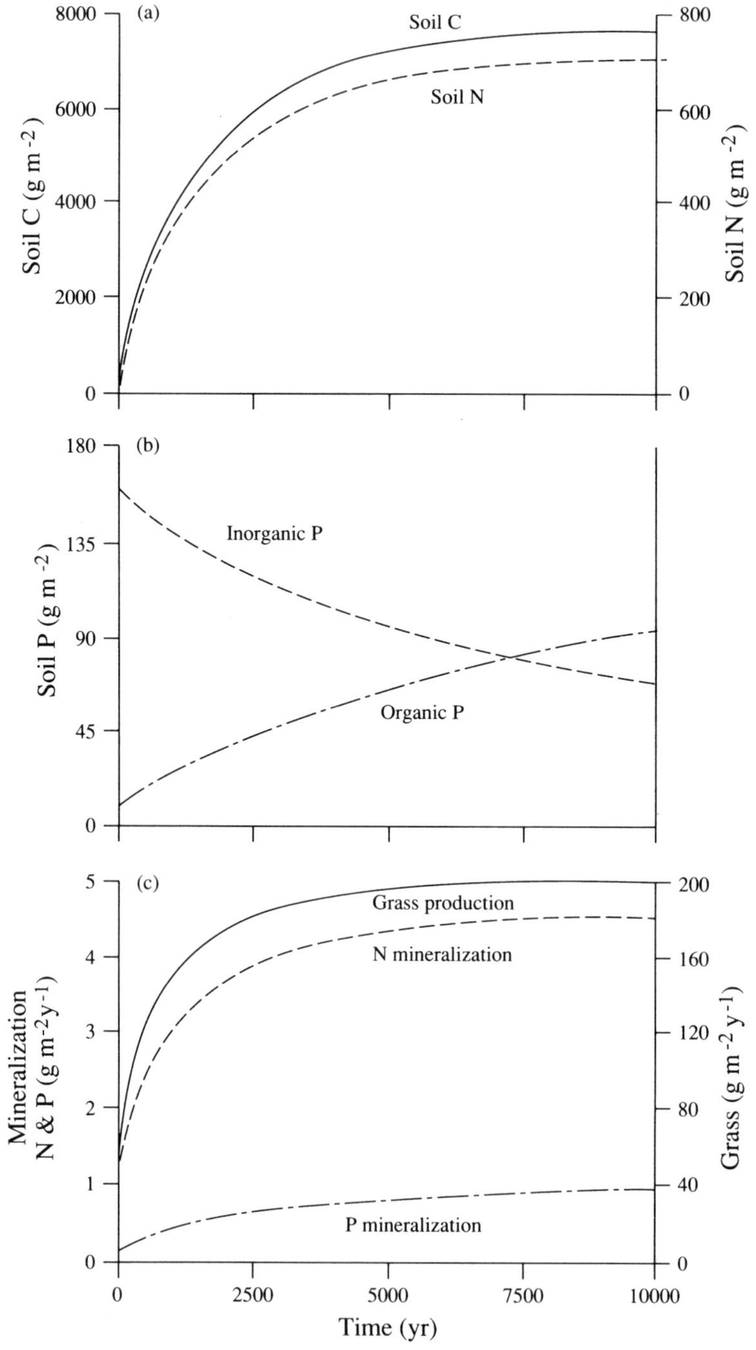

Figure 6.23 Simulated changes in soil C, N, and P during 10,000 years of soil development in a grassland, using the Parton et al. (1988) model.

to P availability during the first 800 years, after which increases in plant production are related to increases in soil N mineralization. Organic P increases throughout the 10,000-year sequence. In simulations of the response of native soils to cultivation, the model predicted a correlated decline in the native levels of organic carbon and nitrogen in the soil, but a relatively small decline in P. Validation of the model is seen in the data of Tiessen et al. (1982), who found declines of 51% for C and 44% for N, but only 30% for P in a silt loam soil cultivated for 90 years in Saskatchewan.

Recommended Readings

Coleman, D. and D.A. Crossley. 1996. *Fundamentals of Soil Ecology*. Academic Press, San Diego.

Dobrovolsky, V.V. 1994. *Biogeochemistry of the World's Land*. CRC Press, Boca Raton, Florida.

Johnson, D.W. and S.E. Lindberg. (eds.). 1992. *Atmospheric Deposition and Nutrient Cycling*. Springer-Verlag, New York.

Likens, G.E. and F.H. Bormann. 1995. *Biogeochemistry of a Forested Ecosystem*. 2nd ed. Springer-Verlag, New York.

Paul, E.A. and F.E. Clark. 1996. *Soil Microbiology and Biochemistry*. 2nd ed. Academic Press, San Diego.

Stevenson, F.J. 1986. *Cycles of Soil*. Wiley, New York.

Swift, M.J., O.W. Heal, and J.M. Anderson. 1979. *Decomposition in Terrestrial Ecosystems*. University of California Press, Berkeley.

7

Biogeochemistry in Freshwater Wetlands and Lakes

Introduction

Oxygen is only sparingly soluble in water and diffuses about 10^4 times more slowly in water than in air. Many organisms inhabiting lakes and flooded soils must survive with relatively low concentrations of oxygen. In some cases, heterotrophic respiration may totally deplete oxygen in wetland environments. For instance, within a few millimeters of depth, the environment of wetland sediments often changes, from one where aerobic metabolism of organic matter (i.e., via the Krebs cycle) is possible, to one where various forms of anaerobic metabolism are required.

Nutrient cycling in lakes and freshwater wetlands is controlled by reduction potential, informally known as *redox,* and by microbial transformations of nutrient elements that occur under conditions in which O_2 is not always abundant. For example, the availability of phosphorus in lakes differs strongly between the surface waters, which are more or less saturated with atmospheric O_2, and deeper waters in which O_2 may be depleted. Anaerobic microbial processes—denitrification, sulfate reduction, and methanogenesis—are responsible for the release of N_2, H_2S, and CH_4 from wetland sediments. Other anaerobic microbial processes are coupled to changes in the oxidation state of iron and manganese in wetland soils. Anaerobic decomposition is often incomplete, so many wetlands store significant amounts of organic carbon—net ecosystem production—in their sediments (Schlesinger 1977). Wet soils contain about 1/3 of all the organic matter stored in soils of the world (Eswaran et al. 1995). Vast deposits of coal represent the net ecosystem production of swamps during the Carboniferous Period (Berner 1984).

Matthews and Fung (1987) estimate that 3.6% of the world's land area is wetland, but the present area of wetlands has been significantly reduced by human activities during the last 100 years. The unique environment of wetlands and their role in chemical transformations mean that their importance to global biogeochemistry is much greater than their proportional surface area on Earth would suggest. Recent environmental legislation recognizes the critical importance of wetlands as wildlife habitat and as an arena for biogeochemistry.

In this chapter we will examine and compare various freshwater wetlands, using the status of oxidation and reduction reactions as a unifying theme. We will begin with a brief review of the concept of redox potential. Then we will discuss the relationship between redox potential and the microbial reactions that occur in the saturated soils and sediments of swamps and lakes. Much of this discussion will also apply to the vast area of arctic tundra, known as muskeg, which is dominated by saturated organic soils known as Histosols. A second section of the chapter treats net primary production and nutrient cycling in lakes. We will compare the nutrient budgets of various wetlands to those of terrestrial ecosystems (Chapter 6).

The biogeochemical reactions in wetlands are linked to the reactions occurring in the surrounding terrestrial environment by the movements of surface runoff and groundwater (Likens and Bormann 1974). Although rivers are the explicit subject of Chapter 8, we will make frequent reference in this chapter to the importance of runoff from upland ecosystems to the nutrient cycles of wetlands. Bogs are an interesting category of wetland with respect to runoff from the surrounding land. True bogs are *ombrotrophic* and depend entirely on the atmosphere for inputs of water and nutrients. Taken literally, ombrotrophic means "to feed on rain" (Du Rietz 1949, Gorham 1957). Other bogs, more correctly termed minerotrophic fens, receive at least a portion of their water and nutrient inputs from groundwater or from surface runoff from the surrounding watershed.

Redox Potential: The Basics

Just as pH expresses the concentration of H^+ in solution, redox potential is used by chemists to express the tendency of an environment to receive or supply electrons. Oxic environments are said to have a high redox potential because O_2 is available as an electron acceptor. For instance, iron (Fe) oxidizes when it shares the electrons of its outer shell with O_2 to become Fe_2O_3 (rust):

$$4Fe + 3O_2 \rightarrow 2Fe_2O_3. \tag{7.1}$$

Heterotrophic organisms in oxic environments capitalize on the use of O_2 as a powerful electron acceptor. Electrons are derived from the metabolism of reduced organic compounds that are obtained from the environment and oxidized to CO_2. In eukaryotic cells aerobic respiration occurs in the mitochondria. Every four electrons that flow across an internal membrane of the mitochondria combine with one O_2 and four H^+ to form two molecules of water—with the internal membrane system allowing an especially efficient capture of energy for biochemistry.

The oxidation state of the environment, or redox potential, is determined by the particular suite of chemical species that is present. Figure 7.1 illustrates a hypothetical situation in which two containers hold iron chloride in different oxidation states, Fe^{2+} and Fe^{3+}. The containers are connected by a wire which passes through a voltmeter and ends in inert platinum electrodes that are placed in each solution. A salt bridge allows Cl^- to diffuse between the containers so as to maintain a neutral charge. One might expect that electrons would flow from left to right until an equilibrium was established:

$$Fe^{2+} - e^- \rightleftarrows Fe^{3+}. \tag{7.2}$$

The voltmeter could then be used to measure the current passing between the containers. For this simple system, we would say that the container on the right is an oxidizing environment, because it draws electrons from Fe^{2+}, the more reduced or electron-rich species on the left.

If the container on the right also contains O_2, a greater voltage would be measured by the voltmeter. When oxygen is present, it acts as a powerful

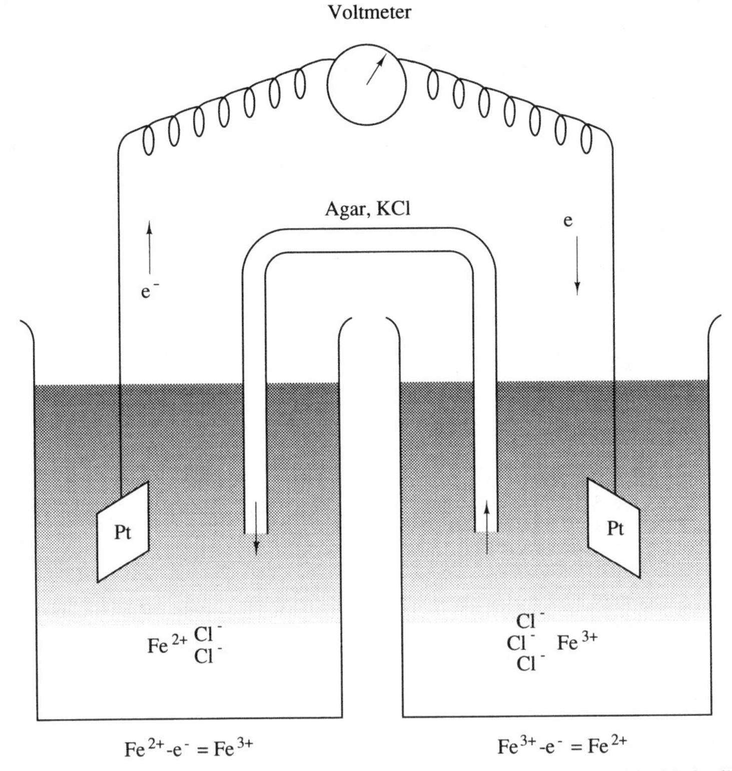

Figure 7.1 A hypothetical chemical cell, connecting two solutions of iron chloride in different oxidation states. The flow of electrons (e^-) can be prevented by the application of 771 mV at the voltmeter. The agar–salt bridge allows Cl^- ions to diffuse between the containers. Modified from Jenny (1980).

electron acceptor, and Fe will precipitate (as Fe^{3+}) in soils and sediments, viz.,

$$4Fe^{3+} + 3O_2 + 12e^- \rightarrow 2Fe_2O_3. \tag{7.3}$$

Thus, Fe^{3+} can accept electrons from more reduced substances, such as Fe^{2+}, but not from O_2, which is more strongly oxidizing. In the absence of strongly oxidizing substances, Fe^{2+} persists in the environment.

Of course, natural environments are not isolated into separate containers, nor do they contain such a simple mixture of constituents. In practice, we measure the redox potential of a natural environment by expressing the disequilibrium of its suite of constituents relative to a standard electrode, which contains H_2 gas overlying a solution of known H^+ concentration. We connect the environment to the standard electrode using an inert platinum electrode, which takes on the potential of the environment, without altering the tendency for electrons to move among chemical constituents. When a voltmeter is placed in this circuit, the redox potential (E_h) is measured as the voltage required to prevent the interconversion of H^+ and H_2 at the standard electrode. In practice, a standard hydrogen electrode is difficult to maintain in the field, so investigators often use other reference electrodes that are calibrated against a hydrogen electrode (Bricker 1982, Faulkner et al. 1989).

When O_2 is present, it accepts electrons at the platinum electrode:

$$O_2 + 4e^- + 4H^+ \rightarrow 2H_2O. \quad (7.4)$$

The electrons are generated at the hydrogen electrode,

$$2H_2 \rightleftarrows 4H^+ + 4e^-, \quad (7.5)$$

and the voltmeter records a high voltage or redox potential (e.g., ca. 1100 mV at pH 2.0). Because Eq. 7.4 is more likely to proceed to the right under acid conditions, a higher redox potential will be found at lower pH, assuming that all other factors are the same. As oxygen is depleted, other constituents, such as Fe^{3+}, may accept electrons, following Eq. 7.2, but a lower voltage will be recorded. An equimolar solution of Fe^{3+} and Fe^{2+} will have a redox potential of +770 mV relative to the standard electrode, when its pH is 2.0.

The pH of the environment affects the redox potential established by Fe^{3+} and other species. In an anoxic environment at pH 5.0, an equilibrium between Fe^{2+} and Fe^{3+} is found at a redox potential of about +400 mV, with the underlying reaction being

$$Fe^{2+} + 3H_2O \rightleftarrows Fe(OH)_3 + 3H^+ + e^-. \quad (7.6)$$

Equation 7.6 is more likely to proceed to the right at higher pH, so Fe^{3+} will prevail in neutral and alkaline environments, while Fe^{2+} is most likely to persist in the acid, anoxic waters of peatbogs. Thus, oxidation proceeds more readily, and at lower redox potentials, in neutral or alkaline environments,[1] and various forms of anaerobic metabolism, such as denitrification, are more likely to occur in acid environments (e.g., Weier and Gilliam 1986).

Much of the recent work expresses redox in units of pe, which is derived from the equilibrium constant of the oxidation- reduction reaction. For any reaction,

$$\text{oxidized species} + e^- + H^+ \rightleftarrows \text{reduced species}, \quad (7.7)$$

[1] This observation accounts for the rapid rust and decay of outdoor grills, which are exposed to alkaline solutions when rainwater mixes with the ash (see Chapter 6).

and the equilibrium constant, K, is determined by

$$\log K = \log[\text{reduced}]\text{-}\log[\text{oxidized}]\text{-}\log[e^-]\text{-}\log[H^+]. \quad (7.8)$$

If we assume that the concentrations of oxidized and reduced species are equal, then

$$pe + pH = \log K. \quad (7.9)$$

Here, pe is the negative logarithm of the electron activity ($-\log[e^-]$), and it expresses the energy of electrons in the system (Bartlett 1986). Because the sum of pe and pH is constant, if one goes up, the other must decline. When a given reaction occurs at lower pH, it will occur at higher redox potential, expressed as pe. Measurements of redox potential that are expressed as voltage, E_h, can be converted to pe following

$$pe = \frac{E_h}{(RT/F)2.3}, \quad (7.10)$$

where R is the universal gas constant (1.987 cal $mole^{-1}$ K^{-1}), F is Faraday's constant (23.06 kcal V^{-1} $mole^{-1}$), T is temperature in Kelvin, and 2.3 is a constant to convert natural to base-10 logarithms.

Environmental chemists use E_h–pH or pe–pH diagrams to predict the oxidation state of various constituents in natural environments (e.g., Fig. 7.2). All diagrams are bounded by two lines. If redox potentials were ever to fall above the upper line, even water would be oxidized, following the reverse of Eq. 7.4. Although the photolysis of water occurs during photosynthesis (Chapter 5) and by exposure of water vapor to ultraviolet light in the upper atmosphere (Chapter 2), we do not normally find such strongly oxidizing conditions in the natural waters at the surface of the earth. An environment dominated by Cl_2 would be more oxidizing than water, so as long as the Earth has contained abundant liquid water, any Cl_2 from volcanic emissions has been dissolved in ocean waters as Cl^- (Bohn et al. 1985):

$$2Cl_2 + 4e^- + 4H^+ \rightarrow 4HCl \quad (7.11)$$

$$2H_2O \rightarrow 4H^+ + 4e^- + O_2. \quad (7.12)$$

Similarly, any conditions above the lower line allow the reaction

$$H_2 + 2OH^- \rightarrow 2H_2O + 2e^-, \quad (7.13)$$

but the reverse of this reaction—the reduction of water—is also rarely seen in the natural environment. Elemental Na reduces water,

$$2Na + 2H_2O \rightarrow 2Na^+ + 2OH^- + H_2, \quad (7.14)$$

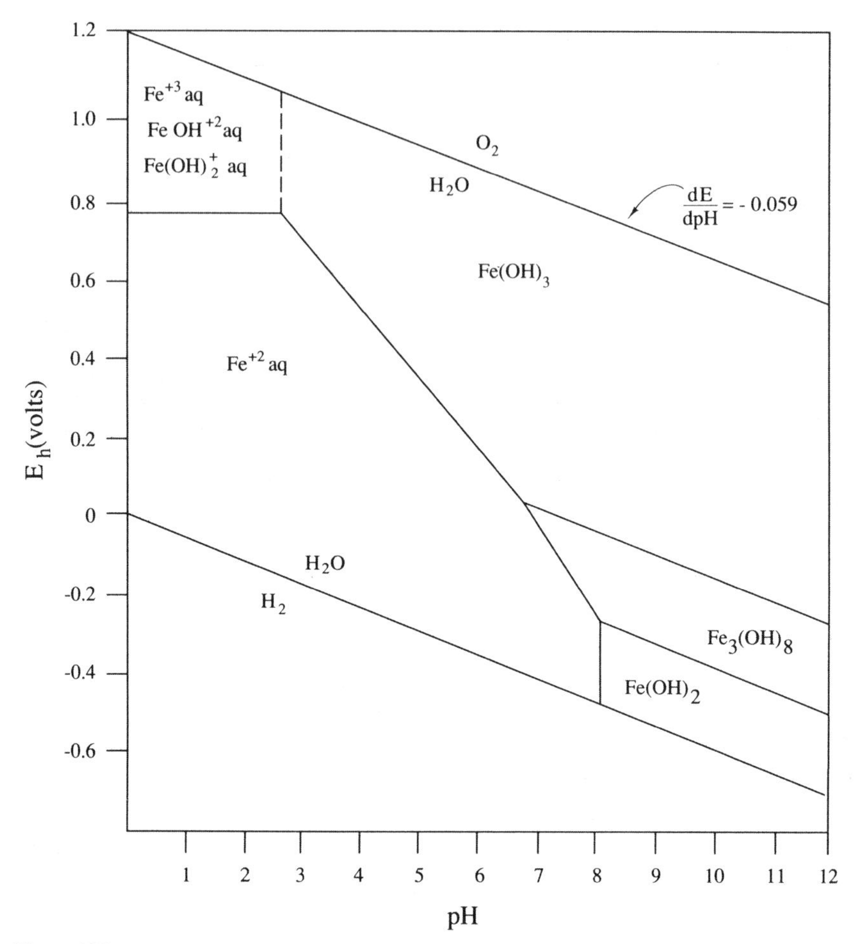

Figure 7.2 The stability of iron and iron hydroxides in soils relative to E_h and pH at 25°C. All conditions refer to 1 mM Fe^{2+} solution. Modified from Ponnamperuma et al. (1967).

which is why sodium exists in ionic form at the surface of the Earth (Bohn et al. 1985). These boundary conditions vary with pH; E_h decreases by 59 mV with each unit of pH increase, reflecting that oxidation requires a higher redox potential under acid conditions.

Figure 7.2 shows the expected forms of iron in natural environments of varying pH and E_h. Note that most transition lines slope downward, indicating that iron is more likely to occur in oxidized form under neutral or alkaline conditions. In interpreting such diagrams, it is important to remember that E_h and pH are properties of the environment, determined by the total suite of chemical species present. Thus, E_h predicts the forms of iron that will be present, but it is not determined solely by the oxidation state of iron in the environment. Important to biogeochemistry, E_h determines what modes of microbial activity are possible in a given environment.

In most cases, organic matter contributes a large amount of "reducing power" that lowers the redox potential in flooded soils and sediments (Bartlett 1986). High concentrations of Fe^{2+} will be found in flooded, low-redox environments, where impeded decomposition leaves undecomposed organic matter in the soil and humic substances impart acidity to the soil solution. Where organic matter is sparse, iron may persist in its oxidized form (Fe^{3+}) even when the soils are flooded (e.g., Couto et al. 1985). The tendency for iron to precipitate in oxidized form at high redox potential or high pH underlies the use of aeration and liming as techniques for ameliorating lakes that are affected by acid mine drainage (Eq. 7.6).

Soils and sediments that resist changes in their redox potential are said to be highly poised. Conceptually, poise is to redox potential as buffer capacity is to pH (Bartlett 1986). As long as soils are exposed to the atmosphere, they will appear to be highly poised, since O_2 will maintain a high redox potential under nearly all conditions. However, in the absence of O_2, these soils may show a rapid decline in redox potential as various weakly oxidizing constituents (e.g., NO_3^-, Mn^{4+}, Fe^{3+}, and SO_4^{2-}) are reduced. Redox potential will fall less rapidly—there is more poise—when concentrations of Mn^{4+} and Fe^{3+} are high (Lovley and Phillips 1988b, Achtnich et al. 1995).

Redox Reactions in Natural Environments

Few oxic environments have redox potentials of less than +600 mV. A progressive decrease in redox potential occurs when soils are flooded (Fig. 7.3). Redox potential drops as heterotrophic respiration of organic carbon depletes the soil of O_2 (Callebaut et al. 1982, Megonigal et al. 1993). The diffusion of oxygen in flooded soils and sediments is so slow that redox potentials also decline rapidly with increasing depth in wetland soils (Stolzy et al. 1981). Where organic carbon is abundant, a strong gradient of redox potential may develop in sediments over a depth as short as 2 mm (e.g., Howeler and Bouldin 1971, Sweerts et al. 1991). The high spatial and temporal variation of redox potential in wetlands accounts for much of the total range of redox potential that has been reported for the surface of the Earth (Fig. 7.4).

The results of many studies suggest that a particular sequence of reactions is expected as progressively lower redox potentials are achieved (Table 7.1; Ponnamperuma 1972, Patrick and Jugsujinda 1992, Achtnich et al. 1995, Peters and Conrad 1996). After O_2 is depleted by aerobic respiration, denitrification begins when the redox falls to +747 mV (at pH 7.0). Denitrifying bacteria use nitrate as an alternative electron acceptor during the oxidation of organic matter (see Eq. 2.18 and Chapter 6). When nitrate is depleted, reduction of Mn^{4+} begins below a redox of +526 mV, followed

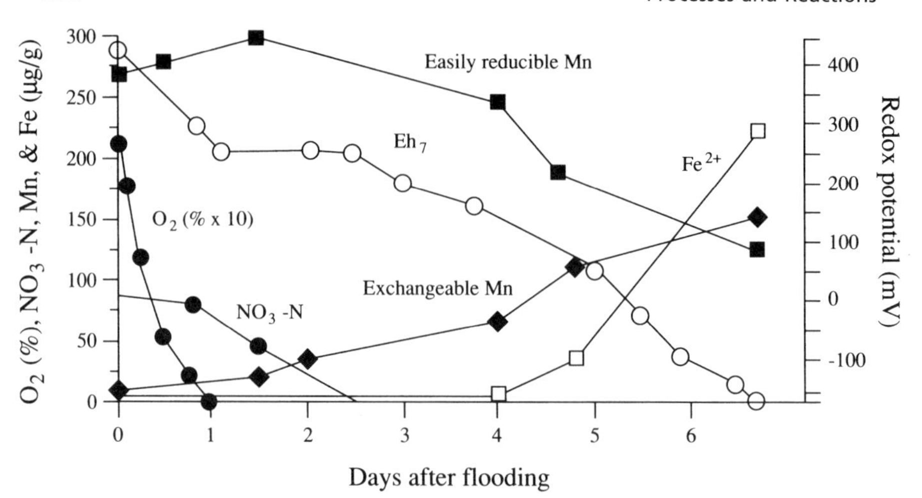

Figure 7.3 Changes in the chemical composition of the waters overlying a flooded soil as a function of time after flooding. Note that the reduction of iron does not begin until fully anaerobic conditions are achieved. Redox potential is expressed at pH 7, that is, E_{h7}. From Turner and Patrick (1968).

by reduction of Fe^{3+} at $E_h < -47$ mV (Lovley 1995). Certain types of bacteria (e.g., *Shewanella putrefaciens*) can couple the reduction of Mn and Fe directly to the oxidation of simple organic substances (Lovley and Phillips 1988a, Nealson and Myers 1992, Caccavo et al. 1992), but usually these reactions are catalyzed by a suite of coexisting bacteria—with some species using fermentation to obtain metabolic energy, while others oxidize the hydrogen, using Mn^{4+} and Fe^{3+} as electron acceptors (Lovley and Phillips 1989), e.g.,

$$C_6H_{12}O_6 \rightarrow CH_3COOH + CH_3CH_2COOH + CO_2 + H_2 \quad (7.15)$$

$$2Fe^{3+} + H_2 \rightarrow 2Fe^{2+} + 2H^+. \quad (7.16)$$

In many cases there appears to be some overlap between the zone of denitrification and the zone of Mn reduction in sediments (e.g., Klinkhammer 1980, Kerner 1993; Fig. 7.3), and most of the microbes in this zone are facultative anaerobes that can tolerate periods of aerobic conditions (Chapter 6). There is little overlap between the zone of Mn reduction and that of Fe reduction, because soil bacteria show an enzymatic preference for Mn^{4+}, and Fe^{3+} reduction will not begin until Mn^{4+} is depleted (Lovley and Phillips 1988b).

Below the zone of Mn^{4+} reduction, most redox reactions are performed by obligate anaerobes. Our earlier emphasis on the redox state of iron (Fig. 7.2) reflects the widespread use of Fe as an index of the transition from mildly oxidizing to strongly reducing conditions. Iron is a convenient

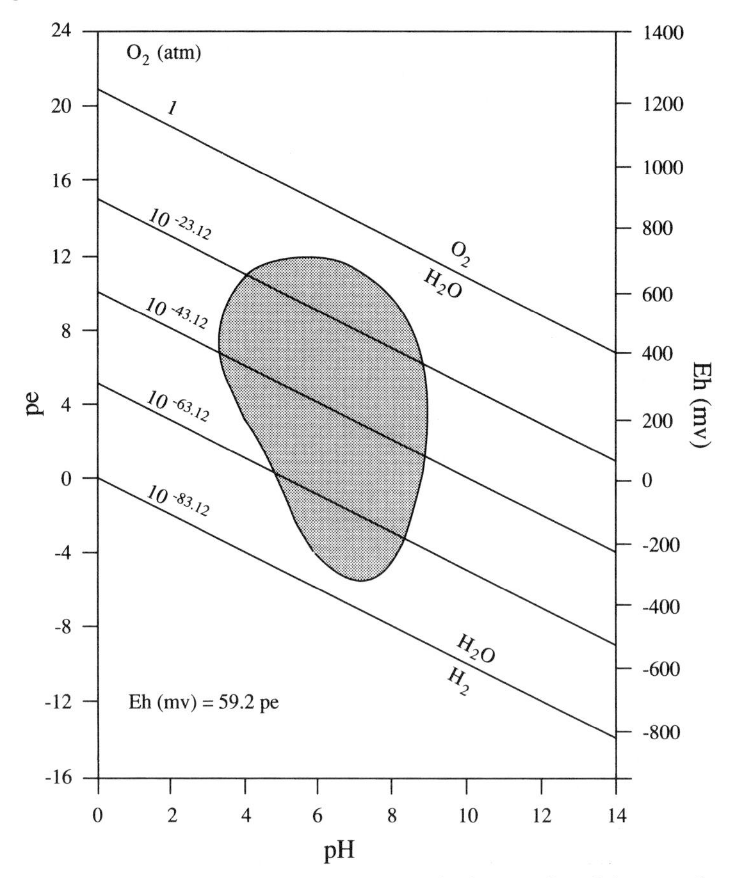

Figure 7.4 The range (shaded) of redox potentials that are found in natural aqueous environments. Modified from Lindsay (1979), based on the original compilation of Baas Becking et al. (1960).

indicator in the field, because oxidized iron is easily recognized in soils by its red color, known as *chroma,* whereas reduced iron is grayish (Evans and Franzmeier 1988). Soil layers with reduced iron are called gley.

Obligate anaerobes such as *Clostridium* use the energy derived from fermentation/Fe^{3+} reduction to engage in nitrogen fixation (Ottow 1971). Such nitrogen fixation is probably essential to augment the meager supplies of nitrogen that result from anaerobic mineralization. Below the depth of iron reduction, the redox potential progressively drops to -221 mV, where sulfate reduction commences, and to -244 mV, where methanogenesis occurs (Lovley and Phillips 1987). These reactions are performed by obligate anaerobic bacteria, some of which also engage in nitrogen fixation (Postgate et al. 1988). A few, highly reducing ($E_h < -700$ mV at pH 7.0) environments appear to allow the production of phosphine gas (PH_3) by

Table 7.1 Thermodynamic Sequence for Reduction of Inorganic Substances by Hydrogen at pH 7.0 and 25°C[a]

Reaction	E_h (V)	ΔG^b
Reduction of O_2		
$O_2 + 4H^+ + 4e^- \rightleftarrows 2H_2O$	0.812	−29.9
Reduction of NO_3^-		
$2NO_3^- + 6H^+ + 6e^- \rightleftarrows N_2 + 3H_2O$	0.747	−28.4
Reduction of Mn^{4+} to Mn^{2+}		
$MnO_2 + 4H^+ + 2e^- \rightleftarrows Mn^{2+} + 2H_2O$	0.526	−23.3
Reduction of Fe^{3+} to Fe^{2+}		
$Fe(OH)_3 + 3H^+ + e^- \rightleftarrows Fe^{2+} + 3H_2O$	−0.047	−10.1
Reduction of SO_4^{2-} to H_2S		
$SO_4^{2-} + 10H^+ + 8e^- \rightleftarrows H_2S + 4H_2O$	−0.221	−5.9
Reduction of CO_2 to CH_4		
$CO_2 + 8H^+ + 8e^- \rightleftarrows CH_4 + 2H_2O$	−0.244	−5.6

[a] Calculated from Stumm and Morgan (1981, p. 459).
[b] Kcal $mole^{-1}$ per e^-, assuming coupling to the oxidation reaction $\frac{1}{4} CH_2O + \frac{1}{4} H_2O \rightarrow \frac{1}{4} CO_2 + H^+ + e^-$ and $\Delta G = -RT \ln(K)$.

microbial reduction of PO_4^{3-} (Gassmann and Glindemann 1993, Dévai et al. 1988).

The environment of flooded soils and sediments exists as a dynamic equilibrium that is maintained by the availability of oxygen at the surface and buried organic carbon as a source of reducing power at depth. A declining yield of metabolic energy determines the order of the anaerobic microbial processes (Stumm and Morgan 1981, Froelich et al. 1979, Patrick and Jugsujinda 1992). Table 7.1 shows that the free energy of reaction (ΔG) is greatest for aerobic respiration (−29.9) and least for methanogenesis (−5.6), and at any redox potential the microbial population conducting metabolism with the greatest energy yield will usually outcompete the rest (Lovley and Klug 1986, Achtnich et al. 1995). Note that the potential energy available to aerobic heterotrophs is only slightly greater than that from denitrifiers, which often coexist in upland soils (Carter et al. 1995, Chapter 6). The low energy yield of reactions that occur at lower redox potential accounts for the inefficiency of anaerobic metabolism and the preservation of organic carbon in sediments (Gale and Gilmour 1988, Albers et al. 1995).

If the surface of a wetland soil is exposed to the air, as might occur with seasonal fluctuations of the water table, the position of each redox reaction will shift downward in the profile (Megonigal et al. 1993). Products of previous reduction reactions become substrates for oxidizing bacteria. For example, the total rate of denitrification is enhanced when seasonal periods of aerobic conditions stimulate the mineralization and nitrification of organic nitrogen, which makes NO_3^- more available for denitrifiers when the water level later rises (Reddy and Patrick 1975, 1976). In continuously flooded soils, nitrate must diffuse downward from aerobic layers supporting nitrification to anaerobic layers supporting denitrification (Patrick and Tusneem 1972). Simultaneously, reduced substances diffuse up from anoxic sediment layers and are reoxidized at the surface (Sweerts et al. 1991). Metabolism can combine unexpected couples of elements in redox reactions; Straub et al (1996) found that microbial reduction of nitrate can be coupled to the oxidation of Fe^{2+} in anaerobic sediments.

Sulfate Reduction

In Chapter 6 we examined the reduction of sulfate that accompanies the uptake, or assimilation, of sulfur by soil microbes and plants. In contrast, *dissimilatory* sulfate reduction in anaerobic soils is analogous to denitrification, in which SO_4^{2-} acts as an alternative electron acceptor during the oxidation of organic matter by bacteria in the genera *Desulfovibrio* and *Desulfotomaculum,* e.g.,

$$2H^+ + SO_4^{2-} + 2(CH_2O) \rightarrow 2CO_2 + H_2S + 2H_2O. \qquad (7.17)$$

These bacteria produce a variety of sulfur gases, including hydrogen sulfide (H_2S), dimethylsulfide [$(CH_3)_2S$)], and carbonyl sulfide (COS) (Cooper et al. 1987, Staubes et al. 1989, Hines et al. 1993, Caron and Kramer 1994). This metabolic pathway evolved at least 2 billion years ago (Chapter 2), and before widespread human industrialization the release of biogenic gases from wetland soils was the dominant source of sulfur gases in the atmosphere (Chapter 13). In an analogous reaction, certain anaerobic bacteria reduce selenium compounds (e.g., SeO_4^{2-}) to Se, which is often toxic to wildlife (Oremland et al. 1989).

The escape of H_2S from wetland soils is often much less than the rate of sulfate reduction at depth as a result of reactions between H_2S and other soil constituents. Brown and MacQueen (1985) found that only 0.3% of the sulfate added to peatland soils was subsequently recovered as H_2S. Hydrogen sulfide can react with Fe^{2+} to precipitate FeS, which gives the characteristic black color to anaerobic soils. FeS may be subsequently converted to pyrite in the reaction

$$FeS + H_2S \rightarrow FeS_2 + 2H^+ + 2e^-. \qquad (7.18)$$

When H_2S diffuses upward through the zone of Fe^{3+}, pyrite (FeS_2) is precipitated following

$$2Fe(OH)_3 + 2H_2S + 2H^+ \rightarrow FeS_2 + 6H_2O + Fe^{2+}. \qquad (7.19)$$

Thus, not all the reduced iron in wetland soils is formed directly by iron-reducing bacteria. In some cases the indirect pathway (Eq. 7.19) may account for most of the total (Canfield 1989a, Jacobson 1994).

A low iron content limits the accumulation of iron sulfides in many wetland sediments (Berner 1984, Rabenhorst and Haering 1989, Giblin et al. 1992); however, the precipitation of iron sulfides is an effective trap for H_2S that might otherwise escape to the atmosphere. During periods of low water, specialized bacteria may reoxidize the iron sulfides (Ghiorse 1984, Jones 1986), releasing SO_4^{2-} that can return by diffusion to the zone of sulfate-reducing bacteria. Thus, high rates of sulfate reduction may be maintained in soils and sediments that have relatively low SO_4^{2-} concentrations, owing to the recycling of sulfur between oxidized and reduced forms (Wieder et al. 1990, Marnette et al. 1992, Urban et al. 1994).

Hydrogen sulfide also reacts with organic matter to form carbon-bonded sulfur that accumulates in peat and lake sediments (Casagrande et al. 1979, Anderson and Schiff 1987). In many cases, the majority of the sulfur in wetland soils is carbon bonded, and only small amounts are found in reduced inorganic forms (H_2S, FeS, and FeS_2; Brown 1985, Wieder and Lang 1988, Krairapanond et al. 1991, Marnette et al. 1992). Apparently carbon-bonded forms—from the original plant debris, from the reaction of H_2S with organic matter, and from the direct immobilization of SO_4 by soil microbes—are relatively stable (Rudd et al. 1986a, Wieder and Lang 1988). Carbon-bonded sulfur accounts for a large fraction of the sulfur in many coals (Casagrande et al. 1977, Altschuler et al. 1983). Organic sediments and coals containing carbon-bonded sulfur that is the result of dissimilatory sulfate reduction show negative values for $\delta^{34}S$ as a result of bacterial discrimination against the rare, heavy isotope $^{34}SO_4^{2-}$ in favor of $^{32}SO_4^{2-}$ during sulfate reduction (Chambers and Trudinger 1979, Hackley and Anderson 1986).

Because H_2S can react with various soil constituents and is oxidized by sulfur bacteria in the overlying sediments and water (Eq. 2.13), many workers once believed that various organic sulfur gases might be the dominant forms escaping from wetland soils. In 1974 Rasmussen was successful in identifying dimethylsulfide as an emission from a temperate pond. However, using new techniques, most investigators now find that H_2S accounts for a large fraction of the emission from wetland soils (Adams et al. 1981, Kelly and Smith 1990). Castro and Dierberg (1987) report a flux of H_2S containing 1 to 110 mg S m^{-2} yr^{-1} from various wetlands in Florida. Nriagu et al. (1987) report a total flux of sulfur gases ranging from 25 to 184 mg m^{-2} yr^{-1} from swamps in Ontario, Canada. The sulfate in rainfall in this region shows a lower $\delta^{34}S$ value during the summer than during the winter (Nriagu et al. 1987). Presumably a portion of the SO_4^{2-} content in this rain is derived from the oxidation of sulfur gases released to the atmosphere by sulfate reduction in these wetlands.

Methanogenesis

The concentration of SO_4^{2-} in most freshwater wetlands is not high, so the zone of sulfate reduction is closely underlain by a zone in which various methanogenic bacteria are active. Methanogenesis can occur via several metabolic pathways (Chapter 2). Methane production in freshwater environments can occur by acetate splitting,

$$CH_3COOH \rightarrow CO_2 + CH_4, \tag{7.20}$$

which produces a $\delta^{13}C$ of -50 to $-65‰$ in CH_4 (Woltemate et al. 1984, Whiticar et al. 1986, Cicerone and Oremland 1988). Acetate-type compounds are produced from cellulose by fermenting bacteria that coexist at the same depths (e.g., Eq. 7.15). Methane is also produced by CO_2 reduction,

$$CO_2 + 4H_2 \rightarrow CH_4 + 2H_2O, \tag{7.21}$$

in which the hydrogen is available as a by-product of fermentation (Eq. 7.15). In this reaction, the CO_2 is found as HCO_3^-, which serves as an electron acceptor in a role analogous to NO_3^- and SO_4^{2-} in denitrification and sulfate reduction. This reaction is also known as dissimilatory CO_2 reduction. Methanogenesis by CO_2 reduction accounts for the limited release of H_2 from wetland soils (Schütz et al. 1988). This methane is highly depleted in ^{13}C, with $\delta^{13}C$ of -60 to $-100‰$ (Whiticar et al. 1986, Lansdown et al. 1992).

Methanogenic bacteria can use only certain organic substrates for acetate splitting, and in many cases there is evidence that sulfate-reducing bacteria are more effective competitors for the same compounds (Schönheit et al. 1982). Sulfate-reducing bacteria also use H_2 as a source of electrons, and they are more efficient in the uptake of H_2 than methanogens engaging in CO_2 reduction (Kristjansson and Schönheit 1983, Achtnich et al. 1995). Thus, in most environments there is little or no overlap between the zone of methanogenesis and the zone of sulfate reduction in sediments. Methanogenesis begins when sulfate is depleted (Lovley and Klug 1986, Kuivila et al. 1989). In a highly productive lake in the Netherlands, Sinke et al. (1992) found that the rate of delivery of organic matter to the sediments during the summer stimulated SO_4 reduction, which depleted SO_4 in the porewaters, allowing methanogens to become active. During other seasons, anaerobic respiration of dead organic matter was completed within the zone of SO_4 reduction, and there was little CH_4 produced. Methanogenesis in marine sediments is inhibited by the high concentrations of SO_4 in seawater, and where methanogenesis occurs in marine environments CO_2 reduction is much more important than acetate splitting, because acetate is entirely depleted within the zone of sulfate-reducing bacteria (Chapter 9).

The flux of methane from wetland soils increases as a function of the height of the water table (Moore and Knowles 1989, Sebacher et al. 1986, Shannon and White 1994), and when soils are flooded the flux increases with soil temperature (Bartlett and Harriss 1993, Roulet et al. 1992). Methanogenesis is also limited by the supply of labile organic matter (Bridgham and Richardson 1992, Cicerone et al. 1992, Valentine et al. 1994, Denier van der Gon and Neue 1995), and CH_4 flux shows a direct correlation to net ecosystem production across a variety of wetland ecosystems (Fig. 7.5). About 3% of the net ecosystem production (Chapter 5) in wetlands escapes to the atmosphere as CH_4.

A variety of bacteria consume methane—methanotrophy (King 1992)—so at any time, the loss of CH_4 from the surface of soils or sediments is determined by the balance between methane production at depth and methane oxidation as it diffuses up through zones of higher redox potential (Fig. 7.6). In some marine sediments, anaerobic methane oxidation is also performed by sulfate-reducing bacteria that use CH_4 as a source of reduced carbon (Reeburgh 1983, Henrichs and Reeburgh 1987, Blair and Aller 1995), viz.,

$$CH_4 + SO_4^{2-} \rightarrow HS^- + HCO_3^- + H_2O. \tag{7.22}$$

In soils, aerobic methanotrophic bacteria can outcompete nitrifying bacteria for O_2 where methane is abundant (Megraw and Knowles 1987). Recent

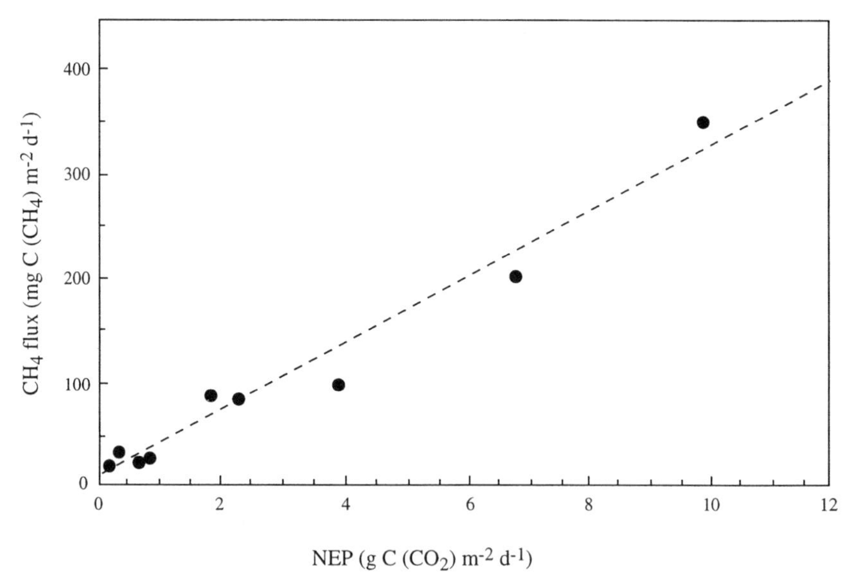

Figure 7.5 Methane flux from the surface of wetland ecosystems as a function of net ecosystem production (NEP), measured as daily net carbon gain by vegetation and soil. Modified from Whiting and Chanton (1993).

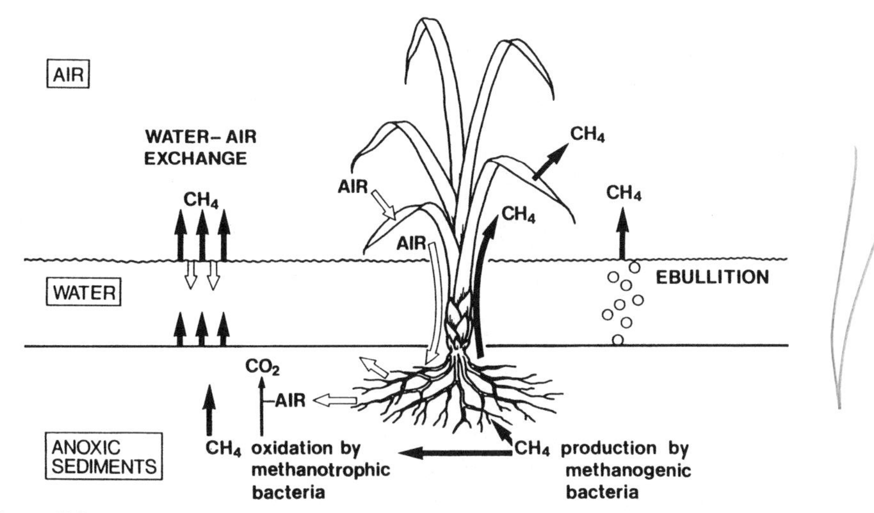

Figure 7.6 Processes of methane production, oxidation, and escape from wetland soils. From Schütz et al. (1991).

studies have employed inhibitors of methane oxidation (e.g., picolinic acid and methylfluoride) to study this process (Bédard and Knowles 1989, Oremland and Culbertson 1992). Methane-oxidizing bacteria also alter the $\delta^{13}C$ of CH_4 escaping to the atmosphere, and comparisons of the isotopic ratio of CH_4 in sediments and surface collections indicate the importance of oxidation (Happell et al. 1993).

In the Florida Everglades, more than 90% of methane production is consumed by methanotrophs before it diffuses to the atmosphere (King et al. 1990). Reeburgh et al. (1993) have estimated that the global rate of methane production in wetlands is about 20% larger than the net release of methane from wetland soils. In Lake Washington, methane oxidation in the upper sediment consumes about half of the methane generated in the deeper sediments, and most of the remainder is oxidized in the water column. About 2% of the carbon entering the lake in net primary production is returned to the atmosphere as CH_4 (Kuivila et al. 1988). In other lakes, oxidation in the water column is thought to dominate the consumption of methane (Rudd and Taylor 1980), and in highly stratified lakes, large concentrations of methane can accumulate in anoxic bottom waters (Tietze et al. 1980).

The loss of methane to the atmosphere is highest when the concentration of methane dissolved in the soil or sediment porewaters exceeds the hydrostatic pressure of the overlying water, allowing bubbles of gas to form. These bubbles may escape to the surface in the process of *ebullition* (Fig. 7.6), which can account for a large fraction of the methane flux to the atmosphere (Devol et al. 1988, J.O. Wilson et al. 1989, Lansdown et al. 1992). When wetland soils and shallow lakes are colonized by vegetation, the plants may act as conduits for the escape of methane to the atmosphere (Dacey 1981, Sebacher et al. 1985, Yavitt and Knapp 1995). The process is enhanced by the tendency for many of these species, including rice, to have hollow stems composed of aerenchymous tissue, which allows O_2 to reach the roots and inadvertently acts as a conduit for CH_4 transport to the surface (Kludze et al. 1993). Bubbles escaping by ebullition may be nearly pure CH_4, whereas bubbles emerging from vegetation are often diluted with N_2 from the atmosphere (Fig. 7.7).

The flux of methane from wetlands shows great spatial variability as a result of differences in soil properties, topography, and vegetation (Pulliam 1993, Bubier et al. 1993, Bartlett et al. 1989, Sass et al. 1994). Harriss et al. (1982) found that methane flux from the Great Dismal Swamp (Virginia) was 1–20 mg CH_4 m^{-2} day^{-1} during the wet season, but the swamp became a sink for atmospheric methane during the dry season, when methane oxidizers were active. Typical values for methane flux range from 10 to 500 mg CH_4 m^{-2} day^{-1} (Bartlett and Harriss 1993, Fig. 7.5). Net regional losses from wetlands are partially balanced by the consumption of atmospheric methane in adjacent upland soils, where it is consumed by methane-oxidizing bacteria (Whalen et al. 1991, Delmas et al. 1992a; see also Chapter 11).

The flux of methane from wetlands accounts for a large portion of the total global flux to the atmosphere (Chapter 11). With the atmospheric concentration increasing at ca. 1%/yr, various workers have asked whether

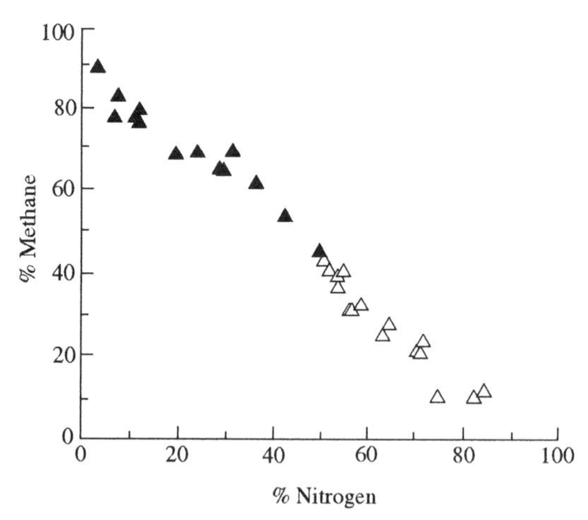

Figure 7.7 Relationship between the N_2 and the CH_4 content of bubbles emitted from the surface of vegetated (open triangles) and bare (closed triangles) wetland soils. From Chanton and Dacey (1991).

changes in ecosystem processes or the area of wetlands might be responsible. Certainly, global methanogenesis has increased with the increasing cultivation of rice, which now accounts for half of the global production of methane from wetlands (Aselmann and Crutzen 1989, Schütz et al. 1989, Matthews et al. 1991, Anastasi et al. 1992). In the future, methane flux may also increase as rising atmospheric CO_2 stimulates the growth of wetland plants, providing a greater supply of organic substances for methane-producing bacteria (Dacey et al. 1994, Hutchin et al. 1995). Changes in the flux of methane to the atmosphere may occur if global warming causes changes in the saturation of peatlands, particularly at northern latitudes. If these soils become warm and dry, the flux of methane may be lower, while the flux of CO_2 to the atmosphere may increase (Whalen and Reeburgh 1990, Freeman et al. 1993, Funk et al. 1994, Moore and Dalva 1993). Methane flux may also be lower in areas impacted by acid rain, where an added deposition of SO_4^{2-} from the atmosphere allows a greater rate of sulfate reduction in the sediments (Wieder et al. 1990, Nedwell and Watson 1995).

Groundwater Environments

Upland ecosystems are underlain by a zone in which the pores of the soil or fractured rock are filled with water—known as groundwater. Water percolating through the zone of unsaturated soils—known as the *vadose* zone—carries dissolved constituents, including organic compounds, to groundwater, where they decompose in anaerobic reactions (Wood and Petraitis 1984, Murphy et al. 1989). Denitrification in groundwater is driven by the low redox potentials and high concentrations of NO_3^- in many aquifers (Korom 1992, Spalding and Parrott 1994). Denitrifying bacteria have been found at a depth of approximately 300 m in coastal South Carolina (Francis et al. 1989), and groundwater denitrification may be a significant source of N_2O in the atmosphere (Ronen et al. 1988). Fe reduction, SO_4 reduction, and methanogenesis are also found in groundwaters (Lovley et al. 1990, Jones et al. 1989). Robertson and Schiff (1994) found that SO_4^{2-} in percolating soil waters was entirely removed by sulfate-reducing bacteria at 15-m depth. Methanogenesis by CO_2 reduction occurs at depths of >100m in the groundwaters of central Iowa (Simpkins and Parkin 1993) and Ontario (Aravena et al. 1995). The rate of microbial activity in groundwater is slow (Chapelle and Lovely 1990, Phelps et al. 1994), but over long periods it can contribute to the weathering of rocks at great depth.

Biomethylations

Microbial reactions in wetland sediments are responsible for the methylation of a wide variety of metals, some of which are toxic to biota and more rapidly assimilated in the methyl form (Craig 1980). We focus on mercury

(Hg), which shows increasing concentration in the global atmosphere (Slemr et al. 1985, Slemr and Langer 1992), apparently due to human industrial activities (Mason et al. 1994). The atmospheric budget is dominated by elemental Hg, which is oxidized to Hg^{2+} and deposited in rainfall (Lindqvist and Rodhe 1985). Mercury deposited in the Greenland ice cap and in lake sediments has increased dramatically in recent years, reflecting an increase in the global emission of Hg by humans (Weiss et al. 1971, Swain et al. 1992).

In wetlands, Hg^{2+} is reduced to Hg, which may return to the atmosphere. Genes for the bacterial reduction of Hg^{2+} are activated by high concentrations of Hg^{2+} in the environment (Nazaret et al. 1994). Hg^{2+} is also converted to methylmercury (CH_3Hg^+), which is the primary form that accumulates with age in fish and other aquatic animals (Bache et al. 1971, Cross et al. 1973, Cabana and Rasmussen 1994). The methylation occurs in a variety of bacteria, especially methanogens, and it usually involves vitamin B-12 (methyl cobalamin) to transfer a reduced methyl group (CH_3^-) to Hg^{2+} (Ridley et al. 1977). The reaction occurs more rapidly under anoxic conditions and at low pH (Miskimmin et al. 1992, Regnell 1994), so it is not surprising that wetlands are the dominant source of methylmercury in the environment (e.g., St. Louis et al. 1994).

Biogeochemistry of "Terrestrial" Wetlands

Swamps, marshes, and bogs are "terrestrial" wetlands, which are important wildlife habitat. Biogeochemical processes in wetland ecosystems are mediated by soils with low redox potential. These areas are often found at the interface between upland and lake ecosystems (Fig. 7.8), and the nutrients received from the adjacent landscapes are often transformed during their passage through wetlands (e.g., Hooper and Morris 1982, Emmett et al. 1994). For example, Devito and Dillon (1993) found that a swamp in Ontario was a sink for NH_4, NO_3, and PO_4, which were converted to organic forms. As seen for many upland ecosystems (Chapter 6), the inputs and outputs of Cl were nearly balanced in this ecosystem, reflecting little involvement of Cl in biogeochemical reactions.

Net primary productivity in wetland ecosystems varies widely, depending upon nutrient supply (Brinson et al. 1981). Because emergent plants dominate the vegetation of these wetlands, net primary production is usually measured using the harvest approaches outlined briefly in Chapter 5. Net primary production is directly related to nutrient inputs in many swamp forests (Brown 1981), and productivity is highest in wetlands that are seasonally dry, allowing for periods of rapid nutrient turnover when soils are exposed to oxygen (Fig. 7.9). In contrast, bogs that receive little or no nutrient input from runoff usually have very low productivity (Schlesinger 1978).

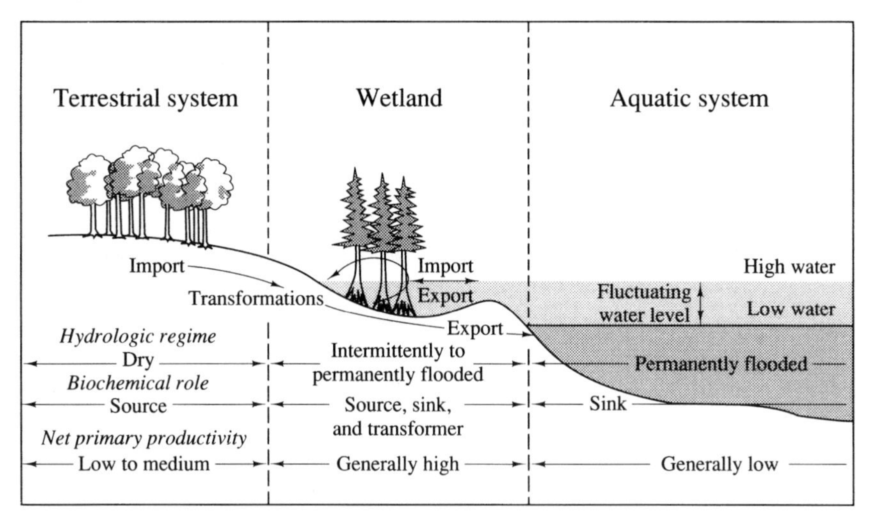

Figure 7.8 A schematic diagram showing the position of wetlands in relation to upland and aquatic ecosystems and the biogeochemical links between these landscape components. From Mitsch and Gosselink (1986).

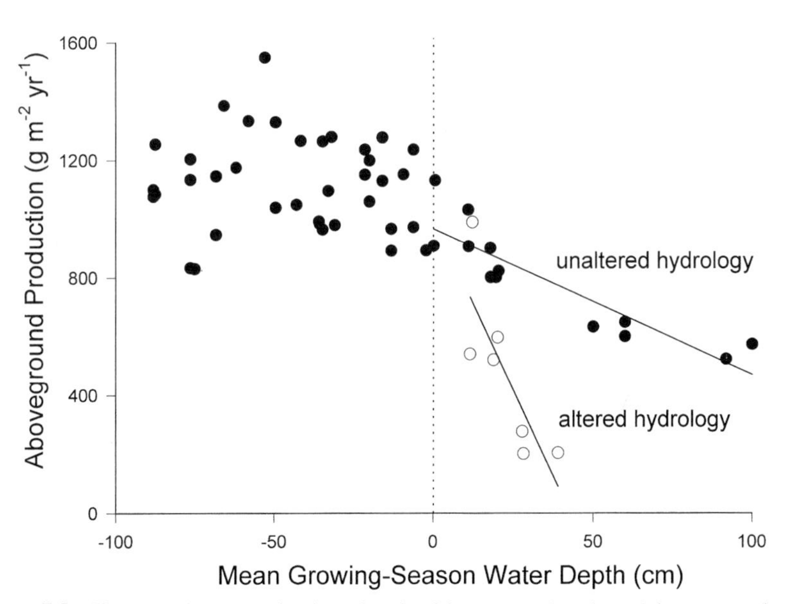

Figure 7.9 The net primary production of wetland forests as a function of the average depth of flooding during the growing season. Modified from Megonigal et al. (1996).

As we have seen, decomposition is impeded in flooded and saturated soils, so freshwater wetlands show large accumulations of soil organic matter (Table 7.2). In bogs, the rate of peat accumulation is determined by the rate of decomposition in the aerobic upper levels (the acrotelm) and in the lower levels (the catotelm) of the deposit. Clymo (1984) proposed a model for peat accumulation, which predicts that peatlands will eventually attain a steady state when the input of detritus from primary production at the peat surface is balanced by the loss of organic matter by decomposition throughout the peat profile. The saturated soils of tundra and boreal forest regions contain about 24% of the total storage of organic matter in soils of the world (Table 5.3). Many of these ecosystems have accumulated soil carbon since the retreat of the last continental glacier (Harden et al. 1992). If these areas are subject to drainage and warmer climatic conditions, the rate of carbon storage will decline, decomposition will increase, and wetlands could become a significant source of CO_2 for the atmosphere (Tate

Table 7.2 Rate of Carbon Accumulation in Some Peatland Ecosystems[a]

Location	Vegetation	Accumulation interval (years)	Accumulation rate[b] (g C m^2/yr)	Reference
Alaska	*Picea* and *Sphagnum*	4790	11–61	Billings (1987)
Alaska	*Eriophorum vaginatum*	7000	26.6	Viereck (1966)
Finland	*Sphagnum-Carex* mire	—	20–28	Francez and Vasander (1995)
Former Soviet Union	Mires, bogs, and fens	3000–7000	12–80	Botch et al. (1995)
Manitoba	*Picea* and *Sphagnum*	2960–7939	13–26	Reader and Stewart (1972)
Ontario	Sphagnum bog	5300	30–32	Belyea and Warner (1996)
Western Canada	*Sphagnum* bogs	9000	13.6–34.9	Kuhry and Vitt (1996)
Wisconsin	*Sphagnum*	8260	17–38	Kratz and DeWitt (1986)
Massachusetts	*Sphagnum*	132	90	Hemond (1980)
North Carolina	Mixed forest	27,700	8	Whitehead (1981)
Georgia	*Taxodium*	6500	22.5	Cohen (1974)
Florida	*Cladium* swamp	25–30	70–105[c]	Craft and Richardson (1993)

[a] Compare to upland ecosystems in Table 5.4.

[b] When data are incomplete, calculated rate assumes a bulk density of 0.1 g/cm^3 and a carbon content of 50%.

[c] Unimpacted sites.

1980, Hutchinson 1980, Armentano and Menges 1986, Gorham 1991, Silvola et al. 1996).

Depending upon landscape position, net primary production in wetlands is often limited by phosphorus or nitrogen. For instance, many bogs receive little or no runoff from the surrounding land, so it is not surprising that phosphorus is in short supply for plant growth and decomposition (Chapin et al. 1978, Damman 1988). By stimulating net primary production of sawgrass, phosphorus inputs to the Florida Everglades tend to increase the rate of peat accretion, despite simultaneously increasing the rate of carbon mineralization at depth (Craft and Richardson 1993, Amador and Jones 1995). As seen for upland terrestrial ecosystems in Chapter 6, the nutrient cycle of bogs is characterized by large nutrient storages in vegetation and peat and relatively low concentrations of available nutrients in the soil. In the tundra of Alaska, Chapin et al. (1978) found that the pool of phosphorus in soil organic matter contained 64% of the total phosphorus in the ecosystem and had a mean residence time of 220 years, while phosphorus available in the soil solution composed 0.3% of the total with a residence time of 10 hr. Based on changes in the concentrations of nitrogen and phosphorus with depth, Damman (1978) suggests that a significant portion of the phosphorus content of peat is mineralized before burial.

In addition to phosphorus, many peatland systems also show shortages of plant-available nitrogen. Nitrogen limits the growth of tundra vegetation, and in a multiple fertilization experiment Shaver et al. (1986) found that the response of *Eriophorum vaginatum* in tussock tundra was greater for N than for P. Low temperatures limit nitrogen mineralization in the tundra (Marion and Black 1987), and as a result of slow decomposition many bogs show a net accumulation of nitrogen in peat (Hemond 1983, Damman 1988, Urban and Eisenreich 1988). Lieffers and MacDonald (1990) found that the nitrogen content of spruce trees in boreal peatlands was higher in areas where drainage diversion has allowed the soils to dry, increasing soil nutrient turnover. Many bog ecosystems show significant amounts of nitrogen fixation (Waughman and Bellamy 1980, Schwintzer 1983, Barsdate and Alexander 1975), which is likely to be in excess of denitrification under field conditions (Bowden 1987, Koerselman et al. 1989). Denitrification is limited by low available nitrate, which is derived from nitrification, requiring oxic conditions (Merrill and Zak 1992, Seitzinger 1994).

Peatland ecosystems that receive drainage from the surrounding uplands (fens) and forests that receive seasonal floodwaters often have relatively high inputs of phosphorus and other elements derived from rock weathering (Mitsch et al. 1979, Waughman 1980, Frangi and Lugo 1985). In these ecosystems, phosphorus and sulfur are retained on iron and aluminum minerals that are constituents of soil organic matter and peat (Richardson 1985, Mowbray and Schlesinger 1988). With greater inputs of nutrient elements from land, net primary production in these systems is more likely to be limited by N than P (e.g., Tilton 1978). Wetlands in low topographic positions are likely to function as effective nutrient sinks (e.g., Verry and Timmons 1982, Urban et al. 1989, Walbridge and Lockaby 1994). Peatlands exposed to decomposition and erosion can also be sources of nutrients to aquatic ecosystems receiving their runoff (e.g., Crisp 1966).

Primary Production and Nutrient Cycling in Lakes

The physical properties of water exert a significant control on net primary productivity and nutrient cycling in lakes. Sunlight warms the surface waters, but light energy is rapidly attenuated by depth. Because fresh water shows its greatest density (1.0 g/cm^3) at 4°C, a stratification of water develops in deep lakes, with warmer surface waters known as the *epilimnion* overlying cooler, deep waters known as the *hypolimnion*. The zone of rapid temperature change is known as the thermocline or metalimnion. Many tropical lakes show permanent stratification (e.g., Kling 1988). In temperate regions, the temperature stratification breaks down seasonally, and lake waters may circulate freely from top to bottom in spring and autumn.

During summer stratification, phytoplankton, the free-floating algae that contribute most of the net production, are confined to the surface layers that contain only a small portion of the total nutrient content of a lake. During these periods, net primary production in the epilimnion depends on direct nutrient inputs to the surface waters (Schindler 1978) and on nutrient regeneration in the epilimnion (Fee et al. 1994). Typically, the surface waters have high redox potential, so organic materials decompose rapidly by aerobic respiration. However, low quantities of available nutrients are found in the epilimnion, as a result of rapid uptake by phytoplankton. During the period of stratification, some dead organic materials sink to the hypolimnion, where their decay leads to the depletion of oxygen, low redox potentials, and greater nutrient availability.

Unlike terrestrial plants, phytoplankton are not bathed in an atmosphere with CO_2. Carbon dioxide dissolves in lake waters according to equilibrium reactions that depend on pH:

$$CO_2 + H_2O \rightleftarrows H^+ + HCO_3^- \rightleftarrows 2H^+ + CO_3^{2-}. \qquad (7.23)$$

At pH <4.3, most carbon dioxide is found as a dissolved gas, between 4.3 and 8.3 as bicarbonate, and >8.3 as carbonate. Together these forms are known as dissolved inorganic carbon (DIC) and abbreviated ΣCO_2. The rate of dissolution of CO_2 in water and the subsequent availability of CO_2 or other forms of inorganic carbon are potential constraints on primary production in lakes.

Net Primary Production

Methods for assessing the net primary production of phytoplankton necessarily must differ from the harvest methods that are used in studies of land vegetation (Chapter 5). Two approaches are common. In the first method, small samples of lake water are confined in glass bottles, clear and opaque, that are resuspended in the water column. After a period of incubation, the O_2 content of the water is measured. An increase in O_2 in the clear bottle is taken as the equivalent of net primary production—i.e., photosynthesis in excess of respiration by the plankton. Net primary production is calculated by assuming a molar equivalent between O_2 production and carbon fixation (Eq. 5.2). Over the same period of incubation, a decrease in O_2 in the dark bottle is taken to be the result of plant respiration. The sum of changes in the light and dark bottles is an estimate of gross primary production.

Many recent studies use variations of the ^{14}C method to measure primary production in freshwaters. This method also uses clear bottles, which are inoculated with DIC containing ^{14}C in a form that is available for phytoplankton. Because the pH of most surface waters lies in the range of 4.3 to 8.3, $NaH^{14}CO_3$ is a frequent choice. The bottles are resuspended in the water column, and during the incubation period photosynthesis is assumed to convert the inorganic ^{14}C to organic forms that accumulate in phytoplankton cells. The bottles are then retrieved and the water filtered. Radiocarbon that is retained on the filter is counted using a scintillation counter and assumed to represent net primary production by the phytoplankton community.

The O_2 and ^{14}C methods have been reviewed exhaustively by Peterson (1980), who examines a number of the sources of error in both methods. The oxygen method is relatively easy and inexpensive to apply to many situations, but it suffers from a number of problems that are created by the artificial environment of the bottles. The bottles contain planktonic bacteria and zooplankton that add to the respiration contributed by the phytoplankton. The sensitivity of most O_2 measurements is relatively low, so long incubations are necessary so that some change in O_2 concentration can be measured. During the incubation, nutrients may be depleted in the bottle, lowering the apparent rate of photosynthesis. Also, during long incubations, increasing O_2 concentrations may increase the rate of photorespiration by phytoplankton.

The artificial nature of the environment confined in bottles also affects the ^{14}C method, but because the technique is more sensitive, the incubations are shorter. A more serious problem stems from the loss of soluble products of photosynthesis (e.g., sugars and amino acids) that are excreted from phytoplankters and lost during filtration. These forms of dissolved organic carbon (DOC) are a component of net primary production. Recently, very small phytoplankton, known as picoplankton, have been found in seawater and some freshwaters (Stockner and Antia 1986). These may also pass through the filters, escaping detection by the ^{14}C method. In practice, the O_2 method often gives higher values for production, particularly when the methods are compared in relatively unproductive environments. Neither method is without potential error, and the best studies often use both approaches.

A compilation of the data from many studies, which showed greater lake productivity in tropical than in temperate or boreal regions, led Brylinsky and Mann (1973) to suggest that available sunlight might control the level of net primary production in lakes. In a similar analysis of a larger data set, Schindler (1978) found no correlation with annual irradiance, but a strong relationship between lake production and the total concentration of nutrients, especially phosphorus (Fig. 7.10). Alternatively, in an evaluation of the pollution impact of phosphorus detergents, a number of workers suggested that lake productivity was limited by the rate at which atmospheric CO_2 could dissolve in surface waters. Subsequent field studies have failed to confirm a CO_2 limitation, except in unusual circumstances of high nutrient availability (Schindler et al. 1972). Søballe and Kimmel (1987) confirmed the importance of phosphorus in a comparison of productivity among 345 rivers and 812 lakes and reservoirs of the United States, and they noted that algal cell density was directly correlated to the availability of phosphorus and the residence or turnover time of the waters. The evidence for a phosphorus limitation of net primary productivity in lakes now appears overwhelming (Vollenweider et al. 1974, Dillon and Rigler 1974, Oglesby 1977). Phosphorus concentration in the epilimnion is directly related to the total chlorophyll content of the water column, which is directly related to net primary productivity (Schindler 1978, Baines and Pace 1994).

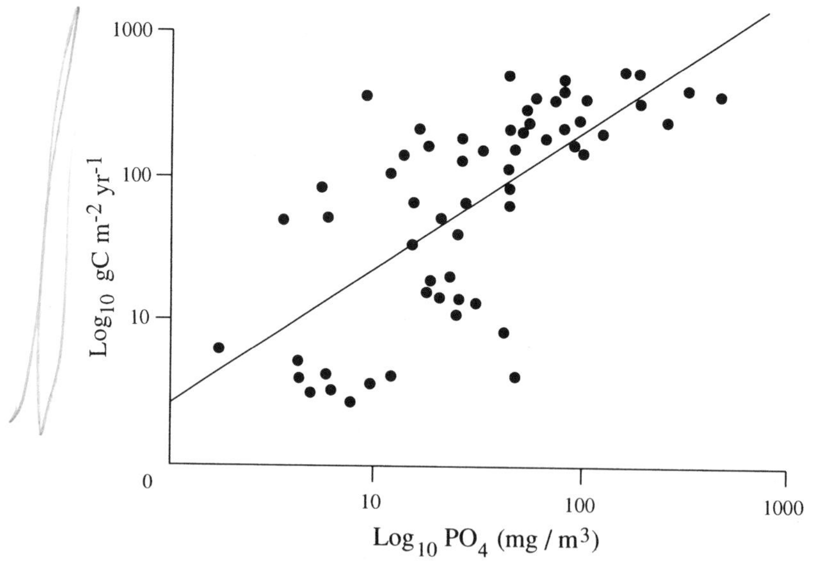

Figure 7.10 Relationship between net primary production and the phosphorus concentration of lakes of the world. From Schindler (1978).

Phosphorus and Nitrogen in Lake Waters

Under most natural conditions the phosphorus inputs to lake ecosystems are relatively small (Ahl 1988). There is little phosphorus in precipitation, and phosphorus is effectively retained in terrestrial watersheds by vegetation (Chapter 6) and by chemical interactions with soil minerals (Chapter 4). Much of the phosphorus entering lakes is carried with soil minerals, which are rapidly sedimented (Sonzogni et al. 1982, Froelich 1988). Available phosphorus may also precipitate with Fe and Mn minerals that are insoluble at the high redox potentials of the surface waters (Figs. 4.4 and 7.2; Mortimer 1941, 1942).

Analyses of lake water typically show that a large proportion of the phosphorus is contained in the plankton biomass and only a small portion is found in available form (Lean 1973). Uptake of phosphorus by phytoplankton is an active process that shows a curvilinear relationship to increasing P concentration, as seen for the roots of land plants (Fig. 6.1; Jansson 1993). Continued net primary production by phytoplankton depends on the rapid cycling of phosphorus between dissolved (e.g., HPO_4^{2-}) and organic forms in the epilimnion (Fee et al. 1994).

Studies of phosphorus cycling have shown that the turnover of phosphorus in the epilimnion is dominated by bacterial decomposition of organic material (Bloesch et al. 1977, Whalen and Cornwell 1985, Levine et al. 1986, Conley et al. 1988). Phytoplankton and bacteria excrete extracellular phosphatases to aid in the mineralization of P (Stewart and Wetzel 1982), and planktonic bacteria may immobilize phosphorus when the C/P ratio of their substrate is high (Vadstein et al. 1993, Caron 1994). Globally, the N/P ratio of freshwater phytoplankton ranges from 20 to 40 (Hecky et al. 1993, Elser and Hassett 1994), and net phosphorus mineralization begins at N/P <16 (Tezuka 1990). Immobilization of N is less common, because the C/N ratio of phytoplankton (8–20) is close to that of bacterial biomass (Tezuka 1990, Hecky et al. 1993, Downing and McCauley 1992, Elser et al. 1995). Nutrient turnover in lakes is enhanced by the activities of larger grazing organisms, which may also be limited by P (Porter 1976, Lehman 1980, Carpenter et al. 1987, Elser and Hassett 1994, Sterner and Hessen 1994).

During a period of stratification, the phosphorus pool in the surface waters is progressively depleted as phytoplankton and other organisms die and sink to the hypolimnion (Levine et al. 1986). Baines and Pace (1994) found that 10 to 50% of the NPP was exported to the deeper waters in 12 lakes of the eastern United States, with a tendency for a greater fractional export in lakes of lower productivity (Fig. 7.11; Cole and Pace 1995). When fecal pellets and dead organisms sink through the thermocline, phosphorus remineralization continues in the lower water column and sediments (Lehman 1988, Gächter et al. 1988, Carignan and Lean 1991). Anoxic hypolim-

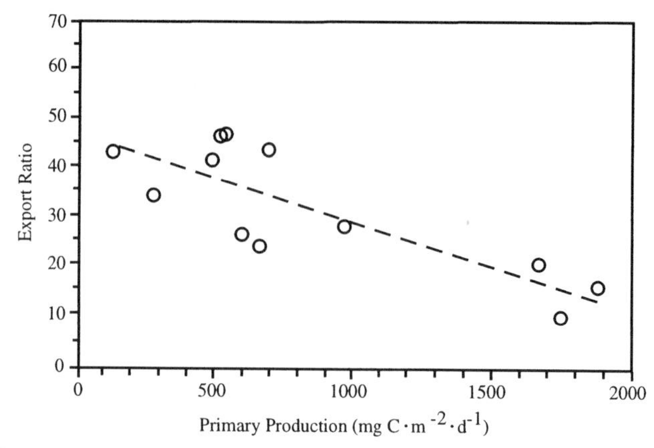

Figure 7.11 The percentage of planktonic primary production that sinks to the hypolimnion (export ratio) in lakes as a function of their net primary productivity. Modified from Baines and Pace (1994).

netic waters often show high concentrations of P, which is returned to the surface during periods of seasonal mixing. Of course, the turnover of phosphorus through the biotic community is incomplete, so some phosphorus is permanently lost to the sediments.

As long as the hypolimnetic waters contain oxygen, a layer of Fe-oxide minerals at the sediment–water interface traps phosphorus that diffuses upward from bacterial decomposition in the sediments or from the dissolution of Fe minerals at lower redox potential at depth. However, when the hypolimnetic waters are anoxic, the Fe minerals are reduced, dissolve, and release P to the overlying waters (Mortimer 1941, 1942, Boström et al. 1982, 1988, Murray 1995). Interactions between elements may be important in determining the release of P from sediments. In most freshwaters, the concentration of SO_4 is low, and P is strongly adsorbed by Fe minerals in the sediment. In the sea, concentrations of SO_4 are higher, and P limitations are less apparent (Chapter 9). Increasing concentrations of SO_4 in lakes affected by acid rain may act through the anion exchange reactions (Eq. 4.13) to drive P into solution, enhancing lake productivity (Caraco et al. 1989, 1993).

In many cases, the dissolution of Fe minerals is limited, and there is little regeneration of phosphorus from sediments (Davison et al. 1982, Levine et al. 1986, Caraco et al. 1990, Golterman 1995). Sedimentary accumulations of undecomposed organic matter and Fe minerals contain P that is lost from the ecosystem (Cross and Rigler 1983).

Despite the limited availability of phosphorus in surface waters, we might expect that, as for land vegetation, processes such as denitrification might make nitrogen the nutrient that is in shortest supply in lakes. Although Goldman (1988) has shown that the primary productivity of some lakes is limited by nitrogen, Schindler (1977) suggests why this is not generally the case. When phytoplankton communities grow in limited supplies of nitrogen, there is a shift in algal dominance from green algae to blue-green algae—which fix nitrogen, adding to its availability and raising the N/P ratio (cf. Fig. 6.3). Among 17 lakes, Smith (1983) found that blue-green algae were common only at N/P ratios <29, and Howarth et al. (1988b) found that significant nitrogen fixation by phytoplankton (blue-green algae) occurred when the N/P ratio was <16. When phosphorus is added as a pollutant, the algal community shifts to species of blue-green algae and total net primary productivity increases. Under such conditions, nitrogen fixation can supply up to 82% of the nitrogen input to the phytoplankton community (Howarth et al. 1988a). When the input of phosphorus ceases, blue-green algae decrease in importance (Edmondson and Lehman 1981, Takamura et al. 1992). Although these shifts in community dominance are not found in all situations (see Canfield et al. 1989), in most cases, the inputs of nitrogen by blue-green algae tend to maintain a phosphorus shortage for the growth and photosynthesis of phytoplankton (Smith 1982). There is no equivalent biogeochemical process that can increase the supply of phosphorus in lakes when it is in short supply.

Other Nutrients

Changes in the dominance of various species of algae are also seen in response to differing availabilities of other nutrient elements. When phosphorus is added to nutrient-poor lakes, the growth of diatoms, which require silicon, may reduce the supply of silicon to low levels, favoring the dominance of other species, such as green algae (Kilham 1971, Schelske et al. 1983, Schelske 1988, Conley et al. 1993). Unlike P, most of the Si in sedimentary organic matter is regenerated to the water column (Schelske 1985, Conley et al. 1988). In many cases, inputs of Si from groundwater link the biogeochemistry of lake ecosystems to processes in the surrounding watershed (Hurley et al. 1985). Titman (1976) showed that subtle differences in the ratio of silicon to phosphorus controlled the dominance shared by two species of diatoms, *Asterionella* and *Cyclotella*. Other studies have shown that net primary production is affected by changes in trace micronutrients, such as B (Subba Rao 1981), Fe (Allen 1972), and Cu (Horne and Goldman 1974). Such changes in the phytoplankton community of lakes are perhaps the best examples of how subtle shifts in the biogeochemistry of the environment can alter the distribution and abundance of species and the productivity of a natural ecosystem.

Lake Budgets

Carbon

Phytoplankton show relatively constant ratios between carbon content and the content of important nutrient elements, such as N and P, so studies of the production and fate of organic carbon are useful in understanding the overall biogeochemistry of lakes. Rich and Wetzel (1978) present a carbon budget for Lawrence Lake, located in southern Michigan (Table 7.3). Net primary production within the ecosystem is known as *autochthonous* production. In this shallow lake, rooted plants contribute 51.3% of the autochthonous net primary productivity, while phytoplankton account for 25.4%. In contrast, Jordan and Likens (1975) report that phytoplankton account for nearly 90% of the production in Mirror Lake, a relatively unproductive lake in New Hampshire with a limited area of shallow water. For Lawrence Lake, DOC that is lost from the rooted plants accounts for 10.6% of the annual input of organic carbon to the lake.

Table 7.3 Origins and Fates of Organic Carbon in Lawrence Lake, Michigan[a]

	g C m^{-2} yr^{-1}	%	% (total)
Net primary productivity (NPP)			
POC			
Phytoplankton	43.3	25.4%	
Epiphytic algae	37.9	22.1%	
Epipelic algae	2.0	1.2%	
Macrophytes	87.9	51.3%	
Total	171.2	100.0%	
DOC			
Littoral	5.5		
Pelagic	14.7		
Total	20.2		
Total NPP	191.4		88.4%
Imports			
POC	4.1	16.3%	
DOC	21.0	83.7%	
Total imports	25.1	100%	11.6%
Total available organic inputs	216.5		100.0%
Respiration			
Benthic	117.5	73.6%	
Water column	42.2	26.4%	
Total respiration	159.7	100.0%	74.2%
Sedimentation	16.8		7.8%
Exports			
POC	2.8	7.3%	
DOC	35.8	92.7%	
Total exports	38.6	100.0%	18.0%
Total removal of carbon	215.1		100.0%

[a] From Rich and Wetzel (1978).

Organic carbon is also derived from streams entering the lake. The inputs of organic carbon from land are known as *allochthonous* production, which is an important source of respirable carbon in lakes (del Giorgio and Peters 1994). Indeed, total respiration exceeds autochthonous production in many unproductive lakes (Fig. 7.12), and lake waters are often supersaturated with CO_2 with respect to the atmosphere (Cole et al. 1994, Quay et al. 1995). Stable isotope ratios, i.e., $\delta^{13}C$, in sediments have been used to estimate the comparative contribution of organic carbon from autochthonous phytoplankton production compared to allochthonous inputs from land vegetation (LaZerte 1983).

Nearly three-fourths of the organic carbon entering Lawrence Lake is respired in the lake, with benthic respiration composing 73.6% of total respiration. In other lakes as much as half of the respiration is due to bacterial decomposition in the water column (e.g., Lehman 1988, Carignan and Lean 1991). Bacterial growth and respiration have proven difficult to study, but new techniques involving the incorporation of ^3H-thymidine into bacterial DNA seems to offer an accurate measurement of heterotrophic bacterial growth in fresh- and seawater (Fuhrman and Azam 1982, Riemann and Bell 1990, Servais 1995).

Only 7.8% of the organic carbon in Lawrence Lake is permanently stored in the sediments composing the net ecosystem production of this ecosystem. The sediment storage of 16.8 g C m^{-2} yr^{-1} is similar to the rate of accumulation of soil organic matter in many terrestrial ecosystems (Table 5.4), but it is derived from a much lower primary production than is typical on land. The greater percentage of net primary production that is stored in aquatic ecosystems speaks for the relative inefficiency of bacterial respiration, often under anaerobic conditions, compared to the importance of aerobic and

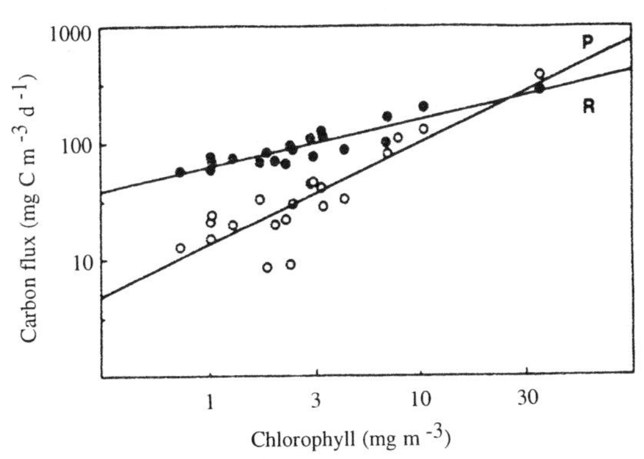

Figure 7.12 Mean summertime plankton respiration (R) and photosynthesis (P) in the surface waters of lakes as a function of chlorophyll concentration—an index of overall lake productivity. From del Giorgio and Peters (1994).

eukaryotic decomposers (fungi) and the associated microfauna in soils (Gale and Gilmour 1988, Benner et al. 1986, Lee 1992).

Examining several lakes, Hutchinson (1938) suggested that the rate of depletion of O_2 in the hypolimnion during seasonal stratification was related to the productivity of the overlying waters. Highly productive waters should contribute large quantities of organic carbon for respiration in the hypolimnion, which is seasonally isolated from sources of oxygen. Nürnberg (1995) found that the concentrations of N and P, determinants of productivity, were directly related to anoxia in the hypolimnion of lakes. Although these relationships seem logical, the generality of Hutchinson's suggestion has been fraught with controversy. Lasenby (1975) found little evidence for the relationship in 14 lakes of southern Ontario, perhaps because a significant amount of oxygen diffuses across the thermocline during periods of stratification (Stauffer 1987). Cornett and Rigler (1979) concluded that "a simple proportionality between biomass in the epilimnion and area hypolimnetic oxygen deficit (AHOD) does not appear to exist." These workers attempted to refine the relationship by examining the role of hypolimnetic volume and water temperature (Cornett and Rigler 1979, 1980). They found that the greatest O_2 consumption occurred in deep lakes with higher water temperatures and a thick hypolimnion. Presumably water temperature controls the rate of bacterial respiration in the water column and sediments. The relationship to hypolimnion thickness was unexpected, because it suggests that the greatest deficits are found in deep lakes with large hypolimnetic volume. Their findings, while not without criticism (Chang and Moll 1980), suggest that the consumption of oxygen in the hypolimnion may be largely the result of respiration in the water column, which is greatest in deep lakes where the transit time for sinking detritus is long (cf. Cole and Pace 1995).

Despite criticisms of the hypolimnetic oxygen deficit theory, it remains as a useful basis for evaluating the mass balance of organic carbon in lakes and the linkage of the carbon and oxygen cycles in freshwaters. In the carbon budget of lakes, aerobic respiration, denitrification, and methanogenesis accounted for 54 to 80% of the summertime mineralization of carbon in the hypolimnion of small lakes in New Hampshire (Mattson and Likens 1993) and southern Ontario (Bédard and Knowles 1991).

Nutrients

Nutrient budgets are constructed by assessing the inputs of nutrients in precipitation, runoff, and N-fixation and the losses of nutrients from lakes due to sedimentation, outflow, and the release of reduced gases. In many cases human impacts now dominate the nutrient budget of lakes (Edmondson and Lehman 1981). Successful attempts to construct nutrient budgets demand an accurate lake water budget. Inputs or losses of nutrients to/

from groundwater are particularly difficult to estimate (Coleman and Deevey 1987, Deevey 1988, Pollman et al. 1991, LaBaugh et al. 1995). The relative turnover or mean residence time of nutrients compared to water indicates the role of biota in geochemical movements.

Most lakes show a net retention of N and P (Cross and Rigler 1983, Galloway et al. 1983; Table 7.4), although in lakes with a substantial annual turnover of water, the net storage of N and P may be relatively small (e.g., Whalen and Cornwell 1985, Windolf et al. 1996). Despite substantial regeneration of Si from the sediments, Schelske (1985) also found net retention of Si in the nutrient budgets for several of the Great Lakes. Many lakes show near-balanced budgets for Mg, Na, and Cl, which are highly soluble and nonlimiting to the growth of phytoplankton (Cole and Fisher 1979, Galloway et al. 1983, Canfield et al. 1984, Jeffries et al. 1988, LeBaugh et al. 1995). Net retention of Ca is seen in lakes where mollusk shells are accumulating in the sediments (Brown et al. 1992) and in some highly productive, alkaline lakes where calcite ($CaCO_3$) may precipitate directly as *marl* during periods when high photosynthetic rates remove CO_2 from the water column (Brunskill 1969, Rosen et al. 1995):

$$Ca^{2+} + 2HCO_3^- \rightarrow CaCO_3 \downarrow + H_2O + CO_2. \qquad (7.24)$$

These lakes will show a net retention of Ca and a relatively short mean residence time for Ca in the water column (Canfield et al. 1984).

Table 7.4 Input–Output Balance (tons/yr) for Cayuga Lake, New York, 1970–1971, and Rawson Lake, Ontario, 1970–1973[a]

Element	Precipitation input	Runoff input	Total input	Discharge output	Percent retained
		Cayuga Lake			
Phosphorus	3	167	170	61	64
Nitrogen	179	2,565	2,744	513	81
Potassium	19	3,480	3,499	3,969	−12
Sulfur	313	24,671	24,984	31,983	−22
		Rawson Lake			
Phosphorus	0.018	0.017	0.035	0.010	71
Nitrogen	0.339	0.346	0.686	0.275	60
Carbon	2.435	19.005	21.440	10.074	53
Potassium	0.059	0.442	0.501	0.434	13
Sulfur	0.055	0.362	0.416	0.331	20

[a] From Likens (1975a).

Despite a water chemistry that would indicate the deposition of marl in Lake Powell (Utah), Reynolds (1978) found that the precipitation of calcite was inhibited by dissolved organic compounds derived from upland soils—linking the biogeochemistry of lakes to that of the surrounding watershed. As seen in terrestrial ecosystems (Chapter 6), when biotic processes lead to the retention or loss of elements in lakes, *bio*geochemical control is exerted on the underlying geochemistry of the surface of the Earth. Lake sediments retain a record of the changes in biogeochemical function through time (Whitehead et al. 1973, Pennington 1981, Brugam 1978, Schelske et al. 1988).

Nitrogen fixation rates in lakes range from 0.1 kg N ha^{-1} yr^{-1} to >90 kg N ha^{-1} yr^{-1} (Howarth et al. 1988a), roughly spanning the range of nitrogen fixation reported for terrestrial ecosystems (Chapter 6). Lakes with high rates of nitrogen fixation show large apparent accumulations of N (Horne and Galat 1985). Doyle and Fisher (1994) found that nitrogen fixation supplied 80% of the N input to a floodplain lake in the Amazon River basin, but N-fixation was reduced when the supply of NO_3 was high (cf. Fig. 6.3). Fewer studies have assessed denitrification and other processes of gaseous loss in lakes. Denitrification can be studied by the application of acetylene-block techniques, and ^{15}N isotopic labels as discussed in Chapter 6 (Seitzinger et al. 1993). The total loss of nitrogen by denitrification exceeds the input of nitrogen by fixation in almost all lakes where both processes have been measured simultaneously (Seitzinger 1988). Molot and Dillon (1993) found that the average rate of denitrification, 9 kg N ha^{-1} yr^{-1}, accounted for 36% of the nitrogen received from the runoff entering seven lakes in Ontario. Gardner et al. (1987) found significant losses of N from Lake Michigan due to denitrification in the sediments. Often denitrification appears limited by the production and availability of NO_3^- in the sediments (Rysgaard et al. 1994), but denitrifiers may also remove NO_3^- that diffuses into the sediment from overlying waters. Yoh et al. (1988) found that both nitrification and denitrification were responsible for the production of N_2O in the water column of several lakes in Japan, but the loss of nitrogen from lakes as N_2 greatly exceeds the loss of N_2O (Seitzinger 1988). Ammonia volatilization may occur in alkaline lakes; Murphy and Brownlee (1981) found that the loss of NH_3 exceeded N inputs by fixation in a highly productive prairie lake (cf. Jellison et al. 1993).

Few studies have examined volatile losses of sulfur as part of the S budget of lakes. Although volatile losses of sulfur occur (Brinkmann and Santos 1974), most H_2S appears to be reoxidized as it passes through the upper sediments (Dornblaser et al. 1994) or the water column (Mazumder and Dickman 1989), so little escapes to the atmosphere. Losses of dimethylsulfide and other volatile sulfides are a minor component of the S budget of most temperate lakes (Nriagu and Holdway 1989, Richards et al. 1991). When reduced forms of sulfur are cycled through reoxidation pathways, high rates of SO_4 reduction can occur in lake sediments, despite low concentrations of SO_4 in lakewater (Urban et al. 1994). Sulfate reduction accounted for 7% of total carbon mineralization in a small New Hampshire lake (Mattson and Likens 1993) and 30% of the carbon mineralization in the sediments of a highly productive lake in Michigan (Smith and Klug 1981).

The input of nutrients relative to lake volume is useful in distinguishing low-productivity *oligotrophic* lakes from high-productivity *eutrophic* lakes. Oligotrophic lakes are often of relatively recent geologic origin (e.g., postglacial) and deep, with cold hypolimnetic waters. Such lakes often show a relatively large ratio between lake area and drainage area, and a long mean residence time for water (Dingman and Johnson 1971). The nutrient input to oligotrophic lakes is dominated by precipitation (Table 7.5). These lakes are nutrient-poor and seldom have productivity >300 mg C m^{-2} day^{-1} (Likens 1975b). In contrast, eutrophic lakes are dominated by nutrient inputs from the surrounding watershed, and they typically have lower N/P ratios as a result of P inputs from runoff (Downing and McCauley 1992). These nutrient-rich lakes are often shallow, with warm, highly productive waters. Of course, sedimentation will eventually convert the physical state of many oligotrophic lakes to shallow, eutrophic conditions, so these concepts have also been used to describe a sequence of lake aging. However, in most cases, nutrient status remains the most useful criterion to distinguish between oligotrophic and eutrophic conditions (Fig. 7.13). Humans may cause rapid "cultural eutrophication" by large nutrient additions in pollutants (Schindler 1974, Vallentyne 1974, Goldman 1988), and the eutrophic conditions can be reversed when pollution controls are implemented (Edmondson and Lehman 1981, Levine and Schindler 1989).

Alkalinity and Acid Rain Effects

Alkalinity is defined as

$$\text{alkalinity} = [2CO_3^{2-} + HCO_3^- + OH^-] - H^+. \quad (7.25)$$

Table 7.5 Sources of Nitrogen and Phosphorus as Percentages of the Total Annual Input to Lake Ecosystems[a]

	Precipitation		Runoff	
	N	P	N	P
Oligotrophic lakes	56	50	44	50
Eutrophic lakes	12	7	88	93

[a] From Likens (1975a).

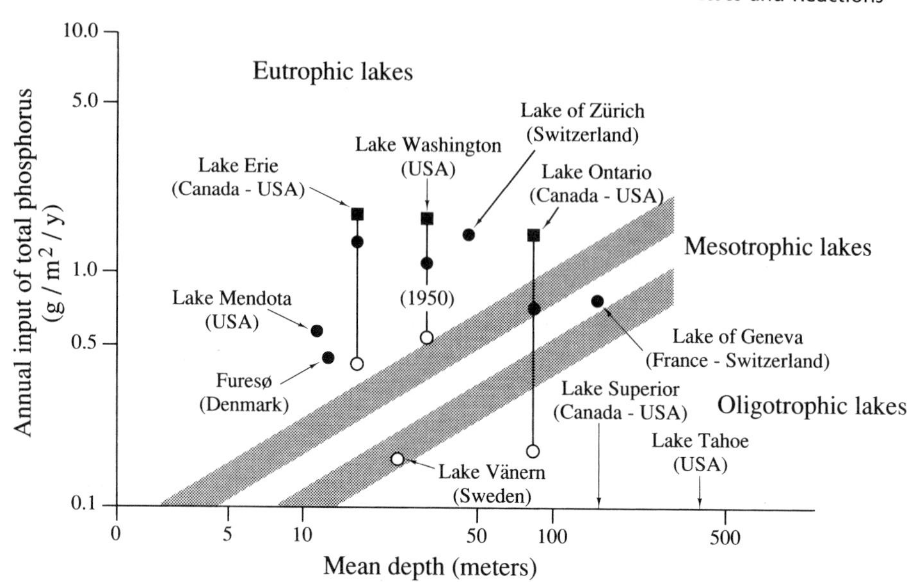

Figure 7.13 The position of important lakes relative to the annual receipt of phosphorus and their mean depth, differentiating oligotrophic and eutrophic lakes. For lakes that have undergone significant pollution, the change from previous conditions (○) to present conditions (●) is shown. From Vollenweider (1968).

It is roughly equivalent to the balance of cations and anions in lake waters, where

$$\text{alkalinity} = [2\text{Ca}^{2+} + 2\text{Mg}^{2+} + \text{Na}^{+} + \text{K}^{+} + \text{NH}_4^{+}] - [2\text{SO}_4^{2-} + \text{NO}_3^{-} + \text{Cl}^{-}]. \tag{7.26}$$

Generally, alkalinity is measured in milliequivalents per liter by titration of a water sample to a pH of 4.3. When present, organic anions, such as humic acids, contribute to the apparent alkalinity of lake waters. Thus, the titration of a water sample to a pH of 4.3 is often said to represent *acid-neutralizing capacity* (ANC), because the consumption of protons by organic anions is not distinguished from H^+ neutralized by HCO_3^- and other inorganic anions. In acid waters, Al^{3+} reduces alkalinity, since it acts as an "acid" cation (Eqs. 4.10–4.12).

Alkalinity is increased by processes that consume SO_4^{2-} or NO_3^- from the water column, including sulfate reduction, sulfate-adsorption on minerals, and denitrification (Rudd et al. 1986b, Anderson and Schiff 1987, Baker et al. 1988, Giblin et al. 1990, Bailey et al. 1995, see also Eq. 7.17). Production of organic carbon and the deposition of calcite by phytoplankton reduce alkalinity by consuming HCO_3^- (Eqs. 7.23 and 7.24). In most lakes, the drainage basin contributes a large amount of alkalinity, because the runoff of cations is usually balanced by HCO_3^- (Chapter 4). Thus, it is not surprising that highly alkaline waters are found in areas underlain by limestone.

When the mean residence time for lake water is < 1 yr, terrestrial sources usually dominate the alkalinity budget of lakes (Shaffer and Church 1989). In lakes with longer mean residence times (4–9 years) the generation of alkalinity by biogeochemical processes within the lake may be greater than the receipt of alkalinity from the surrounding watershed (D.W. Schindler et al. 1986, Wentz et al. 1995). Human-induced changes in land use, such as cultivation and burning, can increase the alkalinity of lakes by increasing the transport of cations and HCO_3^- from land (Renberg et al. 1993). On the other hand, acid rain generally reduces the alkalinity of lakes (L.A. Baker et al. 1991). Acid rain impacts are lower if sulfate reduction increases in response to the sulfate inputs in rainfall (Kilham 1982, D.W. Schindler et al. 1986) and the products of SO_4 reduction are not reoxidized in the lake (Kling et al. 1991, Giblin et al. 1992). We can use changes in the alkalinity status of lakes as a diagnostic tool for the effects of changes in land use and acid rain, analogous to the use of H^+ budgets in terrestrial watersheds (Chapter 6).

Acid rain can affect various other biogeochemical processes in lakes. Recall that nitrification in soil is inhibited at low pH. For two lakes in Ontario, experimental reductions of the pH of lakewater inhibited nitrification in the water column, so that ammonium accumulated (Rudd et al. 1988). Thus, the changing conditions of the atmosphere have the potential to alter ecosystem processes in lakes.

Wetlands and Climatic Change

Wetlands and lakes are vulnerable to global climatic changes that alter patterns of precipitation and evaporation on the world's continents (Chapter 10). Global warming may have particularly severe effects on the fauna of lakes, from which dispersal to new habitats is problematic. Over the last century, the length of the ice-free season has increased by several weeks in several lakes in Wisconsin (Fig. 7.14), and deepening of the thermocline in some lakes has reduced the habitat for coldwater species, such as lake trout, which could disappear entirely from shallow lakes if the warming continues (Schindler et al. 1990). Although we do not know whether or not these changes have been caused by global warming, the effects are indicative of the disruptions that we might expect in lake ecosystems in postulated warmer climates of the future (Carpenter et al. 1992, Melack 1991, Poiani and Johnson 1991).

Summary

Nutrient cycling in wetland ecosystems and lakes is controlled by redox potential and by the microbial transformations of nutrient elements that occur under condi-

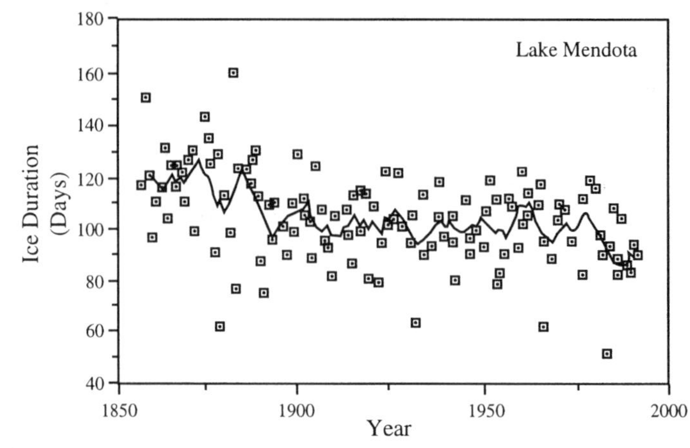

Figure 7.14 The duration of ice in Lake Mendota (Wisconsin) during the last 150 years. The 7-year moving average is plotted through these data. From Robertson et al. (1992).

tions in which O_2 is not always abundant. Under these conditions, decomposition is inhibited and organic carbon accumulates in peat and sediments. Wetland ecosystems are significant to the global cycle of sulfur through the emission of reduced sulfur gases. Wetland ecosystems are also the major source of methane, which is rapidly increasing in the atmosphere (Chapter 11). Depending upon their position at the interface between terrestrial and aquatic ecosystems, wetlands receive varying amounts of runoff from land, which affects their net primary productivity and specific nutrient limitations. Wetland ecosystems are a site of nutrient transformations, and the runoff from wetlands often controls the form of nutrient movement in rivers (Chapter 8).

The physical properties of water control many aspects of nutrient cycling in lakes. Generally most lake ecosystems are stratified into an upper zone where photosynthesis occurs and high redox potentials prevail and a lower zone where oxygen is depleted due to bacterial respiration. The circulation of lake waters, microbial turnover, and redox conditions control the turnover of nutrients in lakes. Net primary productivity in lakes is limited by phosphorus under almost all conditions. Most wetland and lake ecosystems are net sinks for nitrogen and phosphorus that enter from upland ecosystems.

Recommended Readings

Firth, P. and S.G. Fisher. (eds.). 1991. *Global Climate Change and Freshwater Ecosystems.* Springer-Verlag, New York.

Mitsch, W.J. and J.G. Gosselink. 1993. *Wetlands.* 2nd ed. Van Nostrand Reinhold, New York.

Morel, F. 1993. *Principles and Applications of Aquatic Chemistry.* Wiley, New York.

Stumm, W. and J.J. Morgan. 1981. *Aquatic Chemistry.* 2nd ed. Wiley, New York.

Wetzel, R.G. 1983. *Limnology.* 2nd ed. Saunders, Philadelphia.

Zehnder, A.J.B. 1988. *Biology of Anaerobic Microorganisms.* Wiley, New York.

8

Rivers and Estuaries

Introduction

Traditionally geochemists have regarded rivers as conduits leading from land to sea, but it is now clear that this view is too simplistic. Important biogeochemical reactions occur in rivers, transforming chemical elements during downstream transport. Previously we have seen how measurements of streamwater chemistry are useful in calculating weathering rates (Chapter 4) and nutrient losses from terrestrial ecosystems (Chapter 6). In this chapter we will focus on the biogeochemical processes that occur *within* rivers, including transformations of organic carbon, nitrogen, and phosphorus. We will also examine in more detail the factors that control the flow of streamwaters and the origin and concentration of streamwater constituents. We will conclude the chapter with a consideration of the biogeochemistry of salt marshes and estuaries—places where rivers empty into the ocean.

Soil Hydraulics and Stream Hydrology

Vegetation and soil characteristics control the genesis of streamwaters. A canopy of vegetation lowers the initial impact energy of raindrops, reducing their potential to cause soil erosion and allowing greater time for infiltration of water into the soil profile (Lyford and Qashu 1969). In addition, plant roots, earthworms, termites, and other soil organisms promote the downward percolation of moisture through pores in the soil (Beven and Germann 1982). Vegetation increases surface roughness, which slows the rate of runoff from the landscape, relative to that seen from bare soils (Abrahams et al. 1994). There is little surface runoff from most forest ecosystems, but overland flow increases strongly when vegetation is removed (Lull and Sopper 1969). On barren or partially bare land, little precipitation enters the soil, and large amounts of surface runoff are generated even when the rainfall is not intense (Fig. 8.1).

While vegetation promotes the entry of moisture into the soil, it may also reduce the average soil moisture content, owing to the large quantities of soil water that are taken up by roots to support plant transpiration (Chapter 5, Table 8.1). Much of the plant rooting zone is below the depth from which water might otherwise evaporate from the surface (Allison et al. 1983). In some cases trees are known to extract water from soil layers >7 m in depth (Patric et al. 1965, Nepstad et al. 1994). When vegetation is removed, soil water contents increase (Ting and Chang 1985, Schlesinger et al. 1987, Gee et al. 1994), yielding a greater volume of streamflow (Bormann and Likens 1979).

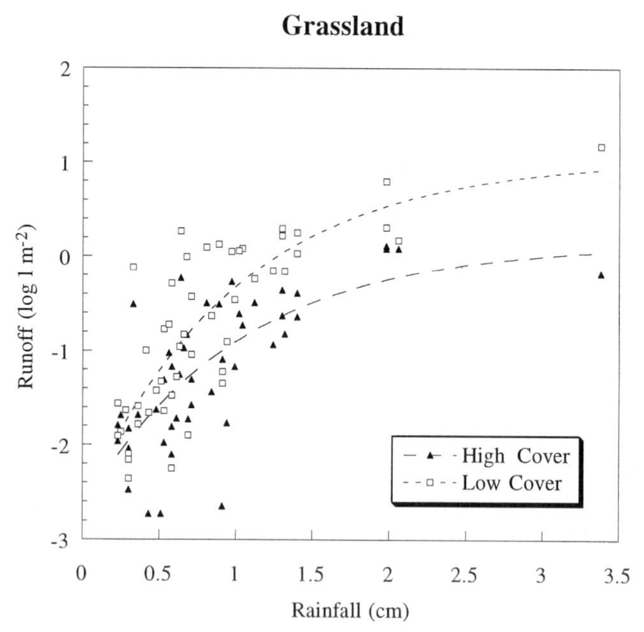

Figure 8.1 Surface runoff from high- and low-cover grassland plots as a function of rainfall in southern New Mexico. Unpublished data of the author.

Table 8.1 Relative Importance of Pathways Leading to the Loss of Water from Terrestrial Ecosystems

Vegetation	Evaporation (%)	Transpiration (%)	Runoff and recharge (%)	Reference
Tropical rainforest	25.6	48.5	25.9	Salati and Vose (1984)
Tropical rainforest	10	40	50	Shuttleworth (1988)
Tropical rainforest	11	56	32	Leopoldo et al. (1995)
Temperate forest	13	32	53	Waring et al. (1981)
Temperate forest[a]	—	54	—	Pražák et al. (1994)
Temperate grassland	35	65	0	Trlica and Biondini (1990)
Temperate grassland	33	67	0	Massman (1992)
Steppe	55	45	0	Floret et al. (1982)
Steppe	56	34	10	Paruelo and Sala (1995)
Desert	28	72	0	Schlesinger et al. (1987)
Desert	20	80	—	Liu et al. (1995)
Desert	73	27	—	Lane et al. (1984)
Desert	65	35	—	Smith et al. (1995)

[a] Growing season only.

Infiltration rates and soil water contents are also affected by soil texture, especially soil porosity. Soil pore volume is related to bulk density,

$$\text{porosity} = 1.00 - \frac{\text{(bulk density)}}{2.66} \times 100\%, \qquad (8.1)$$

so pores compose about 50% of the volume of a soil with a bulk density of 1.33 g/cm^3, which is not unusual for many soils. Under moist conditions, water enters soils with a high proportion of sands (i.e., particles > 2 mm) much faster than those dominated by clays (particles < 0.002 mm), which have a lower porosity. However, as soils dry, clays retain a greater water content at any soil water potential than soils dominated by coarser fractions (Saxton et al. 1986). This effect is due to the high adsorption potential of clays, which tend to retain water on their surface.

The flow of water through terrestrial ecosystems to streamwaters is often modeled using simplified assumptions about the rate of plant uptake and the downward flow of water through the soil profile (Waring et al. 1981, Knight et al. 1985). Downward movement is assumed to occur during any period in which the percolation of water to a particular depth is in excess of the water-holding capacity of that depth and the rate of plant uptake during the interval. Water-holding capacity is commonly called *field capacity*, which is the water content that a soil can retain against the force of gravity.

When excess water drains to the bottom of the profile, it is assumed to be delivered to the stream channel.

In some models, the flow of soil water is calculated by the application of Darcy's law,

$$\text{flux} = kIA, \tag{8.2}$$

where k is the hydraulic conductivity, I is the hydraulic gradient, and A is the cross-sectional area under consideration. Hydraulic conductivity must be determined empirically, usually by observations of the rate of disappearance of water maintained in a ponded cylinder at the surface (Rycroft et al. 1975, Youngs 1987), where

$$\underset{(\text{cm/sec})}{\text{conductance}} = \underset{(\text{g cm}^{-2}\text{ sec}^{-1})}{\text{infiltration rate}} / \underset{(\text{g/cm}^3)}{\text{gradient.}} \tag{8.3}$$

The gradient is the difference in the content of water (g/cm^3) between the surface and some known depth of interest.

Although the large pores in coarse-textured soils conduct water freely when these soils are wet, drainage of these pores causes a rapid decline in hydraulic conductivity with soil drying. Thus, hydraulic conductivity in dry soils is often greater in clays, on which the adsorbed films of water maintain a continuous path for water movement through the soil (e.g., Stuart and Dixon 1973). Darcy's law was originally formulated for use in saturated soils and groundwater, but it is often used successfully in unsaturated soils (Ward 1967). The method is limited because of the large spatial variation in soil properties (Topp et al. 1980, G.V. Wilson et al. 1989) and the effect of channels caused by roots and soil animals (Beven and Germann 1982). More elaborate treatments of flow in unsaturated soils (the vadose zone) are available (e.g., Nielsen et al. 1986), but they are difficult to apply in most field situations.

When precipitation is not occurring, streamflow is largely maintained by the slow drainage of water from the soil profile and from groundwater. This *base flow* declines slowly as the drought period continues. With rainfall, a number of changes are seen in a stream hydrograph, which relates streamflow to time (Fig. 8.2). An immediate increase in flow, known as quick flow or storm flow, may result from surface runoff that enters the stream channel during a storm. At the end of rainfall, the effect of surface runoff disappears rapidly, but base flow is reestablished at a new, higher level, which resumes a slow decline as the soil dries. This increase in base flow is derived from rainwater that infiltrated the soil profile raising the soil moisture content and the amount of water available for drainage.

Long-term observations of streams show that hydrographs are affected by topography, vegetation, and soil characteristics, as well as the pattern

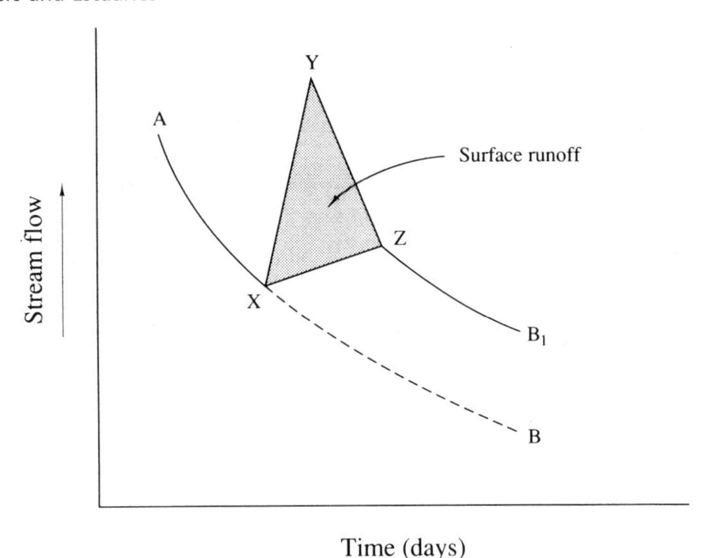

Figure 8.2 A stream hydrograph, showing the effect of a rainstorm at time X on stream runoff, which increases to a peak (Y) during the rainstorm. Streamflow declines rapidly to base flow (Z), which is reestablished at a higher level (B_1) than if the storm had not occurred (B). Modified from Ward (1967).

and intensity of rainfall in individual storms (Ward 1967). Stream hydrographs show what fraction of the flow is derived from surface runoff, which may carry organic debris and soil particles, and what fraction is derived from the drainage of soil water. In a tropical rainforest, base flow can account for 92% of total flow (Lesack 1993), whereas in deserts, virtually all streamflow occurs during storms. As a result, streams draining arid soils carry large amounts of sediment (Langbein and Schumm 1958, Milliman and Meade 1983).

Stream Load

Chemical transport in streams is often divided into two fractions: that carried in the form of dissolved ions and that carried as particulates. The dissolved load is largely derived from rainfall and from the soil solution after it has interacted with soil exchange reactions and the chemical weathering of bedrock (Chapter 4). The particulate load, dominated by the products of mechanical weathering, represents erosion and sediment transport from the surface of the soil. Particulate contents include materials ranging in size from colloidal clays to large boulders and from leaves to floating logs. The particulate load includes material suspended in the water—the suspended load—as well as material that moves along the bottom of the stream channel—the bed load.

Biogeochemical Transformations of C, N, and P

The carbon budget of most small streams is dominated by allochthonous materials, i.e., organic carbon that enters from the surrounding terrestrial ecosystem (Chapter 7). Sometimes this process is easily observed. When leaves fall into a stream, they are shredded and decomposed during downstream movement. In other cases, dissolved organic compounds draining from adjacent soils account for the major proportion of the allochthonous inputs to streams (Fiebig et al. 1990, Dosskey and Bertsch 1994). The compounds of dissolved organic carbon (DOC) include soluble carbohydrates and amino acids, which are leached from decomposing leaves and plant roots (Suberkropp et al. 1976), and humic and fulvic acids from soil organic matter (McDowell and Likens 1988, Chapter 5). Humic acids compose about 60% of the DOC in the Amazon River (Ertel et al. 1986). The movement of dissolved organic carbon into streamwaters is strongly controlled by interactions with clay minerals in the soil (McDowell and Wood 1984, Nelson et al. 1993). When clay minerals are absent, organic compounds move freely from soils to streamwaters, producing deeply stained, blackwater rivers (Beck et al. 1974). Dissolved organic compounds account for 96% of the organic matter in the Ogeechee, a blackwater river draining swamps in southern Georgia (Benke and Meyer 1988).

The carbon budget for Bear Brook in New Hampshire is typical of that for small streams draining temperate forests (Table 8.2). Allochthonous inputs of organic carbon are contributed by dissolved (42%) and particulate (57%) sources, and net primary production by benthic algae and mosses supplies only 0.2% of the organic carbon available for respiration in the

Table 8.2 Yearly Fluxes of Organic Carbon, Nitrogen, and Phosphorus in Bear Brook, New Hampshire[a]

	Organic carbon (g/m^2)	Nitrogen (g/m^2)	Phosphorus (g/m^2)	Atomic ratio C:N:P
Inputs				
Total dissolved	260	56	0.39	1700:320:1
Total fine particulate	12	0.27	0.55	54:1:1
Total coarse particulate	340	8.2	0.70	1300:26:1
Total gaseous	1	<0.1	0	—
Total inputs	620	64	1.6	990:89:1
Outputs				
Total dissolved	260	57	0.29	2300:440:1
Total fine particulate	25	0.43	1.1	59:0.9:1
Total coarse particulate	100	1.8	0.38	720:10:1
Total gaseous	230	?	0	—
Total outputs	620	59	1.8	890:72:1

[a] From Meyer et al. (1981).

stream (Fisher and Likens 1973). As these materials pass through Bear Brook, 230 g C/m^2 (37%) of the organic matter is respired and lost to the atmosphere as CO_2. A considerable amount of the respiration occurs in stream sediments (Hedin 1990, Findlay et al. 1993, Naegeli et al. 1995). In small streams, the ratio of dissolved to particulate organic carbon generally increases as allochthonous materials are degraded during downstream transport (S.G. Fisher 1977).

Larger rivers with more slowly moving waters show a greater importance of autochthonous net primary production by phytoplankton and rooted plants (Naiman and Sedell 1981, Lewis 1988). This production is of special significance, because it is often more easily assimilated by higher trophic levels than is the resistant plant debris derived from land. Despite the large amount of humic material in the Amazon, much of the fish community is supported by the growth of phytoplankton in the river, rather than by organic carbon derived from the rainforest (Araujo-lima et al. 1986). Nevertheless, the rate of plant production in large rivers is often limited by turbidity, so these systems usually retain an overall dominance of allochthonous organic materials (Edwards and Meyer 1987, Cai et al. 1988, Howarth et al. 1992).

Transport of organic materials in the lower reaches of rivers often fuels a large heterotrophic, bacterial population (Edwards and Meyer 1986, Benner et al. 1995). Despite significant internal production in the Amazon, heterotrophic processes dominate total metabolism (Wissmar et al. 1981, Quay et al. 1995). Like the Amazon, most major rivers are supersaturated with CO_2 with respect to the atmosphere (Kempe, 1984, Richey and Victoria 1993). In the Amazon, a portion of the heterotrophic activity is anaerobic, probably occurring in floodplain soils, which release CH_4 to river waters and to the atmosphere (Richey et al. 1988, Tyler et al. 1987, Devol et al. 1988).

In small streams, concentrations of DOC and particulate organic carbon (POC) usually increase with increasing streamflow, as a greater portion of the water is derived from overland flow that carries organic material to the stream channel (Meyer et al. 1988, David et al. 1992, Meybeck 1993, Hornberger et al. 1994). Measurements of small forest streams indicate that streamflow removes about 1 to 5 g C m^{-2} yr^{-1} from the watershed (Moeller et al. 1979, Hope et al. 1994). Remembering that typical forest production is about 400 to 800 g C m^{-2} yr^{-1} (Table 5.2), we find that the metabolism of stream ecosystems is fueled by the transfer of $<1.0\%$ of the net primary production of forests to their runoff waters (Mantoura and Woodward 1983). Somewhat higher values are characteristic of streams draining peatlands and swamps (Mulholland and Kuenzler 1979, Naiman 1982), while lower values characterize grasslands and deserts (Gifford and Busby 1973, Fisher and Grimm 1985, Mulholland and Watts 1982). In many cases, organic materials from the trees of the riparian zone dominate the total input of detritus to the stream (Table 8.3). Large rivers often derive a significant fraction of their organic carbon from their floodplain, especially during seasonal flooding (Cuffney 1988, Grusbaugh and Anderson 1989, Benner et al. 1995).

We can estimate the total riverine transport of organic carbon using several approaches, which provide good examples of the problems of scaling in biogeochemistry (Chapter 1). For example, we might compile studies of small watersheds in which the transfer of organic carbon from land to streamwaters has been carefully measured. These measurements could be multiplied by the area of the globe that is covered by such vegetation, and a global estimate calculated by summation. Alternatively, we might compile studies of large rivers, for which there are measurements of annual riverflow and the load of organic carbon near the mouth. The transport of organic carbon in rivers that have not been studied can be calculated from the regression, allowing us to calculate the total transport for all the major rivers of the world (Fig. 8.3). The large rivers are critical to the second approach, since the 50 largest rivers carry 43% of the total freshwater moving from land to sea. The Amazon alone carries 20%.

The first approach emphasizes the losses of organic carbon from land, while the second approach emphasizes the transfer of organic carbon to the oceans. They will differ by the amount of organic carbon that is metabolized during transport or deposited in floodplains and behind dams (Mulholland 1981, Lal 1995). Regardless of the approach, most estimates of the global riverine transport of organic carbon fall in the range of 0.3 to 0.5×10^{15} g C/yr (Schlesinger and Melack 1981, Meybeck 1988, 1993, Degens et al. 1991, Ludwig et al. 1996). An average concentration of organic carbon in rivers (10 mg C/liter) is derived by dividing the median estimate for global transport by that for river volume (40,000 km^3/yr, Chapter 10).

Most rivers carry low concentrations of dissolved inorganic nitrogen (NH_4^+, NO_3^-) and phosphorus (HPO_4^{2-}), which are actively taken up by plants and soil microbes and retained on land (Chapter 6). The vegetation

Table 8.3 Comparison of Sources of Stream-Exported Organic Carbon from Fourmile Branch Watershed, South Carolina[a]

Source	Area (km^2)	Streamwater, transport			Litterfall input (tons C yr^{-1})	TOC as % of detritus
		POC (tons C yr^{-1})	DOC	TOC		
Watershed	12.6	7.3	21.6	28.9	5810	0.50
Upland forest	11.9	0	2.0	2.0	5545	0.04
Wetland forest	0.7	7.3	19.6	26.9	265	10.2

[a] Total, particulate, and dissolved components are signified by TOC, POC, and DOC, respectively. From Dosskey and Bertsch (1994).

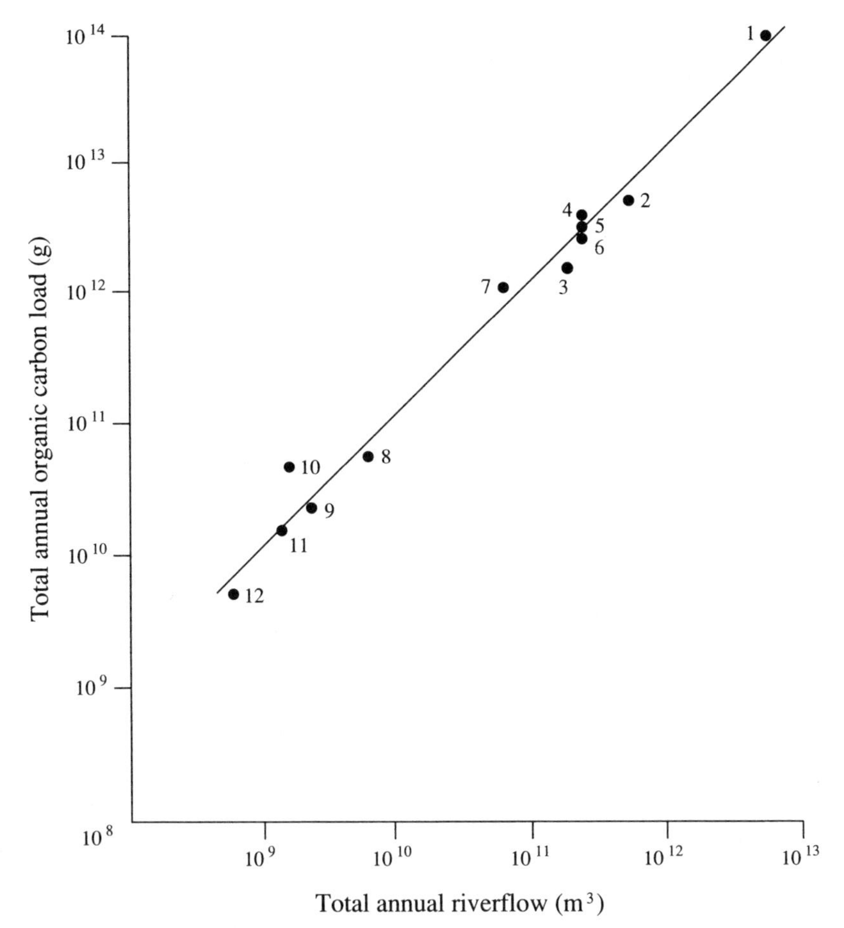

Figure 8.3 Total annual load of organic carbon shown as a logarithmic function of total annual riverflow for major rivers of the world. Rivers 1–7 are among the 50 largest: 1, Amazon; 2, Mississippi; 3, St. Lawrence; 4, MacKenzie; 5, Danube; 6, Volga; and 7, Rhine. From Schlesinger and Melack (1981), with a revision of the data for the St. Lawrence derived from Pocklington and Tan (1987).

in the floodplain of rivers is also an effective trap for nutrients that might otherwise pass from upland ecosystems to the river channel (Jordan et al. 1993, McClain et al. 1994, Brandes et al. 1996). The autochthonous productivity of streams is limited by nutrients, and small streams may shift from net heterotrophy to autotrophy with the addition of phosphorus (Peterson et al. 1985). Decomposition rates are also limited by substrate quality, and experimental additions of P to streams increase the rate of litter decay (Elwood et al. 1981). Decomposition of coarse particulate matter (e.g., leaves) in streams is accompanied by decreases in the C/N and C/P ratios (Table 8.2), just as we see during the decomposition of terrestrial litter (Chapter 6). The decline is due to the immobilization of these essential elements by the microbes involved in litter decay (Meyer 1980, Qualls 1984, Mulholland 1992). During river transport, phosphorus is also adsorbed on sediments and suspended minerals (Meyer 1979, Klotz 1988, Mulholland 1992). Thus, a large fraction of the N and P in streams is carried in particulate and organic forms (Meyer and Likens 1979, Triska et al. 1984, McDowell and Asbury 1994). Floodplain forests, subject to seasonal inundation, receive a significant input of nutrients in deposited sediments that stimulate net primary production (Mitsch et al. 1979, Brown 1981, Walker 1989).

An effective theory for nutrient cycling in stream ecosystems is the concept of nutrient spiraling (Newbold et al. 1983). During downstream transport, dissolved ions are accumulated by bacteria and other stream organisms and converted to organic form. When these organisms die, they are degraded to inorganic forms that are returned to the water, only to be taken up again by organisms that are involved in the further degradation of organic materials. The cycle between inorganic and organic forms may be completed many times while a nutrient atom moves downstream to the ocean. Because the cycle will occur most rapidly when biotic activity is highest, the spiral length or turnover time is an inverse index of ecosystem metabolism (Fig. 8.4). Comparative estimates of spiral length are determined by following the disappearance of isotopic tracers (^{15}N and ^{32}P) from streamwaters. In a small stream in Tennessee, Newbold et al. (1983) found that phosphorus moved downstream at an average velocity of 10.4 m/day, cycling once every 18.4 days, so the average spiral length was about 190 m. Each atom of P spent the majority of its time as a component of particulate matter, but it traveled the greatest distance while in the form of dissolved phosphate.

The Apure River, a tributary of the Orinoco River in Venezuela, transports 0.345 g m^{-2} yr^{-1} of N and 0.068 g m^{-2} yr^{-1} of P from its watershed to the

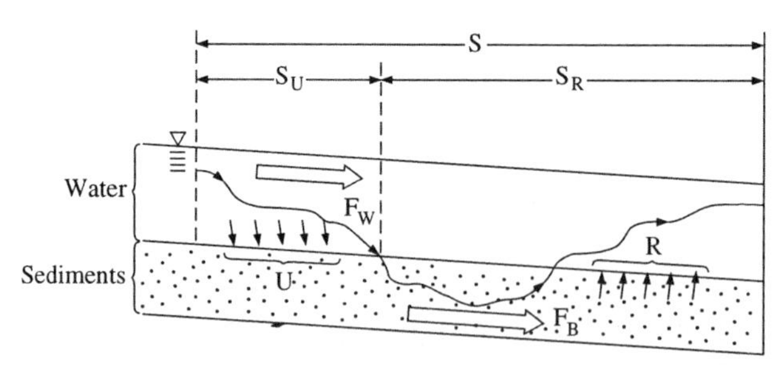

Figure 8.4 Nutrient spiraling in a two-compartment stream. Spiraling length, S, is the sum of the uptake length, S_U, and the remineralization length, S_R. F_w is the downstream flux of dissolved nutrients in the water compartment and F_b is the downward flux in the particulate compartment. Modified from Newbold et al. (1982).

sea (Saunders and Lewis 1988). Dissolved inorganic forms account for only 19% of N and 24% of P. Similarly, on a global basis, rivers transport 33 to 36×10^{12} g/yr of nitrogen, with about 87% in organic forms, and 21×10^{12} g/yr of phosphorus, nearly all in particulates (Meybeck 1982, 1993, Ittekkot and Zhang 1989). Only about 10% of the particulate phosphorus is biologically available; the rest is strongly bound to soil minerals (Ramirez and Rose 1992). The overall C/N ratio in global riverflow is 11.1. Recall that the C/N ratio of fresh plant litter is >100 (Tables 6.5 and 6.7). The lower C/N ratio of particulates in rivers reflects the retention of nitrogen and the respiration of carbon during downstream transport (Meybeck 1982).

The global transport of N and P in rivers has increased greatly as a result of human activities, such as the widespread use of nitrogen fertilizers and phosphorus detergents (Vitousek 1994, Jordan and Weller 1996). Meybeck (1982) estimates that the total riverload of dissolved N has doubled and that for P has tripled over preindustrial levels. These increases are not evenly distributed around the world; they are strongly correlated with human population and energy consumption in the drainage basin (Peierls et al. 1991). However, as a result of the ban on the use of phosphorus-based detergents, total P has declined in rivers of the United States, while total nitrate has continued to increase (Table 8.4). Reflecting the interactions between biogeochemical processes in rivers, von Gunten and Lienert (1993) found that when the emissions of phosphorus to the river Glatt (Switzerland) were reduced, autochthonous net primary production declined, lowering the content of organic matter in sediments, decreasing anaerobic respiration, and reducing the transport of soluble, reduced forms of metals (Mn and Cd) to groundwater.

Table 8.4 Recent Changes in the Delivery of Nitrogen and Phosphorus in Rivers to Coastal Areas of the United States[a]

	Change in load, 1974–1981	
Region	Total nitrate (%)	Total phosphorus (%)
Northeast Atlantic coast	32	−20
Long Island Sound/New York Bight	26	−1
Chesapeake Bay	29	−0.5
Southeast Atlantic coast	20	12
Albemarle/Pamlico Sound	28	0
Gulf Coast	46	55
Great Lakes	36	−7
Pacific Northwest	6	34
California	−5	−5

[a] From Smith et al. (1987). Copyright 1987 by the AAAS.

Other Dissolved Constituents

Variations in the concentration of dissolved ions in rivers can be related to the rate of discharge and to the origin of the waters that contribute to the stream hydrograph (Johnson et al. 1969). In a simple geochemical system, we might expect that streamwater concentrations would be highest at periods of low flow, because most of the water would be derived from drainage of the soil profile where it would be in equilibrium with various rock weathering and ion-exchange reactions (Chapter 4). As streamflow increased, we might expect concentrations to decline as an increasing proportion of the flow is derived from precipitation and surface runoff, with little or no equilibration with the soil mineral phases. This simple geochemical model often explains the behavior of major ions in streamwater (Ca, Mg, Na, Si, Cl, SO_4, and HCO_3), although there are certainly exceptions (Meyer et al. 1988). Most of these ions are easily soluble and nonlimiting to biota.

Jennings (1983) shows dilution of Ca and Mg with increasing discharge in areas of limestone in New Zealand, but the slope of the relationship changes slightly from summer to winter, reflecting a greater weathering of limestone by the more active respiration of roots during the summer (Laudelout and Robert 1994). In contrast, in most areas, the concentration of potassium shows little relation to streamwater discharge, because it is regulated by microbial activity and root uptake (Lewis and Grant 1979, Feller and Kimmins 1979, McDowell and Asbury 1994). At the Hubbard Brook Experimental Forest of New Hampshire, the lowest potassium and nitrate concentrations are found during the low-flow periods of summer, when biotic demands are greatest (Johnson et al. 1969, Likens et al. 1994).

During rainfall or seasonal flooding, the concentrations of dissolved ions are often higher on the rising limb of the stream hydrograph than during the equivalent flows on the declining limb (Whitfield and Schreier 1981, McDiffett et al. 1989). The effect, known as *hysteresis,* is thought to result from an initial flushing of the highly concentrated waters that accumulate in the soil pores during periods of low flow. Not all ions show consistent hysteresis patterns, so to calculate the total annual loss of dissolved ions from a watershed, the streamflow discharge for each day must be multiplied by the concentration measured at that discharge and the products summed for all 365 days.

Even for elements that show lower concentrations at greater discharge, the total removal is greatest during years of high streamflow (Fig. 8.5)—i.e., the increase in flow predominates over the expected dilution of dissolved materials. In comparisons made over large geographic regions, concentrations are greatest in rivers that drain areas with limited runoff; they decline with increasing runoff, but total transport is greater in rivers with greater discharge (e.g., Fig. 8.6; Bluth and Kump 1994).

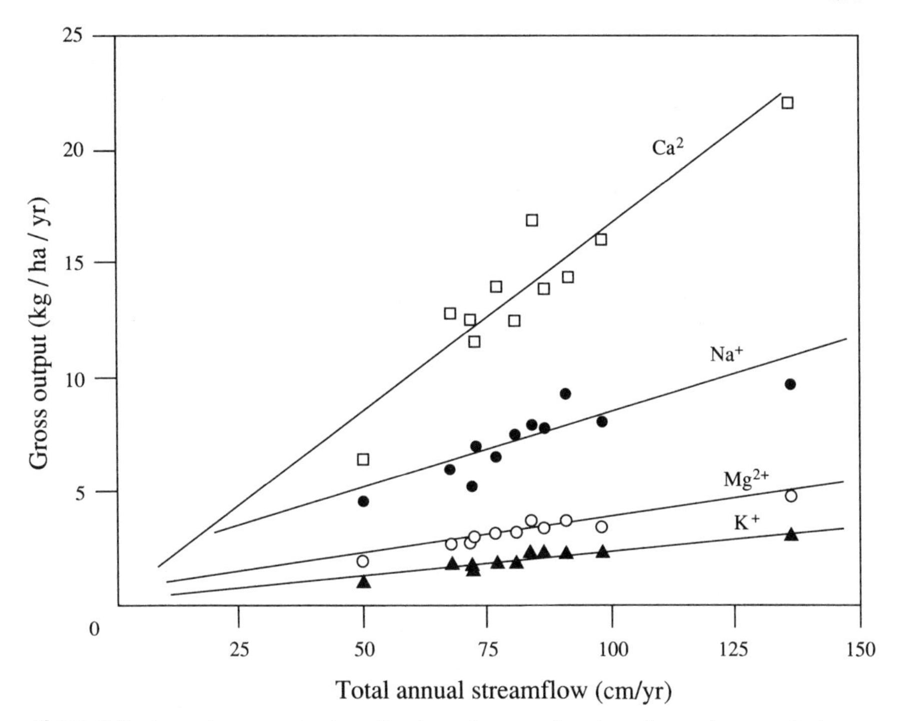

Figure 8.5 Annual streamwater loss of major cations as a function of annual stream discharge in the Hubbard Brook Forest, New Hampshire. From Likens and Bormann (1995).

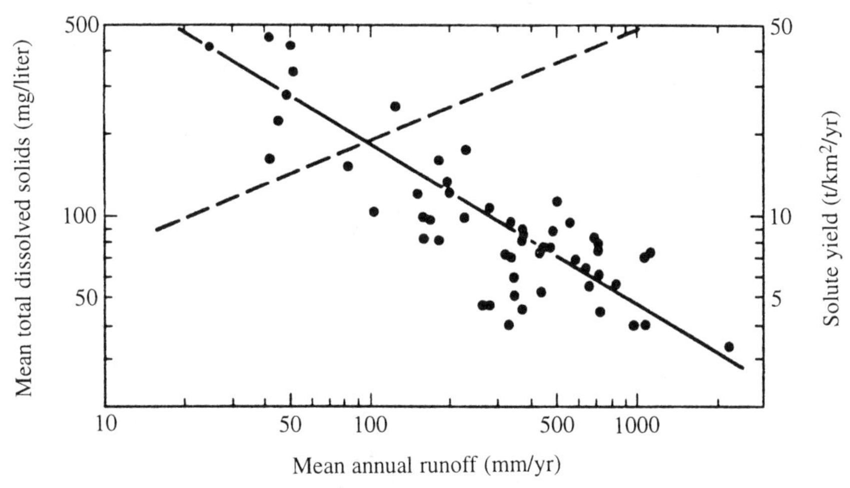

Figure 8.6 Variation of the concentration of total dissolved solids (solid line) and the total annual transport of dissolved substances (dashed line) for various streams in Kenya, as a function of mean annual runoff. From Dunne and Leopold (1978).

Gibbs (1970) used the concentrations of ions in major world rivers to suggest the origins of their dissolved constituents and their waters. Rivers dominated by precipitation show low concentrations of dissolved substances, and a high ratio of Cl to the total of Cl + HCO_3, reflecting the importance of Cl from rainfall (Zone A in Fig. 8.7). Rivers in which the

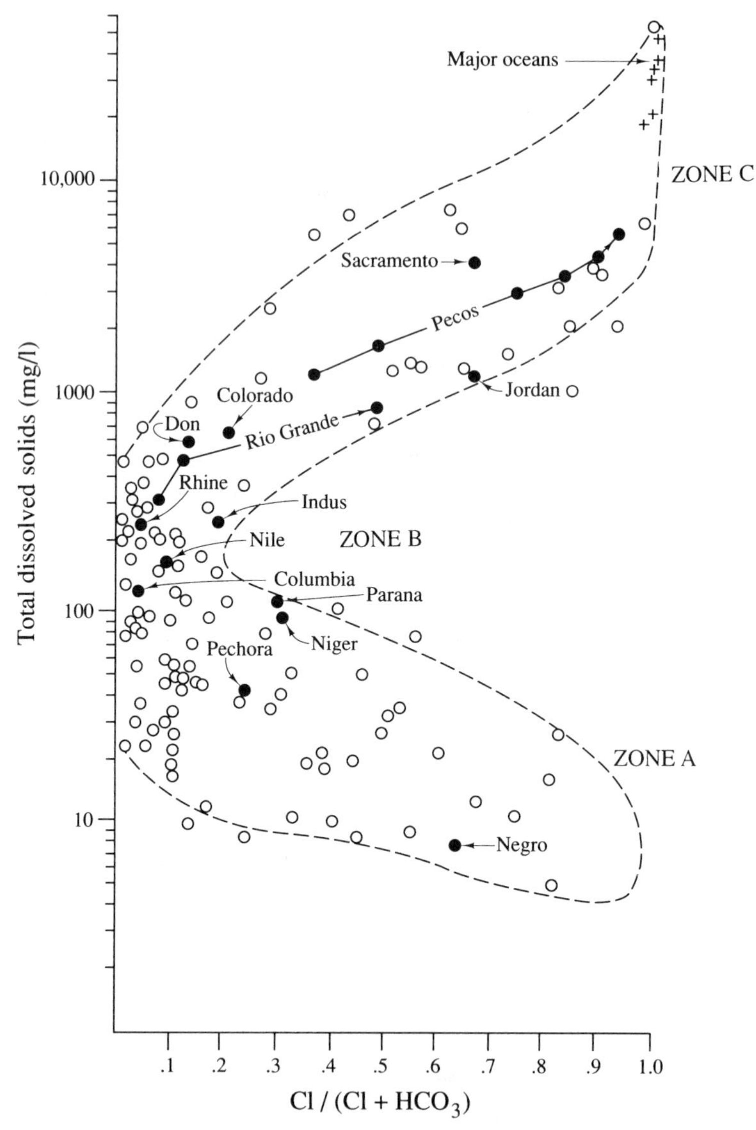

Figure 8.7 Variation in the total dissolved solids in rivers and lakes as a function of the ratio Cl/(Cl + HCO_3) in their waters. From Gibbs (1970).

dissolved load is largely derived from chemical weathering show higher concentrations of dissolved substances, and HCO_3 is the dominant anion (Zone B), reflecting the importance of carbonation weathering in most soils (Chapter 4). Rivers that pass through arid regions lose a significant amount of water to evaporation before reaching the ocean. These rivers (Zone C) show the greatest concentrations of dissolved ions and high ratios of $Cl/(Cl + HCO_3)$, because HCO_3 has been removed by the chemical precipitation of minerals such as $CaCO_3$ on the streambed (Holland 1978). In this scheme, seawater represents the endpoint of the evaporative concentration of river waters. These relationships are also seen when Na and Ca are used to scale the *X* axis, with the relative concentration of Na as an index of rainfall and Ca as an index of chemical weathering.

The composition of "average" riverwater was calculated by Livingstone (1963) from measurements on a large number of rivers (Table 8.5). His estimate of total dissolved transport, 37.6×10^{14} g/yr, is confirmed by more recent work (e.g., see also Tables 4.7 and 9.1). Not all dissolved substances in rivers are derived from rock weathering. A significant fraction of the Na, Cl, and SO_4 in riverflow is derived from marine aerosols ("cyclic salts," Chapter 3) that are deposited on land,[1] and humans have enhanced the atmospheric deposition of NO_3 and SO_4, accounting for the relatively high concentrations of these ions in the runoff from industrialized continents (Table 8.5). Reflecting the importance of carbonation weathering (Chapter 4), about $\frac{2}{3}$ of the HCO_3^- in rivers is derived from the atmosphere, either directly from CO_2 or indirectly via organic decomposition and root respiration that contribute CO_2 to the soil profile (Holland 1978, Meybeck 1987).

Nearly all the Ca, Mg, and K in river water is derived from rock weathering (Table 8.6). Weathering of carbonates is the dominant source for Ca, while silicates are the dominant source for Mg and K (Holland 1978). At least some Na is also derived from weathering, since its content in river water is in excess of the molar equivalent of Cl, which would be expected if seasalt were the sole source. The composition of individual streams may differ strongly from these averages depending upon local conditions. For instance, streams draining areas of carbonate terrain are dominated by Ca and HCO_3 (e.g., Laudelout and Robert 1994), and streamwaters may contain high concentrations of Na, Cl, and SO_4 where evaporite minerals are exposed (e.g., Stallard and Edmond 1983). The river transport of some dissolved ions is also enhanced by human activities, such as mining, which

[1] The amount of Cl in rivers that is derived from the atmosphere is the subject of some controversy (Berner and Berner 1987). Some budgets (e.g., Table 8.6) suggest that a relatively small amount of Cl is "cyclic," based on the observations by Stallard and Edmond (1981, 1983) in the Amazon basin. Other workers have assumed that a larger fraction of the Cl in global riverflow is derived from the sea, with values ranging from 85% (Dobrovolsky 1994, p. 83) to nearly 100% (Möller 1990).

Table 8.5 Mean Composition of Dissolved Ions in River Waters of the World[a]

Continent	HCO_3^-	SO_4^{2-}	Cl^-	NO_3^-	Ca^{2+}	Mg^{2+}	Na^+	K^+	Fe	SiO_2	Sum
North America	68	20	8	1	21	5	9	1.4	0.16	9	142
South America	31	4.8	4.9	0.7	7.2	1.5	4	2	1.4	11.9	69
Europe	95	24	6.9	3.7	31.1	5.6	5.4	1.7	0.8	7.5	182
Asia	79	8.4	8.7	0.7	18.4	5.6	9.3		0.01	11.7	142
Africa	43	13.5	12.1	0.8	12.5	3.8	11		1.3	23.2	121
Australia	31.6	2.6	10	0.05	3.9	2.7	2.9	1.4	0.3	3.9	59
World	58.4	11.2	7.8	1	15	4.1	6.3	2.3	0.67	13.1	120
Anions[b]	0.958	0.233	0.220	0.017							1.428
Cations[b]					0.750	0.342	0.274	0.059			1.425

[a] Livingstone (1963); concentrations in mg/liter.
[b] Millequivalents of strongly ionized components.

Table 8.6 Sources of Major Elements in World River Water (in Percent of Actual Concentrations)[a]

Element	Atmospheric cyclic salt	Weathering: Carbonates	Silicates	Evaporites	Pollution
Ca^{2+}	0.1	65	18	8	9
HCO_3^-	⩽1	61	37	0	2
Na^+	8	0	22	42	28
Cl^-	13	0	0	57	30
SO_4^{2-}[b]	2	0	0	22	43
Mg^{2+}	2	36	54	⩽1	8
K^+	1	0	87	5	7
H_4SiO_4	⩽1	0	99+	0	0

[a] From Berner and Berner (1987).
[b] SO_4^{2-} is also derived from the weathering of pyrite.

accelerate the natural rate of crustal exposure and rock weathering on Earth (Bertine and Goldberg 1971, Martin and Meybeck 1979).

The organic compounds in river water, especially the fulvic acids, are important in the dissolved transport of Fe and Al. These metals form complexes with organic acids (Chapter 4), and they are carried at concentrations well in excess of the solubility of Fe and Al hydroxides in river water (Perdue et al. 1976). The importance of dissolved organic acids in the transport of metals to the sea is a good example of the influence of terrestrial biota over simple geochemical processes that might otherwise determine the movement of materials on the surface of the Earth.

Suspended Load

The products of mechanical weathering and erosion are found in the suspended sediments of river water (Chapter 4). The concentration of suspended sediment often shows a curvilinear relationship with streamflow, increasing exponentially at high flows (Parker and Troutman 1989; Fig. 8.8). At low flows, suspended sediments are dominated by organic materials, but the proportion of organic matter declines as the amount of suspended sediment increases during high flows, when soil erosion is greatest (Meybeck 1982, Ittekkot and Arain 1986, Paolini 1995). Long-term records show that the sediment transport during occasional extreme events often exceeds the total transport during long periods of more normal conditions (Van Sickle 1981, Swanson et al. 1982). Large concentrations of suspended sediments are found during flash floods in deserts (Baker 1977, Fisher and Minckley 1978, Laronne and Reid 1993).

Transport of suspended sediments in world rivers is affected by many factors, including elevation, topographic relief, and runoff from the water-

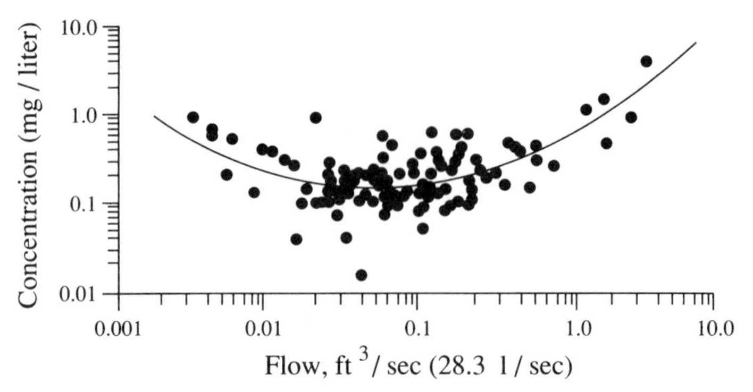

Figure 8.8 Concentration of particulate matter as a function of stream flow in the Hubbard Brook Experimental Forest of New Hampshire. From Bormann et al. (1974).

shed. Although the rivers draining arid regions show high concentrations of suspended sediments, their total flow is limited, so the loss of soil materials per unit of landscape is rather low (Milliman and Meade 1983). Rivers draining southern Asia carry 70% of the global transport of suspended sediments, 13.5×10^{15} g/yr (Fig. 4.14). Large sediment loads in the rivers of China are derived from erosion of massive deposits of wind-derived soils, loess, in their drainage basin. In contrast, the Amazon River carries only about 1.2×10^{15} g of suspended sediment each year, about 9% of the world's total (Meade et al. 1985). Most of the Amazon Basin is situated at low elevations with limited topographic relief, which accounts for its relatively low yield of suspended sediments (Meybeck 1977).

Much of the sediment removed from uplands is deposited in stream channels and floodplains in the lower reaches of rivers (Trimble 1977, Longmore et al. 1983, Gurtz et al. 1988). Thus, the sediment yield per unit area of watershed declines with increasing watershed area (Milliman and Meade 1983). Despite large seasonal variations in volume, the daily sediment transport of the Amazon is rather constant as a result of storage of sediment in the floodplain during periods of rising waters and remobilization during falling waters (Meade et al. 1985).

Erosion increases when vegetation is removed (Bormann et al. 1974), and global land clearing by humans for agriculture and construction has dramatically increased the transport of suspended sediments in many parts of the world (Pimentel et al. 1995). It is likely that the total rate of sediment transport to the sea is now 2× greater than it was several hundred years ago (Gregor 1970, Milliman and Syvitski 1992), but only about 10% of the current rate of soil erosion reaches the ocean (Lal 1995). The remainder is captured in lakes and floodplains and behind dams and other human structures (Dynesius and Nilsson 1994).

Salt Marshes and Estuaries

When large rivers reach sea level, their rate of flow slows, drastically reducing their ability to carry sediment. The load of suspended materials is deposited in the river channel and on the continental shelf. Rivers carrying large sediment loads, such as the Mississippi, may form obvious deltas. The river channel is progressively confined and divided by deposited sediments, which may support broad, flat areas of salt marsh vegetation (Fig. 8.9). The lower reaches of rivers and their salt marshes are subject to daily tidal inundation. An *estuarine* ecosystem consists of the river channel, to the maximum upstream extent of tidal influence, and the adjacent ocean waters, to the maximum seaward extent that they are affected by the addition of freshwater. The estuary also includes any salt marshes that may develop along the shoreline. Estuaries are zones of mixing; within an estuary there is a strong gradient in salinity from land to sea. Thus, estuaries are complicated and dynamic ecosystems that are challenging for studies of biogeochemistry (Kempe 1988, Odum 1988, Morris et al. 1995).

Within estuaries, salt marsh vegetation exists in a dynamic equilibrium between the rate of sediment accumulation and the rate of coastal subsidence or rise in sea level (Frey and Basan 1985). As deposits accumulate, the rate of erosion and the oxidation of organic materials increase, slowing the rate of further accumulation. Conversely, as sea level rises, deposits are inundated more frequently, leading to greater rates of sediment deposition and peat accumulation. Along the Gulf Coast of the United States, the rate of sedimentation has not kept pace with coastal subsidence, and substantial areas of marshland have been lost (DeLaune et al. 1983, Baumann et al. 1984). Further degradation of coastal ecosystems is expected if sea level rises due to global climatic warming (Holligan and Reiners 1992, Moorhead and Brinson 1995).

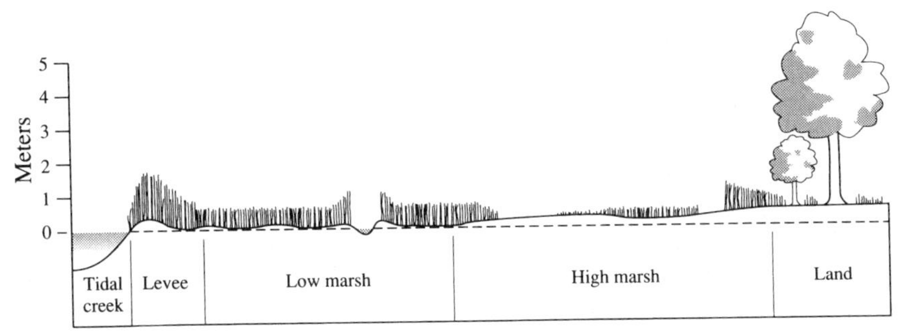

Figure 8.9 Schematic cross section through a salt marsh, showing the relationship between various components of the salt marsh ecosystem and the open waters of the estuary. From Wiegert et al. (1981).

Salt marsh soils are subject to a daily cycle of changing aeration. At high tide, the soils are inundated, and anaerobic conditions may develop throughout the profile. At low tide, the soils drain, allowing high redox potentials in the surface layers (Chapter 7). The transpiration of marsh plants may aid in rapid drying of the soil profile and the return of oxidizing conditions (Dacey and Howes 1984). Tidal fluctuations also affect salinity, which is lowest at low tide, when freshwater runoff from land may flush the soil profile. Tides confer spatial variability on estuaries, since the marsh areas that are closest to the sea are inundated more frequently than those in higher topographic positions.

Many studies make use of the salinity gradient in estuaries for comparative studies of biogeochemical processes (e.g., Fisher et al. 1988). Chloride is much more abundant in seawater than in freshwater, so it is often used as an index of overall salinity. Any position in the estuary or salt marsh can be described by its chloride concentration, which will lie between that of seawater (19,350 mg/liter, Table 9.1) and that of freshwater (ca. 8 mg/liter, Table 8.5). Chloride is particularly useful as an index, because it is very soluble and essentially uninvolved in reactions such as cation exchange, chelation, and precipitation. As such, chloride is known as a *conservative* ion. When river waters mix with seawater, other ions may change in concentration due to biotic uptake or exchange with sediments. Changes in the concentration of these ions in excess of changes that would be expected from simple dilution with seawater—as shown by the concentration of Cl^-—are used to infer biogeochemical processes in the estuary.

Biogeochemistry in Salt Marshes

Due to the complexity of the estuarine ecosystem, many investigators have examined salt marshes as a separate subsystem. These areas are often covered with dense vegetation, of which saltmarsh cordgrass, *Spartina alterniflora,* is the best known in North America. Net primary production in salt marshes is high, ranging from 133 to 1153 g C m^{-2} yr^{-1} aboveground in estuaries of the eastern United States, with a tendency for greater production in warmer climates (Hatcher and Mann 1975). Root growth contributes a substantial fraction of the total net primary production in salt marshes (Gallagher and Plumley 1979, Pomeroy et al. 1981, Howes et al. 1985), and roots show special adaptations for growth in anaerobic sediments in which high concentrations of potentially toxic substances, such as sulfide, are present (Mendelssohn et al. 1981, Carlson and Forrest 1982).

Salt marshes are effective filters and transformers of nutrients (Correll 1981). Salt marshes typically receive NO_3 from rivers and groundwater and a deliver N, usually in organic forms, to the coastal oceans (Valiela and Teal 1979, Childers et al. 1993). Despite long-term storage of organic matter in sediments, most salt marshes are a source of nitrogen and phosphorus for the open waters of their estuaries (Table 8.7).

Table 8.7 Annual Flux of Carbon and Nutrients from Salt Marshes to Coastal Waters[a]

Marsh	Carbon ($g\ C\ m^{-2}\ yr^{-1}$)			Nitrogen ($g\ N\ m^{-2}\ yr^{-1}$)			Phosphorus ($g\ P\ m^{-2}\ yr^{-1}$)	
	DOC	POC	TOC	NH_4^+	NO_3^-	Total	PO_4	TP
Great Sippewissett, Massachusetts		−76[b]		−4.2	−3.8	−24.6	−0.6	
Flax Pond, Long Island, New York	−8.4	+61	+53	−2.0	+1.0		−1.4	−0.3
Canary Creek, Delaware	−38	−62	−100	+0.7	+1.9	−1.2	<−0.1	
Gott's Marsh, Patuxent River, Maryland		−7.3		−0.4	−0.9	−3.7		−0.3
Ware Creek, York River, Virginia	−80	−35	−115	−2.9	+2.3	−2.8	−0.1	+0.7
Carter Creek, York River, Virginia	−25	−116	−142	−0.3	+0.3	−4.0	−0.6	0
Dill Creek, South Carolina		−303					−6.4	
North Inlet, South Carolina			−431					
Barataria Bay, Louisiana	−140	−25	−165					

[a] From Nixon (1980).
[b] Negative values are losses.

The flooded, anaerobic sediments of salt marshes allow significant rates of denitrification, which removes NO_3 from the ecosystem (Valiela and Teal 1979, Smith et al. 1983, Bowden et al. 1991). During a 7-year experiment, White and Howes (1994) found that 16 to 26% of the ^{15}N added to a New England salt marsh was exported to the sea, while 32 to 46% was denitrified. The rate of denitrification varies seasonally depending upon temperature (Kaplan et al. 1979, Jørgensen 1989).

In most salt marshes the dominant form of available nitrogen is NH_4, since nitrification rates are low and denitrifiers remove much of the available NO_3. Compared to upland ecosystems, in which the new inputs of nutrients typically account for about 10% of the annual nutrient availability to plants (Table 6.1), the relative contribution of new inputs and nutrient recycling in salt marshes is about equal (Table 8.8). Nevertheless, salt marsh vegetation is often nitrogen-limited and shows increased growth with nitrogen additions. Nitrogen fixation by blue-green algae and soil bacteria may account for a significant contribution to the nitrogen budget of salt marshes (DeLaune et al. 1989, White and Howes 1994). In a boreal salt marsh, Bazely and Jefferies (1989) found that the growth of blue-green algae was stimulated when the marsh vegetation was grazed by geese. In this case, N-fixation by the algae restored most of the nitrogen that was removed owing to the seasonal migration of the geese through these marshes.

Salt marsh and estuarine sediments show high rates of sulfate reduction (Chapter 7), since they are rich in organic matter, flushed with high concen-

Table 8.8 Nitrogen Budget for the Short *Spartina alterniflora* Areas of Great Sippewissett Marsh, Massachusetts[a]

	Nitrogen ($g\ N\ m^{-2}\ yr^{-1}$)	Percent of annual plant N demand
	Intersystem cycling	
Gains		
N-fixation	12.1	50.0–54.0
Precipitation	0.8	3.0–3.6
Import	0.9	3.7–4.0
Total	13.8	57.0–61.6
Losses		
Denitrification	4.1–5.6	16.9–25.0
Accretion	3.7–4.1	15.3–18.3
Export	2.0–3.2	8.3–14.3
Total	9.8–12.9	40.5–57.6
	Intrasystem cycling	
Remineralization	14.9–16.3	61.6–72.8
Translocation	1.4	5.9–6.3
Total	16.3–17.7	67.4–79.0

[a] Gains and losses are from the vegetated sediment system. From White and Howes (1994).

trations of SO_4 from seawater, and frequently anaerobic. Although the exact magnitude of sulfate reduction is the subject of some controversy (Howes et al. 1984), various investigators have suggested that more than half of the CO_2 released during decomposition of organic matter in salt marshes is associated with sulfate reduction (Howarth 1984, 1993, King 1988, Moeslund et al. 1994).

The initial product of sulfate reduction, H_2S, may be transformed to various reduced sulfur compounds, including pyrite and organic sulfides (Howarth 1979, Giblin 1988, Thode-Andersen and Jørgensen 1989, Kostka and Luther 1995). Hydrogen sulfide may also diffuse upward, where it is oxidized by aerobic bacteria near the soil surface or where it may support photosynthetic sulfur bacteria that oxidize H_2S and fix CO_2 (see Eq. 2.13). In some cases, H_2S, dimethylsulfide [$(CH_3)_2S$], and other reduced S gases escape to the atmosphere (Hines et al. 1993). Despite high rates of loss—averaging about 5 $g\ S\ m^{-2}\ yr^{-1}$ (Goldberg et al. 1981, Steudler and Peterson 1985; cf. wetland emissions p. 236), salt marshes make only a minor contribution to the global flux of reduced sulfur gases, since the total area of salt marshes is not large (Carroll et al. 1986).

Sulfate-reducing bacteria extract only a portion of the energy from the organic carbon compounds they degrade (Table 7.1). In many cases there is substantial reoxidation of organic sulfides and pyrite (Canfield et al. 1993, Moeslund et al. 1994), so the total accumulation of reduced S compounds represents only a small fraction of the total sulfate reduction that has occurred in salt marshes and estuarine sediments (Fig. 8.10). The oxidation process may produce thiosulfate ($S_2O_3^{2-}$), which diffuses downward, where it may be reduced again to sulfide, or upward, where it may be oxidized to sulfate (Luther et al. 1986, Jørgensen 1990). Both pyrite and organic sulfides contribute to the total S in the sediment (Haering et al. 1989, Ferdelman et al. 1991, Brüchert and Pratt 1996), but in a coastal estuary of Denmark, only about 15% of the total sulfide production was permanently buried—mostly as pyrite (Thamdrup et al. 1994). Formation of sulfide minerals is important in the retention of other metals, including some metallic pollutants, in salt marshes (Griffin et al. 1989).

Given the presence of abundant SO_4 from tidal waters, it is not surprising that the rate of methanogenesis in salt marshes is low, because sulfate-reducing bacteria are more effective competitors when SO_4 is abundant (Chapter 7). At a series of sites along the York River in the Chesapeake

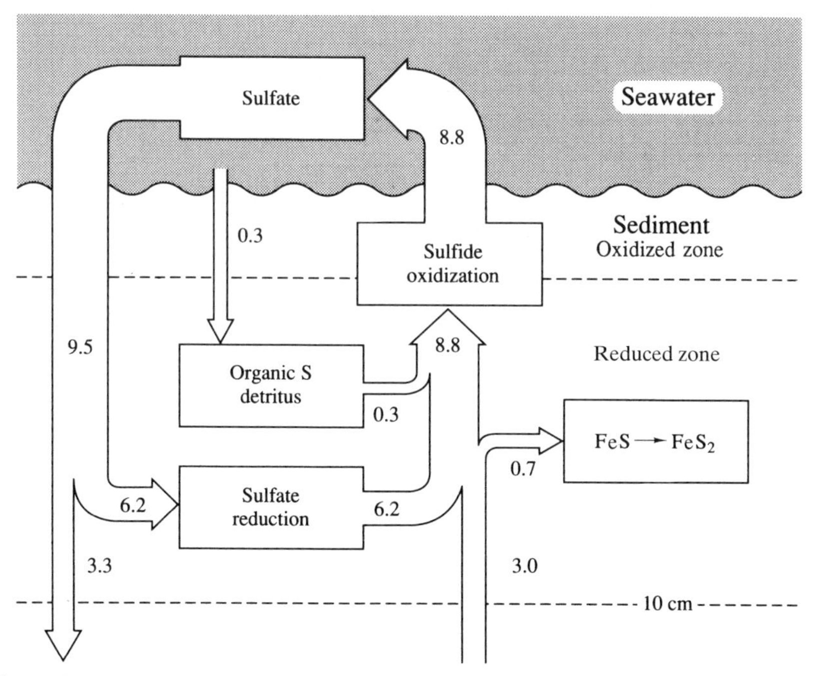

Figure 8.10 Transformations of sulfur in a coastal marine sediment. Note that of 6.2 g S m^{2-} yr^{-1} undergoing sulfate reduction, only 0.7 g S m^{-2} yr^{-1} is permanently stored in the sediment as pyrite or other reduced minerals. From Jørgensen (1977).

Bay estuary, Bartlett et al. (1987) show a gradient of decreasing methanogenesis with increasing salinity, as the SO_4 from seawater progressively inhibits methanogenesis (Fig. 8.11; see also Kelley et al. 1990). Howes et al. (1985) found that only about 0.3% of the total carbon input to the sediments of Sippewissett marsh in Massachusetts was lost through methanogenesis. Slightly higher rates have been reported for the Sapelo Island estuary in Georgia (King and Wiebe 1978), but globally, the methane emissions from saltwater marshes contribute little to the flux of CH_4 to the atmosphere (Chapter 11). Some salt marsh soils also appear to be a small source of phosphine (PH_3) gas to the atmosphere (Dévai and DeLaune 1995).

For many years the large production of fish and shellfish in estuaries was attributed to an abundance of organic carbon flushing from salt marshes to the open water. Indeed, the losses of organic carbon from salt marshes are usually >100 g C m^{-2} yr^{-1}, compared to values of 1 to 5 g C m^{-2} yr^{-1} from uplands (Table 8.7 vs. p. 267). Haines (1977), however, suggested that this paradigm was questionable, because the isotopic ratio of carbon in estuarine animals did not match that of *Spartina*. Using the natural abundance of stable isotopes of both sulfur and carbon, Peterson et al. (1986) showed that the organic carbon in shellfish of Great Sippewissett Marsh is about equally derived from *Spartina* and from phytoplankton production in the

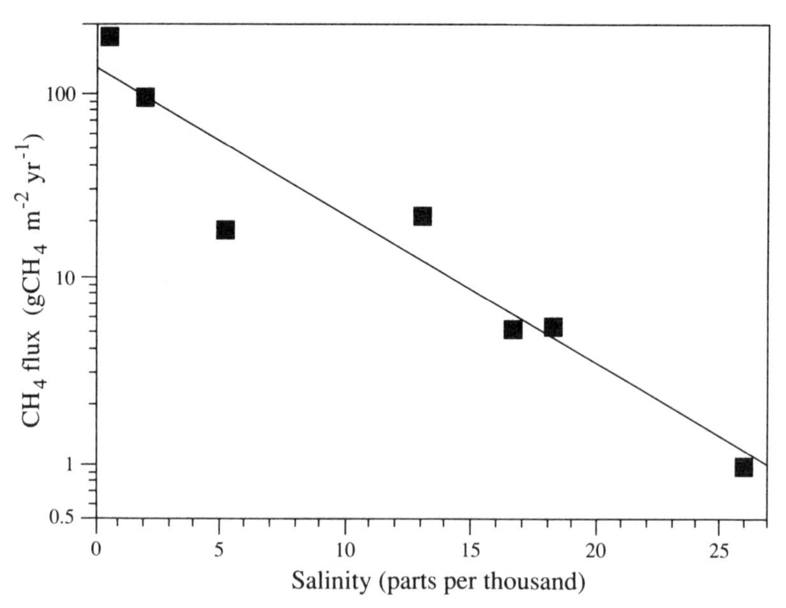

Figure 8.11 Annual methane lost from salt marsh soils as a function of salinity. From Bartlett et al. (1987).

open water (Fig. 8.12). The shellfish show isotopic ratios for C and S that are midway between these sources. Similar results were found in the Sapelo Island marsh (Peterson and Howarth 1987). Carbon from upland, terrestrial vegetation and carbon fixed by sulfur-oxidizing bacteria in salt marsh

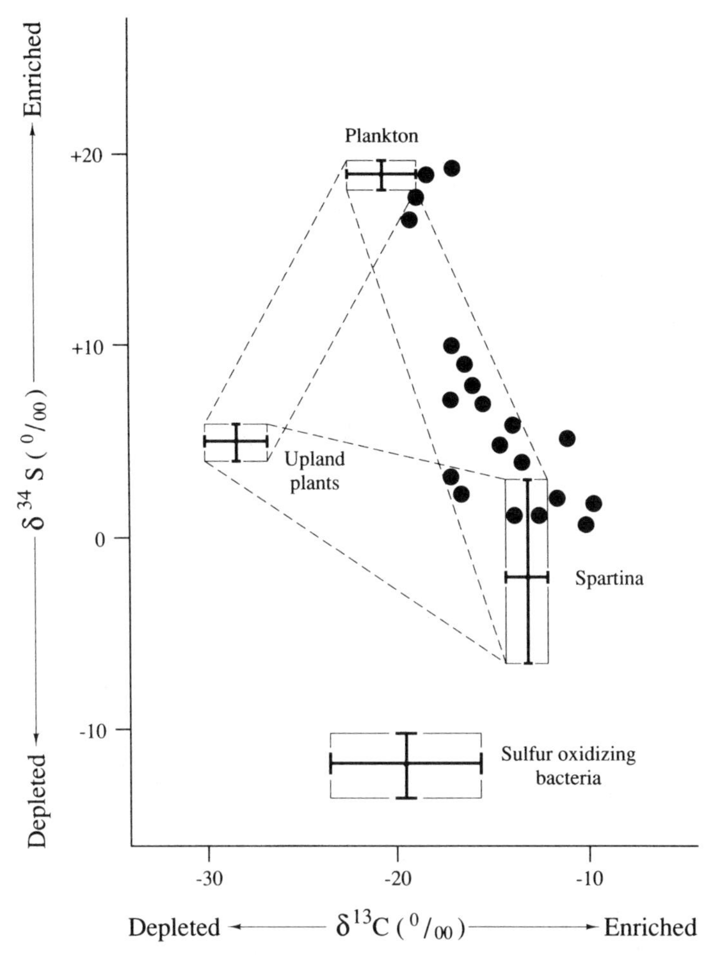

Figure 8.12 In the Great Sippewissett marsh in Massachusetts, the isotope ratio for C and S in estuarine shellfish is shown in relation to the ratios in upland plants, phytoplankton, salt marsh *Spartina,* and sulfur-oxidizing bacteria. The sulfur in sulfur-oxidizing bacteria has an isotope ratio that is very different from that in any of the shellfish, implying that these bacteria are not a major source of food for the higher trophic levels in the estuary. Similarly, the isotope ratio for carbon in terrestrial plants is much more negative than that in the consumers. The isotopic composition of the consumers in the estuary falls between the *Spartina* and phytoplankton, implying that these plant materials are the major sources of food. Modified from Peterson et al. (1986).

soils both play a minor role in supporting the abundant marine life of estuaries.

Open Water Habitats

The mixing of freshwater from rivers and salt water from the sea occurs in the central channel of an estuary. If the estuary is well mixed, the transition from freshwater to seawater is gradual and progressive as one moves downstream. In other cases, inflowing freshwater may extend over a "wedge" of denser saltwater, creating a sharp vertical gradient in salinity throughout much of the estuary (Fig. 8.13). In either case, the transition zone is an arena of rapid biogeochemical transformations and high productivity (Burton 1988).

Seawater has high pH (ca. 8.3), redox potential ($>+200$ mV), and ionic strength (total dissolved ions), relative to freshwater. The mixing of freshwater and seawater causes a rapid precipitation of the dissolved humic compounds carried by rivers. The cations in seawater replace H^+ on the exchange sites of the humic materials (Chapter 4), causing these materials to flocculate and sink to the bottom[2] (Boyle et al. 1977). Most dissolved humic compounds and metals, such as Fe, that are carried with humus substances are precipitated in the estuary or within a short distance of the mouth of the river (Boyle et al. 1974, Sholkovitz 1976, Figueres et al. 1978, Fox 1983). The flocculation of dissolved organic compounds and the deposition of larger plant debris account for a major portion of the organic carbon in estuarine sediments (Hedges et al. 1988), and there is little evidence that organic matter from land contributes much to marine sediments beyond the continental shelf (Hedges and Parker 1976, Prahl et al. 1994). As a result of the removal of terrestrial organic matter, the majority of the organic carbon in estuarine waters is composed of nonhumus substances, presumably from net primary production in the estuary and its salt marshes (Fox 1983, Nixon et al. 1996).

Most river waters are supersaturated with dissolved CO_2, which is derived from the degradation of organic materials during downstream transport. High concentrations of dissolved CO_2 and humic materials cause river waters to be slightly acid. Under these conditions, phosphorus binds to Fe-hydroxide minerals and is transported in the load of suspended sediment (Fig. 4.4, Table 4.8; Eyre 1994). Upon mixing with the higher pH of seawater, phosphorus desorbs from these minerals and contributes to dissolved phosphorus in the estuary (Chase and Sayles 1980, Lebo and Sharp 1992, Berner and Rao 1994, Conley et al. 1995). Seitzinger (1991) found that upon mixing with seawater an increase in pH in the Potomac River estuary caused a release of P from sediments, stimulating a bloom of

[2] A similar reaction between salts, usually $Al_2(SO_4)_3$, and organic matter is frequently used to cleanse sewage waters of dissolved organic compounds.

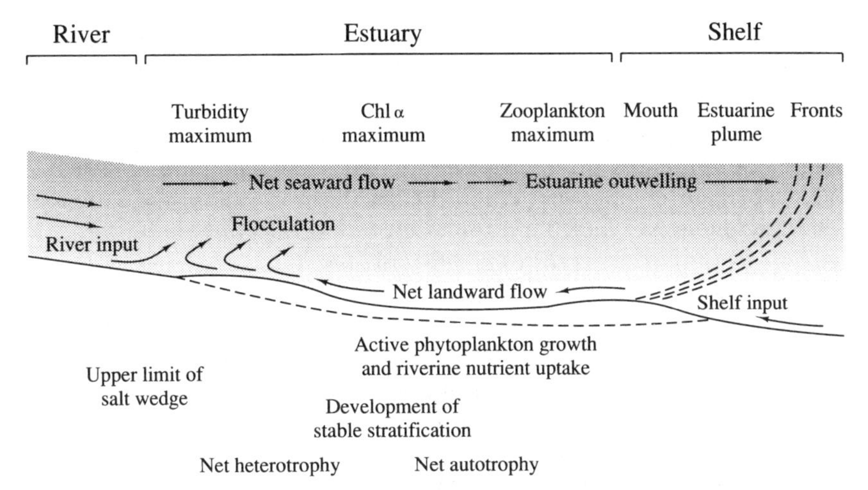

Figure 8.13 Conceptual model of the chemical and biological structure in estuaries. As the suspended load settles from the entering river waters and nutrients are made available, phytoplankton production increases, fueling an increase in zooplankton production and higher trophic levels. From Fisher et al. (1988).

N-fixing blue-green algae—a scenario that is analogous to the shifts in species dominance that are seen in P-polluted lakes (Chapter 7). Phosphorus is often more available in the waters of estuaries than in either freshwater or seawater. De Jonge and Villerius (1989) suggest that the phosphorus bound to carbonate particles in seawater is released as seawater mixes with freshwater and the carbonates dissolve under the acidic conditions of the estuary.

A great deal of effort has been directed toward understanding the nitrogen budget of estuaries. Most river waters do not contain large concentrations of available nitrogen (NO_3 and NH_4), and these forms are removed when the waters pass over coastal salt marshes. Indeed, the filtering action of land and marsh vegetation is so effective that inputs of nitrogen in rain can make a substantial contribution to the nitrogen budget of the central waters of estuaries (Correll and Ford 1982). However, as is the case for terrestrial ecosystems (Chapter 6), most of the nitrogen that supports estuarine productivity is not derived from new inputs but from mineralization and recycling of organic nitrogen within the estuary and its sediments (Stanley and Hobbie 1981).

At the pH and redox potential of seawater, nitrification occurs rapidly in estuarine waters (Billen 1975, Horrigan et al. 1990). Nitrification also occurs in the upper layers of sediment (Admiraal and Botermans 1989). Denitrification in the lower, anaerobic layers of sediment is primarily supported by nitrate diffusing down from the upper sediment (Seitzinger 1988, Kemp et al. 1990), although nitrate in the water column may also diffuse back into the sediments, where it is reduced (Simon 1988, C.S. Law et al.

1991). In Narragansett Bay, Rhode Island, Seitzinger et al. (1980, 1984) found that denitrification removed about 50% of the available NO_3 entering in riverflow and about 35% of that derived from mineralization within the estuary. The major product of denitrification was N_2. In Chesapeake Bay, denitrification leaves the nitrate in the lower water column enriched in $\delta^{15}N$ (Horrigan et al. 1990). When the nitrification rate in the sediments is low, available NO_3 may limit the rate of denitrification, and more NH_4^+ remains to support the growth of phytoplankton in the estuary (Kemp et al. 1990). Storms and tidal currents stir up the sediments in an estuary, releasing large quantities of NH_4 to the water column (Simon 1989).

The net primary productivity of estuaries shows a direct relation to nitrogen inputs (Fig. 8.14). As a result of human inputs of nitrogen in sewage, agricultural runoff, and acid rain, many estuaries show excessive levels of productivity and conditions that resemble the eutrophication of freshwaters (Officer et al. 1984, Nixon et al. 1996). As seen for polluted lakes (Chapter 7), the waters of many estuaries and coastal oceans have high levels of available N and P and evidence of Si limitations of coastal phytoplankton (Justić et al. 1995). Increased autochthonus production in the Chesapeake Bay estuary has led to an increasing occurrence of anoxic conditions in bottom waters and sediments (Cooper and Brush 1991).

The management of polluted estuaries is the subject of much controversy. Some workers argue that an improvement in estuarine conditions will be directly related to efforts to reduce nutrients in inflowing waters (Nixon 1987). Others suggest that the retention of prior inputs and the recircula-

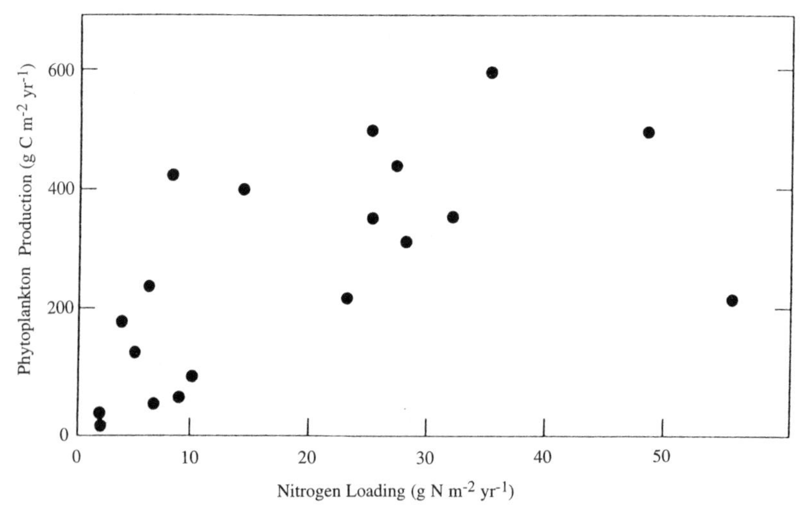

Figure 8.14 Annual phytoplankton productivity in estuaries as a function of the new inputs of nitrogen to their waters. From Boynton et al. (1982).

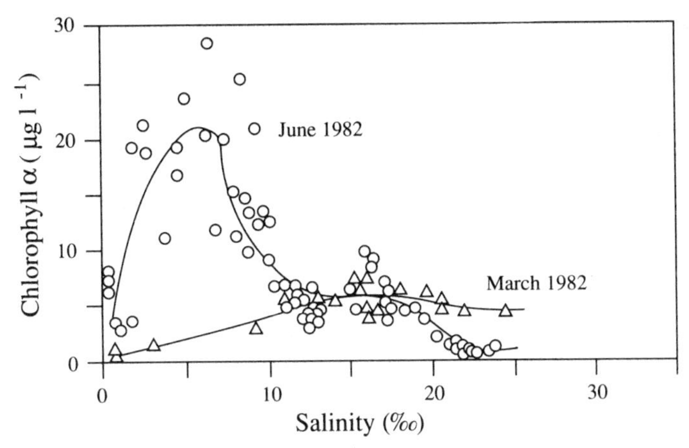

Figure 8.15 Data from Chesapeake Bay, showing the distribution of chlorophyll *a*, as an index of phytoplankton production, relative to salinity in the estuary. From Fisher et al. (1988).

tion of nitrogen within the system mean that efforts to reduce new inputs will not necessarily produce immediate improvements in water quality (Kunishi 1988).

Many estuaries show a peak in net primary productivity at intermediate salinities, reflecting the zone of maximum nutrient availability and phytoplankton abundance (Edmond et al. 1981, Anderson 1986; Fig. 8.15). In other cases, mixing hides any obvious relationship between net primary production and conservative properties, such as salinity, in the estuary (Powell et al. 1989). Phytoplankton productivity and organic matter derived from the surrounding salt marshes fuel the high productivity of fish and shellfish in estuarine waters, but the disruption of estuaries by direct pollution, global sea level rise, and other human perturbations may substantially reduce their potential to contribute to future human food supplies.

Summary

Stream ecosystems are intimately tied to biogeochemical reactions in the surrounding terrestrial ecosystems. The rate of flow and the chemical properties of streamwater are determined by the soil properties and vegetation in the watershed. Biogeochemical transformations of C, N, and P occur in streams. Most stream ecosystems are heterotrophic, showing an excess of respiration over net primary production. During stream transport, available forms of N and P are removed from water and sequestered in organic and inorganic forms. Streams carry a variety of the other ions of biochemistry in dissolved and suspended form, which are largely the products of rock weathering in the watershed. Humans have had dramatic impact on streams and rivers throughout the world, altering the load of dissolved and suspended materials and regulating the flow of water (Dynesius and Nilsson 1994).

The mixing of freshwater and seawater occurs in estuaries, located at the mouth of major rivers. In response to changes in pH, redox potential, and salinity, river waters feed estuaries with a rich solution of available N and P, and high rates of net primary production fuel a productive coastal marine ecosystem. Despite a temporary storage of nutrients in salt marshes and estuarine sediments, river waters are always a net source of nutrients to their estuary and the coastal ocean. As we shall see, rivers are a major source of nutrients in the global budgets of biogeochemical elements in the ocean.

Recommended Readings

Allan, J.D. 1995. *Stream Ecology: Structure and Function of Running Waters.* Chapman and Hall, New York.

Day, J.W., C.A.S. Hall, W.M. Kemp and A. Yanaz-Arancibia. 1989. *Estuarine Ecology.* Wiley, New York.

Degens, E.T., S. Kempe and J.E. Richey. (eds.). 1991. *Biogeochemistry of Major World Rivers.* Wiley, New York.

Drever, J.I. 1988. *The Geochemistry of Natural Waters.* 2nd ed. Prentice-Hall, Englewood Cliffs, New Jersey.

Holland, H.D. 1978. *The Chemistry of the Atmosphere and Oceans.* Wiley, New York.

Knox, G.A. 1986. *Estuarine Ecosystems: A Systems Approach.* 2 vols. CRC Press, Boca Raton, Florida.

Ward, R.C. 1967. *Principles of Hydrology.* McGraw-Hill, London.

9

The Oceans

Introduction

The Earth's waters constitute its hydrosphere. Only small quantities of freshwater contribute to the total; most water resides in the oceans. In this chapter we will examine the biogeochemistry of seawater and the contributions that oceans make to global biogeochemical cycles. We will begin with a brief overview of the circulation of the oceans and the mass balance of the major elements that contribute to the salinity of seawater. Then, we will examine net primary productivity (NPP) in the surface waters and the fate of organic carbon in the sea. Net primary productivity in the oceans is related to the availability of essential nutrient elements, particularly nitrogen and phosphorus. Conversely, biotic processes strongly affect the chemistry of many elements in seawater, including N, P, Si, and a variety of trace metals. We will examine the biogeochemical cycles of essential elements in the sea and the processes that lead to the exchange of gaseous components between the oceans and the atmosphere.

Ocean Circulation

Global Patterns

In Chapter 3 we saw that the circulation of the atmosphere was driven by the receipt of solar energy which heated the atmosphere from the bottom, creating instability in the air column. Unlike the atmosphere, the oceans are heated from the top. Because warm water is less dense than cooler water, the receipt of solar energy conveys stability to the water column, preventing exchange between warm surface waters and deep, cold waters over much of the ocean (Ledwell et al. 1993).

Within the surface layers, seawater is relatively well mixed by the wind (Thorpe 1985, Archer 1995). Depending upon the incident radiation, the surface waters range from 75 to 200 m in depth with a mean temperature of about 18°C. The surface temperature in some tropical seas may reach 30°C. The zone of rapid increase in density between the warm surface waters and the cold deep waters is known as the *pycnocline.* It roughly parallels the gradient in temperature, which is known as the thermocline (Chapter 7). The ocean's deep waters contain about 95% of the volume with a mean temperature of 3°C.

Atmospheric winds (Chapter 3) lead to the formation of currents in the oceans, such as the well-known Gulf Stream in the Atlantic Ocean (Fig. 9.1). In each ocean the trade winds (Fig 3.3a) drive surface currents from east to west along the equator. When these currents encounter land, the waters divide to form currents moving north and south along the eastern borders of the continents. As they move toward the poles, the currents are deflected to the right by the Coriolis force (Fig. 3.3c), so the Gulf Stream crosses the North Atlantic and delivers warm waters to northern Europe. Water returns to the tropical latitudes in cold surface currents that flow along the west side of continents. The cyclic pattern of surface currents in each of the major oceans is called a *gyre.* The global circulation of surface currents transfers heat from the tropics to the polar regions of the Earth (Oort et al. 1994). More than half of the net excess of solar energy received

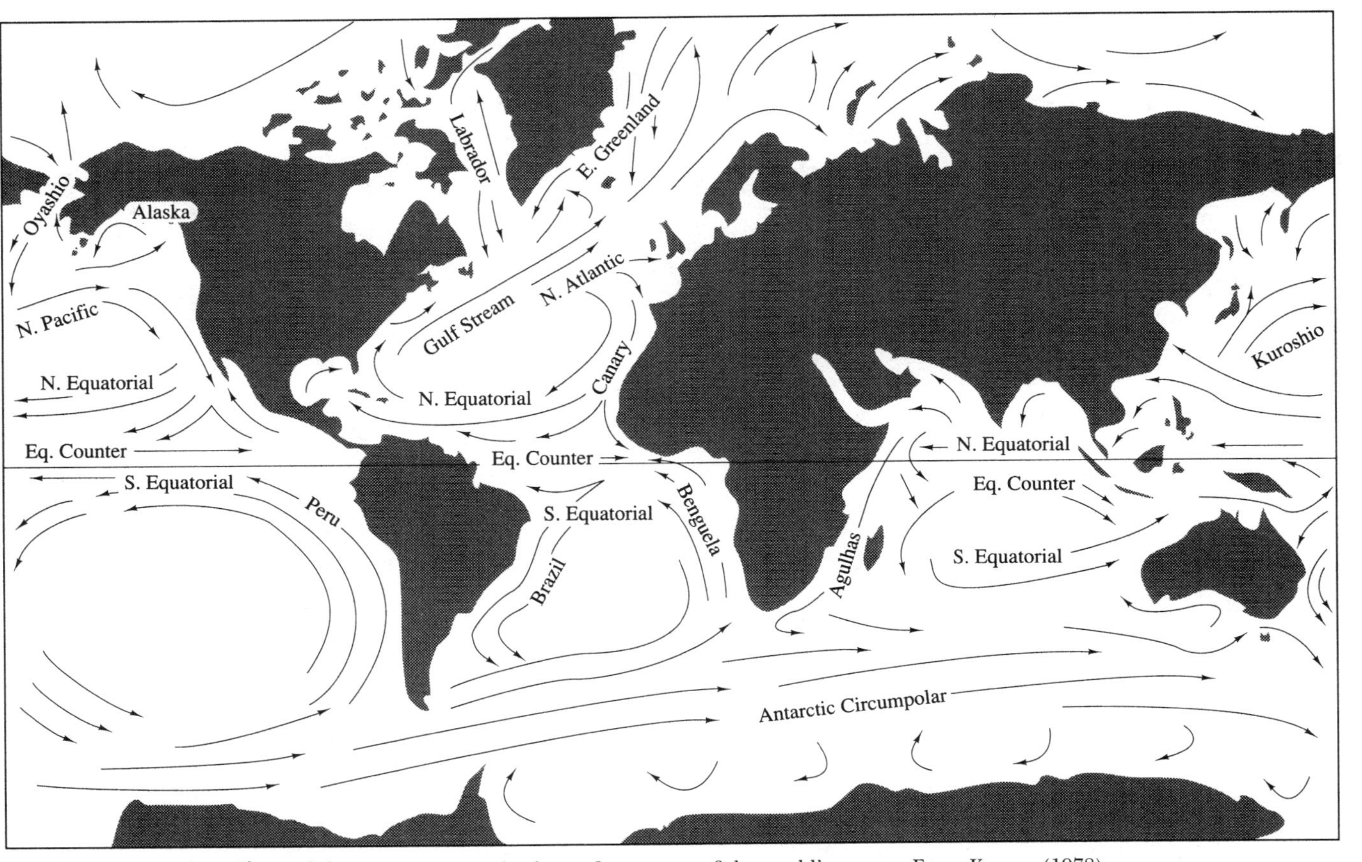

Figure 9.1 Major currents in the surface waters of the world's oceans. From Knauss (1978).

in the tropics is transferred to the poles by ocean circulation; the remainder is transferred through the atmosphere (Vonder Haar and Oort 1973).

With the loss of heat at polar latitudes, the density of seawater increases. Exchange between the surface ocean and the deep waters is possible when the surface waters cool and their density increases to that of the underlying water. Mixing of waters of equal density is known as *isopycnal mixing*. During the winter in the Arctic and Antarctic oceans, the density of some polar waters also increases when fresh water is "frozen out" of seawater and added to the polar ice caps, leaving behind waters of greater salinity that sink to the deep ocean. In contrast, during the summer, the polar oceans have lower surface salinity due to melting from the icecaps. Because the seasonal downwelling of cold polar waters is driven by both temperature and salinity, it is known as *thermohaline circulation*.

Penetration of cold waters to the deep ocean at the poles, which is known as downwelling, creates deep sea currents. For example, North Atlantic deep water (NADW), which forms near Greenland, moves southward through the deep Atlantic, rounding the tip of Africa and entering the Indian and Pacific Oceans. Major zones of upwelling are found in the Pacific Ocean and in the circumpolar southern ocean around 65° S latitude (Toggweiler and Samuels 1993). Deep waters are nutrient rich, so high levels of oceanic productivity are found in zones of upwelling. Upwelling along the western coast of South America yields high levels of net primary production that support the anchovy fishery of Peru.

These patterns of ocean circulation have important implications for biogeochemistry. One might calculate an overall mean residence time of 34,000 years for seawater with respect to river flow (i.e., total ocean volume/annual river flow). In fact, most rivers mix only with the volume of the surface ocean, which has a mean residence time of about 1700 years with respect to river water. If we account for the addition of rain waters and upwelling waters to the volume of the surface ocean, the actual turnover time of the surface waters is even faster. For example, the mean residence time of surface waters in the north Pacific Ocean is about 9–15 years (Michel and Suess 1975). The surface water is also in rapid gaseous equilibrium with the atmosphere. Mean residence time for CO_2 in the surface ocean is about 6 years (Stuiver 1980).

Renewal of the bottom waters is confined to the polar regions. Downward mixing of ^3H_2O produced from the testing of atomic bombs (Fig. 9.2) and downward mixing of anthropogenic chemicals of recent origin (e.g., see Krysell and Wallace 1988) show the rate of entry of surface waters to the deep sea and the movement of deep water toward the equator. The downward transport in the North Atlantic is 13–17 Sv[1]—roughly 10× the annual rate of riverflow to the oceans (Dickson and Brown 1994, MacDonald and Wunsch 1996). Similarly, about 4–5 Sv sink in the Weddell Sea to form

[1] 1 Svedrup (Sv) = 10^6 m^3/sec = 3.2×10^{13} m^3/yr.

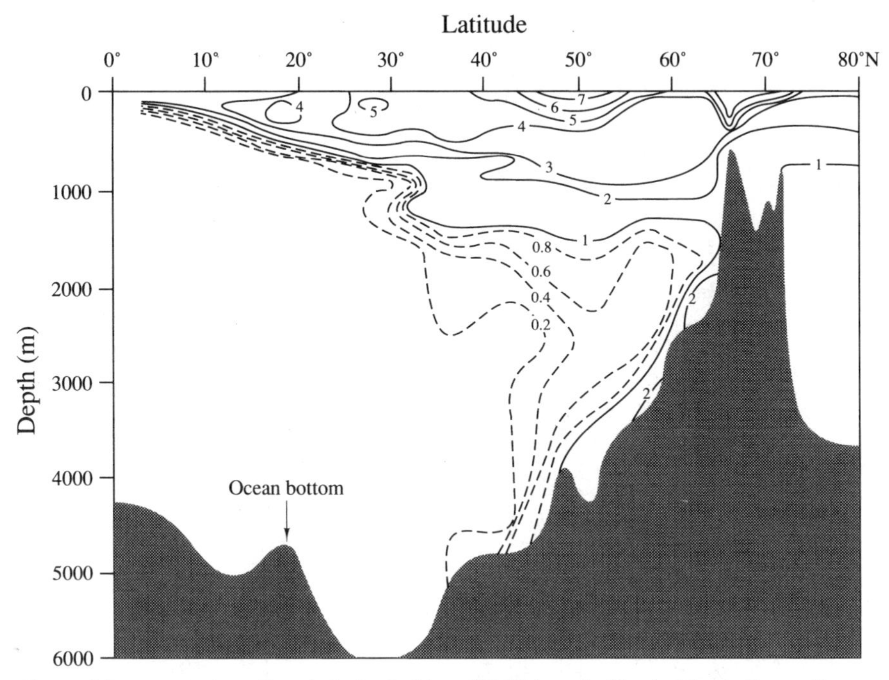

Figure 9.2 Penetration of bomb-derived tritium (^3H_2O) into the North Atlantic Ocean. Data are expressed as the ratio of $^3H/H \times 10^{-18}$ for samples collected in 1972. From Ostlund (1983).

northward-flowing, Antarctic bottom water (Hogg et al. 1982, Schmitz 1995). Because the volume of water entering the deep ocean is much greater than the total annual river flow to the oceans, the mean residence time of the deep ocean is much less than 34,000 years. Estimates of the mean age of bottom waters using ^{14}C dating of dissolved CO_2 range from 275 years for the Atlantic Ocean to 510 years for the Pacific (Stuiver et al. 1983). Thus, the deep waters maintain a historical record of the conditions of the surface ocean several centuries ago.

Deep water currents also transfer seawater between the major ocean basins as a result of the Antarctic circumpolar current. In the Atlantic Ocean, evaporation exceeds the sum of riverflow and precipitation, yielding a higher seawater salinity than in the Pacific (Fig. 9.3). The Atlantic receives a net inflow of less saline waters from the Pacific to restore the water balance. At the same time, dense, saline water flows out of the deep Atlantic to enter the Indian and Pacific Oceans.

Changes in ocean currents, particularly the formation of deep waters, may be associated with changes in global climate. For example, an increase in the rate of downwelling of cold, saline water in the North Atlantic at the start of the last glacial epoch may have led to a decline in atmospheric

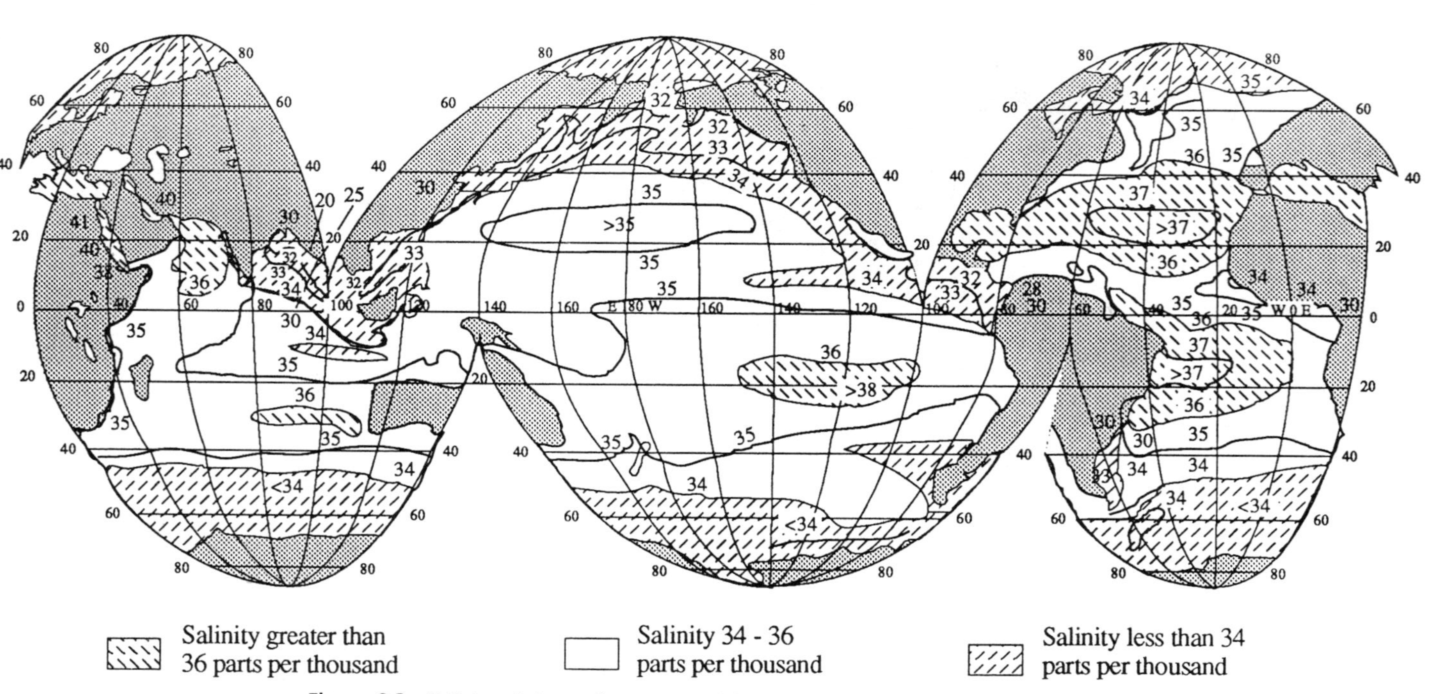

Figure 9.3 Salinity of the surface waters of the world's oceans. From Gross (1977).

CO_2, because CO_2 is more soluble in cold waters (Broecker and Peng 1987). During the last glacial epoch, the concentration of atmospheric CO_2 was about 200 ppm, compared to 280 ppm in the preindustrial atmosphere (Fig. 1.5). However, because the production of deep water is dependent upon the density difference between a warm surface layer and the cold waters that sink beneath it, once the glacial epoch was fully developed, the production of Atlantic deep water was likely to decline, reducing the transport of atmospheric CO_2 to the deep ocean and allowing warmer conditions to return. Climate changes are also likely to have affected the pattern of surface currents; the Gulf Stream appears to have shifted southward, producing a humid climate in southern Europe during the last glacial epoch (Keffer et al. 1988).

El Niño

Ocean currents also show year-to-year variations that affect biogeochemistry and global climate. One of the best known variations in current occurs in the central Pacific Ocean. Under normal conditions, the trade winds drive warm surface waters to the western Pacific, allowing cold bottom waters to upwell along the coast of Peru. Periodically, the surface transport breaks down in an event known as the El Niño–Southern Oscillation (ENSO). During El Niño years, the warm surface waters remain along the coast of Peru, preventing the upwelling of nutrient-rich water. Phytoplankton growth is limited and the fisheries industry collapses (Glynn 1988).

Associated with the warm surface waters in the eastern Pacific are changes in global climate, for example, exceptionally warm winters and greater rainfall in western North America (Molles and Dahm 1990, Swetnam and Betancourt 1990, Redmond and Koch 1991). At the same time the absence of warm surface waters in the western Pacific reduces the intensity of the monsoon rainfalls in southeast Asia and India. Working with atmospheric scientists, oceanographers now recognize that El Niño events are part of a cycle that yields opposite but equally extreme conditions during non-El Nino years. These are known as La Niña conditions (Philander 1989). Although the switch from El Niño to La Niño is poorly understood, it is likely that the conditions at the beginning of each phase reinforce its development, with the cycle averaging between 3 and 5 years between El Niño events. A similar, but less powerful, cyclic pattern of ocean circulation is seen in the Atlantic Ocean (Philander 1989).

The upwelling of cold, deep ocean waters during the La Niña years leads to lower atmospheric temperatures over much of the northern hemisphere. Thus, El Niño–La Niña cycles add variation to the global temperature record, complicating efforts to perceive atmospheric warming that may be due to the greenhouse effect. Moreover, the El Niño–La Niña cycle affects the concentrations of atmospheric CO_2, since the release of CO_2 from cold,

Table 9.1 Major Ion Composition of Seawater, Showing Relationships to Total Salinity and Mean Residence Times for the Elements with Respect to River Water Inputs

Constituent	Concentration in seawater[a] (mg/kg)	Chlorinity ratio[a]	Concentration in river water[b] (mg/kg)	Mean residence time (10^6 yr)
Sodium	10,760	0.5561	5.15	75
Magnesium	1,294	0.0668	3.35	14
Calcium	412	0.0213	13.4	1.1
Potassium	399	0.0206	1.3	11
Strontium	7.9	0.00041	0.03	12
Chloride	19,350	1.0000	5.75	120
Sulfate	2,712	0.1400	8.25	12
Bicarbonate	145	0.0075	52	0.10
Bromide	67	0.0035	.02	100
Boron	4.6	0.00024	0.01	10.0
Fluoride	1.3	0.000067	0.10	0.5
Water				0.034

[a] Holland (1978).
[b] Meybeck (1979) and Holland (1978).

upwelling waters is lower during years of El Niño (Bacastow 1976, Inoue and Sugimura 1992, Wong et al. 1993). During the 1991–1992 El Niño, the east Pacific ocean released 0.3×10^{15} g C as CO_2 to the atmosphere, compared to its normal efflux of 1.0×10^{15} g C (Murray et al. 1995), and the rate of CO_2 increase in the atmosphere slowed for several years (Keeling et al. 1995). In addition to its effects on ocean productivity, El Niño conditions affect other aspects of biogeochemistry in the sea. Lower rates of denitrification in warm El Niño waters may decrease the total marine denitrification rate by as much as 25% over La Niña conditions (Codispoti et al. 1986, Cline et al. 1987). Efforts to understand and predict El Niño events are an important component of global change research.

The Composition of Seawater

Major Ions

Table 9.1 gives the concentration of the major ions in seawater of average salinity, 35‰ (i.e., 35 g of salts per kilogram of water). The mean residence times for these ions are much longer than the mean residence time for water in the oceans, so these elements are uniformly distributed. Although seawater varies slightly in salinity throughout the world (Fig. 9.3), these ions are conservative in the sense that they maintain the same concentrations relative to one another in most ocean waters. Thus, a good estimate of total salinity can be calculated from the concentration of a single ion. Often chloride is used, and the relationship is

$$\text{salinity} = 1.81(\text{Cl}), \tag{9.1}$$

with both values in ‰. Table 9.1 shows the mean ratio between chloride and other major ions in seawater over a wide range of salinity.

Like the atmosphere, the composition of the major elements of seawater has been relatively constant for long periods of time. In the face of continual inputs of new ions in river water, the constant composition of seawater must be maintained by processes that remove ions from the oceans. Table 9.1 shows that the time for rivers to supply the elemental mass in the ocean, the mean residence time, varies from 120 million years for Cl to 1.1 million years for Ca. Biological processes, such as the deposition of calcium carbonate in the shells of animals, are responsible for the relatively rapid cycling of Ca. But even for Cl the mean residence time is much shorter than the age of the oceans.

A number of processes act to remove the major elements from seawater. Earlier, we saw that wind blowing on the ocean surface produces seaspray and marine aerosols that contain the elements of seawater (Chapter 3). A significant portion of the river transport of Cl from land is derived directly from the sea (Table 8.6). The atmospheric transport of these "cyclic salts" removes ions from the sea roughly in proportion to their concentration in seawater. Other processes must act differentially on the major ions, because their concentrations in seawater are much different from the concentrations in rivers. For example, whatever process removes Na from seawater must not be effective until the concentration of Na has built up to high levels (Drever 1988). On the other hand, Ca is the dominant cation in river water (Table 9.1), but it's concentration in seawater is relatively low.

Ions are removed from the oceans when the clays in the suspended sediments of rivers undergo ion exchange with seawater. In rivers, most of the cation exchange sites (Chapter 4) are occupied by Ca. When these clays are delivered to the sea, Ca is released and replaced by other cations, especially Na (Sayles and Mangelsdorf 1977). Some K and Mg may also be taken up by illite and montmorillonite clays that are delivered to coastal oceans by rivers (Chapter 4). Most deep sea clays show higher concentrations of Na, K, and Mg than are found in the suspended matter of river water (Martin and Meybeck 1979). The clays eventually settle to the ocean floor, causing a net loss of these ions from ocean waters.

Other mechanisms of loss occur in ocean sediments. Sediments are porous and the pores contain seawater. Burial of ocean sediments and their porewaters is significant in the removal of Na and Cl, which are the most concentrated ions in seawater. Biological processes are also involved in the burial of elements in sediments. As we will discuss in more detail in a later section, the deposition of $CaCO_3$ by organisms is the major process removing Ca from seawater. Biological processes also cause the removal of

SO_4, which is consumed in sulfate reduction and deposited as pyrite in ocean sediments (see Chapters 7 and 8).

During some periods of the Earth's history, vast deposits of minerals have formed when seawater evaporated from shallow, closed basins. Today, the extensive salt flats, or sabkhas, in the Persian Gulf region are the best examples. Although the area of such seas is limited, the formation of evaporite minerals has been an important mechanism for the removal of Na, Cl, and SO_4 from the oceans in the geologic past (Holland 1978).

So far, the processes that we have discussed for the removal of elements from seawater cannot explain the removal of much of the annual riverflow of Mg and K to the sea. For a time, marine geochemists postulated several reactions of "reverse weathering," whereby silicate minerals were reconstituted in ocean sediments, removing Mg and other cations from the ocean (MacKenzie and Garrels 1966). Although some deep-sea sediments appear to be a small sink for Mg and K (Kastner 1974, Sayles 1981), direct evidence of reverse weathering has proven elusive (Drever 1988). Recently, however, Michalopoulos and Aller (1995) found that alumniosilicate minerals are reconstituted in laboratory cultures of marine sediments from the Amazon River, perhaps sequestering as much as 10% of the annual flux of K to the sea.

In the late 1970s, Corliss et al. (1979) examined the emissions from hydrothermal (volcanic) vents in the sea. One of the best-known hydrothermal systems is found at a depth of 2500 m near the Galapagos Islands in the eastern Pacific Ocean. Hot fluids emanating from these vents are substantially depleted in Mg and SO_4 and enriched in Ca, Li, Rb, Si, and other elements compared to the seawaters that feed the hydrothermal system. Globally the annual sink of Mg in hydrothermal vents, where it leads to the formation of Mg-rich silicate rocks, exceeds the delivery of Mg to the oceans in river water. The flux of Ca to the oceans in rivers, 480×10^{12} g/yr, is incremented by an additional flux of up to 170×10^{12} g/yr from hydrothermal vents (Edmond et al. 1979).

In sum, it appears that most Na and Cl are removed from the sea in porewater burial, sea spray, and evaporites. Magnesium is largely removed in hydrothermal exchange, and calcium and sulfate by the deposition of biogenic sediments. The mass balance of potassium is not well understood, but K appears to be removed by exchange with clay minerals, leading to the formation of illite, and by some reactions with basaltic sediments (Gieskes and Lawrence 1981). Whitfield and Turner (1979) show an indirect correlation between the mean residence time of elements in seawater and their tendency to incorporate into one or more sedimentary forms (Fig. 9.4). Over long periods of time, ocean sediments are subducted to the Earth's mantle, where they are converted into primary silicate minerals, with volatile components being released in volcanic gases (H_2O, CO_2, Cl_2, SO_2, etc.; Fig. 1.4).

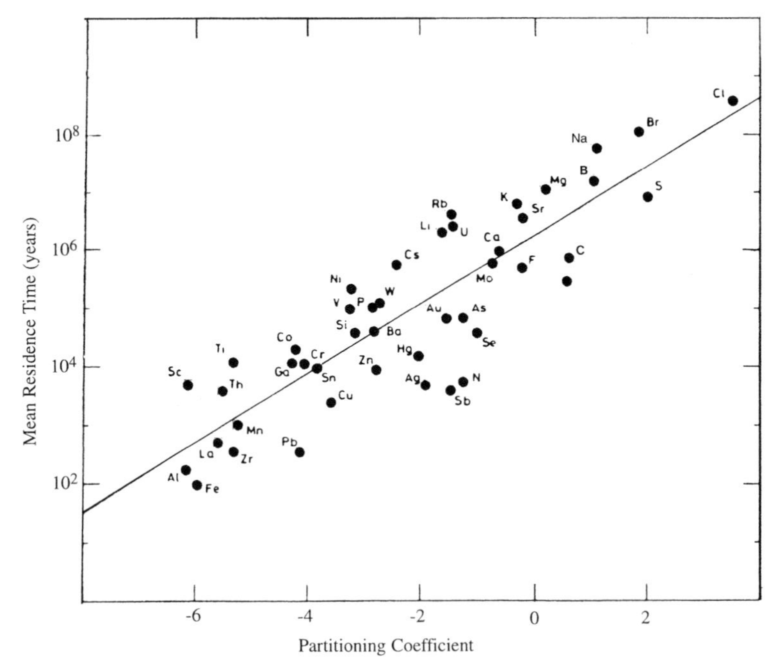

Figure 9.4 Mean residence time of elements in seawater as a function of their concentration in seawater divided by their mean concentration in the Earth's crust—with high values of the index indicating elements that are very soluble. From Whitfield and Turner (1979).

Net Primary Production

Global Patterns

As much as half of the Earth's photosynthesis may occur in the sea (cf. Tables 5.2 and 9.2). Compared to massive forests, the organic carbon produced in the ocean is easy to overlook, because it is largely the result of phytoplankton that are small and ephemeral. Phytoplankton production occurs in the surface mixed layer, in which the distribution of dissolved O_2 is an indirect measure of the rate of photosynthesis (Fig. 9.5). Net primary production in the sea is usually measured using the oxygen-bottle and ^{14}C techniques, as outlined for lake waters in Chapter 7. A new technique based on measurements of the O_2 supersaturation of seawater from photosynthesis has yet to be applied widely in marine environments (Craig and Hayward 1987).

Controversy surrounding the exact magnitude of marine production derives from the tendency for O_2-bottle measurements of NPP to exceed those made using ^{14}C in the same waters (Peterson 1980). Part of the problem can be explained by recent observations of a large biomass of picoplankton, which passes through the filtration steps of the ^{14}C procedure.

Table 9.2 Estimates of Total Marine Primary Productivity and the Proportion That Is New Productivity[a]

Province	% of ocean	Area ($10^{12}m^2$)	Mean production (g C $m^{-2}yr^{-1}$)	Total global production (10^{15}g C yr^{-1})	New production[b] (g C $m^{-2}yr^{-1}$)	Global new production (10^{15}g C yr^{-1})
Open ocean	90	326	130	42	18	5.9
Coastal zone	9.9	36	250	9.0	42	1.5
Upwelling area	0.1	0.36	420	0.15	85	0.03
Total		362		51		7.4

[a] From Knauer (1993).
[b] New productivity defined as C flux at 100 m.

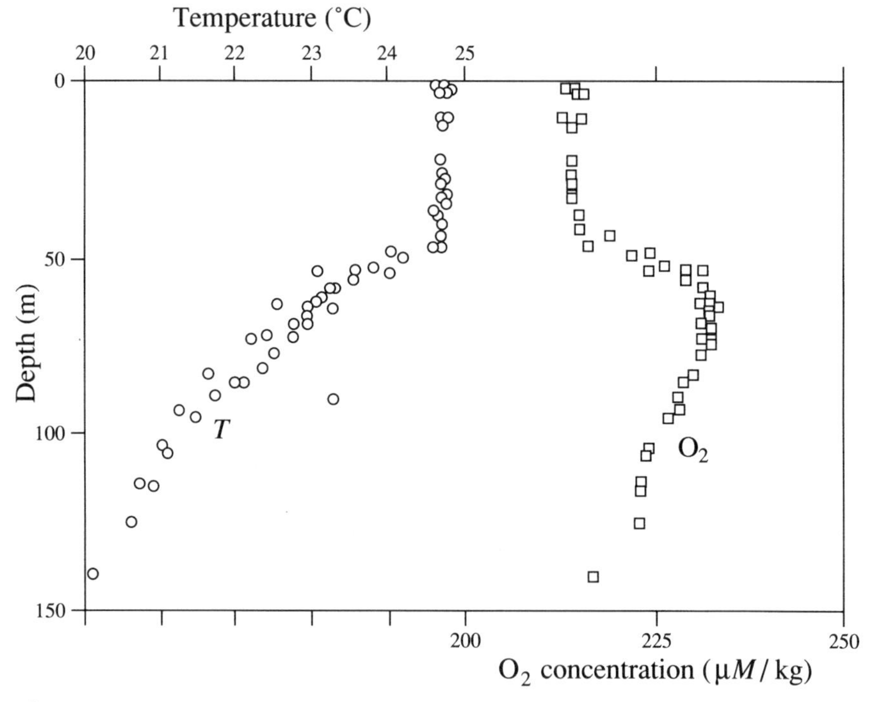

Figure 9.5 Distribution of temperature and O_2 with depth in the North Pacific Ocean. The peak in O_2 at 50 m is not unusual; it reflects the frequent observation that maximum photosynthesis does not occur at the surface, but at a lower level of the euphotic zone where there is maximum nutrient remineralization. From Craig and Hayward (1987). See also Fig. 9.19 for the distribution of O_2 to 1700 m.

In the waters of the eastern tropical Pacific Ocean, Li et al. (1983) found that 25 to 90% of the photosynthetic biomass passes a 1-μm filter, and Stockner and Antia (1986) suggest that such picoplankton may regularly account for up to 50% of ocean production. Marine phytoplankers also release large amounts of dissolved organic carbon to seawater (Baines and Pace 1991), and these compounds—technically a component of NPP—pass through the filtration procedures of the ^{14}C method. Another source of error stems from the possibility that many past studies of NPP have contaminated the seawater samples during application of the ^{14}C technique (Martin et al. 1987).

These methodological problems account for much of the variation among estimates of global marine production. Berger (1989) suggested that global marine NPP was 27×10^{15} g C/yr (Fig. 9.6), but some workers now suggest that marine NPP may be nearly twice that large (Table 9.2). Clearly, biogeochemists must work to improve their estimate of this critical component of the marine biosphere. Despite disagreement on the total value, all workers find that the highest individual values of NPP are measured in coastal regions, where nutrient-rich estuarine waters mix with seawater, and in regions of upwelling, where nutrient- rich deep water reaches the surface (Fig. 9.6). However, as a result of their large area, the open oceans account for about 80% of the total marine NPP, with continental shelf areas accounting for the remainder (Table 9.2). Although massive beds of kelp are found along some coasts, such as the *Macrocystis* kelps of southern California, seaweed accounts for only about 0.1% of marine production globally (Smith 1981, Walsh 1984).

Remote sensing offers significant potential for improving estimates of marine NPP. In 1978 the National Oceanic and Atmospheric Administration (NOAA) launched the Coastal Zone Color Scanner (CZCS) aboard the Nimbus-7 satellite (Hovis et al. 1980, Walsh and Dieterle 1988). The CZCS records the various wavelengths of radiation reflected from the ocean surface. Where ocean waters contain little phytoplankton, there is limited absorption of incident radiation by chlorophyll, and the reflected radiation is blue. Where chlorophyll is abundant, the reflectance contains a greater proportion of green wavelengths (Prézelin and Boczar 1986). The reflected light is indicative of algal biomass in the upper 20–30% of the euphotic zone, where most NPP occurs (Balch et al. 1992). CZCS images show dramatically the distribution of chlorophyll in the coastal ocean (Plate 2). The reflectance data can be used to calculate the concentration of chlorophyll and hence production (Fig. 9.7, Platt and Lewis 1987, Platt and Sathyendranath 1988). Recently, Antonie et al. (1996) used this approach to estimate that marine NPP lies between 36.5 to 45.6×10^{15} g C/yr globally.

A new spectral radiometer (MODIS) being developed by NASA for the Earth Observing System will allow greater satellite coverage of the world's oceans and the potential to monitor suspected long-term trends in oceanic

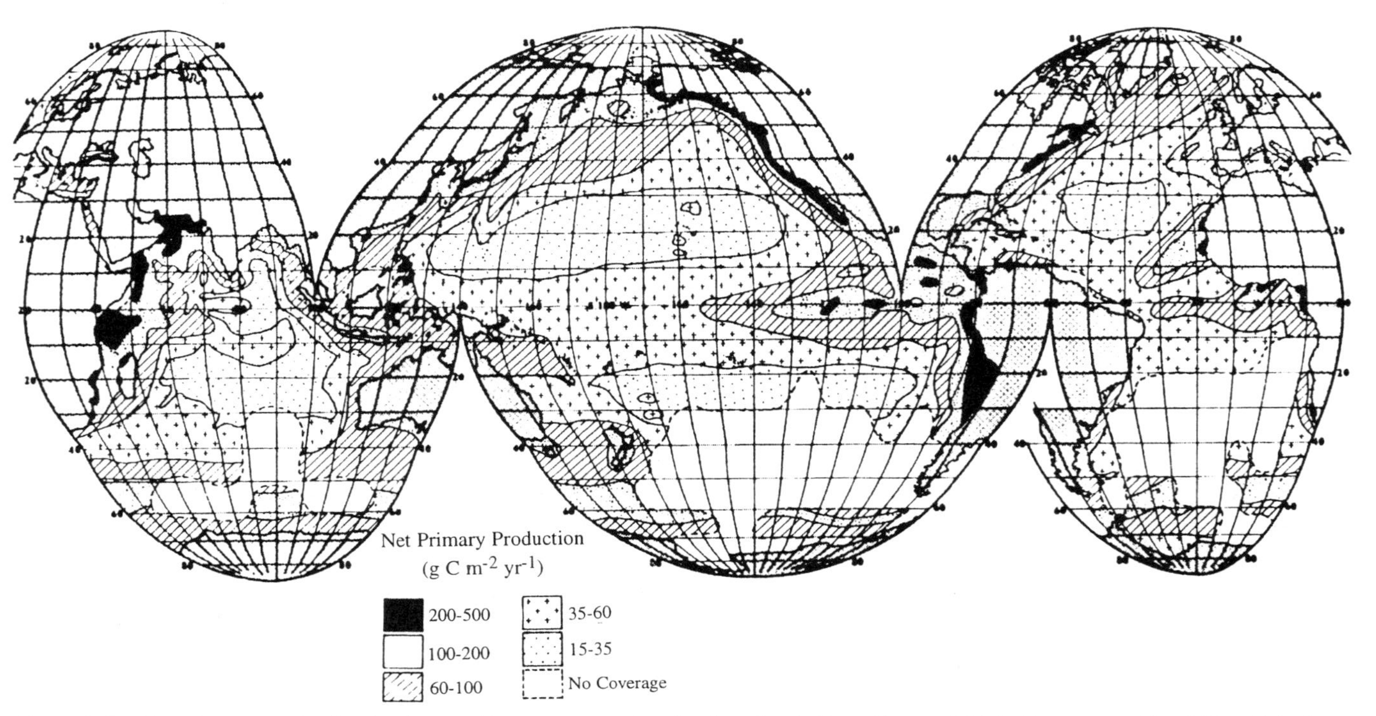

Figure 9.6 Net primary production in the world's oceans in units of g C m^{-2} yr^{-1}. From Berger (1989). Compare to Fig. 5.12, which shows the distribution of net primary production on land.

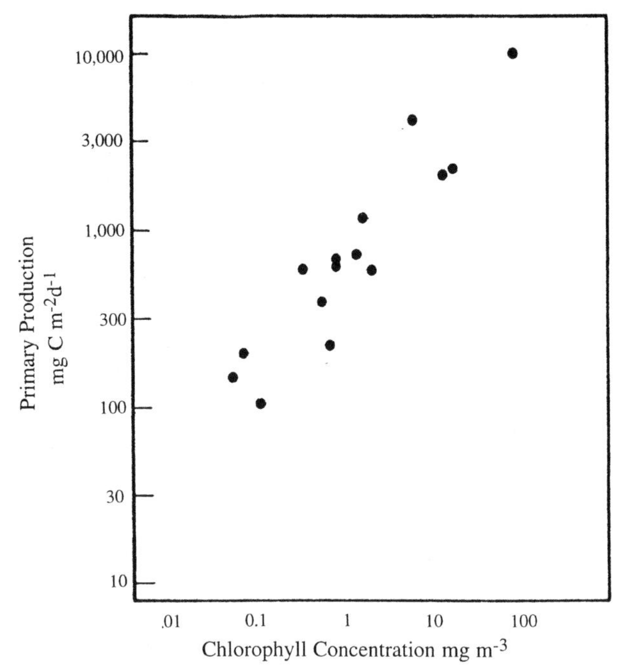

Figure 9.7 Net primary productivity as a function of surface chlorophyll in waters of coastal California. From Eppley et al. (1985).

NPP (e.g., Venrick et al. 1987, Falkowski and Wilson 1992). Future satellites can monitor trends in the marine phytoplankton of the Antarctic ocean (Sullivan et al. 1993), where the loss of stratospheric ozone allows an increasing flux of damaging ultraviolet radiation to the ocean's surface (R.C. Smith et al. 1992).

Fate of Marine Net Primary Production

Most marine NPP is consumed by zooplankton and free-floating bacteria, known as bacterioplankton, in the surface waters. The bacterioplankton also appear to decompose a large fraction of the dissolved organic carbon and organic colloids produced by phytoplankton (Kirchman et al. 1991, Druffel et al. 1992, Moran and Buesseler 1992, Huh and Prahl 1995). Cho and Azam (1988) concluded that bacteria were more important than zooplankton in the consumption of particulate organic carbon in the North Pacific Ocean. Reviewing a large number of studies from marine and freshwater systems, Cole et al. (1988) found that net bacterial growth (production) is about twice that of zooplankton and accounted for the disappearance of 30% of NPP from the photic zone (cf. Ducklow and Carlson 1992). In some areas, gross consumption by bacteria may reach 70% of NPP, especially when NPP is low (Biddanda et al. 1994). Whereas zooplankton represent the first step in a trophic chain that eventually leads to large animals such as fish, bacteria are consumed by a large population of bacteriovores that mineralize nutrients and release CO_2 to the surface waters. Thus, when bacteria are abundant, a large fraction of the carbon fixed by NPP in the sea is not passed to higher trophic levels (Ducklow et al. 1986). In areas where bacterial growth is inhibited by cold waters, more NPP is available to pass to higher trophic levels, including commercial fisheries (Pomeroy and Deibel 1986).

There is general agreement among oceanographers that about 80–90% of the NPP is degraded to inorganic compounds (CO_2, NO_3, PO_4, etc.) in the surface waters, and the remainder sinks below the euphotic zone to the deep ocean. The estimates of sinking are constrained, since greater rates of sinking would remove unreasonably large quantities of nutrients from the surface ocean (Broecker 1974, Eppley and Peterson 1979). The downward flux of organic matter varies seasonally depending upon productivity in the surface water (Deuser et al. 1981, Asper et al. 1992, Sayles et al. 1994). Bacterial degradation continues as particulate organic material (POM) sinks through the water column of the deep ocean. The mean sinking rate is about 350 m/day, so the average particle spends about 10 days in transit to the bottom (Honjo et al. 1982). Bacterial respiration accounts for the consumption of O_2 and the production of CO_2 in the deep water. Honjo et al. (1982) found that respiration rates averaged 2.2 mg C m^{-2} day^{-1} in the deep ocean, where the rate of bacterial respiration is probably limited by cold temperatures. About 95% of the particulate carbon is degraded within a depth of 3000 m and only small quantities reach the sediments of the deep ocean (Suess 1980, Martin et al. 1987, Jahnke 1996). Significant rates of decomposition also continue in the sediments (Emerson et al. 1985, Cole et al. 1987, Bender et al. 1989, K.L. Smith 1992).

If the current, higher estimates of marine NPP are correct, then approximately 7.4×10^{15} g C/yr may sink to the deep waters of the ocean (Knauer 1993). From a compilation of data from sediment cores taken throughout the oceans, Berner (1982) estimates that the rate of incorporation of organic carbon in sediments is 0.157×10^{15} g C/yr. These values suggest that about 98% of the sinking organic debris is degraded in the deep sea (cf. Martin et al. 1991). Degradation of organic carbon continues in marine sediments, and the ultimate rate of burial of organic carbon in the ocean is about 0.085 to 0.126×10^{15} g C/yr (Lein 1984, Berner 1982, Dobrovolsky 1994, p. 168). Even the larger value is much less than 1% of marine NPP.

Maps of the distribution of organic carbon in ocean sediments are similar to maps of the distribution of net primary production in the surface waters (Fig. 9.6), except that a greater fraction of the total burial (83%) occurs on the continental shelf (Premuzic et al. 1982, Berner 1982). Isotopic analyses show that nearly all the sedimentary organic matter in the deep sea is derived from marine production and not from land (Hedges and Parker 1976, Prahl et al. 1994). Indeed, degradation of river-borne organic materials must continue in the ocean, because the total burial of organic carbon in the ocean is less than the global delivery in rivers, 0.4×10^{15} g C/yr (Schlesinger and Melack 1981). This has led to the curious suggestion that the ocean is a net heterotrophic system, because the ratio of total respiration to autochthonous production is >1.0 (Smith and MacKenzie 1987).

Sediment Diagenesis

Organic Diagenesis

Changes in the chemical composition of sediments after deposition are known as *diagenesis*. Many forms of diagenesis are the result of microbial activities that proceed following the order of redox reactions outlined in Chapter 7 (Thomson et al. 1993). Organic marine sediments undergo substantial diagensis after burial as a result of sulfate reduction (Froelich et al. 1979, Berner 1984). In organic-rich sediments, sulfate reduction may begin within a few centimeters of the sediment surface where O_2 is depleted by aerobic respiration (e.g., Thamdrup et al. 1994). Viable anaerobic bacteria extend to depths of >500 m in Pacific Ocean sediments, adding a considerable dimension to the realm of the biosphere on Earth (Parkes et al. 1994).

Globally, Lein (1984) suggests that 14% of sedimentary organic carbon may be oxidized through anaerobic respiration, especially sulfate reduction. In marine environments, sulfate reduction leads to the release of reduced sulfur compounds (e.g., H_2S) and to the deposition of pyrite in sediments (Eqs. 7.17–7.19). The rate of pyrite formation is often limited by the amount of available iron (Boudreau and Westrich 1984), so only a small fraction of the sulfide is retained as pyrite and the remainder escapes to the upper layers of sediment where it is reoxidized (Jørgensen 1977, Thamdrup et al. 1994). When the rate of sulfate reduction is especially high, reduced gases may also escape to the water column.

The importance of sulfate reduction is much greater in organic-rich, near-shore sediments than in sediments of the open ocean (Skyring 1987, Canfield 1989b, 1991). Near-shore environments are characterized by high rates of NPP and a large flux of organic particles to the sediment surface. Sulfate-reduction generally increases with the overall rate of sedimentation, which is greatest near the continents (Canfield 1989b, 1993). Anoxic conditions develop rapidly as organic matter is buried in these sediments. In a marine basin off the coast of North Carolina (USA), Martens and Klump (1984) found that 149 moles C m^{-2} yr^{-1} were deposited, of which 35.6

moles were respired annually. The respiratory pathways included 27% in aerobic respiration, 57% in sulfate reduction leading to CO_2, and 16% in methanogenesis.

In contrast, pelagic (open-ocean) areas have lower NPP, lower downward flux of organic particles, and lower overall rates of sedimentation. The sediments in these areas are generally oxic (Murray and Grundmanis 1980, Murray and Kuivila 1990), so aerobic respiration exceeds sulfate reduction by a large factor (Canfield 1989b), and little organic matter remains to support sulfate reduction at depth (Berner 1984). Among near-shore and pelagic habitats, there is a strong positive correlation between the content of organic carbon and pyrite sulfur in sediments (Fig. 9.8), but it is important to remember that the deposition of pyrite occurs at the expense of organic carbon (Fig. 1.1). Thus, the net ecosystem production of marine environments is represented by the *total* of sedimentary organic carbon + sedimentary pyrite–with the latter resulting from the transformation of organic carbon to reduced sulfur (Eq. 7.17).

The rate of burial of organic carbon depends strongly on the sedimentation rate (Fig. 9.9). Greater preservation of organic matter in near-shore environments is likely to be due to the greater NPP in these regions (Bertrand and Lallier-Vergès 1993), rapid burial (Henrichs and Reeburgh 1987, Canfield 1991), and somewhat less efficient decomposition under anoxic conditions (Canfield 1994, Kristensen et al. 1995). As seen in soils (Chapter 5), the long-term persistence of organic matter in marine sediments is also

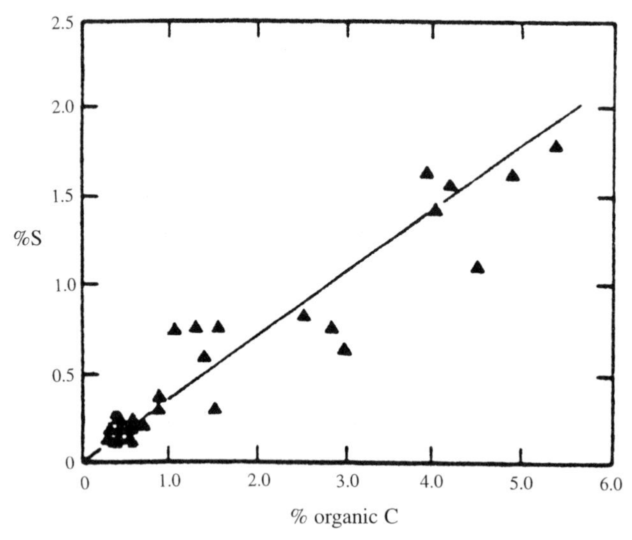

Figure 9.8 Pyrite sulfur content in marine sediments as a function of their organic carbon content. From Berner (1984).

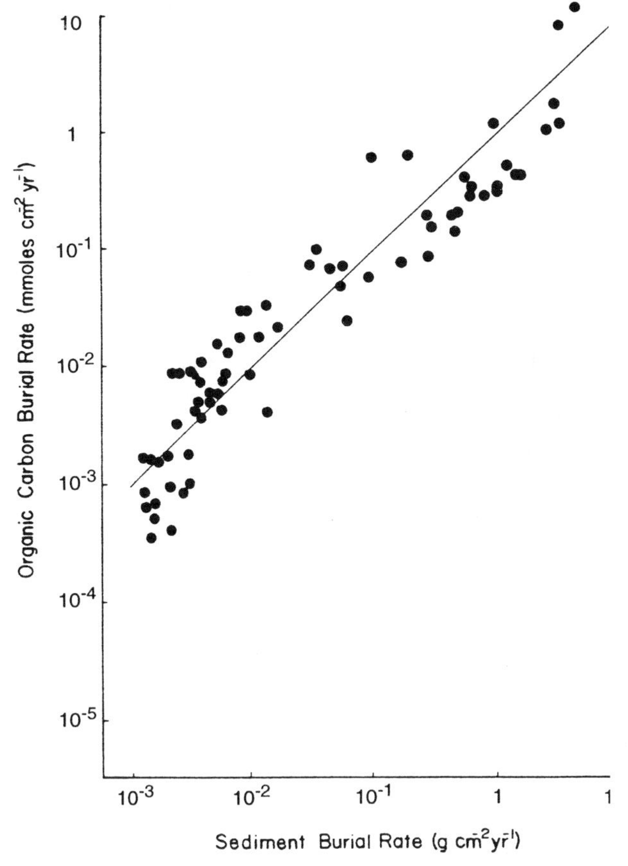

Figure 9.9 Burial of organic carbon in marine sediments as a function of the overall rate of sedimentation. From Berner and Canfield (1989).

enhanced by association with mineral surfaces (Keil et al. 1994, Mayer 1994).

Permanent burial of reduced compounds (organic carbon and pyrite) accounts for the release of O_2 to the atmosphere. The molar ratio is 1.0 for organic carbon, but the burial of 1 mole of reduced sulfur accounts for nearly 2.0 moles of O_2 (Raiswell and Berner 1986, Berner and Canfield 1989; Fig. 1.1). The weight ratio of C/S in most marine shales is about 2.8—equivalent to a molar ratio of 7.5 (Raiswell and Berner 1986). Thus, through geologic time the deposition of reduced sulfur in pyrite may account for about 20% of the O_2 in the atmosphere. As discussed in Chapter 3, the burial of reduced substances through geologic time is thought to regulate the content of O_2 in the atmosphere. During periods of rapid continental uplift, erosion, and sedimentation, large amounts of organic substances were buried and the oxygen content of the atmosphere increased (Des Marais et al. 1992). Rising atmospheric O_2 increases aerobic decomposition in marine sediments, consuming O_2 and limiting the further growth of O_2 in the atmosphere (Walker 1980).

In Chapter 7 we saw that redox potential controls the order of anaerobic metabolism by microbes in sediments. The zone of methanogenesis underlies the zone of sulfate reduction, because the sulfate-reducing bacteria are more effective competitors for reduced substrates. As a result of high concentrations of SO_4 in seawater, methanogenesis in ocean sediments is limited (Oremland and Taylor 1978, Lovley and Klug 1986). Nearly all methanogenesis is the result of CO_2 reduction, because normally acetate is depleted before SO_4 is fully removed from the sediment (Crill and Martens 1986, Whiticar et al. 1986). There is, however, some seasonal variation in the use of CO_2 and acetate that appears to be due to microbial response to temperature (Martens et al. 1986).

Methane is not highly soluble in seawater, and in many areas the surface ocean is supersaturated in CH_4 with respect to the atmosphere (Ward et al. 1987). This methane appears to be due to methanogenesis in decomposing, sinking particles (Scranton and Brewer 1977, Burke et al. 1983, Karl and Tilbrook 1994). Methane released from ocean sediments and hydrothermal vents is easily oxidized by microbes before it reaches the surface (Iversen 1996). The global flux of CH_4 from the oceans to the atmosphere, $<10 \times 10^{12}$ g/yr, is small compared to that from other sources (Liss and Slater 1974, Conrad and Seiler 1988; Table 11.2).

Biogenic Carbonates

A large number of marine organisms precipitate carbonate in their skeletal and protective tissues by the reaction

$$Ca^{2+} + 2HCO_3^- \rightarrow CaCO_3 \downarrow + H_2O + CO_2. \qquad (9.2)$$

Clams, oysters, and other commercial shellfish are the obvious examples, but a vast quantity of $CaCO_3$ is produced by foraminifera, pteropods, and other small zooplankton that are found in the sea (Krumbein 1979, Simkiss and Wilbur 1989). Coccolithophores, a group of marine algae, are responsible for a large amount of $CaCO_3$ deposited on the seafloor of the open ocean.[2] The annual production of $CaCO_3$ by these organisms is much larger than what could be sustained by the supply of Ca to the oceans in river flow (Broecker 1974). However, not all of the $CaCO_3$ produced is stored permanently in the sediment.

[2] Note that the precipitation of carbonate by phytoplankton supplies some of the CO_2 needed for photosynthesis, reducing the net uptake of CO_2 from seawater (Robertson et al. 1994).

Recall that CO_2 is produced in the deep ocean by the degradation of organic materials that sink from the surface waters. Deep ocean waters are supersaturated with CO_2 with respect to the atmosphere as a result of their long isolation from the surface and the progressive accumulation of respiratory CO_2. Carbon dioxide is also more soluble at the low temperatures and high pressures that are found in deep ocean water. (Note that CO_2 effervesces when the pressure of a warm soda bottle is released upon opening). The accumulation of CO_2 makes the deep waters undersaturated with respect to $CaCO_3$, as a result of the formation of carbonic acid:

$$H_2O + CO_2 \rightleftarrows H^+ + HCO_3^- \rightleftarrows H_2CO_3. \quad (9.3)$$

When the skeletal remains of $CaCO_3$-producing organisms sink to the deep ocean, they dissolve:

$$CaCO_3 + H_2CO_3 \rightarrow Ca^{2+} + 2HCO_3^-. \quad (9.4)$$

Their dissolution increases the alkalinity, roughly the concentration of HCO_3^-, in the deep ocean. Small particles may dissolve totally during transit to the bottom, while large particles may survive the journey, and their dissolution occurs as part of sediment diagensis (Honjo et al. 1982, Berelson et al. 1990). The depth at which dissolution begins is called the carbonate lysocline; carbonate dissolution is complete below the *carbonate compensation depth* (CCD). The CCD occurs at roughly 4200–4500 m in the Pacific and 5000 m in the Atlantic Ocean (Kennett 1982). The tendency for a shallower CCD in the Pacific is the result of the longer mean residence time of Pacific deep water, which allows a greater accumulation of respiratory CO_2 (Li et al. 1969).

Dissolution of $CaCO_3$ means that calcareous sediments are found only in shallow ocean basins, and no carbonate sediments are found over much of the pelagic area where the ocean is greater than 4500 m deep. About 5.3×10^{15} g/yr of $CaCO_3$ are produced in the surface layer, and about 3.2×10^{15} g are preserved in shallow, calcareous sediments (Milliman 1993). This recent estimate of carbonate deposition consumes more than the known inputs of Ca to the oceans, suggesting that the Ca budget of the oceans is not now in steady state.

Many studies of carbonate dissolution have employed sediment traps that are anchored at varying depths to capture sinking particles. In most areas, biogenic particles constitute most of the material caught in sediment traps, and most of the $CaCO_3$ is found in the form of calcite. Pteropods, however, deposit an alternative form of $CaCO_3$, known as aragonite, in their skeletal tissues. The downward movement of aragonite has been long overlooked because it is more easily dissolved than calcite and often disappears from sediment traps that are deployed for long periods. As much as

12% of the movement of biogenic carbonate to the deep ocean may occur as aragonite (Berner and Honjo 1981, Betzer et al. 1984).

Geochemists have long puzzled that dolomite [$(Ca,Mg)CO_3$] does not appear to be deposited abundantly in the modern ocean, despite the large concentration of Mg in seawater and the occurrence of massive dolomites in the geologic record. There are few organisms that precipitate Mg calcites in their skeletal carbonates, but thermodynamic considerations would predict that calcite should be converted to dolomite in marine sediments (e.g., Malone et al. 1994). Baker and Kastner (1981) show that the formation of dolomite is inhibited by SO_4^{2-}, but dolomite can form in organic-rich marine sediments in which HCO_3^- is enriched and SO_4^{2-} is depleted by sulfate reduction (Eq. 7.17; Baker and Burns 1985). Dolomite is precipitated in laboratory cultures of the sulfate-reducing bacterium *Desulfovibrio* (Vasconcelos et al. 1995). Thus, the precipitation of dolomite is directly linked to biogeochemical processes in marine sediments. Although dolomite has been a significant sink for marine Mg in the geologic past, its contribution to the removal of Mg from modern seawater is likely to be minor.

Models of Carbon in the Ocean

CO_2 dissolves in seawater as a function of the concentration of CO_2 in the overlying atmosphere. (Recall Henry's Law, Eq. 2.7). The rate of dissolution increases with wind speed, which increases the turbulence of the surface waters[3] and the downward transport of bubbles (Watson et al. 1991, Wanninkhof 1992, Farmer et al. 1993). As it dissolves in water, CO_2 disassociates to form bicarbonate, following Eq. 9.3 (Archer 1995). The solubility of CO_2 in seawater depends on temperature; CO_2 is about twice as soluble at 0°C than at 20°C (Broecker 1974). The temperature of the upper 1 mm of the ocean's surface, the "skin" temperature, is critical to determining the atmosphere-to-ocean flux. Over much of the ocean's surface the skin temperature is about 0.3°C cooler than the underlying waters as a result of evaporation of water from the ocean's surface (Robertson and Watson 1992).

CO_2 enters the deep oceans with the downward flux of cold water at polar latitudes. When cold waters form in equilibrium with an atmosphere of 360 ppm CO_2 (i.e., today), they carry more CO_2 than when they formed in equilibrium with an atmosphere of 280 ppm CO_2—the historical origin of most of today's upwelling waters that are 300 to 500 years old. Brewer et al. (1989) report that North Atlantic deep water now carries a net flux of 0.26×10^{15} g C/yr that is presumably due to the global rise in atmospheric CO_2 during this century.

[3] The term "piston velocity" is often used to describe the mixing of gases with seawater. A piston velocity of 5 m/day for CO_2 implies that atmosphere equilibrates with the upper 5 m of seawater—as if pushed in by a piston—each day.

Although the surface ocean is in theoretical equilibrium with atmospheric CO_2, the surface waters over large areas are often variably undersaturated in CO_2 as a result of photosynthesis (Tans et al. 1990, Watson et al. 1991). Sinking organic materials remove carbon from the surface ocean, and it is replaced by the dissolution of new CO_2 from the atmosphere. Taylor et al. (1992) found that during a 46-day period there was a net downward transport of carbon in the northeast Atlantic Ocean due to the sinking of live (2 g C/m^2) and dead (17 g C/m^2) cells and the downward mixing of living cells by turbulence (3 g C/m^2). Thus, biotic processes act to convert inorganic carbon in the surface waters to organic carbon that is delivered to the deep waters of the ocean. As we have seen, the storage of organic carbon in sediments accounts for $\leqslant 1\%$ of marine NPP; most of the carbon is liberated by bacterial respiration in the deep ocean and released back to the atmosphere as CO_2 in zones of upwelling. Nevertheless, in the absence of a marine biosphere, the atmospheric CO_2 concentration would be much higher than today's—perhaps as high as 470 ppm (Broecker and Peng 1993). A more active "biotic pump" is one postulated explanation for the lower concentrations of atmospheric CO_2 during the last glacial epoch (Broecker 1982, Ganeshram et al. 1995, N. Kumar et al. 1995).

The biosphere also supplies a large quantity of dissolved organic carbon (DOC) to the surface waters of the ocean (Baines and Pace 1991). Most of this is labile and rapidly decomposed in the surface ocean (Kirchman et al. 1991, Druffel et al. 1992). Downwelling waters may carry some DOC to the deep sea, where it is decomposed, adding to the CO_2 content of the deep oceans (Carlson et al. 1994). In this regard, the labile carbohydrates in marine DOC (Benner et al. 1992, Pakulski and Benner 1994) must be distinguished from a large quantity of refractory DOC that is found well mixed throughout the oceans. Most marine DOC shows ^{14}C ages in excess of 6000 years (Williams and Druffel 1987), and some of it may derive from resistant humic substances supplied to seawater by rivers (Moran and Hodson 1994).

Finally, the production and sinking of $CaCO_3$ also delivers carbon to the deep ocean. Most of the Ca^{2+} is derived from weathering on land and is balanced in riverwater by $2HCO_3^-$ (Fig. 1.4). Whether it is preserved in a shallow-water calcareous sediment or sinks to the deep ocean, each molecule of $CaCO_3$ carries the equivalent of one CO_2 and leaves behind the equivalent of one CO_2 in the surface ocean (Eq. 9.2). Globally, the CO_2 sink in sedimentary $CaCO_3$ is about four times larger than the sink in organic sediments (Li 1972). Near-shore environments contain most of the sedimentary storage of $CaCO_3$ and organic carbon; $CaCO_3$ delivered to the deep sea dissolves, producing calcium and bicarbonate that return to the surface waters in zones of upwelling (Eq. 9.4).

Equilibrium with ocean waters controls the concentration of CO_2 in the atmosphere, but the equilibrium can be upset when changes in CO_2 in the atmosphere exceed the rate at which the ocean system can buffer the concentration. The seasonal cycle of terrestrial photosynthesis and the burning of fossil fuels are two processes that affect the concentration of atmospheric CO_2 more rapidly than the ocean can buffer the system. As a result we observe a seasonal oscillation of atmospheric CO_2 and an exponential increase in its concentration in the atmosphere (Fig. 1.3). Given enough time, the oceans could take up nearly all of the CO_2 released from fossil fuels, and the atmosphere would once again show stable concentrations at only slightly higher levels than today (Laurmann 1979). As the oceans take up additional CO_2, the pH of seawater will be buffered at about 8.0 by the dissolution of carbonates (Eqs. 9.3 and 9.4). Already, there is some indication that the concentration of CO_2 dissolved in the surface ocean has increased in response to increasing concentrations in the atmosphere (Fig. 9.10; Inoue et al. 1995), but there is little evidence for greater dissolution of marine carbonates (Broecker et al. 1979).

A large number of models have been developed to explain the response of the ocean to higher concentrations of atmospheric CO_2 (Bacastow and Björkström 1981, Emanuel et al. 1985b, Holligan and Robertson 1996).

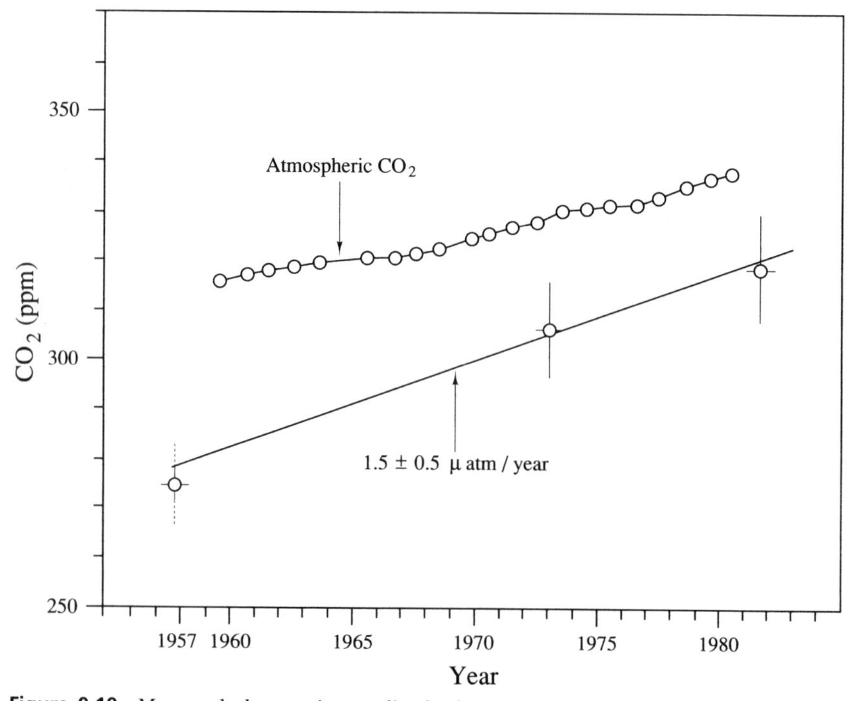

Figure 9.10 Measured changes in CO_2 dissolved in surface waters of the central Atlantic Ocean, showing an increase of 1.5 μatm/yr during recent decades. The trend in atmospheric CO_2 over the same period is shown for comparison. From Takahashi et al. (1983).

Most of these models are constructed to follow parcels of water as they circulate in a simplified ocean basin and to calculate the diffusion of CO_2 between layers that do not mix directly. Figure 9.11 shows a multibox model in which the surface ocean is divided into cold polar waters and warmer waters. In this model, cold waters mix downward to eight layers of the deep ocean, while upwelling returns deep water to the surface, where it releases CO_2 to the atmosphere. The rate of mixing is calculated using oceanographic data for the rate at which ^{14}C and $^{3}H_2O$ from atomic bombs have entered the oceans (Killough and Emanuel 1981) and known constants for the dissolution of CO_2 in water as a function of temperature and pressure

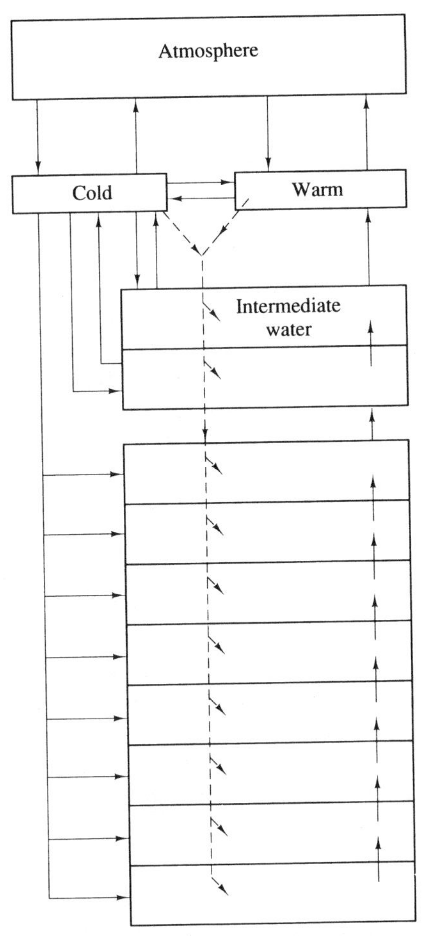

Figure 9.11 A box-diffusion model for the oceans, separating the surface oceans into cold polar waters and warmer waters at other latitudes. Cold polar waters mix with deeper waters as a result of downwelling. Other exchanges are by diffusion. From Emanuel et al. (1985b).

(Sundquist et al. 1979, Archer 1995). The models then adjust the chemistry of the water in each layer according to the carbonate equilibrium reactions given above.

As atmospheric carbon dioxide increases, we would expect an increased dissolution of CO_2 in the oceans, following Henry's Law (Tans et al. 1990). However, the surface ocean provides only a limited volume for CO_2 uptake, and the atmosphere is not in immediate contact with the much larger volume of the deep ocean. In the absence of large changes in NPP, it is the rate of formation of bottom waters in polar regions that limits the rate at which the oceans can take up CO_2. Although most of the ocean models do not yet incorporate the effects of biotic productivity in the sea, nor do they incorporate the full three-dimensional complexity of ocean basins in both hemispheres of the globe, they do allow predictions about future global conditions and hypotheses for further testing (Shaffer 1993, Shaffer and Sarmiento 1995, Semtner 1995).

Nutrient Cycling in the Ocean

Net primary productivity in the sea is limited by nutrients. Production is highest in regions of high nutrient availability—the continental shelf and regions of upwelling (Fig. 9.6)—and in the open ocean the concentrations of available N, P, and Si are normally very low. Nutrients are continuously removed from the surface water by the downward sinking of dead organisms and fecal pellets. Shanks and Trent (1979) found that 4 to 22% of the nitrogen contained in particles (PON) was removed from the surface waters each day. The mean residence time of N, P, and Si in the surface ocean is much less than the mean residence time of water, and there are wide differences in the concentration of these elements between the surface and the deep ocean (Fig. 9.12). These are the nonconservative elements of seawater; their behavior is strongly controlled by *bio*geochemistry.

Nutrients are regenerated in the deep ocean, where the concentrations are much higher than those at the surface. Recalling that the age of deep water in the Pacific Ocean is older than that in the Atlantic, we note that nutrient concentrations are higher in deep Pacific Ocean (Fig. 9.12), because its waters have had a longer time to receive sinking debris which is remineralized at depth. Similarly, in the Atlantic Ocean, nutrient concentrations increase progressively as North Atlantic deep water "ages" during its journey southward (Fig. 9.13).

Internal Cycles

In 1958, Albert Redfield published a paper that has served as a focal point in marine biogeochemistry for the last 40 years. Redfield noted that marine phytoplankters contained N and P in a fairly constant atom ratio to the

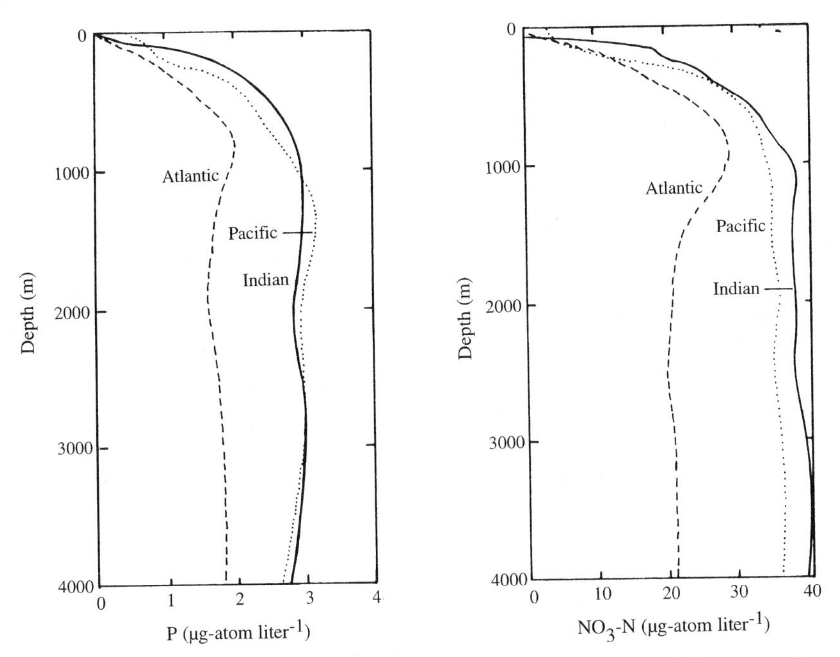

Figure 9.12 Vertical distribution of phosphate and nitrate in the world's oceans. From Svedrup et al. (1942).

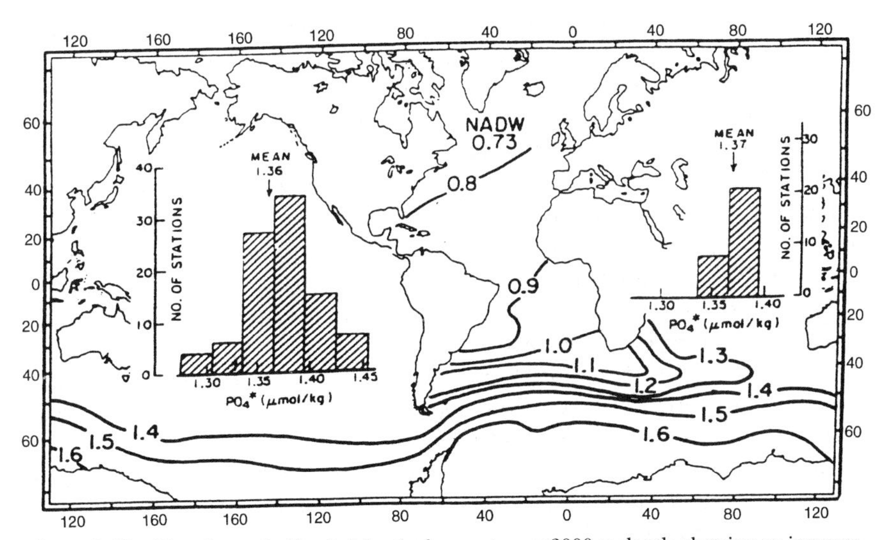

Figure 9.13 Phosphorus in North Atlantic deep waters at 3000-m depth, showing an increase from a mean of 0.73 μM/kg near Greenland to 1.67 μM/kg in the waters that upwell in the southern hemisphere. From Broecker (1991).

content of carbon, 106 C:16 N:1 P (Redfield et al. 1963), as a result of the incorporation of these elements in photosynthesis and growth:

$$106CO_2 + 16NO_3^- + HPO_4^{2-} + 122H_2O + 18H^+ \rightarrow (CH_2O)_{106}(NH_3)_{16}(H_3PO_4) + 138O_2. \quad (9.5)$$

Despite differences in nutrient concentration among the major oceans (Fig. 9.12), upwelling waters contain available C, N, and P (i.e., HCO_3^-, NO_3^-, and HPO_4^{2-}) in the approximate ratio of 800 C:16 N:1 P. Thus, even in the face of the high productivity found in upwelling waters, only about 10% of the HCO_3^- can be consumed by photosynthesis before the N and P are exhausted. In areas of upwelling, the remaining HCO_3^- is lost to the atmosphere as CO_2 (e.g., Murray et al. 1994). Significantly, Redfield (1958) noted that the biota determined the relative concentrations of N and P in the deep sea, and that the biotic demand for N and P was closely matched to the availability of these elements in upwelling waters.

Recognizing that the downward flux of biogenic particles carries $CaCO_3$ as well as organic carbon, Broecker (1974) recalculated Redfield's ratios to include $CaCO_3$. His modified Redfield ratio in sinking particles is 120 C:15 N:1 P:40 Ca. The ratio in upwelling waters is 800 C:15 N:1 P:3200 Ca. Based on these quantities, net production in the surface water could remove all the N and P but only 1.25% of the Ca in upwelling waters. Although biogenic $CaCO_3$ is the main sink for Ca in the ocean, biota exert only a tiny control on the availability of Ca in surface waters. Thus, calcium is a constant, well-mixed, and conservative element in seawater (Table 9.1).

The Redfield ratio allows us to compare the importance of riverflow, upward transport, and internal recycling for their contributions to the net primary production of the surface ocean. To sustain a global marine NPP of 50×10^{15} g C/yr (Table 9.2), phytoplankton must take up about 8.8×10^{15} g N and 1.2×10^{15} g P each year (Table 9.3). Rivers supply about 0.036×10^{15} g N/yr and 0.002×10^{15} g/yr of reactive P to the oceans (Chapter 8). Rivers and vertical movements (upwelling + diffusion) provide only a small fraction (15%) of the total nutrient requirement in the surface ocean (Table 9.3), and nutrient recycling in the surface waters must supply the rest. Rapid turnover of nutrients is consistent with the rapid turnover of organic carbon in the surface ocean.

In the face of nutrient-limited growth and efficient nutrient uptake, phytoplankton maintain very low concentrations of N and P in surface waters (Fig. 9.12). McCarthy and Goldman (1979) showed that much of the nutrient cycling in the surface waters may occur in a small zone, perhaps in a nanoliter (10^{-9} liter) of seawater, which surrounds a dying phytoplankton cell. Growing phytoplankters in the immediate vicinity are able to assimilate the nitrogen as soon as it is released. Often it is difficult to study nutrient cycling on such a small scale, but various workers have applied

Table 9.3 Calculation of the Sources of Nutrients to Sustain a Global Net Primary Production of 50 × 10^{15} g C/yr in the Surface Waters of the Oceans[a]

Flux	Carbon (10^{12} g)	Nitrogen (10^{12} g)	Phosphorus (10^{12} g)
Net primary production[b]	50,000	8838	1219
Amounts supplied			
By rivers[c]		36	2
By atmosphere[d]		45	1
By upwelling		1189	106
Recycling (by difference)		7568	1110

[a] Based on an approach developed by Peterson (1981).
[b] Assuming a Redfield atom ratio of 106:16:1.
[c] Meybeck (1982).
[d] Figure 9.16.

isotopic tracers (e.g., $^{15}NH_4$ and $^{15}NO_3$) to measure nutrient uptake by phytoplankton and bacteria (Glibert et al. 1982, Goldman and Glibert 1982, Harrison et al. 1992, Dickson and Wheeler 1995). Leakage of dissolved organic nitrogen (DON) from phytoplankters may also account for a significant amount of the bacterial uptake and turnover in the surface waters (Kirchman et al. 1994, Bronk et al. 1994, Kroer et al. 1994). During decomposition of organic materials in the surface ocean, nitrogen is mineralized more rapidly than carbon, so that surviving particles carry C/N ratios that are somewhat greater than the Redfield ratio (Sambrotto et al. 1993) and that increase with depth (Honjo et al. 1982, Takahashi et al. 1985, Anderson and Sarmiento 1994).

Nutrient demand by phytoplankers is so great that little of the NH_4 released by mineralization is nitrified in the surface waters, and NH_4 dominates phytoplankton uptake of recycled N (Dugdale and Goering 1967, Harrison et al. 1992, 1996). In contrast, most of the nitrogen mineralized in the deep ocean is converted to NO_3. Nitrate also dominates the nitrogen supply in rivers, so oceanographers can use the fraction of NPP that derives from the uptake of NH_4 versus that derived from NO_3 to estimate the sources of nutrients that sustain NPP in the surface waters (Fig. 9.14). For example, Jenkins (1988) estimated that the upward flux of NO_3 from the deep ocean near Bermuda would support a NPP of about 36 g C m^{-2} yr^{-1}—about 38% of total NPP (Michaels et al. 1994). The remaining production must depend on NH_4 supplied by recycling. The fraction of NPP that is sustained by nutrients delivered from rivers and upwelling is known as *new production*. Globally, new production is about 10–20% of total NPP, but the fraction, f_n, is greatest in areas of cold, upwelling waters (Sathyendranath et al. 1991). To maintain low, steady-state nutrient concentrations in

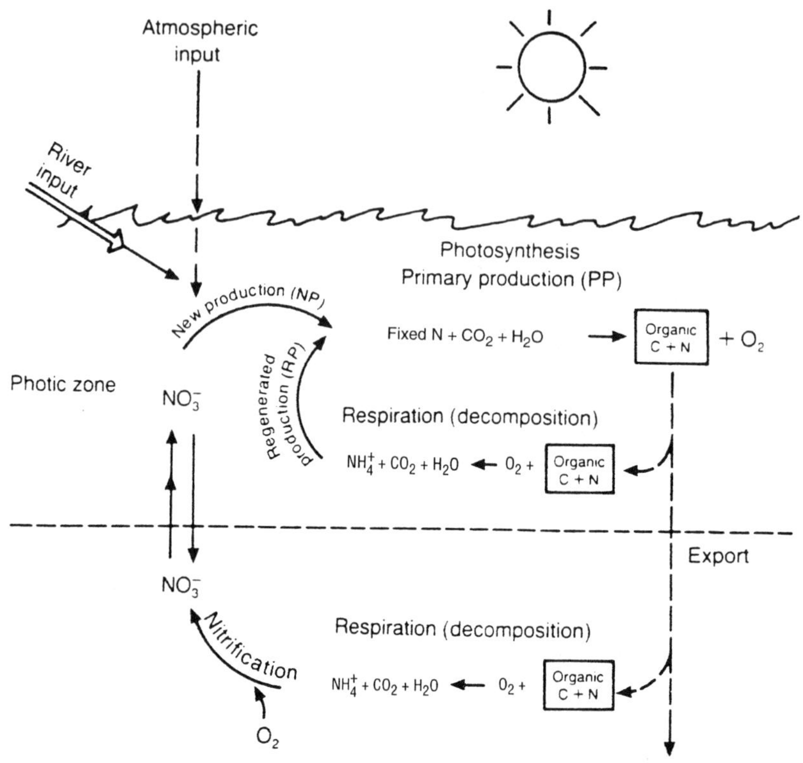

Figure 9.14 Links between the nitrogen and the carbon cycles in the surface ocean. Nitrogen regenerated in the surface waters is largely assimilated by phytoplankters as NH_4, while that diffusing and mixing up from the deep ocean is NO_3. When organic matter sinking to the deep ocean is mineralized, its nitrogen content is initially released as NH_4 and converted to nitrate by nitrifying bacteria. "New production" can be estimated as the fraction of net primary production that is derived from nitrate from rivers, atmospheric deposition, and the deep sea. From Jahnke (1990).

the surface waters of the oceans, the sources of nutrients that sustain "new production" globally are about equal to the annual losses of nutrients in organic debris that sink through the thermocline to the deep sea (Eppley and Peterson 1979).

Nitrogen and Phosphorus Budgets for the Sea

Redfield ratios suggest that the demand for N and P by phytoplankton is closely matched to their concentrations in upwelling waters. Both elements show low concentrations in surface waters (Fig. 9.12), and the concentrations of N and P are correlated with a slope near the Redfield ratio (Holland 1978). These observations suggest that both N and P might simultaneously limit marine productivity, in contrast to the widespread limitation by P in freshwaters (Chapter 7). In fact, NPP in many ocean waters shows a tendency for limitation by available N (Howarth 1988, Vitousek and Howarth 1991). What processes lead to a N limitation in the sea?

In contrast to the high rates of nitrogen fixation by blue-green algae in freshwater habitats, N-fixation in the sea is very limited (Capone and Carpenter 1982, Howarth et al. 1988a, Walsh and Dieterle 1988). Recall that the enzyme of nitrogen fixation, nitrogenase, requires molybdenum and iron in its molecular structure (Chapter 2). Howarth and Cole (1985) showed that the uptake of molybdenum is inhibited by the high concentrations of SO_4 in seawater, and they suggest that the assimilation of molybdenum generally limits N-fixation in the sea (Fig. 9.15). The lower concentration of SO_4 in most lake waters does not inhibit molybdenum uptake (Cole et al. 1993), so blue-green algae dominate lakes with low N/P ratios, adding nitrogen to these ecosystems by nitrogen fixation (Chapter 7).

Paerl et al. (1987) tested this hypothesis in coastal marine waters. They found that additions of Mo, Fe, and P did not affect N-fixation, but additions of organic materials increased the rate significantly. Paulsen et al. (1991) also found that additions of carbohydrates stimulated N-fixation and postulated that these compounds created local zones of active decomposition, where oxygen was depleted and nitrogenase activity is possible. Natural aggregations of organic matter, forming "marine snow," create small microzones of anaerobic conditions in seawater, in which a greater availability of trace micronutrients and low redox potentials could stimulate N-fixation (Alldredge and Cohen 1987, Paerl and Carlton 1988). Anaerobic microzones also develop in bundles of blue-green algae (Paerl and Bebout 1988)

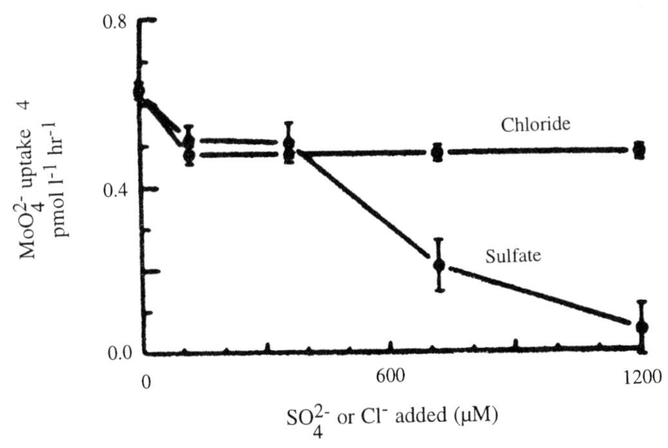

Figure 9.15 Effect of adding SO_4^{2-} and Cl^- on the molybdenum uptake of freshwater phytoplankton. High levels of Cl have no effect, while added SO_4 yields a precipitous decline in Mo uptake. From Howarth and Cole (1985).

and in the endosymbiotic bacteria in diatoms (Martínez et al. 1983), both of which show significant N-fixation in the sea. If nitrogen fixation in the oceans depends on anaerobic microzones in seawater, it may be limited by turbulent conditions that disrupt these microenvironments (Paerl 1985a, but see also Howarth et al. 1993).

Regardless of the mechanism (i.e., Mo deficiency or widespread turbulence) that limits N-fixation in seawater, this nitrogen input contributes significantly to new production only in the oligotrophic waters of the open ocean. For example, fixation by the blue-green algae, *Trichodesmium,* sustains an important component of new production in the tropical Atlantic Ocean, which is distant from sources of upwelling (Carpenter and Romans 1991). Global N-fixation may account for about $10–15 \times 10^{12}$ g N/yr added to the sea (Capone and Carpenter 1982, Walsh and Dieterle 1988, Carpenter and Romans 1991).

The anaerobic microzones created by flocculations of organic matter allow significant rates of denitrification in the oceans, despite the high redox potential of seawater (Alldredge and Cohen 1987). Denitrification in a zone of low O_2 concentration in the eastern Pacific Ocean may result in the loss of 50 to 60×10^{12} g N/yr from the sea (Lui and Kaplan 1989, Codispoti and Christensen 1985, Lipschultz et al. 1990). This denitrification explains the high content of ^{15}N in the residual nitrate in seawater (Lui and Kaplan 1989). As we saw in terrestrial ecosystems (Chapter 6), $^{14}NO_3$ is used preferentially as a substrate in the production of N_2 during denitrification.

Denitrification is also observed in ocean sediments. Christensen et al. (1987) estimate that over 50×10^{12} gN/yr may be lost from the sea by sedimentary denitrification in coastal regions. Devol (1991) found that nitrification occurring within the sediments supplied most of the nitrate for denitrification on the continental shelf of the western United States. Most of the gaseous nitrogen lost from marine environments is N_2—losses of N_2O are less important (Seitzinger et al. 1984, Jørgensen et al. 1984).

Limited inputs of nitrogen in river waters and by nitrogen fixation, and the potential for large losses by denitrification, all reinforce N limitation in the sea. In most areas of the ocean, nitrate is not measurable in surface waters, and phytoplankton respond to nanomolar additions of nitrogen to seawater (Glover et al. 1988). In the open ocean, direct atmospheric deposition of nitrogen in rainfall and dryfall may assume special significance, since these areas are distant from rivers and upwelling. Prospero and Savoie (1989) found that 40 to 70% of the nitrate in the atmosphere over the north Pacific Ocean was derived from soil dusts, presumably from the desert regions of China. The deposition of dust links the NPP of the ocean to the soil biogeochemistry of distant terrestrial ecosystems. An increased deposition of nitrate and organic nitrogen compounds from air pollution may now be responsible for higher marine NPP in areas that are downwind of continental sources (Paerl 1985b, Fanning 1989, Owens et al. 1992, Cornell et al. 1995).

The global model for the N cycle of the oceans (Fig. 9.16) offers a deceptive level of tidiness to our understanding of marine biogeochemistry, and the reader should realize that many fluxes, for example, nitrogen fixation, denitrification, and sedimentary preservation, are not known to better than a factor of two. Nevertheless, the model shows that most NPP is supported by nutrient mineralization in the surface waters, and only

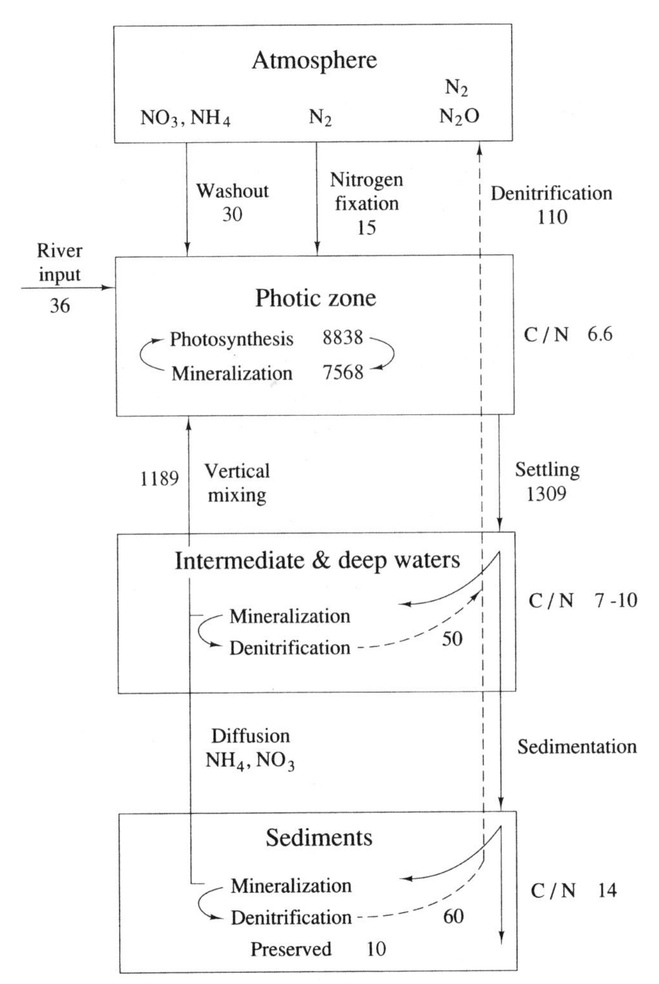

Figure 9.16 Nitrogen budget for the world's oceans, showing major fluxes in units of 10^{12} g N/yr. From an original conception by Wollast (1981), but with newer data added for atmospheric deposition (Duce et al. 1991), nitrogen fixation (Carpenter and Romans 1991), riverflow (Meybeck 1982), denitrification (Christensen et al. 1987), and nutrient regeneration in surface waters (cf. Table 9.3).

small quantities of nutrients are lost to the deep ocean. The mean residence time of available nitrogen in the surface ocean is ≤1 year, while the mean residence time of the *total* pool of N in the surface ocean is about 10 years. Thus, each atom of N cycles through the biota many times. Upon sinking and mineralization in the deep ocean, N enters pools with a mean residence time of about 500 years—largely controlled by the circulation of water through the deep ocean.

Vertical mixing includes both upwelling and upward diffusion from the deep ocean. Upwelling accounts for about half of the global upward flux, and it is centered in coastal areas where the resulting nutrient-rich waters yield high productivity. Away from areas of upwelling, diffusion dominates the upward flux (Table 9.4), but diffusion rates are low (Ledwell et al. 1993), so the total supply of nutrients is limited in most of the open ocean (Lewis et al. 1986, Martin and Gordon 1988). Diffusion appears globally significant only as a result of the large area of open ocean compared to the small area of upwellings.

Although the estimates are subject to large uncertainty, the model of Fig. 9.16 indicates a net loss of nitrogen from the oceans. The overall gaseous losses of nitrogen from the ocean exceed the inputs from rivers and the atmosphere, so that the oceans may be declining in nitrogen content (McElroy 1983, Christensen et al., 1987, Smith and Hollibaugh 1989). Various workers have suggested that, in the absence of denitrification, higher concentrations of NO_3 would be found in the ocean and lower concentrations of N_2 in the atmosphere (Chapters 2 and 12).

As seen in Chapter 8, only a small portion of the total phosphorus transport in rivers (21×10^{12} g P/yr) is carried in dissolved forms; the remainder is adsorbed to Fe and Al oxide minerals that are carried as suspended particles. Some of the adsorbed P is released upon the mixing of freshwater and seawater (Chase and Sayles 1980, Caraco et al. 1989,

Table 9.4 Sources of Fe, PO_4, and NO_3 in Surface Waters of the North Pacific Ocean[a]

Source	Fe	PO_4	NO_3
Concentration at 150 m (μmol m^{-3})	0.075	330	4300
Upwelling (μmol m^{-2} day^{-1})	0.00090	4.0	52
Net upward diffusion (μmol m^{-2} day^{-1})	0.0034	30	400
Atmospheric flux (μmol m^{-2} day^{-1})	0.16	0.102	26
Total fluxes (μmol m^{-2} day^{-1})	0.164	34	480
Percent from advective input	0.5	12	11
Percent from diffusive input	2	88	83
Percent from atmospheric input	98	0	5

[a] From Martin and Gordon (1988).

1990), but most is probably buried with the deposition of river sediments on the continental shelf. The total flux of "bioreactive" P to the oceans is about 2.0×10^{12} g/yr (Ramirez and Rose 1992, Howarth et al. 1995), giving an atom ratio of about 40 for $N/P_{(reactive)}$ in global riverflow.

Deposition of P on the ocean surface, from the dust of deserts, may play a special role in stimulating new production in areas of the open ocean that are distant from rivers and zones of upwelling (Talbot et al. 1986). However, as seen for N, recycling in the surface waters accounts for the vast majority of the P uptake by phytoplankters (Fig. 9.17). Each year a small amount of organic debris, with C/P ratios somewhat greater than the Redfield ratio, sinks through the thermocline to the deep ocean (Honjo et al. 1982). An average of 500 years later, mineralized P (i.e., HPO_4^{2-}) returns to the surface waters in upwelling.

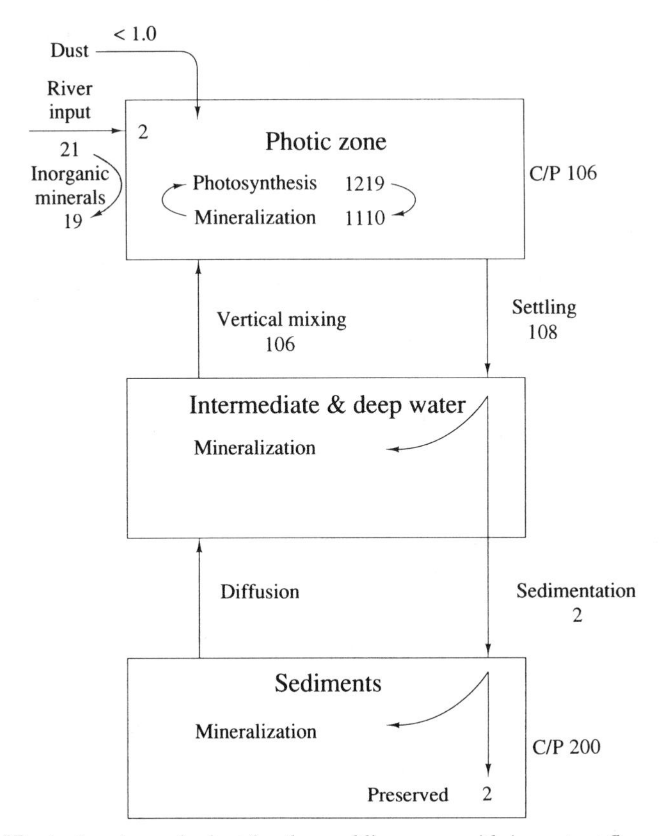

Figure 9.17 A phosphorus budget for the world's oceans, with important fluxes shown in units of 10^{12}g P/yr. From an original conception by Wollast (1981), but with newer data added for dust inputs (Graham and Duce 1979), riverflow (Meybeck 1982), sedimentary preservation (Howarth et al. 1995), and nutrient regeneration in surface waters (cf. Table 9.3).

The C/P ratio in organic matter that is buried in marine sediments is about 200 (Mach et al. 1987, Ingall and van Cappellen 1990, Ramirez and Rose 1992), suggesting that P is mineralized more rapidly than C during the downward transport and sedimentary diagenesis of organic matter in the sea (Honjo et al. 1982, Froelich et al. 1979). Phosphorus release and C/P ratios are greatest in anoxic sediments (Ingall et al. 1993, Ingall and Jahnke 1994). Anoxic environments have lower concentrations of oxidized Fe minerals that can adsorb P as it is mineralized from organic matter (Krom and Berner 1981, Sundby et al. 1992, Berner and Rao 1994). Indeed, Van Cappellen and Ingall (1996) suggest a negative-feedback mechanism by which changes in the concentration of O_2 in the deep ocean control the availability of P, which in turn stablizes the level of O_2 in Earth's atmosphere through geologic time. For instance, if the concentration of O_2 in the atmosphere falls to low levels, available P becomes more plentiful in seawater, marine NPP increases, and O_2 is released to the atmosphere and oceans. This subsequently lowers the abundance of P due to its adsorption on Fe minerals in oxic sediments; marine NPP declines and less O_2 is produced.

In contrast to N, there are no significant gaseous losses of P from the sea. At steady state, the inputs to the sea in river water must be balanced by the burial of phosphorus in ocean sediments. Most of the P carried in suspended sediments is probably deposited near the coast of continents. Burial of biogenic P compounds in sediments of the open ocean is estimated between 1.0 and 2.0×10^{12} g P/yr—similar to the riverflux of bioreactive P (Howarth et al. 1995). Burial occurs with the deposition of organic matter or $CaCO_3$ (Froelich et al. 1982). During sediment diagenesis, organic- and Fe-bound P are coverted to phosphorite (authigenic apatite) and other minerals, which may ultimately dominate the P storage in sediments (Ruttenberg 1993, Filippelli and Delaney 1996, Rasmussen 1996). Phosphorite is formed when PO_4^{3-} produced from the mineralization of organic P combines with Ca and F to form fluorapatite (Ruttenberg and Berner 1993). The F is supplied by inward diffusion from seawater (Froelich et al. 1983, Schuffert et al. 1994). In some areas of the ocean, phosphorite nodules accumulate on the sea floor. These nodules are an enigma; they remain on the surface of the sediment despite growing at rates slower than the rate of sediment accumulation (Burnett et al. 1982).

The mean residence time for reactive P in the oceans, relative to the input in rivers or the loss to sediments, is about 25,000 years (Ruttenberg 1993, Filippelli and Delaney 1996; cf. Fig. 9.4). Thus, each atom of P that enters the sea may complete 50 cycles between the surface and the deep ocean before it is lost to sediments. All forms of buried phosphorus complete a global biogeochemical cycle when geologic processes lift sedimentary rocks above sea level and weathering begins again. Thus, relative to N, the global cycle of P turns very slowly (Chapter 12).

Human Perturbations of Marine Nutrient Cycling

Through the direct release of sewage and indirect losses of fertilizers, the river input of N and P to the oceans has increased in recent years (Meybeck 1982). Fossil fuel pollutants have also increased the atmospheric deposition of N and S on the ocean surface (Whelpdale and Galloway 1994, Cornell et al. 1995). These inputs have probably enhanced the productivity of coastal and estuarine ecosystems (Chapter 8) and perhaps the productivity of the entire ocean. Greater net primary production in the surface ocean should result in a greater transport of particulate carbon to the deep sea, potentially serving as a sink for increasing atmospheric CO_2.

In the open ocean, the net primary production of 42×10^{15} g C/yr (Table 9.2) is supported by nitrogen derived from the atmosphere, from upward diffusion, and from internal recycling (Fig. 9.18). If an additional 17×10^{12} g N/yr is deposited in the surface waters from atmospheric pollution (Galloway et al. 1995), this "excess" nitrogen could result in an increase in the downward flux of organic carbon of about 0.10×10^{15} g/yr, assuming a Redfield atom ratio of 106 C/16 N in new production. Similar calculations using the "excess" flux of N in rivers suggest an increased storage of $<0.30 \times 10^{15}$ g C/yr in coastal zones (Wollast 1991). In the face of a net release of carbon dioxide to the atmosphere of about

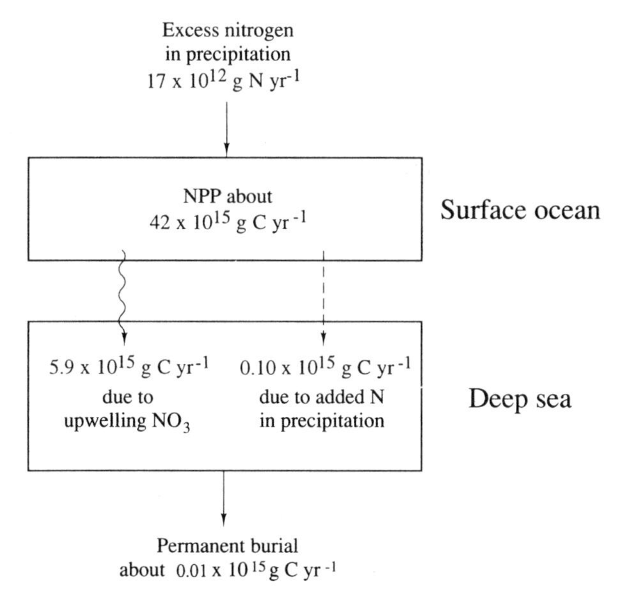

Figure 9.18 Estimated increase in the sedimentation of organic carbon that might be caused by human additions of nitrogen to the world's oceans by precipitation. Updated from an original conception by Peterson and Melillo (1985).

6×10^{15} g C/yr, these ocean sinks are relatively small (MacKenzie et al. 1993). The major ocean sink for CO_2 is found as a result of an increased dissolution of CO_2 in cold waters of the polar oceans (Shaffer 1993). As we discussed earlier, this inorganic sink for CO_2 is limited by the area of polar oceans and the amount of downwelling water.

Human perturbations of marine ecosystems are greatest in estuarine, coastal, and continental shelf waters (Chapter 8). These areas occupy only about 10% of the ocean's surface, but they account for about 18% of ocean productivity (Table 9.2), and 83% of the carbon that is buried in sediments. Globally averaged models (e.g., Figs. 9.16 and 9.17) mask the comparative importance of these regions to the overall biogeochemical cycles of the sea. For example, a significant amount of organic carbon may be transported from the continental shelf to the deep sea (Walsh 1991, Wollast 1991, 1993). If global climate change alters the rate of coastal upwelling (Bakun 1990), significant changes in the ocean's overall biogeochemistry should be expected (Walsh 1984).

Silicon, Iron, and Trace Metals

Diatoms compose a large proportion of the marine phytoplankton, and they require silicon (Si) as a constituent of their cell walls, where it is deposited as opal. As a result of biotic uptake, the concentration of dissolved Si in the surface waters is very low—usually $<2\mu M$. Globally, the annual uptake of Si by diatoms is about 6×10^{15} g (Nelson et al. 1995). Upon the death of diatoms, a large fraction of the opal dissolves, and the Si is recycled in the surface waters. Si concentrations generally increase with depth, but the dissolution of opal is dependent on temperature, so the rate of dissolution in the deep ocean is relatively slow (Honjo et al. 1982). The average concentration in deep waters is about 70 μM.

A mass-balance model for Si in the oceans shows that rivers (1.6×10^{14} g/yr), dust (0.14×10^{14} g/yr), and hydrothermal vents (0.06×10^{14} g/yr) are the main sources, and sedimentation of biogenic opal is the only significant sink (Tréguer et al. 1995). The mean residence time for Si in the oceans is about 15,000 years, which is consistent with its nonconservative behavior in seawater. Most of the Si input is delivered by tropical rivers, as a result of high rates of rock weathering in tropical climates (White and Blum 1995, Chapter 4). Sedimentation in the cold waters of the Antarctic Ocean accounts for 70% of the global sink (Tréguer et al. 1995). Other cold water areas provide most of the remaining sinks (DeMaster 1981), but about 10% of the sink is found in coastal regions, where the growth of diatoms in nutrient-enriched waters may now be limited by silicon (Justić et al. 1995).

Similar to the use of Si by diatoms, marine protists known as acantharians require strontium (Sr). These organisms precipitate celestite ($SrSO_4$) as a skeletal component. As a result of their uptake of Sr in surface waters and the sinking of acantharians to the deep sea, the Sr/Cl ratio in seawater varies from about 392 μg/g in surface waters to >405 μg/g with depth (Bernstein et al. 1987). Thus, biotic processes maintain a slightly nonconservative behavior (i.e., $\pm$1.5%) for Sr in seawater, despite its overall mean residence time of 12,000,000 years (Table 9.1).

All phytoplankton require a suite of micronutrients, for example, iron (Fe), copper (Cu), and zinc (Zn), in their biochemistry. These elements are taken up in surface waters and mineralized when dead organisms sink to the deep ocean. Many of these elements are relatively insoluble in seawater, due to its the high redox potential (Chapter 7). Most metals are found at low concentrations in the surface waters, and concentrations increase with depth in the deep sea (e.g., Fig. 9.19). In response to low concentrations of Fe, some phytoplankton release organic compounds that chelate Fe, increasing its availability in seawater (Wilhelm et al. 1996). In contrast, cyanobacteria avoid toxic levels of Cu by releasing chelators that render Cu less available to biotic uptake (Moffett and Brand 1996).

Near the continents, the concentration of Fe in seawater is normally adequate to support phytoplankton growth, but in the central Pacific Ocean, Martin and Gordon (1988) found that internal sources of Fe could sustain only a small percentage of the observed NPP. They suggested that as much as 98% of the new production in this area is supported by Fe derived from dust deposited from the atmosphere (Table 9.4). Most of the dust is probably transported from the deserts of central China (Duce and Tindale 1991). Growth of phytoplankton appears to be limited by iron, so small quantities of NO_3 and PO_4 remain in surface seawaters even during periods of peak production (Fig. 9.19). Some workers have even suggested that fertilizing the oceans with Fe might be an effective way to stimulate new production and lower atmospheric CO_2.

During 1993 and 1995, the "iron hypothesis" was tested by fertilizing patches of the Pacific Ocean with $FeSO_4$. During this experiment, the rate of photosynthesis by phytoplankton increased significantly in the surface waters (Martin et al. 1994, Kolber et al. 1994). When Fe was supplied continuously, the fertilization produced a massive bloom of phytoplankton, which removed a large portion of the dissolved CO_2 from the surface water (Coale et al. 1996). Iron fertilization is probably not a cure for rising atmospheric CO_2, but it is interesting to speculate that a greater global dispersal of iron-rich dust during the last glacial period may have stimulated marine production, drawing down atmospheric CO_2 (Martin 1990, N. Kumar et al. 1995).

Zinc (Zn) is an essential component of carbonic anhydrase—the enzyme that allows phytoplankton to convert HCO_3^- in seawater to CO_2 for photosynthesis (Morel et al. 1994). Low concentrations of Zn in surface waters can limit the growth of phytoplankton in different marine environments (Brand

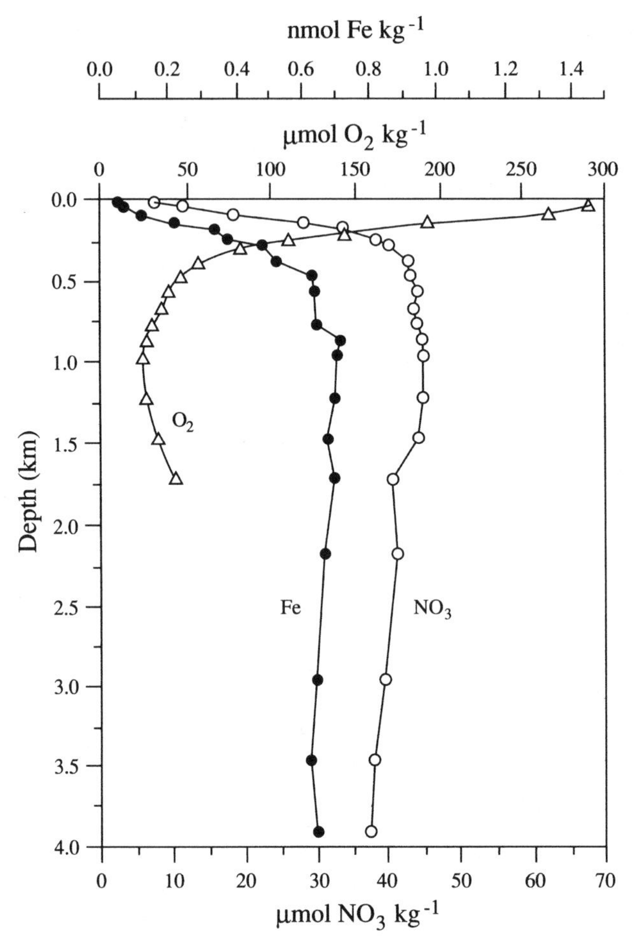

Figure 9.19 Vertical distribution of Fe, NO_3, and O_2 in the central North Pacific Ocean. From Martin et al. (1989).

et al. 1983, Sunda and Huntsman 1992). Like Fe, the concentrations of Zn increase with depth in the deep sea (Bruland 1989). Among samples of surface and deep waters, the concentrations of Fe, Zn, and other metals are often well correlated to those of N, P, and Si, suggesting that biological processes control the distribution of these elements in seawater. For example, Zn is correlated to Si in the northeast Pacific (Bruland et al. 1978a), and Ni is correlated to P in most seawater (Sclater et al. 1976). Nickel appears to be an essential element for diatoms that use urea as a nitrogen source (Price and Morel 1991).

Uptake and accumulation of trace metals also extends to the tendency for some nonessential, toxic metals, such as cadmium (Cd) and mercury

(Hg), to accumulate in phytoplankton (Table 9.5). Cadmium appears to substitute for zinc in biochemical molecules, allowing diatoms to maintain growth in zinc-deficient seawater (Price and Morel 1990, Lee et al. 1995). Cadmium (Cd) is well correlated with available P in waters of the Pacific ocean (Fig. 9.20; Boyle et al. 1976), and the concentration of Cd in marine sediments is often taken as an index of the availability of P in seawater of the geologic past (Hester and Boyle 1982). When marine phosphate rock is used as a fertilizer, cadmium is often an undesirable trace contaminant.

Mercury enters the ocean by deposition from the atmosphere, where the concentrations are increasing by >1%/yr (Slemr and Langer 1992). Most of the mercury entering the oceans is revolatilized, so the oceans do not constitute a large sink in the global cycle of mercury (Mason et al. 1994). However, in areas of seawater where oxygen is depleted, inorganic mercury is converted to methylmercury (Mason and Fitzgerald 1993) by processes similar to those in freshwater wetlands (Chapter 7). Methylmercury is the major form accumulating in fish (Cross et al. 1973), leading to concerns for human health.

When nonessential elements (e.g., Al, Ti, Ba, Hg, and Cd) and essential elements (e.g., Si and P) show similar variations in concentration with depth, it is tempting to suggest that both are affected by biotic processes, but the correlation does not indicate whether the association is active or passive. Organisms actively accumulate essential micronutrients by enzymatic uptake, whereas other elements may show passive accumulations, as a result of coprecipitation or adsorption on dead, sinking particles. Titanium (Ti) shows nonconservative behavior is seawater, with concentrations ranging from 10 μM at the surface to >200 μM at depth (Orians et al.

Table 9.5 Ratio of the Concentration of Elements in Phytoplankton to the Concentration of Elements in Seawater[a]

Element	Ratio
Al	25,000
Cd	910
Cu	17,000
Fe	87,000
Mg	0.59
Mn	9,400
N	19,000
Na	0.14
N:	1700
P	15,000
Zn	65,000

[a] From Bowen (1966).

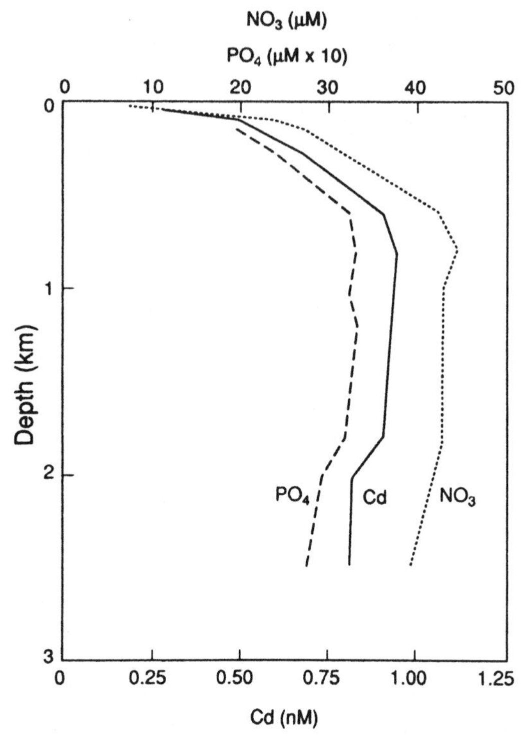

Figure 9.20 Depth distribution of nitrate, phosphate, and cadmium in the coastal waters of California. Bruland et al. (1978b).

1990). Widespread observations of nonconservative behavior of barium (Ba) in seawater do not appear to result from direct biotic uptake. $BaSO_4$ precipitates on dead, sinking phytoplankton, especially diatoms, as a result of the high concentrations of SO_4 that surround these organisms during decomposition (Bishop 1988).

In the Mediterranean Sea, aluminum shows a concentration minimum at a depth of 60 m, where Si and NO_3 are also depleted. MacKenzie et al. (1978) suggested that this distribution is the result of biotic activity, and active uptake has been confirmed in laboratory studies (Moran and Moore 1988). Other workers have found that organic particles carry Al to the deep ocean, but that the association is passive (Hydes 1979, Deuser et al. 1983). High Al in surface waters is due to atmospheric inputs of dust (Orians and Bruland 1985, 1986). Aluminum declines in concentration with depth as a result of scavenging by organic particles and by sedimentation of mineral particles.

Manganese (Mn), an essential element for photosynthesis (Chapter 5), is found at higher concentrations in the surface waters (0.1 μg/liter) than in the deep waters (0.02 μg/liter) of the ocean. Calculating a Mn budget for the oceans, Bender et al. (1977) attribute the high surface concentrations to the input of dust to the ocean surface (Guieu et al. 1994). Manganese appears less limiting than Fe and Zn for the growth of marine phytoplankton in surface waters (Brand et al. 1983). As in the case of Al, the deposition of Mn in dust must exceed the rate of biotic uptake, downward transport, and remineralization of Mn in the deep sea.

The Mn budget of the ocean has long puzzled oceanographers, who recognized that the Mn concentration in ocean sediments greatly exceeds that found in the average continental rock (Broecker 1974, Martin and Meybeck 1979). Other sources of Mn are found in riverflow and in releases from hydrothermal vents (Edmond et al. 1979). Various deep-sea bacteria appear to concentrate Mn by oxidizing Mn^{2+} in seawater to Mn^{4+} that is deposited in sediment (Krumbein 1971, Ehrlich 1975, 1982). The most impressive sedimentary accumulations are seen in Mn nodules that range in diameter from 1 to 15 cm and cover large portions of the seafloor (Broecker 1974, McKelvey 1980). As we discussed for phosphorus nodules, the rate of growth of Mn nodules, about 1 to 300 mm/million years (Odada 1992), is slower than the mean rate of sediment accumulation, yet they remain on the surface of the seafloor. Various hypotheses invoking sediment stirring by biota have been suggested to explain the enigma, but none is proven. In addition to a high concentration of Mn (15–25%), these nodules also contain high concentrations of Fe, Ni, Cu, and Co and are a potential economic mineral resource.

These diverse observations suggest that the geochemistry of many trace elements in seawater is controlled directly and indirectly by biota. Cherry et al. (1978) show that the mean residence time for 14 trace elements in ocean water is inversely related to their concentration in sinking fecal pellets (Fig. 9.21). Some of these elements are mineralized in the deep ocean, but the fate for many trace constituents is downward transport in organic particles and burial in the sediments of the deep sea (Turekian 1977, Lal 1977, Li 1981). Elements with less interaction with biota remain as the major constituents of seawater (Table 9.1).

Biogeochemistry of Hydrothermal Vent Communities

At a depth of 2500 m a remarkable community of organisms is found in association with hydrothermal vents in the east Pacific Ocean. Discovered in 1977, this community consists of bacteria, tube worms, mollusks, and other organisms, many of which are recognized as new species (Corliss et al. 1979, Grassle 1985). Similar communities are found at hydrothermal vents in the Gulf of Mexico and other areas. In total darkness, these communities are supported by bacterial chemosynthesis, in which hydrogen sulfide (H_2S) from the hydrothermal emissions is metabolized using O_2 and CO_2

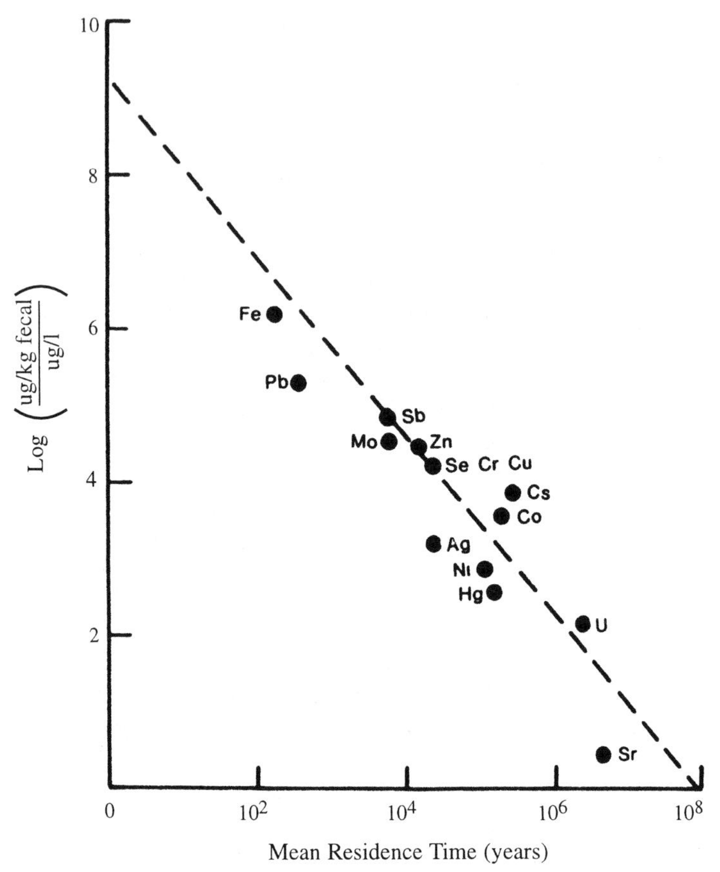

Figure 9.21 The ratio between the concentration of an element in sinking fecal pellets (μg/kg) and its concentration in seawater (μg/liter), plotted as a function of its mean residence time in the ocean. From Cherry et al. (1978).

from the deep sea waters to produce carbohydrate (Jannasch and Wirsen 1979, Jannasch and Mottl 1985):

$$O_2 + 4H_2S + CO_2 \rightarrow CH_2O + 4S\downarrow + 3H_2O. \quad (9.6)$$

Consumption of H_2S by chemosynthetic bacteria is correlated with declines in O_2 when seawater mixes with hydrothermal water (K.S. Johnson et al. 1986). At first glance the reaction would appear to result in the production of organic matter without photosynthesis. We must remember, however, that the dependence of this reaction on O_2 links chemosynthesis in the deep sea to photosynthesis occurring in other locations on Earth. Other bacteria at hydrothermal vents employ chemosynthetic reactions based on methane, hydrogen, and reduced metals that are emitted in conjunction with H_2S (Jannasch and Mottl 1985).

On the basis of the chemosynthetic reactions, bacterial growth feeds the higher organisms found in the hydrothermal communities (Grassle 1985). Some of the bacteria are symbiotic in higher organisms. Symbiotic bacteria in the tube worm *Riftia* deposit elemental sulfur, leading to the rapid growth of tubular columns of sulfur up to 1.5 m long (Cavanaugh et al. 1981, Lutz et al. 1994). Filter-feeding clams up to 30 cm in diameter occur in dense mats near the vents. These communities are dynamic; a particular vent may be active for only about 10 years. Because they are below the carbonate compensation depth, the clam shells slowly dissolve when the vent activity ceases (Grassle 1985). The offspring of these organisms must continually disperse to colonize new vent systems.

Various metallic elements and silicon are soluble in the hot, low redox conditions of hydrothermal vents. Upon mixing with seawater, the precipitation of metallic sulfides removes about 96×10^{12} g S/yr from the ocean (Edmond et al. 1979, Jannasch 1989). Mn and Fe are also deposited as insoluble oxides (MnO_2, FeO) and nodules on the sea floor. The iron oxides act to scavenge vanadium (V) and other elements from seawater and may remove 25% of the annual riverine input of V to the ocean each year (Trefry and Metz 1989).

Hydrothermal vents attain global significance for their effect on the Ca, Mg, and SO_4 budgets of the oceans, but these bizzare chemosynthetic communities speak strongly for the potential for life to exist in unusual locations where oxidized and reduced substances are brought together by biogeochemical cycles.

The Marine Sulfur Cycle and Global Climate

Sulfur is abundant in the oceans, where it is found as SO_4^{2-}. Rivers and atmospheric deposition are the major sources of SO_4 in the sea (Fig. 9.22), but most of the atmospheric deposition is derived from seasalt aerosols that are simply redeposited on the ocean's surface. Metallic sulfides precipitated at hydrothermal vents and biogenic pyrite in sediments are the major marine sinks. Sulfate shows a highly conservative behavior in seawater, with a mean residence time of about 10 to 12 million years relative to inputs from rivers (Table 9.1).

Our understanding of the marine sulfur cycle has developed rapidly within the last 20 years. Of greatest significance, the oceans are now recognized as a major source of dimethylsulfide [$(CH_3)_2S$] in the atmosphere. Trace quantities of this gas impart the "odor of the sea" to coastal regions (Andreae 1986). Dimethylsulfide (DMS) is produced during the decomposition of dimethylsulfoniopropionate (DMSP) from dying phytoplankton cells (Andreae and Barnard 1984, Andreae 1990,

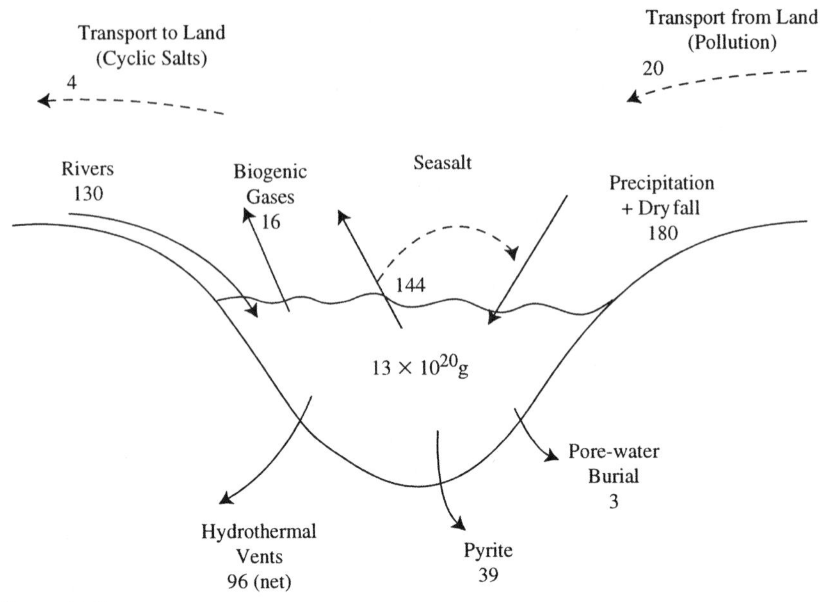

Figure 9.22 Sulfur budget for the world's oceans, showing important fluxes in units of 10^{12} g S/yr. Riverflux from Husar and Husar (1985), gaseous outputs from Bates et al. (1992), hydrothermal flux from Jannasch (1989), and pyrite deposition from Berner (1982). See also Fig. 13.1.

Kiene 1990). The reaction is mediated by an enzyme, dimethyl sulfoxide (DMSO) reductase, which contains Mo (Stiefel 1996). Grazing by zooplankton seems to be important to the release of DMS to seawater (Dacey and Wakeham 1986). Only a small percent of the total production of DMS is lost to the atmosphere; the rest is degraded by microbes in the surface waters (Kiene and Bates 1990). The mean residence time of DMS in seawater is about 2 days.

In an effort to balance the global sulfur cycle, DMS was first proposed as a major gaseous output of the sea by Lovelock et al. (1972). In 1977, Maroulis and Bandy were able to measure DMS as an atmospheric constituent along the eastern coast of the United States. It is now widely recognized as a trace constituent in seawater and in the marine atmosphere, and the diffusion gradient of DMS across the sea–air interface indicates a global flux of at least 15×10^{12} g S/yr to the atmosphere (Erickson et al. 1990, Bates et al. 1992). This is the largest *natural* emission of a sulfur gas to the atmosphere globally (Möller 1984, Spiro et al. 1992).

In the atmosphere, DMS is rapidly oxidized by OH radicals, forming sulfate aerosols that are deposited in precipitation (Chapter 3). Nearly 80% of the non-seasalt sulfate in the atmosphere over the North Pacific Ocean appears to be derived from DMS, with the soil dust and pollution contributing the rest (Savoie and Prospero 1989). Marine DMS is estimated to contribute up to 10% of the atmospheric sulfur over industrial Europe (Tarrasón et al. 1995).

In contrast to terrestrial and freshwater wetland environments, where H_2S dominates the losses of gaseous sulfur, the oceans emit only small quantities of H_2S (T.W. Andreae et al. 1991). The oceans are also a source of carbonyl sulfide (COS) in the atmosphere, but the flux of COS is only a small component of the marine sulfur budget (about 0.2×10^{12} g S/yr) (Chapter 13). Thus, dimethylsulfide is the major form of gaseous sulfur lost from the sea. Dimethylsulfide is also an important sulfur gas emitted from salt marshes (Steudler and Peterson 1985, Hines et al. 1993). Iverson et al. (1989) show that the production of DMS increases as a function of increasing salinity, as river water mixes with seawater in estuaries of the eastern United States.

In addition to helping balance the marine sulfur budget, dimethylsulfide attains global significance for its potential effects on climate. Charlson et al. (1987) recognized that the oxidation of DMS to sulfate aerosols would increase the abundance of cloud condensation nuclei in the atmosphere, leading to greater cloudiness (Bates et al. 1987). Clouds over the sea reflect incoming sunlight, leading to global cooling (Chapter 3). The production of DMS is directly related to the growth of marine phytoplankton (Andreae and Barnard 1984, Turner et al. 1988, T.W. Andreae et al. 1994), so an increase in marine NPP increases the production of DMS. If higher NPP is associated with warmer sea surface temperatures, then the flux of DMS would have the potential to act as a negative feedback on global warming that might otherwise occur by the greenhouse effect.

This hypothesis for a biotic regulation on global temperature is intriguing, for it may be responsible for the moderation of global climate throughout geologic time. Given the strong arguments in favor of global warming by increased atmospheric CO_2, the potential negative feedbacks of DMS are the subject of intense scientific scrutiny and debate. The flux of DMS from the sea is greater in summer than winter, as a result of greater sea surface temperature (Prospero et al. 1991, Tarrasón et al. 1995). The concentration of DMS in seawater is well correlated to that in the air over the North Pacific Ocean (Watanabe et al. 1995). Cloud condensation nuclei also appear to be well correlated to the atmospheric burden of DMS in nonpolluted areas (Ayers and Gras 1991, Putaud et al. 1993, Andreae et al. 1995), and the ice-core record of methanesulphonate (MSA)—a DMS degradation product—suggests higher concentrations during the last glacial epoch than today (Legrand et al. 1991). Certainly the ocean's temperature was lower during the last glacial, but if, for other reasons (e.g., a greater deposition of iron-rich dust), marine productivity was higher during the last glacial, an increased flux of DMS may have reinforced the global cooling of Earth's climate (Turner et al. 1996).

Schwartz (1988) argued that anthropogenic emissions of SO_2 should have the same effect as natural emissions of DMS, because SO_2 is also oxidized to produce condensation nuclei in the atmosphere. Using a general circulation model for global climate, Wigley (1989) suggested that climatic cooling by SO_2 may have offset some of the temperature change expected from the greenhouse effect. Because SO_2 has a short atmospheric lifetime (Chapter 3), its effect is regional and centered on areas of industry (Kiehl and Briegleb 1993, Falkowski et al. 1992, Langner et al. 1992). Nevertheless, it is possible that an increased flux of both SO_2 and DMS will act to dampen the greenhouse effect during the next century.

The Sedimentary Record of Biogeochemistry

Marine sediments contain a record of the conditions of the oceans through geologic time. Sediments and sedimentary rocks rich in $CaCO_3$ (calcareous ooze) show the past location of shallow, productive seas, where foraminifera and coccolithopores were abundant. Sediments deposited in the deep sea are dominated by silicate clay minerals, with high concentrations of Fe and Mn (red clays). Opal indicates the past environment of diatoms, whereas sediments with abundant organic carbon are associated with near-shore areas, where burial of organic materials is rapid (Fig. 9.9). Direct identification of preserved organisms and changes in their species composition have also been used to infer patterns of ocean climate, circulation, and productivity during the geologic past (Weyl 1978, Corliss et al. 1986).

Calcareous sediments contain a record of paleotemperature. When the continental ice caps grew during glacial periods, the water they contained was depleted in $H_2^{18}O$, relative to ocean water, because $H_2^{16}O$ evaporates more readily from seawater and subsequently contributes more to continental rainfall and snowfall. When large quantities of water were lost from the ocean and stored in ice, the waters that remained in the ocean were enriched in $H_2^{18}O$ compared to today. Because carbonates precipitate in an equilibrium reaction with seawater (Eq. 9.2), an analysis of changes in the ^{18}O content of sedimentary carbonates is an indication of past changes in ocean volume and temperature (Fig. 9.23).

The history of the Sr content of seawater is also of particular interest to geochemists, because its isotopic ratio changes as a result of changes in the rate of rock weathering on land (Dia et al. 1992). Most strontium is ultimately removed from the oceans by coprecipitation with $CaCO_3$ (Kinsman 1969, Pingitore and Eastman 1986). During periods of extensive weathering, the ^{87}Sr content of seawater increases as a result of the high content of that isotope in continental rocks. Thus, changes in the ^{87}Sr content of marine carbonate rocks offer an index of the relative rate of rock weathering over long periods (Richter et al. 1992).

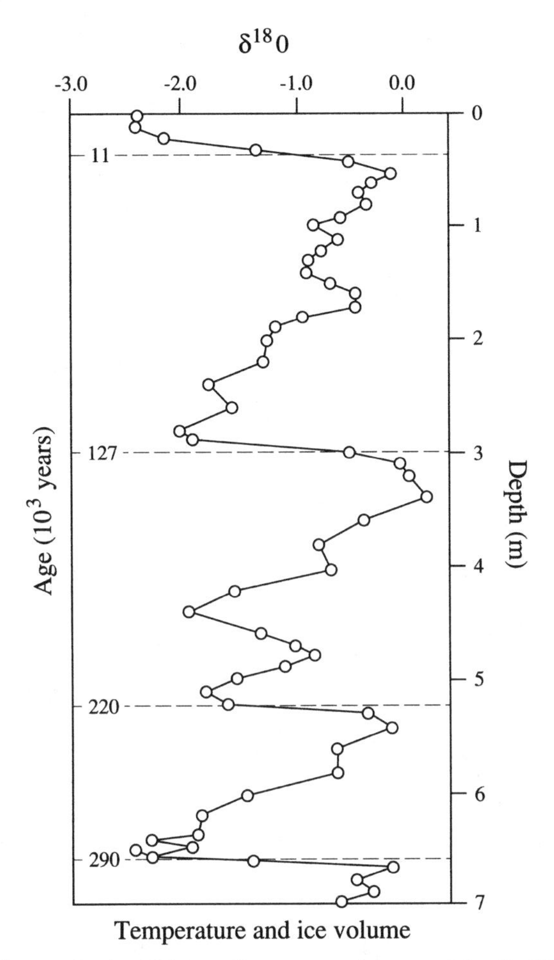

Figure 9.23 Changes in the $\delta^{18}O$ in sedimentary carbonates of the Caribbean Sea during the last 300,000 years. Enrichment of $\delta^{18}O$ during the last glacial epoch (20,000 ybp) is associated with lower sea levels and a greater proportion of $H_2^{18}O$ in seawater. From Broecker (1973).

Carbonates are also a major sink (20%) of boron in the oceans (Vengosh et al. 1991), and the isotopic ratio of boron in carbonate varies as a function of seawater pH. The isotopic ratio of boron in sedimentary foraminifera of the Miocene (21 million years ago) indicates that seawater pH was lower (7.4) than that of today (8.2), consistent with suggestions of higher atmospheric CO_2 during that period (Spivack et al. 1993). Similarly, the boron isotope ratios of sedimentary carbonate indicate a higher seawater pH during the last glacial, when atmospheric CO_2 was low (Sanyal et al. 1995).

The sedimentary record of ^{13}C in organic matter and in $CaCO_3$ contains a record of the biotic productivity of Earth. Recall that photosynthesis discriminates against $^{13}CO_2$ relative to $^{12}CO_2$ (Chapter 5), slightly enriching plant materials in ^{12}C compared to the atmosphere. When large amounts of organic matter are stored on land and in ocean sediments, $^{13}CO_2$ accumulates in the atmosphere and the ocean (i.e., $^{13}HCO_3$). Arthur et al. (1988) suggest that the relatively high ^{13}C content of marine carbonates during the late Cretaceous reflects a greater storage of organic carbon from photosynthesis. Similar changes are seen in the ^{13}C of coal age (Permian) brachiopods (Brand 1989). When the storage of organic carbon is greater, there is the potential for an increase in atmospheric O_2, as postulated for the Permian (Berner and Canfield 1989).

Summary

Biogeochemistry in the sea offers striking contrasts to that on land. The environment on land is spatially heterogeneous; within short distances there are great variations in soil characteristics, including redox potential and nutrient turnover. In contrast, the sea is relatively well mixed. Large, long-lived plants dominate the primary production on land, versus small, ephemeral phytoplankton in the sea. A fraction of the organic matter in the sea escapes decomposition and accumulates in sediments, whereas soils contain little permanent storage of organic matter.

Through its buffering of atmospheric composition and temperature, the oceans exert enormous control over the climate of Earth. At a pH of 8.2 and a redox potential of +200 mV, seawater sets the conditions for biogeochemistry on the 71% of the Earth's surface that is covered by water. Most of the major ions in the oceans have long mean residence times and their concentration in seawater has been constant for nearly all of geologic time. All of this reinforces the traditional, and unfortunate, view that the ocean is a constant body that offers nearly infinite dilution potential for the effluents of modern society.

Looking at the sedimentary record, however, we see that the ocean is subject to large changes in volume and productivity, due to changes in global climate and nutrient flux. Already, we have strong reason to suspect that the productivity of coastal waters is affected by human inputs of N and P. Changes in the temperature and productivity of the central ocean basins may well indicate that global climate change is affecting the oceans as a whole (Venrick et al. 1987, Strong 1989, Polovina et al. 1995). Humans extract a large harvest of fish and shellfish from the oceans—amounting to 8% of marine net primary productivity (Pauly and Christensen 1995). Recent declines in the populations of important commercial fishes suggest that it is doubtful that this harvest is sustainable for future generations.

Recommended Readings

Berger, W.H., V.S. Smetacek, and G. Wefer. (eds.). 1989. *Productivity of the Ocean: Present and Past.* Wiley, New York.

Broecker. W.S. 1974. *Chemical Oceanography*. Harcourt Brace Jovanovich, New York.
Drever, J.I. 1988. *The Geochemistry of Natural Waters*. Prentice Hall, Englewood Cliffs, New Jersey.
Holland, H.D. 1978. *The Chemistry of the Atmosphere and Oceans*. Wiley, New York.
Libes, S.M. 1992. *An Introduction to Marine Biogeochemistry*. Wiley, New York.

PART II

Global Cycles

10

The Global Water Cycle

Introduction

The annual circulation of water is the largest movement of a chemical substance at the surface of the Earth. Through evaporation and precipitation, water transfers much of the heat energy received by the Earth from the tropics to the poles, just as a steam heating system transfers heat from the furnace to the rooms of a house. Movements of water through the atmosphere determine the distribution of rainfall on Earth, and the annual availability of water on land is the single most important factor that determines the growth of plants (Kramer 1982). Where precipitation exceeds evapotranspiration on land, there is runoff. Runoff carries the products of mechanical and chemical weathering to the sea.

In this chapter we will examine a general outline of the global hydrologic cycle and then look briefly at some indications of past changes in the hydrologic cycle and global water balance. Finally, we will look, somewhat speculatively, at future changes in the water cycle that may accompany global climate change. These changes would have direct effects on global patterns of plant growth, the rate of rock weathering, and biogeochemical cycles. Thus, changes in the water cycle have strong implications for the future of agricultural productivity and for the social and economic well-being of human society. Significantly, widespread drought seems associated with the collapse of the early Mesopotamian civilization in the Middle East around 2200 B.C. (Weiss et al. 1993) and the disappearance of the Maya civilization in Mexico around 900 A.D. (Hodell et al. 1995).

The Global Water Cycle

The quantities of water in the global hydrologic cycle are so large that it is traditional to describe the pools and transfers in units of km^3 (Fig. 10.1). Each km^3 contains 10^{12} liters and weighs 10^{15} g. The flux of water in the water cycle may also be expressed in units of average depth. For example, if all the rainfall on land were spread evenly over the surface, each weather station would record a depth of about 70 cm/yr. Units of depth can also be used to express runoff and evaporation (e.g., Fig. 8.5). Annual evaporation from the oceans removes the equivalent of 100 cm of water each year from the surface area of the sea.

Not surprisingly, the oceans are the dominant pool in the global water cycle (Fig. 10.1). Seawater composes over 97% of all the water at the surface of the Earth. The equivalent depth of seawater is 3500 m—the mean depth of the oceans (Chapter 9). Water held in polar ice caps and in continental glaciers is the next largest contributor to the global pool. Soils contain 121,800 km^3 of water, of which about 58,100 km^3 is within the rooting zone of plants (Webb et al. 1993). Human society depends on a relatively small

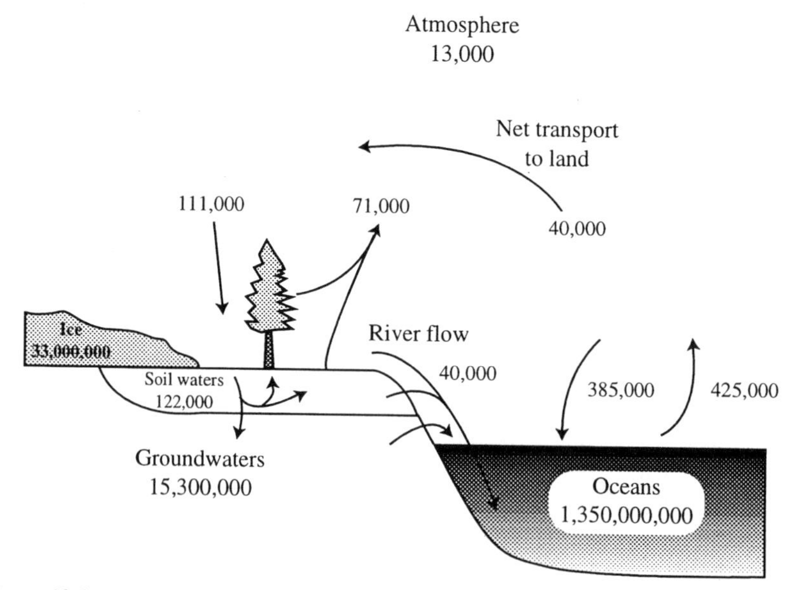

Figure 10.1 The global water cycle. Pools (km^3) and flux (km^3/yr) are mostly from Lvovitch (1973) and Chahine (1992), with some newer values as derived from the text.

pool of liquid freshwater in lakes and rivers. The large pool of freshwater below the vadose zone is known as groundwater (Chapter 7). Global estimates of the volume of groundwater are poorly constrained—4,200,000 to 15,300,000 km^3—but, except as a result of human activities, groundwater is largely inaccessible to the biosphere. The pool of water in the atmosphere is tiny, equivalent to about 3 cm of rainfall at any given time (Eq. 3.4). Nevertheless, enormous quantities of water move through the atmosphere each year.

Evaporation removes about 425,000 km^3 of water from the world's oceans each year. Thus, the mean residence time of ocean water with respect to the atmosphere is about 3100 years. Only about 385,000 km^3/yr of this water returns to the oceans in rainfall; the rest contributes to precipitation on land, which totals 111,000 km^3/yr. Plant transpiration and evaporation from soils return 71,000 km^3/yr to the atmosphere. Thus, with respect to precipitation inputs or evapotranspiration losses, the mean residence time of soil water is about 1 year. Owing to the excess of precipitation over evapotranspiration on land, about 40,000 km^3/yr becomes runoff.

These global average values obscure large regional differences in the water cycle. Evaporation from the oceans is not uniform, but ranges from 4 mm/day in tropical latitudes to <1 mm/day at the poles (Mitchell 1983). Although much precipitation falls at tropical latitudes, an excess of evaporation over precipitation in the tropics provides a net regional flux of water vapor to the atmosphere. Net evaporative loss accounts for the high salinity in tropical oceans (Fig. 9.3), and the movement of water vapor in the atmosphere carries latent heat to polar regions (Vonder Haar and Oort 1973).

On land the relative balance of precipitation and evaporation differs strongly between regions. In tropical rainforests, precipitation may greatly exceed evapotranspiration. Shuttleworth (1988) calculates that 50% of the rainfall becomes runoff in the Amazon rainforests (Table 8.1). In desert regions, precipitation and evapotranspiration are essentially equal, so there is no runoff and only limited recharge of groundwater (e.g., Phillips et al. 1988). As a global average, rivers carry about 1/3 of the precipitation from land to the sea. Less than 10% of precipitation becomes groundwater (Zektser and Loaiciga 1993), so the mean residence time of groundwater is over 1000 years.

The concept of *potential evapotranspiration* (PET), developed by hydrologists, expresses the maximum evapotranspiration that would be expected to occur under the climatic conditions of a particular site, assuming that water is always present in the soil and plant cover is 100%. Potential evapotranspiration is greater than the evaporation from an open pond, as a result of the plant uptake of water from the deep soil and a leaf area index >1.0 in many plant communities (Chapter 5). In tropical rainforests, PET and actual evapotranspiration (AET) are about equal (Vörösmarty et al. 1989).

In deserts, PET greatly exceeds actual AET, owing to long periods when the soils are dry. In southern New Mexico, precipitation averages about 21 cm/year, but the receipt of solar energy could potentially evaporate over 200 cm/yr from the soil (Phillips et al. 1988).

Actual evapotranspiration is often useful as a predictor of net primary production (Webb et al. 1978), decomposition (Fig. 5.15), and soil activity (Fig. 4.3). Changes in climate that affect rainfall and AET would have a dramatic effect on the biosphere. Annual variability in AET is greatest in ecosystems with low AET, reflecting large year-to-year variations in both rainfall and net primary production in deserts (Frank and Inouye 1994). Actual evapotranspiration is more constant in tundra and boreal forest ecosystems, where wet soils do not constrain the supply of water to plants. The net primary productivity of land plants (60×10^{15}g C/yr) and the actual evapotranspiration of water from land (71×10^{18} g/yr) indicate that the global average water-use efficiency of vegetation is about 1.28 mmol CO_2 fixed per mole of water lost (Eq. 5.3)—well within the range measured by physiologists studying individual leaves (Chapter 5).

The sources of water contributing to precipitation also differ greatly in different regions of the Earth. Nearly all the rainfall over the oceans is derived from the oceans. On land, much of the rainfall in maritime and monsoonal climates is also derived from evaporation from the sea. In contrast, 25–50% of the water falling in the Amazon Basin is derived from evapotranspiration within the basin, with the rest derived from long-distance atmospheric transport (Salati and Vose 1984, Eltahir and Bras 1994). Evapotranspiration in Amazon forests is maximized by deep-rooted plants (Nepstad et al. 1994), and the regional importance of evapotranspiration in the Amazon basin speaks strongly for the long-term implications of forest destruction in that region. Using a general circulation model of the Earth's climate, Lean and Warrilow (1989) show that a replacement of the Amazon rainforest by a savanna would decrease regional evaporation and precipitation and increase surface temperatures (cf. Shukla et al. 1990). Similarly, in semiarid regions, precipitation may decline as a result of the removal of vegetation, leading to soil warming (Balling 1989) and increasing desertification (Schlesinger et al. 1990, Chahine 1995, Dirmeyer and Shukla 1996). Thus, the transpiration of land plants is an important factor determining the movement of water in the hydrologic cycle and Earth's climate (Shukla and Mintz 1982, Chahine 1992).

Estimates of global riverflow range from 33,500 km^3/yr to 47,000 km^3/yr (Lvovitch 1973, Speidel and Agnew 1982). Most recent workers assume a value of about 40,000 km^3/yr (Fig. 10.1). The distribution of flow among rivers is highly skewed. The 50 largest rivers carry about 43% of the total riverflow, so reasonable estimates of the global transport of organic carbon, inorganic nutrients, and suspended sediments can be based on data from a few large rivers (e.g., Fig. 8.3).

As a result of the positions of the continents and their surface topography, relative to global climatic patterns, there are large regional differences in the delivery of runoff to the sea. The average runoff from North America is about 32 cm/yr, whereas, the average runoff from Australia, which has a large area of internal drainage and deserts, is only 4 cm/yr (Tamrazyan 1989). Thus, the delivery of dissolved and suspended sediment to the oceans varies greatly between rivers draining different continents (Table 4.7, Fig. 4.14). In the northern hemisphere, 77% of the water discharge comes from rivers in which the flow is now regulated by dams and other human structures (Dynesius and Nilsson 1994) which strongly affect the sediment transport to the sea. Postel et al. (1996) calculate that humans now use 54% of the volume of rivers globally, converting a large portion of it to water vapor as a result of irrigated agriculture.

The mean residence time of the oceans with respect to riverflow is about 34,000 years, which is 10× less dynamic than the exchange with the atmosphere. The mean residence times differ among ocean basins. The mean residence time for the Pacific Ocean with respect to riverflow is 43,700 years—significantly longer than that for the Atlantic (9600 years). This is consistent with the greater accumulation of nutrients in deep Pacific waters and a shallower carbonate compensation depth in the Pacific Ocean (Chapter 9). Despite the enormous riverflow of the Amazon, which carries about 20% of the annual freshwater delivered to the sea, continental runoff to the Atlantic ocean is less than the loss of water through evaporation. Thus, the Atlantic Ocean has a net water deficit, which accounts for its greater salinity (Fig. 9.3). Conversely, the Pacific ocean receives a greater proportion of the total freshwater returning to the sea each year. Ocean currents carry water from the Pacific and Indian oceans to the Atlantic ocean to restore the balance (Chapter 9).

Models of the Hydrologic Cycle

A variety of models have been developed to predict the movement of water through terrestrial ecosystems. Watershed models follow the fate of water received in precipitation and calculate runoff after subtraction of losses due to plant uptake (Waring et al. 1981, Moorhead et al. 1989, Ostendorf and Reynolds 1993). In these models, the soil is considered as a collection of small boxes, in which the annual input and output of water must be equal. Water entering the soil in excess of its water holding capacity is routed to the next lower soil layer or to the next downslope soil unit on the landscape via subsurface flow (Chapter 8). Models of water movement in the soil can be coupled to models of soil chemistry to predict the loss of elements in runoff (e.g., Nielsen et al. 1986, Knight et al. 1985).

A major difficulty in building these models is the calculation of plant uptake and transpiration loss. This flux is usually computed using a formulation of the basic diffusion law, in which the loss of water is determined by the gradient, or vapor pressure deficit, between plant leaves and the atmosphere. The loss is mediated by a resistance term, which includes stomatal conductance and wind speed (Chapter 5). In a model of forest hydrology, Running et al. (1989) assume that canopy conductance decreases to zero when air temperatures fall below 0°C or soil water potential declines below −1.6 MPa. Their model appears to give an accurate regional prediction of evapotranspiration and primary productivity for a variety of forest types in western Montana.

Larger scale models have been developed to assess the contribution of continental land areas to the global hydrologic cycle. For example, Vörösmarty et al. (1989) divided South America into 5700 boxes, each $1/2° \times 1/2°$ in size. Large-scale maps of each nation were used to characterize the vegetation and soils in each box, and data from local weather stations were used to characterize the climate. A model (Fig. 10.2) is used to calculate the water balance in each unit. During periods of rainfall, soil moisture storage is allowed to increase up to a maximum water-holding capacity determined by soil texture. During dry periods, water is lost to evapotranspiration, with the rate becoming a declining fraction of PET as the soil dries.

This type of model can be coupled to other models, including general circulation models of the Earth's climate (Chapter 3), to predict global biogeochemical phenomena. For example, a monthly prediction of soil moisture content for the South American continent can be used with known relationships between soil moisture and denitrification (Potter et al. 1996; Chapter 6) to predict the loss of N_2O and the total loss of gaseous nitrogen from soils to the atmosphere. The excess water in the water balance model is routed to stream channels, where it can be used to predict the flow of the major rivers draining the continent (Russell and Miller 1990). Changes in land use and the destruction of vegetation are easily added to these models, allowing a prediction of future changes in continental-scale hydrology and biogeochemistry.

The History of the Water Cycle

As we learned in Chapter 2, water was delivered to the primitive Earth by planetesimals, meteors, and comets. The accretion of the Earth was largely complete by 3.8 billion years ago (bya). Water was released from the Earth's crust in volcanic eruptions (i.e., degassing), and as long as the Earth's temperature was >100°C, the water vapor remained in the atmosphere. When the Earth cooled <100°C, nearly all the water condensed to form the oceans. Even then, a small amount of water vapor and CO_2 remaining in the Earth's atmosphere was enough to raise the temperature of the

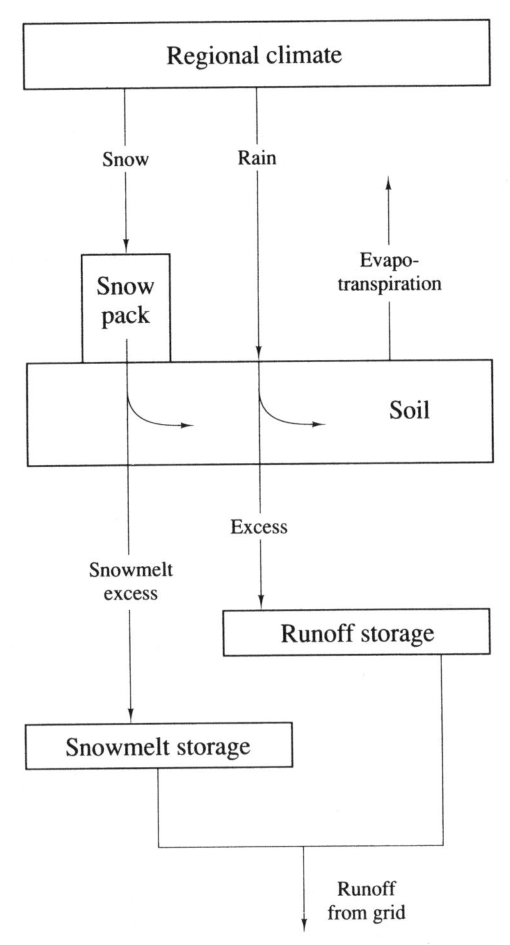

Figure 10.2 Components of a model for the hydrologic cycle of South America. From Vörösmarty et al. (1989).

Earth above freezing. Without this greenhouse effect the Earth might have become a frozen ball of ice—like Mars today.

There is good evidence of liquid oceans on Earth as early as 3.8 bya, and it is likely that the volume of water in the hydrologic cycle has not changed appreciably since that time. Owing to the low content of water vapor in the atmosphere, only 0.1% of the water on Earth appears to have been lost by the photolysis of H_2O in the stratosphere (Walker 1977). Much larger quantities appear to have been lost from Venus, where all water remained as vapor (Chapter 2). The total inventory of volatiles on Earth (Table 2.2) suggests that about 155×10^{22} g of water was degassed from its crust. The difference between this value and the total of the pools in Fig. 10.1 is largely contained in sedimentary rocks. In addition, the accumulation of O_2 in the atmosphere and in oxidized minerals of the Earth's crust suggests that about 2% of the Earth's water has been consumed by net photosynthesis through geologic time (Table 2.2).

Throughout the Earth's history, changes in relative sea level have accompanied periods of tectonic activity that increase (or decrease) the volume of submarine mountains. Changes in sea level also accompany changes in global temperature that lead to glaciations (Degens et al. 1981). The geologic record shows large changes in ocean volume during the 16 continental glaciations that occurred during the Pleistocene Epoch, extending to 2 million years ago. During the most recent glaciation, which reached a peak 18,000 years ago, $42{,}000 \times 10^3$ km^3 of seawater was sequestered in the polar ice caps (Starkel 1989). This represents 3% of the ocean volume, and it lowered the sea level about 120 m from that of present day (Fairbanks 1989). As we saw in Chapter 9, the Pleistocene glaciations are recorded in calcareous marine sediments. During periods of glaciation, the ocean was relatively rich in $H_2^{18}O$, which evaporates more slowly than $H_2^{16}O$. Calcium carbonate precipitated in these oceans shows higher values of $\delta^{18}O$, which can be used as an index of paleotemperature (Fig. 9.23).

Although many causes have been suggested, most workers now believe that ice ages are related to small variations in the Earth's orbit around the Sun (Harrington 1987). These variations lead to differences in the receipt of solar energy, particularly in polar regions. Once polar ice begins to accumulate, the cooling accelerates, because snow has a high reflectivity or albedo to incoming solar radiation. Proponents of this theory believe that the low concentrations of atmospheric CO_2 (Fig. 1.5) and the high concentrations of sulfate aerosols (Legrand et al. 1991) and atmospheric dust (Petit et al. 1990) during the last ice age were probably an effect, rather than a cause, of global cooling. These changes in the atmosphere may have reinforced the rate of cooling (Harvey 1988). At the present time, the Earth is unusually warm; we are about halfway through an interglacial period, which should end about 12,000 A.D.

Continental glaciations represent a major disruption—a loss of steady-state conditions—in Earth's water cycle. These changes in global climate appear to have affected the circulation of the oceans and the interaction of the oceans with the atmosphere (Chapter 9). Global cooling yields lower rates of evaporation, reducing the circulation of moisture through the atmosphere and reducing precipitation. One model of global climate suggests that 18,000 years ago, total precipitation was 14% lower than that of today (Gates 1976). Throughout most of the world, the area of deserts expanded, and total net primary productivity and plant biomass on land may have been much lower than today's (Shackleton 1977, J.M. Adams et al. 1990, Friedlingstein et al. 1995). Greater wind erosion of desert soils contributed to the accumulation of dust in ocean sediments, polar ice caps, and loess deposits (Chapter 3; Yung et al. 1996). The southwestern United States appears to have been an exception. Over most of this desert area, the climate of 18,000 years ago was wetter than today (Van Devender and Spaulding 1979, Wells 1983, Marion et al. 1985).

Changes in the rate of global river flow produce changes in the delivery of dissolved and suspended matter to the sea. Broecker (1982) suggests that erosion of exposed continental shelf sediments during the glacial sea-level minimum may have led to a greater nutrient content of seawater and higher marine net primary productivity in glacial times. Worsley and Davies (1979) show that deep-sea sedimentation rates throughout geologic time have been greatest during periods of relatively low sea level, when a greater area of continents is displayed.

The Water Cycle under Scenarios of Future Climate

It is widely believed that global warming could cause a melting of the polar ice caps, leading to a rise in sea level and a flooding of coastal areas during the next century. Using a variety of methods, most workers measure a rise in sea level of 1 to 2 mm/yr during the last 100 years (Gornitz 1995; Fig. 10.3). Observations of sea-level rise are complicated by the ongoing isostatic adjustments of continental elevations in response to the melting of ice

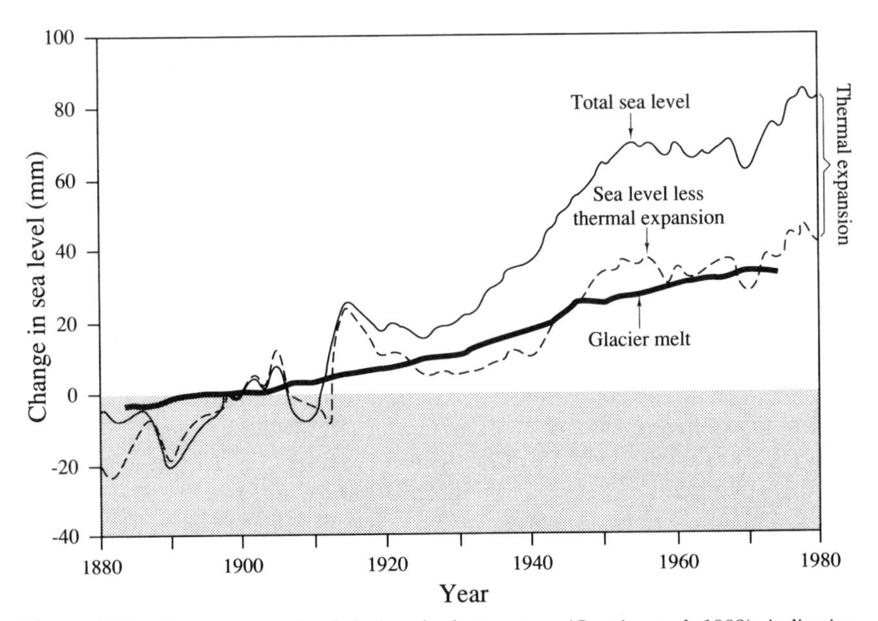

Figure 10.3 Changes in sea level during the last century (Gornitz et al. 1982), indicating the proportion due to thermal expansion of the oceans and that due to melting of glaciers. From Jacobs (1986), after Meier (1984). Copyright 1984 by the AAAS.

from the last continental glaciation. After removing this factor, Peltier and Tushingham (1989) found a rise of 2.4 mm/yr from 1920 to 1970, which they suggest indicates global warming. Although a longer record is clearly needed, recent measurements by the TOPEX/POSEIDON satellite suggest that relative sea level rose at a rate of 3.9 mm/yr globally during 1993 and 1994 (Nerem 1995).

Sea surface temperatures (Strong 1989, Parrilla et al. 1994) have risen over the last 100 years, so at least some of the rise in sea level must be attributed to the thermal expansion of water at warmer temperatures. Some rise in sea level may also stem from human activities, including the extraction of groundwater that is delivered to the sea by rivers (Sahagian et al. 1994). The remaining rise in sea level is likely due to the melting of mountain glaciers throughout the world—an indication of a global warming trend (Oerlemans 1994). At present it is difficult to ascertain whether the volume of the massive ice sheets on Greenland and Antarctica is changing (Jacobs 1992), but we can anticipate that improved measurements of ice volume by satellite remote sensing will soon aid our understanding of this portion of the water cycle (Zwally et al. 1989).

Just as the volume of a glass of water is not affected by ice cubes that may melt within it, sea level is not affected by changes in the area or volume of ice, known as *sea ice,* that is floating on the oceans' surface. Nevertheless, changes sea ice are a useful index of trends in climate that may ultimately affect the hydrologic cycle. Repeated measurements by submarines show no trend in ice thickness at the North Pole during 1977-1990 (McLaren et al. 1992), but satellite measurements using microwave sounding show that the area of Arctic sea ice is shrinking (Johannessen et al. 1995). In contrast, Antarctic sea ice appeared to show little change from 1978 to 1987 (Gloersen and Campbell 1991), although several of the ice shelves on the Antarctic Peninsula appear to have retreated during the past 50 years (Vaughan and Doake 1996).

In response to global warming, most climate models predict a more humid world, in which the movements of water in the hydrologic cycle through evaporation, precipitation, and runoff are enhanced (Neilson and Marks 1994, Loaiciga et al. 1995). Increased cloudiness may moderate the degree of warming, but a new steady state in Earth's temperature would be found at a higher value than that of today (Raval and Ramanathan 1989). Not all areas of the land will be affected equally. Most of the anticipated temperature change is confined to high latitudes, and Manabe and Wetherald (1986) show that large areas of the central United States and Asia may experience a reduction in soil moisture, leading to more arid conditions. Due to the thermal buffer capacity of water, the oceans may warm more slowly than the land surface. Because most precipitation is generated from the oceans, land areas may experience severe drought during the transient period of global warming (Rind et al. 1990, Dirmeyer and Shulka 1996).

Such changes in precipitation and temperature will lead to large-scale adjustments in the distribution of vegetation and global net primary production (Emanuel et al. 1985a, T.M. Smith et al. 1992, Neilson and Marks 1994).

Are any observed changes in the hydrologic cycle consistent with these predictions of global warming? Oltmans and Hofmann (1995) note an increase in stratospheric water vapor over Boulder, Colorado, from 1981 to 1994. A portion of this increase may be due to increasing atmospheric concentrations of methane, some of which is destroyed in the stratosphere producing water vapor (Eq. 3.21). However, the observed increase in water vapor appears to exceed that derived from CH_4, perhaps indicating an ongoing global warming trend.

Analyzing the rainfall records of 1487 weather stations, Bradley et al. (1987) found an increase in precipitation over most of the midlatitudes in the northern hemisphere in the last 30 to 40 years—consistent with changes expected for a warmer planet. Their data also show a decrease in precipitation over North Africa and the Middle East—consistent with the increasing occurrence of drought in the Sahel. In many areas precipitation also seems to be becoming more variable—droughts are more frequent—consistent with the predictions of several general circulation models of future climate (Tsonis 1996). Over much of the world, the historical record of precipitation is scanty, and we must hope that global estimates of precipitation will improve dramatically with the application of satellite remote sensing (Petty 1995). Because water vapor absorbs microwave energy, the relative transmission of microwave radiation through the atmosphere is related to water vapor content and rainfall, and satellite remote sensing of the microwave emission from Earth can measure the rainfall over large areas (e.g., Weng et al. 1994).

Greater precipitation should lead to greater runoff from land (Miller and Russell 1992). Probst and Tardy (1987) found a 3% increase in streamflow in major world rivers over the last 65 years. This increased streamflow may be an indication of global climate change, but it may also relate to the human destruction of vegetation leading to greater runoff (Chapter 8). We might also speculate that greater streamflow is expected due to greater water-use efficiency by vegetation growing in a high-CO_2 atmosphere (Chapter 5; Idso and Brazel 1984). Finally, greater runoff may be due to the surprising decline in evaporation rates reported over a large proportion of the United States and Russia during the last several decades (Peterson et al. 1995).

The historical pattern of runoff for each continent and for the world as a whole shows a cyclic pattern (Fig. 10.4). The cycles for the continents are not synchronous, so the cycles in the global record are "damped," relative to those on each continent. In sum, the recent increases in water vapor, precipitation, and streamflow are consistent with predicted changes in the water cycle with global warming, but such observations must be

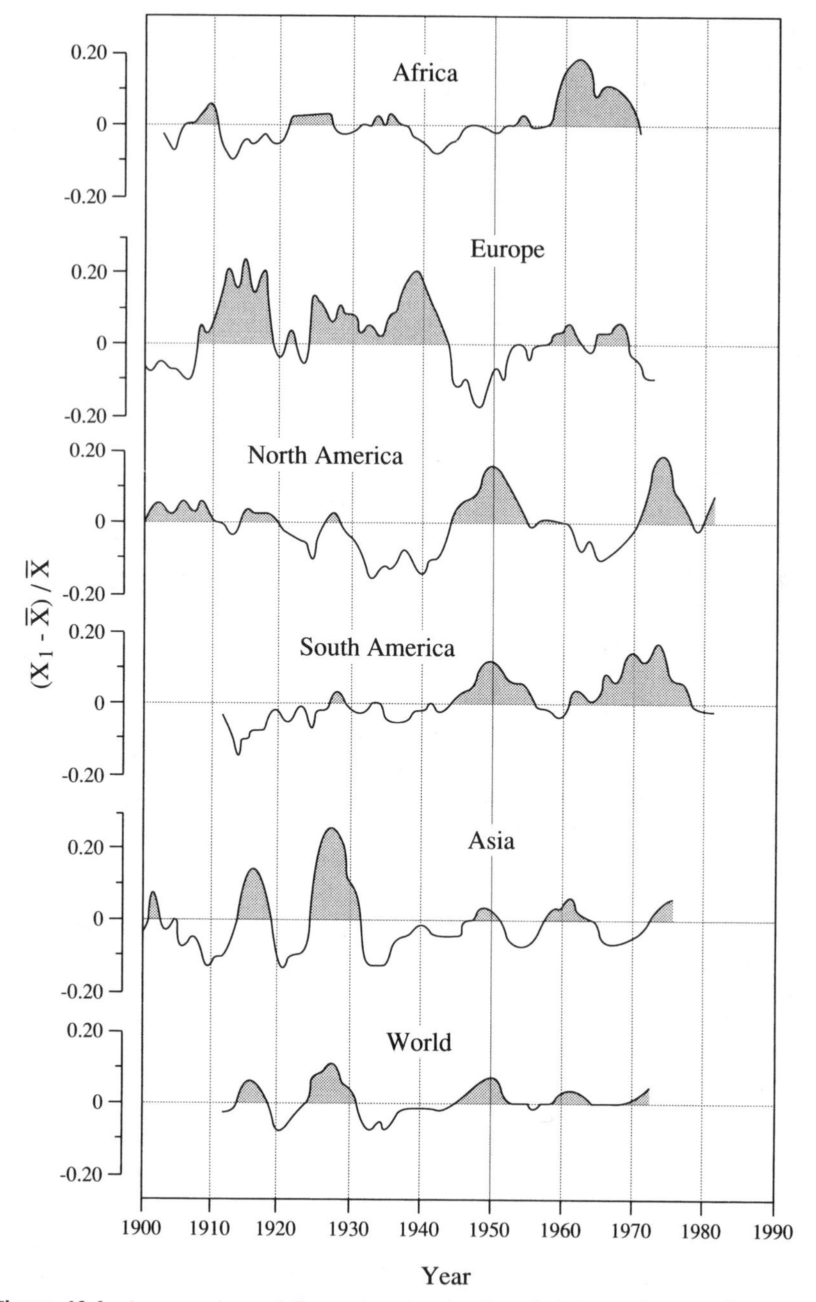

Figure 10.4 A comparison of fluctuations in riverflow draining various continents and averaged for the world. Variation is expressed as the difference between an annual value and the long-term mean, as a fraction of the long-term mean. From Probst and Tardy (1987).

evaluated in the context of long-term cycles in climate that have occurred through geologic time.

Summary

Through evaporation and precipitation the hydrologic cycle transfers water and heat throughout the global system. Receipt of water in precipitation is one of the primary factors controlling net primary production on land. Changes in the hydrologic cycle through geologic time are associated with changes in global temperature. All evidence suggests that movements in the hydrologic cycle were slower in glacial time, and that they would be likely to increase with global warming. Movements of water on the surface of the Earth affect the rate of rock weathering and other biogeochemical phenomena.

Recommended Readings

Baumgartner, A. and E. Reichel. 1975. *The World Water Balance.* R. Olenburg, Munich.

Oliver, H.R. and S.A. Oliver. 1995. *The Role of Water and the Hydrologic Cycle in Global Change.* Springer-Verlag, Berlin.

Raschke, E. and D. Jacob. 1993. *Energy and Water Cycles in the Climate System.* Springer-Verlag, Berlin.

Sumner, G. 1988. *Precipitation: Process and Analysis.* Wiley, New York.

Van der Leeden, F., F.L. Troise, and D.K. Todd. 1990. *The Water Encyclopedia.* Lewis Publishers, Chelsea, Michigan.

Ward, R.C. 1967. *The Principles of Hydrology.* McGraw-Hill, New York.

11

The Global Carbon Cycle

Introduction

The carbon cycle is of central interest to biogeochemistry. Living tissue is composed primarily of carbon, so estimates of the global production and destruction of organic carbon give us an overall index of the health of the biosphere—both past and present. Photosynthetic organisms capture sunlight energy in organic compounds that fuel the biosphere and account for the presence of molecular O_2 in our atmosphere. Thus, the carbon and oxygen cycles on Earth are inextricably linked, and the presence of O_2 in Earth's atmosphere sets the redox potential for organic metabolism in most habitats. Through oxidation and reduction, organisms transform the other important elements of life (e.g., N, P, and S) in reactions that capitalize on the presence of organic carbon and oxygen on Earth. Finally, there is good evidence that through the burning of fossil fuels and other activities, humans have altered the global cycle of carbon, causing the atmospheric concentration of CO_2 to rise to levels that have never been experienced during the evolutionary history of most species that now occupy our planet (Berner and Lasaga 1989).

In this chapter we will consider a simple model for the carbon cycle on the Earth and for assessing human impacts on that cycle. We will then consider the magnitude of past fluctuations in the carbon cycle to gain some perspective of the current human impact. We will look briefly at t budgets of methane (CH_4) and carbon monoxide (CO) in the atmosphe... Because increasing concentrations of carbon dioxide and methane are associated with global warming through the greenhouse effect (Fig. 3.2), the global carbon cycle is directly linked to considerations of global climate change and to international efforts to combat global warming. Finally, we will examine the linkage of the carbon and oxygen cycles on Earth, as a means of "cross-checking" our estimated budgets for these elements at the global level.

The Modern Carbon Cycle

The Earth contains about 10^{23} g of carbon (Table 2.2). All but a small portion is buried in sedimentary rocks, where it is found in organic compounds (1.56×10^{22} g C; Des Marais et al. 1992) and carbonate (6.5×10^{22} g C; Li 1972). The sum of the active pools near the Earth's surface is about 40×10^{18} g C (Fig. 11.1), and extractable fossil fuels are estimated

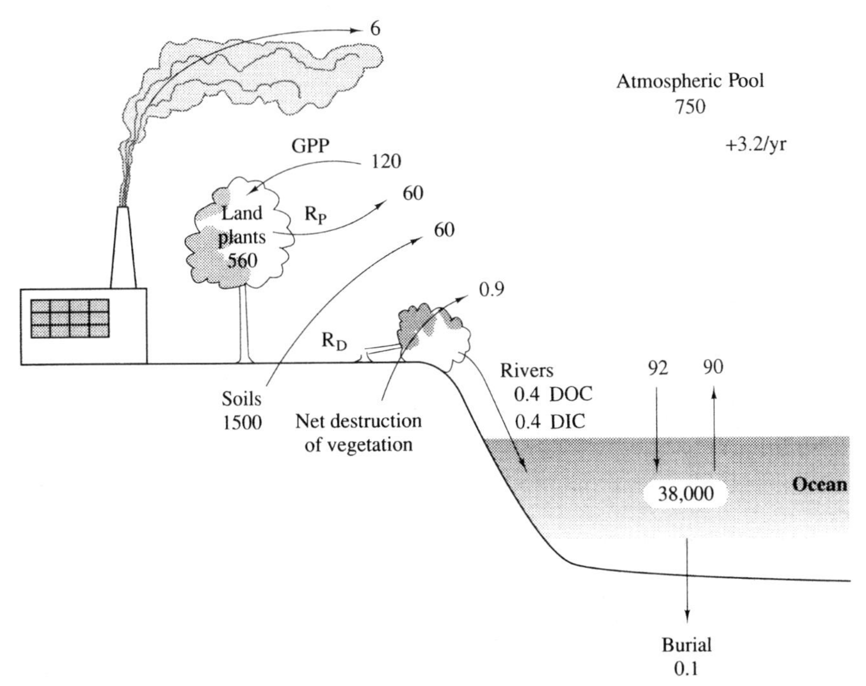

Figure 11.1 The present-day global carbon cycle. All pools are expressed in units of 10^{15} g C and all annual fluxes in units of 10^{15} g C/yr, averaged for the 1980s. Most of the values are from Schimel et al. (1995); others are derived in the text.

at 4×10^{18} g C. Dissolved inorganic carbon in the ocean is the largest near-surface pool, which has an enormous capacity to buffer changes in the atmosphere via Henry's Law (Fig. 2.7). At equilibrium, the sea contains about 56X as much carbon as the atmosphere. The largest pool of carbon on land is contained in soils (Table 5.3). Surprisingly, at the present time, the atmosphere contains more carbon than all of the Earth's living vegetation (Table 5.2).

The largest fluxes of the global carbon cycle are those that link atmospheric carbon dioxide to land vegetation and to the sea (Fig. 11.1). Global net primary production on land is estimated at 60×10^{15} g C/yr (Chapter 5). Considering land vegetation alone, we find that each molecule of CO_2 in the atmosphere has the potential to be captured in net primary production in about 12.5 years. The annual exchange of CO_2 with the oceans is somewhat greater, so the overall mean residence time of CO_2 in the atmosphere is about 5 years.[1] Because this mean residence time is only slightly longer than the mixing time for the atmosphere, CO_2 shows minor regional and seasonal variations that are superimposed on its global average concentration of 360 ppm in 1996 (Fig. 3.6).

Oscillations in the atmospheric content of CO_2 are the result of the seasonal uptake of CO_2 by photosynthesis in each hemisphere and seasonal differences in the use of fossil fuels and in the exchange of CO_2 with the oceans. The oscillation in CO_2 is mirrored by slight oscillations in atmospheric O_2, which has a much longer mean residence time in the atmosphere and a much larger pool size (Keeling and Shertz 1992, R.F. Keeling et al. 1996). Globally, about two-thirds of the terrestrial vegetation occurs in regions with seasonal periods of growth, and the remainder occurs in the moist tropics, where growth occurs throughout the year (Box 1988). The seasonal effect of photosynthesis on atmospheric CO_2 is most pronounced in the northern hemisphere (Fig. 3.6), which contains most of the world's continental area. At high, northern latitudes, vegetation accounts for about 50% of the annual variation in atmospheric CO_2 (D'Arrigo et al. 1987). In the southern hemisphere, smaller fluctuations in atmospheric CO_2 are seasonally reversed relative to the northern hemisphere, and they appear to be dominated by exchange with ocean waters (Keeling et al. 1984). The oscillation at Mauna Loa, Hawaii, located at 19° N latitude, is about 6 ppm/yr (Fig. 1.3), equivalent to a transfer of about 13×10^{15} g C/yr to and from the atmosphere. This value is less than the annual net

[1] This calculation is based on the traditional definition of mean residence time for a steady-state pool—i.e., mass/input. The recent report of the Intergovernmental Panel on Climate Change (Houghton et al. 1995) indicates a mean residence time of 50 to 200 years for atmospheric CO_2 (Table 3, p. 25). This is actually the time that it would take for the current human perturbation of the atmosphere to disappear into other pools at the Earth's surface (e.g., the ocean), if the use of fossil fuels were to cease. Thus, it would take several centuries to return steady-state conditions to the carbon cycle on Earth.

primary productivity of land plants (Table 5.2), owing to the asynchrony of terrestrial photosynthesis and respiration throughout the globe and the buffering of atmospheric CO_2 concentrations by exchange with the oceans.

The release of CO_2 in fossil fuels, currently about 6×10^{15} g C/yr, is one of the best-known values in the global carbon cycle (Marland and Boden 1993, Subak et al. 1993). If all this CO_2 accumulated in the atmosphere, the annual increment would be about 0.8%/yr. In fact, the atmospheric increase is about 0.4%/yr (1.5 ppm), because only 56% of the fossil fuel release remains in the atmosphere (Keeling et al. 1995). This constitutes the "airborne fraction." Where is the rest?

Using models of ocean circulation and CO_2 dissolution in seawater, oceanographers believe that about 33% (2×10^{15} g C/yr) of the CO_2 released from fossil fuels enters the oceans each year (Quay et al. 1992, Siegenthaler and Sarmiento 1993). Thus, Fig. 11.1 shows an annual uptake by the oceans (92×10^{15} g/yr) that is slightly greater than the return of CO_2 to the atmosphere (90×10^{15} g C/yr). Following Henry's Law (Eq. 2.7), excess CO_2 dissolves in seawater, where it is buffered by the dissolution of marine carbonates (Eqs. 9.3 and 9.4). Owing to the low levels of nutrients in seawater, changes in marine NPP are believed to be relatively unimportant to the current oceanic uptake of anthropogenic CO_2 (Shaffer 1993).

Remembering that the exchange of CO_2 between the atmosphere and the oceans takes place only in the surface waters (Chapter 9), we can calculate the mean residence time of CO_2 in the surface ocean—about 11 years—by dividing the pool of carbon in surface waters (1020×10^{15} g C) by the rate of influx (92×10^{15} g C/yr). A similar mixing time is calculated from the distribution of ^{14}C in the surface ocean (Chapter 9). Turnover of carbon in the entire ocean is much slower, about 350 years—consistent with the age of deep ocean waters. Thus, the uptake of CO_2 by the oceans is constrained by mixing of surface and deep waters—not by the rate of dissolution of CO_2 across the surface (Chapter 9). Significantly, if the oceans were well mixed, they might take up as much as 6×10^{15} g C/yr (Keeling 1983), indicating that it is the rate of release from fossil fuels relative to the rate at which the oceans can take up carbon that accounts for the current increase of CO_2 in the atmosphere. If the release of CO_2 from fossil fuels were curtailed, nearly all the CO_2 that has accumulated in the atmosphere would eventually dissolve in the oceans and the global carbon cycle would return to a steady state.

Taken alone, the atmospheric increase and oceanic uptake of CO_2 account for 89% of the annual emissions from fossil fuels. Considering the errors associated with these global estimates, it would seem that we have a fairly tidy picture of the global carbon cycle. However, many terrestrial ecologists believe that there have also been substantial releases of CO_2 from terrestrial vegetation and soils, caused by the destruction of forest vegetation in favor of agriculture, especially in the tropics (Chapter 5). If their calculations are accurate, then the atmospheric budget is unbalanced, and a large amount of carbon dioxide that ought to be in the atmosphere is "missing" (Table 11.1).

The net release of CO_2 from vegetation is difficult to estimate globally (Chapter 5). At any time, some land is being cleared, while agriculture is abandoned in other areas that are allowed to regrow. Historical changes in the ^{13}C and ^{14}C isotopic ratios in atmospheric CO_2 show unequivocal evidence of a net release from the biosphere early in this century (Chapter 5, Wilson 1978). Indeed, until about 1960, the release from land clearing may have exceeded the release from fossil fuel combustion (Houghton et al. 1983). In 1990, the estimate of CO_2 released from deforestation in the tropics (1.6×10^{15} g C) was partially balanced by the estimate of carbon captured by the regrowth of forests in temperate regions (0.7×10^{15} g C; Dixon et al. 1994). Forests in the United States—mostly in New England—now appear to accumulate 0.1 to 0.2×10^{15} g C/yr (Birdsey et al. 1993, Turner et al. 1995), and a similar amount of forest regrowth is estimated for Europe (Houghton et al. 1987, Kauppi et al. 1992). Global estimates of changes in the carbon held in vegetation and soils will improve with the application of remote sensing by satellites, which have already suggested that the current rate of deforestation in the tropics is lower than what was widely believed just a few years ago (Skole and Tucker 1993). Still, any net release of carbon from land complicates our ability to balance a carbon dioxide budget for the atmosphere (Table 11.1).

We might reconcile the carbon dioxide budget of the atmosphere if we find evidence that the pool of carbon in land vegetation and soils has increased as a result of a global stimulation of plant growth by higher concentrations of atmospheric CO_2 (Chapter 5). Despite widespread forest destruction, enhanced uptake of CO_2 in areas of undisturbed vegetation could add to the pool of carbon on land. A recent estimate, using eddy-correlation measurements (Chapter 5), shows net CO_2 uptake in an undisturbed tropical rainforest, which may be indicative of a CO_2-fertilization effect (Grace et al. 1995, 1996). Excess deposition of nitrogen from the atmosphere may also stimulate plant growth in some regions (Townsend et al. 1996). The overall stimulation of terrestrial photosynthesis by human activities is informally known as the "beta" factor in models of the global carbon cycle. Beta is defined as the change in NPP that would derive from a doubling of atmospheric CO_2 con-

Table 11.1 Sources and Sinks of CO_2 in the Atmosphere in Units of 10^{15}g C/yr

		Net emissions	=	Net changes in the carbon cycle				
Fossil fuel emission	+	Net destruction of vegetation	=	Atmospheric increase	+	Ocean uptake	+	Unknown sink
6	+	0.9	=	3.2	+	2.0	+	1.7

centration. In controlled experiments with tree seedlings, the beta factor usually lies in a range of 32 to 41% as a result of CO_2 fertilization (Poorter 1993, Wullschleger et al. 1995, Norby 1996).

The historical record of CO_2 in the atmosphere offers several indirect approaches for estimating changes in global net primary production and the potential for a significant, positive beta factor. For example, in the record of atmospheric CO_2 (Fig. 1.3), the seasonal decline each summer is largely due to photosynthesis, while the seasonal upswing derives from decomposition. An increasing *amplitude* of the CO_2 oscillation, after the removal of fossil fuel and El Niño effects, implies a greater activity of the terrestrial biosphere. Such a trend is evident in an analysis of the Mauna Loa record of CO_2 (Fig. 11.2), in which the amplitude has increased by about 0.54%/yr since 1958 (Bacastow et al. 1985, Keeling 1993). The increase at northern latitudes has been about 40% since 1960, perhaps as a result of climatic warming over the same period (C.D. Keeling et al. 1996). Although an increasing annual oscillation in the concentration of atmospheric CO_2 suggests that biospheric processes have been stimulated, we should not necessarily assume that a greater amount of carbon is being stored on land. Greater rates of decomposition may simply balance increased rates of photosynthesis (Houghton 1987, C.D. Keeling et al. 1996).

There is a small difference (about 4 ppm) in the concentration of atmospheric CO_2 between the northern and the southern hemispheres, owing to the predominant use of fossil fuel in the northern hemisphere (Keeling 1993). This observed latitudinal gradient in atmospheric CO_2 can be compared to the gradient that would be expected due to the mixing of the atmosphere, as calculated from general circulation models (Chapter 3). In fact, many of these models suggest that the concentration in the northern hemisphere, particularly at temperate latitudes, should be even greater than what is actually observed in the atmosphere, implying that there is

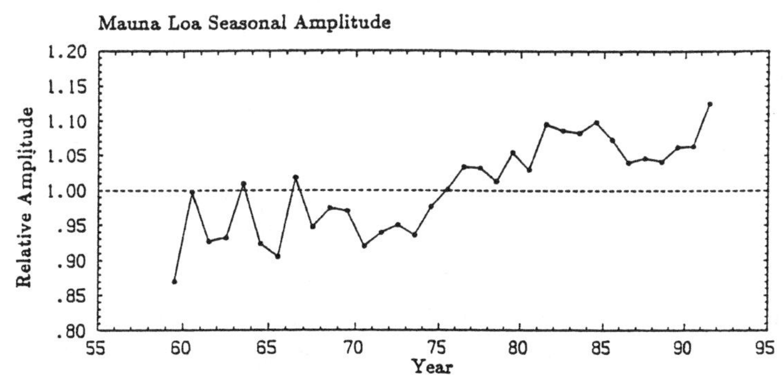

Figure 11.2 Increasing amplitude of the oscillations in atmospheric CO_2 at Mauna Loa, Hawaii. From Keeling (1993).

significant uptake, perhaps by land plants (Tans et al. 1990, Denning et al. 1995). The relative importance of land plants (versus uptake by the ocean) is also indicated by a latitudinal gradient in the isotopic ratio of atmospheric CO_2 (i.e., $\delta^{13}C$), which is fractionated by photosynthesis, but not by dissolution in seawater (Ciais et al. 1995).

Despite substantial theoretical and indirect evidence that it *should* occur, there is little direct evidence (e.g., changes in tree-ring width) that an enhanced growth of land plants currently acts as a sink for atmospheric CO_2 (Chapter 5). The response of land plants may be limited by inadequate supplies of nutrients and water over most of the Earth's land surface (Comins and McMurtrie 1993). Growth of plants at elevated CO_2 produces tissues and leaf litterfall with a high C/N ratio, slowing decomposition in soils and exacerbating nutrient deficiencies (Coûteaux et al. 1991, Cotrufo et al. 1994, Chapter 6). In response, plants often appear to allocate increasing amounts of photosynthate to the growth of roots and to root metabolic activities (Norby et al. 1992, Zak et al. 1993, D.W. Johnson et al. 1994, Rouhier et al. 1994, DeLucia et al. 1996). Parallel changes in soil organic matter, however, are difficult to measure due to the enormous natural variability of soils.

Our knowledge of the response of the terrestrial biosphere to high CO_2 is limited to a few studies that have examined entire ecosystems, including both vegetation and soils (Ceulemans and Mousseau 1994). Oechel et al. (1994) found that wet tundra ecosystems in Alaska showed a complete physiological adjustment and no net carbon storage in response to a 3-year exposure to elevated CO_2. However, carbon storage in these ecosystems increased when both CO_2 and temperature were maintained at elevated levels. It is possible that the warmer soil temperatures enhanced decomposition, improving the supply of nutrients for plant growth (Van Cleve et al. 1990, Peterjohn et al. 1994). Because the C/N ratio of soil organic matter (12–15) is lower than the C/N ratio of plant tissues (160, Table 6.5), a small amount of additional nitrogen mineralization in soils could yield a large enhancement of net production and carbon sequestration in plants (Rastetter et al. 1992, McGuire et al. 1992). Using a model for carbon cycling in grassland ecosystems (Fig. 6.22), Parton et al. (1995) found that global warming stimulated CO_2 loss from grassland soils, but about half of the loss was recaptured by greater rates of plant growth under high CO_2.

Many areas of the world receive an excess atmospheric deposition of nitrogen derived from anthropogenic emissions of NO_x and NH_3 (Chapters 3 and 12). In some areas the N input is so extreme that symptoms of forest decline are observed (Chapter 6), but in other areas the added nitrogen has the potential to act as a fertilizer stimulating plant growth (Townsend et al. 1996). If the global human production of fixed N (100×10^{12} g/yr; Chapter 12) were all stored in the woody tissues of plants with a C/N ratio of 160 (Table 6.5), then as much as 16×10^{15} g C/yr might be stored in terrestrial ecosystems. Such a large storage is unlikely, because nitrogen is also removed from the land by runoff and denitrification (Chapters 6, 8, and 12). In some areas, the fate of the excess nitrogen deposited from the atmosphere is unclear—some may accumulate in soil organic matter (Nadelhoffer et al. 1995, Mäkipää 1995); however, commercial nitrogen fertilization of forests often produces a significant increase in plant growth and only a small increase in soil organic compounds (e.g., Neilsen et al. 1992, Harding and Jokela 1994).

Over longer periods of time, changes in the distribution of vegetation as a result of global climate change could also affect the concentration of atmospheric CO_2 (Chapter 5). Coupled to models of climate change, most models of the global carbon cycle suggest an increase in the carbon content of vegetation and soils, when vegetation is in equilibrium with a warmer and wetter world of the future (T.M. Smith et al. 1992). If the adjustment of vegetation to climate occurs over 100 years, these models suggest that the net uptake by the terrestrial biosphere could be as high as 1.8×10^{15} g C/yr—mostly in vegetation. However, other models suggest that changes in vegetation and soils during the transition in climate may yield the opposite effect. Smith and Shugart (1993) estimate large losses of carbon from vegetation during the transient period of drought that is likely to accompany global warming over the next century (Chapter 10).

In sum, whole ecosystem response will be determined by various factors—CO_2, nutrient availability, and global patterns of temperature and rainfall—which are all affected by human activities. Although it seems unlikely that enhanced growth by terrestrial vegetation will ultimately stem the rise of CO_2 that is derived from fossil fuels (cf. Idso and Kimball 1993 vs. Amthor 1995), the response of the terrestrial biosphere could have a dramatic impact on the future composition of the atmosphere (Fig. 11.3).

In our view of the global carbon cycle, it is important to recognize that the annual movements of carbon, rather than the amount stored in various reservoirs, are most important. The ocean contains the largest pool of carbon near the surface of the Earth ($38,000 \times 10^{15}$ g), but most of that pool is not involved with rapid exchange with the atmosphere. Similarly, desert soil carbonates contain more carbon (930×10^{15} g) than land vegetation, but the exchange between desert soils and the atmosphere is tiny (0.023×10^{15} g C/yr), yielding a turnover time of 85,000 years in that pool (Schlesinger 1982, 1985b). In soils, the small, active pool of organic matter near the surface is more likely to respond rapidly to carbon additions than the larger pool of refractory carbon at depth (Trumbore 1993, Harrison et al. 1993).

All explanations for the increasing concentrations of CO_2 in the atmosphere must rely on documented, recent changes in the global carbon cycle. A flux that has not changed in recent times, no matter how large, is not likely to affect the concentration of atmospheric CO_2 (Houghton et

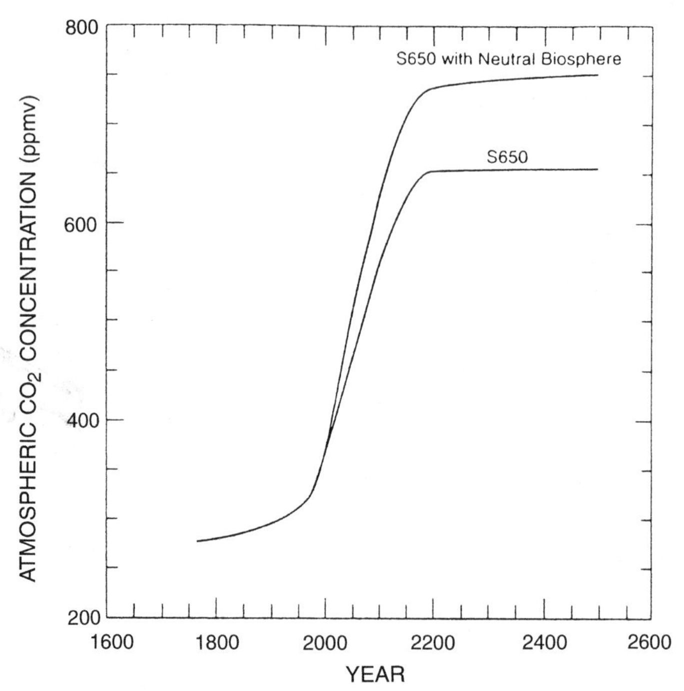

Figure 11.3 The effect of an "active" terrestrial biosphere on a model-derived trajectory for the future course of atmospheric CO_2. The lower curve shows the trajectory of CO_2 concentration that would result from an international agreement to limit fossil fuels emissions, so as to achieve a concentration of 650 ppm CO_2. The upper curve shows the concentrations that are reached—750 ppm—with the same policy but no interaction of the terrestrial biosphere. From Schimel (1995).

al. 1983). For example, the release of CO_2 in forest fires is of no consequence to changes in atmospheric CO_2 unless the frequency or area of forest fires has changed in recent times (Adams et al. 1977, Auclair and Carter 1993, Kasischke et al. 1995). The carbon flux in rivers or sinking pteropods cannot serve as a net sink for anthropogenic CO_2 in the ocean, unless the flux in these pathways is greater as a result of human activities. Similarly, the storage of carbon in peatland soils is not a sink for fossil fuel CO_2, unless the rate of storage in these areas, which has occurred throughout the Holocene, has increased significantly during the Industrial Revolution (Harden et al. 1992).

On the other hand, relatively small changes in large pools of carbon can have a dramatic impact on the carbon dioxide content of the atmosphere, especially if they are not balanced by simultaneous changes in other components of the carbon cycle. A 1% increase in the rate of decomposition on land, as a result of global warming, would release nearly 0.6×10^{15} g C/yr to the atmosphere. Schimel et al. (1994) estimated that the soil carbon pool could lose 0.7% of its content (11×10^{15} g C) for every degree of global warming during the next century (cf. Kirschbaum 1995, Trumbore et al. 1996). On the other hand, a 0.2%/yr-increment to the biomass of carbon on land, as a result of a greater storage of NPP, could balance the CO_2 budget of the atmosphere (Table 11.1). We can speculate that this increment should be first realized in vegetation, because such a small percentage of NPP that enters the soil survives to become a component of soil organic matter (Schlesinger 1990).

The largest global pool of carbon is found in sedimentary rocks, including the fossil fuels. Storage of organic carbon in these deposits accounts for the accumulation of O_2 in the atmosphere through geologic time (Chapter 2). In the absence of human perturbations, the exchange between the fossil pool and the atmosphere could be ignored in global models. Humans affect the global system by creating a large biogeochemical flux where none existed before.

Temporal Perspectives of the Carbon Cycle

Studies of the biogeochemistry of carbon on Earth must begin with a consideration of the origin of carbon as an element and with theories that explain its differential abundance on the planets of our solar system (Chapter 2). During the early development of Earth, the carbon cycle was decidedly non-steady state: the carbon content of the planet grew with the receipt of planetesimals and meteorites, and the atmospheric content increased as volcanoes released CO_2. The oldest geologic sediments suggest that atmospheric CO_2 may have been as high as 3% on the primitive Earth, providing a substantial greenhouse effect during a time of low solar output (Walker 1985, Rye et al. 1995). Even today, 360 ppm of CO_2 and substantial concentrations of other greenhouse gases in our atmosphere raise the surface temperature of the Earth above freezing—obviously an essential condition for the persistence of the biosphere (Ramanathan 1988).

As discussed Chapter 4, CO_2 in atmosphere interacts with the crust of the Earth, causing rock weathering (Eq. 4.3). Carbon dioxide is removed from the atmosphere and transferred via rivers to the oceans, where it is eventually deposited on the seafloor in carbonate rocks, adding to the Earth's crust (Fig. 1.4). As early as 1918, Arrhenius speculated that the consumption of CO_2 by rock weathering might eventually cool the planet through a loss of its natural "greenhouse effect":

> As the crust grew thicker, the supply of this gas [CO_2] diminished and was further used up in the process of disintegration [weathering]. As a consequence the temperature slowly decreased, although decided fluctuations occurred with the changing volcanic activity during different periods. Supply and consumption of carbon dioxide fairly

balanced as disintegration ran parallel with the proportion of this gas in the air. (Svante Arrhenius, 1918, *The Destinies of the Stars,* p. 177)

Fortunately, CO_2 is also released from the crust of the Earth as a result of tectonic activity. In the complete geochemical cycle of carbon (Fig. 1.4), subduction of the oceanic crust carries carbonate minerals deposited on the seafloor to the interior of the Earth, where CO_2 and other volatile elements are once again released by hydrothermal and volcanic emissions. S.N. Williams et al. (1992) and Bickle (1994) estimate that between 0.02 and 0.05×10^{15} g C/yr are released as CO_2 from volcanoes around the world. Presumably, most of the CO_2 being vented by volcanoes today has made at least one previous trip through this cycle. If this cycle were not complete, rock weathering would deplete CO_2 from the atmosphere, and dissolved CO_2 in the oceans, in about 1,000,000 years.

On Earth, this geochemical cycle has helped to maintain the concentration of atmospheric CO_2 below 1% for the last 100 million years (Berner and Lasaga 1989). On Mars, where this cycle has slowed or stopped, the atmosphere contains a small amount of CO_2, and the planet is very cold (Chapter 2). On Venus, where CO_2 cannot react with crustal minerals, the atmosphere contains a large amount of CO_2, and the planet is very hot (Nozette and Lewis 1982). During periods of extensive volcanism, the atmospheric concentration of CO_2 on Earth may have been greater than today's, leading to warmer climates (Owen and Rea 1985); however, a continuous geologic record of liquid oceans on Earth indicates that CO_2 and other greenhouse gases have always remained at levels that produce relatively moderate surface temperatures.

Despite their long-term significance in buffering atmospheric CO_2, the annual transfers of carbon in the geochemical cycle are relatively small. The massive quantities of CO_2 that are now tied up in the carbonate minerals of the Earth's crust are the result of a slow accumulation of these materials over long periods of Earth's history. Today rivers carry about 500×10^{12} g yr of Ca^{2+} (Milliman 1993) and 0.40×10^{15} g of carbon as HCO_3^- (Sarmiento and Sundquist 1992, Suchet and Probst 1995) to the sea. For seawater to maintain fairly constant concentrations of Ca, an equivalent amount of Ca must be deposited as $CaCO_3$ in ocean sediments, carrying 0.15×10^{15} g C/yr to the oceanic crust. Dividing the mass of carbonate rocks by their annual rate of formation, we find that each atom of carbon sequestered in marine carbonate spends more than 400 million years in that reservoir.

With the appearance of life, a *bio*geochemical cycle was added on top of the underlying geochemical cycle of carbon on Earth. Models of the modern biogeochemical cycle of carbon must focus on the large annual transfer of CO_2 from the atmosphere to plants as a result of photosynthesis and the large return of CO_2 to the atmosphere as a result of decomposition (Fig. 11.1). Today, the fluxes of carbon in the biogeochemical cycle of carbon, mostly expressed in units of 10^{15} to 10^{17} g C/yr, dwarf the fluxes of the underlying geochemical cycle of carbon, where the movements are typically 10^{13} to 10^{14} g C/yr.

During Earth's history, at times when the production of organic carbon by photosynthesis has exceeded its decomposition, organic carbon has accumulated in geologic sediments. The earliest organic carbon is present in rocks from 3.8 bya, with the pool increasing to 1.56×10^{22} g by about 540 mya (Des Marais et al. 1992). During that interval, between 10 and 20% of all the carbon buried in marine sediments was organic—similar to the ratio found in modern marine sediments (Li 1972, Dobrovolsky 1994, p. 163). During the Carboniferous (300 mya), large deposits of organic carbon were stored in freshwater environments, leading to modern economic deposits of coal. During the Tertiary, the precursors to modern deposits of petroleum were added to marine sediments. Net storage of organic carbon in sediments has varied between about 0.04 and 0.07×10^{15} g C/yr during the last 300 million years (Fig. 11.4; Berner and Raiswell 1983); a rate of 0.10×10^{15} g C/yr is estimated for the present (Chapter 9).

Life also stimulated some of the reactions in the underlying geochemical cycle of carbon. Various marine organisms enhance the deposition of calcareous sediments, which now cover more than half of the ocean's seafloor (Kennett 1982). Land plants, by maintaining high concentrations of CO_2 in the soil pore space, raise the rate of carbonation weathering, speeding the reaction of CO_2 with the Earth's crust (Chapter 4). Land plants and soil microbes also excrete a variety of organic compounds, by-products of photosynthesis, that enhance rock weathering. Various models developed and summarized by Berner (1992, 1994) suggest that the atmospheric concentration of CO_2 declined precipitously as land plants gained dominance about 350 mya.

The appearance of life on Earth also allowed interactions among the carbon and sulfur cycles, providing further biogeochemical mechanisms

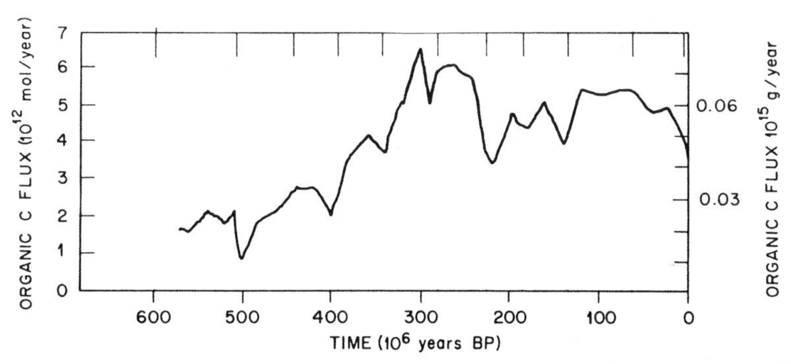

Figure 11.4 Burial of organic carbon on Earth over the last 600 million years. From Olson et al. (1985).

that have buffered atmospheric CO_2 within narrow limits over the last several million years. As illustrated by the model of Garrels and Lerman (1981) (Fig. 1.1), if atmospheric CO_2 were to rise, greater rates of photosynthesis and storage of organic carbon in marine sediments should follow. When the storage of organic carbon increases, the pool of sedimentary pyrite decreases, consuming the oxygen that is released by photosynthesis and forming gypsum. The fact that neither gypsum nor Mg carbonates (dolomites) are significant components of modern marine sediments or evaporites speaks to the efficacy of various processes controlling atmospheric CO_2 within narrow limits. Although the model does not consider nutrient limitations in the ocean, it suggests that fluctuations in atmospheric CO_2 should be small and short-lived, because the atmosphere is in rapid exchange with other compartments of the global carbon cycle.

Collections of gas trapped in ice cores from the Antarctic provide a historical record of atmospheric CO_2 for the last 220,000 years (Fig. 1.5). Until the Industrial Revolution, concentrations varied between 200 and 280 ppm, with the lowest values found in layers of ice that were deposited during the most recent ice age. Although the exact magnitude is controversial, the mass of carbon stored in vegetation and soils was also lower during the last glacial, as a result of the advance of continental ice sheets and widespread desertification of land habitats (J.M. Adams et al. 1990, Servant 1994, Bird et al. 1994, Friedlingstein et al. 1995, Prentice and Sykes 1995, Crowley 1995). Thus, glacial conditions must have produced changes in the oceans that allowed a large uptake of CO_2 (Faure 1990, Sundquist 1993). Increased marine NPP or an increased efficiency of the marine "biotic pump" (Chapter 9) seem unlikely (Leuenberger et al. 1992), but a decrease in the amount of carbon stored as carbonates could lead to a greater retention of CO_2 in the oceans as HCO_3^- (Eqs. 9.3 and 9.4). Most of the $CaCO_3$ dissolution in marine sediments is driven by CO_2 released during the decomposition of organic matter (Berelson et al. 1990), so Archer and Maier-Reimer (1944) suggest that an increase in the ratio of organic carbon to carbonate in sinking particles could lead to a greater dissolution of carbonate in marine sediments. Recently, Sanyal et al. (1995) report evidence for a higher pH of the glacial ocean, consistent with a greater dissolution of carbonates in the deep sea.

At the end of the last glacial, 17,000 years ago, atmospheric CO_2 rose to about 280 ppm, where it remained with minor variations until the beginning of the Industrial Revolution (Fig. 11.5; Etheridge et al. 1996). The increase in concentration from 280 ppm to today's value of 360 ppm represents a global change of 30% in less than 200 years! Although the current level of CO_2 is not unprecedented in the geologic record, our concern is the speed at which a basic characteristic of the planet has changed to levels not previously experienced during human history or during the evolution

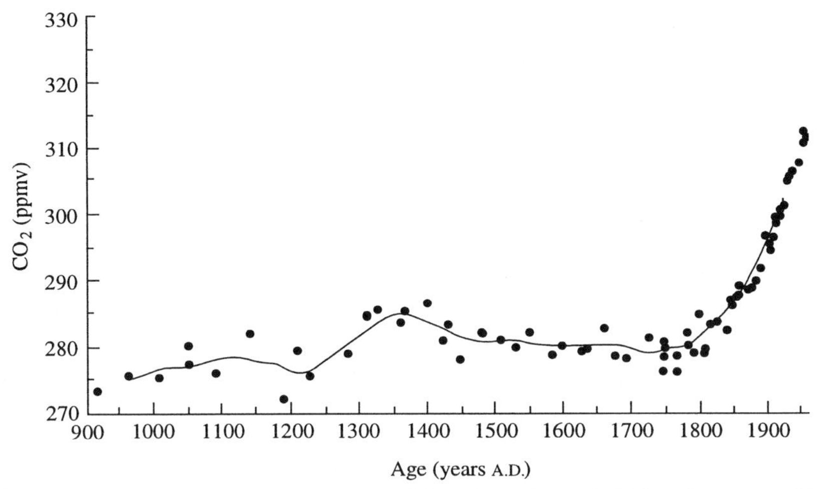

Figure 11.5 Concentrations of atmospheric CO_2 estimated from bubbles of gas trapped in ice cores from Antarctica. From Barnola et al. (1995).

of current ecosystems. If global temperature and atmospheric CO_2 are related, then we are destined for significant global warming during the near future.

These perspectives of the global carbon cycle extend from processes that occur on a time scale of 10^9 years to those that occur annually. The global carbon cycle is composed of large, rapid transfers in the *bio*geochemical cycle superimposed on the underlying, small, slow transfers of the geologic cycle. Buffering of atmospheric CO_2 over geologic time involves small net changes in carbon storage that occur relatively slowly. Thus, an increase in rock weathering as a result of high CO_2 and rising global temperature is not likely to be an effective buffer to the rapid release of CO_2 from fossil fuels. In contrast, the current exchange of CO_2 between the atmosphere and the biosphere is about 150×10^{15} g C/yr, so the biosphere is more likely to buffer the rise of CO_2 as a result of human activities. The current increase in atmospheric CO_2 results from our ability to change the flux of CO_2 to the atmosphere by an amount that is significant relative to the biogeochemical reactions that buffer the system over long periods of time.

Atmospheric Methane

At first glance, the annual flux of methane would seem to be only a minor component of the global carbon cycle. All sources of methane in the atmosphere are in the range of 10^{12} to 10^{14} g C/yr, which is several orders of magnitude lower than the values for CO_2 shown in Fig. 11.1. Globally, the atmospheric methane concentration is 1.75 ppm, versus 360 ppm for CO_2 (Table 3.1). However, over the last several decades the concentration of methane in the atmosphere has increased at an average rate of about 1%/yr (Fig. 11.6), which is much faster than the rate of CO_2 increase over the same interval (Fig. 11.5). Each molecule of methane in the atmosphere has the potential to contribute about 25× as much greenhouse warming as each molecule of CO_2 over the next century (Lashof and Ahuja 1990, Albritton et al. 1995). Because the absorbance of infrared radiation by carbon dioxide may eventually approach 100% in the wavebands in which CO_2 is effective, increases in atmospheric methane may be relatively more important to global climate change during the next century (Dickinson and Cicerone 1986).

The current increase in atmospheric methane stems from a net accumulation of about 30 × 10^{12} g/yr in a global pool of about 4.96 × 10^{15} g. The cause of the increase is not obvious, because a wide variety of sources contribute to the total annual production of about 535 × 10^{12} g/yr (Table 11.2). The sum of anthropogenic sources is about 2× the sum of natural sources, so it is perhaps surprising that the annual increase of methane in the atmosphere is not more than 1%/yr. The estimate of total flux is fairly well constrained, because it yields a mean residence time for atmospheric

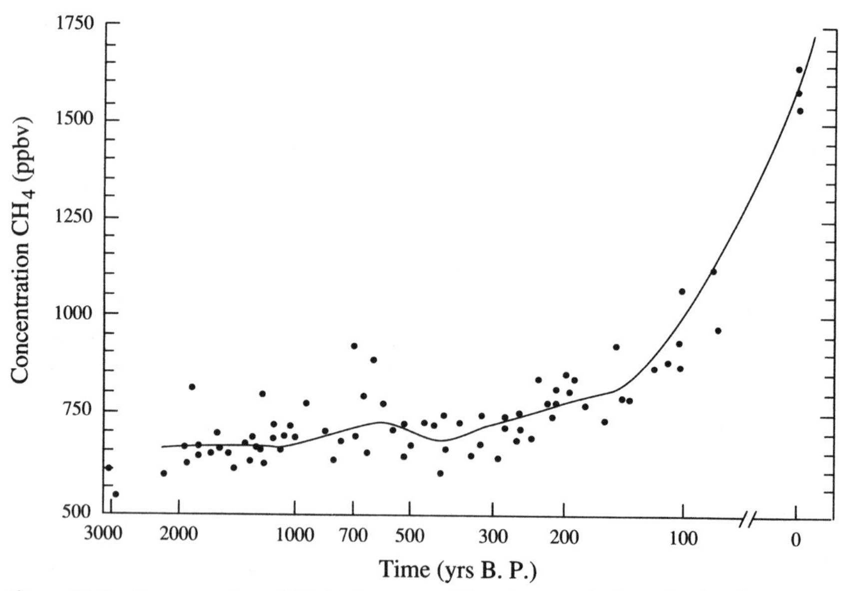

Figure 11.6 Concentration of CH_4 in air extracted from ice cores in Greenland and Antarctica and from contemporary air samples. From Cicerone and Oremland (1988).

Table 11.2 Estimated Sources and Sinks of Methane in the Atmosphere in Units of 10^{12}g CH_4/yr[a]

Sources	Range	Likely
	Natural	
Wetlands		
Tropics	30–80	65
Northern latitude	20–60	40
Others	5–15	10
Termites	10–50	20
Ocean	5–50	10
Freshwater	1–25	5
Geological	5–15	10
Total		160
	Anthropogenic	
Fossil fuel related		
Coal mines	15–45	30
Natural gas	25–50	40
Petroleum industry	5–30	15
Coal combustion	5–30	15
Waste management system		
Landfills	20–70	40
Animal waste	20–30	25
Domestic sewage treatment	15–80	25
Enteric fermentation	65–100	85
Biomass burning	20–80	40
Rice paddies	20–100	60
Total		375
Total sources		535
Sinks		
Reaction with OH	330–560	445
Removal in stratosphere	25–55	40
Removal by soils	15–45	30
Total sinks		515
Atmospheric increase	30–35	30

[a] From Prather et al. (1995). Similar budgets are given by Cicerone and Ormeland (1988), Servant (1991), and Khalil and Shearer (1993).

CH_4 of about 9 years, which is consistent with independent calculations based on methane consumption (Cicerone and Oremland 1988, Khalil and Rasmussen 1990, Prinn et al. 1995) and with the spatial variation in CH_4 concentration in the atmosphere (Fig. 3.5). The concentration of CH_4 is slightly higher in the northern hemisphere (Fig. 3.4), suggesting that it is the location of major emissions.

Surprisingly, the concentration of methane in the atmosphere of the northern hemisphere oscillates, showing a minimum concentration in midsummer (Steele et al. 1987, Khalil et al. 1993b, Dlugokencky et al. 1994). While methane emissions from wetlands are greatest during warm periods, the summer is also the time of the most rapid destruction of atmopheric methane by OH radicals (Khalil et al. 1993c).

Methanogenesis in wetland habitats is the dominant natural source of atmospheric methane (Chapter 7). Matthews and Fung (1987) estimated 110×10^{12} g/yr stem from anaerobic decomposition in natural wetlands globally. The rate of production is higher in tropical wetlands than in boreal wetlands (Schütz et al. 1991, Bartlett and Harriss 1993, Cao et al. 1996), reflecting the positive relationship between temperature, net ecosystem production, and the rate of methanogenesis in many wetland ecosystems (Chapter 7). Because tropical wetlands cover a large area of the world, they dominate the methane flux from wetlands globally (Aselmann and Crutzen 1989, Fung et al. 1991, Bartlett and Harriss 1993).

Changes in the global distribution of wetlands may be related to the increase in atmospheric methane over the last century. Although many wetlands have been drained, Harriss et al. (1988) found that the current management of wetland areas in southern Florida has potentially enhanced the flux of methane to the atmosphere. A large portion of the current increase in atmospheric methane may derive from an increase in the worldwide area of rice cultivation. Because most rice paddies are found in warm climates, they often yield a large CH_4 flux, which is enhanced by the upward transport of CH_4 through the hollow stems of rice (Chapter 7). Matthews et al. (1991) provide maps of the global distribution of CH_4 production from rice cultivation, which is likely to increase by more than 1%/yr during the next several decades (Anastasi et al. 1992).

Many grazing animals and termites maintain a population of anaerobic microbes that conduct fermentation at low redox potentials in their digestive tract. Digestion in these animals provides the functional equivalent of a mobile wetland soil! The flatulence of grazing animals makes a significant contribution to the global sources of methane (Table 11.2). About 78×10^{12} g/yr of CH_4 are derived from domestic and wild animals, with humans contributing 1×10^{12} g/yr (Crutzen et al. 1986, Lerner et al. 1988). Anastasi and Simpson (1993) suggest that larger herds of grazing animals may increase the global flux of methane from animals by about 1×10^{12} g/yr during the next several decades—only a small portion of the current rate of increase of methane in the atmosphere. Some termites and other insects also make a small, but significant, contribution to atmospheric methane as a result of anaerobic decomposition in their hindgut (Khalil et al. 1990, Brauman et al. 1992, Hackstein and Stumm 1994). It is not likely, however, that the flux of CH_4 from termites has increased significantly in recent years.

Forest fires produce methane as a product of incomplete combustion. We know little about the annual area of burning in the preindustrial world, but it is likely that the current flux of CH_4 from forest fires has increased as a result of high, recent rates of biomass burning in the tropics (Andreae 1991). Kaufman et al. (1990) used remote sensing of fires in Brazil to calculate a loss of 7×10^{12} g CH_4/yr in that region in 1987, and Delmas et al. (1991) found a flux of 9.2×10^{12} g CH_4 from burning of African savannas. CH_4 typically accounts for 1% of the total carbon lost by fire (Levine et al. 1993).

Humans contribute directly to atmospheric methane during the production and use of fossil fuels and due to the disposal of wastes. The flux of methane from landfills increases linearly with the amount of material buried, where it presumably decomposes under anoxic conditions (Thorneloe et al. 1993). Inadvertent releases of fossil CH_4 during the mining and use of coal and natural gas must account for about 15–20% of the total annual flux of CH_4 to the atmosphere, based on the ^{14}C age of atmospheric methane (Ehhalt 1974, Wahlen et al. 1989, Quay et al. 1991). Biomass burning and releases of natural gas appear to have increased the $\delta^{13}C$ of atmospheric methane from a preindustrial value of ca. −50‰ to the value of −47‰ that is observed today (Craig et al. 1988, Quay et al. 1988). Increasing emissions from wetlands cannot be responsible, because methanogenesis by acetate or CO_2 reduction yields CH_4 that is more depleted in ^{13}C (Chapter 7). Indeed, an increased release of CH_4 by biomass burning would seem to be the only single source that is consistent with recent observed changes in both atmospheric $^{13}CH_4$ (Craig et al. 1988) and $^{14}CH_4$ (Wahlen et al. 1989). However, a combination of changing contributions from various source and sink reactions is also possible (Whiticar 1993).

The major sink for atmospheric methane is reaction with hydroxyl radical in the atmosphere (Chapter 3). Each year about 445×10^{12} g is removed from the troposphere by this process. As a result of its mean atmospheric lifetime of 9 years, about 40×10^{12} g of CH_4 mixes into the stratosphere, where it is destroyed by similar reactions, producing CO_2 and water vapor. Some workers have suggested that the current increase in atmospheric methane is derived from a reduction in the sink strength offered by hydroxyl radicals, which react more rapidly with CO, which is also increasing in atmosphere (Khalil and Rasmussen 1985). Although this mechanism cannot be dismissed, it is inconsistent with indirect observations that the concentration of hydroxyl radicals has not decreased, and may have even increased, in the atmosphere during the last decade (Prinn et al. 1995).

A small amount of methane diffuses from the atmosphere into upland soils, where it is oxidized by methanotrophic bacteria (King 1992). In desert soils where the supply of labile organic matter is limited, soil bacteria consume an average of 0.66 mg CH_4 m^{-2} day^{-1}, with the greatest rates observed after rainstorms (Striegl et al. 1992). Consumption of CH_4 in temperate and tropical forest soils typically ranges from 1.0 to 5.0 mg CH_4 m^{-2} day^{-1} (Crill 1991, Bartlett and Harriss 1993, Adamsen and King 1993), with lower values after rainstorms, which tend to retard the diffusion of O_2 and CH_4 in clay-rich soils (Koschorreck and Conrad 1993, Castro et al. 1994a, 1995). Methanotrophic bacteria remain active at extremely low CH_4 concentrations (Conrad 1994), so the global significance of this process appears limited by the rate of diffusion of methane into the soil (Born et al. 1990, King and Adamsen 1992).

Some of the methanotrophic activity in soils derives from the activities of nitrifying bacteria, which can use CH_4 as an alternative substrate to NH_4^+ (Jones and Morita 1983, Hyman and Wood 1983, Bédard and Knowles 1989). Steudler et al. (1989) suggested that the consumption of CH_4 by nitrifying bacteria may be lower in forests that currently receive a large atmospheric deposition of NH_4^+, because the NH_4^+/CH_4 ratio has greatly increased in these regions. Both Castro et al. (1994b) and Mosier et al. (1991) found reduced methane uptake when forest and grassland soils were fertilized with nitrogen. Methane uptake by soils is also lower after land clearing, which stimulates nitrification (Hütsch et al. 1994, Keller and Reiners 1994). With fertilization or land clearing, ammonium oxidation produces small amounts of nitrite (NO^{2-}_2), which may cause a persistent inhibition of methanotrophic bacteria in soils (King and Schnell 1994, Schnell and King 1994).

Over large regions the sink for methane in upland soils consumes only a small fraction of the production of methane in adjacent, wet lowland soils (e.g., Whalen et al. 1991, Delmas et al. 1992b). The global estimate of the sink for atmospheric methane in soils is about 30×10^{12} g/yr. Given this relatively small value, it is unlikely that changes in this process by human activities can account for the current increase in atmospheric CH_4 globally (e.g., Willison et al. 1995).

Ice-core records of atmospheric methane show that concentrations were about 400 ppb during the last glacial period, increasing abruptly to the preindustrial value of 700 ppb as the glaciers melted (Chappellaz et al. 1990, Raynaud et al. 1993). The increase during deglaciation seems to have occurred while many northern wetlands were still covered with ice, suggesting that changes in tropical wetlands may have caused the initial methane increase, which reinforced the global warming during deglaciation (Chappellaz et al. 1993). Concentrations of atmospheric methane showed minor variation during the Holocene (±15%; Blunier et al. 1995), but beginning about 200 years ago the concentration began to increase rapidly (Fig. 11.6). The atmospheric concentration of methane has doubled during this period. The rise in atmospheric methane is paralleled by a rise in formaldehyde—a methane oxidation product (Eq. 3.21)—in polar ice cores (Staffelbach et al. 1991). Recently measured increases in stratospheric water vapor are also linked to an increasing transport of CH_4 to the stratosphere where it is oxidized (Thomas et al. 1989, Oltmans and Hofmann 1995).

Although the annual increase in methane in the atmosphere averaged about 1%/yr during the 1980s, the rate of increase slowed considerably during the early 1990s (Steele et al. 1992, Dlugokencky et al. 1994). A variety of explanations have been offered: Some workers suggest that less natural gas was leaking from gas fields of the former Soviet Union, as a result of slower economic growth and efforts to patch pipeline leaks (Law and Nisbet 1996). Other workers believe that the slower growth of atmospheric CH_4, as well as CO_2 and N_2O, in the early 1990s was related to the global cooling that followed the volcanic eruption of Mt. Pinatubo in June 1991. Bekki et al. (1994) suggested that an increasing depletion of stratospheric ozone allowed greater amounts of uvB radiation to penetrate to the troposphere, where it produces OH radicals that oxidize CH_4. The rise in atmospheric methane returned to its historic growth rate of 1%/yr by the mid-1990s.

Future changes in the global methane budget may accompany global warming, which could shift the balance between aerobic and anaerobic decomposition in wetlands and increase the ratio of CO_2 to CH_4 emitted from these ecosystems (Whalen and Reeburgh 1990, Funk et al. 1994, Moore and Dalva 1993). On the other hand, an increasing flux of CH_4 may accompany a CO_2-induced stimulation of plant growth in wetlands (Dacey et al. 1994, Hutchin et al. 1995). Methanogenic bacteria show a greater positive response to temperature than methane-oxidizing bacteria, suggesting that the flux of methane from wetland soils could increase with global warming (King and Adamsen 1992, Dunfield et al. 1993). Catastrophic release of methane from marine sediments, where it is held as methane hydrate (clathrate) could also yield a large increase in atmospheric methane and greenhouse warming in the future (Revelle 1983, MacDonald 1990). Given methane's potential as a greenhouse gas and indications that increasing concentrations of CH_4 may have preceded the global warming 10,000 years ago, a better understanding the global methane budget is paramount if biogeochemists are to contribute to the development of effective international policy to combat global warming (Nisbet and Ingham 1995).

Carbon Monoxide

Carbon monoxide has a low concentration (45 to 250 ppb) and a short lifetime (2 months) in the atmosphere (Table 3.4). The short lifetime is consistent with wide regional and seasonal variations in its concentration (Fig. 3.5); the concentration of CO in the northern hemisphere is typically 3X larger than that in the southern hemisphere (Dianov-Klokov et al. 1989, Novelli et al. 1992, 1994). The budget for CO is dominated by anthropogenic sources (Table 11.3), which are concentrated in the northern hemisphere. Human activities account for more than 2/3 of the annual production of CO. Over the past several decades, the concentration of CO has increased at a rate of >1%/yr (Khalil and Rasmussen 1988, Dianov-Klokov et al. 1989), presumably as a result of increasing emissions from fossil fuel combustion and an increasing production of CO as a methane oxidation product (Chapter 3). Surprisingly, CO concentrations began to decline during the early 1990s, in conjunction with the slower growth rate of other trace gases in the atmosphere (Novelli et al. 1994, Khalil and Rasmussen 1994). The decline in CO may be related to the slower growth rate of atmospheric CH_4, which produces CO as an oxidation product.

Some CO is taken up by vegetation, but the dominant sink for CO is oxidation by hydroxyl radical in the atmosphere (Eqs. 3.29 to 3.34). Because it is oxidized to CO_2 so rapidly, carbon monoxide is normally included as a component of the CO_2 flux in most accounts of the global carbon cycle (e.g., Fig. 11.1). Actually, the direct release of CO may account for about 5% of the total carbon emitted during fossil fuel combustion and perhaps

Table 11.3 Estimated Sources and Sinks of Carbon Monoxide in the Atmosphere[a]

Sources	Flux (10^{12}g CO/yr)
Fossil fuels	300–550
Biomass burning	300–700
Vegetation	60–160
Oceans	20–200
Methane oxidation	400–1000
NMHC oxidation	200–600
Total	1800–2700
Sinks	
OH reaction	1400–2600
Stratospheric destruction	~100
Soil uptake	250–640
Total	2100–3000

[a] From Prather et al. (1995). An emission of 13 × 10^{12} g/yr from the ocean is now suggested by Bates et al. (1995).

as much as 15% of the carbon released during biomass burning (Andreae 1991).

Carbon monoxide shows limited absorption of infrared radiation. Its main effect on the greenhouse warming of Earth is probably indirect—by slowing the destruction of methane in the atmosphere (Lashof and Ahuja 1990). More importantly, carbon monoxide plays a major role in atmospheric chemistry by controlling levels of tropospheric ozone (Chapter 3). High concentrations of atmospheric ozone over the tropical regions of South America and Africa appear to be related to the production of CO by forest burning, followed by the reaction of CO with OH radicals to produce ozone (Fig. 3.8). Crutzen et al. (1985) estimate that up to 3% of net primary production may be lost as CO or as volatile hydrocarbons that are oxidized to CO in the atmosphere.

Synthesis: Linking the Carbon and Oxygen Cycles

Even on a lifeless Earth, photolysis of water vapor in the atmosphere might produce small amounts of O_2, as it has on other planets (Chapter 2). However, during the geologic history of Earth, significant amounts of atmospheric O_2 appeared only following the advent of autotrophic photosynthesis, and O_2 began to accumulate to its present level when the annual production exceeded the reaction of O_2 with reduced crustal minerals, especially FeS_2. The current atmospheric pool of O_2 is only a small fraction of the total O_2 produced over geologic time (Fig. 2.7). The net production of O_2 over geologic time is balanced stoichiometrically by the storage of reduced organic carbon (1.56×10^{22} g) and sedimentary pyrite (4.97×10^{21} g S, Table 13.1) in the Earth's crust. Oxygen is likely to have accumulated most rapidly during periods when large amounts of organic sediments were buried (Des Marais et al. 1992). We have little evidence of historical variations in atmospheric O_2, but geochemical models suggest that the concentrations may have ranged from 15 to 35% during the past 500 million years (Berner and Canfield 1989). The highest values would be expected in the Carboniferous and Permian, when a large amount of organic matter was buried in sediments (Fig. 11.4).

Large fluctuations of O_2 would have dramatic implications for the physiology, morphology, and evolution of most organisms (Graham et al. 1995). Fortunately, the pool of atmospheric O_2 is well buffered over geologic time because increases in O_2 expand the area and depth of aerobic respiration in marine sediments, leading to a greater consumption of O_2 which may stem any further increase in atmospheric O_2 concentration (Chapters 3 and 9). The small amount of organic matter that escapes oxidation in the sea is balanced over geologic time by the uplift and weathering of organic carbon in sedimentary rocks (Fig. 11.7). Higher levels of O_2 may also increase the adsorption of P to iron minerals in marine sediments, subse-

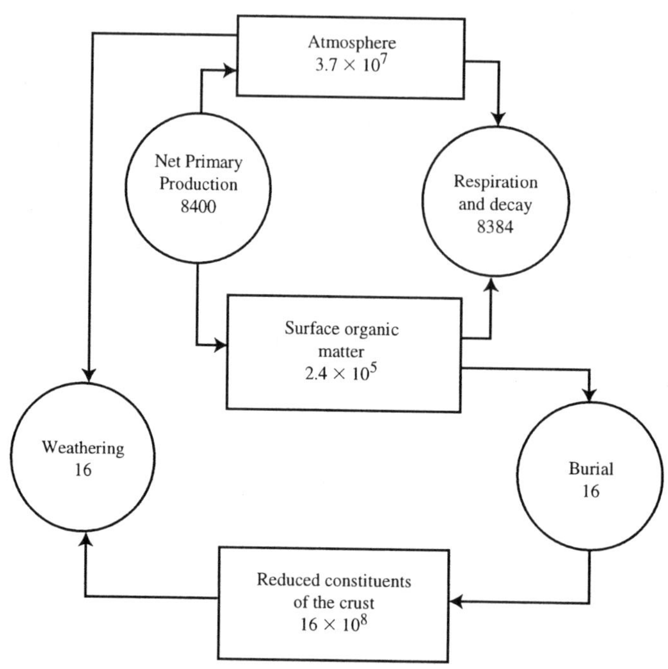

Figure 11.7 A simple model for the global biogeochemical cycle of O_2. Data are expressed in units of 10^{12} moles of O_2 per year or the equivalent amount of reduced compounds. Note that a small misbalance in the ratio of photosynthesis to respiration can result in a net storage of reduced organic materials in the crust and an accumulation of O_2 in the atmosphere. Modified from an original conception by Walker (1980) to reflect values derived in this text.

quently lowering nutrient availability and NPP in the sea (Van Cappellen and Ingall 1996). This interaction potentially provides a negative-feedback mechanism, buffering the concentration of O_2 in Earth's atmosphere (Chapter 9).

Like the carbon cycle, the modern oxygen cycle is composed of a set of large, annual fluxes superimposed on the smaller, slow fluxes of the geologic cycle (Walker 1980). The current atmospheric pool of O_2 is maintained in a dynamic equilibrium between the production of O_2 by photosynthesis and its consumption in respiration (Fig. 11.7). The annual fluctuation of O_2 in the atmosphere due to photosynthesis and respiration is about $\pm 0.0020\%$ in an average background concentration of 20.946% (Keeling and Shertz 1992). The mean residence time of O_2 in the atmosphere is about 4000 years—significantly shorter than what would be predicted merely by the reaction of O_2 with the Earth's crust.

Given the large amount of O_2 in the atmosphere, it is difficult to measure the annual production and consumption of O_2 as an independent check on our estimates of photosynthesis and decay. However, an examination of the isotopic composition of atmospheric O_2 (i.e., $\delta^{18}O$) allows us to constrain the atmospheric O_2 budget within certain limits. Photosynthesis does not discriminate among the oxygen isotopes of water—the O_2 released has an isotopic composition that is identical to that in the seawater or the soil water in which the plant is growing. Respiration discriminates among oxygen isotopes, consuming $^{16}O_2$ in preference to $^{18}O_2$. The $\delta^{18}O$ of atmospheric O_2 (+23.5‰) suggests that gross primary production must be $>180 \times 10^{15}$ g C/yr on land and about 140×10^{15} g C/yr in the oceans (Bender et al. 1994). Assuming that net primary production is one-half of gross primary production, in both cases, these values indicate that NPP is somewhat higher than what we have estimated independently for land (Table 5.2) and marine (Table 9.2) habitats globally. Nevertheless, these values offer an upper limit for NPP, which helps constrain our estimates for the global carbon cycle (Fig. 11.1).

The oxygen cycle is directly linked to other biogeochemical cycles. For example, assuming that about half the annual circulation of N on land (1200×10^{12}g) and about 15% of the N cycle in the oceans (8000×10^{12} g; Fig. 9.16) are derived from the plant uptake of NO_3 and the return of NH_4, then about 3% of the annual production of O_2 by photosynthesis is used to oxidize NH_4 in the nitrification reactions. The formation and oxidation of sedimentary pyrite, through sulfate reduction, also affects the concentration of O_2 in the atmosphere. For every mole of pyrite-S oxidized, nearly 2 moles of O_2 are consumed from the atmosphere (Fig. 1.1). Currently, the annual burial of pyrite in marine sediments accounts for about 20% of the oxygen in our atmosphere (Chapter 9).

Methanogenesis in freshwater sediments returns CH_4 to the atmosphere, where it is oxidized (Henrichs and Reeburgh 1987). Methane oxidation in the atmosphere accounts for about 1% of the total consumption of atmospheric O_2 each year. In the absence of methanogenesis, the burial of organic carbon in freshwater sediments might be greater and the atmospheric content of O_2 might be slightly higher. Thus, methanogenesis acts as a negative feedback in the regulation of atmospheric O_2 (Watson et al. 1978, Kump and Garrels 1986).

It is perhaps entertaining to speculate whether the carbon cycle on Earth drives the oxygen cycle, or vice versa. Over geologic time, the answer is obvious: the conditions on our neighboring planets provide ample evidence that O_2 is derived from life. Now, however, the carbon and oxygen cycles are inextricably linked, and the discussion seems philosophical. The metabolism of eukaryotic organisms, including humans, depends on the flow of electrons from reduced organic molecules to oxygen.

Recommended Readings

Dobrovolsky, V.V. 1994. *Biogeochemistry of the World's Land.* CRC Press, Boca Raton, Florida.

Heimann, M. (ed.). 1993. *The Global Carbon Cycle.* Springer-Verlag, New York.
Houghton, J.T., L.G. Meira Filho, J. Bruce, H. Lee, B.A. Callander, E. Haites, N. Harris, and K. Maskell. (eds.). 1995. *Climate Change 1994.* Cambridge University Press, Cambridge.
Khalil, M.A.K. (ed.). 1993. *Atmospheric Methane: Sources, Sinks, and Role in Global Change.* Springer-Verlag, New York.
Woodwell, G.M. and F.T. MacKenzie. (eds). 1995. *Biotic Feedbacks in the Global Climatic System.* Oxford University Press, New York.

12

The Global Cycles of Nitrogen and Phosphorus

Introduction

The availability of nitrogen and phosphorus controls many aspects of local ecosystem function and global biogeochemistry. Nitrogen often limits the rate of net primary production on land and in the sea (Vitousek and Howarth 1991). In living tissues, nitrogen is an integral part of enzymes, which mediate the biochemical reactions in which carbon is reduced (i.e., photosynthesis) or oxidized (respiration). Phosphorus is an essential component of DNA, ATP, and the phospholipid molecules of cell membranes. Changes in the availability of N and P are likely to have controlled the size and activity of the biosphere through geologic time.

A large number of biochemical transformations of nitrogen are possible, since nitrogen is found at valence states ranging from −3 (in NH_3) to +5 (in NO_3^-). A variety of microbes capitalize on the potential for transformations of N among these states and use the energy released by the changes in redox potential to maintain their life processes (Rosswall 1982). Collecely, these microbial reactions drive the cycle of nitrogen (Fig. 12.1). In contrast, whether it occurs in soils or in biochemistry, phosphorus is almost always found in combination with oxygen (i.e., as PO_4^{3-}). Most metabolic activity is associated with the synthesis or destruction of high-energy bonds between a phosphate ion and various organic molecules, but in nearly all cases the phosphorus atom remains at a valence of +5 in these reactions.

The most abundant form of nitrogen at the surface of the Earth, N_2, is the least reactive species. Nitrogen fixation converts atmospheric N_2 to one of the forms of "fixed" (or "odd," Chapter 3) nitrogen that can be used by biota. Nitrogen-fixing species are most abundant in nitrogen-poor habitats, where their activity increases the availability of nitrogen for the biosphere (Eq. 2.12). At the same time, denitrifying bacteria return N_2 to the atmosphere (Eq. 2.18), lowering the overall stock of nitrogen available for life on Earth.

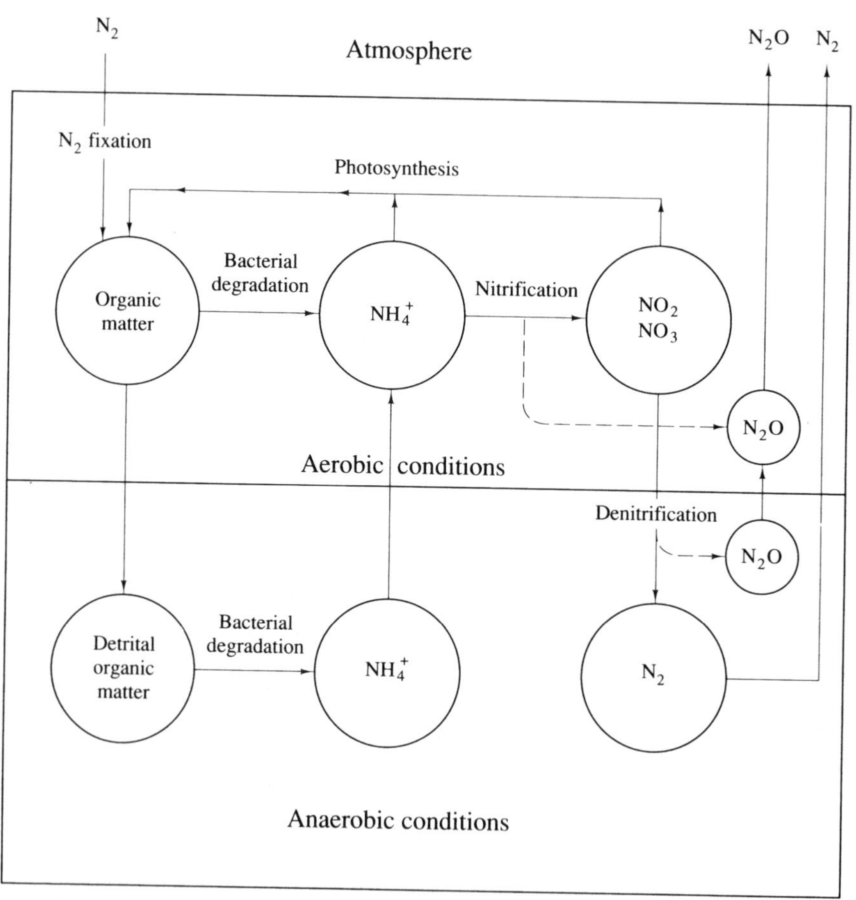

Figure 12.1 Microbial transformations in the nitrogen cycle. From Wollast (1981).

Rocks of the continental crust hold the reservoir of phosphorus that becomes available to the biosphere through rock weathering. Land plants can increase the rate of rock weathering in P-deficient habitats (Chapter 4), but in nearly all cases the phosphorus content of rocks is relatively low. Subsequent reactions between dissolved P and other minerals reduce the availability of P in soil solutions or seawater. Thus, in most habitats—both on land and in the sea—the availability of P is controlled by the degradation of organic forms of P (e.g., Fig. 6.17). This *bio*geochemical cycle temporarily retains and recycles some P from the unrelenting flow of P from weathered rock to ocean sediments. The global P cycle is complete only when sedimentary rocks are lifted above sea level and the weathering begins again.

In this chapter we will examine our current understanding of the global cycles of N and P. We will attempt to balance N and P budgets for the world's land area and the sea. For N, the balance between N-fixation and denitrification through geologic time determines the nitrogen available to biota and the global nitrogen cycle. One of the by-products of nitrification and denitrification is N_2O (nitrous oxide), which is both a greenhouse gas and a cause of ozone destruction in the stratosphere (Chapter 3). We will formulate a tentative budget for N_2O based on our current, limited understanding of the sources of this gas in the atmosphere.

The Global Nitrogen Cycle

Land

Figure 12.2 presents the global nitrogen cycle, showing the linkage between the atmosphere, the land, and the oceans. The atmosphere contains the largest pool (3.9×10^{21} g N; Table 3.1). Relatively small amounts of N are found in terrestrial biomass (3.5×10^{15} g) and in soil organic matter (95 to 140×10^{15} g; Post et al. 1985; Batjes 1996). The mean C/N ratios for terrestrial biomass and soil organic matter are about 160 and 15, respectively. At any time, the pool of inorganic nitrogen, NH_4^+ and NO_3^-, in soils is very small. The uptake of N by organisms is so rapid that little nitrogen remains in inorganic forms, despite the large annual flux through this pool (Chapter 6).

The nitrogen that bathes the terrestrial biosphere is not available to most organisms, because the great strength of the triple bond in N_2 makes this molecule practically inert. All nitrogen that is available to biota was originally derived from nitrogen fixation—either by lightning or by free-living and symbiotic microbes (Chapter 6). The rate of nitrogen fixation by lightning, which produces momentary conditions of high pressure and temperature allowing N_2 and O_2 to combine, is poorly known, but relatively small. Most recent, global estimates are $<3 \times 10^{12}$ g N/yr (Borucki and Chameides 1984, Lawrence et al. 1994, P. P. Kumar et al. 1995, Ridley et al. 1996), and the calculated total annual deposition of oxidized N (NO_x) from the preindustrial atmosphere precludes an estimate higher than about

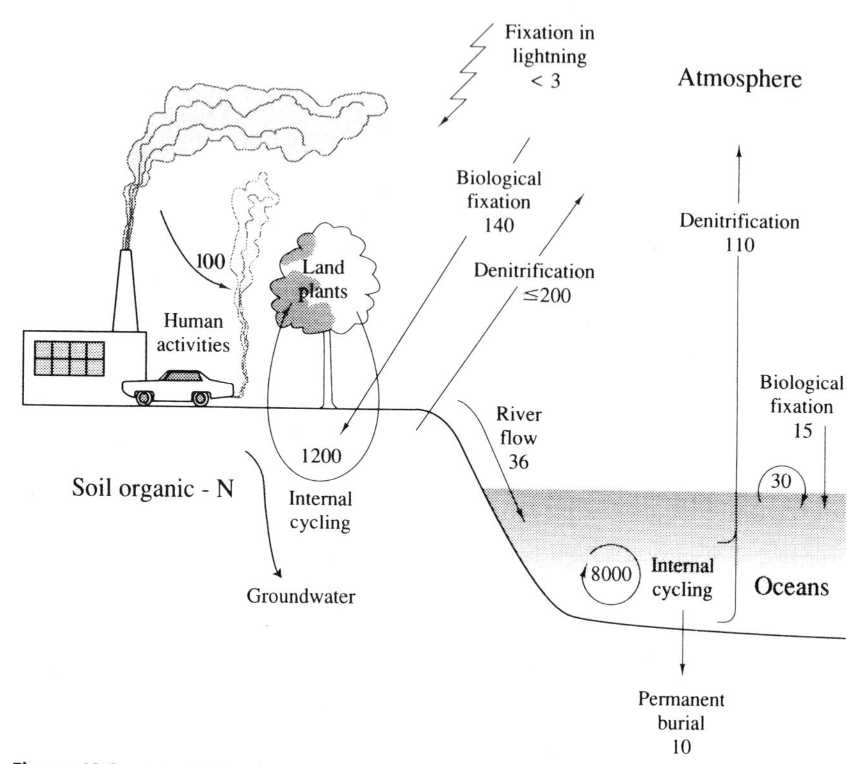

Figure 12.2 The global nitrogen cycle. Each flux is shown in units of 10^{12} g N/yr. Values are derived in the text.

20×10^{12} g N/yr (Logan 1983, Lyons et al. 1990). Assuming that lightning is distributed uniformly over land and sea, a liberal estimate for the deposition of N fixed by lightning over land would be 5×10^{12} g N/yr. The present-day deposition of oxidized nitrogen on land is about 30×10^{12} g/yr, owing to the additional NO_x that is emitted from soils, biomass burning, and human activities (Penner et al. 1991, Kasibhatla et al. 1993).

A widely cited estimate of total biological nitrogen fixation on land is 140×10^{12} g N/yr (Burns and Hardy 1975). This is equivalent to about 10 kg N/yr for each hectare of the Earth's land surface. Most studies of asymbiotic nitrogen fixation on land report values ranging from 1 to 5 kg ha^{-1} yr^{-1} (Chapter 6). A value of 3 kg N ha^{-1} yr^{-1} multiplied by the world's land area suggests that asymbiotic fixation contributes about 44×10^{12} g N/yr to the global total. The remainder is assumed to come from symbiotic fixation in higher plants. This flux is not distributed uniformly among natural ecosystems; the greatest values are often found in areas of disturbed or successional vegetation (Chapter 6). About 40×10^{12} g N/yr of N fixation occurs in agricultural fields as a result of the cultivation of N-fixing crops (e.g., soybeans) (Burns and Hardy 1975, Galloway et al. 1995). In any case, in the modern world, the various forms of biotic N-fixation dwarf abiotic fixation by lightning as the source of fixed N on land. Taking all forms of N-fixation as the only source, the mean residence time of nitrogen in the terrestrial biosphere is about 700 years (i.e., pool/input).

Assuming that the estimates of terrestrial net primary production, 60×10^{15} g C/yr, are roughly correct and that the mean C/N ratio of net primary production is about 50, the nitrogen requirement of land plants is about 1200×10^{12} g/yr (Chapter 6).[1] Thus, nitrogen fixation supplies only about 12% of the nitrogen that is assimilated by land plants each year. The remaining nitrogen must be derived from internal recycling and the decomposition of dead materials in the soil (Chapter 6). When the turnover in the soil is calculated with respect to the input of dead plant materials, the mean residence time of nitrogen in soil organic matter is >100 years. Thus, the mean residence time of N exceeds that of C in both land vegetation and soils (Chapters 5 and 11).

Humans have a dramatic impact on the global N cycle. In addition to planting N-fixing species for crops, humans produce nitrogen fertilizers through the Haber process, viz.,

$$3CH_4 + 6H_2O \rightarrow 3CO_2 + 12H_2 \tag{12.1}$$

$$4N_2 + 12H_2 \rightarrow 8NH_3, \tag{12.2}$$

in which natural gas is burned to produce hydrogen, which is combined with N_2 to form ammonia under conditions of high temperature and pressure. Fertilizer production supplies $>80 \times 10^{12}$ g N/yr to agricultural ecosystems (Schlesinger and Hartley 1992, Matthews 1994), and the subsequent loss of NH_3 from these areas carries fixed N to adjacent natural ecosystems where it is deposited (Draaijers et al. 1989, Hesterberg et al. 1996).

Fossil fuel combustion also releases about 20×10^{12} g of fixed N (viz., NO_x) annually (Logan 1983, Hameed and Dignon 1988, Müller 1992). Some of this is derived from the organic nitrogen contained in fuels (Bowman 1991), but it is best regarded as a source of new, fixed N for the biosphere because in the absence of human activities, this N would remain inaccessible in the Earth's crust (Galloway et al. 1995). Owing to the short residence time of NO_x in the atmosphere (Eq. 3.27), most of this nitrogen is deposited by precipitation over land, where it enters biogeochemical cycles. Only a small portion of NO_x undergoes long-distance transport in

[1] Most primary production consists of short-lived tissues with a C/N ratio that is much lower than that of wood (160), which composes most of the terrestrial biomass.

the troposphere, accounting for the rising levels of NO_3^- deposited in Greenland snow (Fig. 12.3). Forest ecosystems downwind of major population centers now receive enormous nitrogen inputs that may be related to their decline (Aber et al. 1989, Schulze 1989).

In total, about 240×10^{12} g of newly fixed N is delivered from the atmosphere to the Earth's land surface each year—40% by natural and 60% by human-derived sources. In the absence of processes removing nitrogen, a very large pool of nitrogen would be found on land in a relatively short time. Each year rivers carry about 36×10^{12} g N from land to the sea (Chapter 8), and some workers believe that this estimate may be too low as a result of an underestimate of the transport of N in particulate materials (Wollast 1993). Humans may account for as much as half of the present-day riverine transport of N in rivers (Galloway et al. 1995). Human additions of fixed nitrogen to the terrestrial biosphere have also resulted in marked increases in the nitrogen content of groundwaters, especially in many agricultural areas (Spalding and Exner 1993). For example, the loss of nitrate to groundwater was the largest single fate of nitrogen added to the fields of a dairy farm in Ontario (Barry et al. 1993). The global sink in groundwaters may approach 11×10^{12} g N/yr—calculated from an estimate of the

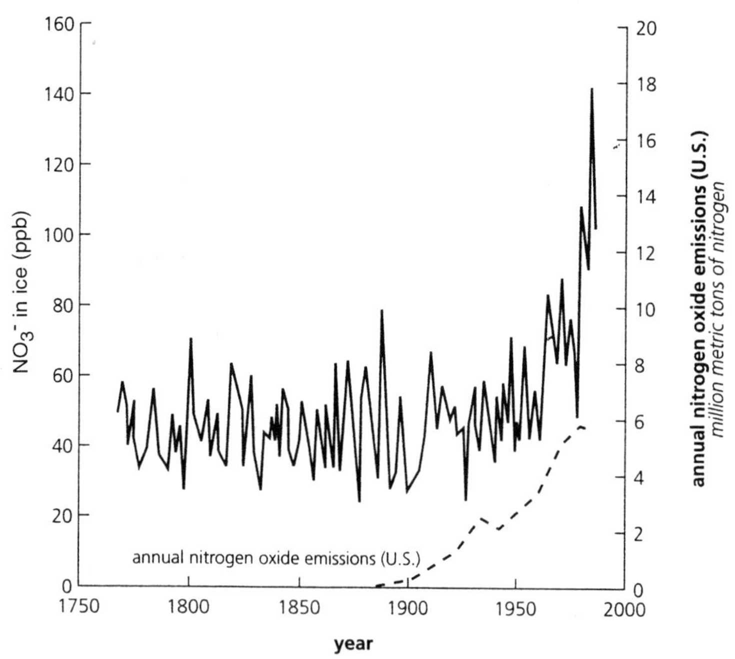

Figure 12.3 The 200-year record of nitrate in layers of the Greenland ice pack and the annual production of nitric oxides by fossil fuel combustion in the United States. Modified from Mayewski et al. (1990).

annual flux of groundwater (11,000 km^3/yr; Zektser and Loaiciga 1993) and a recent increase of 1 mg N/liter in these waters. Despite these large transports, riverflow and groundwater cannot account for all of the nitrogen that is deposited on land. The remaining nitrogen is assumed to be lost by denitrification in terrestrial soils (Chapter 6) and in wetlands (Chapter 7), and during forest fires (Chapter 6).

Estimates of global denitrification in terrestrial ecosystems range from 13 to 233 $\times 10^{12}$ g N/yr (Bowden 1986). At least half of the denitrification on land occurs in wetlands (Bowden 1986). If N-fixation and denitrification were once in balance, then a terrestrial denitrification rate of $>70 \times 10^{12}$ g N/yr was most likely in the preindustrial world (i.e., fixation minus riverflow) (Delwiche 1970). Most of the loss occurs as N_2, but the small fraction that is lost as N_2O during nitrification and denitrification (Chapter 6) contributes significantly to the global budget of this gas. Indeed, the current rise in atmospheric N_2O can be used to estimate the overall increase in global denitrification as a result of human activities. If we assume that the N_2/N_2O ratio in denitrification is about 22 (Chapter 6) and that the recent increase of N_2O in the atmosphere (nearly 4×10^{12} g N/yr) all derives from increased denitrification, then it is possible that the overall loss of N_2 from denitrification has increased by as much as 90×10^{12} g N/yr, helping to balance the present-day N budget on land (Fig. 12.2).

Nitrogen in biomass is volatilized as NH_3, NO_x, and N_2 during fires—the latter constituting a form of *pyrodenitrification* (Chapter 6). About 30% of the nitrogen in fuel is converted to N_2, and globally biomass burning may return as much as 50×10^{12} g N/yr to the atmosphere as N_2 (Kuhlbusch et al. 1991). To the extent that the rate of biomass burning has increased in recent years, this form of denitrification may have increased as well.

In balancing the terrestrial N cycle, we concentrate on processes that affect the net production or loss of fixed nitrogen. We do not include processes that recycle N that was fixed at an earlier time. Thus, NH_3 volatilization from biomass burning (Table 12.1) and the natural emission of NO_x from soils (Table 12.2) can be ignored to the extent that these forms are redeposited on land in precipitation. Ammonia and NO_x have relatively short atmospheric lifetimes, so they are usually deposited in precipitation and dryfall near their point of origin (Chapter 3).

Sea

The world's oceans receive about 36×10^{12} g N/yr in dissolved forms in rivers (Chapter 8), about 15×10^{12} g N/yr via biological N-fixation (Chapter 9), and about 30×10^{12} g N in precipitation (Duce et al. 1991). Note that while the flux in rivers is a rather small component of the terrestrial cycle, it contributes about 40% of the total nitrogen delivered annually to the sea (Fig. 12.2). Much of the precipitation flux is NH_4^+ that is derived from

Table 12.1 A Budget for Atmospheric NH_3[a]

Sources	"Best" estimate (10^{12}g N/yr)	Potential range (10^{12}g N/yr)
Domestic animals	32	24–40
Sea surface	13	8–18
Undisturbed soils	10	6–45
Fertilizers	9	5–10
Biomass burning	5	1–9
Human excrement	4	
Coal combustion	2	
Automobiles	0.2	
Total inputs	75	50–128
Sinks		
Wet deposition on land	30	
Dry deposition on land	10	
Wet deposition on sea surface	16	
Reaction with OH radical	1	
Total outputs	57	

[a] From Schlesinger and Hartley (1992). An alternative budget showing emissions of 45 Tg N/yr is given by Dentener and Crutzen (1994).

Table 12.2 Estimated Global Emissions of NO_x Typical of the Last Decade (Tg N/yr)[a]

Sources	Magnitude 10^{12}g N/yr
Fossil fuel combustion	24[b]
Soil release (natural and anthropogenic)	12[c]
Biomass burning	8[d]
Lightning	5
NH_3 oxidation	3[e]
Aircraft	0.4
Transport from stratosphere	0.1 (0.6 total NO_y)

[a] From Prather et al. (1995).

[b] Müller (1992) estimates 21 Tg N/yr from this source.

[c] Davidson (1991) suggests an emission of 20 Tg N/yr from soils, whereas Yienger and Levy (1995) give 5.5 Tg N/yr.

[d] Andreae (1991).

[e] Warneck (1988) suggests 1 Tg N/yr (cf. Table 12.1).

NH_3 volatilized from the sea (Schlesinger and Hartley 1992). Cornell et al. (1995) suggest that the atmospheric deposition on the ocean surface may be somewhat greater than 30×10^{12} g N/yr, because most workers have focused only on NH_4^+ and NO_3^- and ignored significant inputs of dissolved organic N.

The riverflux of N assumes its greatest importance in coastal seas and estuaries, whereas N inputs from the atmosphere are most important in the open oceans. In the surface ocean, the pool of inorganic nitrogen is very small. As we have shown for terrestrial ecosystems, most of the net primary production in the sea is supported by nitrogen recycling in the water column (Table 9.3). The deep ocean contains a large pool of inorganic nitrogen (570×10^{15} g N) derived from the decomposition of organic matter. Permanent burial of organic nitrogen in sediments is small, so most of the nitrogen input to the oceans must be returned to the atmosphere as N_2 by denitrification (Fig. 9.16). Important areas of denitrification are found in the anaerobic deep waters of the eastern tropical Pacific Ocean and the Arabian Sea (Chapter 9). Globally, marine denitrification may account for the return of 110×10^{12} g N/yr to the atmosphere as N_2.

Temporal Variations in the Global Nitrogen Cycle

The earliest atmosphere on Earth is thought to have been dominated by nitrogen, since N is abundant in volcanic emissions and only sparingly soluble in seawater (Chapter 2). Before the origin of life, nitrogen was fixed by lightning and in the shock waves of meteors, which create local conditions of high temperature and pressure in the atmosphere (Mancinelli and McKay 1988). The rate of N-fixation was very low, perhaps about 6% of the present-day rate, because abiotic fixation in an atmosphere dominated by N_2 and CO_2 is much slower than in an atmosphere of N_2 and O_2 (Kasting and Walker 1981). The best estimates of abiotic fixation suggest that it had a limited effect on the content of atmospheric nitrogen, but it provided a small but important supply of fixed nitrogen, largely NO_3^-, to the waters of the primitive Earth (Kasting and Walker 1981, Mancinelli and McKay 1988). The limited supply of fixed nitrogen in the primitive oceans is likely to have led to the early evolution of N-fixation in marine biota (Chapter 2).

With respect to N-fixation by lightning, the mean residence time of N_2 in the atmosphere is about 1.3 billion years. The mean residence time of atmospheric nitrogen decreases to about 20,000,000 years when biological nitrogen fixation is included. This is much shorter than the history of life on Earth, and it speaks strongly for the importance of denitrification in returning N_2 to the atmosphere over geologic time. Denitrification closes the global biogeochemical cycle of nitrogen, but it also means that nitrogen remains in short supply for the biosphere. In the absence of denitrification, most nitrogen on Earth would be found as NO_3^- in seawater, and the oceans would be quite acidic (Sillén 1966).

The origin of denitrification is uncertain. Mancinelli and McKay (1988) argue for its appearance before the advent of atmospheric O_2, suggesting that a facultative tolerance of O_2 evolved later (cf. Castresana and Saraste 1995). Others suggest that denitrification is more recent (Broda 1975, Betlach 1982). These investigators point out that denitrifying bacteria are facultative anaerobes, switching from simple heterotrophic respiration to NO_3 respiration under anaerobic conditions. This view is supported by the observation that the denitrification enzymes are somewhat tolerant of low concentrations of O_2, allowing denitrifying bacteria to persist in environments with fluctuations in redox potential (Bonin et al. 1989, McKenney et al. 1994, Carter et al. 1995).

Requiring oxygen as a reactant, nitrification clearly arose after photosynthesis and the development of an O_2-rich atmosphere. Today, the rate of denitrification is controlled by the rate of nitrification, which supplies NO_3^- as a substrate (Fig. 6.12). In any case, the major microbial reactions in the nitrogen cycle (Fig. 12.1) are all likely to have been in place at least 1 billion years ago.

Because NO_3^- is very soluble, there is little reliable record of changes in the content of NO_3^- in seawater through geologic time. Only changes in the deposition of organic nitrogen are recorded in sediments. Recently, Altabet and Curry (1989) suggested that the $^{15}N/^{14}N$ record in sedimentary foraminifera may be useful in reconstructing the past record of ocean N chemistry. The isotope ratio in sedimentary organic matter increases when high rates of denitrification remove NO_3^- from the oceans, leaving the residual pool of nitrate enriched in ^{15}N (Altabet et al. 1995, Ganeshram et al. 1995).

Assuming a steady state in the ocean nitrogen cycle, the mean residence time for an atom of fixed N in the sea is about 8000 years. During this time, this atom will make several trips through the deep ocean, each lasting 200 to 500 years (Chapter 9). Because the turnover of N is much longer than the mixing time for ocean water, NO_3 shows a relatively uniform distribution in deep ocean water. In a provocative paper, McElroy (1983) suggested that the oceans are not presently in steady state; the rate of denitrification exceeds known inputs (cf. Fig. 12.2). He argues that the oceans received a large input of nitrogen during the continental glaciation 20,000 years ago, and they have been recovering from this input ever since (Christensen et al. 1987). His suggestion is consistent with sedimentary evidence of greater net primary production in the oceans during the last ice age (Broecker 1982), and with the low ratio of $^{15}N/^{14}N$ in sedimentary organic matter of glacial age (Ganeshram et al. 1995).

McElroy's paper serves to remind us of another important aspect of global biogeochemistry: although the assumption of a steady state is useful in the construction of global models, such as Fig. 12.2, it is sometimes not realistic. The Earth has experienced large fluctuations in its biogeochemical function through geologic time. Changes in the global distribution and circulation of nitrogen may have accompanied climatic changes—just as the rise in CO_2 concentrations at the end of the last glacial indicates a period of non-steady-state conditions in the global carbon cycle (Chapter 11).

Human activities have now disrupted steady-state conditions in the nitrogen cycle (Delwiche 1970, Galloway et al. 1995). Humans have greatly accelerated the natural rate of N-fixation; the current production of N-fertilizers is increasing at an exponential rate (Fig. 12.4). Enrichments of nitrogen in terrestrial ecosystems, stimulating the rates of nitrification and denitrification, are likely to account for the rapid rise in the atmospheric content of N_2O (Vitousek 1994). These changes in the global nitrogen cycle have important implications for potential greenhouse warming of Earth's atmosphere and for the persistence of species diversity in impacted ecosystems (Schlesinger 1994).

Nitrous Oxide

Currently, biogeochemists are devoting a large research effort toward understanding the global budget of nitrous oxide, N_2O. This trace atmospheric

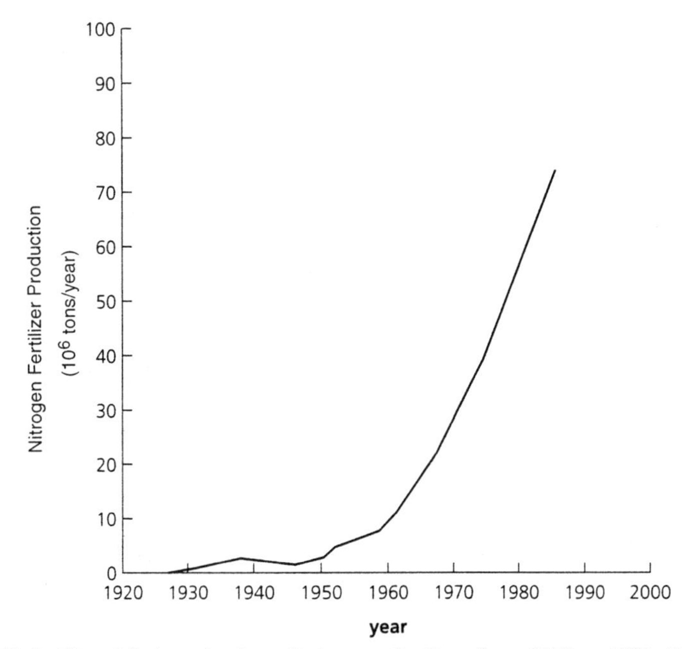

Figure 12.4 The global production of nitrogen fertilizer from 1920 to 1985. From Smil (1991).

constituent has a mean concentration of 311 ppb, which indicates a global pool of 2.4×10^{15} g N_2O or 1.5×10^{15} g N in the atmosphere (Table 3.1). The concentration of N_2O is increasing at an annual rate of 0.3% (Prinn et al. 1990, Khalil and Rasmussen 1992a). Each molecule of N_2O has the potential to contribute about 300× to the greenhouse effect relative to each molecule of CO_2, so the current increase in the atmosphere has the potential to impact global climate over the next century (Lashof and Ahuja 1990, Albritton et al. 1995).

The only significant sink for N_2O—stratospheric destruction (Eqs. 3.43 and 3.44)—consumes about 12×10^{12} g N as N_2O per year (Minschwander et al. 1993). A few soils also appear to consume N_2O (Ryden 1981, Cicerone 1989, Donoso et al. 1993), but the global sink in soils is unknown and probably very small (Blackmer and Bremner 1976, Conrad 1994). The mean residence time for N_2O in the atmosphere is about 120 years, consistent with observations of a relatively uniform (311 ± 1 ppb) concentration of N_2O around the world (Fig. 3.5). Unfortunately, estimates of sources—particularly sources that have changed greatly in recent years—are poorly constrained (Table 12.3).

The oceans appear to be a source of N_2O to the atmosphere as a result of nitrification in the deep sea (Cohen and Gordon 1979, Oudot et al. 1990). Some N_2O may subsequently be *consumed* by denitrification in areas of low O_2 (Cohen and Gordon 1978, Kim and Craig 1990), but in many areas seawater is supersaturated in N_2O with respect to the atmosphere. Specifically, the waters of the northwest Indian Ocean, a local zone of upwelling, may account for 20% of the total flux of N_2O from the oceans to the atmosphere (Law and Owens 1990). Based on the belief that the N_2O supersaturation of seawater was widespread, calculated emissions from the ocean dominated the earliest global estimates of N_2O sources (Liss and Slater 1974, Hahn 1974). When more extensive sampling showed that the areas of supersaturation were limited, these workers substantially lowered their estimate of N_2O production in marine ecosystems (Hahn 1981, Liss 1983, Butler et al. 1989). The most extensive survey of ocean waters suggests a flux of about 4×10^{12} g N/yr, emitted as N_2O to the atmosphere (Nevison et al. 1995). A large portion of this may derive from coastal waters (Bange et al. 1996).

Soil emissions from nitrification and denitrification (Chapter 6) are now thought to compose the largest global source of N_2O (Table 12.3). Particularly large emissions of N_2O are found from tropical soils (Matson and Vitousek 1990, Bouwman et al. 1993). Conversion of tropical forests to cultivated lands and pasture results in greater N_2O emissions (Matson and Vitousek 1990, Keller and Reiners 1994), and the flux of N_2O increases when agricultural lands and forests are fertilized or manured (Mosier et al. 1991, Castro et al. 1994b, Bouwman et al. 1995, Nevison et al. 1996). Presumably the increased flux of N_2O from disturbed and fertilized soils

Table 12.3 Estimated Sources and Sinks of N_2O Typical of the Last Decade (10^{12}g N/yr)[a]

Sources	
Natural	
Oceans	4
Tropical soils	
Wet forests	3
Dry savannas	1
Temperate soils	
Forests	1
Grasslands	1
Total identified natural sources	10
Anthropogenic	
Cultivated soils	3.5
Biomass burning	0.5
Industrial sources	1.3
Cattle and feed lots	0.4
Total identified anthropogenic sources	5.7
Total identified sources	15.7
Sinks	
Stratospheric destruction	12.3
Soil microbial activity	?
Atmospheric increase	3.9
Total identified sinks	16.2

[a] From Prather et al. (1995), except ocean flux (Nevison et al. 1995).

stems from higher rates of nitrification that makes NO_3 available to denitrifying bacteria (Chapter 6). Globally the flux of N_2O from fertilizer use is about 0.7×10^{12} g N/yr (Eichner 1990), although this estimate is poorly constrained (0.03–2.0 Tg N/yr; Matthews 1994). Downward leaching of fertilizer nitrate also has the potential to stimulate denitrification in groundwaters, and Ronen et al. (1988) suggest that groundwater may be an important source of N_2O to the atmosphere—up to 1×10^{12} g N/yr.

Relatively small emissions of N_2O result from the combustion of fossil fuels or biomass (Muzio and Kramlich 1988, Linak et al. 1990, Cofer et al. 1991, Andreae 1991, Khalil and Rasmussen 1992b, Berges et al. 1993), but the industrial production of nylon (Thiemens and Trogler 1991) and other chemicals results in a significant—1.3×10^{12} g N/yr—flux of N_2O to the atmosphere (Table 12.3). Disposal of human sewage may also represent a large source of N_2O in the atmosphere (Kaplan et al. 1978). Anthropogenic sources of N_2O are more than enough to explain its rate of increase in the atmosphere, but the total compilation of sources—both natural and anthropogenic—is slightly less than the known sinks, including the rate of N_2O accumulation in the atmosphere (Table 12.3). Clearly, more research is needed to refine the budget for atmospheric N_2O.

Cores extracted from the Antarctic ice cap show that the concentration of N_2O was much lower during the last glacial period (Leuenberger and Siegenthaler 1992). At the end of the Pleistocene, concentrations rose from 185 ppb to about 280 ppb and remained fairly constant until the Industrial Revolution, when they increased to the present-day value of about 311 ppb (Fig. 12.5; Zardini et al. 1989). Relatively high rates of N_2O emission have been observed in a variety of wetlands (Bowden 1986), and deglaciation uncovered large areas of boreal peatland, perhaps accounting for the increase in atmospheric N_2O at the end of the last glaciation. Concentrations of atmospheric N_2O may now be increasing as global warming affects wetlands in tundra and boreal regions (Khalil and Rasmussen 1989, 1992a).

The Global Phosphorus Cycle

The global cycle of P is unique among the cycles of the major biogeochemical elements in having no significant gaseous component (Fig. 12.6). The redox potential of most soils is too high to allow for the production of

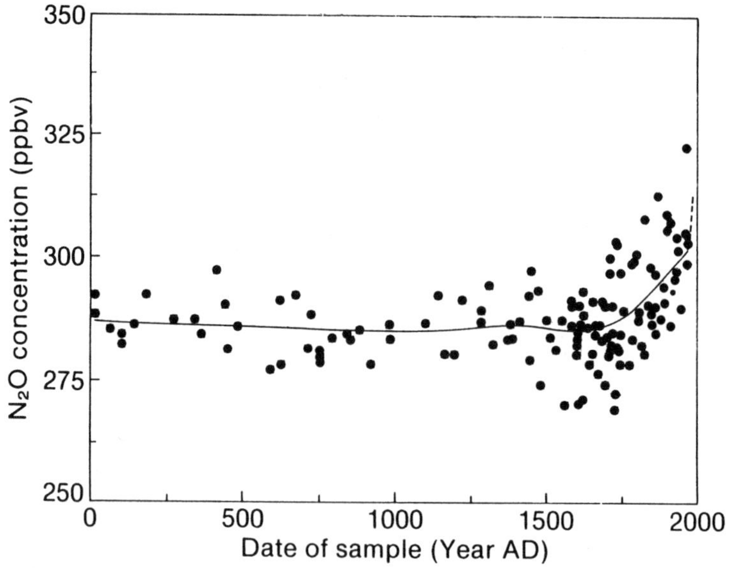

Figure 12.5 Nitrous oxide measurements from ice-core samples, as compiled by Watson et al. (1990).

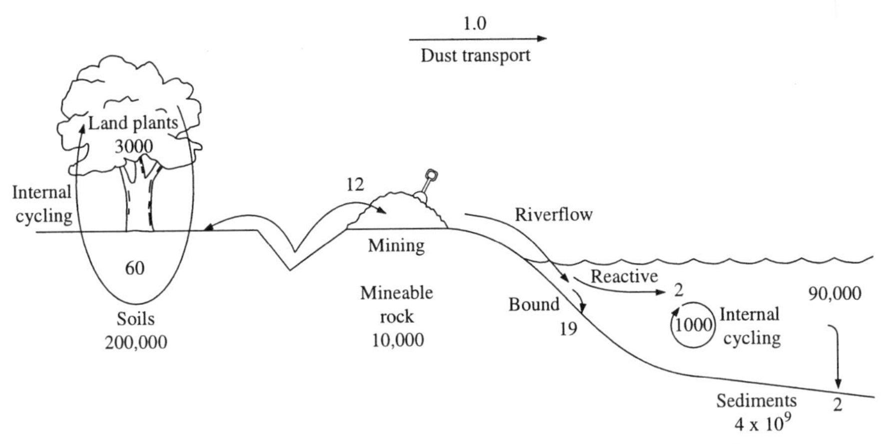

Figure 12.6 The global phosphorus cycle. Each flux is shown in units of 10^{12} g P/yr. Values are derived from the text and from Jahnke (1992). See also Fig. 9.17.

phosphine gas (PH_3; Bartlett 1986), except under very specialized, local conditions (e.g., Dévai et al. 1988, Dévai and DeLaune 1995). The global flux of P in phosphine is probably $<0.04 \times 10^{12}$g/yr (Gassmann and Glindemann 1993). The flux of P through the atmosphere in soil dust and seaspray (1×10^{12} g P/yr) is also much smaller than other transfers in the global P cycle (Graham and Duce 1979), although it is known to make a significant contribution to the supply of available P when it is deposited in some tropical forests (Swap et al. 1992, Newman 1995) and in the open ocean (Talbot et al. 1986).

Unlike transfers in the global nitrogen cycle, the major source of reactive P in the global P cycle is not provided by microbial reactions. Nearly all the phosphorus in terrestrial ecosystems is originally derived from the weathering of calcium phosphate minerals, especially apatite [$Ca_5(PO_4)_3OH$]. Root exudates and mycorrhizae may increase the rate of rock weathering on land (Chapter 4), but there is no process, equivalent to N-fixation, that can produce dramatic increases in phosphorus availability for plants in P-deficient habitats. The phosphorus content of most rocks is not large, and in most soils only a small fraction of the total P is available to biota (Chapter 4). Thus, on both land and at sea, the biota persist as a result of a well-developed recycling of phosphorus in organic forms (Figs. 6.17 and 9.17).

The main flux of P in the global cycle is carried by rivers, which transport about 21×10^{12} g P/yr to the sea (Meybeck 1982). Only about 10% of this flux is potentially available to marine biota; the remainder is strongly bound to soil particles that are rapidly sedimented on the continental shelf (Chapter 9). The solubility product of apatite is only about 10^{-58} (Lindsay and Vlek 1977), so at a seawater pH of 8.0, the phosphorus concentration in equilibrium with apatite would be about 10^{-8} molar (cf. Fig. 4.4). In fact, organic and colloidal forms of P maintain its concentration in excess of that in equilibrium with respect to apatite; the average content of P in deep ocean water is about 3×10^{-6} M (Figs. 9.12 and 9.20). The concentration of PO_4^{3-} in the surface oceans is low, but the large volume of the deep sea accounts for a substantial pool of P (Fig. 12.6). The overall mean residence time for reactive P in the sea is about 25,000 years (Chapter 9).

The turnover of P through the organic pools in the surface ocean is only a few days. Nearly 90% of the phosphorus taken up by marine biota is regenerated in the surface ocean, and most of the rest is mineralized in the deep sea (Fig. 9.17). Eventually, however, phosphorus is deposited in ocean sediments, which contain the largest phosphorus pool near the surface of the Earth. About 2×10^{12} g P/yr are added to sediments of the open ocean—roughly equivalent to the delivery of reactive P to the oceans by rivers (Howarth et al. 1995). On a time scale of hundreds of millions of years, these sediments are uplifted and subject to rock weathering, completing the global cycle. Today, most of the phosphorus in rivers is derived from the weathering of sedimentary rocks, and it represents P that has made at least one complete journey through the global cycle (Griffith et al. 1977).

In many areas humans have enhanced the availability of P by mining phosphate rocks that can be used as fertilizer. Most of the economic deposits of phosphate are found in sedimentary rocks of marine origin, so the mining activity directly enhances the turnover of the global P cycle. In the United States, some of the largest deposits of phosphate rock are found in Florida and North Carolina. In many areas, the flux of P in rivers is significantly higher than it was in prehistoric times as a result of erosion, pollution, and fertilizer runoff (Howarth et al. 1995).

Linking the Global Cycles of C, N, and P

The cycles of important biogeochemical elements are linked at many levels. Stock et al. (1990) describe how P is used to activate a transcriptional protein, stimulating nitrogen fixation in bacteria when nitrogen is in short supply. In this case, an understanding of the interaction between these elements is gained through the study of molecular biology. In Chapter 5, we saw that the photosynthetic rate of land plants is related to the N and P content of their leaves, linking the availability of these elements in plant cells to the net production of organic C for plant growth. In marine ecosystems, net primary productivity is often calculated from the Redfield ratio of C:N:P in phytoplankton biomass (Chapter 9). New production is predicted by estimating the mass of upwelling water and its nutrient content. Whatever our viewpoint—from molecules to whole ecosystems—the movements of N, P, and C are strongly linked in biogeochemistry (Reiners 1986).

Nitrogen fixation by free-living bacteria appears inversely related to the N/P ratio in soil (Fig. 6.3), and the rate of accumulation of N is greatest in soils with high P content (Walker and Adams 1958). Similarly, N/P ratios <29 appear to stimulate N-fixation in freshwater ecosystems (Chapter 7). One might speculate that the high demand for P by N-fixing organisms links the global cycles of N and P, with P being the ultimate limit on nitrogen availability and net primary production. Indeed, in many soils the accumulation of organic carbon is correlated to available P (Chapter 6). Despite these theoretical arguments for a phosphorus limitation of the biosphere through geologic time, net primary production in most terrestrial and marine ecosystems usually shows an immediate response to additions of N (Fig. 12.7). Denitrification appears to maintain small supplies of N in most ecosystems.

Summary

For both N and P, a small biogeochemical cycle with relatively rapid turnover is coupled to a large global pool with relatively slow turnover. For P, the large pool is found in unweathered rock and soil. For N, the major pool is found in the atmosphere.

The biogeochemical cycle of N begins with the fixation of atmospheric nitrogen, which transfers a small amount of inert N_2 to the biosphere. This transfer is balanced by denitrification, which returns N_2 to the atmosphere. The balance of these processes maintains a steady-state concentration of N_2 in the atmosphere with a turnover time of 10^7 years. In the absence of denitrification, most of the N inventory on the Earth would eventually be sequestered in the ocean and in organic sediments. Denitrification closes the global nitrogen cycle, and it causes nitrogen to cycle more rapidly than phosphorus, which has no gaseous phase. The mean residence time of phosphorus in sedimentary rocks is measured in 10^8 yr, and the phosphorus cycle is complete only as a result of tectonic movements of the Earth's crust.

Once within the biosphere, the movements of N and P are more rapid than in their global cycles, showing turnover times ranging from hours (for soluble P in the soil) to hundreds of years (for N in biomass). In response to nutrient limitations, biotic recycling in terrestrial and marine habitats allows much greater rates of net primary production than rates of N-fixation and rock weathering alone would otherwise support (Tables 6.1 and 9.3). The high efficiency of nutrient recycling may explain why, in the face of widespread nitrogen limitation, only about 2.5% of global net primary production is diverted to nitrogen fixation (Gutschick 1981).

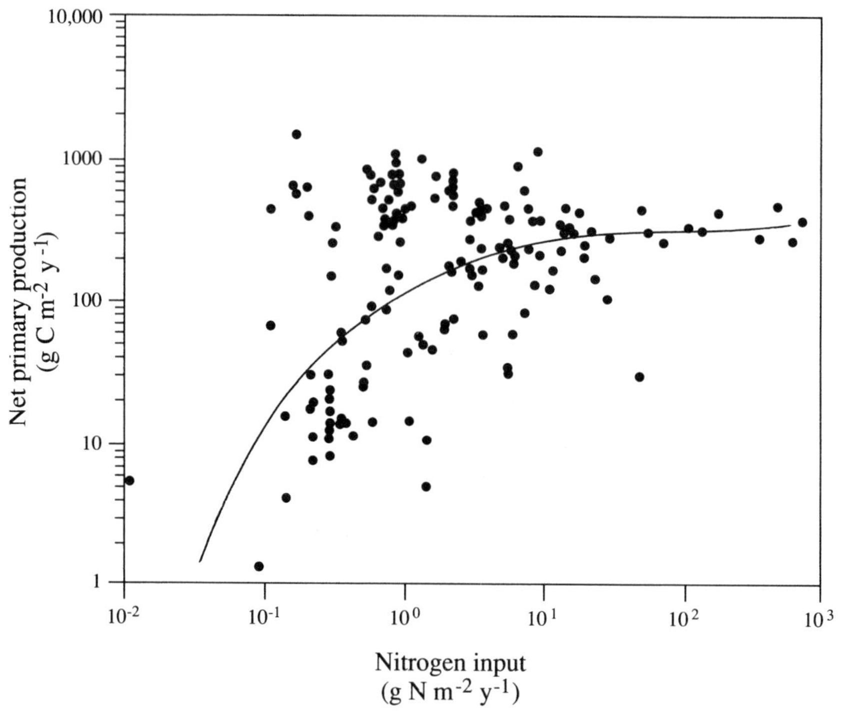

Figure 12.7 Net primary productivity versus nitrogen inputs to terrestrial, aquatic, and marine ecosystems. Net primary production increases in direct response to added nitrogen up to inputs of about 10 g N m^{-1} yr^{-1}. Inputs in excess of that level are rarely found in natural ecosystems, but are seen in polluted environments and agricultural soils. Modified from Levin (1989).

Human perturbations of the global nitrogen and phosphorus cycles are widespread and dramatic. Through the production of fertilizers, humans have doubled the rate at which nitrogen enters the biogeochemical cycle on land. It is unclear how rapidly denitrification will respond to this global increase in nitrogen availability, but the rising concentrations of atmospheric N_2O are perhaps one indication of an ongoing biotic response (Vitousek 1994). Increasing nitrogen availability has led to the local extinction of species from polluted ecosystems (Schlesinger 1994) and shifted the limitation of net primary production in these systems from N to P (e.g., Mohren et al. 1986). Increasing transport of N and P in rivers has shifted many estuarine and coastal ecosystems to a condition of Si deficiency (Justić et al. 1995). All these changes indicate the effect of a single species—the human—in upsetting a steady state in global nutrient cycling.

Recommended Readings

Bolin, B. and R.B. Cook (eds.). 1983. *The Major Biogeochemical Cycles and Their Interactions.* Wiley, New York.

Porter, R. and D.W. Fitzsimons. 1978. *Phosphorus in the Environment: Its Chemistry and Biochemistry*. Elsevier, Amsterdam.

Sprent, J.I. 1988. *The Ecology of the Nitrogen Cycle*. Cambridge University Press, Cambridge.

Tiessen, H. (ed.). 1995. *Phosphorus Cycling in Terrestrial and Aquatic Ecosystems*. Wiley, Chichester, U.K.

13

The Global Sulfur Cycle

Introduction

Sulfur is found in valence states ranging from +6 in SO_4^{2-} to −2 in sulfides. The original pool of sulfur on Earth was held in igneous rocks, largely as igneous pyrite (FeS_2). Degassing and weathering of the crust under an atmosphere containing O_2 transferred a large amount of S to the oceans, where it is now found as SO_4^{2-}. When SO_4^{2-} is assimilated by organisms, it is reduced and converted into organic sulfur, which is an essential component of protein. However, the live biosphere contains relatively little sulfur. Today, the major global pools of S are found in sedimentary pyrite, seawater, and evaporites derived from ocean water (Table 13.1).

As in the case of nitrogen, microbial transformations between valence states drive the global S cycle. Under anoxic conditions, SO_4 is a substrate for sulfate reduction, which may lead to the release of reduced gases to the atmosphere and to the deposition of sedimentary pyrite (Chapters 7 and 9). Anoxic environments can also support sulfur-based photosynthesis, which is likely to have been one of the first forms of photosynthesis on Earth (Chapter 2). On the other hand, in the presence of oxygen, reduced sulfur compounds are oxidized by microbes. In some cases, the oxidation of S is coupled to the reduction of CO_2, in the reactions of S-based chemosynthesis (Eq. 2.15).

Understanding the biogeochemistry of S has enormous economic significance. Many metals are mined from sulfide minerals in hydrothermal

Table 13.1 Active Reservoirs of Sulfur Near the Surface of the Earth

Reservoir	10^{18} g S
Atmosphere	0.0000028
Seawater	1280.
Sedimentary rocks	
Evaporites	2470.
Shales	4970.
Land plants	0.0085
Soil organic matter	0.0155
Total	8720

From Holser et al. (1989) and Dobrovolsky (1994).

ore deposits (Meyer 1985). Microbial reactions involving sulfur bacteria are increasingly being used to remove metals from relatively low-grade ore (Lundgren and Silver 1980). Sulfur is an important constituent of coal and oil, and SO_2 is emitted to the atmosphere when these fuels are burned. A large amount of SO_2 is also emitted during the smelting of copper ores (Cullis and Hirschler 1980, Oppenheimer et al. 1985). An understanding of the relative importance of natural sulfur compounds in the atmosphere compared to anthropogenic SO_2 is essential to evaluate the causes of acid rain and the impact of acid rain on natural ecosystems.

In this chapter we will review the global sulfur cycle. As for carbon (Chapter 11), nitrogen, and phosphorus (Chapter 12), we will attempt to establish a budget for S on land and in the atmosphere. We will couple these to the budget for marine S (Fig. 9.22) to form an overall, global picture of the S cycle. The biogeochemical cycle of S has varied through Earth's history as a result of the appearance of new metabolic pathways and changes in their importance. We will review the history of the S cycle as it is told by sedimentary rocks. Finally, we will evaluate human impact on the S cycle and the global production of acidic sulfur substances in acid rain.

The Global Sulfur Cycle

No sulfur gas is a long-lived or major constituent of the atmosphere. Thus, all attempts to model the global S cycle must explain the fate of the large annual input of sulfur compounds to the atmosphere. The short mean residence time for atmospheric sulfur compounds, as a result of their oxidation to SO_4, allows us to express all the fluxes in the global budget in units of 10^{12} g of S, regardless of the original form of emission. Despite the small atmospheric content of S compounds, the total annual flux of S compounds through the atmosphere (about 300×10^{12} g S/yr) rivals the movements of N in the global nitrogen cycle (compare Fig. 13.1 to Fig. 12.2).

In 1960, Eriksson examined the potential origins of SO_4 in Swedish rainfall, and hence, indirectly, sources of SO_4 in the atmosphere. He reasoned that if the Cl^- in rainfall was derived from the ocean, then seaspray should also carry SO_4 roughly in proportion to the ratio of SO_4^{2-} to Cl^- in seawater. His calculation suggested that about 4×10^{12} g S/yr deposited on land must be derived from the sea. At about the same time, however, Junge (1960) was evaluating the SO_4 content of rainfall, and he calculated that about 73×10^{12} g S/yr was deposited on land. Clearly, there were

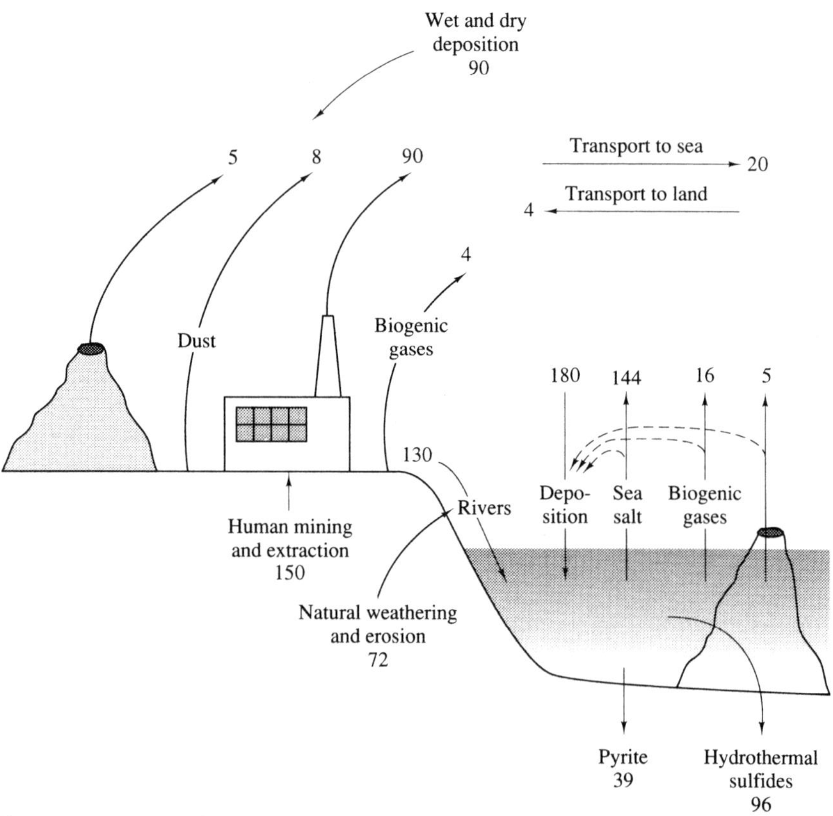

Figure 13.1 The global S cycle with annual flux shown in units of 10^{12} g S/yr. The derivation of most values is described in the text, with the marine values taken from Fig. 9.22. The net flux from land to sea is extrapolated from Whelpdale and Galloway (1994).

other sources of SO_4 in the atmosphere and in rainfall. Junge's maps showed that SO_4 was abundant in the rainfall of industrial regions and in areas downwind of deserts (Fig. 3.15c). Desert soils are a source of gypsum ($CaSO_4 \cdot 2H_2O$) in atmospheric dust (Reheis and Kihl 1995), and the burning of fossil fuels in industrial regions contributes SO_2 to air pollution (Langner et al. 1992, Spiro et al. 1992). In the intervening years, new sources of S in the atmosphere have been recognized, and global flux estimates have been revised repeatedly. Nevertheless, our understanding of the global S cycle is primitive, and most of the estimates illustrated in Fig. 13.1 are subject to considerable uncertainty.

Episodic events, including volcanic eruptions and dust storms, contribute to the global biogeochemical cycle of S. Sulfur emissions from volcanoes are especially difficult to quantify. Legrand and Delmas (1987) used the deposition of SO_4 in the Antarctic ice pack to estimate the contribution of volcanoes to the global S cycle during the last 220 years. The Tambora eruption of 1815 was the largest, releasing 50×10^{12} g S to the atmosphere. Typically, major eruptions, such as that of Mt. Pinatubo (15 June 1991), release $5–10 \times 10^{12}$ g S each (Bluth et al. 1993). When the volcanic emissions are averaged over many years, the annual global flux is about 10×10^{12} g S/yr (Stoiber et al. 1987, Berresheim and Jaeschke 1983, Bates et al. 1992). About 70% of this leaks passively from volcanoes and the remainder is derived from periodic, explosive events (Bluth et al. 1993, Allard et al. 1994).

The movement of S in soil dust is also episodic and poorly understood. Many of the large particles are deposited locally, while smaller particles may undergo long-range transport in the atmosphere (Chapter 3). Savoie et al. (1987) found that dust from the deserts of the Middle East contributed SO_4 to the waters of the northwest Indian Ocean. Ivanov (1983) suggests a global flux of 8×10^{12} g S/yr owing to dust transport in the troposphere—about 10% of the fossil fuel release.

Estimates of the flux of biogenic sulfur gases from land differ by a factor of 10, with recent values being lower than those of earlier studies (cf. Adams et al. 1981, versus Goldan et al. 1987, Lamb et al. 1987b). The dominant sulfur gas emitted from freshwater wetlands and anoxic soils is H_2S, with dimethylsulfide and carbonyl sulfide (COS) playing lesser roles (Chapter 7). Emissions from plants are poorly understood and deserving of further study (Chapter 6). Most recent estimates suggest that the total flux of biogenic sulfur from land is $<1 \times 10^{12}$ g S/yr (Bates et al. 1992). Forest fires emit an additional 3×10^{12} g S/yr (Andreae 1991).

It seems certain that direct emissions from human industrial activities are the largest sources of S gases in the atmosphere. Estimates of this flux have ranged from 50 to 100×10^{12} g S/yr, globally (Möller 1984, Hameed and Dignon 1988, Spiro et al. 1992, Müller 1992), with slightly lower values in more recent years as pollution abatement has become more widespread. Ice cores from Greenland show a large increase in the deposition of SO_4

from the atmosphere since the beginning of the Industrial Revolution (Herron et al. 1977, Mayewski et al. 1986, 1990).

Owing to the reactivity of S gases in the atmosphere, most of the anthropogenic emission of SO_2 is deposited locally in precipitation and dryfall. Total deposition of S on land may be as high as 120×10^{12} g S/yr (Andreae and Jaeschke 1992), but a value of 90×10^{12} g S/yr balances the global S cycle shown in Fig. 13.1. Deposition in dryfall and the direct absorption of SO_2 are poorly understood, so this global estimate is subject to revision. The estimate of atmospheric deposition on land accounts for a large fraction of the total emissions from land. The remainder undergoes long-distance transport in the atmosphere and accounts for a net transfer of S from land to sea (Whelpdale and Galloway 1994).

Human activities also affect the transport of S in rivers. Berner (1971) estimated that at least 28% of the SO_4 content of rivers is derived from air pollution, mining, erosion, and other human activities, whereas, Husar and Husar (1985) suggest that the current river transport of about 131×10^{12} g S/yr is roughly double that of preindustrial conditions. Other workers indicate that human activities have raised the current river transport to $>200 \times 10^{12}$ g S/yr (Brimblecombe et al. 1989). A small fraction of the natural river load of SO_4 is derived from rainfall, which includes cyclic salts that are carried through the atmosphere from the ocean (4×10^{12} g S/yr in Fig. 13.1). Weathering of pyrite and gypsum also contributes to the SO_4 content of river water (Table 8.6).

The marine portion of the global S cycle is largely taken from Fig. 9.22. The ocean is a large source of aerosols that contain SO_4, but most of these are redeposited in the ocean in precipitation and dryfall. Dimethylsulfide [$(CH_3)_2S$ or DMS] is the major biogenic gas emitted from the sea. There is a wide range of estimates of marine DMS flux (Andreae and Jaeschke 1992), but Erickson et al. (1990) suggest that the global flux of DMS from the sea may be only slightly greater than 15×10^{12} g S/yr—a value that has been adopted by most recent workers (Langner and Rodhe 1991, Spiro et al. 1992, Bates et al. 1992). Thus, DMS is the largest natural source of sulfur gases in the atmosphere (Ferek et al. 1986). The mean residence time of DMS in the atmosphere is about 1 day (Table 3.4) as a result of its oxidation to SO_4. Thus, most of the sulfur from DMS is redeposited in the oceans.

Although they are subject to great revision, the current estimates of inputs to the ocean are slightly in excess of the estimate of total sinks, implying that the oceans are increasing in SO_4 by over 10^{13} g S/yr. Such an increase will be difficult to document, because the content in the oceans is 1.28×10^{21}g S. As calculated in Chapter 9, the mean residence time for SO_4 in seawater is over 10,000,000 years, with respect to the current inputs from rivers.

Temporal Perspectives of the Global Sulfur Cycle

During the accretion of the primordial Earth, sulfur was among the gases that were released by crustal outgassing to form the secondary atmosphere (Chapter 2). Even today, volcanic emissions contain appreciable concentrations of SO_2 and H_2S (Table 2.1). When the oceans condensed on Earth, the atmosphere was essentially swept clear of S gases, owing to their high solubility in water. On Venus, where there is no ocean, crustal degassing has resulted in a large concentration of SO_2 in the atmosphere (Oyama et al. 1979). The dominant form of S in the earliest seas on Earth is likely to have been SO_4^{2-}; high concentrations of Fe^{2+} in the primitive ocean would have precipitated any sulfides, which are insoluble under anoxic conditions (Walker and Brimblecombe 1985). The SO_4 content of the oceans apparently increased until about 400,000,000 years ago and then decreased slightly to the amount found today (Zehnder and Zinder 1980). The total inventory of S compounds on the surface of the Earth (nearly 10^{22} g S) represents the total crustal outgassing of S through geologic time (Table 2.2).

The ratio of ^{32}S to ^{34}S in the total S inventory on Earth is thought to be similar to the ratio of 22.22 measured in the Canyon Diablo Troilite (CDT), a meteorite collected in California. The sulfur isotope ratio in this rock is accepted as an international standard and assigned a value of 0.00. In other samples, deviations from this ratio are expressed as $\delta^{34}S$, with the units of parts per thousand parts (‰)—a convention that we also used for the isotopes of carbon (Chapter 5) and nitrogen (Chapter 6). Presumably the $\delta^{34}S$ isotope ratio in the earliest oceans was 0.00, because there is no reason to expect any discrimination between the isotopes of S during crustal degassing. When evaporite minerals precipitate from seawater, there is little differentiation among the isotopes of sulfur, so geologic deposits of gypsum and barite ($BaSO_4$) carry a record of the isotopic composition of S in seawater. In the earliest sedimentary rocks, dating to 3.8 bya, $\delta^{34}S$ is close to 0.00 (Schidlowski et al. 1983).

Dissimilatory sulfate reduction by bacteria strongly differentiates among the isotopes of sulfur, as a result of a more rapid enzymatic reaction with $^{32}SO_4$. By itself, sulfate metabolism results in an isotopic depletion of −18‰ ^{34}S in sedimentary sulfides relative to the ratio in seawater (Canfield and Teske 1996). Stronger fractionations, sometimes as much as −50‰, occur as a result of repeated cycles of oxidation and reduction (Canfield and Thamdrup 1994, Canfield and Teske 1996). The evolution of sulfate reduction dates to 2.4 to 2.7 bya, based on the first occurrence of sedimentary rocks with depletion of ^{34}S (Cameron 1982, Schidlowski et al. 1983, Habicht and Canfield 1996). Today, the average $\delta^{34}S$ in sedimentary sulfides in all of the Earth's crust is about −10 to −12‰ (Holser and Kaplan 1966, Migdisov et al. 1983).

Figure 13.2 shows a three-box model for the S cycle, in which marine SO_4 and sedimentary sulfides are connected through microbial oxidation and reduction reactions, which discriminate between the sulfur isotopes. During periods of Earth's history when large amounts of sedimentary pyrite were formed as a result of sulfate reduction, seawater SO_4 became enriched in residual $^{34}SO_4$. Currently, about 50% of the pool of S near the surface of the Earth is found in reduced form (Li 1972, Holser et al. 1989), and the $\delta^{34}S$ of seawater is +21‰ (Kaplan 1975, Rees et al. 1978). Because there is little differentiation among isotopes of S during the precipitation of evaporites, the sedimentary record of evaporites indicates changes in

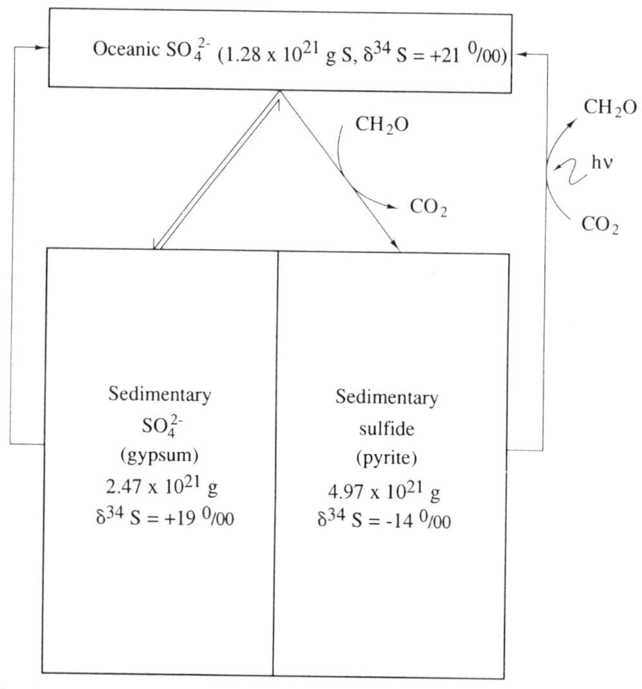

Figure 13.2 A model for the global sulfur cycle showing the linkage and partitioning of S between oxidized and reduced pools near the surface of the Earth. Transfers of S from seawater to pyrite involve a major fractionation between ^{34}S and ^{32}S isotopes, whereas exchange between seawater SO_4 and sedimentary SO_4 (largely gypsum) involves only minor fractionation. The sum of all pools, nearly 10^{22} g, represents the total outgassing of S from the crust (cf. Table 2.2). About 15% now resides in the ocean. Estimates of the pool of S in sedimentary sulfides show a wide range of values; the value here, from Holser et al. (1989), is close to that estimated from the pool of sedimentary organic carbon (1.56 × 10^{22} g; Des Marais et al. 1992) divided by the mean C/S ratio in marine sediments (2.8; Raiswell and Berner 1986). Isotope ratios in seawater and gypsum are taken from Holser et al. (1989). The isotope ratio of S in sedimentary sulfides is derived by mass balance to yield $\delta^{34}S$ of +4.2 in the global inventory.

the $\delta^{34}S$ of seawater over geologic time. Changes in the isotopic composition of seawater indicate changes in the relative size of the reservoir of sedimentary pyrite, which occur as a result of changes in the net global balance between sulfate reduction and the oxidation of sedimentary sulfides. By contrast, the annual uptake of S by plants [assimilatory reduction (590×10^{12} g S/yr on land and 1320×10^{12} g S/yr in the sea); Dobrovolsky 1994] and the decomposition of plant detritus have little effect on the isotopic composition of the major reservoirs of the global S cycle.

During the last 600,000,000 years, seawater SO_4 has varied between +10 and +30‰ in $\delta^{34}S$ (Fig. 13.3), with an average value close to that of today. Seawater sulfate shows a marked positive excursion (+32‰) in $\delta^{34}S$ during the Cambrian (550 mya), when the deposition of pyrite must have been greatly in excess of the oxidation of sulfide minerals exposed on land. Seawater sulfate was less concentrated in ^{34}S, i.e., $\delta^{34}S$ of +10‰, during the Carboniferous and Permian, when a large proportion of the Earth's net primary production occurred in freshwater swamps, where SO_4, sulfate reduction, and pyrite deposition are less important (Berner 1984). Presum-

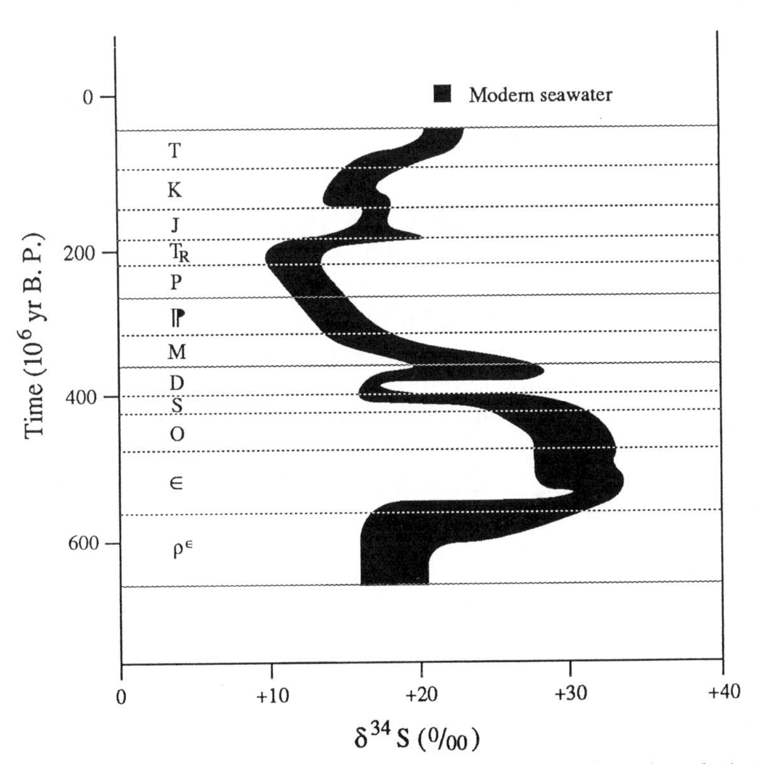

Figure 13.3 Variations in the isotopic composition of seawater SO_4 through geologic time. From Kaplan (1975).

ably the concentration of SO_4 in seawater was also greater during that interval, because the rate of pyrite formation was depressed.

Although the sulfur cycle has shown shifts between net sulfur oxidation and net sulfur reduction in the geologic past, the current human impact is probably unprecedented in the geologic record. As for the carbon cycle, the present day cycle of S is not in steady state. Human activities have added a large flux of gaseous sulfur to the atmosphere, some of which is transported globally. Humans are mining coal and extracting petroleum from the Earth's crust at a rate that mobilizes 150×10^{12} g S/yr, more than double the rate of 100 years ago (Brimblecombe et al. 1989). The net effect of these processes is to increase the pool of oxidized sulfur (SO_4) in the global cycle, at the expense of the storage of reduced sulfur in the Earth's crust. Human activities cause only a tiny change in the global pools of S, but they produce massive changes in the annual flux of S through the atmosphere.

Various workers have attempted to use measurements of $\delta^{34}S$ to deduce the origin of the SO_4 in rainfall and the extent of human impact on the movement of S in the atmosphere (Grey and Jensen 1972, Nriagu et al. 1991). Unfortunately, the potential sources of SO_4 show a wide range of values for $\delta^{34}S$, making the identification of specific sources difficult (Nielsen 1974). For example, the sulfur in coal may be depleted in $\delta^{34}S$ if it is found as pyrite or enriched in $\delta^{34}S$ if it is derived from the sulfur that was originally assimilated by the plants forming coal (Hackley and Anderson 1986). Thus, coals show a wide range in $\delta^{34}S$. Similarly, petroleum shows a range of -10.0 to $+25‰$ in ^{34}S (Krouse and McCready 1979). Desert dusts containing SO_4 range in $\delta^{34}S$ from $-35‰$ to $+17‰$, depending on the parent material of soil formation (Schlesinger and Peterjohn 1988). In the eastern United States, $\delta^{34}S$ of rainfall varies seasonally between $+6.4‰$ in winter and $+2.9‰$ in summer, consistent with any of these sources or a combination of them (Nriagu and Coker 1978). The lower values of summer are thought to reflect the influence of biogenic sulfur derived from sulfate reduction in wetlands (Nriagu et al. 1987).

When SO_2 is emitted as an air pollutant, it forms sulfuric acid through heterogeneous reactions with water in the atmosphere (Chapter 3). As a strong acid that is completely dissociated in water, H_2SO_4 suppresses the disassociation of natural, weak acids in rainfall. For example, in the absence of strong acids, the dissolution of CO_2 in water will form a weak solution of carbonic acid, H_2CO_3, and rainfall pH will be about 5.6:

$$CO_2 + H_2O \rightleftarrows H^+ + HCO_3^-. \qquad (13.1)$$

In the presence of strong acids that lower the pH below 4.3, this reaction moves to the left, and carbonic acid makes no contribution to free acidity. In many industrialized areas, free acidity in precipitation is almost wholly determined by the concentration of the strong acid anions, SO_4^{2-} and NO_3^- (Table 13.2). Rock weathering that was primarily driven by carbonation weathering in the preindustrial age is now driven by anthropogenic H^+ (Johnson et al. 1972).

It is interesting to estimate the global sources of acidity in the atmosphere. In this analysis, we are interested only in reactions that are net sources of H^+, so we can ignore the movements of soil dusts and seaspray, because the strong-acid anions they contain, NO_3^- and SO_4^{2-}, are largely balanced by cations (especially Ca and Na) that are emitted at the same time. If the pH of all rainfall on Earth was 5.6 as a result of an equilibrium with atmospheric CO_2, the total deposition of H^+ ions would be 1.24×10^{12} moles/year. The production of NO by lightning produces additional acidity, because NO dissolves in rainwater, forming HNO_3. Globally, N- fixation by lightning contributes 0.21×10^{12} moles of H^+/yr, and other natural sources of NO_x (soils and forest fires) contribute 2×10^{12} moles of H^+/yr. In the years following massive eruptions, SO_2 from volcanoes is distributed globally and dominates the atmospheric deposition of S (Mayewski et al. 1990, Langway et al. 1995). On the average, however, volcanic emanations of SO_2 contribute 0.63×10^{12} moles H^+/yr, and the oxidation of biogenic S gases produces 1.2×10^{12} moles H^+/yr. Thus, the total, natural production of H^+ in the atmosphere is normally about 5.3×10^{12} moles/yr. In contrast, the anthropogenic production of NO_x and SO_2 results in about 4.0×10^{12} moles of H^+/year—nearly as much as all natural sources of acidity combined.

Table 13.2 Sources of Acidity in Acid Rainfall Collected in Ithaca, New York, on July 11, 1975 (Ambient pH 3.84)[a]

		Contribution to	
Component	Concentration in precipitation (mg/liter)	Free acidity at pH 3.84 (μeq/liter)	Total acidity in a titration to pH 9.0 (μeq/liter)
H_2CO_3	0.62	0	20
Clay	5	0	5
NH_4^+	0.53	0	29
Dissolved Al	0.050	0	5
Dissolved Fe	0.040	0	2
Dissolved Mn	0.005	0	0.1
Total organic acids	0.43	2	5.7
HNO_3	2.80	40	40
H_2SO_4	5.60	102	103
Total		144	210

[a] From Galloway et al. (1976). Copyright 1976 by the AAAS.

The only net source of alkalinity in the atmosphere comes from the reaction of NH_3 with the strong acids H_2SO_4 and HNO_3 to form aerosols, $(NH_4)_2SO_4$ and NH_4NO_3 (Eq. 3.5). However, the "natural" global emission of NH_3, about 28×10^{12} g N/yr (Table 12.1), reduces the "natural" production of H^+ by only about 2.0×10^{12} moles/year, or 37% (cf. Savoie et al. 1993). Thus, even though the current acidity of the atmosphere is much higher as a result of human activities, the atmosphere has always acted as an acidic medium with respect to the Earth's crust throughout geologic time.

The Atmospheric Budget of Carbonyl Sulfide (COS)

Showing an average concentration of about 500 parts per trillion, carbonyl sulfide (COS; alternatively OCS) is the most abundant sulfur gas in the atmosphere (Table 3.1). The pool in the atmosphere contains about 2.8×10^{12} g S (Chin and Davis 1995). Based on the global budget of Table 13.3, the mean residence time for COS in the atmosphere is about 5 years. Our understanding of COS is primitive. Estimates of sources of COS have been revised downward during the last few decades (cf. Khalil and Rasmussen 1984, Servant 1989, Chin and Davis 1993), and the budget of COS in the atmosphere is not balanced—showing an excess of sources over sinks. Nevertheless, there is no strong indication that COS is increasing in the atmosphere (Hofmann 1990, Rinsland et al. 1992).

Table 13.3 Global Budget for Carbonyl Sulfide (COS) in the Atmosphere[a]

Source or sink	COS (10^{12} g S/yr)
Sources	
Oceans	0.17
Soils	0.14
Biomass burning	0.07
Fossil fuels	0.02
Volcanoes	0.01
Oxidation of CS_2	0.18
Oxidation of DMS	0.09
Total sources	0.68
Sinks	
Oxidation by OH	0.17
Stratospheric photolysis	0.02
Vegetation uptake	0.23
Total sinks	0.42

[a] From Chin and Davis (1993), except for DMS source (Barnes et al. 1994).

The major source of COS appears to be the ocean, where it is produced by a photochemical reaction with dissolved organic matter (Ferek and Andreae 1984). Andreae and Ferek (1992) suggest a global marine flux of 0.41×10^{12} g S—substantially higher than the value in Table 13.3. The lower value used here is consistent with recent measurements that suggest that the open ocean is not a source, and a perhaps even a small sink, for COS (Weiss et al. 1995). Other sources of COS include biomass burning (Nguyen et al. 1995), fossil fuel combustion, and the oxidation of CS_2—largely produced in industry—by OH radicals in the atmosphere (Table 13.3). Early indications of a large source of COS from upland soils (Adams et al. 1981) have been reduced by more recent measurements (Goldan et al. 1987), and the global emission of COS from salt marshes is limited by the small extent of salt marsh vegetation (Steudler and Peterson 1985, Carroll et al. 1986).

Some COS is oxidized in the troposphere via OH radicals, but the major tropospheric sink for COS, first reported by Goldan et al. (1988), appears to be uptake by vegetation. Berresheim and Vulcan (1992) found no evidence of net COS uptake by a loblolly pine (*Pinus taeda*) forest in Georgia (USA), but Kesselmeier and Merk (1993) found that a variety of crop plants took up COS whenever the ambient concentration was greater than 150 ppt. Uptake by vegetation may account for 55% (Table 13.3) to 89% (Servant 1989) of the total annual destruction of COS globally.

A small amount of COS is mixed into the stratosphere, where it is destroyed by a photochemical reaction involving the OH radical, producing SO_4 (Chapter 3). In fact, aside from the periodic eruptions of large volcanoes, COS appears to be the main source of SO_4 aerosols in the stratosphere (Hofmann and Rosen 1983, Servant 1986). There is some evidence that these aerosols have increased in recent years (Hofmann 1990). In the absence of a discernible trend in COS, some workers have suggested that SO_2 from high-altitude aircraft may be responsible (Hofmann 1991). These aerosols affect the amount of solar radiation entering the troposphere, and they are an important component of the radiation budget of the Earth (Turco et al. 1980). Humans appear to make large contributions to the budget of COS (Table 13.3), and any increase in COS-derived aerosols in the stratosphere has potential consequences for predictive models of future global warming (Hofmann and Rosen 1980). In return, global warming and an increasing flux of uvB radiation penetrating to the Earth's surface may enhance the production of COS in ocean waters (Najjar et al. 1995).

Summary

The major pool of S in the global cycle is found in the crustal minerals gypsum and pyrite. Additional S is found dissolved in ocean water. Thus, with respect to pools, the global S cycle resembles the global cycle of phosphorus (Chapter 12). In contrast, the largest pool of the global N cycle is found in the atmosphere.

In other respects, however, there are strong similarities between the global cycles of N and S. In both cases, the major annual movement of the element is through the atmosphere, and under natural conditions a large portion of the movement is through the production of reduced gases of N and S by biota. These gases return N and S to the atmosphere, providing a closed global cycle with a relatively rapid turnover. In contrast, the ultimate fate for P is incorporation into ocean sediments, where the cycle of P is complete only as a result of long-term sedimentary uplift.

Biogeochemistry exerts a major influence on the global S cycle. The largest pool of S near the surface of the Earth is found in sedimentary pyrite, as a result of sulfate reduction. The sedimentary record shows that the relative extent of sulfate reduction has varied through geologic time. In the absence of sulfate-reducing bacteria, the concentration of SO_4 in seawater would be higher and that of O_2 in the atmosphere would be lower than present day values.

Current human perturbation of the sulfur cycle is extreme—roughly doubling the annual mobilization of sulfur from the crust of the Earth. As a result of fossil fuel combustion, areas that are downwind of industrial regions now receive massive amounts of acidic deposition from the atmosphere. This excess acidity is likely to lead to changes in rock weathering (Chapter 4), forest growth (Chapter 6), and ocean productivity (Chapter 9).

Recommended Readings

Brimblecombe, P. and A.Y. Lein (eds.). 1989. *Evolution of the Global Biogeochemical Sulphur Cycle.* Wiley, Chichester, U.K.

Howarth, R.W., J.W.B. Stewart, and M.V. Ivanov. (eds.). 1992. *Sulphur Cycling on the Continents.* Wiley, Chichester,U.K.

Ivanov, M.V. and J.R. Freney (eds.). 1983. *The Global Biogeochemical Sulphur Cycle.* Wiley, Chichester, U.K.

14

A Perspective

There is an underlying theme to this book: the biota control the basic chemical conditions on the surface of the Earth. With an atmosphere containing 21% oxygen, the chemical environment on Earth stands in stark contrast to that on our sterile planetary neighbors—Mars and Venus. The metabolic activities of organisms, which link oxidation and reduction reactions, produce the relatively stable conditions on Earth that are conducive to the persistence of life. The system is fueled by the capture of sunlight energy by green plants. Chemical elements circulate between the land, the sea, and the atmosphere in linked, global biogeochemical cycles. All this occurs in a zone about 50 km wide—a thin skin at the surface of our planet's 6371-km radius.

Today, humans are carrying the Earth into a new realm. Our species now uses or controls up to 40% of the net primary productivity on land, and directly or indirectly, we harvest nearly 8% of the productivity of the sea. Planetary conditions that were relatively constant for the developmental history of human social and economic systems are now changing rapidly—upsetting steady-state conditions in chemistry, if not in climate. Our population is increasing exponentially, so that we may double our number in the next 50 years. As our population increases, we will leave much less natural habitat for other species in the biosphere (Fig. 14.1). Our future is uncertain, but as we come close to controlling all of the Earth's terrestrial productivity, it is certain that the prognosis for most other species on our planet is not rosy.

A second theme of biogeochemistry is the recognition that the integrated diversity of the natural biosphere maintains the stable planetary conditions upon which all life depends. Biochemical elements move between the crust, the atmosphere, and the oceans—in global cycles—as a result of the metabolic diversity of life. Reductions in species diversity are soon followed by lower net primary production, increasing losses of nutrient elements in streamwater, and other indices of ecosystem degradation (Woodwell 1970, Rapport et al. 1985, Tilman et al. 1996). Thus, human impoverishment of

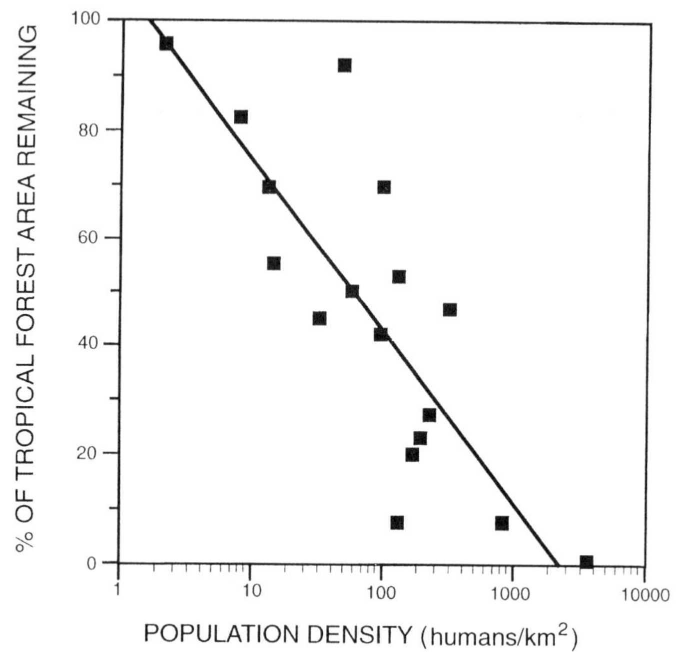

Figure 14.1 The relationship between the percentage of undisturbed tropical rainforest and human population density in 17 countries of southeast Asia. From the data of Collins et al. (1991), as plotted by Sinclair et al. (1995).

the diversity of the biosphere threatens the stability of our planet's condition.

• • •

The biosphere faces a dilemma: each person on Earth wishes to achieve the highest possible standard of living, and our numbers are increasing exponentially. The human pursuit of a better life and the by-products of this quest now foul the atmosphere and the waters of our planet, denude its vegetation, and erode its soils. Industrialization has markedly altered the chemical environment in which we live. For example, we have *doubled* the annual supply of fixed nitrogen to the Earth's soils and the flux of oxidized sulfur gases to its atmosphere.

Luckily, we have within our grasp the ability to control some of the environmental consequences of our lifestyle. Most of the by-products of human society can be contained and cleansed with the application of appropriate technology and human interest in doing so. Our ingenuity has offered increasing supplies of fresh water, mineral resources, and food for our increasing population. As economic incentives demand it, we can use energy efficiently and recycle many of the waste products of modern society.

In these endeavors, the need for collective, political action is paramount, for the motivation of individuals is often lacking (Hardin 1968).

Thus, I believe that the greatest problem facing human society is our rising population. I was born into a world with 2.5 billion people; today there are more than double that number, and each student reading this book is likely to live to see the human population rise to 10 billion individuals. This rate of population growth masks all efforts to use resources wisely and clean up the by-products of modern society. There are few basic axioms of ecology, but one of the most fundamental is that which predicts the ultimate collapse of a population showing exponential growth in a closed environment.

In a variety of experiments during the 1930s, Gause (1934) showed that stable populations of the freshwater protozoan *Paramecium* could be maintained in small aquaria only if supplies of food and water were replenished continuously. Initially the population grew rapidly, but without fresh water, toxic wastes accumulated and the population perished. Now, I am not at all hopeful that our planet will receive an interplanetary delivery of fresh resources. We live in a closed chemical system and the persistence of life in that system demands that we manage it well—for ourselves and for the myriad of other species that help determine its stable condition. Changes in atmospheric chemistry suggest that we are failing in our planetary stewardship. Today it is true that more people live at a higher standard of living than ever before, but our rapid increase in population leaves an increasing number of people at the edge of tolerable existence and it pushes us ever closer to a day of planetary reckoning. Our focus for maintaining life and quality of life on this planet should be on controlling human population growth.

References

Aber, J.D. and J.M. Melillo. 1980. Litter decomposition: Measuring relative contributions of organic matter and nitrogen to forest soils. Canadian Journal of Botany 58: 416–421.

Aber, J.D., K.J. Nadelhoffer, P. Steudler, and J.M. Melillo. 1989. Nitrogen saturation in northern forest ecosystems. BioScience 39: 378–386.

Aber, J.D., A. Magill, R. Boone, J.M. Melillo, P. Steudler, and R. Bowden. 1993. Plant and soil responses to chronic nitrogen additions at the Harvard Forest, Massachusetts. Ecological Applications 3: 156–166.

Abrahams, A.D., A.J. Parsons, and J. Wainwright. 1994. Resistance to overland flow on semiarid grassland and shrubland hillslopes, Walnut Gulch, southern Arizona. Journal of Hydrology 156: 431–446.

Achtnich, C., F. Bak, and R. Conrad. 1995. Competition for electron donors among nitrate reducers, ferric iron reducers, sulfate reducers, and methanogens in anoxic paddy soil. Biology and Fertility of Soils 19: 65–72.

Ackerman, S.A. and H. Chung. 1992. Radiative effects of airborne dust on regional energy budgets at the top of the atmosphere. Journal of Applied Meteorology 31: 223–233.

Adams, D.F., S.O. Farwell, E. Robinson, M.R. Pack, and W.L. Bamesberger. 1981. Biogenic sulfur source strengths. Environmental Science and Technology 15: 1493–1498.

Adams, J.A.S., M.S.M. Mantovani, and L.L. Lundell. 1977. Wood versus fossil fuel as a source of excess carbon dioxide in the atmosphere: A preliminary report. Science 196: 54–56.

Adams, J.M., H. Faure, L. Faure-Denard, J.M. McGlade, and F.I. Woodward. 1990. Increases in terrestrial carbon storage from the Last Glacial Maximum to the present. Nature 348: 711–714.

Adams, M.A. and P.M. Attiwill. 1982. Nitrate reductase activity and growth response of forest species to ammonium and nitrate sources of nitrogen. Plant and Soil 66: 373–381.

Adams, R.M., C. Rosenzweig, P.M. Peart, J.T. Ritchie, B.A. McCarl, J.D. Glyer, R.B. Curry, J.W. Jones, K.J. Boote, and L.H. Allen. 1990. Global climate change and US agriculture. Nature 345: 219–224.

Adams, W.A., M.A. Raza, and L.J. Evans. 1980. Relationships between net redistribution of Al and Fe and extractable levels in podzolic soils derived from lower Palaeozoic sedimentary rocks. Journal of Soil Science 31: 533–545.

Adamsen, A.P.S. and G.M. King. 1993. Methane consumption in temperate and subarctic forest soils: Rates, vertical zonation, and responses to water and nitrogen. Applied and Environmental Microbiology 59: 485–490.

Admiraal, W. and Y.J.H. Botermans. 1989. Comparison of nitrification rates in three branches of the lower river Rhine. Biogeochemistry 8: 135–151.

Ae, N., J. Arihara, K. Okada, T. Yoshihara, and C. Johansen. 1990. Phosphorus uptake by pigeon pea and its role in cropping systems of the Indian subcontinent. Science 248: 477–480.

Aerts, R. 1996. Nutrient resorption from senescing leaves of perennials: are there general patterns? Journal of Ecology 84: 597–608.

Agee, C.B. 1990. A new look at differentiation of the Earth from melting experiments on the Allende meteorite. Nature 346: 834–837.

Aguilar, R. and R.D. Heil. 1988. Soil organic carbon, nitrogen, and phosphorus quantities in northern Great Plains rangeland. Soil Science Society of America Journal 52: 1076–1081.

Ahl, T. 1988. Background yield of phosphorus from drainage area and atmosphere: An empirical approach. Hydrobiologia 170: 35–44.

Albers, B.P., F. Beese, and A. Hartmann. 1995. Flow-microcalorimetry measurements of aerobic and anaerobic soil microbial activity. Biology and Fertility of Soils 19: 203–208.

Albritton, D.L., R.G. Derwent, I.S.A. Isaksen, M. Lal, and D.J. Wuebbles. 1995. Trace gas radiative forcing indices. pp. 205–231. In J.T. Houghton, L.G. Meira Filho, J. Bruce, H. Lee, B.A. Callander, E. Haites, N. Harris, and K. Maskell (eds.). *Climate Change 1994.* Cambridge University Press, Cambridge.

Alexander, E.B. 1988. Rates of soil formation: Implications for soil-loss tolerance. Soil Science 145: 37–45.

Allan, C.J. and N.T. Roulet. 1994. Solid phase controls of dissolved aluminum within upland Precambrian shield catchments. Biogeochemistry 26: 85–114.

Allan, C.J., N.T. Roulet, and A.R Hill. 1993. The biogeochemistry of pristine headwater Precambrian shield watersheds: An analysis of material transport within a heterogeneous landscape. Biogeochemistry 22: 37–79.

Allard, P., J. Carbonnelle, N. Métrich, H. Loyer, and P. Zettwoog. 1994. Sulphur output and magma degassing budget of Stromboli volcano. Nature 368: 326–330.

Alldredge, A.L. and Y. Cohen. 1987. Can microscale chemical patches persist in the sea? Microelectrode study of marine snow, fecal pellets. Science 235: 689–691.

Allègre, C.J., G. Manhès, and C. Göpel. 1995. The age of the Earth. Geochimica et Cosmochimica Acta 59: 1445–1456.

Allen, H.L. 1972. Phytoplankton photosynthesis, micronutrient interactions, and inorganic carbon availability in a soft-water Vermont lake. pp. 63–80. In G.E. Likens (ed.), *Nutrients and Eutrophication.* American Society of Limnology and Oceanography, Lawrence, Kansas.

Allen, L.H. 1990. Plant responses to rising carbon dioxide and potential interactions with air pollutants. Journal of Environmental Quality 19: 15–34.

Allen, M.F. 1992. *Mycorrhizas: Plant-fungus Relationships.* Chapman and Hall, New York.

Allison, G.B., G.J. Barnes, and M.W. Hughes. 1983. The distribution of deuterium and ^{18}O in dry soils. 2. Experimental. Journal of Hydrology 64: 377–397.

Altabet, M.A. and W.B. Curry. 1989. Testing models of past ocean chemistry using foraminifera $^{15}N/^{14}N$. Global Biogeochemical Cycles 3: 107–119.

Altabet, M.A., R. Francois, D.W. Murray, and W.L. Prell. 1995. Climate-related variations in denitrification in the Arabian Sea from sediment $^{15}N/^{14}N$ ratios. Nature 373: 506–509.

Altschuler, Z.S., M.M. Schnepfe, C.C. Silber, and F.O. Simon. 1983. Sulfur diagenesis in Everglades peat and origin of pyrite in coal. Science 221: 221–227.

Altshuller, A.P. 1991. The production of carbon monoxide in the homogeneous NO_x-induced photooxidation of volatile organic compounds in the troposphere. Journal of Atmospheric Chemistry 13: 155–182.

Alvarez, E., A. Martinez, and R. Calvo. 1992. Geochemical aspects of aluminum in forest soils in Galicia (N.W. Spain). Biogeochemistry 16: 167–180.

Amador, J.A. and R.D. Jones. 1995. Carbon mineralization in pristine and phosphorus-enriched peat soils of the Florida Everglades. Soil Science 159: 129–141.

Ambrose, S.H. and N.E. Sikes. 1991. Soil carbon isotope evidence for Holocene habitat change in the Kenya Rift Valley. Science 253: 1402–1405.

Ambus, P. and S. Christensen. 1994. Measurement of N_2O emission from a fertilized grassland: An analysis of spatial variability. Journal of Geophysical Research 99: 16549–16555.

Ames, R.N., C.P.P. Reid, L.K. Porter, and C. Cambardella. 1983. Hyphal uptake and transport of nitrogen from two ^{15}N-labelled sources by *Glomus mosseae,* a vesicular-arbuscular mycorrhizal fungus. New Phytologist 95: 381–396.

Amthor, J.S. 1984. The role of maintenance respiration in plant growth. Plant, Cell and Environment 7: 561–569.

Amthor, J.S. 1989. *Respiration and Crop Productivity.* Springer-Verlag, New York.

Amthor, J.S. 1995. Terrestrial higher-plant response to increasing atmospheric [CO_2] in relation to the global carbon cycle. Global Change Biology 1:243–274.

Amundson, R.G. and E.A. Davidson. 1990. Carbon dioxide and nitrogenous gases in the soil atmosphere. Journal of Geochemical Exploration 38: 13–41.

Amundson, R.G. and H. Jenny. 1991. The place of humans in the state factor theory of ecosystems and their soils. Soil Science 151: 99–109.

Amundson, R.G., O.A. Chadwick, and J.M. Sowers. 1989. A comparison of soil climate and biological activity along an elevational gradient in the eastern Mojave Desert. Oecologia 80: 395–400.

Anastasi, C. and V.J. Simpson. 1993. Future methane emissions from animals. Journal of Geophysical Research 98: 7181–7186.

Anastasi, C., M. Dowding, and V.J. Simpson. 1992. Future CH_4 emissions from rice production. Journal of Geophysical Research 97: 7521–7525.

Anbar, A.D., Y.L. Yung, and F.P. Chavez. 1996. Methyl bromide: Ocean sources, ocean sinks, and climate sensitivity. Global Biogeochemical Cycles 10: 175–190.

Anbeek, C. 1993. The effect of natural weathering on dissolution rates. Geochimica et Cosmochimica Acta 57: 4963–4975.

Anders, E. 1989. Pre-biotic organic matter from comets and asteroids. Nature 342: 255–257.

Anders, E. and N. Grevesse. 1989. Abundances of the elements: Meteoritic and solar. Geochimica et Cosmochimica Acta 53: 197–214.

Anders, E. and T. Owen. 1977. Mars and Earth: Origin and abundance of volatiles. Science 198: 453–465.

Anderson, D.W. 1977. Early stages of soil formation on glacial till mine spoils in a semi-arid climate. Geoderma 19: 11–19.

Anderson, D.W. and E.A. Paul. 1984. Organo-mineral complexes and their study by radiocarbon dating. Soil Science Society of America Journal 48: 298–301.

Anderson, G.F. 1986. Silica, diatoms and a freshwater productivity maximum in Atlantic coastal plain estuaries, Chesapeake Bay. Estuarine, Coastal and Shelf Science 22: 183–197.

Anderson, I.C., J.S. Levine, M.A. Poth, and P.J. Riggan. 1988. Enhanced biogenic emissions of nitric oxide and nitrous oxide following surface biomass burning. Journal of Geophysical Research 93: 3893–3898.

Anderson, J.M., P. Ineson, and S.A. Huish. 1983. Nitrogen and cation mobilization by soil fauna feeding on leaf litter and soil organic matter from deciduous woodlands. Soil Biology and Biochemistry 15: 463–467.

Anderson, J.P.E. and K.H. Domsch. 1980. Quantities of plant nutrients in the microbial biomass of selected soils. Soil Science 130: 211–216.

Anderson, L.A. and J.L. Sarmiento. 1994. Redfield ratios of remineralization determined by nutrient data analysis. Global Biogeochemical Cycles 8: 65–80.

Anderson, R.F. and S.L. Schiff. 1987. Alkalinity generation and the fate of sulfur in lake sediments. Canadian Journal of Fisheries and Aquatic Sciences 44(Suppl.): 188–193.

Andreae, M.O. 1986. The oceans as a source of biogenic gases. Oceanus 29(4):27–35.

Andreae, M.O. 1990. Ocean-atmosphere interactions in the global biogeochemical sulfur cycle. Marine Chemistry 30: 1–29.

Andreae, M.O. 1991. Biomass burning: Its history, use, and distribution and its impact on environmental quality and global climate. pp. 3–21. In J.S. Levine (ed.), *Global Biomass Burning.* MIT Press, Cambridge, Massachusetts.

Andreae, M.O. and T.W. Andreae. 1988. The cycle of biogenic sulfur compounds over the Amazon basin. I. Dry season. Journal of Geophysical Research 93: 1487–1497.

Andreae, M.O. and W.R. Barnard. 1984. The marine chemistry of dimethylsulfide. Marine Chemistry 14: 267–279.

Andreae, M.O. and R.J. Ferek. 1992. Photochemical production of carbonyl sulfide in seawater and its emission to the atmosphere. Global Biogeochemical Cycles 6: 175–183.

Andreae, M.O. and W.A. Jaeschke. 1992. Exchange of sulphur between biosphere and atmosphere over temperate and tropical regions. pp. 27–61. In R.W. Howarth, J.W.B. Stewart, and M.V. Ivanov (eds.), *Sulphur Cycling on the Continents.* Wiley, New York.

Andreae, M.O., R.J. Charlson, F. Bruynseels, H. Storms, R. Van Grieken, and W. Maenhaut. 1986. Internal mixture of sea salt, silicates, and excess sulfate in marine aerosols. Science 232: 1620–1623.

Andreae, M.O., H. Berresheim, H. Bingemer, D.J. Jacob, B.L. Lewis, S.-M. Li, and R.W. Talbot. 1990. The atmospheric sulfur cycle over the Amazon Basin. 2. Wet season. Journal of Geophysical Research 95: 16813–16824.

Andreae, M.O., B.E. Anderson, D.R. Blake, J.D. Bradshaw, J.E. Collins, G.L. Gregory, G.W. Sachse, and M.C. Shipham. 1994. Influence of plumes from biomass burning on atmospheric chemistry over the equatorial and tropical South Atlantic during CITE 3. Journal of Geophysical Research 99: 12793–12808.

Andreae, M.O., W. Elbert, and S.J. de Mora. 1995. Biogenic sulfur emissions and aerosols over the tropical South Atlantic. 3. Atmospheric dimethylsulfide, aerosols, and cloud condensation nuceli. Journal of Geophysical Research 100: 11335–11356.

Andreae, T.W., G.A. Cutter, N. Hussain, J. Radford-Knoery and M.O. Andreae. 1991. Hydrogen sulfide and radon in and over the western North Atlantic ocean. Journal of Geophysical Research 96: 18753–18760.

Andreae, T.W., M.O. Andreae, and G. Schebeske. 1994. Biogenic sulfur emissions and aerosols over the tropical South Atlantic. I. Dimethylsufide in seawater and in the atmospheric boundary layer. Journal of Geophysical Research 99: 22819–22829.

Andrews, M. 1986. The partitioning of nitrate assimilation between root and shoot of higher plants. Plant, Cell and Environment 9: 511–519.

Antibus, R.K., J.G. Croxdale, O.K. Miller, and A.E. Linkins. 1981. Ecotomycorrhizal fungi of *Salix rotundifolia.* III. Resynthesized mycorrhizal complexes and their surface phosphatase activities. Canadian Journal of Botany 59: 2458–2465.

Antoine, D., J-M. André, and A. Morel. 1996. Oceanic primary production 2. Estimation at global scale from satellite (coastal zone color scanner) chlorophyll. Global Biogeochemical Cycles 10: 57–69.

Antweiler, R.C. and J.I. Drever. 1983. The weathering of a late Tertiary volcanic ash: Importance of organic solutes. Geochimica et Cosmochimica Acta 47: 623–629.

Appenzeller, C. and H.C. Davies. 1992. Structure of stratospheric intrusions into the troposphere. Nature 358: 570–572.

April, R. and D. Keller. 1990. Mineralogy of the rhizosphere in forest soils of the eastern United States: Mineralogic studies of the rhizosphere. Biogeochemistry 9: 1–18.

April, R., R. Newton, and L.T. Coles. 1986. Chemical weathering in two Adirondack watersheds: Past and present-day rates. Geological Society of America Bulletin 97: 1232–1238.

Araujo-lima, C.A.R.M., B.R. Forsberg, R. Victoria, and L. Martinelli. 1986. Energy sources for detritivorous fishes in the Amazon. Science 234: 1256–1258.

Aravena, R., L.I. Wassenaar, and J.F. Barker. 1995. Distribution and isotopic characterization of methane in a confined aquifer in southern Ontario, Canada. Journal of Hydrology 173: 51–70.

Archer, D. 1995. Upper ocean physics as relevant to ecosystem dynamics: A tutorial. Ecological Applications 5: 724–739.

Archer, D. and E. Maier-Reimer. 1994. Effect of deep-sea sedimentary calcite preservation on atmospheric CO_2 concentration. Nature 367: 260–263.

Arianoutsou, M. and N.S. Margaris. 1981. Fire-induced nutrient losses in a Phryganic (East Mediterranean) ecosystem. International Journal of Biometeorology 25: 341–347.

Arkley, R.J. 1963. Calculation of carbonate and water movement in soil from climatic data. Soil Science 92: 239–248.

Arkley, R.J. 1967. Climates of some great soil groups of the western United States. Soil Science 103: 389–400.

Armentano, T.V. and E.S. Menges. 1986. Patterns of change in the carbon balance of organic soil-wetlands of the temperate zone. Journal of Ecology 74: 755–774.

Arnold, F. and T. Bührke. 1983. New H_2SO_4 and HSO_3 vapour measurements in the stratosphere—evidence for a volcanic influence. Nature 301: 293–295.

Arrhenius, S. 1918. *The Destinies of the Stars*. G.P. Putnam's Sons, New York.

Art, H.W., F.H. Bormann, G.K. Voigt, and G.M. Woodwell. 1974. Barrier island forest ecosystem: The role of meteorologic inputs. Science 184: 60–62.

Arthur, M.A., W.E. Dean, and L.M. Pratt. 1988. Geochemical and climatic effects of increased marine organic carbon burial at the Cenomanian/Turonian boundary. Nature 335: 714–717.

Aselmann, T. and P.J. Crutzen. 1989. Global distribution of natural freshwater wetlands and rice paddies, their net primary productivity, seasonality and possible methane emissions. Journal of Atmospheric Chemistry 8: 307–358.

Asper, V.L., W.G. Deuser, G.A. Knauer, and S.E. Lohrenz. 1992. Rapid coupling of sinking particle fluxes between surface and deep ocean waters. Nature 357: 670–672.

Atkin, O.K. 1996. Reassessing the nitrogen relations of arctic plants: A mini-review. Plant, Cell, and Environment 19: 695–704.

Auclair, A.N.D. and T.B. Carter. 1993. Forest wildfires as a recent source of CO_2 at northern latitudes. Canadian Journal of Forest Research 23: 1528–1536.

Aumann, G.D. 1965. Microtine abundance and soil sodium levels. Journal of Mammalogy 46: 594–604.

Autry, A. and J.W. Fitzgerald. 1993. Saturation potentials for sulfate adsorption by field-moist forest soils. Soil Biology and Biochemistry 25: 833–838.

Avnimelech, Y. and J.R. McHenry. 1984. Enrichment of transported sediments with organic carbon, nutrients, and clay. Soil Science Society of America Journal 48: 259–266.

Awasthi, O.P., E. Sharma, and L.M.S. Palni. 1995. Stemflow: A source of nutrients in some naturally growing epiphytic orchids of the Sikkim Himalaya. Annals of Botany 75: 5–11.

Awramik, S.M., J.W. Schopf, and M.R. Walter. 1983. Filamentous fossil bacteria from the Archean of western Australia. Precambrian Research 20: 357–374.

Axelsson, B. 1981. Site differences in yield differences in biological production or in redistribution of carbon within trees. Department of Ecological and Environmental Research, Research Report 9, Swedish University of Agricultural Science, Uppsala.

Ayers, G.P. and J.L. Gras. 1991. Seasonal relationship between cloud condensation nuceli and aerosol methanesulphonate in marine air. Nature 353: 834–835.

Ayers, G.P., S.A. Penkett, R.W. Gillett, B. Bandy, I.E. Galbally, C.P. Meyer, C.M. Elsworth, S.T. Bentley, and B.W. Forgan. 1992. Evidence for photochemical control of ozone concentrations in unpolluted marine air. Nature 360: 446–449.

Azevedo, J. and D.L. Morgan. 1974. Fog precipitation in coastal California forests. Ecology 55: 1135–1141.

Baas Becking, L.G.M., I.R. Kaplan, and D. Moore. 1960. Limits of the natural environment in terms of pH and oxidation-reduction potentials. Journal of Geology 68: 243–384.

Bacastow, R.B. 1976. Modulation of atmospheric carbon dioxide by the southern oscillation. Nature 261: 116–118.

Bacastow, R.B. and A. Björkström. 1981. Comparison of ocean models for the carbon cycle. pp. 29–79. In B. Bolin (ed.), *Carbon Cycle Modelling*. Wiley, New York.

Bacastow, R.B., C.D. Keeling, and T.P. Whorf. 1985. Seasonal amplitude increase in atmospheric CO_2 concentration at Mauna Loa, Hawaii, 1959–1982. Journal of Geophysical Research 90: 10529–10540.

Bache, B.W. 1984. The role of calcium in buffering soils. Plant, Cell and Environment 7: 391–395.

Bache, C.A., W.H. Gutenmann, and D.J. Lisk. 1971. Residues of total mercury and methylmercuric salts in lake trout as a function of age. Science 172: 951–952.

Bachmann, P.A., P.L. Luisi, and J. Lang. 1992. Autocatalytic self-replicating micelles as models for prebiotic structures. Nature 357: 57–59.

Baethgen, W.E. and M.M. Alley. 1987. Nonexchangeable ammonium nitrogen contribution to plant available nitrogen. Soil Science Society of America Journal 51: 110–115.

Bailey, S.W., C.T. Driscoll, and J.W. Hornbeck. 1995. Acid-base chemistry and aluminum transport in an acidic watershed and pond in New Hampshire. Biogeochemistry 28: 69–91.

Baines, S.B. and M.L. Pace. 1991. The production of dissolved organic matter by phytoplankton and its importance to bacteria: Patterns across marine and freshwater systems. Limnology and Oceanography 36: 1078–1090.

Baines, S.B. and M.L. Pace. 1994. Relationships between suspended and particulate matter and sinking flux along a trophic gradient and implications for the fate of planktonic primary production. Canadian Journal of Fisheries and Aquatic Sciences 51: 25–36.

Baker, L.A., C.D. Pollman, and J.M. Eilers. 1988. Alkalinity regulation of softwater Florida lakes. Water Resources Research 24: 1069–1082.

Baker, L.A., A.T. Herlihy, P.R. Kaufmann, and J.M. Eilers. 1991. Acidic lakes and streams in the United States: The role of acidic deposition. Science 252: 1151–1154.

Baker, P.A. and S.J. Burns. 1985. Occurrence and formation of dolomite in organic-rich continental margin sediments. American Association of Petroleum Geologists Bulletin 69: 1917–1930.

Baker, P.A. and M. Kastner. 1981. Constraints on the formation of sedimentary dolomite. Science 213: 214–216.

Baker, V.R. 1977. Stream-channel response to floods, with examples from central Texas. Geological Society of America Bulletin 88: 1057–1071.

Baker, V.R., R.G. Strom, V.C. Gulick, J.S. Kargel, G. Komatsu, and V.S. Kale. 1991. Ancient oceans, ice sheets and the hydrological cycle on Mars. Nature 352: 589–594.

Bakun, A. 1990. Global climate change and intensification of coastal ocean upwelling. Science 247: 198–201.

Balch, W., R. Evans, J. Brown, G. Feldman, C. McClain, and W. Esaias. 1992. The remote sensing of ocean primary productivity: Use of a new data compilation to test satellite algorithms. Journal of Geophysical Research 97: 2279–2294.

Baldocchi, D.D., S.B. Verma, and D.E. Anderson. 1987. Canopy photosynthesis and water-use efficiency in a deciduous forest. Journal of Applied Ecology 24: 251–260.

Baldock, J.A., J.M. Oades, A.G. Waters, X. Peng, A.M. Vassallo, and M.A. Wilson. 1992. Aspects of the chemical structure of soil organic materials as revealed by solid-state ^{13}C NMR spectroscopy. Biogeochemistry 16: 1–42.

Balling R.C. 1989. The impact of summer rainfall on the temperature gradient along the United States-Mexico border. Journal of Applied Meteorology 28: 304–308.

Bange, H.W., S. Rapsomanikis, and M.O. Andreae. 1996. Nitrous oxide in coastal waters. Global Biogeochemical Cycles 10: 197–207.

Banin, A. and J. Navrot. 1975. Origin of life: Clues from relations between chemical compositions of living organisms and natural environments. Science 189: 550–551.

Barber, S.A. 1962. A diffusion and mass-flow concept of soil nutrient availability. Soil Science 93: 39–49.

Barnes, I., K.H. Becker, and I. Patroescu. 1994. The tropospheric oxidation of dimethyl sulfide: A new source of carbonyl sulfide. Geophysical Research Letters 21: 2389–2392.

Barnola, J.M., M. Anklin, J. Porcheron, D. Raynaud, J. Schwander, and B. Stauffer. 1995. CO_2 evolution during the last millennium as recorded by Antarctic and Greenland ice. Tellus 47B: 264–272.

Barrett, L.R. and R.J. Schaetzl. 1992. An examination of podzolization near Lake Michigan using chronofunctions. Canadian Journal of Soil Science 72: 527–541.

Barrie, L.A. and J.M. Hales. 1984. The spatial distributions of precipitation acidity and major ion wet deposition in North America during 1980. Tellus 36B: 333–355.

Barry, D.A.J., D. Goorahoo, and M.J. Goss. 1993. Estimation of nitrate concentrations in groundwater using a whole farm nitrogen budget. Journal of Environmental Quality 22: 767–775.

Barsdate, R.J. and V. Alexander. 1975. The nitrogen balance of Arctic tundra: Pathways, rates, and environmental implications. Journal of Environmental Quality 4: 111–117.

Bartel-Ortiz, L.M. and M.B. David. 1988. Sulfur constituents and transformations in upland and floodplain forest soils. Canadian Journal of Forest Research 18: 1106–1112.

Bartlett, D.S., K.B. Bartlett, J.M. Hartman, R.C. Harriss, D.I. Sebacher, R. Pelletier-Travis, D.D. Dow, and D.P. Brannon. 1989. Methane emissions from the Florida Everglades: Patterns of variability in a regional wetland ecosystem. Global Biogeochemical Cycles 3: 363–374.

Bartlett, K.B. and R.C. Harriss. 1993. Review and assessment of methane emissions from wetlands. Chemosphere 26: 261–320.

Bartlett, K.B., D.S. Bartlett, R.C. Harriss, and D.I. Sebacher. 1987. Methane emissions along a salt marsh salinity gradient. Biogeochemistry 4: 183–202.

Bartlett, R.J. 1986. Soil redox behavior. pp. 179–207. In D.L. Sparks (ed.), *Soil Physical Chemistry*. CRC Press, Boca Raton, Florida.

Bates, T.R. and J.P. Lynch. 1996. Stimulation of root hair elongation in *Arabidopsis thaliana* by low phosphorus availability. Plant, Cell, and Environment 19: 529–538.

Bates, T.S., R.J. Charlson, and R.H. Gammon. 1987. Evidence for the climatic role of marine biogenic sulphur. Nature 329: 319–321.

Bates, T.S., B.K. Lamb, A. Guenther, J. Dignon, and R.E. Stoiber. 1992. Sulfur emissions to the atmosphere from natural sources. Journal of Atmospheric Chemistry 14: 315–337.

Bates, T.S., K.C. Kelly, J.E. Johnson, and R.H. Gammon. 1995. Regional and seasonal variations in the flux of oceanic carbon monoxide to the atmosphere. Journal of Geophysical Research 100: 23093–23101.

Batjes, N.H. 1996. Total carbon and nitrogen in the soils of the world. European Journal of Soil Science 47: 151–163.

Baumann, R.H., J.W. Day, and C.A. Miller. 1984. Mississippi deltaic wetland survival: Sedimentation versus coastal submergence. Science 224: 1093–1095.

Baumgärtner, M. and R. Conrad. 1992. Effects of soil variables and season on the production and consumption of nitric oxide in oxic soils. Biology and Fertility of Soils 14: 166–174.

Bazely, D.R. and R.L. Jefferies. 1989. Lesser snow geese and the nitrogen economy of a grazed salt marsh. Journal of Ecology 77: 24–34.

Bazzaz, F.A. 1990. The response of natural ecosystems to the rising global CO_2 levels. Annual Review of Ecology and Systematics 21: 167–196.

Beadle, N.C.W. 1966. Soil phosphate and its role in molding segments of the Australian flora and vegetation, with special reference to xeromorphy and sclerophylly. Ecology 47: 992–1007.

Beck, K.C., J.H. Reuter, and E.M. Perdue. 1974. Organic and inorganic geochemistry of some coastal plain rivers of the southeastern United States. Geochimica et Cosmochimica Acta 38: 341–364.

Beckwith, S.V.W. and A.I. Sargent. 1996. Circumstellar disks and the search for neighbouring planetary systems. Nature 383: 139–144.

Bédard, C. and R. Knowles. 1989. Physiology, biochemistry, and specific inhibitors of CH_4, NH_4^+, and CO oxidation by methanotrophs and nitrifiers. Microbiological Reviews 53: 68–84.

Bédard, C. and R. Knowles. 1991. Hypolimnetic O_2 consumption, denitrification, and methanogeneis in a thermally stratified lake. Canadian Journal of Fisheries and Aquatic Sciences 48: 1048–1054.

Beerling, D.J. and W.G. Chaloner. 1993. Stomatal density responses of Egyptian *Olea europaea* L. leaves to CO_2 change since 1327 B.C. Annals of Botany 71: 431–435.

Beilke, S. and D. Lamb. 1974. On the absorption of SO_2 in ocean water. Tellus 26: 268–271.

Beke, G.J. 1990. Soil development in a 100-year old dike near Grant Pré, Nova Scotia. Canadian Journal of Soil Science 70: 683–692.

Bekki, S., K.S. Law, and J.A. Pyle. 1994. Effect of ozone depletion on atmospheric CH_4 and CO concentrations. Nature 371: 595–597.

Belillas, C.M. and F. Rodà. 1991. Nutrient budgets in a dry heathland watershed in northeastern Spain. Biogeochemisty 13: 137–157.

Bell, D.R. and G.R. Rossman. 1992. Water in Earth's mantle: The role of nominally anhydrous minerals. Science 255: 1391–1397.

Bell, R.A. 1993. Cryptoendolithic algae of hot semiarid lands and deserts. Journal of Phycology 29: 133–139.

Bellini, G., M.E. Summer, D.E. Radcliffe, and N.P. Qafoku. 1996. Anion transport through columns of highly weathered acid soil: Adsorption and retardation. Soil Science Society of America Journal 60: 132–137.

Belser, L.W. and E.L. Mays. 1980. Specific inhibition of nitrite oxidation by chlorate and its use in assessing nitrification in soils and sediments. Applied and Environmental Microbiology 39: 505–510.

Belyea, L.R. and B.G. Warner. 1996. Temporal scale and the accumulation of peat in a *Sphagnum bog.* Canadian Journal of Botany 74: 366–377.

Bender, M., R. Jahnke, R. Weiss, W. Martin, D.T. Heggie, J. Orchardo, and T. Sowers. 1989. Organic carbon oxidation and benthic nitrogen and silica dynamics in San Clemente Basin, a continental borderland site. Geochimica et Cosmochimica Acta 53: 685–697.

Bender, M., T. Sowers, and L. Labeyrie. 1994. The Dole effect and its variations during the last 130,000 years as measured in the Vostok ice core. Global Biogeochemical Cycles 8: 363–376.

Bender, M.L., G.P. Klinkhammer, and D.W. Spencer. 1977. Manganese in seawater and the marine manganese balance. Deep Sea Research 24: 799–812.

Benke, A.C. and J.L. Meyer. 1988. Structure and function of a blackwater river in the southeastern U.S.A. Verhandelingen–Internationale Vereinigung fuer Theoretische und Angewandte Limnologie 23: 1209–1218.

Benner, R., M.A. Moran, and R.E. Hodson. 1986. Biogeochemical cycling of lignocellulosic carbon in marine and freshwater ecosystems: Relative contributions of procaryotes and eucaryotes. Limnology and Oceanography 31: 89–100.

Benner, R., J.D. Pakulski, M. McCarthy, J.I. Hedges, and P.G. Hatcher. 1992. Bulk chemical characteristics of dissolved organic matter in the ocean. Science 255: 1561–1564.

Benner, R., S. Opsahl, G. Chin-Leo, J.E. Richey, and B.R. Forsberg. 1995. Bacterial carbon metabolism in the Amazon River system. Limnology and Oceanography 40: 1262–1270.

Bennett, P.C., M.E. Melcer, D.I. Siegel, and J.P. Hassett. 1988. The dissolution of quartz in dilute aqueous solutions of organic acids at 25°C. Geochimica et Cosmochimica Acta 52: 1521–1530.

Ben-Shahar, R. and M.J. Coe. 1992. The relationships between soil factors, grass nutrients and the foraging behaviour of wildebeest and zebra. Oecologia 90: 422–428.

Benzing, D.H. and A. Renfrow. 1974. The nutritional status of *Encyclia tampense* and *Tillandsia circinata* on *Taxodium ascendens* and the availability of nutrients to epiphytes on this host in South Floria. Bulletin of the Torrey Botanical Club 101: 191–197.

Berelson, W.M., D.E. Hammond, and G.A. Cutter. 1990. *In situ* measurements of calcium carbonate dissolution rates in deep-sea sediments. Geochimica et Cosmochimica Acta 54: 3013–3020.

Berg, B. 1988. Dynamics of nitrogen (^{15}N) in decomposing Scots pine (*Pinus sylvestris*) needle litter: Long-term decomposition in a Scots pine forest. VI. Canadian Journal of Botany 66: 1539–1546.

Berg, B., M.P. Berg, P. Bottner, E. Box, A. Breymeyer, R. Calvo de Anta, M. Couteaux, A. Escudero, A. Gallardo, W. Kratz, M. Madeira, E. Mälkönen, C. McClaugherty, V. Meentemeyer, F. Muñoz, P. Piussi, J. Remacle, and A. Virzo de Santo. 1993. Litter mass loss rates in pine forests of Europe and eastern United States: Some relationships with climate and litter quality. Biogeochemistry 20: 127–159.

Berg, P., K. Klemedtsson, and T. Rosswall. 1982. Inhibitory effect of low partial pressures of acetylene on nitrification. Soil Biology and Biochemistry 14: 301–303.

Berger, A.L. 1978. Long-term variations of caloric insolation resulting from the Earth's orbital elements. Quaternary Research 9: 139–167.

Berger, T.W. and G. Glatzel. 1994. Deposition of atmospheric constituents and its impact on nutrient budgets of oak forests (*Quercus petraea* and *Quercus robur*) in lower Austria. Forest Ecology and Management 70: 183–193.

Berger, W.H. 1989. Global maps of ocean productivity. pp. 429–455. In W.H. Berger, V.S. Smetacek, and G. Wefer (eds.), *Productivity of the Oceans: Present and Past.* Wiley, New York.

Berges, M.G.M., R.M. Hofmann, D. Scharffe, and P.J. Crutzen. 1993. Nitrous oxide emissions from motor vehicles in tunnels and their global extrapolation. Journal of Geophysical Research 98: 18527–18531.

Berggren, D. and J. Mulder. 1995. The role of organic matter in controlling aluminum solubility in acidic mineral soil horizons. Geochimica et Cosmochimica Acta 59: 4167–4180.

Berkner, L.V. and L.C. Marshall. 1965. On the origin and rise of oxygen concentration in the Earth's atmosphere. Journal of the Atmospheric Sciences 22: 225–261.

Berliner, R., B. Jacoby, and E. Zamski. 1986. Absence of *Cistus incanus* from basaltic soils in Israel: Effect of mycorrhizae. Ecology 67: 1283–1288.

Berner, E.K. and R.A. Berner. 1987. *The Global Water Cycle.* Prentice Hall, Englewood Cliffs, New Jersey.

Berner, R.A. 1971. Worldwide sulfur pollution of rivers. Journal of Geophysical Research 76: 6597–6600.

Berner, R.A. 1982. Burial of organic carbon and pyrite sulfur in the modern ocean: Its geochemical and environmental significance. American Journal of Science 282: 451–473.

Berner, R.A. 1984. Sedimentary pyrite formation: An update. Geochimica et Cosmochimica Acta 48: 605–615.

Berner, R.A. 1992. Weathering, plants, and the long-term carbon cycle. Geochimica et Cosmochimica Acta 56: 3225–3231.

Berner, R.A. 1994. GEOCARB II: A revised model of atmospheric CO_2 over Phanerozoic time. American Journal of Science 294: 56–91.

Berner, R.A. and D.E. Canfield. 1989. A new model for atmospheric oxygen over Phanerozoic time. American Journal of Science 289: 333–361.

Berner, R.A. and S. Honjo. 1981. Pelagic sedimentation of aragonite: Its geochemical significance. Science 211: 940–942.

Berner, R.A. and A.C. Lasaga. 1989. Modeling the geochemical carbon cycle. Scientific American 260(3): 74–81.

Berner, R.A. and R. Raiswell. 1983. Burial of organic carbon and pyrite sulfur in sediments over Phanerozoic time: A new theory. Geochimica et Cosmochimica Acta 47: 855–862.

Berner, R.A. and J.-L. Rao. 1994. Phosphorus in sediments of the Amazon River and estuary: Implications for the global flux of phosphorus to the sea. Geochimica et Cosmochimica Acta 58: 2333–2339.

Bernier, B. and M. Brazeau. 1988a. Foliar nutrient status in relation to sugar maple dieback and decline in the Quebec Appalachians. Canadian Journal of Forest Research 18: 754–761.

Bernier, B. and M. Brazeau. 1988b. Magnesium deficiency symptoms associated with sugar maple dieback in a Lower Laurentians site in southeastern Quebec. Canadian Journal of Forest Research 18: 1265–1269.

Bernstein, R.E., P.R. Betzer, R.A. Feeley, R.H. Byrne, M.F. Lamb, and A.F. Michaels. 1987. Acantharian fluxes and strontium to chlorinity ratios in the North Pacific Ocean. Science 237: 1490–1494.

Berresheim, H. and W. Jaeschke. 1983. The contribution of volcanoes to the global atmospheric sulfur budget. Journal of Geophysical Research 88: 3732–3740.

Berresheim, H. and V.D. Vulcan. 1992. Vertical distributions of COS, CS_2, DMS and other volatile sulfur compounds in a loblolly pine forest. Atmospheric Environment 26A: 2031–2036.

Bertine, K.K. and E.D. Goldberg. 1971. Fossil fuel combustion and the major sedimentary cycle. Science 173: 233–235.

Bertrand, P. and E. Lallier-Vergès. 1993. Past sedimentary organic matter accumulation and degradation controlled by productivity. Nature 364: 786–788.

Betlach, M.R. 1982. Evolution of bacterial denitrification and denitrifier diversity. Antonie van Leeuwenhoek Journal of Microbiology 48: 585–607.

Betzer, P.R., R.H. Byrne, J.G. Acker, C.S. Lewis, R.R. Jolly, and R.A. Feely. 1984. The oceanic carbonate system: A reassessment of biogenic controls. Science 226: 1074–1077.

Beven, K. and P. Germann. 1982. Macropores and water flow in soils. Water Resources Research 18: 1311–1325.

Beyer, L., H.-R. Schulten, R. Fruend, and U. Irmler. 1993. Formation and properties of organic matter in a forest soil, as revealed by its biological activity, wet chemical analysis, CPMAS ^{13}C-NMR spectroscopy and pyrolysis-field ionization mass spectrometry. Soil Biology and Biochemistry 25: 587–596.

Bickle, M.J. 1994. The role of metamorphic decarbonation reactions in returning strontium to the silicate sediment mass. Nature 367: 699–704.

Biddanda, B., S. Opsahl, and R. Benner. 1994. Plankton respiration and carbon flux through bacterioplankton on the Louisiana shelf. Limnology and Oceanography 39: 1259–1275.

Billen, G. 1975. Nitrification in the Scheldt estuary (Belgium and the Netherlands). Estuarine and Coastal Marine Science 3: 79–89.

Billings, W.D. 1950. Vegetation and plant growth as affected by chemically altered rocks in the western Great Basin. Ecology 31: 62–74.

Billings, W.D. 1987. Carbon balance of Alaskan tundra and taiga ecosystems: past, present and future. Quaternary Science Reviews 6: 165–177.

Billings, W.D., J.O. Luken, D.A. Mortensen, and K.M. Peterson. 1982. Arctic tundra: A source or sink for atmospheric carbon dioxide in a changing environment? Oecologia 53: 7–11.

Billings, W.D., K.M. Peterson, J.O. Luken, and D.A. Mortensen. 1984. Interaction of increasing atmospheric carbon dioxide and soil nitrogen on the carbon balance of tundra microcosms. Oecologia 65: 26–29.

Binkley, D. 1986. *Forest Nutrition Management.* Wiley, New York.

Binkley, D. and S.C. Hart. 1989. The components of nitrogen availability assessments in forest soils. Advances in Soil Science 10: 57–112.

Binkley, D. and D.D. Richter. 1987. Nutrient cycles and H^+ budgets of forest ecosystems. Advances in Ecological Research 16: 1–51.

Bird, M.I., J. Lloyd, and G.D. Farquhar. 1994. Terrestrial carbon storage at the LGM. Nature 371: 566.

Birdsey, R.A., A.J. Plantinga, and L.S. Heath. 1993. Past and prospective carbon storage in United States forests. Forest Ecology and Management 58: 33–40.

Birk, E.M. and P.M. Vitousek. 1986. Nitrogen availability and nitrogen use efficiency in loblolly pine stands. Ecology 67: 69–79.

Birkeland, P.W. 1984. *Soils and Geomorphology*. Oxford University Press, Oxford.

Bishop, J.K.B. 1988. The barite-opal-organic carbon association in oceanic particulate matter. Nature 332: 341–343.

Blackmer, A.M. and J.M. Bremner. 1976. Potential of soil as a sink for atmospheric nitrous oxide. Geophysical Research Letters 3: 739–742.

Blair, N.E. and R.C. Aller. 1995. Anaerobic methane oxidation on the Amazon shelf. Geochimica et Cosmochimica Acta 59: 3707–3715.

Blaise, T. and J. Garbaye. 1983. Effets de la fertilisation minérale sur les ectomycorhizes d'une hêtraie. Acta Oecologica Oceologia [Series]: Plantarum 4: 165–169.

Blatt, H. and R.L. Jones. 1975. Proportions of exposed igneous, metamorphic, and sedimentary rocks. Geological Society of America Bulletin 86: 1085–1088.

Blaustein, A.R., P.D. Hoffman, D.G. Hokit, J.M. Kiesecker, S.C. Walls, and J.B. Hays. 1994. UV repair and resistance to solar uV-B in amphibian eggs: A link to population declines? Proceedings of the National Academy of Sciences, U.S.A. 91: 1791–1795.

Bloesch, J., P. Stadelmann, and H. Bührer. 1977. Primary production, mineralization, and sedimentation in the euphotic zone of two Swiss lakes. Limnology and Oceanography 22: 511–526.

Bloom, A.J., S.S. Sukrapanna, and R.L. Warner. 1992. Root respiration associated with ammonium and nitrate absorption and assimilation by barley. Plant Physiology 92: 1294–1301.

Bloom, A.J., L.E. Jackson, and D.R. Smart. 1993. Root growth as a function of ammonium and nitrate in the root zone. Plant, Cell and Environment 16: 199–206.

Bloomfield, C. 1972. The oxidation of iron sulphides in soils in relation to the formation of acid sulphate soils, and of ochre deposits in field drains. Journal of Soil Science 23: 1–16.

Blunier, T., J. Chappellaz, J. Schwander, B. Stauffer, and D. Raynaud. 1995. Variations in atmospheric methane concentration during the Holocene epoch. Nature 374: 46–49.

Bluth, G.J.S. and L.R. Kump. 1991. Phanerozoic paleogeology. American Journal of Science 291: 284–308.

Bluth, G.J.S. and L.R. Kump. 1994. Lithologic and climatologic controls on river chemistry. Geochimica et Cosmochimica Acta 58: 2341–2359.

Bluth, G.J.S., C.C. Schnetzler, A.J. Krueger, and L.S. Walter. 1993. The contribution of explosive volcanism to global atmospheric sulphur dioxide concentrations. Nature 366: 327–329.

Bockheim, J.G. 1980. Solution and use of chronofunctions in studying soil development. Geoderma 24: 71–85.

Boerner, R.E.J. 1984. Foliar nutrient dynamics and nutrient use efficiency of four deciduous tree species in relation to site fertility. Journal of Applied Ecology 21: 1029–1040.

Bohn, H.L., B.L. McNeal, and G.A. O'Connor. 1985. *Soil Chemistry*. 2nd ed. Wiley, New York.

Bojkov, R.D. and V.E. Fioletov. 1995. Estimating the global ozone characteristics during the last 30 years. Journal of Geophysical Research 100: 16537–16551.

Bolan, N.S. 1991. A critical review on the role of mycorrhizal fungi in the uptake of phosphorus by plants. Plant and Soil 134: 189–207.

Bolan, N.S., A.D. Robson, N.J. Barrow, and L.A.G. Aylmore. 1984. Specific activity of phosphorus in mycorrhizal and non-mycorrhizal plants in relation to the availability of phosphorus to plants. Soil Biology and Biochemistry 16: 299–304.

Bolte, M. and C.J. Hogan. 1995. Conflict over the age of the Universe. Nature 376: 399–402.

Bonan, G.B. 1993. Physiological derivation of the observed relationship between net primary production and mean annual air temperature. Tellus 45B: 397–408.

Bondietti, E.A., C.F. Baes, and S.B. McLaughlin. 1989. Radial trends in cation ratios in tree rings as indicators of the impact of atmospheric deposition on forests. Canadian Journal of Forest Research 19: 586–594.

Bonin, P., M. Gilewicz, and J.C. Bertrand. 1989. Effects of oxygen on each step of denitrification on *Pseudomonas nautica*. Canadian Journal of Microbiology 35: 1061–1064.

Boring, L.R., C.D. Monk, and W.T. Swank. 1981. Early regeneration of a clear-cut southern Appalachian forest. Ecology 62: 1244–1253.

Boring, L.R., W.T. Swank, J.B. Waide, and G.S. Henderson. 1988. Sources, fates, and impacts of nitrogen inputs to terrestrial ecosystems: Review and synthesis. Biogeochemistry 6: 119–159.

Bormann, B.T. and J.C. Gordon. 1984. Stand density effects in young red alder plantations: Productivity, photosynthate partitioning, and nitrogen fixation. Ecology 65: 394–402.

Bormann, F.H. and G.E. Likens. 1979. *Pattern and Process in a Forested Ecosystem.* Springer-Verlag, New York.

Bormann, F.H., G.E. Likens, T.G. Siccama, R.S. Pierce, and J.S. Eaton. 1974. The export of nutrients and recovery of stable conditions following deforestation at Hubbard Brook. Ecological Monographs 44: 255–277.

Born, M., H. Dörr, and I. Levin. 1990. Methane consumption in aerated soils of the temperate zone. Tellus 42B: 2–8.

Borucki, W.J. and W.L. Chameides. 1984. Lightning: Estimates of the rates of energy dissipation and nitrogen fixation. Reviews of Geophysics and Space Physics 22: 363–372.

Boss, A.P. 1988. High temperatures in the early solar nebula. Science 241: 565–567.

Boström, B., M. Jansson, and C. Forsberg. 1982. Phosphorus release from lake sediments. Archiv für Hydrobiologie Beiheft Ergebnisse der Limnologie 18: 5–59.

Boström, B., J.M. Andersen, S. Fleischer, and M. Jansson. 1988. Exchange of phosphorus across the sediment-water interface. Hydrobiologia 170: 229–244.

Botch, M.S., K.I. Kobak, T.S. Vinson, and T.P. Kolchugina. 1995. Carbon pools and accumulation in peatlans of the former Soviet Union. Global Biogeochemical Cycles 9: 37–46.

Botkin, D.B. and C.R. Malone. 1968. Efficiency of net primary production based on light intercepted during the growing season. Ecology 49: 438–444.

Botkin, D.B. and L.G. Simpson. 1990. Biomass of North American boreal forest: A step toward accurate global measures. Biogeochemistry 9: 161–174.

Botkin, D.B., L.G. Simpson, and R.A. Nisbet. 1993. Biomass and carbon storage of the North American deciduous forest. Biogeochemistry 20: 1–17.

Botkin, D.B., P.A. Jordan, A.S. Dominski, H.S. Lowendorf, and G.E. Hutchinson. 1973. Sodium dynamics in a northern ecosystem. Proceedings of the National Academy of Sciences, U.S.A. 70: 2745–2748.

Boudreau, B.P. and J.T. Westrich. 1984. The dependence of bacterial sulfate reduction on sulfate concentration in marine sediments. Geochimica et Cosmochimica Acta 48: 2503–2516.

Boutron, C.F., U. Görlach, J.-P. Candelone, M.A. Bolshov, and D.J. Delmas. 1991. Decrease in anthropogenic lead, cadmium and zinc in Greenland snows since the late 1960s. Nature 353: 153–156.

Boutron, C.F., J-P. Candelone, and S. Hong. 1994. Past and recent changes in the large-scale tropospheric cycles of lead and other heavy metals as documented in Antarctic and Greenland snow and ice: A review. Geochimica et Cosmochimica Acta 58: 3217–3225.

Bouwman, A.F., I. Fung, E. Matthews, and J. John. 1993. Global analysis of the potential for N_2O production in natural soils. Global Biogeochemical Cycles 7: 557–597.

Bouwman, A.F., K.W. Van der Hoek, and J.G.J. Olivier. 1995. Uncertainties in the global source distribution of nitrous oxide. Journal of Geophysical Research 100: 2785–2800.

Bowden, R.D., K.J. Nadelhoffer, R.D. Boone, J.M. Melillo, and J.B. Garrison. 1993. Contributions of aboveground litter, belowground litter, and root respiration to total soil respiration in a temperate mixed hardwood forest. Canadian Journal of Forest Research 23: 1402–1407.

Bowden, W.B. 1986. Gaseous nitrogen emissions from undisturbed terrestrial ecosystems: An assessment of their impacts on local and global nitrogen budgets. Biogeochemistry 2: 249–279.

Bowden, W.B. 1987. The biogeochemistry of nitrogen in freshwater wetlands. Biogeochemistry 4: 313–348.

Bowden, W.B., C.J. Vörösmarty, J.T. Morris, B.J. Peterson, J.E. Hobbie, P.A. Steudler, and B. Moore. 1991. Transport and processing of nitrogen in a tidal freshwater wetland. Water Resources Research 27: 389–408.

Bowen, G.D. and S.E. Smith. 1981. The effects of mycorrhizas on nitrogen uptake by plants. pp. 237–247. In F.E. Clark and T. Rosswall (eds.), *Terrestrial Nitrogen Cycles.* Swedish Natural Science Research Council, Stockholm.

Bowen, H.J.M. 1966. *Trace Elements in Biochemistry.* Academic Press, New York.

Bowman, C.T. 1991. Chemistry of gaseous pollutant formation and destruction. pp. 215–260. In W. Bartok and A.F. Sarofim (eds.), *Fossil Fuel Combustion: A Source Book.* Wiley, New York.

Bowring, S.A. and T. Housh. 1995. The Earth's early evolution. Science 269: 1535–1540.

Box, E.O. 1988. Estimating the seasonal carbon source-sink geography of a natural, steady-state terrestrial biosphere. Journal of Applied Meteorology 27: 1109–1124.

Box, E.O., B.N. Holben, and V. Kalb. 1989. Accuracy of the AVHRR vegetation index as a predictor of biomass, primary productivity, and net CO_2 flux. Vegetatio 80: 71–89.

Boyle, E.A., R. Collier, A.T. Dengler, J.M. Edmond, A.C. Ng, and R.F. Stallard. 1974. On the chemical mass-balance in estuaries. Geochimica et Cosmochimica Acta 38: 1719–1728.

Boyle, E.A., F. Sclater, and J.M. Edmond. 1976. On the marine geochemistry of cadmium. Nature 263: 42–44.

Boyle, E.A., J.M. Edmond, and E.R. Sholkovitz. 1977. The mechanism of iron removal in estuaries. Geochimica et Cosmochimica Acta 41: 1313–1324.

Boyle, J.R. and G.K. Voigt. 1973. Biological weathering of silicate minerals. Implications for tree nutrition and soil genesis. Plant and Soil 38: 191–201.

Boynton, W.R., W.M. Kump, and C.W. Keefe. 1982. A comparative analysis of nutrients and other factors influencing estuarine phytoplankton production. pp. 69–90. In V.S. Kennedy (ed.), *Estuarine Comparisons*. Academic Press, New York.

Bradley, R.S., H.F. Diaz, J.K. Eischeid, P.D. Jones, P.M. Kelly, and C.M. Goodess. 1987. Precipitation fluctuations over northern hemisphere land areas since the mid-19th century. Science 237: 171–175.

Brady, P.V. 1991. The effect of silicate weathering on global temperature and atmospheric CO_2. Journal of Geophysical Research 96: 18101–18106.

Brady, P.V. and S.A. Carroll. 1994. Direct effects of CO_2 and temperature on silicate weathering: Possible implications for climate control. Geochimica et Cosmochimica Acta 58: 1853–1856.

Bramley, R.G.V. and R.E. White. 1990. The variability of nitrifying activity in field soils. Plant and Soil 126: 203–208.

Brand, L.E., W.G. Sunda, and R.R.L. Guillard. 1983. Limitation of marine phytoplankton reproductive rates by zinc, manganese, and iron. Limnology and Oceanography 28: 1182–1198.

Brand, U. 1989. Biogeochemistry of Late Paleozoic North American brachiopods and secular variation of seawater composition. Biogeochemistry 7: 159–193.

Brandes, J.A., M.E. McClain, and T.P. Pimentel. 1996. ^{15}N evidence for the origin and cycling of inorganic nitrogen in a small Amazonian catchment. Biogeochemistry 34: 45–56.

Brasseur, G. and C. Granier. 1992. Mount Pinatubo aerosols, chlorofluorocarbons, and ozone depletion. Science 257: 1239–1242.

Brasseur, G. and M.H. Hitchman. 1988. Stratospheric response to trace gas perturbations: Changes in ozone and temperature distributions. Science 240: 634–637.

Brauman, A., M.D. Kane, M. Labat, and J.A. Breznak. 1992. Genesis of acetate and methane by gut bacteria of nutritionally diverse termites. Science 257: 1384–1387.

Bravard, S. and D. Righi. 1989. Geochemical differences in an Oxisol-Spodosol toposequence of Amazonia, Brazil. Geoderma 44: 29–42.

Bray, J.R. and E. Gorham. 1964. Litter production in forests of the world. Advances in Ecological Research 2: 101–157.

Bremner, J.M. and A.M. Blackmer. 1978. Nitrous oxide: Emission from soils during nitrification of fertilizer nitrogen. Science 199: 295–296.

Brewer, P.G., C. Goyet, and D. Dyrssen. 1989. Carbon dioxide transport by ocean currents at 25°N Latitude in the Atlantic Ocean. Science 246: 477–479.

Bricker, O.P. 1982. Redox potential: Its measurement and importance in water systems. pp. 55–83. In R.A. Minear and L.H. Kieth (eds.), *Water Analysis*. Vol. 1. Academic Press, New York.

Bridgham, S.D. and C.J. Richardson. 1992. Mechanisms controlling soil respiration (CO_2 and CH_4) in southern peatlands. Soil Biology and Biochemistry 24: 1089–1099.

Brimblecombe, P. and G.A. Dawson. 1984. Wet removal of highly soluble gases. Journal of Atmospheric Chemistry 2: 95–107.

Brimblecombe, P., C. Hammer, H. Rodhe, A. Ryaboshapko, and C.F. Boutron. 1989. Human influence on the sulphur cycle. pp. 77–121. In P. Brimblecombe and A.Y. Lein (eds.), *Evolution of the Global Biogeochemical Sulphur Cycle*. Wiley, New York.

Brimhall, G.H., O.A. Chadwick, C.J. Lewis, W. Compston, I.S. Williams, K.J. Danti, W.E. Dietrich, M.E. Power, D. Hendricks, and J. Bratt. 1991. Deformational mass transport and invasive processes in soil evolution. Science 255: 695–702.

Brinkmann, W.L.F. and U. De M. Santos. 1974. The emission of biogenic hydrogen sulfide from Amazonian floodplain lakes. Tellus 26: 261–267.

Brinson, M.M., A.E. Lugo, and S. Brown. 1981. Primary productivity, decomposition and consumer activity in freshwater wetlands. Annual Review of Ecology and Systematics 12: 123–161.

Brock, T.D. 1985. Life at high temperatures. Science 230: 132–138.

Broda, E. 1975. The history of inorganic nitrogen in the biosphere. Journal of Molecular Evolution 7: 87–100.

Broecker, W.S. 1973. Factors controlling CO_2 content in the oceans and atmosphere. pp. 32–50. In G.M. Woodwell and E.V. Pecan (eds.), *Carbon and the Biosphere*. CONF 720510. National Technical Information Service, Washington, D.C.

Broecker, W.S. 1974. *Chemical Oceanography*. Harcourt Brace Jovanovich, New York.

Broecker, W.S. 1982. Ocean chemistry during glacial time. Geochimica et Cosmochimica Acta 46: 1689–1705.

Broecker, W.S. 1991. The great ocean conveyor. Oceanography 4: 79–89.

Broecker, W.S. and T.-H. Peng. 1987. The oceanic salt pump: Does it contribute to the glacial-interglacial difference in atmospheric CO_2 content? Global Biogeochemical Cycles 1: 251–259.

Broecker, W.S. and T.-H. Peng. 1993. What caused the glacial to interglacial CO_2 change? pp. 95–115. In M. Heimann (ed.), *The Global Carbon Cycle*. Springer-Verlag, New York.

Broecker, W.S., T. Takahashi, H.J. Simpson, and T.-H. Peng. 1979. Fate of fossil fuel carbon dioxide and the global carbon budget. Science 206: 409–418.

Bronk, D.A., P.M. Glibert, and B.B. Ward. 1994. Nitrogen uptake, dissolved organic nitrogen release, and new production. Science 265: 1843–1846.

Bronson, K.F. and A.R. Mosier. 1993. Effect of nitrogen fertilizer and nitrification inhibitors on methane and nitrous oxide fluxes in irrigated corn. pp. 278–289. In R.S. Oremland (ed.), *Biogeochemistry and Global Change*. Chapman and Hall, New York.

Brook, G.A., M.E. Folkoff, and E.O. Box. 1983. A world model of soil carbon dioxide. Earth Surface Processes and Landforms 8: 79–88.

Brooks, P.C., D.S. Powlson, and D.S. Jenkinson. 1982. Measurement of microbial biomass phosphorus in soil. Soil Biology and Biochemistry 14: 319–329.

Brooks, P.C., D.S. Powlson, and D.S. Jenkinson. 1984. Phosphorus in the soil microbial biomass. Soil Biology and Biochemistry 16: 169–175.

Brooks, P.C., A. Landman, G. Pruden, and D.S. Jenkinson. 1985. Chloroform fumigation and the release of soil nitrogen: A rapid direct extraction method to measure microbial biomass nitrogen in soil. Soil Biology and Biochemistry 17: 837–842.

Brooks, R.R. 1973. Biogeochemical parameters and their significance for mineral exploration. Journal of Applied Ecology 10: 825–836.

Brown, B.E., J.L. Fassbender, and R. Winkler. 1992. Carbonate production and sediment transport in a marl lake of southeastern Wisconsin. Limnology and Oceanography 37: 184–191.

Brown, K.A. 1985. Sulphur distribution and metabolism in waterlogged peat. Soil Biology and Biochemistry 17: 39–45.

Brown, K.A. and J.F. MacQueen. 1985. Sulphate uptake from surface water by peat. Soil Biology and Biochemistry 17: 411–420.

Brown, K.R. 1991. Carbon dioxide enrichment accelerates the decline in nutrient status and relative growth rate of *Populus tremuloides* Michx. seedlings. Tree Physiology 8: 161–173.

Brown, S. 1981. A comparison of the structure, primary productivity, and transpiration of cypress ecosystems in Florida. Ecological Monographs 51: 403–427.

Brown, S. and A.E. Lugo. 1982. The storage and production of organic matter in tropical forests and their role in the global carbon cycle. Biotropica 14: 161–187.

Brown, S. and A.E. Lugo. 1984. Biomass of tropical forests: A new estimate based on forest volumes. Science 223: 1290–1293.

Brownlee, D.E. 1992. The origin and early evolution of the Earth. pp. 9–20. In S.S. Butcher, R.J. Charlson, G.H. Orians, and G.V. Wolfe (eds.), *Global Biogeochemical Cycles*. Academic Press, San Diego.

Brüchert, V. and L.M. Pratt. 1996. Contemporaneous early diagenetic formation of organic and inorganic sulfur in estuarine sediments from St. Andrew Bay, Florida, USA. Geochimica et Cosmochimica Acta 60: 2325–2332.

Brugam, R.B. 1978. Human disturbance and the historical development of Linsley pond. Ecology 59: 19–36.

Bruland, K.W. 1989. Complexation of zinc by natural organic ligands in the central North Pacific. Limnology and Oceanography 34: 269–285.

Bruland, K.W., G.A. Knauer, and J.H. Martin. 1978a. Zinc in north-east Pacific water. Nature 271: 741–743.

Bruland, K.W., G.A. Knauer, and J.H. Martin. 1978b. Cadmium in northeast Pacific waters. Limnology and Oceanography 23: 618–625.

Brune, W.H., J.G. Anderson, D.W. Toohey, D.W. Fahey, S.R. Kawa, R.L. Jones, D.S. McKenna, and L.R. Poole. 1991. The potential for ozone depletion in the Arctic polar stratosphere. Science 252: 1260–1266.

Brunskill, G.J. 1969. Fayetteville Green Lake, New York. II. Precipitation and sedimentation of calcite in a meromictic lake with laminated sediments. Limnology and Oceanography 14: 830–847.

Brylinsky, M. and K.H. Mann. 1973. An analysis of factors governing productivity in lakes and reservoris. Limnology and Oceanography 18: 1–14.

Bubier, J.L., T.R. Moore, and N.T. Roulet. 1993. Methane emissions from wetlands in the midboreal region of northern Ontario, Canada. Ecology 74: 2240–2254.

Bundy, L.G. and J.M. Bremner. 1973. Inhibition of nitrification in soils. Soil Science Society of America Proceedings 37: 396–398.

Burbidge, E.M., G.R. Burbidge, W.A. Fowler, and F. Hoyle. 1957. Synthesis of the elements in stars. Reviews of Modern Physics 29: 547–650.

Burford, J.R. and J.M. Bremner. 1975. Relationships between the denitrification capacities of soils and total, water-soluble and readily decomposable soil organic matter. Soil Biology and Biochemistry 7: 389–394.

Burke, I.C. 1989. Control of nitrogen mineralization in a sagebrush steppe landscape. Ecology 70: 1115–1126.

Burke, I.C., W.K. Lauenroth, and D.P. Coffin. 1995. Soil organic matter recovery in semiarid grasslands: Implications for the Conservation Reserve Program. Ecological Applications 5: 793–801.

Burke, M.K. and D.J. Raynal. 1994. Fine root growth phenology, production, and turnover in a northern hardwood forest ecosystem. Plant and Soil 162: 135–146.

Burke, R.A., D.F. Reid, J.M. Brooks, and D.M. Lavoie. 1983. Upper water column methane geochemistry in the eastern tropical North Pacific. Limnology and Oceanography 28: 19–32.

Burnett, W.C., M.J. Beers, and K.K. Roe. 1982. Growth rates of phosphate nodules from the continental margin off Peru. Science 215: 1616–1618.

Burns, R.C. and R.W.F. Hardy. 1975. *Nitrogen Fixation in Bacteria and Higher Plants.* Springer-Verlag, New York.

Burns, R.G. 1982. Enzyme activity in soil: Location and a possible role in microbial ecology. Soil Biology and Biochemistry 14: 423–427.

Burton, D.L. and E.C. Beauchamp. 1984. Field techniques using the acetylene blockage of nitrous oxide reduction to measure denitrification. Canadian Journal of Soil Science 64: 555–562.

Burton, J.D. 1988. Riverborne materials and the continent-ocean interface. pp. 299–321. In A. Lerman and M. Meybeck (eds.), *Physical and Chemical Weathering in Geochemical Cycles.* Kluwer Academic Publishers, Dordrecht, The Netherlands.

Butler, J.H., J.W. Elkins, T.M. Thompson, and K.B. Egan. 1989. Tropospheric and dissolved N_2O of the West Pacific and East Indian Oceans during the El Niño Southern Oscillation event of 1987. Journal of Geophysical Research 94: 14865–14877.

Buyanovsky, G.A. and G.H. Wagner. 1983. Annual cycles of carbon dioxide level in soil air. Soil Science Society of America Journal 47: 1139–1145.

Buyanovsky, G.A., M. Aslam, and G.H. Wagner. 1994. Carbon turnover in soil physical fractions. Soil Science Society of America Journal 58: 1167–1173.

Byrne, R., J. Michaelsen, and A. Soutar. 1977. Fossil charcoal as a measure of wildfire frequency in southern California: A preliminary analysis. pp. 361–367. In H.A. Mooney and C.E. Conrad (eds.), *Symposium on the Environmental Consequences of Fire and Fuel Management in Mediterranean Ecosystems.* U.S. Forest Service, Washington, D.C.

Cabana, G. and J.B. Rasmussen. 1994. Modelling food chain structure and contaminant bioaccumulation using stable nitrogen isotopes. Nature 372: 255–257.

Caccavo, F., R.P. Blakemore, and D.R. Lovley. 1992. A hydrogen-oxidizing, Fe(III)-reducing microorganism from the Great Bay Estuary, New Hampshire. Applied and Environmental Microbiology 58: 3211–3216.

Cachier, H. and J. Ducret. 1991. Influence of biomass burning on equatorial African rains. Nature 352: 228–230.

Cachier, H., M.-P. Brémond, and P. Buat-Ménard. 1989. Carbonaceous aerosols from different tropical biomass burning scenarios. Nature 340: 371–373.

Cahoon, D.R., B.J. Stocks, J.S. Levine, W.R. Cofer, and K.P. O'Neill. 1992. Seasonal distribution of African savanna fires. Nature 359: 812–815.

Cai, D.-L., F.C. Tan, and J.M. Edmond. 1988. Sources and transport of particulate organic carbon in the Amazon River and estuary. Estuarine, Coastal and Shelf Science 26: 1–14.

Cairns-Smith, A.G. 1985. The first organisms. Scientific American 252(6): 90–100.

Caldeira, K. and J.F. Kasting. 1992. The life span of the biosphere revisited. Nature 360: 721–723.

Caldwell, M.M. and L.B. Camp. 1974. Belowground productivity of two cool desert communities. Oecologia 17: 123–130.

Callaway, R.M., E.H. DeLucia, and W.H. Schlesinger. 1994. Biomass allocation of montane and desert ponderosa pine: An analog for response to climate change. Ecology 75: 1474–1481.

Callebaut, F., D. Gabriels, W. Winjauw, and M. De Boodt. 1982. Redox potential, oxygen diffusion rate, and soil gas composition in relation to water table level in two soils. Soil Science 134: 149–156.

Cambardella, C.A. and E.T. Elliott. 1994. Carbon and nitrogen dynamics of soil organic matter fractions from cultivated grassland soils. Soil Science Society of America Journal 58: 123–130.

Cameron, E.M. 1982. Sulphate and sulphate reduction in early Precambrian oceans. Nature 296: 145–148.

Campbell, C.A., E.A. Paul, D.A. Rennie, and K.J. McCallum. 1967. Factors affecting the accuracy of the carbon-dating method in soil humus studies. Soil Science 104: 81–85.

Campos, M.L.A.M., P.D. Nightingale, and T.D. Jickells. 1996. A comparison of methyl iodide emissions from seawater and wet depositional fluxes of iodine over the southern North Sea. Tellus 48B: 106–114.

Canfield, D.E. 1989a. Reactive iron in marine sediments. Geochimica et Cosmochimica Acta 53: 619–632.

Canfield, D.E. 1989b. Sulfate reduction and oxic respiration in marine sediments: Implications for organic carbon preservation in euxinc environments. Deep Sea Research 36: 121–138.

Canfield, D.E. 1991. Sulfate reduction in deep-sea sediments. American Journal of Science 291: 177–188.

Canfield, D.E. 1993. Organic matter oxidation in marine sediments. pp. 333–363. In R. Wollast, F.T. MacKenzie, and L. Chou (eds.), *Interactions of C, N, P and S in Biogeochemical Cycles and Global Change*. Springer-Verlag, New York.

Canfield, D.E. 1994. Factors influencing organic carbon preservation in marine sediments. Chemical Geology 114: 315–329.

Canfield, D.E. and A. Teske. 1996. Late Proterozoic rise in atmospheric oxygen concentration inferred from phylogenetic and sulphur-isotope studies. Nature 382: 127–132.

Canfield, D.E. and B. Thamdrup. 1994. The production of ^{34}S-depleted sulfide during bacterial disproportionation of elemental sulfur. Science 266: 1973–1975.

Canfield, D.E., W.J. Green, T.J. Gardner, and T. Ferdelman. 1984. Elemental residence times in Acton Lake, Ohio. Archiv für Hydrobiologie 100: 501–509.

Canfield, D.E., E. Philips, and C.M. Duarte. 1989. Factors influencing the abundance of blue-green algae in Florida lakes. Canadian Journal of Fisheries and Aquatic Sciences 46: 1232–1237.

Canfield, D.E., B. Thamdrup, and J.W. Hansen. 1993. The anaerobic degradation of organic matter in Danish coastal sediments: Iron reduction, manganese reduction, and sulfate reduction. Geochimica et Cosmochimica Acta 57: 3867–3883.

Cannon, H.L. 1960. Botanical prospecting for ore deposits. Science 132: 591–598.

Cao, M., S. Marshall, and K. Gregson. 1996. Global carbon exchange and methane emissions from natural wetlands: Application of a process-based model. Journal of Geophysical Research 101: 14399–14414.

Capone, D.G. and E.J. Carpenter. 1982. Nitrogen fixation in the marine environment. Science 217: 1140–1142.

Caraco, N.F., J.J. Cole, and G.E. Likens. 1989. Evidence for sulphate-controlled phosphorus release from sediments of aquatic systems. Nature 341: 316–318.

Caraco, N.F., J.J. Cole, and G.E. Likens. 1990. A comparison of phosphorus immobilization in sediments of freshwater and coastal marine systems. Biogeochemistry 9: 277–290.

Caraco, N.F., J.J. Cole, and G.E. Likens. 1993. Sulfate control of phosphorus availability in lakes. Hydrobiologia 253: 275–280.

Carignan, R. and D.R.S. Lean. 1991. Regeneration of dissolved substances in a seasonally anoxic lake: The relative importance of processes occurring in the water column and in the sediments. Limnology and Oceanography 36: 683–707.

Carlisle, A., A.H.F. Brown, and E.J. White. 1966. The organic matter and nutrient elements in the precipitation beneath a sessile oak (*Quercus petraea*) canopy. Journal of Ecology 54: 87–98.

Carlson, C.A., H.W. Ducklow, and A.F. Michaels. 1994. Annual flux of dissolved organic carbon from the euphotic zone in the northwestern Sargasso sea. Nature 371: 405–408.

Carlson, P.R. and J. Forrest. 1982. Uptake of dissolved sulfide by *Spartina alterniflora:* Evidence from natural sulfur isotope abundance ratios. Science 216: 633–635.

Carlyle, J.C. and D.C. Malcolm. 1986. Larch litter and nitrogen availability in mixed larch-spruce stands. I. Nutrient withdrawal, redistribution, and leaching loss from larch foliage at senescence. Canadian Journal of Forest Research 16: 321–326.

Caron, D.A. 1994. Inorganic nutrients, bacteria, and the microbial loop. Microbial Ecology 28: 295–298.

Caron, F. and J.R. Kramer. 1994. Formation of volatile sulfides in freshwater environments. Science of the Total Environment 153: 177–194.

Carpenter, E.J. and K. Romans. 1991. Major role of the cyanobacterium *Trichodesmium* in nutrient cycling in the North Atlantic ocean. Science 254: 1356–1358.

Carpenter, S.R., J.F. Kitchell, J.R. Hodgson, P.A. Cochran, J.J. Elser, M.M. Elser, D.M. Lodge, D. Kretchmer, X. He, and C.N. von Ende. 1987. Regulation of lake primary productivity by food web structure. Ecology 68: 1863–1876.

Carpenter, S.R., S.G. Fisher, N.B. Grimm, and J.F. Kitchell. 1992. Global change and freshwater ecosystems. Annual Review of Ecology and Systematics 23: 119–139.

Carr, M.H. 1987. Water on Mars. Nature 326: 30–35.

Carroll, M.A., L.E. Heidt, R.J. Cicerone, and R.G. Prinn. 1986. OCS, H_2S, and CS_2 fluxes from a salt water marsh. Journal of Atmospheric Chemistry 4: 375–395.

Carter, J.P., Y.H. Hsiao, S. Spiro, and D.J. Richardson. 1995. Soil and sediment bacteria capable of aerobic nitrate respiration. Applied and Environmental Microbiology 61: 2852–2858.

Casagrande, D.J., K. Siefert, C. Berschinski, and N. Sutton. 1977. Sulfur in peat-forming systems of the Okefenokee Swamp and Florida Everglades: Origins of sulfur in coal. Geochimica et Cosmochimica Acta 41: 161–167.

Casagrande, D.J., G. Idowu, A. Friedman, P. Rickert, K. Siefert, and D. Schlenz. 1979. H_2S incorporation in coal precursors: Origins of organic sulphur in coal. Nature 282: 599–600.

Casey, W.H., H.R. Westrich, J.F. Banfield, G. Ferruzzi, and G.W. Arnold. 1993. Leaching and reconstruction at the surfaces of dissolving chain-silicate minerals. Nature 366: 253–256.

Castelle, A.J. and J.N. Galloway. 1990. Carbon dioxide dynamics in acid forest soils in Shenandoah National Park, Virginia. Soil Science Society of America Journal 54: 252–257.

Castresana, J. and M. Saraste. 1995. Evolution of energetic metabolism: the respiration-early hypothesis. Trends in Biochemical Science 20: 443–448.

Castro, M.S. and F.E. Dierberg. 1987. Biogenic hydrogen sulfide emissions from selected Florida wetlands. Water, Air and Soil Pollution 33: 1–13.

Castro, M.S., J.M. Melillo, P.A. Steudler, and J.W. Chapman. 1994a. Soil moisture as a predictor of methane uptake by temperate forest soils. Canadian Journal of Forest Research 24: 1805–1810.

Castro, M.S., W.T. Peterjohn, J.M. Melillo, P.A. Steudler, H.L. Gholz, and D. Lewis. 1994b. Effects of nitrogen fertilization on the fluxes of N_2O, CH_4, and CO_2 from soils in a Florida slash pine plantation. Canadian Journal of Forest Research 24: 9–13.

Castro, M.S., P.A. Steudler, J.M. Melillo, J.D. Aber, and R.D. Bowden. 1995. Factors controlling atmospheric methane consumption by temperate forest soils. Global Biogeochemical Cycles 9: 1–10.

Cavanaugh, C.M., S.L. Gardiner, M.J. Jones, H.W. Jannasch, and J.B. Waterbury. 1981. Prokaryotic cells in the hydrothermal vent tube worm *Riftia pachyptila* Jones: Possible chemoautotrophic symbionts. Science 213: 340–342.

Cebrián, J. and C.M. Duarte. 1995. Plant growth-rate dependence of detrital carbon storage in ecosystems. Science 268: 1606–1608.

Cejudo, F.J., A. de La Torre, and A. Paneque. 1984. Short-term ammonium inhibition of nitrogen fixation in *Azotobacter*. Biochemical and Biophysical Research Communications 123: 431–437.

Ceulemans, R. and M. Mousseau. 1994. Effects of elevated atmospheric CO_2 on woody plants. New Phytologist 127: 425–446.

Chadwick, O.A., E.F. Kelley, D.M. Merritts, and R.G. Amundson. 1994. Carbon dioxide consumption during soil development. Biogeochemistry 24: 115–127.

Chadwick, O.A., W.D. Nettleton, and G.J. Staidl. 1995. Soil polygenesis as a function of Quaternary climate change, northern Great Basin, USA. Geoderma 68: 1–26.

Chahine, M.T. 1992. The hydrologic cycle and its influence on climate. Nature 359: 373–380.

Chahine, M.T. 1995. Observations of local cloud and moisture feedbacks over high ocean and desert surface temperatures. Journal of Geophysical Research 100: 8919–8927.

Chambers, L.A. and P.A. Trudinger. 1979. Microbiological fractionation of stable sulfur isotopes: A review and critique. Geomicrobiology Journal 1: 249–293.

Chameides, W.L. and D.D. Davis. 1982. Chemistry in the troposphere. Chemical and Engineering News 60(40): 38–52.

Chameides, W.L., R.W. Lindsay, J. Richardson, and C.S. Kiang. 1988. The role of biogenic hydrocarbons in urban photochemical smog: Atlanta as a case study. Science 241: 1473–1475.

Chameides, W.L., F. Fehsenfeld, M.O. Rodgers, C. Cardelino, J. Martinez, D. Parrish, W. Lonneman, D.R. Lawson, R.A. Rasmussen, P. Zimmerman, J. Greenberg, P. Middleton, and T. Wang. 1992. Ozone precursor relationships in the ambient atmosphere. Journal of Geophysical Research 97: 6037–6055.

Chandler, A.S., T.W. Choularton, G.J. Dollard, A.E.J. Eggleton, M.J. Gay, T.A. Hill, B.M.R. Jones, B.J. Tyler, B.J. Bandy, and S.A. Penkett. 1988. Measurements of H_2O_2 and SO_2 in clouds and estimates of their reaction rate. Nature 336: 562–565.

Chang, S., D. Des Marais, R. Mack, S.L. Miller, and G.E. Strathearn. 1983. Prebiotic organic syntheses and the origin of life. pp. 53–92. In J.W. Schopf (ed.), *Earth's Earliest Biosphere.* Princeton University Press, Princeton, New Jersey.

Chang, W.Y.B. and R.A. Moll. 1980. Prediction of hypolimnetic oxygen deficits: Problems of interpretation. Science 209: 721–722.

Chanton, J.P. and J.W.H. Dacey. 1991. Effects of vegetation on methane flux, reservoirs, and carbon isotopic composition. pp. 65–92. In T.D. Sharkey, E.A. Holland, and H.A. Mooney (eds.), *Trace Gas Emissions by Plants.* Academic Press, San Diego.

Chapelle, F.H. and D.R. Lovley. 1990. Rates of microbial metabolism in deep coastal plain aquifers. Applied and Environmental Microbiology 56: 1865–1874.

Chapin, D.M., L.C. Bliss, and L.J. Bledsoe. 1991. Environmental regulation of nitrogen fixation in a high arctic lowland ecosystem. Canadian Journal of Botany 69: 2744–2755.

Chapin, F.S. 1974. Morphological and physiological mechanisms of temperature compensation in phosphate absorption along a latitudinal gradient. Ecology 55: 1180–1198.

Chapin, F.S. 1980. The mineral nutrition of wild plants. Annual Review of Ecology and Systematics 11: 233–260.

Chapin, F.S. 1988. Ecological aspects of plant mineral nutrition. Advances in Mineral Nutrition 3: 161–191.

Chapin, F.S. and R.A. Kedrowski. 1983. Seasonal changes in nitrogen and phosphorus fractions and autumn retranslocation in evergreen and deciduous taiga trees. Ecology 64: 376–391.

Chapin, F.S. and L. Moilanen. 1991. Nutritional controls over nitrogen and phosphorus resorption from Alaskan birch leaves. Ecology 72: 709–715.

Chapin, F.S. and W.C. Oechel. 1983. Photosynthesis, respiration, and phosphate absorption by *Carex aquatilis* ecotypes along latitudinal and local environmental gradients. Ecology 64: 743–751.

Chapin, F.S., R.J. Barsdate, and D. Barèl. 1978. Phosphorus cycling in Alaskan coastal tundra: A hypothesis for the regulation of nutrient cycling. Oikos 31: 189–199.

Chapin, F.S., P.M. Vitousek, and K. Van Cleve. 1986a. The nature of mineral limitation in plant communities. American Naturalist 127: 48–58.

Chapin, F.S., G.R. Shaver, and R.A. Kedrowski. 1986b. Environmental controls over carbon, nitrogen and phosphorus fractions in *Eriophorum vaginatum* in Alaskan tussock tundra. Journal of Ecology 74: 167–195.

Chapin, F.S., L. Moilanen, and K. Kielland. 1993. Preferential use of organic nitrogen for growth by a non-mycorrhizal arctic sedge. Nature 361: 150–153.

Chapman, D.J. and J.W. Schopf. 1983. Biological and biochemical effects of the development of an aerobic environment. pp. 302–320. In J.W. Schopf (ed.), *Earth's Earliest Biosphere.* Princeton University Press, Princeton, New Jersey.

Chapman, S.B., J. Hibble, and C.R. Rafarel. 1975. Net aerial production by *Calluna vulgaris* on lowland heath in Britain. Journal of Ecology 63: 233–258.

Chappellaz, J., J.M. Barnola, D. Raynaud, Y.S. Korotkevich and C. Lorius. 1990. Ice-core record of atmospheric methane over the past 160,000 years. Nature 345: 127–131.

Chappellaz, J., T. Blunier, D. Raynaud, J.M. Barnola, J. Schwander, and B. Stauffer. 1993. Synchronous changes in atmospheric CH_4 and Greenland climate between 40 and 8 kyr BP. Nature 366: 443–445.

Charlson, R.J., J.E. Lovelock, M.O. Andreae, and S.G. Warren. 1987. Oceanic phytoplankton, atmospheric sulphur, cloud albedo and climate. Nature 326: 655–661.

Charlson, R.J., S.E. Schwartz, J.M. Hales, R.D. Cess, J.A. Coakley, J.E. Hansen and D.J. Hofmann. 1992. Climate forcing by anthropogenic aerosols. Science 255: 423–430.

Chase, E.M. and F.L. Sayles. 1980. Phosphorus in suspended sediments of the Amazon river. Estuarine and Coastal Marine Science 11: 383–391.

Chaussidon, M. and F. Robert. 1995. Nucleosynthesis of ^{11}B-rich boron in the presolar cloud recorded in meteoritic chondrules. Nature 374: 337–339.

Chen, D.L., P.M. Chalk, and J.R. Freney. 1995. Distribution of reduced products of ^{15}N-labelled nitrate in anaerobic soils. Soil Biology and Biochemistry 27: 1539–1545.

Cherry, R.D., J.J.W. Higgo, and S.W. Fowler. 1978. Zooplankton fecal pellets and element residence times in the ocean. Nature 274: 246–248.

Chesworth, W. and F. Macias-Vasquez. 1985. pe, pH, and podzolization. American Journal of Science 285: 128–146.

Chesworth, W., J. Dejou, and P. Larroque. 1981. The weathering of basalt and relative mobilities of the major elements at Belbex, France. Geochimica et Cosmochimica Acta 45: 1235–1243.

Chevalier, R.A. and C.L. Sarazin. 1987. Hot gas in the Universe. American Scientist 75: 609–618.

Childers, D.L., N.H. McKellar, R.F. Dame, F.H. Sklar, and E.R. Blood. 1993. A dynamic nutrient budget of subsystem interactions in a salt marsh estuary. Estuarine, Coastal and Shelf Science 36: 105–131.

Chin, M. and D.D. Davis. 1993. Global sources and sinks of OCS and CS_2 and their distributions. Global Biogeochemical Cycles 7: 321–337.

Chin, M. and D.D. Davis. 1995. A reanalysis of carbonyl sulfide as a source of stratospheric background sulfur aerosol. Journal of Geophysical Research 100: 8993–9005.

Cho, B.C. and F. Azam. 1988. Major role of bacteria in biogeochemical fluxes in the ocean's interior. Nature 332: 441–443.

Chorover, J., P.M. Vitousek, D.A. Everson, A.M. Esperanza, and D. Turner. 1994. Solution chemistry profiles of mixed-conifer forests before and after fire. Biogeochemistry 26: 115–144.

Christensen, J.P., J.W. Murray, A.H. Devol, and L.A. Codispoti. 1987. Denitrification in continental shelf sediments has major impact on the oceanic nitrogen budget. Global Biogeochemical Cycles 1: 97–116.

Christensen, N.L. 1973. Fire and the nitrogen cycle in California chaparral. Science 181: 60–68.

Christensen, N.L. 1977. Fire and soil-plant nutrient relations in a pine-wiregrass savanna on the coastal plain of North Carolina. Oecologia 31: 27–44.

Christensen, N.L. and T. MacAller. 1985. Soil mineral nitrogen transformations during succession in the Piedmont of North Carolina. Soil Biology and Biochemistry 17: 675–681.

Christensen-Dalsgaard, J., W. Däppen, S.V. Ajukov, E.R. Anderson, H.M. Antia, S. Basu, V.A. Baturin, G. Berthomieu, B. Chaboyer, S.M. Chitre, A.N. Cox, P. Demarque, J. Donatowicz, W.A. Dziembowski, M. Gabriel, D.O. Gough, D.B. Guenther, J.A. Guzik, J.W. Harvey, F. Hill, G. Houdek, C.A. Iglesias, A.G. Kosovichev, J.W. Leibacher, P. Morel, C.R. Proffitt, J. Provost, J. Reiter, E.J. Rhodes, F.J. Rogers, I.W. Roxburgh, M.J. Thompson, and R.K. Ulrich. 1996. The current state of solar modeling. Science 272: 1286–1292.

Chyba, C.F. 1987. The cometary contribution to the oceans of primitive Earth. Nature 330: 632–635.

Chyba, C.F. 1990a. Impact delivery and erosion of planetary oceans in the early inner solar system. Nature 343: 129–133.

Chyba, C.F. 1990b. Extraterrestrial amino acids and terrestrial life. Nature 348: 113–114.

Chyba, C.F. and C. Sagan. 1992. Endogenous production, exogenous delivery and impact-shock synthesis of organic molecules: An inventory for the origins of life. Nature 355: 125–132.

Ciais, P., P.P. Tans, J.W.C. White, M. Trolier, R.J. Francey, J.A. Berry, D.R. Randall, P.J. Sellers, J.C. Collatz, and D.S. Schimel. 1995. Partitioning of ocean and land uptake of CO_2 as inferred by $\delta^{13}C$ measurements from the NOAA Climate Monitoring and Diagnostics Laboratory global air sampling network. Journal of Geophysical Research 100: 5051–5070.

Cicerone, R.J. 1987. Changes in stratospheric ozone. Science 237: 35–42.

Cicerone, R.J. 1989. Analysis of sources and sinks of atmospheric nitrous oxide (N_2O). Journal of Geophysical Research 94: 18265–18271.

Cicerone, R.J. and R.S. Oremland. 1988. Biogeochemical aspects of atmospheric methane. Global Biogeochemical Cycles 2: 299–327.

Cicerone, R.J., C.C. Delwiche, S.C. Tyler, and P.R. Zimmerman. 1992. Methane emissions from California rice paddies with varied treatments. Global Biogeochemical Cycles 6: 233–248.

Clark, J.S. 1990. Fire and climate change during the last 750 yr in northwestern Minnesota. Ecological Monographs 60: 135–159.

Clark, J.S. and P.D. Royall. 1994. Pre-industrial particulate emissions and carbon sequestration from biomass burning in North America. Biogeochemistry 23: 35–51.

Clarkson, D.T. and J.B. Hanson. 1980. The mineral nutrition of higher plants. Annual Review of Plant Physiology 31: 239–298.

Clayton, H., J.R.M. Arah, and K.A. Smith. 1994. Measurement of nitrous oxide emissions from fertilized grassland using closed chambers. Journal of Geophysical Research 99: 16599–16607.

Clayton, J.L. 1976. Nutrient gains to adjacent ecosystems during a forest fire: An evaluation. Forest Science 22: 162–166.

Clemens, S.C., J.W. Farrell, and L.P. Gromet. 1993. Synchronous changes in seawater strontium isotope composition and global climate. Nature 363: 607–610.

Cline, J.D., D.P. Wisegarver, and K. Kelly-Hansen. 1987. Nitrous oxide and vertical mixing in the equatorial Pacific during the 1982-1983 El Niño. Deep Sea Research 34: 857–873.

Cloud, P. 1973. Paleoecological significance of the banded iron formation. Economic Geology 68: 1135–1143.

Clymo, R.S. 1984. The limits to peat bog growth. Philosophical Transactions of the Royal Society of London 303B: 605–654.

Coale, K.H., K.S. Johnson, S.E. Fitzwater, R.M. Gordon, S. Tanner, F.P. Chavez, L. Ferioli, C. Sakamoto, P. Rogers, F. Millero, P. Steinberg, P. Nightingale, D. Cooper, W.P. Cochlan, M.R. Landry, J. Constantinou, G. Rollwagen, A. Trasvina, and R. Kudela. 1996. A massive phytoplankton bloom induced by an ecosystem-scale iron fertilization experiment in the equatorial Pacific Ocean. Nature 383: 495–501.

Cochran, M.F. and R.A. Berner. 1992. The quantitative role of plants in weathering. pp. 473–476. In Y.K. Kharaka and A.S. Maest (eds.), *Water-Rock Interactions*. Balkema, Rotterdam, The Netherlands.

Codispoti, L.A. and J.P. Christensen. 1985. Nitrification, denitrification and nitrous oxide cycling in the eastern tropical South Pacific Ocean. Marine Chemistry 16: 277–300.

Codispoti, L.A., G.E. Friederich, T.T. Packard, H.E. Glover, P.J. Kelley, R.W. Spinrad, R.T. Barber, J.W. Elkins, B.B. Ward, F. Lipschultz, and N. Lostaunau. 1986. High nitrite levles off northern Peru: A signal of instability in the marine denitrification rate. Science 233: 1200–1202.

Cofer, W.R., J.S. Levine,, E.L. Winstead, and B.J. Stocks. 1990. Gaseous emissions from Canadian boreal forest fires. Atmospheric Environment 24A: 1653–1659.

Cofer, W.R., J.S. Levine, E.L. Winstead, and B.J. Stocks. 1991. New estimates of nitrous oxide emissions from biomass burning. Nature 349: 689–691.

Cogbill, C.V. and G.E. Likens. 1974. Acid precipitation in the northeastern United States. Water Resources Research 10: 1133–1137.

Cohen, A.D. 1974. Petrography and paleoecology of Holocene peats from the Okefenokee swamp-marsh complex of Georgia. Journal of Sedimentary Petrology 44: 716–726.

Cohen, Y. and L.I. Gordon. 1978. Nitrous oxide production in the oxygen minimum of the eastern tropical North Pacific: Evidence for its consumption during denitrification and possible mechanisms for its production. Deep Sea Research 25: 509–524.

Cohen, Y. and L.I. Gordon. 1979. Nitrous oxide production in the ocean. Journal of Geophysical Research 84: 347–353.

Cole, C.V. and S.R. Olsen. 1959. Phosphorus solubility in calcareous soils. I. Dicalcium phosphate activities in equilibrium solutions. Soil Science Society of America Proceedings 23: 116–118.

Cole, C.V., G.S. Innis, and J.W.B. Stewart. 1977. Simulation of phosphorus cycling in semiarid grasslands. Ecology 58: 1–15.

Cole, C.V., E.T. Elliott, H.W. Hunt, and D.C. Coleman. 1978. Trophic interactions in soils as they affect energy and nutrient dynamics. V. Phosphorus transformations. Microbial Ecology 4: 381–387.

Cole, D.W. and M. Rapp. 1981. Element cycling in forest ecosystems. pp. 341–409. In D.E. Reichle (ed.), *Dynamic Properties of Forest Ecosystems*. Cambridge University Press, Cambridge.

Cole, J.J. and S.G. Fisher. 1979. Nutrient budgets of a temporary pond ecosystem. Hydrobiologia 63: 213–222.

Cole, J.J. and M.L. Pace. 1995. Bacterial secondary production in oxic and anoxic freshwaters. Limnology and Oceanography 40: 1019–1027.

Cole, J.J., S. Honjo, and J. Erez. 1987. Benthic decomposition of organic matter at a deep-water site in the Panama basin. Nature 327: 703–704.

Cole, J.J., S. Findlay, and M.L. Pace. 1988. Bacterial production in fresh and saltwater ecosystems: A cross-system overview. Marine Ecology: Progress Series 43: 1–10.

Cole, J.J., J.M. Lane, R. Marino, and R.W. Howarth. 1993. Molybdenum assimilation by cyanobacteria and phytoplankton in freshwater and salt water. Limnology and Oceanography 38: 23–35.

Cole, J.J., N.F. Caraco, G.W. Kling, and T.K. Kratz. 1994. Carbon dioxide supersaturation in the surface waters of lakes. Science 265: 1568–1570.

Coleman, J.M. and E.S. Deevey. 1987. Lacustrine sediment/groundwater nutrient dynamics. Biogeochemistry 4: 3–14.

Coley, P.D., J.P. Bryant, and F.S. Chapin. 1985. Resource availability and plant antiherbivore defense. Science 230: 895–899.

Collins, N.M., J.A. Sayer, and T.C. Whitmore. (eds.). 1991. *The Conservation Atlas of Tropical Forests: Asia and the Pacific.* Macmillan, London.

Comins, H.N. and R.E. McMurtrie. 1993. Long-term response of nutrient-limited forests to CO_2 enrichment: Equilibrium behavior of plant-soil models. Ecological Applications 3: 666–681.

Conley, D.J., M.A. Quigley, and C.L. Schelske. 1988. Silica and phosphorus flux from sediments: Importance of internal recycling in Lake Michigan. Canadian Journal of Fisheries and Aquatic Sciences 45: 1030–1035.

Conley, D.J., C.L. Schelske, and E.F. Stoermer. 1993. Modification of the biogeochemical cycle of silica with eutrophication. Marine Ecology: Progress Series 101: 179–192.

Conley, D.J., W.M. Smith, J.C. Cornwell, and T.R. Fisher. 1995. Transformation of particle-bound phosphorus at the land-sea interface. Estuarine, Coastal and Shelf Science 40: 161–176.

Connell, M.J., R.J. Raison, and P.K. Khanna. 1995. Nitrogen mineralization in relation to site history and soil properties for a range of Australian forest soils. Biology and Fertility of Soils 20: 213–220.

Conrad, R. 1994. Compensation concentration as critical variable for regulating the flux of trace gases between soil and atmosphere. Biogeochemistry 27: 155–170.

Conrad, R. and W. Seiler. 1988. Methane and hydrogen in seawater (Atlantic Ocean). Deep Sea Research 35: 1903–1917.

Conrad, R., W. Seiler, and G. Bunse. 1983. Factors influencing the loss of fertilizer nitrogen into the atmosphere as N_2O. Journal of Geophysical Research 88: 6709–6718.

Conway, T.J., P. Tans, I.S. Waterman, K.W. Thoning, K.A. Masarie, and R.H. Gammon. 1988. Atmospheric carbon dioxide measurements in the remote global troposphere, 1981-1984. Tellus 40B: 81–115.

Cook, E.A., L.R. Iverson, and R.L. Graham. 1989. Estimating forest productivity with thematic mapper and biogeographical data. Remote Sensing of Environment 28: 131–141.

Cooke, R.U. and A. Warren. 1973. *Geomorphology in Deserts.* University of California Press, Berkeley.

Cooper, S.R. and G.S. Brush. 1991. Long-term history of Chesapeake Bay anoxia. Science 254: 992–996.

Cooper, W.J., D.J. Cooper, E.S. Saltzman, W.Z. De Mello, D.L. Savoie, R.G. Zika, and J.M. Prospero. 1987. Emissions of biogenic sulphur compounds from several wetland soils in Florida. Atmospheric Environment 21: 1491–1495.

Copi, C.J., D.N. Schramm, and M.S. Turner. 1995. Big-bang nucleosynthesis and the baryon density of the Universe. Science 267: 192–199.

Corliss, B.H., D.G. Martinson, and T. Keffer. 1986. Late Quaternary deep-ocean circulation. Geological Society of America Bulletin 97: 1106-1121.

Corliss, J.B., J. Dymond, L.I. Gordon, R.P. von Herzen, R.D. Ballard, K. Green, D. Williams, A. Bainbridge, K. Crane, and T.H. Van Andel. 1979. Submarine thermal springs on the Galápagos Rift. Science 203: 1073–1083.

Cornell, S., A. Rendell, and T. Jickells. 1995. Atmospheric inputs of dissolved organic nitrogen to the oceans. Nature 376: 243–246.

Cornett, R.J. and F.H. Rigler. 1979. Hypolimnetic oxygen deficits: Their prediction and interpretation. Science 205: 580–581.

Cornett, R.J. and F.H. Rigler. 1980. The areal hypolimnetic oxygen deficit: An empirical test of the model. Limnology and Oceanography 25: 672–679.

Correll, D.L. 1981. Nutrient mass balances for the watershed, headwaters, intertidal zone, and basin of the Rhode River estuary. Limnology and Oceanography 26: 1142–1149.

Correll, D.L. and D. Ford. 1982. Comparison of precipitation and land runoff as sources of estuarine nitrogen. Estuarine, Coastal and Shelf Science 15: 45–56.

Correll, D.L., C.O. Clark, B. Goldberg, V.R. Goodrich, D.R. Hayes, W.H. Klein, and W.D. Schecher. 1992. Spectral ultraviolet-B radiation fluxes at the Earth's surface: Long-term variations at 39°N, 77°W. Journal of Geophysical Research 97: 7579–7591.

Cosby, B.J., G.M. Hornberger, J.N. Galloway, and R.F. Wright. 1985. Modeling the effects of acid deposition: Assessment of a lumped parameter model of soil water and streamwater chemistry. Water Resources Research 21: 51–63.

Cosby, B.J., G.M. Hornberger, R.F. Wright, and J.N. Galloway. 1986. Modeling the effects of acid deposition: Control of long-term sulfate dynamics by soil sulfate adsorption. Water Resources Research 22: 1283–1291.

Cotrufo, M.F., P. Ineson, and A.P. Rowland. 1994. Decomposition of tree leaf litters grown under elevated CO_2: Effect of litter quality. Plant and Soil 163: 121–130.

Courchesne, F. and W.H. Hendershot. 1989. Sulfate retention in some podzolic soils of the southern Laurentians, Quebec. Canadian Journal of Soil Science 69: 337–350.

Coûteaux, M.-M., M. Mousseau, M.-L. Célérier, and P. Bottner. 1991. Increased atmospheric CO_2 and litter quality: Decomposition of sweet chestnut leaf litter with animal food webs of different complexities. Oikos 61: 54–64.

Couto, W., C. Sanzonowicz, and A. de O. Barcellos. 1985. Factors affecting oxidation-reduction processes in an Oxisol with a seasonal water table. Soil Science Society of America Journal 49: 1245–1248.

Covington, W.W. and S.S. Sackett. 1992. Soil mineral nitrogen changes following prescribed burning in ponderosa pine. Forest Ecology and Management 54: 175–191.

Cox, P.A. 1989. *The Elements.* Oxford University Press, Oxford.

Cox, T.L., W.F. Harris, B.S. Ausmus, and N.T. Edwards. 1978. The role of roots in biogeochemical cycles in an eastern deciduous forest. Pedobiologia 18: 264–271.

Craft, C.B. and C.J. Richardson. 1993. Peat accretion and phosphorus accumulation along a eutrophication gradient in the northern Everglades. Biogeochemistry 22: 133–156.

Craig, H. and T. Hayward. 1987. Oxygen supersaturation in the ocean: Biological versus physical contributions. Science 235: 199–202.

Craig, H., C.C. Chou, J.A. Welhan, C.M. Stevens, and A. Engelkemeir. 1988. The isotopic composition of methane in polar ice cores. Science 242: 1535–1539.

Craig, P.J. 1980. Metal cycles and biological methylation. pp. 169-227. In O. Hutzinger (ed.), *The Handbook of Environmental Geochemistry.* Vol. 1, Part A. *The Natural Environment and the Biogeochemical Cycles.* Springer-Verlag, New York.

Crane, B.R., L.M. Siegel, and E.D. Getzoff. 1995. Sulfite reductase structure at 1.6 Å: Evolution and catalysis for reduction of inorganic anions. Science 270: 59–67.

Crawley, J.L., R.C. Burruss, and H.D. Holland. 1969. Chemical weathering in central Iceland: An analog of pre-Silurian weathering. Science 165: 391–392.

Crews, T.E., K. Kitayama, J.H. Fownes, R.H. Riley, D.A. Herbert, D. Mueller-Dombois, and P.M. Vitousek. 1995. Changes in soil phosphorus fractions and ecosystem dynamics across a long chronosequence in Hawaii. Ecology 76: 1407–1424.

Crill, P.M. 1991. Seasonal patterns of methane uptake and carbon dioxide release by a temperate woodland soil. Global Biogeochemical Cycles 5: 319–334.

Crill, P.M. and C.S. Martens. 1986. Methane production from bicarbonate and acetate in an anoxic marine sediment. Geochimica et Cosmochimica Acta 50: 2089–2097.

Crisp, D.T. 1966. Input and output of minerals for an area of Pennine moorland: The importance of precipitation, drainage, peat erosion and animals. Journal of Applied Ecology 3: 327–348.

Cromack, K., P. Sollins, W.G. Graustein, K. Speidel, A.W. Todd, G. Spycher, C.Y. Li, and R.L. Todd. 1979. Calcium oxalate accumulation and soil weathering in mats of the hypogeous fungus *Hysterangium crassum.* Soil Biology and Biochemistry 11: 463–468.

Cronan, C.S. 1980. Solution chemistry of a New Hampshire subalpine ecosystem: A biogeochemical analysis. Oikos 34: 272–281.

Cronan, C.S. and G.R. Aiken. 1985. Chemistry and transport of soluble humic substances in forest watersheds of the Adirondack Park, New York. Geochimica et Cosmochimica Acta 49: 1697–1705.

Cronan, C.S. and D.F. Grigal. 1995. Use of calcium/aluminum ratios as indicators of stress in forest ecosystems. Journal of Environmental Quality 24: 209–226.

Cronan, C.S., C.T. Driscoll, R.M. Newton, J.M. Kelly, C.L. Schofield, R.J. Bartlett, and R. April. 1990. A comparative analysis of aluminum biogeochemistry in a northeastern and a southeastern forested watershed. Water Resources Research 26: 1413–1430.

Cross, A.F. and W.H. Schlesinger 1995. A literature review and evaluation of the Hedley fractionation: Applications to the biogeochemical cycle of soil phosphorus in natural ecosystems. Geoderma 64: 197–214.

Cross, F.A., L.H. Hardy, N.Y. Jones, and R.T. Barber. 1973. Relation between total body weight and concentrations of manganese, iron, copper, zinc, and mercury in white muscle of bluefish (*Pomatomus saltatrix*) and a bathyl-demersal fish *Antimora rostrata*. Journal of the Fisheries Research Board of Canada. 30: 1287–1291.

Cross, P.M. and F.H. Rigler. 1983. Phosphorus and iron retention in sediments measured by mass budget calculations and directly. Canadian Journal of Fisheries and Aquatic Sciences 40: 1589–1597.

Crowley, T.J. 1995. Ice age terrestrial carbon changes revisited. Global Biogeochemical Cycles 9: 377–389.

Crutzen, P.J. 1983. Atmospheric interactionshomogeneous gas reactions of C, N, and S containing compounds. pp. 67–114. In B. Bolin and R.B. Cook (eds.), *The Major Biogeochemical Cycles and Their Interactions*. Wiley, New York.

Crutzen, P.J. 1988. Variability in atmospheric-chemical systems. pp. 81–108. In T. Rosswall, R.G. Woodmansee, and P.G. Risser (eds.), *Scales and Global Change*. Wiley, New York.

Crutzen, P.J. and M.O. Andreae. 1990. Biomass burning in the tropics: Impact on atmospheric chemistry and biogeochemical cycles. Science 250: 1669–1678.

Crutzen, P.J. and P.H. Zimmermann. 1991. The changing photochemistry of the troposphere. Tellus 43: 136–151.

Crutzen, P.J., L.E. Heidt, J.P. Krasnec, W.H. Pollock, and W. Seiler. 1979. Biomass burning as a source of atmospheric gases CO, H_2, N_2O, NO, CH_3Cl, and COS. Nature 282: 253–256.

Crutzen, P.J., A.C. Delany, J. Greenberg, P. Haagenson, L. Heidt, R. Lueb, W. Pollock, W. Seiler, A. Wartburg, and P. Zimmerman. 1985. Tropospheric chemical composition measurements in Brazil during the dry season. Journal of Atmospheric Research 2: 233–256.

Crutzen, P.J., I. Aselmann, and W. Seiler. 1986. Methane production by domestic animals, wild ruminants, other herbivorous fauna, and humans. Tellus 38B: 271–284.

Cuevas, E. and E. Medina. 1986. Nutrient dynamics within Amazonian forest ecosystems. I. Nutrient flux in fine litter fall and efficiency of nutrient utilization. Oecologia 68: 466–472.

Cuevas, E. and E. Medina. 1988. Nutrient dynamics within Amazonian forests. II. Fine root growth, nutrient availability and leaf litter decomposition. Oecologia 76: 222–235.

Cuffney, T.F. 1988. Input, movement and exchange of organic matter within a subtropical coastal blackwater river-floodplain system. Freshwater Biology 19: 305–320.

Cullis, C.F. and M.M. Hirschler. 1980. Atmospheric sulphur: Natural and man-made sources. Atmospheric Environment 14: 1263–1278.

Curtis, P.S. 1996. A meta-analysis of leaf gas exchange and nitrogen in trees grown under elevated carbon dioxide. Plant, Cell, and Environment 19: 127–137.

Cushon, G.H. and M.C. Feller. 1989. Asymbiotic nitrogen fixation and denitrification in a mature forest in coastal British Columbia. Canadian Journal of Forest Research 19: 1194–1200.

Cyr, H. and M.L. Pace. 1993. Magnitude and patterns of herbivory in aquatic and terrestrial ecosystems. Nature 361: 148–150.

Dacey, J.W.H. 1981. Pressurized ventilation in the yellow waterlily. Ecology 62: 1137–1147.

Dacey, J.W.H. and B.L. Howes. 1984. Water uptake by roots controls water table movement and sediment oxidation in short *Spartina* marsh. Science 224: 487–489.

Dacey, J.W.H. and S.G. Wakeham. 1986. Oceanic dimethylsulfide: Production during zooplankton grazing on phytoplankton. Science 233: 1314–1316.

Dacey, J.W.H., B.G. Drake, and M.J. Klug. 1994. Stimulation of methane emission by carbon dioxide enrichment of marsh vegetation. Nature 370: 47–49.

Dahlgren, R.A. and F.C. Ugolini. 1989. Effects of tephra addition on soil processes in Spodosols in the Cascade Range, Washington, U.S.A. Geoderma 45: 331–355.

Dahlgren, R.A. and W.J. Walker. 1993. Aluminum release rates from selected Spodosol Bs horizons: Effect of pH and solid-phase aluminum pools. Geochimica et Cosmochimica Acta 57: 57–66.

Dalal, R.C. and R.J. Mayer. 1986a. Long-term trends in fertility of soils under continuous cultivation and cereal cropping in southern Queensland. III. Distribution and kinetics of soil organic carbon in particle-size fractions. Australian Journal of Soil Research 24: 293–300.

Dalal, R.C. and R.J. Mayer. 1986b. Long-term trends in fertility of soils under continuous cultivation and cereal cropping in southern Queensland. IV. Loss of organic carbon from different density fractions. Australian Journal of Soil Research 24: 301–309.

Damman, A.W.H. 1978. Distribution and movement of elements in ombrotrophic peat bogs. Oikos 30: 480–495.

Damman, A.W.H. 1988. Regulation of nitrogen removal and retention in sphagnum bogs and other peatlands. Oikos 51: 291–305.

Daniels, W.L., L.W. Zelazny, and C.J. Everett. 1987. Virgin hardwood forest soils of the southern Appalachian mountains: II. Weathering, mineralogy, and chemical properties. Soil Science Society of America Journal 51: 730–738.

Darling, M.S. 1976. Interpretation of global differences in plant calorific values: The significance of desert and arid woodland vegetation. Oecologia 23: 127–139.

D'Arrigo, R., G.C. Jacoby, and I.Y. Fung. 1987. Boreal forests and atmosphere-biosphere exchange of carbon dioxide. Nature 329: 321–323.

David, M.B., M.J. Mitchell, and J.P. Nakas. 1982. Organic and inorganic sulfur constituents of a forest soil and their relationship to microbial activity. Soil Science Society of America Journal 46: 847–852.

David, M.B., J.O. Reuss, and P.M. Walthall. 1988. Use of a chemical equilibrium model to understand soil chemical processes that influence soil solution and surface water alkalinity. Water, Air, and Soil Pollution 38: 71–83.

David, M.B., G.F. Vance, and W.J. Fasth. 1991. Forest soil response to acid and salt additions of sulfate. II. Aluminum and base cations. Soil Science 151: 208–219.

David, M.B., C.F. Vance, and J.S. Kahl. 1992. Chemistry of dissolved organic carbon and organic acids in two streams draining forested watersheds. Water Resources Research 28: 389–396.

Davidson, E.A. 1991. Fluxes of nitrous oxide and nitric oxide from terrestrial ecosystems. pp. 219–235. In J.E. Rogers and W.B. Whitman (eds.), *Microbial Production and Consumption of Greenhouse Gases.* American Society for Microbiology, Washington, D.C.

Davidson, E.A. and I.L. Ackerman. 1993. Changes in soil carbon inventories following cultivation of previously untilled soils. Biogeochemistry 20: 161–193.

Davidson, E.A. and W.T. Swank. 1987. Factors limiting denitrification in soils from mature and disturbed southeastern hardwood forests. Forest Science 33: 135–144.

Davidson, E.A., W.T. Swank, and T.O. Perry. 1986. Distinguishing between nitrification and denitrification as sources of gaseous nitrogen production in soil. Applied and Environmental Microbiology 52: 1280–1286.

Davidson, E.A., J.M. Stark, and M.K. Firestone. 1990. Microbial production and consumption of nitrate in an annual grassland. Ecology 71: 1968–1975.

Davidson, E.A., S.C. Hart, C.A. Shanks, and M.K. Firestone. 1991a. Measuring gross nitrogen mineralization, immobilization, and nitrification by ^{15}N isotopic pool dilution in intact soil cores. Journal of Soil Science 42: 335–349.

Davidson, E.A., P.M. Vitousek, P.A. Matson, R. Riley, G. García-Méndez, and J.M. Maass. 1991b. Soil emissions of nitric oxide in a seasonally dry tropical forest of Mexico. Journal of Geophysical Research 96: 15439–15445.

Davidson, E.A., S.C. Hart, and M.K. Firestone. 1992. Internal cycling of nitrate in soils of a mature coniferous forest. Ecology 73: 1148–1156.

Davidson, E.A., P.A. Matson, P.M. Vitousek, R. Riley, K. Dunkin, G. García-Méndez, and J.M. Maass. 1993. Processes regulating soil emissions of NO and N_2O in a seasonally dry tropical forest. Ecology 74: 130–139.

Davies, W.G. 1972. *Introduction to Chemical Thermodynamics*. Saunders, Philadelphia.

Davis, W.L. and C.P. McKay. 1996. Origins of life: A comparison of theories and application to Mars. Origins of Life and Evolution of the Biosphere 26: 61–73.

Davison, W., C. Woof, and E. Rigg. 1982. The dynamics of iron and manganese in a seasonally anoxic lake: Direct measurement of fluxes using sediment traps. Limnology and Oceanography 27: 987–1003.

Dean, J.M. and A.P. Smith. 1978. Behavioral and morphological adaptations of a tropical plant to high rainfall. Biotropica 10: 152–154.

De Angelis, M., N.I. Barkov, and V.N. Petrov. 1987. Aerosol concentrations over the last climatic cycle (160 kyr) from an Antarctic ice core. Nature 325: 318–321.

DeBano, L.F. and C.E. Conrad. 1978. The effects of fire on nutrients in a chaparral ecosystem. Ecology 59: 489–497.

DeBano, L.F. and J.M. Klopatek. 1988. Phosphorus dynamics of pinyon-juniper soils following simulated burning. Soil Science Society of America Journal 52: 271–277.

DeBell, D.S. and C.W. Ralston. 1970. Release of nitrogen by burning light forest fuels. Soil Science Society of America Proceedings 34: 936–938.

de Bergh, C., B. Bézard, T. Owen, D. Crisp, J.-P. Maillard, and B.L. Lutz. 1991. Deuterium on Venus: Observations from Earth. Science 251: 547–549.

de Duve, C. 1995. The beginnings of life on Earth. American Scientist 83: 428–437.

Deevey, E.S. 1970a. Mineral cycles. Scientific American 223(3): 148–158.

Deevey, E.S. 1970b. In defense of mud. Bulletin of the Ecological Society of America 51(1): 5–8.

Deevey, E.S. 1988. Estimation of downward leakage from Florida lakes. Limnology and Oceanography 33: 1308–1320.

DeFries, R.S., C.B. Field, I. Fung, C.O. Justice, S. Los, P.A. Matson, E. Matthews, H.A. Mooney, C.S. Potter, K. Prentice, P.J. Sellers, J.R.G. Townshend, C.J. Tucker, S.L. Ustin, and P.M. Vitousek. 1995. Mapping the land surface for global atmosphere-biosphere models: Towards continuous distributions of vegetation's functional properties. Journal of Geophysical Research 100: 20867–20882.

Degens, E.T., H.K. Wong, and S. Kempe. 1981. Factors controlling global climate of the past and the future. pp. 3–24. *In* G.E. Likens (ed.), *Some Perspectives of the Major Biogeochemical Cycles*. Wiley, New York.

Degens, E.T., S. Kempe, and J.E. Richey. 1991. Summary: Biogeochemistry of major world rivers. pp. 323–347. In E.T. Degens, S. Kempe, and J.E. Richey (eds.), *Biogeochemistry of Major World Rivers*. Wiley, New York.

De Jonge, V.N. and L.A. Villerius. 1989. Possible role of carbonate dissolution in estuarine phosphate dynamics. Limnology and Oceanography 34: 332–340.

De Kimpe, C.R. and Y.A. Martel. 1976. Effects of vegetation on the distribution of carbon, iron, and aluminum in the B horizons of Northern Appalachian Spodosols. Soil Science Society of America Journal 40: 77–80.

del Arco, J.M., A. Escudero, and M.V. Garrido. 1991. Effects of site characteristics on nitrogen retranslocation from senescing leaves. Ecology 72: 701–708.

DeLaune, R.D., R.H. Baumann, and J.G. Gosselink. 1983. Relationships among vertical accretion, coastal submergence, and erosion in a Louisiana Gulf Coast marsh. Journal of Sedimentary Petrology 53: 147–157.

DeLaune, R.D., T.C. Feijtel, and W.H. Patrick. 1989. Nitrogen flows in Louisiana Gulf Coast salt marsh: Spatial considerations. Biogeochemistry 8: 25–37.

Delcourt, H.R. and W.F. Harris. 1980. Carbon budget of the southeastern U.S. biota: Analysis of historical change in trend from source to sink. Science 210: 321–323.

del Giorgio, P.A. and R.H. Peters. 1994. Patterns in planktonic P:R ratios in lakes: Influence of lake trophy and dissolved organic carbon. Limnology and Oceanography 39: 772–787.

Delmas, R. and J. Servant. 1983. Atmospheric balance of sulphur above an equatorial forest. Tellus 35B: 110–120.

Delmas, R.A., A. Marenco, J.P. Tathy, B. Cros, and J.G.R. Baudet. 1991. Sources and sinks of methane in the African savanna: CH_4 emissions from biomass burning. Journal of Geophysical Research 96: 7287–7299.

Delmas, R.A., J. Servant, J.P. Tathy, B. Cros, and M. Labat. 1992a. Sources and sinks of methane and carbon dioxide exchanges in mountain forest in equatorial Africa. Journal of Geophysical Research 97: 6169–6179.

Delmas, R.A., J.P. Tathy, and B. Cros. 1992b. Atmospheric methane budget in Africa. Journal of Atmospheric Chemistry 14: 395–409.

Delmas, R.A., J.P. Lacaux, J.C. Menaut, L. Abbadie, X. Le Roux, G. Helas, and J. Lobert. 1995. Nitrogen compound emission from biomass burning in tropical African savanna FOS/DECAFE 1991 Experiment (Lamto, Ivory Coast). Journal of Atmospheric Chemistry 22: 175–193.

Delsemme, A.H. 1992. Cometary origin of carbon, nitrogen and water on the Earth. Origins of Life and Evolution of the Biosphere 21: 279–298.

DeLuca, T.H. and D.R. Keeney. 1993. Glucose-induced nitrate assimilation in prairie and cultivated soils. Biogeochemistry 21: 167–176.

DeLucia, E.H. and W.H. Schlesinger. 1991. Resource-use efficiency and drought tolerance in adjacent Great Basin and Sierran plants. Ecology 72: 51–58.

DeLucia, E.H. and W.H. Schlesinger. 1995. Photosynthetic rates and nutrient-use efficiency among evergreen and deciduous shrubs in Okefenokee Swamp. International Journal of Plant Science 156: 19–28.

DeLucia, E.H., W.H. Schlesinger, and W.D. Billings. 1988. Water relations and the maintenance of Sierran conifers on hydrothermally altered rock. Ecology 69: 303–311.

DeLucia, E.H., R.M. Callaway, E.M. Thomas, and W.H. Schlesinger. 1996. Mechanisms of P acquisition for ponderosa pine seedlings under different climatic regimes. Annals of Botany, in press.

Delwiche, C.C. 1970. The nitrogen cycle. Scientific American 223(3): 136–146.

DeMaster, D.J. 1981. The supply and accumulation of silica in the marine environment. Geochimica et Cosmochimica Acta 45: 1715–1732.

Deng, S.P. and M.A. Tabatabai. 1994. Cellulase activity of soils. Soil Biology and Biochemistry 26: 1347–1354.

Denier van der Gon, H.A.C. and H.U. Neue. 1995. Influence of organic matter incorporation on the methane emission from a wetland rice field. Global Biogeochemical Cycles 9: 11–22.

Denmead, O.T., J.R. Freney, and J.R. Simpson. 1976. A closed ammonia cycle within a plant canopy. Soil Biology and Biochemistry 8: 161–164.

Denning, A.S., I.Y. Fung, and D. Randall. 1995. Latitudinal gradient of atmospheric CO_2 due to seasonal exchange with land biota. Nature 376: 240–243.

Dentener, F.J. and P.J. Crutzen. 1994. A three-dimensional model of the global ammonia cycle. Journal of Atmospheric Chemistry 19: 331–369.

Deshler, T., D.J. Hoffman, J.V. Hereford, and C.B. Sutter. 1990. Ozone and temperature profiles over McMurdo station Antarctica in the spring of 1989. Geophysical Research Letters 17: 151–154.

Des Marais, D.J., H. Strauss, R.E. Summons, and J.M. Hayes. 1992. Carbon isotope evidence for the stepwise oxidation of the Proterozoic environment. Nature 359: 605–609.

Dethier, D.P., S.B. Jones, T.P. Feist, and J.E. Ricker. 1988. Relations among sulfate, aluminum, iron, dissolved organic carbon, and pH in upland forest soils of northwestern Massachusetts. Soil Science Society of America Journal 52: 506–512.

Detwiler, R.P. 1986. Land use change and the global carbon cycle: The role of tropical soils. Biogeochemistry 2: 67–93.

Deuser, W.G., E.H. Ross, and R.F. Anderson. 1981. Seasonality in the supply of sediment to the deep Sargasso sea and implications for the rapid transfer of matter to the deep ocean. Deep Sea Research 28: 495–505.

Deuser, W.G., P.G. Brewer, T.D. Jickells, and R.F. Commeau. 1983. Biological control of the removal of abiogenic particles from the surface ocean. Science 219: 388–391.

Dévai, I., and R.D. DeLaune. 1995. Evidence for phosphine production and emission from Louisiana and Florida marsh soils. Organic Geochemistry 23: 277–279.

Dévai, I., L. Felföldy, I. Wittner, and S. Plósz. 1988. Detection of phosphine: New aspects of the phosphorus cycle in the hydrosphere. Nature 333: 343–345.

Devito, K.J. and P.J. Dillon. 1993. The influence of hydrologic conditions and peat oxia on the phosphorus and nitrogen dynamics of a conifer swamp. Water Resources Research 29: 2675–2685.

Devol, A.H. 1991. Direct measurement of nitrogen gas fluxes from continental shelf sediments. Nature 349: 319–321.

Devol, A.H., J.E. Richey, W.A. Clark, S.L. King, and L.A. Martinelli. 1988. Methane emissions to the troposphere from the Amazon floodplain. Journal of Geophysical Research 93: 1583–1592.

Dhamala, B.R. and M.J. Mitchell. 1995. Sulfur speciation, vertical distribution, and seasonal variation in a northern hardwood forest soil, U.S.A. Canadian Journal of Forest Research 25: 234–243.

Dia, A.N., A.S. Cohen, R.K. O'Nions, and N.J. Shackleton. 1992. Seawater Sr isotope variation over the past 300 kyr and influence of global climate cycles. Nature 356: 786–788.

Dianov-Klokov, V.I., L.N. Yurganov, E.I. Grechko, and A.V. Dzhola. 1989. Spectroscopic measurements of atmospheric carbon monoxide and methane. I. Latitudinal distribution. Journal of Atmospheric Chemistry 8: 139–151.

Diáz-Raviña, M., M.J. Acea, and T. Carballas. 1993. Microbial biomass and its contribution to nutrient concentrations in forest soils. Soil Biology and Biochemistry 25: 25–31.

Dickerson, R.E. 1978. Chemical evolution and the origin of life. Scientific American 239(3): 70–86.

Dickerson, R.R., G.J. Huffman, W.T. Luke, L.J. Nunnermacker, K.E. Pickering, A.C.D. Leslie, C.G. Lindsey, W.G.N. Slinn, T.J. Kelley, P.H. Daum, A.C. Delany, J.P. Greenberg, P.R. Zimmerman, J.F. Boatman, J.D. Ray, and D.H. Stedman. 1987. Thunderstorms: An important mechanism in the transport of air pollutants. Science 235: 460–465.

Dickerson, R.R., B.G. Doddridge, P. Kelley, and K.P. Rhoads. 1995. Large-scale pollution of the atmosphere over the remote Atlantic ocean: Evidence from Bermuda. Journal of Geophysical Research 100: 8945–8952.

Dickinson, R.E. and R.J. Cicerone. 1986. Future global warming from atmospheric trace gases. Nature 319: 109–115.

Dickson, M.-L. and P.A. Wheeler. 1995. Nitrate uptake rates in a coastal upwelling regime: A comparison of PN-specific, absolute, and Chl *a*-specific rates. Limnology and Oceanography 40: 533–543.

Dickson, R.R. and J. Brown. 1994. The production of North Atlantic deep water: Sources, rates, and pathways. Journal of Geophysical Research 99: 12319–12341.

Dillon, P.J. and F.H. Rigler. 1974. The phosphorus-chlorophyll relationship in lakes. Limnology and Oceanography 19: 767–773.

Dingman, S.L. and A.H. Johnson. 1971. Pollution potential of some New Hampshire lakes. Water Resources Research 7: 1208–1215.

Dinkelaker, B. and H. Marschner. 1992. In vivo demonstration of acid phosphatase activity in the rhizosphere of soil-grown plants. Plant and Soil 144: 199–205.

Dirmeyer, P.A. and J. Shukla. 1996. The effect on regional and global climate of expansion of the world's deserts. Quarterly Journal of the Royal Meteorological Society 122: 451–482.

Dixon, K.W., J.S. Pate, and W.J. Bailey. 1980. Nitrogen nutrition of the tuberous sundew *Drosera erythrorhiza* Lindl. with special reference to catch of arthropod fauna by its gladular leaves. Australian Journal of Botany 28: 283–297.

Dixon, R.K., S. Brown, R.A. Houghton, A.M. Solomon, M.C. Trexler, and J. Wisniewski. 1994. Carbon pools and flux in global forest ecosystems. Science 263: 185–190.

Dlugokencky, E.J., L.P. Steele, P.M. Lang, and K.A. Masarie. 1994. The growth rate and distribution of atmospheric methane. Journal of Geophysical Research 99: 17021–17043.

Dobrovolsky, V.V. 1994. *Biogeochemistry of the World's Land.* CRC Press, Boca Raton, Florida.

Dodd, J.C., C.C. Burton, R.G. Burns, and P. Jeffries. 1987. Phosphatase activity associated with the roots and the rhizosphere of plants infected with vesicular-arbuscular mycorrhizal fungi. New Phytologist 107: 163–172.

Donahue, T.M., J.H. Hoffman, R.R. Hodges, and A.J. Watson. 1982. Venus was wet: A measurement of the ratio of deuterium to hydrogen. Science 216: 630–633.

Donoso, L., R. Santana, and E. Sanhueza. 1993. Seasonal variation of N_2O fluxes at a tropical savanna site: Soil consumption of N_2O during the dry season. Geophysical Research Letters 20: 1379–1382.

Doolittle, R.F., D.-F. Feng, S. Tsang, G. Cho, and E. Little. 1996. Determining divergence times of the major kingdoms of living organisms with a protein clock. Science 271: 470–477.

Dormaar, J.F. 1979. Organic matter characteristics of undisturbed and cultivated chernozemic and solonetzic A horizons. Canadian Journal of Soil Science 59: 349–356.

Dorn, H.-P., J. Callies, U. Platt, and D.H. Ehhalt. 1988. Measurement of tropospheric OH concentrations by laser long-path absorption spectroscopy. Tellus 40B: 437–445.

Dornblaser, M., A.E. Giblin, B. Fry, and B.J. Peterson. 1994. Effects of sulfate concentration in the overlying water on sulfate reduction and sulfur storage in lake sediments. Biogeochemistry 24: 129–144.

Dörr, H. and K.O. Münnich. 1989. Downward movement of soil organic matter and its influence on trace-element transport (^{210}Pb, ^{137}Cs) in the soil. Radiocarbon 31: 655–663.

Dosskey, M.G. and P.M. Bertsch. 1994. Forest sources and pathways of organic matter transport to a blackwater stream: A hydrologic approach. Biogeochemistry 24: 1–19.

Downing, J.A. and E. McCauley. 1992. The nitrogen:phosphorus relationship in lakes. Limnology and Oceanography 37: 936–945.

Downs, M.R., K.J. Nadelhoffer, J.M. Melillo, and J.D. Aber. 1996. Immobilization of a ^{15}N-labeled nitrate addition by decomposing forest litter. Oecologia 105: 141–150.

Doyle, R.D. and T.R. Fisher. 1994. Nitrogen fixation by periphyton and plankton on the Amazon floodplain at Lake Calado. Biogeochemistry 26: 41–66.

Draaijers, G.P.J., W.P.M.F. Ivens, M.M. Bos, and W. Bleuten. 1989. The contribution of ammonia emissions from agriculture to the deposition of acidifying and eutrophying compounds onto forests. Environmental Pollution 60: 55–66.

Dregne, H.E. 1976. *Soils of Arid Regions.* Elsevier Scientific Publishers, Amsterdam.

Drever, J.I. 1988. *The Geochemistry of Natural Waters.* 2nd ed. Prentice-Hall, Englewood Cliffs, New Jersey.

Drever, J.I. 1994. The effect of land plants on weathering rates of silicate minerals. Geochimica et Cosmochimica Acta 58: 2325–2332.

Drever, J.I. and C.L. Smith. 1978. Cyclic wetting and drying of the soil zone as an influence on the chemistry of ground water in arid terrains. American Journal of Science 278: 1448–1454.

Drever, J.I. and J. Zobrist. 1992. Chemical weathering of silicate rocks as a function of elevation in the southern Swiss Alps. Geochimica et Cosmochimica Acta 56: 3209–3216.

Driscoll, C.T., N. van Breemen, and J. Mulder. 1985. Aluminum chemistry in a forested Spodosol. Soil Science Society of America Journal 49: 437–444.

Druffel, E.R.M., P.M. Williams, J.E. Bauer, and J.R. Ertel. 1992. Cycling of dissolved and particulate organic matter in the open ocean. Journal of Geophysical Research 97: 15639–15659.

Drury, C.F., R.P. Voroney, and E.G. Beauchamp. 1991. Availability of NH_4^+-N to microorganisms and the soil internal N cycle. Soil Biology and Biochemistry 23: 165–169.

Drury, C.F., D.J. McKenney, and W.I. Findlay. 1992. Nitric oxide and nitrous oxide production from soil: Water and oxygen effects. Soil Science Society of America Journal 56: 766–770.

Duce, R.A. and N.W. Tindale. 1991. Atmospheric transport of iron and its deposition in the ocean. Limnology and Oceanography 36: 1715–1726.

Duce, R.A., C.K. Unni, B.J. Ray, J.M. Prospero, and J.T. Merrill. 1980. Long-range atmospheric transport of soil dust from Asia to the tropical North Pacific: Temporal variability. Science 209: 1522–1524.

Duce, R.A., P.S. Liss, J.T. Merrill, E.L. Atlas, P. Buat-Menard, B.B. Hicks, J.M. Miller, J.M. Prospero, R. Arimoto, T.M. Church, W. Ellis, J.N. Galloway, L. Hansen, T.D. Jickells, A.H. Knap, K.H. Reinhardt, B. Schneider, A. Soudine, J.J. Tokos, S. Tsunogai, R. Wollast, and M. Zhou. 1991. The atmospheric input of trace species to the world ocean. Global Biogeochemical Cycles 5: 193–259.

Ducklow, H.W. and C.A. Carlson. 1992. Oceanic bacterial production. Advances in Microbial Ecology 12: 113–181.

Ducklow, H.W., D.A. Purdie, P.J. LeB. Williams, and J.M. Davies. 1986. Bacterioplankton: A sink for carbon in a coastal marine plankton community. Science 232: 865–867.

Duff, S.M.G., G. Sarath, and W.C. Plaxton, 1994. The role of acid phosphatases in plant phosphorus metabolism. Physiologia Plantarum 90: 791–800.

Dugdale, R.C. and J.J. Goering. 1967. Uptake of new and regenerated forms of nitrogen in primary productivity. Limnology and Oceanography 12: 196–206.

Duggin, J.A., G.K. Voigt, and F.H. Bormann. 1991. Autotrophic and heterotrophic nitrification in response to clear-cutting northern hardwood forest. Soil Biology and Biochemistry 23: 779–787.

Dunfield, P., R. Knowles, R. Dumont, and T.R. Moore. 1993. Methane production and consumption in temperate and subarctic peat soils: Response to temperature and pH. Soil Biology and Biochemistry 25: 321–326.

Dunn, P.H., L.F. DeBano, and G.E. Eberlein. 1979. Effects of burning on chaparral soils: II. Soil microbes and nitrogen mineralization. Soil Science Society of America Journal 43: 509–514.

Dunne, T. and L.B. Leopold. 1978. *Water in Environmental Planning.* Freeman, San Francisco.

Du Rietz, G.E. 1949. Huvudenheter och Huvudgränser i Svensk Myrvegetation. Svensk Botanisk Tidskrift 43: 274–309.

Durka, W., E.-D. Schulze, G. Gebauer, and S. Voerkelius. 1994. Effects of forest decline on uptake and leaching of deposited nitrate determined from ^{15}N and ^{18}O measurements. Nature 372: 765–767.

Duyzer, J. and D. Fowler. 1994. Modelling land atmospheric exchange of gaseous oxides of nitrogen in Europe. Tellus 46B: 353–372.

Dynesius, M. and C. Nilsson. 1994. Fragmentation and flow regulation of river systems in the northern third of the world. Science 266: 753–762.

Dyrness, C.T., K. Van Cleve, and J.D. Levison. 1989. The effect of wildfire on soil chemistry in four forest types in interior Alaska. Canadian Journal of Forest Research 19: 1389–1396.

Edmond, J.M., C. Measures, R.E. McDuff, L.H. Chan, R. Collier, B. Grant, L.I. Gordon, and J.B. Corliss. 1979. Rdige crest hydrothermal activity and the balances of the major and minor elements in the ocean: The Galapagos data. Earth and Planetary Science Letters 46: 1–18.

Edmond, J.M., E.A. Boyle, B. Grant, and R.F. Stallard. 1981. The chemical mass balance in the Amazon plume. I: The nutrients. Deep Sea Research 28: 1339–1374.

Edmondson, W.T. and J.T. Lehman. 1981. The effect of changes in the nutrient income on the condition of Lake Washington. Limnology and Oceanography 26: 1–29.

Edwards, N.T. 1975. Effects of temperature and moisture on carbon dioxide evolution in a mixed deciduous forest floor. Soil Science Society of America Proceedings 39: 361–365.

Edwards, N.T. and W.F. Harris. 1977. Carbon cycling in a mixed deciduous forest floor. Ecology 58: 431–437.

Edwards, N.T. and P. Sollins. 1973. Continuous measurement of carbon dioxide evolution from partitioned forest floor components. Ecology 54: 406–412.

Edwards, P.J. 1977. Studies of mineral cycling in a montane rain forest in New Guinea. II. The production and disappearance of litter. Journal of Ecology 65: 971–992.

Edwards, P.J. 1982. Studies of mineral cycling in a montane rain forest in New Guinea. V. Rates of cycling in throughfall and litter fall. Journal of Ecology 70: 807–827.

Edwards, P.J. and P.J. Grubb. 1982. Studies of mineral cycling in a montane rain forest in New Guinea. IV. Soil characteristics and the division of mineral elements between the vegetation and soil. Journal of Ecology 70: 649–666.

Edwards, R.T. and J.L. Meyer. 1986. Production and turnover of planktonic bacteria in two southeastern blackwater rivers. Applied and Environmental Microbiology 52: 1317–1323.

Edwards, R.T. and J.L. Meyer. 1987. Metabolism of a sub-tropical low gradient blackwater river. Freshwater Biology 17: 251–263.

Eghbal, M.K., R.J. Southard, and L.D. Whittig. 1989. Dynamics of evaporite distribution in soils on a fan-playa transect in the Carrizo Plain, California. Soil Science Society of America Journal 53: 898–903.

Ehhalt, D.H. 1974. The atmospheric cycle of methane. Tellus 26: 58–70.

Ehleringer, J.R. and R.K. Monson. 1993. Evolutionary and ecological aspects of photosynthetic pathway variation. Annual Review of Ecology and Systematics 24: 411–439.

Ehrlich, H.L. 1975. The formation of ores in the sedimentary environment of the deep sea with microbial participation: The case for ferromanganese concretions. Soil Science 119: 36–41

Ehrlich, H.L. 1982. Enhanced removal of Mn^{2+} from seawater by marine sediments and clay minerals in the presence of bacteria. Canadian Journal of Microbiology 28: 1389–1395.

Eichner, M.J. 1990. Nitrous oxide emissions from fertilized soils: Summary of available data. Journal of Environmental Quality 19: 272–280.

Eisele, K.A., D.S. Schimel, L.A. Kapustka, and W.J. Parton. 1989. Effects of available P and N:P ratios on non-symbiotic dinitrogen fixation in tallgrass prairie soils. Oecologia 79: 471–474.

Elkins, J.W., T.W. Thompson, T.H. Swanson, J.H. Butler, B.D. Hall, S.O. Cummings, D.A. Fisher, and A.G. Raffo. 1993. Decrease in the growth rates of atmospheric chlorofluorocarbons 11 and 12. Nature 364: 780–783.

Elliott, E.T. 1986. Aggregate structure and carbon, nitrogen, and phosphorus in native and cultivated soils. Soil Science Society of America Journal 50: 627–633.

Ellsworth, D.S., R. Oren, C. Huang, N. Phillips, and G.R. Hendrey. 1995. Leaf and canopy responses to elevated CO_2 in a pine forest under free-air CO_2 enrichment. Oecologia 104:139–146.

Elser, J.J. and R.P. Hassett. 1994. A stoichiometric analysis of the zooplankton-phytoplankton interaction in marine and freshwater ecosystems. Nature 370: 211–213.

Elser, J.J., T.H. Chrzanowski, R.W. Sterner, J.H. Schampel, and D.K. Foster. 1995. Elemental ratios and the uptake and release of nutrients by phytoplankton and bacteria in three lakes of the Canadian Shield. Microbial Ecology 29: 145–162.

Elsgaard, L., M.F. Isaksen, B.B. Jørgensen, A-M. Alayse, and H.W. Jannasch. 1994. Microbial sulfate reduction in deep-sea sediments at the Guaymas Basin hydrothermal vent area: Influence of temperature and substrates. Geochimica et Cosmochimica Acta 58: 3335–3343.

Eltahir, E.A.B. and R.L. Bras. 1994. Precipitation recycling in the Amazon basin. Quarterly Journal of the Royal Meteorological Society 120: 861–880.

Elwood, J.W., J.D. Newbold, A.F. Trimble, and R.W. Stark. 1981. The limiting role of phosphorus in a woodland stream ecosystem: Effects of P enrichment on leaf decomposition and primary production. Ecology 62: 146–158.

Emanuel, W.R., H.H. Shugart, and M.P. Stevenson. 1985a. Climatic change and the broad-scale distribution of terrestrial ecosystem complexes. Climatic Change 7: 29–43.

Emanuel, W.R., I.Y.-S. Fung, G.G. Killough, B. Moore, and T.-H. Peng. 1985b. Modeling the global carbon cycle and changes in the atmospheric carbon dioxide levels. pp. 141–173. In J.R. Trabalka (ed.), *Atmospheric Carbon Dioxide and the Global Carbon Cycle.* U.S. Department of Energy, Washington, D.C.

Emerson, S., K. Fischer, C. Reimers, and D. Heggie. 1985. Organic carbon dynamics and preservation in deep-sea sediments. Deep Sea Research 32: 1–21.

Emmett, B.A., J.A. Hudson, P.A. Coward, and B. Reynolds. 1994. The impact of a riparian wetland on streamwater quality in a recently afforested upland catchment. Journal of Hydrology 162: 337–353.

Enriquez, S., C.M. Duarte, and K. Sand-Jensen. 1993. Patterns in decomposition rates among photosynthetic organisms: The importance of detritus C:N:P content. Oecologia 94: 457–471.

Eppley, R.W. and B.J. Peterson. 1979. Particulate organic matter flux and planktonic new production in the deep ocean. Nature 282: 677–680.

Eppley, R.W., E. Stewart, M.R. Abbott, and U. Heyman. 1985. Estimating ocean primary production from satellite chlorophyll: Introduction to regional differences and statistics for the southern California Bight. Journal of Plankton Research 7: 57–70.

Epstein, C.B. and M. Oppenheimer. 1986. Empirical relation between sulphur dioxide emissions and acid deposition derived from monthly data. Nature 323: 245–247.

Erickson, D.J., J.S. Ghan, and J.E. Penner. 1990. Global ocean-to-atmosphere dimethyl sulfide flux. Journal of Geophysical Research 95: 7543–7552.

Eriksson, E. 1960. The yearly circulation of chloride and sulfur in nature; meteorological, geochemical and pedological implications. Part II. Tellus 12: 63–109.

Ertel, J.R., J.I. Hedges, A.H. Devol, J.E. Richey, and M.N.G. Ribeiro. 1986. Dissolved humic substances of the Amazon river system. Limnology and Oceanography 31: 739–754.

Escudero, A., J.M. del Arco, I.C. Sanz, and J. Ayala. 1992. Effects of leaf longevity and retranslocation efficiency on the retention time of nutrients in the leaf biomass of different woody species. Oecologia 90: 80–87.

Esser, B.K. and K.K. Turekian. 1988. Accretion rate of extraterrestrial particles determined from osmium isotope systematics of Pacific pelagic clay and manganese nodules. Geochimica et Cosmochimica Acta 52: 1383–1388.

Esser, G., I. Aselmann, and H. Lieth. 1982. Modelling the carbon reservoir in the system compartment 'litter.' Mitteilungen aus dem Geologisch-Paläontologischen Institut der Universität Hamburg 52: 39–58.

Eswaran, H., E. Van Den Berg, and P. Reich. 1993. Organic carbon in soils of the world. Soil Science Society of America Journal 57: 192–194.

Eswaran, H., E. Van Den Berg, P. Reich, and J. Kimble. 1995. Global soil carbon resources. pp. 27–43. In R. Lal, J. Kimble, E. Levine, and B.A. Stewart (eds.), *Soils and Global Change.* CRC Lewis Publishers, Boca Raton, Florida.

Etheridge, D.M., L.P. Steele, R.L. Langenfelds, R.J. Francy, J.-M. Barnola, and V.I. Morgan. 1996. Natural and anthropogenic changes in atmospheric CO_2 over the last 1000 years from air in Antarctic ice and firn. Journal of Geophysical Research 101: 4115–4128.

Evans, C.V. and D.P. Franzmeier. 1988. Color index values to represent wetness and aeration in some Indiana soils. Geoderma 41: 353–368.

Evans, J.R. 1989. Photosynthesis and nitrogen relationships in leaves of C_3 plants. Oecologia 78: 9–19.

Evans, L.J. 1980. Podzol development north of Lake Huron in relation to geology and vegetation. Canadian Journal of Soil Science 60: 527–539.

Evans, R.D. and J.R. Ehleringer. 1993. A break in the nitrogen cycle in aridlands? Evidence from $\delta^{15}N$ of soils. Oecologia 94: 314–317.

Eyre, B. 1994. Nutrient biogeochemistry in the tropical Moresby River estuary system North Queensland, Australia. Estuarine, Coastal and Shelf Science 39: 15–31.

Fahey, T.J. 1983. Nutrient dynamics of aboveground detritus in lodgepole pine (*Pinus contorta* ssp. *latifolia*) ecosystems, southeastern Wyoming. Ecological Monographs 53: 51–72.

Fahey, T.J. and J.W. Hughes. 1994. Fine root dynamics in a northern hardwood forest ecosystem, Hubbard Brook Experimental Forest, N.H. Journal of Ecology 82: 533–548.

Fairbanks, R.G. 1989. A 17,000-year glacio-eustatic sea level record: Influence of glacial melting rates on the Younger Dryas event and deep-ocean circulation. Nature 342: 637–642.

Falkengren-Grerup, U. 1995. Interspecies differences in the preference of ammonium and nitrate in vascular plants. Oecologia 102: 305–311.

Falkowski, P.G. and C. Wilson. 1992. Phytoplankton productivity in the North Pacific Ocean since 1900 and implications for absorption of anthropogenic CO_2. Nature 358: 741–743.

Falkowski, P.G., Y. Kim, Z. Kolber, C. Wilson, C. Wirick, and R. Cess. 1992. Natural versus anthropogenic factors affecting low-level cloud albedo over the North Atlantic. Science 256: 1311–1313.

Fanale, F.P. 1971. A case for catastrophic early degassing of the Earth. Chemical Geology 8: 79–105.

Fanning, K.A. 1989. Influence of atmospheric pollution on nutrient limitation in the ocean. Nature 339: 460–463.

Farman, J.C., B.G. Gardiner, and J.D. Shanklin. 1985. Large losses of total ozone in Antarctic reveal seasonal ClO_x/NO_x interaction. Nature 315: 207–210.

Farmer, D.M., C.L. McNeil, and B.D. Johnson. 1993. Evidence for the importance of bubbles in increasing air-sea gas flux. Nature 361: 620–623.

Farquhar, G.D., R. Wetselaar, and P.M. Firth. 1979. Ammonia volatilization from senescing leaves of maize. Science 203: 1257–1258.

Farquhar, G.D., K.T. Hubick, A.G. Condon, and R.A. Richards. 1989. Carbon isotope fractionation and plant water-use efficiency. pp. 21–40. In P.W. Rundel, J.R. Ehleringer, and K.A. Nagy (eds.), *Stable Isotopes in Ecological Research.* Springer-Verlag, New York.

Farrar, J.F. 1985. The respiratory source of CO_2. Plant, Cell and Environment 8: 427–438.

Fassbinder, J.W.E., H. Stanjek, and H. Vali. 1990. Occurrence of magnetic bacteria in soil. Nature 343: 161–163.

Faulkner, S.P., W.H. Patrick, and R.P. Gambrell. 1989. Field techniques for measuring wetland soil parameters. Soil Science Society of America Journal 53: 883–890.

Faure, H. 1990. Changes in the global continental reservoir of carbon. Palaeography, Palaeoclimatology, and Palaeoecology 82: 47–52.

Fearnside, P.M., N. Leal, and F.M. Fernandes. 1993. Rainforest burning and the global carbon budget: Biomass, combustion efficiency, and charcoal formation in the Brazilian Amazon. Journal of Geophysical Research 98: 16733–16743.

Federer, C.A. 1983. Nitrogen mineralization and nitrification: Depth variation in four New England forest soils. Soil Science Society of America Journal 47: 1008–1014.

Federer, C.A. and J.W. Hornbeck. 1985. The buffer capacity of forest soils in New England. Water, Air, and Soil Pollution 26: 163–173.

Fee, E.J., R.E. Hecky, G.W. Regehr, L.L. Hendzel, and P. Wilkinson. 1994. Effects of lake size on nutrient availability in the mixed layer during summer stratification. Canadian Journal of Fisheries and Aquatic Sciences 51: 2756–2768.

Feller, M.C. and J.P. Kimmins. 1979. Chemical characteristics of small streams near Haney in southwestern British Columbia. Water Resources Research 15: 247–258.

Ferdelman, T.G., T.M. Church, and G.W. Luther. 1991. Sulfur enrichment of humic substances in a Delaware salt marsh sediment core. Geochimica et Cosmochimica Acta 55: 979–988.

Ferek, R.J. and M.O. Andreae. 1984. Photochemical production of carbonyl sulphide in marine surface waters. Nature 307: 148–150.

Ferek, R.J., R.B. Chatfield, and M.O. Andreae. 1986. Vertical distribution of dimethylsulphide in the marine atmosphere. Nature 320: 514–516.

Ferris, J.P., A.R. Hill, R. Liu, and L.E. Orgel. 1996. Synthesis of long prebiotic oligomers on mineral surfaces. Nature 381: 59–61.

Fiebig, D.M., M.A. Lock, and C. Neal. 1990. Soil water in the riparian zone as a source of carbon for a headwater stream. Journal of Hydrology 116: 217–237.

Field, C. and H.A. Mooney. 1986. The photosynthesis-nitrogen relationship in wild plants. pp. 25–55. In T.J. Givnish (ed.), *On the Economy of Plant Form and Function.* Cambridge University Press, Cambridge.

Field, C., J. Merino, and H.A. Mooney. 1983. Compromises between water-use efficiency and nitrogen-use efficiency in five species of California evergreens. Oecologia 60: 384–389.

Field, C.B., J.T. Randerson, and C.M. Malmström. 1995. Global net primary production: Combining ecology and remote sensing. Remote Sensing of Environment 51: 74–88.

Fife, D.N. and E.K.S. Nambiar. 1984. Movement of nutrients in Radiata pine needles in relation to the growth of shoots. Annals of Botany 54: 303–314.

Figueres, G., J.-M. Martin, and M. Meybeck. 1978. Iron behaviour in the Zaire estuary. Netherlands Journal of Sea Research 12: 329–337.

Filippelli, G.M. and M.L. Delaney. 1996. Phosphorus geochemistry of equatorial Pacific sediments. Geochimica et Cosmochimica Acta 60: 1479–1495.

Findlay, S., D. Strayer, C. Goumbala, and K. Gould. 1993. Metabolism of streamwater dissolved organic carbon in the shallow hyporheic zone. Limnology and Oceanography 38: 1493–1499.

Firestone, M.K. 1982. Biological denitrification. pp. 289–326. In F.J. Stevenson (ed.), *Nitrogen in Agricultural Soils.* American Society of Agronomy, Madison, Wisconsin.

Firestone, M.K. and E.A. Davidson. 1989. Microbiological basis of NO and N_2O production and consumption in soil. pp. 7–21. In M.O. Andreae and D.S. Schimel (eds.), *Exchange of Trace Gases Between Terrestrial Ecosystems and the Atmosphere.* Wiley, New York.

Firestone, M.K., R.B. Firestone, and J.M. Tiedje. 1980. Nitrous oxide from soil denitrification: Factors controlling its biological production. Science 208: 749–751.

Fisher, R.F. 1972. Spodosol development and nutrient distribution under *Hydnaceae* fungal mats. Soil Science Society of America Proceedings 36: 492–495.

Fisher, R.F. 1977. Nitrogen and phosphorus mobilization by the fairy ring fungus, *Marasmius oreades* (Bolt.) Fr. Soil Biology and Biochemistry 9: 239–241.

Fisher, S.G. 1977. Organic matter processing by a stream-segment ecosystem: Fort River, Massachusetts, U.S.A. Internationale Revue der Gesamten Hydrobiologie 62: 701–727.

Fisher, S.G. and N.B. Grimm. 1985. Hydrologic and material budgets for a small Sonoran desert watershed during three consecutive cloudburst floods. Journal of Arid Environments 9: 105–118.

Fisher, S.G. and G.E. Likens. 1973. Energy flow in Bear Brook, New Hampshire: An integrative approach to stream ecosystem metabolism. Ecological Monographs 43: 421–439.

Fisher, S.G. and W.L. Minckley. 1978. Chemical characteristics of a desert stream in flash flood. Journal of Arid Environments 1: 25–33.

Fisher, T.R., L.W. Harding, D.W. Stanley, and L.G. Ward. 1988. Phytoplankton, nutrients, and turbidity in the Chesapeake, Delaware, and Hudson estuaries. Estuarine, Coastal and Shelf Science 27: 61–93.

Fishman, J., C.E. Watson, J.C. Larsen, and J.A. Logan. 1990. Distribution of tropospheric ozone determined from satellite data. Journal of Geophysical Research 95: 3599–3617.

Fitzgerald, J.W., T.L. Andrew, and W.T. Swank. 1984. Availability of carbon-bonded sulfur for mineralization in forest soils. Canadian Journal of Forest Research 14: 839–843.

Fitzgerald, J.W., T.C. Strickland, and J.T. Ash. 1985. Isolation and partial characterization of forest floor and soil organic sulfur. Biogeochemistry 1: 155–167.

Flaig, W., H. Beutelspacher, and E. Rietz. 1975. Chemical composition and physical properties of humic substances. pp. 1–211. In J.E. Gieseking (ed.), *Soil Components.* Vol. 1. *Organic Components.* Springer-Verlag, New York.

Fleisher, Z., A. Kenig, I. Ravina, and J. Hagin. 1987. Model of ammonia volatilization from calcareous soils. Plant and Soil 103: 205–212.

Flint, R.F. 1971. *Glacial and Quaternary Geology.* Wiley, New York.

Floret, C., R. Pontanier, and S. Rambal. 1982. Measurement and modelling of primary production and water use in a south Tunisian steppe. Journal of Arid Environments 5: 77–90.

Fomenkova, M.N., S. Chang, and L.M. Mukhin. 1994. Carbonaceous components in the comet Halley dust. Geochimica et Cosmochimica Acta 58: 4503–4512.

Foster, R.C. 1981. Polysaccharides in soil fabrics. Science 214: 665–667.

Fowells, H.A. and R.W. Krauss. 1959. The inorganic nutrition of loblolly pine and virginia pine with special reference to nitrogen and phosphorus. Forest Science 5: 95–111.

Fowler, W.A. 1984. The quest for the origin of the elements. Science 226: 922–935.

Fox, G.E., E. Stackebrandt, R.B. Hespell, J. Gibson, J. Maniloff, T.A. Dyer, R.S. Wolfe, W.E. Balch, R.S. Tanner, L.J. Magrum, L. B. Zablen, R. Blakemore, R. Gupta, L. Bonen, B.J. Lewis, D.A. Stahl, K.R. Luehrsen, K.N. Chen, and C.R. Woese. 1980. The phylogeny of prokaryotes. Science 209: 457–463.

Fox, L.E. 1983. The removal of dissolved humic acid during estuarine mixing. Estuarine, Coastal and Shelf Science 16: 431–440.

Fox, T.R. and N.B. Comerford. 1992a. Influence of oxalaate loading on phosphorus and aluminum solubility in Spodosols. Soil Science Society of America Journal 56: 290–294.

Fox, T.R. and N.B. Comerford. 1992b. Rhizosphere phosphatase activity and phosphatase hydrolysable organic phosphorus in two forested Spodosols. Soil Biology and Biochemistry 24: 579–583.

Francez, A.-J. and H. Vasander. 1995. Peat accumulation and peat decomposition after human disturbance in French and Finnish mires. Acta Oecologica [Series] Oecologia Plantarum 16 599–608.

Francis, A.J., J.M. Slater, and C.J. Dodge. 1989. Denitrification in deep subsurface sediments. Geomicrobiology Journal 7: 103–116.

François, L.M. and J.-C. Gérard. 1986. Reducing power of ferrous iron in the Archean ocean. 1. Contribution of photosynthetic oxygen. Paleoceanography 1: 355–368.

Frangi, J.L. and A.E. Lugo. 1985. Ecosystem dynamics of a subtropical floodplain forest. Ecological Monographs 55: 351–369.

Frank, D.A. and R.S. Inouye. 1994. Temporal variation in actual evapotranspiration of terrestrial ecosystems: Patterns and ecological implications. Journal of Biogeography 21: 401–411.

Frank, D.A., R.S. Inouye, N. Huntly, G.W. Minshall, and J.E. Anderson. 1994. The biogeochemistry of a north-temperate grassland with native ungulates: Nitrogen dynamics at Yellowstone National Park. Biogeochemistry 26: 163–188.

Freedman, W.L., B.F. Madore, J.R. Mould, R. Hill, L. Ferrarese, R.C. Kennicutt, A. Saha, P.B. Stetson, J.A. Graham, H. Ford, J.G. Hoessel, J. Huchra, S.M. Hughes, and G.D. Illingworth. 1994. Distance to the Virgo cluster glaxy M100 from Hubble Space Telescope observations of Cepheids. Nature 371: 757–762.

Freeland, W.J., P.H. Caicott, and D.P. Geiss. 1985. Allelochemicals, minerals and herbivore population size. Biochemical Systematics and Ecology 13: 195–206.

Freeman, C., M.A. Lock, and B. Reynolds. 1993. Fluxes of CO_2, CH_4, and N_2O from a Welsh peatland following simulation of water table draw-down: Potential feedback to climatic change. Biogeochemistry 19: 51–60.

Frey, R.W. and P.B. Basan. 1985. Coastal salt marshes. pp. 225–301. In R.A. Davis (ed.), *Coastal Sedimentary Environments.* 2nd ed. Springer-Verlag, New York.

Fridovich, I. 1975. Superoxide dismutases. Annual Review of Biochemistry 44: 147–159.

Friedland, A.J. and A.H. Johnson. 1985. Lead distribution and fluxes in a high-elevation forest in northern Vermont. Journal of Environmental Quality 14: 332–336.

Friedlander, G., J.W. Kennedy, and J.M. Miller. 1964. *Nuclear and Radiochemistry.* 2nd ed. Wiley, New York.

Friedlingstein, P., K.C. Prentice, I.Y. Fung, J.G. John, and G.P. Brasseur. 1995. Carbon-biosphere-climate interactions in the last glacial maximum climate. Journal of Geophysical Research 100: 7203–7221.

Friedmann, E.I. 1982. Endolithic microorganisms in the Antarctic cold desert. Science 215: 1045–1053.

Froelich, P.N. 1988. Kinetic control of dissolved phosphate in natural rivers and estuaries: A primer on the phosphate buffer mechanism. Limnology and Oceanography 33: 649–668.

Froelich, P.N., G.P. Klinkhammer, M.L. Bender, N.A. Luedtke, G.R. Heath, D. Cullen, P. Dauphin, D. Hammond, B. Hartman, and V. Maynard. 1979. Early oxidation of organic matter in pelagic sediments of the eastern equatorial Atlantic: Suboxic diagenesis. Geochimica et Cosmochimica Acta 43: 1075–1090.

Froelich, P.N., M.L. Bender, N.A. Luedtke, G.R. Heath, and T. DeVries. 1982. The marine phosphorus cycle. American Journal of Science 282: 474–511.

Froelich, P.N., K.H. Kim, R. Jahnke, W.C. Burnett, A. Soutar, and M. Deakin. 1983. Pore water fluoride in Peru continental margin sediments: Uptake from seawater. Geochimica et Cosmochimica Acta 47: 1605–1612.

Fruchter, J.S., D.E. Robertson, J.C. Evans, K.B. Olsen, E.A. Lepel, J.C. Laul, K.H. Abel, R.W. Sanders, P.O. Jackson, N.S. Wogman, R.W. Perkins, H.H. van Tuyl, R.H. Beauchamp, J.W. Shade, J.L. Daniel, R.L. Erikson, G.A. Sehmel, R.N. Lee, A.V. Robinson, O.R. Moss, J.K. Briant, and W.C. Cannon. 1980. Mount St. Helens ash from the 18 May 1980 eruption: Chemical, physical, mineralogical, and biological properties. Science 209: 1116–1125.

Fuhrman, J.A. and F. Azam. 1982. Thymidine incorporation as a measure of heterotrophic bacterioplankton production in marine surface waters: Evaluation and field results. Marine Biology 66: 109–120.

Fung, I.Y., C.J. Tucker, and K.C. Prentice. 1987. Application of advanced very high resolution radiometer vegetation index to study atmosphere-biosphere exchange of CO_2. Journal of Geophysical Research 92: 2999–3015.

Fung, I.Y., J. John, J. Lerner, E. Matthews, M. Prather, L.P. Steele, and P.J. Fraser. 1991. Three-dimensional model synthesis of the global methane cycle. Journal of Geophysical Research 96: 13033–13065.

Funk, D.W., E.R. Pullman, K.M. Peterson, P.M. Crill, and W.D. Billings. 1994. Influence of water table on carbon dioxide, carbon monoxide, and methane fluxes from taiga bog ecosystems. Global Biogeochemical Cycles 8: 271–278.

Furrer, G., J. Westall, and P. Sollins. 1989. The study of soil chemistry through quasi-steady state models: I. Mathematical definition of the model. Geochimica et Cosmochimica Acta 53: 595–601.

Gächter, R., J.S. Meyer, and A. Mares. 1988. Contribution of bacteria to release and fixation of phosphorus in lake sediments. Limnology and Oceanography 33: 1542–1558.

Gale, P.M. and J.T. Gilmour. 1988. Net mineralization of carbon and nitrogen under aerobic and anaerobic conditions. Soil Science Society of America Journal 52: 1006–1010.

Gallagher, J.L. and F.G. Plumley. 1979. Underground biomass profiles and productivity in Atlantic coastal marshes. American Journal of Botany 66: 156–161.

Gallardo, A. and W.H. Schlesinger. 1992. Carbon and nitrogen limitations of soil microbial biomass in desert ecosystems. Biogeochemistry 18: 1–17.

Gallardo, A. and W.H. Schlesinger. 1994. Factors limiting microbial biomass in the mineral soil and forest floor of a warm-temperate forest. Soil Biology and Biochemistry 26: 1409–1415.

Galloway, J.N. and G.E. Likens. 1979. Atmospheric enhancement of metal deposition in Adirondack lake sediments. Limnology and Oceanography 24: 427–433.

Galloway, J.N. and D.M. Whelpdale. 1987. WATOX-86 overview and western North Atlantic ocean S and N atmospheric budgets. Global Biogeochemical Cycles 1: 261–281.

Galloway, J.N., G.E. Likens, and E.S. Edgerton. 1976. Acid precipitation in the northeastern United States: pH and acidity. Science 194: 722–724.

Galloway, J.N., C.L. Schofield, N.E. Peters, G.R. Hendrey, and E.R. Altwicker. 1983. Effect of atmospheric sulfur on the composition of three Adirondack lakes. Canadian Journal of Fisheries and Aquatic Sciences 40: 799–806.

Galloway, J.N., G.E. Likens, and M.E. Hawley. 1984. Acid precipitation: Natural versus anthropogenic components. Science 226: 829–831.

Galloway, J.N., W.H. Schlesinger, H. Levy, A. Michaels, and J.L. Schnoor. 1995. Nitrogen fixation: Anthropogenic enhancement-environmental response. Global Biogeochemical Cycles 9: 235–252.

Ganeshram, R.S., T.F. Pedersen, S.E. Calvert, and J.W. Murray. 1995. Large changes in oceanic nutrient inventories from glacial to interglacial periods. Nature 376: 755-758.

Gardner, L.R. 1981. Element mass balances for South Carolina coastal plain watersheds. Water, Air, and Soil Pollution 15: 271–284.

Gardner, L.R. 1990. The role of rock weathering in the phosphorus budget of terrestrial watersheds. Biogeochemistry 11: 97–110.

Gardner, L.R., I. Kheoruenromne, and H.S. Chen. 1978. Isovolumetric geochemical investigation of a buried granite saprolite near Columbia, S.C. USA. Geochimica et Cosmochimica Acta 42: 417–424.

Gardner, W.S., T.F. Nalepa, and J.M. Malczyk. 1987. Nitrogen mineralization and denitrification in Lake Michigan sediments. Limnology and Oceanography 32: 1226–1238.

Garrels, R.M. and A. Lerman. 1981. Phanerozoic cycles of sedimentary carbon and sulfur. Proceedings of the National Academy of Sciences, U.S.A. 78: 4652–4656.

Garrels, R.M. and F.T. MacKenzie. 1971. *Evolution of Sedimentary Rocks.* W.W. Norton, New York.

Garten, C.T. 1976. Correlations between concentrations of elements in plants. Nature 261: 686–688.

Garten, C.T. 1990. Foliar leaching, translocation, and biogenic emission of ^{35}S in radiolabeled loblolly pines. Ecology 71: 239–251.

Garten, C.T. 1993. Variation in foliar ^{15}N abundance and the availability of soil nitrogen on Walker Branch watershed. Ecology 74: 2098–2113.

Garten, C.T. and H. van Miegroet. 1994. Relationships between soil nitrogen dynamics and natural ^{15}N abundance in plant foliage from Great Smoky Mountains National Park. Canadian Journal of Forest Research 24: 1636–1645.

Garten, C.T., E.A. Bondietti, and R.D. Lomax. 1988. Contribution of foliar leaching and dry deposition to sulfate in net throughfall below deciduous trees. Atmospheric Environment 22: 1425–1432.

Gassmann, G. and D. Glindemann. 1993. Phosphane (PH_3) in the biosphere. Angewandte Chemie, International Edition in English 32: 761–763.

Gates, W.L. 1976. Modeling the ice-age climate. Science 191: 1138–1144.

Gatz, D.F. and A.N. Dingle. 1971. Trace substances in rain water: Concentration variations during convective rains, and their interpretation. Tellus 23: 14–27.

Gaudichet, A., F. Echalar, B. Chatenet, J.P. Quisefit, G. Malingre, H. Cachier, P. Buat-Menard, P. Artaxo, and W. Maenhaut. 1995. Trace elements in tropical African savanna biomass burning aerosols. Journal of Atmospheric Chemistry 22: 19–39.

Gause, G.F. 1934. *The Struggle for Existence.* Hafner, New York.

Gebhart, D.L., H.B. Johnson, H.S. Mayeux, and H.W. Polley. 1994. The CRP increases soil organic carbon. Journal of Soil and Water Conservation 49: 488–492.

Gee, G.W., P.J. Wierenga, B.J. Andraski, M.H. Young, M.J. Fayer, and M.L. Rockhold. 1994. Variations in water balance and recharge potential at three western desert sites. Soil Science Society of America Journal 58: 63–72.

Gensel, P.G. and H.N. Andrews. 1987. The evolution of early land plants. American Scientist 75: 478–489.

Georgiadis, M.M., H. Komiya, P. Chakrabarti, D. Woo, J.J. Kornuc, and D.C. Rees. 1992. Crystallographic structure of the nitrogenase iron protein from *Azotobacter vinelandii.* Science 257: 1653–1659.

Georgii, H.-W. and D. Wötzel. 1970. On the relationship between drop size and concentration of trace elements in rainwater. Journal of Geophysical Research 75: 1727–1731.

Gersper, P.L. and N. Holowaychuk. 1971. Some effects of stem flow from forest canopy trees on chemical properties of soils. Ecology 52: 691–702.

Ghiorse, W.C. 1984. Biology of iron- and manganese-depositing bacteria. Annual Review of Microbiology 38: 515–550.

Gholz, H.L. 1982. Environmental limits on aboveground net primary production, leaf area, and biomass in vegetation zones of the Pacific Northwest. Ecology 63: 469–481.

Gholz, H.L., R.F. Fisher, and W.L. Pritchett. 1985. Nutrient dynamics in slash pine plantation ecosystems. Ecology 66: 647–659.

Gibbs, R.J. 1970. Mechanisms controlling world water chemistry. Science 170: 1088–1090.

Giblin, A.E. 1988. Pyrite formation in marshes during early diagenesis. Geomicrobiology Journal 6: 77–97.

Giblin, A.E., G.E. Likens, D. White, and R.W. Howarth. 1990. Sulfur storage and alkalinity generation in New England lake sediments. Limnology and Oceanography 35: 852–869.

Giblin, A.E., G.E. Likens, and R.W. Howarth. 1992. The importance of reduced inorganic sulfur to the sulfur cycle of lakes. Mitteilungen aus dem Geologisch-Paläontologischen Institute Universität Hamburg 72: 233–244.

Gieskes, J.M. and J.R. Lawrence. 1981. Alteration of volcanic matter in deep sea sediments: Evidence from the chemical composition of interstitial waters from deep sea drilling cores. Geochimica et Cosmochimica Acta 45: 1687–1703.

Gifford, G.F. and F.E. Busby. 1973. Loss of particulate organic materials from semiarid watersheds as a result of extreme hydrologic events. Water Resources Research 9: 1443–1449.

Gijsman, A.J. 1990. Nitrogen nutrition of Douglas-fir (*Pseudotsuga menziesii*) on strongly acid sandy soil. I. Growth, nutrient uptake, and ionic balance. Plant and Soil 126: 53–61.

Gile, L.H., F.F. Peterson, and R.B. Grossman. 1966. Morphological and genetic sequences of carbonate accumulation in desert soils. Soil Science 101: 347–360.

Gillespie, A.R. and P.E. Pope. 1990. Rhizosphere acidification increases phosphorus recovery of black locust: I. Induced acidification and soil response. Soil Science Society of America Journal 54: 533–537.

Gillette, D.A., G.J. Stensland, A.L. Williams, W. Barnard, D. Gatz, P.C. Sinclair, and T.C. Johnson. 1992. Emissions of alkaline elements calcium, magnesium, potassium, and sodium from open sources in the contiguous United States. Global Biogeochemical Cycles 6: 437–457.

Gilmore, A.R., G.Z. Gertner, and G.L. Rolfe. 1984. Soil chemical changes associated with roosting birds. Soil Science 138: 158–163.

Glass, S.J. and M.J. Matteson. 1973. Ion enrichment in aerosols dispersed from bursting bubbles in aqueous salt solutions. Tellus 25: 272–280.

Gleason, J.F., P.K. Bhartia, J.R. Herman, R. McPeters, R. Newman, R.S. Stolarski, L. Flynn, G. Labow, D. Larko, C. Seftor, C. Wellemeyer, W.D. Komhyr, A.J. Miller, and W. Planet. 1993. Record low global ozone in 1992. Science 260: 523–526.

Glibert, P.W., F. Lipschultz, J.J. McCarthy, and M.A. Altabet. 1982. Isotope dilution models of uptake and remineralization of ammonium by marine plankton. Limnology and Oceanography 27: 639–650.

Gloersen, P. and W.J. Campbell. 1991. Recent variations in Arctic and Antarctic sea-ice covers. Nature 352: 33–36.

Glover, H.E., B.B. Prézelin, L. Campbell, M. Wyman, and C. Garside. 1988. A nitrate-dependent *Synechococcus* bloom in surface Sargasso sea water. Nature 331: 161–163.

Glynn, P.W. 1988. El Niño - Southern Oscillation 1982–1983: Nearshore population, community, and ecosystem responses. Annual Review of Ecology and Systematics 19: 309–345.

Godbold, D.L., E. Fritz, and A. Hüttermann. 1988. Aluminum toxicity and forest decline. Proceedings of the National Academy of Sciences, U.S.A. 85: 3888–3892.

Godden, J.W., S. Turley, D.C. Teller, E.T. Adman, M.Y. Liu, W.J. Payne, and J. LeGall. 1991. The 2.3 Angstrom X-ray structure of nitrite reductase from *Achromobacter cycloclastes*. Science 253: 438–442.

Goldan, P.D., W.C. Kuster, D.L. Albritton, and F.C. Fehsenfeld. 1987. The measurement of natural sulfur emissions from soils and vegetation: Three sites in the eastern United States revisited. Journal of Atmospheric Chemistry 5: 439–467.

Goldan, P.D., R. Fall, W.C. Kuster, and F.C. Fehsenfeld. 1988. Uptake of COS by growing vegetation: A major tropospheric sink. Journal of Geophysical Research 93: 14186–14192.

Goldberg, A.B., P.J. Maroulis, L.A. Wilner, and A.R. Bandy. 1981. Study of H_2S emissions from a salt water marsh. Atmospheric Environment 15: 11–18.

Goldberg, D.E. 1982. The distribution of evergreen and deciduous trees relative to soil type: An example from the Sierra Madre, Mexico, and a general model. Ecology 63: 942–951.

Goldberg, D.E. 1985. Effects of soil pH, competition, and seed predation on the distributions of two tree species. Ecology 66: 503–511.

Goldich, S.S. 1938. A study in rock-weathering. Journal of Geology 46: 17–58.

Goldman, C.R. 1988. Primary productivity, nutrients, and transparency during the early onset of eutrophication in ultra-oligotrophic Lake Tahoe, California-Nevada. Limnology and Oceanography 33: 1321–1333.

Goldman, J.C. and P.M. Glibert. 1982. Comparative rapid ammonium uptake by four species of marine phytoplankton. Limnology and Oceanography 27: 814–827.

Golley, F.B. 1972. Energy flux in ecosystems. pp. 69-90. In J.A. Wiens (ed.), *Ecosystem Structure and Function*. Oregon State University Press, Corvallis.

Golterman, H.L. 1995. The role of the ironhydroxide-phosphate-sulphide system in the phosphate exchange between sediments and overlying water. Hydrobiologia 297: 43–54.

Gorham, E. 1957. The development of peat lands. Quarterly Review of Biology 32: 145–166.

Gorham, E. 1961. Factors influencing supply of major ions to inland waters, with special reference to the atmosphere. Geological Society of America Bulletin 72: 795–840.

Gorham, E. 1991. Northern peatlands: Role in the carbon cycle and probable responses to climatic warming. Ecological Applications 1: 182–195.

Gorham, E., P.M. Vitousek, and W.A. Reiners. 1979. The regulation of chemical budgets over the course of terrestrial ecosystem succession. Annual Review of Ecology and Systematics 10: 53–84.

Gorham, E., F.B. Martin, and J.T. Litzau. 1984. Acid rain: Ionic correlations in the eastern United States, 1980-1981. Science 225: 407–409.

Gornitz, V. 1995. Monitoring sea level changes. Climatic Change 31: 515–544.

Gornitz, V., S. Lebedeff, and J. Hansen. 1982. Global sea level trend in the past century. Science 215: 1611–1614.

Gosz, J.R. 1981. Nitrogen cycling in coniferous ecosystems. pp. 405–426. In F.E. Clark and T. Rosswall (eds.), *Terrestrial Nitrogen Cycles.* Swedish Natural Science Research Council, Stockholm.

Gosz, J.R., R.T. Holmes, G.E. Likens, and F.H. Bormann. 1978. The flow of energy in a forest ecosystem. Scientific American 238(3): 92–102.

Goulden, M.L., J.W. Munger, S.-M. Fan, B.C. Daube, and S.C. Wofsy. 1996. Exchange of carbon dioxide by a deciduous forest: Response to interannual climate variability. Science 271: 1576–1578.

Goward, S.N., C.J. Tucker, and D.G. Dye. 1985. North American vegetation patterns observed with the NOAA-7 advanced very high resolution radiometer. Vegetatio 64: 3–14.

Gower, S.T., K.A. Vogt, and C.C. Grier. 1992. Carbon dynamics of Rocky Mountain douglas-fir: Influence of water and nutrient availability. Ecological Monographs 62: 43–65.

Gower, S.T., S. Pongracic, and J.J. Landsberg. 1996. A global trend in belowground carbon allocation: Can we use the relationship at smaller scales? Ecology 77: 1750–1755.

Grace, J., J. Lloyd, J. McIntyre, A.C. Miranda, P. Meir, H.S. Miranda, C. Nobre, J. Moncrieff, J. Massheder, Y. Malhi, I. Wright, and J. Gash. 1995. Carbon dioxide uptake by an undisturbed tropical rain forest in Southwest Amazonia, 1992 to 1993. Science 270: 778–780.

Grace, J., Y. Malhi, J. Lloyd, J. McIntyre, A.C. Miranda, P. Meir, and H.S. Miranda. 1996. The use of eddy covariance to infer the net carbon dioxide uptake of Brazilian rain forest. Global Change Biology 2: 209–217.

Graedel, T.E. and P.J. Crutzen. 1993. *Atmospheric Change.* Freeman, New York.

Graedel, T.E. and W.C. Keene. 1995. Tropospheric budget of reactive chlorine. Global Biogeochemical Cycles 9: 47–77.

Graham, J.B., R. Dudley, N.M. Aguilar, and C. Gans. 1995. Implications of the late Palaeozoic oxygen pulse for physiology and evolution. Nature 375: 117–120.

Graham, N.E. 1995. Simulation of recent global temperature trends. Science 267: 666–671.

Graham, R.C. and E. Franco-Vizcaíno. 1992. Soils on igneous and metavolcanic rocks in the Sonoran desert of Baja California, Mexico. Geoderma 54: 1–21.

Graham, W.F. and R.A. Duce. 1979. Atmospheric pathways of the phosphorus cycle. Geochimica et Cosmochimica Acta 43: 1195–1208.

Grandstaff, D.E. 1986. The dissolution rate of forsteritic olivine from Hawaiian beach sand. pp. 41-59. In S.M. Colman and D.P. Dethier (eds.), *Rates of Chemical Weathering of Rocks and Minerals.* Academic Press, Orlando, Florida.

Granhall, U. 1981. Biological nitrogen fixation in relation to environmental factors and functioning of natural ecosystems. pp. 131–144. In F.E. Clark and T. Rosswall (eds.), *Terrestrial Nitrogen Cycles.* Swedish Natural Science Research Council, Stockholm.

Grasman, B.T. and E.C. Hellgren. 1993. Phosphorus nutrition in white-tailed deer: Nutrient balance, physiological responses, and antler growth. Ecology 74: 2279–2296.

Grassle, J.F. 1985. Hydrothermal vent animals: Distribution and biology. Science 229: 713–717.

Graumlich, L.J. 1991. Subalpine tree growth, climate, and increasing CO_2: An assessment of recent growth trends. Ecology 72: 1–11.

Graustein, W.C., K. Cromack, and P. Sollins. 1977. Calcium oxalate: Occurrence in soils and effect on nutrient and geochemical cycles. Science 198: 1252–1254.

Graveland, J., R. van der Wal, J.H. Van Balen, and A.J. van Noordwijk. 1994. Poor reproduction in forest passerines from decline of snail abundance on acidified soils. Nature 368: 446–448.

Gray, J.T. 1982. Community structure and productivity in *Ceanothus megacarpus* chaparral and coastal sage scrub of southern California. Ecological Monographs 52: 415–435.

Gray, J.T. 1983. Nutrient use by evergreen and deciduous shrubs in southern California. I. Community nutrient cycling and nutrient-use efficiency. Journal of Ecology 71: 21–41.

Gray, J.T. and W.H. Schlesinger. 1981. Nutrient cycling in Mediterranean type ecosystems. pp. 259–285. In P.C. Miller (ed.), *Resource Use by Chaparral and Matorral.* Springer-Verlag, New York.

Graybill, D.A. and S.B. Idso. 1993. Detecting the aerial fertilization effect of atmospheric CO_2 enrichment in tree-ring chronologies. Global Biogeochemical Cycles 7: 81–95.

Greeley, R. and B.D. Schneid. 1991. Magma generation on Mars: Amounts, rates, and comparisons with Earth, Moon, and Venus. Science 254: 996–998.

Green, C.J., A.M. Blackmer, and N.C. Yang. 1994. Release of fixed ammonium during nitrification in soils. Soil Science Society of America Journal 58: 1411–1415.

Greenberg, J.P. and P.R. Zimmerman. 1984. Nonmethane hydrocarbons in remote tropical, continental, and marine atmospheres. Journal of Geophysical Research 89: 4767–4778.

Greenland, D.J. 1971. Interactions between humic and fulvic acids and clays. Soil Science 111: 34–41.

Gregor, B. 1970. Denudation of the continents. Nature 228: 273–275.

Gressel, N., J.G. McColl, C.M. Preston, R.H. Newman, and R.F. Powers. 1996. Linkages between phosphorus transformations and carbon decomposition in a forest soil. Biogeochemistry 33:97–123.

Grey, D.C. and M.L. Jensen. 1972. Bacteriogenic sulfur in air pollution. Science 177: 1099–1100.

Grier, C.C. and S.W. Running. 1977. Leaf area of mature northwestern coniferous forests: Relation to site water balance. Ecology 58: 893–899.

Griffin, T.M., M.C. Rabenhorst, and D.S. Fanning. 1989. Iron and trace metals in some tidal marsh soils of the Chesapeake Bay. Soil Science Society of America Journal 53: 1010–1019.

Griffith, E.J., C. Ponnamperuma, and N.W. Gabel. 1977. Phosphorus, A key to life on the primitive Earth. Origins of Life 8: 71–85.

Griffiths, R.P., M.E. Harmon, B.A. Caldwell, and S.E. Carpenter. 1993. Acetylene reduction in conifer logs during early stages of decomposition. Plant and Soil 148: 53–61.

Groffman, P.M. and J.M. Tiedje. 1989. Denitrification in north temperate forest soils: Spatial and temporal patterns at the landscape and seasonal scales. Soil Biology and Biochemistry 21: 613–620.

Gross, M.G. 1977. *Oceanography: A View of the Earth.* 2nd ed. Prentice-Hall, Englewood Cliffs, New Jersey.

Grotzinger, J.P. and J.F. Kasting. 1993. New constraints on Precambrian ocean composition. Journal of Geology 101: 235–243.

Grusbaugh, J.W. and R.V. Anderson. 1989. Upper Mississippi River: Seasonal and floodplain forest influences on organic matter transport. Hydrobiologia 174: 235–244.

Gu, B., J. Schmitt, Z. Chen, L. Liang, and J.F. McCarthy. 1995. Adsorption and desorption of different organic matter fractions on iron oxide. Geochimica et Cosmochimica Acta 59: 219–229.

Guadalix, M.E. and M.T. Pardo. 1991. Sulphate sorption by variable charge soils. Journal of Soil Science 42: 607–614.

Guenther, A.B., P. Zimmerman, and M. Wildermuth. 1994. Natural volatile organic compound emission rate estimates for U.S. woodland landscapes. Atmospheric Environment 28: 1197–1210.

Guenther, A.B., C.N. Hewitt, D. Erickson, R. Fall, C. Geron, T. Graedel, P. Harley, L. Klinger, M. Lerdau, W.A. McKay, T. Pierce, B. Scholes, R. Steinbrecher, R. Tallamraju, J. Taylor, and P. Zimmerman. 1995. A global model of natural volatile organic compound emissions. Journal of Geophysical Research 100: 8873–8892.

Guieu, C., R. Duce, and R. Arimoto. 1994. Dissolved input of manganese to the ocean: Aerosol source. Journal of Geophysical Research 99: 18789–18800.

Gurtz, M.E., G.R. Marzolf, K.T. Killingbeck, D.L. Smith, and J.V. McArthur. 1988. Hydrologic and riparian influences on the import and storage of coarse particulate organic matter in a prairie stream. Canadian Journal of Fisheries and Aquatic Sciences 45: 655–665.

Guthrie, R.L. and J.E. Witty. 1982. New designation for soil horizons and layers and the new Soil Survey Manual. Soil Science Society of America Journal 46: 443–444.

Gutschick, V.P. 1981. Evolved strategies in nitrogen acquisition by plants. American Naturalist 118: 607–637.

Habicht, K.S. and D.E. Canfield. 1996. Sulphur isotope fractionation in modern microbial mats and the evolution of the sulphur cycle. Nature 382: 342–343.

Hackley, K.C. and T.F. Anderson. 1986. Sulfur isotopic variations in low-sulfur coals from the Rocky Mountain region. Geochimica et Cosmochimica Acta 50: 1703–1713.

Hackstein, J.H.P. and C.K. Stumm. 1994. Methane production in terrestrial arthropods. Proceedings of the National Academy of Sciences, U.S.A. 91: 5441–5445.

Haering, K.C., M.C. Rabenhorst, and D.S. Fanning. 1989. Sulfur speciation in some Chesapeake Bay tidal marsh soils. Soil Science Society of America Journal 53: 500–505.

Hahn, J. 1974. The North Atlantic Ocean as a source of atmospheric N_2O. Tellus 26: 160–168.

Hahn, J. 1980. Organic constituents of natural aerosols. Annals of the New York Academy of Sciences 338: 359–376.

Hahn, J. 1981. Nitrous oxide in the oceans. pp. 191–277. In C.C. Delwiche (ed.), *Denitrification, Nitrification, and Atmospheric Nitrous Oxide.* Wiley, New York.

Haines, B., M. Black, and C. Bayer. 1989. Sulfur emissions from roots of the rain forest tree *Stryphnodendron excelsum*. pp. 58–69. In E.S. Saltzman and W.J. Cooper (eds.), *Biogenic Sulfur in the Environment*. American Chemical Society, Washington, D.C.

Haines, E.B. 1977. The origins of detritus in Georgia salt marsh estuaries. Oikos 29: 254–260.

Hall, D.T., D.F. Strobel, P.D. Feldman, M.A. McGrath, and H.A. Weaver. 1995. Detection of an oxygen atmosphere on Jupiter's moon Europa. Nature 373: 677–679.

Hameed, S. and J. Dignon. 1988. Changes in the geographical distributions of global emissions of NO_x and SO_x from fossil fuel combustion between 1966 and 1980. Atmospheric Environment 22: 441–449.

Han, T.-M. and B. Runnegar. 1992. Megascopic eukaryotic algae from the 2.1-billion-year-old Negaunee iron-formation, Michigan. Science 257: 232–235.

Handley, L.L. and J.A. Raven. 1992. The use of natural abundance of nitrogen isotopes in plant physiology and ecology. Plant, Cell and Environment 15: 965–985.

Hanks, T.C. and D.L. Anderson. 1969. The early thermal history of the Earth. Physics of the Earth and Planetary Interiors 2: 19–29.

Hansen, J., D. Johnson, A. Lacis, S. Lebedeff, P. Lee, D. Rind, and G. Russell. 1981. Climatic impact of increasing atmospheric carbon dioxide. Science 213: 957–966.

Hao, W.M. and M.-H. Liu. 1994. Spatial and temporal distribution of tropical biomass burning. Global Biogeochemical Cycles 8: 495–503.

Happell, J.D., J.P. Chanton, G.J. Whiting, and W.J. Showers. 1993. Stable isotopes as tracers of methane dynamics in Everglades marshes with and without active populations of methane oxidizing bacteria. Journal of Geophysical Research 98: 14771–14782.

Harden, J.W. 1988. Genetic interpretations of elemental and chemical differences in a soil chronosequence, California. Geoderma 43: 179–193.

Harden, J.W., E.M. Taylor, C. Hill, R.K. Mark, L.D. McFadden, M.C. Reheis, J.M. Sowers, and S.G. Wells. 1991. Rates of soil development from four soil chronosequences in the southern Great Basin. Quaternary Research 35: 383–399.

Harden, J.W., E.T. Sundquist, R.F. Stallard, and R.K. Mark. 1992. Dynamics of soil carbon during deglaciation of the Laurentide ice sheet. Science 258: 1921–1924.

Hardin, G. 1968. The tragedy of the commons. Science 162: 1243–1248.

Harding, R.B. and E.J. Jokela. 1994. Long-term effects of forest fertilization on site organic matter and nutrients. Soil Science Society of America Journal 58: 216–221.

Harley, J.L. and S.E. Smith. 1983. *Mycorrhizal Symbiosis*. Academic Press, New York.

Harmon, M.E., J.F. Franklin, F.J. Swanson, P. Sollins, S.V. Gregory, J.D. Lattin, N.H. Anderson, S.P. Cline, N.G. Aumen, J.R. Sedell, G.W. Lienkaemper, K. Cromack, and K.W. Cummins. 1986. Ecology of coarse woody debris in temperate ecosystems. Advances in Ecological Research 15: 133–302.

Harmon, M.E., W.K. Ferrell, and J.F. Franklin. 1990. Effects on carbon storage of conversion of old-growth forests to young forests. Science 247: 699–702.

Harper, C.L. and S.B. Jacobsen. 1996. Noble gases and Earth's accretion. Science 273: 1814–1818.

Harrington, J.B. 1987. Climatic change: A review of causes. Canadian Journal of Forest Research 11: 1313–1339.

Harris, W.G., K.A. Hollien, T.L. Yuan, S.R. Bates, and W.A. Acree. 1988. Nonexchangeable potassium associated with hydroxy-interlayered vermiculite from coastal plain soils. Soil Science Society of America Journal 52: 1486–1492.

Harrison, A.F. 1982. ^{32}P-method to compare rates of mineralization of labile organic phosphorus in woodland soils. Soil Biology and Biochemistry 14: 337–341.

Harrison, K.G., W.S. Broecker, and G. Bonani. 1993. A strategy for estimating the impact of CO_2 fertilization on soil carbon storage. Global Biogeochemical Cycles 7: 69–80.

Harrison, M.J. and M.L. van Buuren. 1995. A phosphate transporter from the mycorrhizal fungus *Glomus versiforme.* Nature 378: 626–629.

Harrison, R.B., D.W. Johnson, and D.E. Todd. 1989. Sulfate adsorption and desorption reversibility in a variety of forest soils. Journal of Environmental Quality 18: 419–426.

Harrison, W.G., L.R. Harris, D.M. Karl, G.A. Knauer, and D.G. Redalje. 1992. Nitrogen dynamics at the VERTEX time-series site. Deep Sea Research 39: 1535–1552.

Harrison, W.G., L.R. Harris, and B.D. Irwin. 1996. The kinetics of nitrogen utilization in the oceanic mixed layer: Nitrate and ammonium interactions at nanomolar concentrations. Limnology and Oceanography 41: 16–32.

Harriss, R.C., D.I. Sebacher, and F.P. Day. 1982. Methane flux in the Great Dismal Swamp. Nature 297: 673–674.

Harriss, R.C., D.I. Sebacher, K.B. Bartlett, D.S. Bartlett, and P.M. Crill. 1988. Sources of atmospheric methane in the South Florida environment. Global Biogeochemical Cycles 2: 231–243.

Harvey, L.D.D. 1988. Climatic impact of ice-age aerosols. Nature 334: 333–335.

Hatcher, B.G. and K.H. Mann. 1975. Above-ground production of marsh cordgrass (*Spartina alterniflora*) near the northern end of its range. Journal of the Fisheries Research Board of Canada 32: 83–87.

Haynes, R.J. and K.M. Goh. 1978. Ammonium and nitrate nutrition of plants. Biological Reviews 53: 465–510.

Heckathorn, S.A. and E.H. DeLucia. 1995. Ammonia volatilization during drought in perennial C_4 grasses of tallgrass prairie. Oecologia 101: 361–365.

Hecky, R.E., P. Campbell, and L.L. Hendzel. 1993. The stoichiometry of carbon, nitrogen, and phosphorus in particulate matter of lakes and oceans. Limnology and Oceanography 38: 709–724.

Hedges, J.I. and P.L. Parker. 1976. Land-derived organic matter in surface sediments from the Gulf of Mexico. Geochimica et Cosmochimica Acta 40: 1019–1029.

Hedges, J.I., W.A. Clark, and G.L. Cowie. 1988. Organic matter sources to the water column and surficial sediments of a marine bay. Limnology and Oceanography 33: 1116–1136.

Hedin, L.O. 1990. Factors controlling sediment community respiration in woodland stream ecosystems. Oikos 57: 94–105.

Hedin, L.O., L. Granat, G.E. Likens, T.A. Buishand, J.N. Galloway, T.J. Butler, and H. Rodhe. 1994. Steep declines in atmospheric base cations in regions of Europe and North America. Nature 367: 351–354.

Hedin, L.O., J.J. Armesto, and A.H. Johnson. 1995. Patterns of nutrient loss from unpolluted, old-growth temperate forests: Evaluation of biogeochemical theory. Ecology 76: 493–509.

Hedley, M.J., P.H. Nye, and R.E. White. 1982a. Plant-induced changes in the rhizosphere of rape (*Brassica napus* var. Emerald) seedlings. II. Origin of the pH change. New Phytologist 91: 31–44.

Hedley, M.J., J.W.B. Stewart, and B.S. Chauhan. 1982b. Changes in inorganic and organic soil phosphorus fractions induced by cultivation practices and by laboratory incubations. Soil Science Society of America Journal 46: 970–976.

Heintzenberg, J. 1989. Fine particles in the global troposphere: A review. Tellus 41B: 149–160.

Hemond, H.F. 1980. Biogeochemistry of Thoreau's Bog, Concord, Massachusetts. Ecological Monographs 50: 507–526.

Hemond, H.F. 1983. The nitrogen budget of Thoreau's Bog. Ecology 64: 99-109.

Henderson, G.S., W.T. Swank, J.B. Waide, and C.C. Grier. 1978. Nutrient budgets of Appalachian and Cascade Region watersheds: A comparison. Forest Science 24: 385–397.

Henderson-Sellers, A. and K. McGuffie. 1987. *A Climate Modelling Primer.* Wiley, New York.

Henrichs, S.M., and W.S. Reeburgh. 1987. Anaerobic mineralization of marine sediment organic matter: Rates and the role of anaerobic processes in the oceanic carbon economy. Geomicrobiology Journal 5: 191–237.

Herman, J.R., R. McPeters, and D. Larko. 1993. Ozone depletion at northern and southern latitudes derived from January 1979 to December 1991 Total Ozone Mapping Spectrometer data. Journal of Geophysical Research 98: 12783–12793.

Herron, M.M., C.C. Langway, H.W. Weiss, and J.H. Cragin. 1977. Atmospheric trace metals and sulfate in the Greenland ice sheet. Geochimica et Cosmochimica Acta 41: 915–920.

Hess, S.L., R.M. Henry, C.B. Leovy, J.A. Ryan, J.E. Tillman, T.E. Chamberlain, H.L. Cole, R.G. Dutton, G.C. Greene, W.E. Simon, and J.L. Mitchell. 1976. Preliminary meteorological results on Mars from the Viking I Lander. Science 193: 788–791.

Hester, K. and E. Boyle. 1982. Water chemistry control of cadmium content in recent benthic foraminifera. Nature 298: 260–262.

Hesterberg, R., A. Blatter, M. Fahrni, M. Rosset, A. Neftel, W. Eugster, and H. Wanner. 1996. Deposition of nitrogen-containing compounds to an extensively managed grassland in central Switzerland. Environmental Pollution 91: 21–34.

Hewitt, C.M. and R.M. Harrison. 1985. Tropospheric concentrations of the hydroxyl radicala review. Atmospheric Environment 19: 545–554.

Hidy, G.M. 1970. Theory of diffusive and impactive scavenging. pp. 355–371. In R.J. Englemann and W.G.N. Slinn (eds.), *Precipitation Scavenging (1970).* U.S. Atomic Energy Commission, Division of Technical Information, Oak Ridge, Tennessee.

Hines, M.E., R.E. Pelletier, and P.M. Crill. 1993. Emissions of sulfur gases from marine and freshwater wetlands of the Florida Everglades: Rates and extrapolation using remote sensing. Journal of Geophysical Research 98: 8991–8999.

Hingston, F.J., R.J. Atkinson, A.M. Posner, and J.P. Quirk. 1967. Specific adsorption of anions. Nature 215: 1459–1461.

Hodell, D.A., J.H. Curtis, and M. Brenner. 1995. Possible role of climate in the collapse of Classic Maya civilization. Nature 375: 391–394.

Hoff, T., B.M. Stummann, and K.W. Henningsen. 1992. Structure, function and regulation of nitrate reductase in higher plants. Physiologia Plantarum 84: 616–624.

Hofmann, D.J. 1990. Increase in the stratospheric background sulfuric acid aerosol mass in the past 10 years. Science 248: 996–1000.

Hofmann, D.J. 1991. Aircraft sulphur emissions. Nature 349: 659.

Hofmann, D.J. and J.M. Rosen. 1983. Sulfuric acid droplet formation and growth in the stratosphere after the 1982 eruption of El Chichón. Science 222: 325–327.

Högberg, P. 1991. Development of ^{15}N enrichment in a nitrogen-fertilized forest soil-plant system. Soil Biology and Biochemistry 23: 335–338.

Hogg, N.G., P. Biscaye, W. Gardner, and W.J. Schmitz. 1982. On the transport and modification of Antarctic Bottom Water in the Vema Channel. Journal of Marine Research 40(Suppl.): 231–263.

Højberg, O., N.P. Revsbech, and J.M. Tiedje. 1994. Denitrification in soil aggregates analyzed with microsensors for nitrous oxide and oxygen. Soil Science Society of America Journal 58: 1691–1698.

Hole, F.D. 1981. Effects of animals on soil. Geoderma 25: 75–112.

Holland, H.D. 1978. *The Chemistry of the Atmosphere and Oceans.* Wiley, New York.

Holland, H.D. 1984. *The Chemical Evolution of the Atmosphere and Oceans.* Princeton University Press, Princeton, New Jersey.

Holland, H.D., B. Lazar, and M. McCaffrey. 1986. Evolution of the atmosphere and oceans. Nature 320: 27–33.

Holland, H.D., C.R. Feakes, and E.A. Zbinden. 1989. The Flin Flon paleosol and the composition of the atmosphere 1.8 Bybp. American Journal of Science 289: 362–389.

Holligan, P.M. and W.A. Reiners. 1992. Predicting the responses of the coastal zone to global change. Advances in Ecological Research 22: 211–255.

Holligan, P.M. and J.E. Robertson. 1996. Significance of ocean carbonate budgets for the global carbon cycle. Global Change Biology 2: 85–95.

Hollinger, D.Y. 1986. Herbivory and the cycling of nitrogen and phosphorus in isolated California oak trees. Oecologia 70: 291–297.

Hollinger, D.Y., F.M. Kelliher, J.N. Byers, J.E. Hunt, T.M. McSeveny, and P.L. Weir. 1994. Carbon dioxide exchange between an undisturbed old-growth temperate forest and the atmosphere. Ecology 75: 134–150.

Holser, W.T. and I.R. Kaplan. 1966. Isotope geochemistry of sedimentary sulfates. Chemical Geology 1: 93–135.

Holser, W.T., J.B. Maynard, and K.M. Cruikshank. 1989. Modelling the natural cycle of sulphur through Phanerozoic time. pp. 21–56. In P. Brimblecombe and A.Y. Lein (eds.), *Evolution of the Global Biogeochemical Sulphur Cycle.* Wiley, New York.

Holton, J.R., P.H. Haynes, M.E. McIntyre, A.R. Douglass, R.B. Rood, and L. Pfister. 1995. Stratosphere-troposphere exchange. Reviews of Geophysics 33: 403–439.

Honda, M., I. McDougall, D.B. Patterson, A. Doulgeris, and D.A. Clague. 1991. Possible solar noble-gas component in Hawaiian basalts. Nature 349: 149–151.

Honeycutt, C.W., R.D. Heil, and C.V. Cole. 1990. Climatic and topographic relations of three Great Plains soils. I. Soil morphology. Soil Science Society of America Journal 54: 469–475.

Hong, J.-I., Q. Feng, V. Rotello, and J. Rebek. 1992. Competition, cooperation, and mutation: Improving a synthetic replicator by light irradiation. Science 255: 848–850.

Honjo, S., S.J. Manganini, and J.J. Cole. 1982. Sedimentation of biogenic matter in the deep ocean. Deep Sea Research 29: 609–625.

Hooper, F.F. and L.S. Morris. 1982. Mat-water phosphorus exchange in an acid bog lake. Ecology 63: 1411–1421.

Hooper, P.R., I.W. Herrick, E.R. Laskowski, and C.R. Knowles. 1980. Composition of the Mount St. Helens ashfall in the Moscow-Pullman area on 18 May 1980. Science 209: 1125–1126.

Hope, D., M.F. Billett, and M.S. Cresser. 1994. A review of the export of carbon in river water: Fluxes and processes. Environmental Pollution 84: 301–324.

Hornberger, G.M., K.E. Bencala, and D.M. McKnight. 1994. Hydrological controls on dissolved organic carbon during snowmelt in the Snake River near Montezuma, Colorado. Biogeochemistry 25: 147–165.

Horne, A.J. and D.L. Galat. 1985. Nitrogen fixation in an oligotrophic, saline desert lake: Pyramid Lake, Nevada. Limnology and Oceanography 30: 1229–1239.

Horne, A.J. and C.R. Goldman. 1974. Suppression of nitrogen fixation by blue-green algae in a eutrophic lake with trace additions of copper. Science 183: 409–411.

Horodyski, R.J. and L.P. Knauth. 1994. Life on land in the Precambrian. Science 263: 494–498.

Horowitz, N.H. 1977. The search for life on Mars. Scientific American 237(5): 52–61.

Horrigan, S.G., J.P. Montoya, J.L. Nevins, and J.J. McCarthy. 1990. Natural isotopic composition of dissolved inorganic nitrogen in the Chesapeake Bay. Estuarine, Coastal and Shelf Science 30: 393–410.

Hosker, R.P. and S.E. Lindberg. 1982. Review: Atmospheric deposition and plant assimilation of gases and particles. Atmospheric Environment 16: 889–910.

Hough, A.M. and R.G. Derwent. 1990. Changes in the global concentration of tropospheric ozone due to human activities. Nature 344: 645–648.

Houghton, J.T. 1986. *The Physics of Atmospheres.* 2nd ed. Cambridge University Press, Cambridge.

Houghton, J.T., G.J. Jenkins, and J.J. Ephraums. (eds.). 1990. *Climate Change: The IPCC Scientific Assessment.* Cambridge University Press, Cambridge.

Houghton, J.T., L.G. Meira Filho, J. Bruce, H. Lee, B.A. Callander, E. Haites, N. Harris, and K. Maskell. 1995. *Climate Change 1994.* Cambridge University Press, Cambridge.

Houghton, R.A. 1987. Biotic changes consistent with the increased seasonal amplitude of atmospheric CO_2 concentrations. Journal of Geophysical Research 92: 4223–4230.

Houghton, R.A. 1993. The role of the world's forests in global warming. pp. 21–58. In K. Ramakrishna and G.M. Woodwell (eds.), *World Forests for the Future.* Yale University Press, New Haven, Connecticut.

Houghton, R.A. 1995. Land-use change and the carbon cycle. Global Change Biology 1: 275–287.

Houghton, R.A. and D.L. Skole. 1990. Carbon. pp. 393–408. In B.L. Turner, W.C. Clark, R.W. Kates, J.F. Richards, J.T. Mathews, and W.B. Meyer (eds.), *The Earth as Transformed by Human Action.* Cambridge University Press, Cambridge.

Houghton, R.A., J.E. Hobbie, J.M. Melillo, B. Moore, B.J. Peterson, G.R. Shaver, and G.M. Woodwell. 1983. Changes in the carbon content of terrestrial biota and soils between 1860 and 1980: A net release of CO_2 to the atmosphere. Ecological Monographs 53: 235–262.

Houghton, R.A., R.D. Boone, J.R. Fruci, J.E. Hobbie, J.M. Melillo, C.A. Palm, B.J. Peterson, G.R. Shaver, G.M. Woodwell, B. Moore, D.L. Skole, and N. Myers. 1987. The flux of carbon from terrestrial ecosystems to the atmosphere in 1980 due to changes in land use: Geographic distribution of the global flux. Tellus 39B: 122–139.

Houle, D. and R. Carignan. 1992. Sulfur speciation and distribution in soils and aboveground biomass of a boreal coniferous forest. Biogeochemistry 16: 63–82.

Hovis, W.A., D.K. Clark, F. Anderson, R.W. Austin, W.H. Wilson, E.T. Baker, D. Ball, H.R. Gordon, J.L. Mueller, S.Z. El-Sayed, B. Sturm, R.C. Wrigley, and C.S. Yentsch. 1980. Nimbus-7 Coastal Zone Color Scanner: System description and initial imagery. Science 210: 60–63.

Howard, J.A. and C.W. Mitchell. 1985. *Phytogeomorphology.* Wiley, New York.

Howarth, R.W. 1979. Pyrite: Its rapid formation in a salt marsh and its importance in ecosystem metabolism. Science 203: 49–51.

Howarth, R.W. 1984. The ecological significance of sulfur in the energy dynamics of salt marsh and coastal marine sediments. Biogeochemistry 1: 5–27.

Howarth, R.W. 1988. Nutrient limitation of net primary production in marine ecosystems. Annual Review of Ecology and Systematics 19: 89–110.

Howarth, R.W. 1993. Microbial processes in salt-marsh sediments. pp. 239–259. In T.E. Ford (ed.), *Aquatic Microbiology: An Ecological Approach.* Blackwell Scientific Publishers, Oxford.

Howarth, R.W. and J.J. Cole. 1985. Molybdenum availability, nitrogen limitation, and phytoplankton growth in natural waters. Science 229: 653–655.

Howarth, R.W., R. Marino, J. Lane, and J.J. Cole. 1988a. Nitrogen fixation in freshwater, estuarine, and marine ecosystems. I. Rates and importance. Limnology and Oceanography 33: 669–687.

Howarth, R.W., R. Marino, and J.J. Cole. 1988b. Nitrogen fixation in freshwater, estuarine, and marine ecosystems. 2. Biogeochemical controls. Limnology and Oceanography 33: 688–701.

Howarth, R.W., R. Marino, R. Garritt, and D. Sherman. 1992. Ecosystem respiration and organic carbon processing in a large tidally influenced river: The Hudson River. Biogeochemistry 16: 83–102.

Howarth, R.W., T. Butler, K. Lunde, D. Swaney, and C.R. Chu. 1993. Turbulence and planktonic nitrogen fixation: A mesocosm experiment. Limnology and Oceanography 38: 1696–1711.

Howarth, R.W., H.S. Jensen, R. Marino, and H. Postma. 1995. Transport to and processing of P in near-shore and oceanic waters. pp. 323–345. In H. Tiessen (ed.), *Phosphorus in the Global Environment.* Wiley, New York.

Howeler, R.H. and D.R. Bouldin. 1971. The diffusion and consumption of oxygen in submerged soils. Soil Science Society of America Proceedings 35: 202–208.

Howell, D.G. and R.W. Murray. 1986. A budget for continental growth and denudation. Science 233: 446–449.

Howes, B.L., J.W.H. Dacey, and G.M. King. 1984. Carbon flow through oxygen and sulfate reduction pathways in salt marsh sediments. Limnology and Oceanography 29: 1037–1051.

Howes, B.L., J.W.H. Dacey, and J.M. Teal. 1985. Annual carbon mineralization and belowground production of *Spartina alterniflora* in a New England salt marsh. Ecology 66: 595–605.

Huang, P.M. 1988. Ionic factors affecting aluminum transformations and the impact on soil and environmental sciences. Advances in Soil Science 8: 1–78.

Huber, R., M. Kurr, H.W. Jannasch, and K.O. Stetter. 1989. A novel group of abyssal methanogenic archaebacteria (*Methanopyrus*) growing at 110°C. Nature 342: 833–836.

Hudson, B.D. 1995. Reassessment of Polynov's ion mobility series. Soil Science Society of America Journal 59: 1101–1103.

Hue, N.V. 1991. Effects of organic acids/anions on P sorption and phytoavailability in soils with different mineralogies. Soil Science 152: 463–471.

Huenneke, L.F., S.P. Hamburg, R. Koide, H.A. Mooney, and P.M. Vitousek. 1990. Effects of soil resources on plant invasion and community structure in Californian serpentine grassland. Ecology 71: 478–491.

Huh, C.-A. and F.G. Prahl. 1995. Role of colloids in upper ocean biogeochemistry in the northeast Pacific Ocean elucidated from ^{238}U-^{234}Th disequilibria. Limnology and Oceanography 40: 528–532.

Hulett, L.D., A.J. Weinberger, K.J. Northcutt, and M. Ferguson. 1980. Chemical species in fly ash from coal-burning power plants. Science 210: 1356–1358.

Humayun, M. and R.N. Clayton. 1995. Potassium isotope cosmochemistry: Genetic implications of volatile element depletion. Geochimica et Cosmochimica Acta 59: 2131–2148.

Hunten, D.M. 1993. Atmospheric evolution of the terrestrial planets. Science 259: 915–920.

Hurley, J.P., D.E. Armstrong, G.J. Kenoyer, and C.J. Bowser. 1985. Ground water as a silica source for diatom production in a precipitation-dominated lake. Science 227: 1576–1578.

Hurst, D.F., D.W.T. Griffith, and G.D. Cook. 1994. Trace gas emissions from biomass buring in tropical Australian savannas. Journal of Geophysical Research 99: 16441–16456.

Husar, R.B. and J.D. Husar. 1985. Regional river sulfur runoff. Journal of Geophysical Research 90: 1115–1125.

Huss-Danell, K. 1986. Nitrogen in shoot litter, root litter, and root exudates from nitrogen-fixing *Alnus incana*. Plant and Soil 91: 43–49.

Huston, M. 1993. Biological diversity, soils, and economics. Science 262: 1676–1680.

Hutchin, P.R., M.C. Press, J.A. Lee, and T.W. Ashenden. 1995. Elevated concentrations of CO_2 may double methane emissions from mires. Global Change Biology 1: 125–128.

Hutchins, K.S. and B.M. Jakosky. 1996. Evolution of Martian atmospheric argon: Implications for sources of volatiles. Journal of Geophysical Research 101: 14933–14949.

Hutchinson, G.E. 1938. On the relation between oxygen deficit and the productivity and typology of lakes. Internationale Revue der Gesamten Hydrobiologie und Hydrographie 36: 336–355.

Hutchinson, G.E. 1943. The biogeochemistry of aluminum and of certain related elements (concluded). Quarterly Review of Biology 18: 331–363.

Hutchinson, G.L., W.D. Guenzi, and G.P. Livingston. 1993. Soil water controls on aerobic soil emission of gaseous nitrogen oxides. Soil Biology and Biochemistry 25: 1–9.

Hutchinson, J.N. 1980. The record of peat wastage in the East Anglian fenlands at Holme Post, 1948-1978 A.D. Journal of Ecology 68: 229–249.

Hütsch, B.W., C.P. Webster, and D.S. Powlson. 1994. Methane oxidation in soil as affected by land use, soil pH and N fertilization. Soil Biology and Biochemistry 26: 1613–1622.

Hydes, D.J. 1979. Aluminum in seawater: Control by inorganic processes. Science 205: 1260–1262.

Hyman, M.R. and P.M. Wood. 1983. Methane oxidation by *Nitrosomonas europaea*. Biochemical Journal 212: 31–37.

Idso, K.E. and S.B. Idso. 1994. Plant responses to atmospheric CO_2 enrichment in the face of environmental constraints: A review of the past 10 years' research. Agricultural and Forest Meteorology 69: 153–203.

Idso, S.B. and A.J. Brazel. 1984. Rising atmospheric carbon dioxide concentrations may increase streamflow. Nature 312: 51–53.

Idso, S.B. and B.A. Kimball. 1993. Tree growth in carbon dioxide enriched air and its implications for global carbon cycling and maximum levels of atmospheric CO_2. Global Biogeochemical Cycles 7: 535–555.

Illmer, P., A. Barbato, and F. Schinner. 1995. Solubilization of hardly-soluble $AlPO_4$ with P-solubilizing microorganisms. Soil Biology and Biochemistry 27: 265–270.

Ingall, E. and R. Jahnke. 1994. Evidence for enhanced phosphorus regeneration from marine sediments overlain by oxygen depleted waters. Geochimica et Cosmochimica Acta 58: 2571–2575.

Ingall, E.D. and P. van Cappellen. 1990. Relation between sedimentation rate and burial of organic phosphorus and organic carbon in marine sediments. Geochimica et Cosmochimica Acta 54: 373–386.

Ingall, E.D., R.M. Bustin, and P. van Cappellen. 1993. Influence of water column anoxia on the burial and preservation of carbon and phosphorus in marine shales. Geochimica et Cosmochimica Acta 57: 303–316.

Ingestad, T. 1979a. Nitrogen stress in birch seedlings. II. N, K, P, Ca, and Mg nutrition. Physiologia Plantarum 45: 149–157.

Ingestad, T. 1979b. Mineral nutrient requirements of *Pinus silvestris* and *Picea abies* seedlings. Physiologia Plantarum 45: 373–380.

Ingestad, T. 1982. Relative addition rate and external concentration: Driving variables used in plant nutrition research. Plant, Cell and Environment 5: 443–453.

Ingham, R.E. and J.K. Detling. 1990. Effects of root-feeding nematodes on aboveground net primary production in a North American grassland. Plant and Soil 121: 279–281.

Inoue, H.Y. and Y. Sugimura. 1992. Variations and distributions of CO_2 in and over the equatorial Pacific during the period from the 1986/88 El Niño event to the 1988/89 La Niña event. Tellus 44B: 1–22.

Inoue, H.Y., H. Matsueda, M. Ishii, K. Fushimi, M. Hirota, I. Asanuma, and Y. Takasugi. 1995. Long-term trend of the partial pressure of carbon dioxide (pCO_2) in surface waters of the western North Pacific, 1984-1993. Tellus 47B: 391–413.

Insam, H. 1990. Are the soil microbial biomass and basal respiration governed by the climatic regime? Soil Biology and Biochemistry 22: 525–532.

Inskeep, W.P. and P.R. Bloom. 1986. Kinetics of calcite precipitation in the presence of water-soluble organic ligands. Soil Science Society of America Journal 50: 1167–1172.

Irwin, J.G. and M.L. Williams. 1988. Acid rain: Chemistry and transport. Environmental Pollution 50: 29–59.

Isaksen, I.S.A., and O. Hov. 1987. Calculation of trends in the tropospheric concentration of O_3, OH, CO, CH_4, and NO_x. Tellus 39B: 271–285.

Ittekkot, V. and R. Arain. 1986. Nature of particulate organic matter in the river Indus, Pakistan. Geochimica et Cosmochimica Acta 50: 1643–1653.

Ittekkot, V. and S. Zhang. 1989. Pattern of particulate nitrogen transport in world rivers. Global Biogeochemical Cycles 3: 383–391.

Ivanov, M.V. 1983. Major fluxes of the global biogeochemical cycle of sulphur. pp. 449-463. In M.V. Ivanov and J.R. Freney (eds.), *The Global Biogeochemical Sulphur Cycle*. Wiley, New York.

Iversen, N. 1996. Methane oxidation in coastal marine environments. pp. 51–68. In J.C. Murrell and D.P. Kelly (eds.), *Microbiology of Atmospheric Trace Gases*. Springer-Verlag, Berlin.

Ivens, W., P. Kauppi, J. Alcamo, and M. Posch. 1990. Sulfur deposition onto European forests: Throughfall data and model estimates. Tellus 42B: 294–303.

Iverson, R.L., F.L. Nearhoof, and M.O. Andreae. 1989. Production of dimethylsulfonium propionate and dimethylsulfide by phytoplankton in estuarine and coastal waters. Limnology and Oceanography 34: 53–67.

Jackson, L.E., J.P. Schimel, and M.K. Firestone. 1989. Short-term partitioning of ammonium and nitrate between plants and microbes in an annual grassland. Soil Biology and Biochemisty 21: 409–415.

Jackson, R.B., J.H. Manwaring, and M.M. Caldwell. 1990. Rapid physiological adjustment of roots to localized soil enrichment. Nature 344: 58–60.

Jacob, D.J. and S.C. Wofsy. 1990. Budgets of reactive nitrogen, hydrocarbons, and ozone over the Amazon forest during the wet season. Journal of Geophysical Research 95: 16737–16754.

Jacob, D.J., J.A. Logan, G.M. Gardner, R.M. Yevich, C.M. Spivakovsky, S.C. Wofsy, S. Sillman, and M.J. Prather. 1993. Factors regulating ozone over the United States and its export to the global atmosphere. Journal of Geophysical Research 98: 14817–14826.

Jacobs, S.S. 1986. The polar ice sheets: A wild card in the deck? Oceanus 29(4): 50–54.

Jacobs, S.S. 1992. Is the Antarctic ice sheet growing? Nature 360: 29–33.

Jacobson, M.E. 1994. Chemical and biological mobilization of Fe(III) in marsh sediments. Biogeochemistry 25: 41–60.

Jacoby, G.C. and R.D. D'Arrigo. 1995. Tree ring width and density evidence of climatic and potential forest change in Alaska. Global Biogeochemical Cycles 9: 227–234.

Janke, R.A. 1990. Ocean flux studies: A status report. Reviews of Geophysics 28: 381–398.

Jahnke, R.A. 1992. The phosphorus cycle. pp. 301–315. In S.S. Butcher, R.J. Charlson, G.H. Orians, and G.V. Wolfe. (eds.), *Global Biogeochemical Cycles*. Academic Press, London.

Jahnke, R.A. 1996. The global ocean flux of particulate organic carbon: Areal distribution and magnitude. Global Biogeochemical Cycles 10: 71–88.

James, B.R. and S.J. Riha. 1986. pH buffering in forest soil organic horizons: Relevance to acid precipitation. Journal of Environmental Quality 15: 229–234.

James, P.B., H.H. Kieffer, and D.A. Paige. 1992. The seasonal cycle of carbon dioxide on Mars. pp. 934–968. In H. Kieffer, B.M. Jakosky, C.W. Snyder, and M.A. Matthews (eds.), *Mars*. University of Arizona Press, Tucson.

Jannasch, H.W. 1989. Sulphur emission and transformations at deep sea hydrothermal vents. pp. 181–190. In P. Brimblecombe and A.Y. Lein (eds.), *Evolution of the Global Biogeochemical Sulphur Cycle*. Wiley, New York.

Jannasch, H.W. and M.J. Mottl. 1985. Geomicrobiology of deep-sea hydrothermal vents. Science 229: 717–725.

Jannasch, H.W. and C.O. Wirsen. 1979. Chemosynthetic primary production at East Pacific sea floor spreading centers. BioScience 29: 592–598.

Janos, D.P. 1980. Vesicular-arbuscular mycorrhizae affect lowland tropical rain forest plant growth. Ecology 61: 151–162.

Jansson, M. 1993. Uptake, exchange, and excretion of orthophosphate in phosphate-starved *Scenedesmus quadricauda* and *Pseudomonas* K7. Limnology and Oceanography 38: 1162–1178.

Jaramillo, V.J. and J.K. Detling. 1988. Grazing history, defoliation, and competition: Effects on shortgrass production and nitrogen accumulation. Ecology 69: 1599–1608.

Jarrell, W.M. and R.B. Beverly. 1981. The dilution effect in plant nutrition studies. Advances in Agronomy 34: 197–224.

Jeffries, D.L., J.M. Klopatek, S.O. Link, and H. Bolton. 1992. Acetylene reduction by cryptogamic crusts from a blackbrush community as related to resaturation and dehydration. Soil Biology and Biochemistry 24: 1101–1105.

Jeffries, D.S., R.G. Semkin, R. Neureuther, and M. Seymour. 1988. Ion mass budgets for lakes in the Turkey Lakes watershed, June 1981-May 1983. Canadian Journal of Fisheries and Aquatic Sciences 45(Suppl.): 47–58.

Jellison, R., L.G. Miller, J.M. Melack, and G.L. Dana. 1993. Meromixis in hypersaline Mono Lake, California. 2. Nitrogen fluxes. Limnology and Oceanography 38: 1020–1039.

Jenkins, W.J. 1988. Nitrate flux into the euphotic zone near Bermuda. Nature 331: 521–523.

Jenkinson, D.S. 1990. The turnover of organic carbon and nitrogen in soil. Philosophical Transactions of the Royal Society of London 329B: 361–368.

Jenkinson, D.S. and D.S. Powlson. 1976. The effects of biocidal treatments on metabolism in soil. I. Fumigation with chloroform. Soil Biology and Biochemistry 8: 167–177.

Jenkinson, D.S. and J.H. Rayner. 1977. The turnover of soil organic matter in some of the Rothamsted classical experiments. Soil Science 123: 298–305.

Jenkinson, D.S., D.E. Adams, and A. Wild. 1991. Model estimates of CO_2 emissions from soil in response to global warming. Nature 351: 304–306.

Jennings, J.N. 1983. Karst landforms. American Scientist 71: 578–586.

Jenny, H. 1980. *The Soil Resource*. Springer-Verlag, New York.

Jersak, J., R. Amundson, and G. Brimhall. 1995. A mass balance analysis of podzolization: Examples from the northeastern United States. Geoderma 66: 15–42.

Jiang, Y. and Y.L. Yung. 1996. Concentrations of tropospheric ozone from 1979 to 1992 over tropical Pacific South America from TOMS data. Science 272: 714–716.

Joergensen, R.G. 1996. Quantification of the microbial biomass by determining ninhydrin-reactive N. Soil Biology and Biochemistry 28: 301–306.

Joergensen, R.G. and T. Mueller. 1996. The fumigation-extraction method to estimate soil microbial biomass: Calibration of the K_{en} value. Soil Biology and Biochemistry 28: 33–37.

Johannessen, O.M., M. Miles, and E. Bjørgo. 1995. The Arctic's shrinking sea ice. Nature 376: 126–127.

Johnson, A.H., S.B. Andersen, and T.G. Siccama. 1994. Acid rain and soils of the Adirondacks. I. Changes in pH and available calcium, 1930-1984. Canadian Journal of Forest Research 24: 39–45.

Johnson, D.B., M.A. Ghauri and M.F. Said. 1992. Isolation and characterization of an acidophilic, heterotrophic bacterium capable of oxidizing ferrous iron. Applied and Environmental Microbiology 58: 1423–1428.

Johnson, D.W. 1984. Sulfur cycling in forests. Biogeochemistry 1: 29–43.

Johnson, D.W. and D.W. Cole. 1980. Anion mobility in soils: Relevance to nutrient transport from forest ecoystems. Environment International 3: 79–90.

Johnson, D.W. and D.E. Todd. 1983. Relationships among iron, aluminum, carbon, and sulfate in a variety of forest soils. Soil Science Society of America Journal 47: 792–800.

Johnson, D.W., D.W. Cole, S.P. Gessel, M.J. Singer, and R.V. Minden. 1977. Carbonic acid leaching in a tropical, temperate, subalpine, and northern forest soil. Arctic and Alpine Research 9: 329–343.

Johnson, D.W., G.S. Henderson, and D.E. Todd. 1981. Evidence of modern accumulations of adsorbed sulfate in an east Tennessee forested Ultisol. Soil Science 132: 422–426.

Johnson, D.W., G.S. Henderson, D.D. Huff, S.E. Lindberg, D.D. Richter, D.S. Shriner, D.E. Todd, and J. Turner. 1982. Cycling of organic and inorganic sulphur in a chestnut oak forest. Oecologia 54: 141–148.

Johnson, D.W., D.W. Cole, H. Van Miegroet, and F.W. Horng. 1986. Factors affecting anion movement and retention in four forest soils. Soil Science Society of America Journal 50: 776–783.

Johnson, D.W., G.S. Henderson, and D.E. Todd. 1988. Changes in nutrient distribution in forests and soils of Walker Branch watershed, Tennessee, over an eleven-year period. Biogeochemistry 5: 275–293.

Johnson, D.W., D. Geisinger, R. Walker, J. Newman, J. Vose, K. Elliot, and T. Ball. 1994. Soil pCO_2, soil respiration, and root activity in CO_2-fumigated and nitrogen-fertilized ponderosa pine. Plant and Soil 165: 129–138.

Johnson, K.S., C.L. Beehler, C.M. Sakamoto-Arnold, and J.J. Childress. 1986. In situ measurements of chemical distributions in a deep-sea hydrothermal vent field. Science 231: 1139–1141.

Johnson, N.M. 1971. Mineral equilibria in ecosystem geochemistry. Ecology 52: 529–531.

Johnson, N.M., G.E. Likens, F.H. Bormann, and R.S. Pierce. 1968. Rate of chemical weathering of silicate minerals in New Hampshire. Geochimica et Cosmochimica Acta 32: 531–545.

Johnson, N.M., G.E. Likens, F.H. Bormann, D.W. Fisher, and R.S. Pierce. 1969. A working model for the variation in stream water chemistry at the Hubbard Brook Experimental Forest, New Hampshire. Water Resources Research 5: 1353–1363.

Johnson, N.M., R.C. Reynolds, and G.E. Likens. 1972. Atmospheric sulfur: Its effect on the chemical weathering of New England. Science 177: 514–516.

Johnson, N.M., C.T. Driscoll, J.S. Eaton, G.E. Likens, and W.H. McDowell. 1981. 'Acid rain,' dissolved aluminum and chemical weathering at the Hubbard Brook Experimental Forest, New Hampshire. Geochimica et Cosmochimica Acta 45: 1421–1437.

Johnston, D.A. 1980. Volcanic contribution of chlorine to the stratosphere: More significant to ozone than previously estimated? Science 209: 491–493.

Jonas, P.R., R.J. Charlson, and H. Rodhe. 1995. Aerosols. pp. 127–162. In J.T. Houghton, L.G. Meira Filho, J. Bruce, H. Lee, B.A. Callander, E. Haites, N. Harris, and K. Maskell (eds.), *Climate Change 1994.* Cambridge University Press, Cambridge.

Jonasson, S., J.P. Bryant, F.S. Chapin, and M. Andersson. 1986. Plant phenolics and nutrients in relation to variations in climate and rodent grazing. American Naturalist 128: 394–408.

Jones, A.E. and J.D. Shanklin. 1995. Continued decline of total ozone over Halley, Antarctica, since 1985. Nature 376: 409–411.

Jones, J.G. 1986. Iron transformations by freshwater bacteria. Advances in Microbial Ecology 9: 149–185.

Jones, M.J. 1973. The organic matter content of the savanna soils of west Africa. Journal of Soil Science 24: 42–53.

Jones, R.D. and R.Y. Morita. 1983. Methane oxidation by *Nitrosococcus oceanus* and *Nitrosomonas europaea.* Applied and Environmental Microbiology 45: 401–410.

Jones, R.E., R.E. Beeman, and J.M. Suflita. 1989. Anaerobic metabolic processes in the deep terrestrial subsurface. Geomicrobiology Journal 7: 117–130.

Jones, R.L. and H.C. Hanson. 1985. *Mineral Licks, Geophagy, and Biogeochemistry of North American Ungulates.* Iowa State University Press, Ames.

Jordan, C.F. 1971. A world pattern in plant energetics. American Scientist 59: 425–433.

Jordan, M. and G.E. Likens. 1975. An organic carbon budget for an oligotrophic lake in New Hampshire, U.S.A. VerhandlungenInternationale Vereinigung fuer Theoretische und Angewandte Limnologie 19: 994–1003.

Jordan, T.E. and D.E. Weller. 1996. Human contributions to terrestrial nitrogen flux. BioScience 46: 655–664.

Jordan, T.E., D.L. Correll, and D.E. Weller. 1993. Nutrient interception by a riparian forest receiving inputs from adjacent cropland. Journal of Environmental Quality 22: 467–473.

Jørgensen, B.B. 1977. The sulfur cycle of a coastal marine sediment (Limfjorden, Denmark). Limnology and Oceanography 22: 814–832.

Jørgensen, B.B. 1990. A thiosulfate shunt in the sulfur cycle of marine sediments. Science 249: 152–154.

Jørgensen, B.B., M.F. Isaksen, and H.W. Jannasch. 1992. Bacterial sulfate reduction ahove 100°C in deep-sea hydrothermal vent sediments. Science 258: 1756–1757.

Jorgensen, J.R., C.G. Wells, and L.J. Metz. 1980. Nutrient changes in decomposing loblolly pine forest floor. Soil Science Society of America Journal 44: 1307–1314.

Jørgensen, K.S. 1989. Annual pattern of denitrification and nitrate ammonification in estuarine sediment. Applied and Environmental Microbiology 55: 1841–1847.

Jørgensen, K.S., H.B. Jensen, and J. Sørensen. 1984. Nitrous oxide production from nitrification and denitrification in marine sediment at low oxygen concentrations. Canadian Journal of Microbiology 30: 1073–1078.

Jouzel, J., N.I. Barkov, J.M. Barnola, M. Bender, J. Chappellaz, C. Genthon, V.M. Kotlyakov, V. Lipenkov, C. Lorius, J.R. Petit, D. Raynaud, G. Raisbeck, C. Ritz, T. Sowers, M. Stievenard, F. Yiou, and P. Yiou. 1993. Extending the Vostok ice-core record of palaeoclimate to the penultimate glacial period. Nature 364: 407–412.

Juang, F.H.T. and N.M. Johnson. 1967. Cycling of chlorine through a forested watershed in New England. Journal of Geophysical Research 72: 5641–5647.

Junge, C.E. 1960. Sulfur in the atmosphere. Journal of Geophysical Research 65: 227–237.

Junge, C.E. 1974. Residence time and variability of tropospheric trace gases. Tellus 26: 477–488.

Junge, C.E. and R.T. Werby. 1958. The concentration of chloride, sodium, potassium, calcium, and sulfate in rain water over the United States. Journal of Meteorology 15: 417–425.

Jurgensen, M.F. 1973. Relationship between nonsymbiotic nitrogen fixation and soil nutrient statusA review. Journal of Soil Science 24: 512–522.

Jurinak, J.J., L.M. Dudley, M.F. Allen, and W.G. Knight. 1986. The role of calcium oxalate in the availability of phosphorus in soils of semiarid regions: A thermodynamic study. Soil Science 142: 255–261.

Justić, D., N.N. Rabalais, R.E. Turner, and Q. Dortch. 1995. Changes in nutrient structure of river-dominated coastal waters: Stoichiometric nutrient balance and its consequences. Estuarine, Coastal and Shelf Science 40: 339–356.

Kadeba, O. 1978. Organic matter status of some savanna soils of northern Nigeria. Soil Science 125: 122–127.

Kahn, R. 1985. The evolution of CO_2 on Mars. Icarus 62: 175–190.

Kaplan, I.R. 1975. Stable isotopes as a guide to biogeochemical processes. Proceedings of the Royal Society of London 189B: 183–211.

Kaplan, W., I. Valiela, and J.M. Teal. 1979. Denitrification in a salt marsh ecosystem. Limnology and Oceanography 24: 726–734.

Kaplan, W.A., J.W. Elkins, C.E. Kolb, M.B. McElroy, S.C. Wofsy, and A.P. Duran. 1978. Nitrous oxide in fresh water systems: An estimate of the yield of atmospheric N_2O associated with disposal of human waste. Pure and Applied Geophysics 116: 424–438.

Kaplan, W.A., S.C. Wofsy, M. Keller, and J.M. Da Costa. 1988. Emission of NO and deposition of O_3 in a tropical forest system. Journal of Geophysical Research 93: 1389–1395.

Karl, D.M. and B.D. Tilbrook. 1994. Production and transport of methane in oceanic particulate organic matter. Nature 368: 732–734.

Karltun, E. and J.P. Gustafsson. 1993. Interference by organic complexation of Fe and Al on the SO_4^{2-} adsorption in spodic B horizons in Sweden. Journal of Soil Science 44: 625–632.

Kasibhatla, P.S., H. Levy, and W.J. Moxim. 1993. Global NO_x, NHO_3, PAN, and NO_y distributions from fossil fuel combustion emissions: A model study. Journal of Geophysical Research 98: 7165–7180.

Kasischke, E.S., L.L. Bourgeau-Chavez, N.L. Christensen, and E. Haney. 1994. Observations on the sensitivity of ERS-1 SAR image intensity to changes in aboveground biomass in young loblolly pine forests. International Journal of Remote Sensing 15: 3–16.

Kasischke, E.S., N.L. Christensen, and B.J. Stocks. 1995. Fire, global warming, and the carbon balance of boreal forests. Ecological Applications 5: 437–451.

Kass, D.M. and Y.L. Yung. 1995. Loss of atmosphere from Mars due to solar wind-induced sputtering. Science 268: 697–699.

Kasting, J.F. 1993. Earth's early atmosphere. Science 259: 920–926.

Kasting, J.F. and J.C.G. Walker. 1981. Limits on oxygen concentration in the prebiological atmosphere and the rate of abiotic fixation of nitrogen. Journal of Geophysical Research 86: 1147–1158.

Kasting, J.F., O.B. Toon, and J.B. Pollack. 1988. How climate evolved on the terrestrial planets. Scientific American 258(2): 90–97.

Kastner, M. 1974. The contribution of authigenic feldspars to the geochemical balance of alkalic metals. Geochimica et Cosmochimica Acta 38: 650–653.

Kauffman, J.B., R.L. Sanford, D.L. Cummings, I.H. Salcedo, and E.V.S.B. Sampaio. 1993. Biomass and nutrient dynamics associated with slash pine fires in neotropical dry forests. Ecology 74: 140–151.

Kaufman, Y.J., C.J. Tucker, and I. Fung. 1990. Remote sensing of biomass burning in the tropics. Journal of Geophysical Research 95: 9927–9939.

Kauppi, P.E., K. Mielikäinen, and K. Kuusela. 1992. Biomass and carbon budget of European forests, 1971 to 1990. Science 256: 70–74.

Keeling, C.D. 1983. The global carbon cycle: What we know and could know from atmospheric, biospheric, and oceanic observations. pp. II.3–62. In *Proceedings: Carbon Dioxide Research Conference: Carbon Dioxide, Science and Consensus.* CONF 820970. U.S. Department of Energy, Washington, D.C.

Keeling, C.D. 1993. Global observations of atmospheric CO_2. pp. 1–29. In M. Heimann (ed.), *The Global Carbon Cycle.* Springer-Verlag, New York.

Keeling, C.D. and T.P. Whorf. 1994. Atmospheric CO_2 from sites in the SIO air sampling network. pp. 16–26 *In* T.A. Boden, D.P. Kaiser, R.J. Sepanski, and F.W. Stoss (eds.), *Trends '93: A Compendium of Data on Global Change.* Carbon Dioxide Information Analysis Center, Oak Ridge National Laboratory, Oak Ridge, Tennessee.

Keeling, C.D., A.F. Carter, and W.G. Mook. 1984. Seasonal, latitudinal, and secular variations in the abundance and isotopic ratios of atmospheric CO_2. 2. Results from oceanographic crusies in the tropical Pacific ocean. Journal of Geophysical Research 89: 4615–4628.

Keeling, C.D., S.C. Piper, and M. Heimann. 1989. A three-dimensional model of atmospheric CO_2 transport based on observed winds: 4. Mean annual gradients and interannual variations. pp. 305-363. In D.H. Peterson (ed.), *Aspects of Climate Variability in the Pacific and Western Americas.* American Geophysical Union, Washington, D.C.

Keeling, C.D., T.P. Whorf, M. Wahlen, and J. van der Plicht. 1995. Interannual extremes in the rate of rise of atmospheric carbon dioxide since 1980. Nature 375: 666–670.

Keeling, C.D., J.F.S. Chin, and T.P. Whorf. 1996. Increased activity of northern vegetation inferred from atmospheric CO_2 measurements. Nature 382: 146–149.

Keeling, R.F. and S.R. Shertz. 1992. Seasonal and interannual variations in atmospheric oxygen and implications for the global carbon cycle. Nature 358: 723–727.

Keeling, R.F., S.C. Piper, and M. Heimann. 1996. Global and hemispheric CO_2 sinks deduced from changes in atmospheric O_2 concentration. Nature 381: 218–221.

Keeney, D.R. 1980. Prediction of soil nitrogen availability in forest ecosystems: A literature review. Forest Science 26: 159–171.

Keffer, T., D.G. Martinson, and B.H. Corliss. 1988. The position of the Gulf Stream during Quaternary glaciations. Science 241: 440–442.

Keil, R.G., D.B. Montiuçon, F.G. Prahl, and J.I. Hedges. 1994. Sorptive preservation of labile organic matter in marine sediments. Nature 370: 549–552.

Keller, C.K. and B.D. Wood. 1993. Possibility of chemical weathering before the advent of vascular land plants. Nature 364: 223–225.

Keller, M. and W.A. Reiners. 1994. Soil-atmosphere exchange of nitrous oxide, nitric oxide, and methane under secondary succession of pasture to forest in the Atlantic lowlands of Costa Rica. Global Biogeochemical Cycles 8: 399–409.

Keller, M., W.A. Kaplan, S.C. Wofsy, and J.M. Da Costa. 1988. Emission of N_2O from tropical forest soils: Response to fertilization with NH_4^+, NO_3^-, and PO_4^{3-}. Journal of Geophysical Research 93: 1600–1604.

Kelley, C.A., C.S. Martens, and J.P. Chanton. 1990. Variations in sedimentary carbon remineralization rates in the White Oak River estuary, North Carolina. Limnology and Oceanography 35: 372–383.

Kellogg, W.W. 1992. Aerosols and global warming. Science 256: 598.

Kelly, D.P. and N.A. Smith. 1990. Organic sulfur compounds in the environment. Advances in Microbial Ecology 11: 345–385.

Kemp, W.M., P. Sampou, J. Caffrey, M. Mayer, K. Henriksen, and W.R. Boynton. 1990. Ammonium recycling versus denitrification in Chesapeake Bay sediments. Limnology and Oceanography 35: 1545–1563.

Kempe, S. 1984. Sinks of the anthropogenically enhanced carbon cycle in surface fresh waters. Journal of Geophysical Research 89: 4657–4676.

Kempe, S. 1988. Estuariestheir natural and anthropogenic changes. pp. 251-285. In T. Rosswall, R.G. Woodmansee, and R.G. Risser (eds.), *Scales and Global Change.* Wiley, London

Kennett, J. 1982. *Marine Geology*. Prentice Hall, Englewood Cliffs, New Jersey.

Kerner, M. 1993. Coupling of microbial fermentation and respiration processes in an intertidal mudflat of the Elbe Estuary. Limnology and Oceanography 38: 314–330.

Kerr, J.B. and C.T. McElroy. 1993. Evidence for large upward trends of ultraviolet-B radiation linked to ozone depletion. Science 262: 1032–1034.

Kesselmeier, J. and L. Merk. 1993. Exchanges of carbonyl sulfide (COS) between agricultural plants and the atmosphere: Studies on the deposition of COS to peas, corn and rapeseed. Biogeochemistry 23: 47–60.

Kesselmeier, J., F.X. Meixner, U. Hofmann, A.-L. Ajavon, S. Leimbach, and M.O. Andreae. 1993. Reduced sulfur compound exchange between the atmosphere and tropical tree species in southern Cameroon. Biogeochemistry 23: 23–45.

Khalil, M.A.K. and R.A. Rasmussen. 1984. Global sources, lifetimes, and mass balances of carbonyl sulfide (OCS) and carbon disulfide (CS_2) in the Earth's atmosphere. Atmospheric Environment 18: 1805–1813.

Khalil, M.A.K. and R.A. Rasmussen. 1985. Causes of increasing atmospheric methane: Depletion of hydroxyl radicals and the rise of emissions. Atmospheric Environment 19: 397–407.

Khalil, M.A.K. and R.A. Rasmussen. 1988. Carbon monoxide in the Earth's atmosphere: Indications of a global increase. Nature 332: 242–245.

Khalil, M.A.K. and R.A. Rasmussen. 1989. Climate-induced feedbacks for the global cycles of methane and nitrous oxide. Tellus 41B: 554–559.

Khalil, M.A.K. and R.A. Rasmussen. 1990. Constraints on the global sources of methane and an analysis of recent budgets. Tellus 42B: 229–236.

Khalil, M.A.K. and R.A. Rasmussen. 1992a. The global sources of nitrous oxide. Journal of Geophysical Research 97: 14651–14660.

Khalil, M.A.K and R.A. Rasmussen. 1992b. Nitrous oxide from coal-fired power plants: Experiments in the plumes. Journal of Geophysical Research 97: 14645–14649.

Khalil, M.A.K. and R.A. Rasmussen. 1994. Global decrease in atmospheric carbon monoxide concentration. Nature 370: 639–641.

Khalil, M.A.K. and M.J. Shearer. 1993. Sources of methane: An overview. pp. 180–198. In M.A.K. Khalil (ed.), *Atmospheric Methane: Sources, Sinks, and Role in Global Change.* Springer-Verlag, New York.

Khalil, M.A.K., R.A. Rasmussen, J.R.J. French, and J.A. Holt. 1990. The influence of termites on atmospheric trace gases: CH_4, CO_2, $CHCl_3$, N_2O, CO, H_2, and light hydrocarbons. Journal of Geophysical Research 95: 3619–3634.

Khalil, M.A.K., R.A. Rasmussen, and R. Gunawardena. 1993a. Atmospheric methyl bromide: Trends and global mass balance. Journal of Geophysical Research 98: 2887–2896.

Khalil, M.A.K., R.A. Rasmussen, and F. Moraes. 1993b. Atmospheric methane at Cape Meares: Analysis of a high-resolution data base and its environmental implications. Journal of Geophysical Research 98: 14753–14770.

Khalil, M.A.K., M.J. Shearer, and R.A. Rasmussen. 1993c. Methane sinks and distribution. pp. 168-179. In M.A.K. Khalil (ed.), *Atmospheric Methane: Sources, Sinks, and Role in Global Change.* Springer-Verlag, Berlin.

Kieffer, H.H. 1976. Soil and surface temperatures at the Viking landing sites. Science 194: 1344–1346.

Kiehl, J.T. and B.P. Briegleb. 1993. The relative roles of sulfate aerosols and greenhouse gases in climate forcing. Science 260: 311–314.

Kielland, K. 1994. Amino acid absorption by arctic plants: Implications for plant nutrition and nitrogen cycling. Ecology 75: 2373–2383.

Kiene, R.P. 1990. Dimethyl sulfide production from dimethylsulfoniopropionate in coastal seawater samples and bacterial cultures. Applied and Environmental Microbiology 56: 3292–3297.

Kiene, R.P. and T.S. Bates. 1990. Biological removal of dimethyl sulphide from sea water. Nature 345: 702–705.

Kilham, P. 1971. A hypothesis concerning silica and the freshwater planktonic diatoms. Limnology and Oceanography 16: 10–18.

Kilham, P. 1982. Acid precipitation: Its role in the alkalization of a lake in Michigan. Limnology and Oceanography 27: 856–867.

Killingbeck, K.T. 1985. Autumnal resorption and accretion of trace metals in gallery forest trees. Ecology 66: 283–286.

Killingbeck, K.T. 1996. Nutrients in senesced leaves: Keys to the search for potential resorption and resorption proficiency. Ecology 77: 1716–1727.

Killough, G.G. and W.R. Emanuel. 1981. A comparison of several models of carbon turnover in the ocean with respect to their distributions of transit time and age, and responses to atmospheric CO_2 and ^{14}C. Tellus 33: 274–290.

Kim, J. and D.C. Rees. 1992. Structural models for the metal centers in the nitrogenase molybdenum-rich protein. Science 257: 1677–1682.

Kim, J., S.B. Verma, and R.J. Clement. 1992. Carbon dioxide budget in a temperate grassland ecosystem. Journal of Geophysical Research 97: 6057–6063.

Kim, K.-R. and H. Craig. 1990. Two-isotope characterization of N_2O in the Pacific ocean and constraints on its origin in deep water. Nature 347: 58–61.

King, G.M. 1988. Patterns of sulfate reduction and the sulfur cycle in a South Carolina salt marsh. Limnology and Oceanography 33: 376–390.

King, G.M. 1992. Ecological aspects of methane oxidation, a key determinant of global methane dynamics. Advances in Microbial Ecology 12: 431–468.

King, G.M. and A.P.S. Adamsen. 1992. Effects of temperature on methane consumption in a forest soil and pure cultures of the methanotroph *Methylomonas rubra.* Applied and Environmental Microbiology 58: 2758–2763.

King, G.M. and S. Schnell. 1994. Effect of increasing atmospheric methane concentration on ammonium inhibition of soil methane consumption. Nature 370: 282–284.

King, G.M. and W.J. Wiebe. 1978. Methane release from soils of a Georgia salt marsh. Geochimica et Cosmochimica Acta 42: 343–348.

King, G.M., P. Roslev, and H. Skovgaard. 1990. Distribution and rate of methane oxidation in sediments of the Florida Everglades. Applied and Environmental Microbiology 56: 2902–2911.

Kinsman, D.J.J. 1969. Interpretation of Sr^{2+} concentrations in carbonate minerals and rocks. Journal of Sedimentary Petrology 39: 486–508.

Kira, T. and T. Shidei. 1967. Primary production and turnover of organic matter in different forest ecosystems of the western Pacific. Japanese Journal of Ecology 17: 70–87.

Kirchman, D.L., Y. Suzuki, C. Garside, and H.W. Ducklow. 1991. High turnover rates of dissolved organic carbon during a spring phytoplankton bloom. Nature 352: 612–614.

Kirchman, D.L., H.W. Ducklow, J.J. McCarthy, and C. Garside. 1994. Biomass and nitrogen uptake by heterotrophic bacteria during the spring phytoplankton bloom in the North Atlantic Ocean. Deep Sea Research 41: 879–895.

Kirchner, J.W. 1989. The Gaia hypothesis: Can it be tested? Reviews of Geophysics 27: 223–235.

Kirschbaum, M.U.F. 1995. The temperature dependence of soil organic matter decomposition, and the effect of global warming on soil organic C storage. Soil Biology and Biochemistry 27: 753–760.

Klappa, C.F. 1980. Rhizoliths in terrestrial carbonates: Classification, recognition, genesis and significance. Sedimentology 27: 613–629.

Klein, H.P. 1979. The Viking mission and the search for life on Mars. Reviews of Geophysics and Space Physics 17: 1655–1662.

Kleinman, L., Y.-N. Lee, S.R. Springston, L. Nunnermacker, X. Zhou, R. Brown, K. Hallock, P. Klotz, D. Leahy, J.H. Lee, and L. Newman. 1994. Ozone formation at a rural site in the southeastern United States. Journal of Geophysical Research 99: 3469–3482.

Kling, G.W. 1988. Comparative transparency, depth of mixing, and stability of stratification in lakes of Cameroon, West Africa. Limnology and Oceanography 33: 27–40.

Kling, G.W., A.E. Giblin, B. Fry, and B.J. Peterson. 1991. The role of seasonal turnover in lake alkalinity dynamics. Limnology and Oceanography 36: 106–122.

Klinkhammer, G.P. 1980. Early diagensis in sediments from the eastern equatorial Pacific. II. Pore water metal results. Earth and Planetary Science Letters 49: 81–101.

Klopatek, J.M. 1987. Nitrogen mineralization and nitrification in mineral soils of pinyon-juniper ecosystems. Soil Science Society of America Journal 51: 453–457.

Klotz, R.L. 1988. Sediment control of soluble reactive phosphorus in Hoxie Gorge Creek, New York. Canadian Journal of Fisheries and Aquatic Sciences 45: 2026–2034.

Kludze, H.K., R.D. DeLaune, and W.H. Patrick. 1993. Aerenchyma formation and methane and oxygen exchange in rice. Soil Science Society of America Journal 57: 386–391.

Knauer, G.A. 1993. Productivity and new production of the oceanic systems. pp. 211-231. In R. Wollast, F.T. MacKenzie, and L. Chou (eds.), *Interactions of C, N, P and S Biogeochemical Cycles and Global Change.* Springer-Verlag, New York.

Knauss, J.A. 1978. *Introduction to Physical Oceanography.* Prentice-Hall, Englewood Cliffs, New Jersey.

Knight, D.H., T.J. Fahey, and S.W. Running. 1985. Water and nutrient outflow from contrasting lodgepole pine forests in Wyoming. Ecological Monographs 55: 29–48.

Knoll, A.H. 1992. The early evolution of eukaryotes: A geological perspective. Science 256: 622–627.

Knoll, M.A. and W.C. James. 1987. Effect of the advent and diversification of vascular land plants on mineral weathering through geologic time. Geology 15: 1099–1102.

Knowles, R. 1982. Denitrification. Microbiological Reviews 46: 43–70.

Kodama, H. and M. Schnitzer. 1977. Effect of fulvic acid on the crystallization of Fe(III) oxides. Geoderma 19: 279–291.

Kodama, H. and M. Schnitzer. 1980. Effect of fulvic acid on the crystallization of aluminum hydroxides. Geoderma 24: 195–205.

Koerselman, W., H. De Caluwe, and W.H. Kieskamp. 1989. Denitrification and dinitrogen fixation in two quaking fens in the Vechtplassen area, The Netherlands. Biogeochemistry 8: 153–165.

Koide, R.T. 1991. Nutrient supply, nutrient demand and plant response to mycorrhizal infection. New Phytologist 117: 365–386.

Kolber, Z.S., R.T. Barber, K.H. Coale, S.E. Fitzwater, R.M. Greene, K.S. Johnson, S. Lindley, and P.G. Falkowski. 1994. Iron limitation of phytoplankton photosynthesis in the equatorial Pacific Ocean. Nature 371: 145–149.

Koopmans, C.J., A. Tietema, and A.W. Boxman. 1996. The fate of ^{15}N enriched throughfall in two coniferous forest stands at different nitrogen deposition levels. Biogeochemistry 34: 19–44.

Körner, C. 1989. The nutritional status of plants from high altitudes: A worldwide comparison. Oecologia 81: 379–391.

Körner, C. 1993. CO_2 fertilization: The great uncertainty in future vegetation development. pp. 53–70. In A.M. Solomon and H.H. Shugart (eds.), *Vegetation Dyanmics and Global Change.* Chapman and Hall, New York.

Korom, S.F. 1992. Natural denitrification in the saturated zone: A review. Water Resources Research 28: 1657–1668.

Koschorreck, M. and R. Conrad. 1993. Oxidation of atmospheric methane in soil: Measurements in the field, in soil cores, and in soil samples. Global Biogeochemical Cycles 7: 109–121.

Kostka, J.E. and G.W. Luther. 1995. Seasonal cycling of Fe in saltmarsh sediments. Biogeochemistry 29: 159–181.

Krairapanond, N., R.D. DeLaune, and W.H. Patrick. 1991. Sulfur dynamics in Louisiana coastal freshwater marsh soils. Soil Science 151: 261–273.

Kramer, P.J. 1981. Carbon dioxide concentration, photosynthesis, and dry matter production. BioScience 31: 29–33.

Kramer, P.J. 1982. Water and plant productivity of yield. pp. 41–47. In M. Rechcigl (ed.), *Handbook of Agricultural Productivity.* CRC Press, Boca Raton, Florida.

Kratz, T.K. and C.B. DeWitt. 1986. Internal factors controlling peatland-lake ecosystem development. Ecology 67: 100–107.

Krause, H.H. 1982. Nitrate formation and movement before and after clear-cutting of a monitored watershed in central New Brunswick, Canada. Canadian Journal of Forest Research 12: 922–930.

Kristensen, E., S.I. Ahmed, and A.H. Devol. 1995. Aerobic and anaerobic decomposition of organic matter in marine sediment: Which is fastest? Limnology and Oceanography 40: 1430–1437.

Kristjansson, J.K. and P. Schönheit. 1983. Why do sulfate-reducing bacteria outcompete methanogenic bacteria for substrates? Oecologia 60: 264–266.

Kroehler, C.J. and A.E. Linkins. 1991. The absorption of inorganic phosphate from 32-P labeled inositol hexaphosphate by *Eriophorum vaginatum.* Oecologia 85: 424–428.

Kroer, N., N.O.G. Jørgensen, and R.B. Coffin. 1994. Utilization of dissolved nitrogen by heterotrophic bacterioplankton: A comparison of three ecosystems. Applied and Environmental Microbiology 60: 4116–4123.

Krom, M.D. and R.A. Berner. 1981. The diagenesis of phosphorus in a nearshore marine sediment. Geochimica et Cosmochimica Acta 45: 207–216.

Krouse, H.R. and R.G.L. McCready. 1979. Reductive reactions in the sulfur cycle. pp. 315–368. In P.A. Trudinger and D.J. Swaine (eds.), *Biogeochemical Cycling of Mineral-Forming Elements.* Elsevier Scientific Publishers, Amsterdam.

Krumbein, W.E. 1971. Manganese-oxidizing fungi and bacteria in recent shelf sediments of the Bay of Biscay and the North Sea. Naturwissenschaften 58: 56–57.

Krumbein, W.E. 1979. Calcification by bacteria and algae. pp. 47–68. In P.A. Trudinger and D.J. Swaine (eds.), *Biogeochemical Cycling of Mineral-Forming Elements.* Elsevier Scientific Publishers, Amsterdam.

Krysell, M. and D.W.R. Wallace. 1988. Arctic Ocean ventilation studied with a suite of anthropogenic halocarbon tracers. Science 242: 746–749.

Kuhlbusch, T.A., J.M. Lobert, P.J. Crutzen, and P. Warneck. 1991. Molecular nitrogen emissions from denitrification during biomass burning. Nature 351: 135–137.

Kuhry, P. and D.H. Vitt. 1996. Fossil carbon/nitrogen ratios as a measure of peat decomposition. Ecology 77: 271–275.

Kuivila, K.M., J.W. Murray, A.H. Devol, M.E. Lidstrom, and C.E. Reimers. 1988. Methane cycling in the sediments of Lake Washington. Limnology and Oceanography 33: 571–581.

Kuivila, K.M., J.W. Murray, A.H. Devol, and P.C. Novelli. 1989. Methane production, sulfate reduction and competition for substrates in the sediments of Lake Washington. Geochimica et Cosmochimica Acta 53: 409–416.

Kumar, N., R.F. Anderson, R.A. Mortlock, P.N. Froelich, P. Kubik, B. Dittrich-Hannen, and M. Suter. 1995. Increased biological productivity and export production in the glacial Southern Ocean. Nature 378: 675–680.

Kumar, P.P., G.K. Manohar, and S.S. Kandalgaonkar. 1995. Global distribution of nitric oxide produced by lightning and its seasonal variation. Journal of Geophysical Research 100: 11203–11208.

Kump, L.R. and R.M. Garrels. 1986. Modeling atmospheric O_2 in the global sedimentary redox cycle. American Journal of Science 286: 337–360.

Kump, L.R. and H.D. Holland. 1992. Iron in Precambrian rocks: Implications for the global oxygen budget of the ancient Earth. Geochimica et Cosmochimica Acta 56: 3217–3223.

Kunishi, H.M. 1988. Sources of nitrogen and phosphorus in an estuary of the Chesapeake Bay. Journal of Environmental Quality 17: 185–188.

Kuramoto, K. and T. Matsui. 1996. Partitioning of H and C between the mantle and core during the core formation in the Earth: Its implications for the atmospheric evolution and redox state of early mantle. Journal of Geophysical Research 101: 14909–14932.

Kvenvolden, K., J. Lawless, K. Pering, E. Peterson, J. Flores, C. Ponnamperuma, I.R. Kaplan, and C. Moore. 1970. Evidence for extraterrestrial amino-acids and hydrocarbons in the Murchison meteorite. Nature 228: 923–926.

LaBaugh, J.W., D.O. Rosenbergy, and T.C. Winter. 1995. Groundwater contribution to the water and chemical budgets of Williams Lake, Minnesota, 1980-1991. Canadian Journal of Fisheries and Aquatic Sciences 52: 754–767.

Ladd, J.N., J.M. Oades, and M. Amato. 1981. Microbial biomass formed from ^{14}C, ^{15}N-labelled plant material decomposing in soils in the field. Soil Biology and Biochemistry 13: 119–126.

Lagage, P.O. and E. Pantin. 1994. Dust depletion in the inner disk of *B* Pictoris as a possible indicator of planets. Nature 369: 628–630.

Laj, P., J.M. Palais, and H. Sigurdsson. 1992. Changing sources of impurities to the Greenland ice sheet over the last 250 years. Atmospheric Environment 26A: 2627–2640.

Lajtha, K. 1987. Nutrient reabsorption efficiency and the response to phosphorus fertilization in the desert shrub *Larrea tridentata* (DC.) Cov. Biogeochemistry 4: 265–276.

Lajtha, K. and S.H. Bloomer. 1988. Factors affecting phosphate sorption and phosphate retention in a desert ecosystem. Soil Science 146: 160–167.

Lajtha, K. and W.H. Schlesinger. 1986. Plant response to variations in nitrogen availability in a desert shrubland community. Biogeochemistry 2: 29–37.

Lajtha, K. and W.H. Schlesinger. 1988. The biogeochemistry of phosphorus cycling and phosphorus availability along a desert soil chronosequence. Ecology 69: 24–39.

Lajtha, K. and W.G. Whitford. 1989. The effect of water and nitrogen amendments on photosynthesis, leaf demography, and resource-use efficiency in *Larrea tridentata,* a desert evergreen shrub. Oecologia 80: 341–348.

Lal, D. 1977. The oceanic microcosm of particles. Science 198: 997–1009.

Lal, R. 1995. Global soil erosion by water and carbon dynamics. pp. 131–142. In R. Lal, J. Kimble, E. Levine, and B.A. Stewart (eds.), *Soils and Global Change.* CRC Lewis Publishers, Boca Raton, Florida.

Lamb, B., A. Guenther, D. Gay, and H. Westberg. 1987a. A national inventory of biogenic hydrocarbon emissions. Atmospheric Environment 21: 1695–1705.

Lamb, B., H. Westberg, G. Allwine, L. Bamesberger, and A. Guenther. 1987b. Measurement of biogenic sulfur emissions from soils and vegetation: Application of dynamic enclosure methods with Natusch filter and CG/FPD analysis. Journal of Atmospheric Chemistry 5: 469–491.

Lamb, D. 1985. The influence of insects on nutrient cycling in eucalypt forests: A beneficial role? Australian Journal of Ecology 10: 1–5.

Lambert, R.L., G.E. Lang, and W.A. Reiners. 1980. Loss of mass and chemical change in decaying boles of a subalpine balsam fir forest. Ecology 61: 1460–1473.

Lane, L.J., E.M. Romney, and T.E. Hakonson. 1984. Water balance calculations and net production of perennial vegetation in the northern Mojave Desert. Journal of Range Management 37: 12–18.

Lang, G.E. and R.T.T. Forman. 1978. Detrital dynamics in a mature oak forest: Hutcheson Memorial Forest, New Jersey. Ecology 59: 580–595.

Langbein, W.B. and S.A. Schumm. 1958. Yield of sediment in relation to mean annual precipitation. Transactions of the American Geophysical Union 39: 1076–1084.

Langford, A.O. and F.C. Fehsenfeld. 1992. Natural vegetation as a source or sink for atmospheric ammonia: A case study. Science 255: 581–583.

Langford, A.O., F.C. Fehsenfeld, J. Zachariassen, and D.S. Schimel. 1992. Gaseous ammonia fluxes and background concentrations in terrestrial ecosystems of the United States. Global Biogeochemical Cycles 6: 459–483.

Langner, J. and H. Rodhe. 1991. A global three-dimensional model of the tropospheric sulfur cycle. Journal of Atmospheric Chemistry 13: 225–263.

Langner, J., H. Rodhe, P.J. Crutzen, and P. Zimmerman. 1992. Anthropogenic influence on the distribution of tropospheric sulphate aerosol. Nature 359: 712–716.

Langway, C.C., K. Osada, H.B. Clausen, C.U. Hammer, H. Shoji, and A. Mitani. 1994. New chemical stratigraphy over the last millennium for Byrd Station, Antarctica. Tellus 46B: 40–51.

Langway, C.C., K. Osada, H.B. Clausen, C.U. Hammer, and H. Shoji. 1995. A 10-century comparison of prominent bipolar volcanic events in ice cores. Journal of Geophysical Research 100: 16241–16247.

Lansdown, J.M., P.D. Quay, and S.L. King. 1992. CH_4 production via CO_2 reduction in a temperate bog: A source of ^{13}C-depleted CH_4. Geochimica et Cosmochimica Acta 56: 3493–3503.

Lantzy, R.J. and F.T. MacKenzie. 1979. Atmospheric trace metals: Global cycles and assessment of man's impact. Geochimica et Cosmochimica Acta 43: 511–525.

Lapeyrie, F., G.A. Chilvers, and C.A. Bhem. 1987. Oxalic acid synthesis by the mycorrhizal fungus *Paxillus involutus* (Batsch. ex. Fr.) Fr. New Phytologist 106: 139–146.

Laronne, J.B. and I. Reid. 1993. Very high rates of bedload sediment transport by ephemeral desert rivers. Nature 366: 148–150.

Lasenby, D.C. 1975. Development of oxygen deficits in 14 southern Ontario lakes. Limnology and Oceanography 20: 993–999.

Lashof, D.A. and D.R. Ahuja. 1990. Relative contributions of greenhouse gas emissions to global warming. Nature 344: 529–531.

Laskowski, R., M. Niklińska, and M.J. Maryański. 1995. The dynamics of chemical elements in forest litter. Ecology 76: 1393–1406.

Laudelout, H. and M. Robert. 1994. Biogeochemistry of calcium in a broad-leaved forest ecosystem. Biogeochemistry 27: 1–22.

Lauenroth, W.K. and W.C. Whitman. 1977. Dynamics of dry matter production in a mixed-grass prairie in western North Dakota. Oecologia 27: 339–351.

Laurmann, J.A. 1979. Market penetration characteristics for energy production and atmospheric carbon dioxide growth. Science 205: 896–898.

Laursen, K.K., P.V. Hobbs, L.F. Radke, and R.A. Rasmussen. 1992. Some trace gas emissions from North American biomass fires with an assessment of regional and global fluxes from biomass burning. Journal of Geophysical Research 97: 20687–20701.

Law, C.S. and N.J.P. Owens. 1990. Significant flux of atmospheric nitrous oxide from the northwest Indian Ocean. Nature 346: 826–828.

Law, C.S., A.P. Rees, and N.J.P. Owens. 1991. Temporal variability of denitrification in estuarine sediments. Estuarine, Coastal and Shelf Science 33: 37–56.

Law, K.R., H.W. Nesbitt, and F.J. Longstaffe. 1991. Weathering of granitic tills and the genesis of a podzol. American Journal of Science 291: 940–976.

Law, K.S. and E.G. Nisbet. 1996. Sensitivity of the CH_4 growth rate to changes in CH_4 emissions from natural gas and coal. Journal of Geophysical Research 101: 14387–14397.

Lawless, J.G. and N. Levi. 1979. The role of metals ions in chemical evolution: Polymerization of alanine and glycine in a cation-exchanged clay environment. Journal of Molecular Evolution 13: 281–286.

Lawrence, W.G., W.L. Chameides, P.S. Kasibhatla, H. Levy, and W. Moxim. 1994. Lightning and atmospheric chemistry: The rate of atmospheric NO production. pp. 189–202. In H. Volland (ed.), *The Handbook of Atmospheric Electrodynamics.* CRC Press, Boca Raton, Florida.

Lawson, D.R. and J.W. Winchester. 1979. A standard crustal aerosol as a reference for elemental enrichment factors. Atmospheric Environment 13: 925–930.

LaZerte, B.D. 1983. Stable carbon isotope ratios: Implications for the source of sediment carbon and for phytoplankton carbon assimilation in Lake Memphremagog, Quebec. Canadian Journal of Fisheries and Aquatic Sciences 40: 1658–1666.

Leahey, A. 1947. Characteristics of soils adjacent to the MacKenzie River in the Northwest Territories of Canada. Soil Science Society of America Proceedings 12: 458–461.

Lean, D.R.S. 1973. Phosphorus dynamics in lake water. Science 179: 678–680.

Lean, J. and D.A. Warrilow. 1989. Simulation of the regional climatic impact of Amazon deforestation. Nature 342: 411–413.

Leavitt, S.W. and A. Long. 1988. Stable carbon isotope chronologies from trees in the southwestern United States. Global Biogeochemical Cycles 2: 189–198.

Lebo, M.E. and J.H. Sharp. 1992. Modeling phosphorus cycling in a well-mixed coastal plain estuary. Estuarine, Coastal and Shelf Science 35: 235–252.

Lechowicz, M.J. and G. Bell. 1991. The ecology and genetics of fitness in forest plants. II. Microspatial heterogeneity of the edaphic environment. Journal of Ecology 79: 687–696.

Ledwell, J.R., A.J. Watson, and C.S. Law. 1993. Evidence for slow mixing across the pycnocline from an open-ocean tracer-release experiment. Nature 364: 701–703.

Lee, C. 1992. Controls on organic carbon preservation: The use of stratified water bodies to compare intrinsic rates of decomposition in oxic and anoxic systems. Geochimica et Cosmochimica Acta 56: 3323–3335.

Lee, D.H., J.R. Granja, J.A. Martinez, K. Severin, and M.R. Ghadiri. 1996. A self-replicating peptide. Nature 382: 525–528.

Lee, J., S.B. Roberts, and F.M.M. Morel. 1995. Cadmium: A nutrient for the marine diatom *Thalassiosira weissflogii*. Limnology and Oceanography 40: 1056–1063.

Lee, J.A. and G.R. Stewart. 1978. Ecological aspects of nitrogen assimilation. Advances in Botanical Research 6: 1–43.

Legrand, M. and R.J. Delmas. 1987. A 220-year continuous record of volcanic H_2SO_4 in the Antarctic ice sheet. Nature 327: 671–676.

Legrand, M., C. Feniet-Saigne, E.S. Saltzman, C. Germain, N.I. Barkov, and V.N. Petrov. 1991. Ice-core record of oceanic emissions of dimethylsulfide during the last climate cycle. Nature 350: 144–146.

Legrand, M., M. De Angelis, T. Staffelbach, A. Neftel, and B . Stauffer. 1992. Large perturbations of ammonium and organic acids content in the Summit-Greenland ice core. Fingerprint from forest fires? Geophysical Research Letters 19: 473–475.

Lehman, J.T. 1980. Release and cycling of nutrients between planktonic algae and herbivores. Limnology and Oceanography 25: 620–632.

Lehman, J.T. 1988. Hypolimnetic metabolism in Lake Washington: Relative effects of nutrient load and food web structure on lake productivity. Limnology and Oceanography 33: 1334–1347.

Lein, A.Y. 1984. Anaerobic consumption of organic matter in modern marine sediments. Nature 312: 148–150.

Leisman, G.A. 1957. A vegetation and soil chronosequence on the Mesabi iron range spoil banks, Minnesota. Ecological Monographs 27: 221–245.

Lelieveld, J. and P.J. Crutzen. 1990. Influences of cloud photochemical processes on tropospheric ozone. Nature 343: 227–233.

Lelieveld, J. and P.J. Crutzen. 1991. The role of clouds in tropospheric photochemistry. Journal of Atmospheric Chemistry 12: 229–267.

Leopoldo, P.R., W.K. Franken, and N.A. Villa Nova. 1995. Real evapotranspiration and transpiration through a tropical rain forest in central Amazonia as estimated by the water balance method. Forest Ecology and Management 73: 185–195.

Lerner, J., E. Matthews, and I. Fung. 1988. Methane emission from animals: A global high-resolution data base. Global Biogeochemical Cycles 2: 139–156.

Lesack, L.F.W. 1993. Export of nutrients and major ionic solutes from a rain forest catchment in the central Amazon basin. Water Resources Research 29: 743–758.

Lesack, L.F.W. and J.M. Melack. 1991. The deposition, composition, and potential sources of major ionic solutes in rain of the central Amazon basin. Water Resources Research 27: 2953–2977.

Leschine, S.B., K. Holwell, and E. Canale-Parola. 1988. Nitrogen fixation by anaerobic cellulolytic bacteria. Science 242: 1157–1159.

Lescop-Sinclair, K. and S. Payette. 1995. Recent advance of the arctic treeline along the eastern coast of Hudson Bay. Journal of Ecology 83: 929–936.

Leuenberger, M. and U. Siegenthaler. 1992. Ice-age atmospheric concentration of nitrous oxide from an Antarctic ice core. Nature 360: 449–451.

Leuenberger, M., U. Siegenthaler, and C.C. Langway. 1992. Carbon isotope composition of atmospheric CO_2 during the last ice age from an Antarctic ice core. Nature 357: 488–490.

Levin, S.A. 1989. Challenges in the development of a theory of community and ecosystem structure and function. pp. 242–255. In J. Roughgarden, R.M. May, and S.A. Levin (eds.), *Perspectives in Ecological Theory*. Princeton University Press, Princeton, New Jersey.

Levine, J.S., W.R. Cofe, D.I. Sebacher, E.L. Winstead, S. Sebacher, and P.J. Boston. 1988. The effects of fire on biogenic soil emissions of nitric oxide and nitrous oxide. Global Biogeochemical Cycles 2: 445–449.

Levine, J.S., W.R. Cofer, and J.P. Pinto. 1993. Biomass burning. pp. 299–313. In M.A.K. Khalil (ed.), *Atmospheric Methane: Sources, Sinks and Role in Global Change*. Springer-Verlag, New York

Levine, S.N. and D.W. Schindler. 1989. Phosphorus, nitrogen, and carbon dynamics of Experimental Lake 303 during recovery from eutrophication. Canadian Journal of Fisheries and Aquatic Sciences 46: 2–10.

Levine, S.N., M.P. Stainton, and D.W. Schindler. 1986. A radiotracer study of phosphorus cycling in a eutrophic Canadian Shield Lake, Experimental Lake 227, northwestern Ontario. Canadian Journal of Fisheries and Aquatic Sciences 43: 366–378.

Lewis, M.R., W.G. Harrison, N.S. Oakey, D. Hebert, and T. Platt. 1986. Vertical nitrate fluxes in the oligotrophic ocean. Science 234: 870–873.

Lewis, W.M. 1974. Effects of fire on nutrient movement in a South Carolina pine forest. Ecology 55: 1120–1127.

Lewis, W.M. 1981. Precipitation chemistry and nutrient loading by precipitation in a tropical watershed. Water Resources Research 17: 169–181.

Lewis, W.M. 1986. Nitrogen and phosphorus runoff losses from a nutrient-poor tropical moist forest. Ecology 67: 1275–1282.

Lewis, W.M. 1988. Primary production in the Orinoco River. Ecology 69: 679–692.

Lewis, W.M. and M.C. Grant. 1979. Relationships between stream discharge and yield of dissolved substances from a Colorado Mountain watershed. Soil Science 128: 353–363.

Lewis, W.M., S.K. Hamilton, S.L. Jones, and D.D. Runnels. 1987. Major element chemistry, weathering and elements yields for the Caura River drainage, Venezuela. Biogeochemistry 4: 159–181.

Li, W.K.W., D.V. Subba Rao, W.G. Harrison, J.C. Smith, J.J. Cullen, B. Irwin, and T. Platt. 1983. Autotrophic picoplankton in the tropical ocean. Science 219: 292–295.

Li, Y.-H. 1972. Geochemical mass balance among lithosphere, hydrosphere, and atmosphere. American Journal of Science 272: 119–137.

Li, Y.-H. 1981. Geochemical cycles of elements and human perturbation. Geochimica et Cosmochimica Acta 45: 2073–2084.

Li, Y.-H., T. Takahashi, and W.S. Broecker. 1969. The degree of saturation of $CaCO_3$ in the oceans. Journal of Geophysical Research 74: 5507–5525.

Lieffers, V.J. and S.E. MacDonald. 1990. Growth and foliar nutrient status of black spruce and tamarack in relation to depth of water table in some Alberta peatlands. Canadian Journal of Forest Research 20: 805–809.

Lieth, H. 1975. Modeling the primary productivity of the world. pp. 237–263. In H. Lieth and R.H. Whittaker (eds.), *Primary Productivity of the Biosphere.* Springer-Verlag, New York.

Lightfoot, D.C. and W.G. Whitford. 1987. Variation in insect densities on desert creosotebush: Is nitrogen a factor? Ecology 68: 547–557.

Likens, G.E. 1975a. Nutrient flux and cycling in freshwater ecosystems. pp. 314–348. In F.G. Howell, J.B. Gentry, and M.H. Smith (eds.), *Mineral Cycling in Southeastern Ecosystems.* National Technical Information Service, Springfield, Virginia.

Likens, G.E. 1975b. Primary production of inland aquatic ecosystems. pp. 185–202. In H. Lieth and R.H. Whittaker (eds.), *Primary Productivity of the Biosphere.* Springer-Verlag, New York.

Likens, G.E. and F.H. Bormann. 1974. Linkages between terrestrial and aquatic ecosystems. BioScience 24: 447–456.

Likens, G.E. and F.H. Bormann. 1995. *Biogeochemistry of a Forested Ecosystem.* 2nd ed. Springer-Verlag, New York.

Likens, G.E., F.H. Bormann, N.M. Johnson, D.W. Fisher, and R.S. Pierce. 1970. Effects of forest cutting and herbicide treatment on nutrient budgets in the Hubbard Brook watershed-ecosystem. Ecological Monographs 40: 23–47.

Likens, G.E., F.H. Bormann, and N.M. Johnson. 1981. Interactions between major biogeochemical cycles in terrestrial ecosystems. pp. 93–112. In G.E. Likens (ed.), *Some Perspectives of the Major Biogeochemical Cycles.* Wiley, New York.

Likens, G.E., F.H. Bormann, R.S. Pierce, J.S. Eaton, and R.E. Munn. 1984. Long-term trends in precipitation chemistry at Hubbard Brook, New Hampshire. Atmospheric Environment 18: 2641–2647.

Likens, G.E., F.H. Bormann, L.O. Hedin, C.T. Driscoll, and J.S. Eaton. 1990. Dry deposition of sulfur: A 23-year record for the Hubbard Brook forest ecosystem. Tellus 42B: 319–329.

Likens, G.E., C.T. Driscoll, D.C. Buso, T.G. Siccama, C.E. Johnson, G.M. Lovett, D.F. Ryan, T. Fahey, and W.A. Reiners. 1994. The biogeochemistry of potassium at Hubbard Brook. Biogeochemistry 25: 61–125.

Likens, G.E., C.T. Driscoll, and D.C. Buso. 1996. Long-term effects of acid rain: Response and recovery of a forest ecosystem. Science 272: 244–246.

Linak, W.P., J.A. McSorley, R.E. Hall, J.V. Ryan, R.K. Srivastava, J.O.L. Wendt, and J.B. Mereb. 1990. Nitrous oxide emissions from fossil fuel combustion. Journal of Geophysical Research 95: 7533–7541.

Lindberg, S.E. and C.T. Garten. 1988. Sources of sulphur in forest canopy throughall. Nature 336: 148–151.

Lindberg, S.E. and G.M. Lovett. 1985. Field measurements of particle dry deposition rates to foliage and inert surfaces in a forest canopy. Environmental Science and Technology 19: 238–244.

Lindberg, S.E. and R.R. Turner. 1988. Factors influencing atmospheric deposition, stream export, and landscape accumulation of trace metals in forested watersheds. Water, Air, and Soil Pollution 39: 123–156.

Lindberg, S.E., G.M. Lovett, D.D. Richter, and D.W. Johnson. 1986. Atmospheric deposition and canopy interactions of major ions in a forest. Science 231: 141–145.

Lindeboom, H.J. 1984. The nitrogen pathway in a penguin rookery. Ecology 65: 269–277.

Lindqvist, O. and H. Rodhe. 1985. Atmospheric mercurya review. Tellus 37B: 136–159.

Lindsay, W.L. 1979. *Chemical Equilibria in Soils.* Wiley, New York.

Lindsay, W.L. and E.C. Moreno. 1960. Phosphate phase equilibria in soils. Soil Science Society of America Proceedings 24: 177–182.

Lindsay, W.L. and P.L.G. Vlek. 1977. Phosphate minerals. pp. 639–672. In J.B. Dixon and S.B. Weed (eds.), *Minerals in Soil Environments.* Soil Science Society of America, Madison, Wisconsin.

Linkins, A.E., R.L. Sinsabaugh, C.A. McClaugherty, and J.M. Melillo. 1990. Cellulase activity on decomposing leaf litter in microcosms. Plant and Soil 123: 17–25.

Lipschultz, F., S.C. Wofsy, B.B. Ward, L.A. Codispoti, G. Friedrich, and J.W. Elkins. 1990. Bacterial transformations of inorganic nitrogen in the oxygen-deficient waters of the eastern tropical Pacific Ocean. Deep Sea Research 37: 1513–1541.

Liss, P.S. 1983. The exchange of biogeochemically important gases across the air-sea interface. pp. 411–426. In B. Bolin and R.B. Cook (eds.), *The Major Biogeochemical Cycles and Their Interactions.* Wiley, New York.

Liss, P.S. and P.G. Slater. 1974. Flux of gases across the air-sea interface. Nature 247: 181–184.

Littke, W.R., C.S. Bledsoe, and R.L. Edmonds. 1984. Nitrogen uptake and growth *in vitro* by *Hebeloma crustuliniforme* and other Pacific Northwest mycorrhizal fungi. Canadian Journal of Botany 62: 647–652.

Liu, B., F. Phillips, S. Hoines, A.R. Campbell, and P. Sharma. 1995. Water movement in desert soil traced by hydrogen and oxygen isotopes, chloride, and chlorine-36, southern Arizona. Journal of Hydrology 168: 91–110.

Liu, T.S., X.F. Gu, Z.S. An, and Y.X. Fan. 1981. The dust fall in Beijing, China on April 18, 1980. pp. 149–157. In T.L. Pewe (ed.), *Desert Dust: Origin, Characteristics, and Effects on Man.* Geological Society of America, Special Paper 186, Boulder, Colorado.

Livingston, G.P., P.M. Vitousek, and P.A. Matson. 1988. Nitrous oxide flux and nitrogen transformations across a landscape gradient in Amazonia. Journal of Geophysical Research 93: 1593–1599.

Livingstone, D.A. 1963. Chemical composition of rivers and lakes. U.S. Geological Survey, Professional Paper 440G, Washington, D.C.

Llewellyn, M. 1975. The effects of the lime aphid (*Eucallipterus tiliae* L.) (Aphididae) on the growth of lime *Tilia* × *vulgaris* Hayne). II. The primary production of saplings and mature trees, the energy drain imposed by the aphid populations, and revised standard deviations of aphid population energy budgets. Journal of Applied Ecology 12: 15–23.

Lloyd, J. and G.D. Farquhar. 1994. ^{13}C discrimination during CO_2 assimilation by the terrestrial biosphere. Oecologia 99: 201–215.

Loaiciga, H.A., J.B. Valdes, R. Vogel, J. Garvey, and H. Schwarz. 1995. Global warming and the hydrologic cycle. Journal of Hydrology 174: 82–127.

Lobert, J.M., D.H. Scharffe, W.M. Hao, and P.J. Crutzen. 1990. Importance of biomass burning in the atmospheric budgets of nitrogen-containing gases. Nature 346: 552–554.

Lobert, J.M., J.H. Butler, S.A. Montzka, L.S. Geller, R.C. Myers, and J.W. Elkins. 1995. A net sink for atmospheric CH_3Br in the East Pacific ocean. Science 267: 1002–1006.

Lockwood, P.V., J.M. McGarity, and J.L. Charley. 1995. Measurement of chemical weathering rates using natural chloride as a tracer. Geoderma 64: 215–232.

Logan, J.A. 1983. Nitrogen oxides in the troposphere: Global and regional budgets. Journal of Geophysical Research 88: 10785–10807

Logan, J.A. 1985. Tropospheric ozone: Seasonal behavior, trends, and anthropogenic influence. Journal of Geophysical Research 90: 10463–10482.

Logan, J.A., M.J. Prather, S.C. Wofsy, and M.B. McElroy. 1981. Tropospheric chemistry: A global perspective. Journal of Geophysical Research 86: 7210–7254.

Lohrmann, R. and L.E. Orgel. 1973. Prebiotic activation processes. Nature 244: 418–420.

Longmore, M.E., B.M. O'Leary, C.W. Rose, and A.L. Chandica. 1983. Mapping soil erosion and accumulation with the fallout isotope caesium-137. Australian Journal of Soil Research 21: 373–385.

Lonsdale, W.M. 1988. Predicting the amount of litterfall in forests of the world. Annals of Botany 61: 319–324.

Love, S.G. and D.E. Brownlee. 1993. A direct measurement of the terrestrial mass accretion rate of cosmic dust. Science 262: 550–553.

Lovelock, J.E. 1979. *Gaia: A New Look at Life on Earth.* Oxford University Press, Oxford.

Lovelock, J.E. and M. Whitfield. 1982. Life span of the biosphere. Nature 296: 561–563.

Lovelock, J.E., R.J. Maggs, and R.A. Rasmussen. 1972. Atmospheric dimethyl sulphide and the natural sulphur cycle. Nature 237: 452–453.

Lovett, G.M. 1994. Atmospheric deposition of nutrients and pollutants in North America: An ecological perspective. Ecological Applications 4: 629–650.

Lovett, G.M. and S.E. Lindberg. 1986. Dry deposition of nitrate to a deciduous forest. Biogeochemistry 2: 137–148.

Lovett, G.M. and S.E. Lindberg. 1993. Atmospheric deposition and canopy interactions of nitrogen in forests. Canadian Journal of Forest Research 23: 1603–1616.

Lovett, G.M., W.A. Reiners, and R.K. Olson. 1982. Cloud droplet deposition in subalpine balsam fir forests: Hydrological and chemical inputs. Science 218: 1303–1304.

Lovley, D.R. 1995. Microbial reduction of iron, manganese, and other metals. Advances in Agronomy 54: 175–231.

Lovley, D.R. and M.J. Klug. 1986. Model for the distribution of sulfate reduction and methanogenesis in freshwater sediments. Geochimica et Cosmochimica Acta 50: 11–18.

Lovley, D.R. and E.J.P. Phillips. 1987. Competitive mechanisms for inhibition of sulfate reduction and methane production in the zone of ferric iron reduction in sediments. Applied and Environmental Microbiology 53: 2636–2641.

Lovley, D.R. and E.J.P. Phillips. 1988a. Novel model of microbial energy metabolism: Organic carbon oxidation coupled to dissimilatory reduction of iron or manganese. Applied and Environmental Microbiology 54: 1472–1480.

Lovley, D.R. and E.J.P. Phillips. 1988b. Manganese inhibition of microbial iron reduction in anaerobic sediments. Geomicrobiology Journal 6: 145–155.

Lovley, D.R. and E.J.P. Phillips. 1989. Requirement for a microbial consortium to completely oxidize glucose in Fe(III)-reducing sediments. Applied and Environmental Microbiology 55: 3234–3236.

Lovley, D.R., F.H. Chapelle, and E.J.P. Phillips. 1990. Fe(III)-reducing bacteria in deeply buried sediments of the Atlantic coastal plain. Geology 18: 954–957.

Lowry, B., D. Lee, and C. Hébant. 1980. The origin of land plants: A new look at an old problem. Taxon 29: 183–197.

Lucas, Y., F.J. Luizão, A. Chauvel, J. Rouiller, and D. Hahon. 1993. The relation between biological activity of the rain forest and mineral composition of soils. Science 260: 521–523.

Ludwig, W., J.-L. Probst, and S. Kempe. 1996. Predicting the oceanic input of organic carbon by continental erosion. Global Biogeochemical Cycles 10: 23–41.

Lugo, A.E. and S. Brown. 1986. Steady state terrestrial ecosystems and the global carbon cycle. Vegetatio 68: 83–90.

Lugo, A.E., M.J. Sanchez, and S. Brown. 1986. Land use and organic carbon content of some subtropical soils. Plant and Soil 96: 185–196.

Lui, K.-K. and I.R. Kaplan. 1989. The eastern tropical Pacific Ocean as a source of ^{15}N-enriched nitrate in seawater off southern California. Limnology and Oceanography 34: 820–830.

Luizão, F., P. Matson, G. Livingston, R. Luizão, and P. Vitousek. 1989. Nitrous oxide flux following tropical land clearing. Global Biogeochemical Cycles 3: 281–285.

Luke, W.T., R.R. Dickerson, W.F. Ryan, K.E. Pickering, and L.J. Nunnermacker. 1992. Tropospheric chemistry over the lower Great Plains of the United States: 2. Trace gas profiles and distribution. Journal of Geophysical Research 97: 20647–20670.

Lull, H.W. and W.E. Sopper. 1969. Hydrologic effects from urbanization of forested watersheds in the Northeast. U.S. Department of Agriculture, Northeast Forest Experiment Station Research Paper NE-146, Upper Darby, Pennsylvania.

Lundgren, D.G. and M. Silver. 1980. Ore leaching by bacteria. Annual Review of Microbiology 34: 263–283.

Lundström, U.S. 1993. The role of organic acids in the soil solution chemistry of a podzolized soil. Journal of Soil Science 44: 121–133.

Lunine, J.I. 1989. Origin and evolution of outer solar system atmospheres. Science 245: 141–147.

Lupton, J.E. and H. Craig. 1981. A major helium-3 source at 15°S on the east Pacific Rise. Science 214: 13–18.

Luther, F.M. and R.D. Cess. 1985. Review of the recent carbon dioxide-climate controversy. pp. 321–335. In M.C. MacCraken and F.M. Luther (eds.), *Projecting the Climatic Effects of Increasing Carbon Dioxide.* U.S. Department of Energy, Er-0237, Washington, D.C.

Luther, G.W., T.M. Church, J.R. Scudlark, and M. Cosman. 1986. Inorganic and organic sulfur cycling in salt-marsh pore waters. Science 232: 746–749.

Lutz, R.A., T.M. Shank, D.J. Fornari, R.M. Haymon, M.D. Lilley, K.L. von Damm, and D. Desbruyeres. 1994. Rapid growth at deep-sea vents. Nature 371: 663–664.

Luxmoore, R.J., T. Grizzard, and R.H. Strand. 1981. Nutrient translocation in the outer canopy and understory of an eastern deciduous forest. Forest Science 27: 505–518.

Lvovitch, M.I. 1973. The global water balance. EOS 54: 28–42.

Lyford, F.P. and H.K. Qashu. 1969. Infiltration rates as affected by desert vegetation. Water Resources Research 5: 1373–1376.

Lyons, W.B., P.A. Mayewski, M.J. Spencer, and M.S. Twickler. 1990. Nitrate concentrations in snow from remote areas: Implication for the global NO_x flux. Biogeochemistry 9: 211–222.

Maathuis, F.J.M. and D. Sanders. 1994. Mechanism of high-affinity potassium uptake in roots of *Arabidopsis thaliana*. Proceedings of the National Academy of Sciences USA 91: 9272–9276.

MacCracken, M.C. 1985. Carbon dioxide and climate change: Background and overview. pp. 1–23. In M.C. MacCraken and F.M. Luther (eds.), *Projecting the Climatic Effects of Increasing Carbon Dioxide.* U.S. Department of Energy, Er-0237, Washington, D.C.

MacDonald, A.M. and C. Wunsch. 1996. An estimate of global ocean circulation and heat fluxes. Nature 382: 436–439.

MacDonald, G.J. 1990. Role of methane clathrates in past and future climates. Climatic Change 16: 247–281.

MacDonald, G.M., T.W.D. Edwards, K.A. Moser, R. Pienitz, and J.P. Smol. 1993. Rapid response of treeline vegetation and lakes to past climate warming. Nature 361: 243–246.

MacDonald, N.W. and J.B. Hart. 1990. Relating sulfate adsorption to soil properties in Michigan forest soils. Soil Science Society of America Journal 54: 238–245.

Mach, D.L., A. Ramirez, and H.D. Holland. 1987. Organic phosphorus and carbon in marine sediments. American Journal of Science 287: 429–441.

MacIntyre, F. 1974. The top millimeter of the ocean. Scientific American 230(5): 62–77.

MacKenzie, F.T. and R.M. Garrels. 1966. Chemical mass balance between rivers and oceans. American Journal of Science 264: 507–525.

MacKenzie, F.T., M. Stoffyn, and R. Wollast. 1978. Aluminum in seawater: Control by biological activity. Science 199: 680–682.

MacKenzie, F.T., L. May Ver, C. Sabine, M. Lane, and A. Lerman. 1993. C, N, P, and S in global biogeochemical cycles and modeling of global change. pp. 1–61. In R. Wollast, F.T. MacKenzie, and L. Chou (eds.), *Interactions of C, N, P and S in Biogeochemical Cycles and Global Change.* Springer-Verlag, New York.

Mackney, D. 1961. A podzol development sequence in oakwoods and heath in central England. Journal of Soil Science 12: 23–40.

Madsen, H.B. and P. Nørnberg. 1995. Mineralogy of four sandy soils developed under heather, oak, spruce and grass in the same fluvioglacial deposit in Denmark. Geoderma 64: 233–256.

Mäkipää, R. 1995. Effect of nitrogen input on carbon accumulation of boreal forest soils and ground vegetation. Forest Ecology and Management 79: 217–226.

Malaney, R.A. and W.A. Fowler. 1988. The transformation of matter after the big bang. American Scientist 76: 472–477.

Malcolm, R.E. 1983. Assessment of phosphatase activity in soils. Soil Biology and Biochemistry 15: 403–408.

Malone, M.J., P.A. Baker, and S.J. Burns. 1994. Recrystallization of dolomite: Evidence from the Monterey Formation (Miocene), California. Sedimentology 41: 1223–1239.

Manabe, S. and R.T. Wetherald. 1980. On the distribution of climate change resulting from an increase in CO_2 content of the atmosphere. Journal of the Atmospheric Sciences 37: 99–118.

Manabe, S. and R.T. Wetherald. 1986. Reduction in summer soil wetness induced by an increase in atmospheric carbon dioxide. Science 232: 626–628.

Mancinelli, R.L. and C.P. McKay. 1988. The evolution of nitrogen cycling. Origins of Life and Evolution of the Biosphere 18: 311–325.

Mankin, W.G. and M.T. Coffey. 1984. Increased stratospheric hydrogen chloride in the El Chichón cloud. Science 226: 170–172.

Mankin, W.G., M.T. Coffey, and A. Goldman. 1992. Airborne observations of SO_2, HCl, and O_3 in the stratospheric plume of the Pinatubo volcano in July 1991. Geophysical Research Letters 19: 179–182.

Manley, S.L., K. Goodwin, and W.J. North. 1992. Laboratory production of bromoform, methylene bromide, and methyl iodide by macroalgae and distribution in nearshore southern California waters. Limnology and Oceanography 37: 1652–1659.

Mann, L.K. 1986. Changes in soil carbon storage after cultivation. Soil Science 142: 279–288.

Manney, G.L., L. Froidevaux, J.W. Waters, R.W. Zurek, W.G. Read, L.S. Elson, J.B. Kumer, J.L. Mergenthaler, A.E. Roche, A. O'Neill, R.S. Harwood, I. MacKenzie, and R. Swinbank. 1994. Chemical depletion of ozone in the Arctic lower stratosphere during winter 1992-93. Nature 370: 429–434.

Manö, S. and M.O. Andreae. 1994. Emission of methyl bromide from biomass burning. Science 263: 1255–1257.

Mantoura, R.F.C. and E.M.S. Woodward. 1983. Conservative behaviour of riverine dissolved organic carbon in the Severn Estuary: Chemical and geochemical implications. Geochimica et Cosmochimica Acta 47: 1293–1309.

Marcano-Martinez, E. and M.B. McBride. 1989. Calcium and sulfate retention by two Oxisols of the Brazilian cerrado. Soil Science Society of America Journal 53: 63–69.

Marion, G.M. and C.H. Black. 1987. The effect of time and temperature on nitrogen mineralization in Arctic tundra soils. Soil Science Society of America Journal 51: 1501–1508.

Marion, G.M. and C.H. Black. 1988. Potentially available nitrogen and phosphorus along a chaparral fire cycle chronosequence. Soil Science Society of America Journal 52: 1155–1162.

Marion, G.M. and W.H. Schlesinger. 1994. Quantitative modeling of soil forming processes in deserts: The CALDEP and CALGYP models. pp. 129–145. In R.B. Bryant and R.W. Arnold (eds.), *Quantitative Modeling of Soil Forming Processes,* Soil Science Society of America, Special Publication 39, Madison, Wisconsin.

Marion, G.M., W.H. Schlesinger, and P.J. Fonteyn. 1985. CALDEP: A regional model for soil $CaCO_3$ (caliche) deposition in southwestern deserts. Soil Science 139: 468–481.

Markewich, H.W. and M.J. Pavich. 1991. Soil chronosequence studies in temperate to subtropical, low-latitude, low-relief terrain with data from the eastern United States. Geoderma 51: 213–239.

Markewich, H.W., M.J. Pavich, M.J. Mausbach, R.G. Johnson, and V.M. Gonzalez. 1989. A guide for using soil and weathering profile data in chronosequence studies of the Coastal Plain of the eastern United States. U.S. Geological Survey, Bulletin 1589D, Washington, D.C.

Marks, P.L. and F.H. Bormann. 1972. Revegetation following forest cutting: Mechanisms for return to steady-state nutrient cycling. Science 176: 914–915.

Marland, G. and T. Boden. 1993. The magnitude and distribution of fossil-fuel-related carbon releases. pp. 117-138. In M. Heimann (ed.), *The Global Carbon Cycle.* Springer-Verlag, New York.

Marnette, E.C., C. Hordijk, N. van Breemen, and T. Cappenberg. 1992. Sulfate reduction and S-oxidation in a moorland pool sediment. Biogeochemistry 17: 123–143.

Maroulis, P.J. and A.R. Bandy. 1977. Estimate of the contribution of biologically produced dimethyl sulfide to the global sulfur cycle. Science 196: 647–648.

Marquis, R.J. and C.J. Whelan. 1994. Insectivorous birds increase growth of white oak through consumption of leaf-chewing insects. Ecology 75: 2007–2014.

Marrs, R.H., J. Protor, A. Heaney, and M.D. Mountford. 1988. Changes in soil nitrogen-mineralization and nitrification along an altitudinal transect in tropical rain forest in Costa Rica. Journal of Ecology 76: 466–482.

Martens, C.S. and J.V. Klump. 1984. Biogeochemical cycling in an organic-rich coastal marine basin. 4. An organic carbon budget for sediments dominated by sulfate reduction and methanogenesis. Geochimica et Cosmochimica Acta 48: 1987–2004.

Martens, C.S., N.E. Blair, C.D. Green, and D.J. Des Marais. 1986. Seasonal variations in the stable carbon isotopic signature of biogenic methane in a coastal sediment. Science 233: 1300–1303.

Martens, R. 1995. Current methods for measuring microbial biomass C in soil: Potentials and limitations. Biology and Fertility of Soils 19: 87–99.

Martin, C.W. and R.D. Harr. 1988. Precipitation and streamwater chemistry from undisturbed watersheds in the Cascade Mountains of Oregon. Water, Air, and Soil Pollution 42: 203–219.

Martin, H.W. and D.L. Sparks. 1985. On the behavior of nonexchangeable potassium in soils. Communications in Soil Science and Plant Analysis 16: 133–162.

Martin, J.H. 1990. Glacial-interglacial CO_2 change: The iron hypothesis. Paleoceanography 5: 1–13.

Martin, J.H. and R.M. Gordon. 1988. Northeast Pacific iron distributions in relation to phytoplankton productivity. Deep Sea Research 35: 177–196.

Martin, J.H., G.A. Knauer, D.M. Karl, and W.W. Broenkow. 1987. VERTEX: Carbon cycling in the northeast Pacific. Deep Sea Research 34: 267–285.

Martin, J.H., R.M. Gordon, S. Fitzwater, and W.W. Broenkow. 1989. VERTEX: Phytoplankton/iron studies in the Gulf of Alaska. Deep Sea Research 36: 649–680.

Martin, J.H., K.H. Coale, K.S. Johnson, S.E. Fitzwater, R.M. Gordon, S.J. Tanner, C.N. Hunter, V.A. Elrod, J.L. Wowicki, T.L. Coley, R.T. Barber, S. Lindley, A.J. Watson, K. Van Scoy, C.S. Law, M.I. Liddicoat, R. Ling, T. Stanton, J. Stockel, C. Collins, A. Anderson, R. Bidigare, M. Ondrusek, M. Latasa, F.J. Millero, K. Lee, W. Yao, J.Z. Zhang, G. Friederich, C. Sakamoto, F. Chavez, K. Buck, Z. Kolber, R. Greene, P. Falkowski, S.W. Chisholm, F. Hoge, R. Swift, J. Yungel, S. Turner, P. Nightingale, A. Hatton, P. Liss, and N.W. Tindale. 1994. Testing the iron hypothesis in ecosystems of the equatorial Pacific Ocean. Nature 371: 123–129.

Martin, J.-M. and M. Meybeck. 1979. Elemental mass-balance of material carried by major world rivers. Marine Chemistry 7: 173–206.

Martin, W.R., M. Bender, M. Leinen, and J. Orchardo. 1991. Benthic organic carbon degradation and biogenic silica dissolution in the central equatorial Pacific. Deep Sea Research 38A: 1481–1516.

Martínez, L.A., M.W. Silver, J.M. King, and A.L. Allredge. 1983. Nitrogen fixation by floating diatom mats: A source of new nitrogen to oligotrophic ocean waters. Science 221: 152–154.

Marty, B. 1995. Nitrogen content of the mantle inferred from N_2-Ar correlation in oceanic basalts. Nature 377: 326–329.

Marumoto, T., J.P.E. Anderson, and K.H. Domsch. 1982. Mineralization of nutrients from soil microbial biomass. Soil Biology and Biochemistry 14: 469–475.

Marx, D.H., A.B. Hatch, and J.F. Mendicino. 1977. High soil fertility decreases sucrose content and susceptibility of loblolly pine roots to ectomycorrhizal infection by *Pisolithus tinctorius*. Canadian Journal of Botany 55: 1569–1574.

Mason, R.P. and W.F. Fitzgerald. 1993. The distribution and biogeochemical cycling of mercury in the equatorial Pacific Ocean. Deep Sea Research 40: 1897–1924.

Mason, R.P., W.F. Fitzgerald, and F.M.M. Morel. 1994. The biogeochemical cycling of elemental mercury: Anthropogenic influences. Geochimica et Cosmochimica Acta 58: 3191–3198.

Massman, W.J. 1992. A surface energy balance method for partitioning evapotranspiration data into plant and soil components for a surface with partial cover. Water Resources Research 28: 1723–1732.

Matson, P.A. and P.M. Vitousek. 1987. Cross-system comparisons of soil nitrogen transformations and nitrous oxide flux in tropical ecosystems. Global Biogeochemical Cycles 1: 163–170.

Matson, P.A. and P.M. Vitousek. 1990. Ecosystem approach to a global nitrous oxide budget. BioScience 40: 667–672.

Matson, P.A., P.M. Vitousek, J.J. Ewel, M.J. Mazzarino, and G.P. Robertson. 1987. Nitrogen transformations following tropical forest felling and burning on a volcanic soil. Ecology 68: 491–502.

Matson, P.A., C. Volkmann, K. Coppinger, and W.A. Reiners. 1991. Annual nitrous oxide flux and soil nitrogen characteristics in sagebrush steppe ecosystems. Biogeochemistry 14: 1–12.

Matson, P.A., L. Johnson, C. Billow, J. Miller, and R. Pu. 1994. Seasonal patterns and remote spectral estimation of canopy chemistry across the Oregon transect. Ecological Applications 4: 280–298.

Matthews, E. 1994. Nitrogenous fertilizers: Global distribution of consumption and associated emissions of nitrous oxide and ammonia. Global Biogeochemical Cycles 8: 411–439.

Matthews, E. and I. Fung. 1987. Methane emission from natural wetlands: Global distribution, area, and environmental characteristics of sources. Global Biogeochemical Cycles 1: 61–86.

Matthews, E., I. Fung, and J. Lerner. 1991. Methane emission from rice cultivation: Geographic and seasonal distribution of cultivated areas and emissions. Global Biogeochemical Cycles 5: 3–24.

Mattson, M.D. and G.E. Likens. 1993. Redox reactions of organic matter decomposition in a soft water lake. Biogeochemistry 19: 149–172.

Mattson, W.J. 1980. Herbivory in relation to plant nitrogen content. Annual Review of Ecology and Systematics 11: 119–161.

Mattson, W.J. and N.D. Addy. 1975. Phytophagous insects as regulators of forest primary production. Science 190: 515–522.

Mayer, B., K.H. Feger, A. Giesemann, and H.-J. Jäger. 1995. Interpretation of sulfur cycling in two catchments in the Black forest (Germany) using stable sulfur and oxygen isotope data. Biogeochemistry 30: 31–58.

Mayer, L.M. 1994. Relationships between mineral surfaces and organic carbon concentrations in soils and sediments. Chemical Geology 114: 347–363.

Mayewski, P.A., W.B. Lyons, M.J. Spencer, M. Twickler, W. Dansgaard, B. Koci, C.I. Davidson, and R.E. Honrath. 1986. Sulfate and nitrate concentrations from a south Greenland ice core. Science 232: 975–977.

Mayewski, P.A., W.B. Lyons, M.J. Spencer, M.S. Twickler, C.F. Buck, and S. Whitlow. 1990. An ice-core record of atmospheric response to anthropogenic sulphate and nitrate. Nature 346: 554–556.

Mayor, M. and D. Queloz. 1995. A Jupiter-mass companion to a solar-type star. Nature 378: 355–359.

Mazumder, A. and M.D. Dickman. 1989. Factors affecting the spatial and temporal distribution of phototrophic sulfur bacteria. Archive de Hydrobiologia 116: 209–226.

McCarthy, J.J. and J.C. Goldman. 1979. Nitrogenous nutrition of marine phytoplankton in nutrient-depleted waters. Science 203: 670–672.

McCarty, G.W., J.J. Meisinger, and F.M.M. Jenniskens. 1995. Relationships between total-N, biomass-N and active-N in soil under different tillage and N fertilizer treatments. Soil Biology and Biochemistry 27: 1245–1250.

McClain, M.E., J.E. Richey, and T.P. Pimentel. 1994. Groundwater nitrogen dynamics at the terrestrial-lotic interface of a small catchment in the central Amazon basin. Biogeochemistry 27: 113–127.

McColl, J.G. and D.F. Grigal. 1975. Forest fire: Effects on phosphorus movement to lakes. Science 188: 1109–1111.

McCormick, M.P., L.W. Thomason, and C.R. Trepte. 1995. Atmospheric effects of the Mt. Pinatubo eruption. Nature 373: 399–404.

McDiffett, W.F., A.W. Beidler, T.F. Dominick, and K.D. McCrea. 1989. Nutrient concentration-stream discharge relationships during storm events in a first-order stream. Hydrobiologia 179: 97–102.

McDowell, L.R. 1992. *Minerals in Animal and Human Nutrition.* Academic Press, Orlando, Florida.

McDowell, W.H. and C.E. Asbury. 1994. Export of carbon, nitrogen, and major ions from three tropical montane watersheds. Limnology and Oceanography 39: 111–125.

McDowell, W.H. and G.E. Likens. 1988. Origin, composition, and flux of dissolved organic carbon in the Hubbard Brook Valley. Ecological Monographs 58: 177–195.

McDowell, W.H. and T. Wood. 1984. Podzolization: Soil processes control dissolved organic carbon concentrations in stream water. Soil Science 137: 23–32.

McElroy, M.B. 1983. Marine biological controls on atmospheric CO_2 and climate. Nature 302: 328–329.

McElroy, M.B. and R.J. Salawitch. 1989. Changing composition of the global stratosphere. Science 243: 763–770.

McElroy, M.B., Y.L. Yung, and A.O. Nier. 1976. Isotopic composition of nitrogen: Implications for the past history of Mars' atmosphere. Science 194: 70–72.

McElroy, M.B., M.J. Prather, and J.M. Rodriguez. 1982. Escape of hydrogen from Venus. Science 215: 1614–1615.

McFadden, L.D. and D.M. Hendricks. 1985. Changes in the content and composition of pedogenic iron oxyhydroxides in a chronosequence of soils in southern California. Quaternary Research 23: 189–204.

McGill, W.B. and C.V. Cole. 1981. Comparative aspects of cycling of organic C, N, S and P through soil organic matter. Geoderma 26: 267–286.

McGuire, A.D., J.M. Melillo, L.A. Joyce, D.W. Kicklighter, A.L. Grace, B. Moore, and C.J. Vörösmarty. 1992. Interactions between carbon and nitrogen dynamics in estimating net primary productivity for potential vegetation in North America. Global Biogeochemical Cycles 6: 101–124.

McKay, C.P., O.B. Toon, and J.F. Kasting. 1991. Making Mars habitable. Nature 352: 489–496.

McKay, D.S., E.K. Gibson, K.L. Thomas-Keprta, H. Vali, C.S. Romanek, S.J. Clemett, X.D.F. Chillier, C.R. Maechling, and R.N. Zare. 1996. Search for past life on Mars: Possible relic biogenic activity in Martian meteorite ALH84001. Science 273: 924–930.

McKelvey, V.E. 1980. Seabed minerals and the law of the sea. Science 209: 464–472.

McKenney, D.J., C.F. Drury, W.I. Findlay, B. Mutus, T. McDonnell, and C. Gajda. 1994. Kinetics of denitrification by *Pseudomonas fluorescenes*: Oxygen effects. Soil Biology and Biochemistry 26: 901–908.

McLaren, A.S., J.E. Walsh, R.H. Bourke, R.L. Weaver, and W. Wittmann. 1992. Variability in sea-ice thickness over the North Pole from 1977 to 1990. Nature 358: 224–226.

McLaughlin, S.B. and G.E. Taylor. 1981. Relative humidity: Important modifier of pollutant uptake by plants. Science 211: 167–169.

McNaughton, S.J. 1988. Mineral nutrition and spatial concentrations of African ungulates. Nature 334: 343–345.

McNaughton, S.J. 1990. Mineral nutrition and seasonal movements of African migratory ungulates. Nature 345: 613–615.

McNaughton, S.J. and F.S. Chapin. 1985. Effects of phosphorus nutrition and defoliation on C_4 graminoids from the Serengeti plains. Ecology 66: 1617–1629.

McNaughton, S.J., M. Oesterheld, D.A. Frank, and K.J. Williams. 1989. Ecosystem-level patterns of primary productivity and herbivory in terrestrial habitats. Nature 341: 142–144.

McSween, H.Y. 1989. Chondritic meteorites and the formation of planets. American Scientist 77: 146–153.

Meade, C., J.A. Reffner, and E. Ito. 1994. Synchrotron infrared absorbance measurements of hydrogen in $MgSiO_3$ perovskite. Science 264: 1558–1560.

Meade, R.H., T. Dunne, J.E. Richey, U. De M. Santos, and E. Salati. 1985. Storage and remobilization of suspended sediment in the lower Amazon River of Brazil. Science 228: 488–490.

Meentemeyer, V. 1978a. Climatic regulation of decomposition rates of organic matter in terrestrial ecosystems. pp. 779–789. In D.C. Adriano and I.L. Brisbin (eds.). *Environmental Chemistry and Cycling Processes.* CONF 760429. National Technical Information Service, Springfield, Virginia.

Meentemeyer, V. 1978b. Macroclimate and lignin control of litter decomposition rates. Ecology 59: 465–472.

Meentemeyer, V., E.O. Box, and R. Thompson. 1982. World patterns and amounts of terrestrial plant litter production. BioScience 32: 125–128.

Megonigal, J.P., W.H. Patrick, and S.P. Faulkner. 1993. Wetland identification in seasonally flooded forest soils: Soil morphology and redox dynamics. Soil Science Society of America Journal 57: 140–149.

Megonigal, J.P., W.H. Conner, S. Kroeger, and R.R. Sharitz. 1997. Aboveground production in southeastern floodplain forests: A test of the subsidy-stress hypothesis. Ecology 78: 370–384.

Megraw, S.R. and R. Knowles. 1987. Active methanotrophs suppress nitrification in a humisol. Biology and Fertility of Soils 4: 205–212.

Meier, M.F. 1984. Contribution of small glaciers to global sea level. Science 226: 1418–1421.

Melack, J.M. 1991. Reciprocal interactions among lakes, large rivers, and climate. pp. 68–87. In P. Firth and S.G. Fisher (eds.), *Global Climate Change and Freshwater Ecosystems.* Springer-Verlag, New York.

Melamed, R., J.J. Jurinak, and L.M. Dudley. 1994. Anion exclusion-pore water velocity interaction affecting transport of bromine through an Oxisol. Soil Science Society of America Journal 58: 1405–1410.

Melillo, J.M., J.D. Aber, and J.F. Muratore. 1982. Nitrogen and lignin control of hardwood leaf litter decomposition dynamics. Ecology 63: 621–626.

Melillo, J.M., J.D. Aber, P.A. Steudler, and J.P. Schimel. 1983. Denitrification potentials in a successional sequence of northern hardwood forest stands. pp. 217–228. In R. Hallberg (ed.), *Environmental Biogeochemistry.* Swedish Natural Science Research Council, Stockholm.

Melillo, J.M., A.D. McGuire, D.W. Kicklighter, B. Moore, C.J. Vorosmarty, and A.L. Schloss. 1993. Global climate change and terrestrial net production. Nature 363: 234–240.

Melin, J., H. Nômmik, U. Lohm, and J. Flower-Ellis. 1983. Fertilizer nitrogen budget in a Scots pine ecosystem attained by using root-isolated plots and ^{15}N tracer technique. Plant and Soil 74: 249–263.

Melosh, H.J. and A.M. Vickery. 1989. Impact erosion of the primordial atmosphere of Mars. Nature 338: 487–489.

Mendelssohn, I.A., K.L. McKee, and W.H. Patrick. 1981. Oxygen deficiency in *Spartina alterniflora* roots: Metabolic adaptation to anoxia. Science 214: 439–441.

Mengel, K. and H.W. Scherer. 1981. Release of nonexchangeable (fixed) soil ammonium under field conditions during the growing season. Soil Science 131: 226–232.

Merrill, A.G. and D.R. Zak. 1992. Factors controlling denitrification rates in upland and swamp forests. Canadian Journal of Forest Research 22: 1597–1604.

Meybeck, M. 1977. Dissolved and suspended matter carried by rivers: Composition, time and space variations, and world balance. pp. 25–32. In H.L. Golterman (ed.), *Interactions between Sediments and Freshwater.* Dr. W. Junk Publishers, Amsterdam.

Meybeck, M. 1979. Concentrations des eaux fluviales en éléments majeurs et apports en solution aux oceans. Revue de Geologie Dynamique et de Geographe Physique 21: 215–246.

Meybeck, M. 1980. Pathways of major elements from land to ocean through rivers. pp. 18–30. In J.-M. Martin, J.D. Burton, and D. Eisma (eds.), *Proceedings of the Review and Workshop on River Inputs to Ocean Systems.* Food and Agriculture Organization, Rome.

Meybeck, M. 1982. Carbon, nitrogen, and phosphorus transport by world rivers. American Journal of Science 282: 401–450.

Meybeck, M. 1987. Global chemical weathering of surficial rocks estimated from river dissolved loads. American Journal of Science 287: 401–428.

Meybeck, M. 1988. How to establish and use world budgets of riverine materials. pp. 247–272. In A. Lerman and M. Meybeck (eds.), *Physical and Chemical Weathering in Geochemical Cycles.* Kluwer Academic Publishers, Dordrecht, The Netherlands.

Meybeck, M. 1993. C, N, P and S in rivers: From sources to global inputs. pp. 163–193. In R. Wollast, F.T. MacKenzie and L. Chou (eds.), *Interactions of C, N, P and S in Biogeochemical Cycles and Global Change.* Springer-Verlag, Berlin.

Meyer, C. 1985. Ore metals through geologic history. Science 227: 1421–1428.

Meyer, J. 1979. The role of sediments and bryophytes in phosphorus dynamics in a headwater stream ecosystem. Limnology and Oceanography 24: 365–375.

Meyer, J. 1980. Dynamics of phosphorus and organic matter during leaf decomposition in a forest stream. Oikos 34: 44–53.

Meyer, J.L. and G.E. Likens. 1979. Transport and transformation of phosphorus in a forest stream ecosystem. Ecology 60: 1255–1269.

Meyer, J.L., G.E. Likens, and J. Sloane. 1981. Phosphorus, nitrogen, and organic carbon flux in a headwater stream. Archiv für Hydrobiologie 91: 28–44.

Meyer, J.L., W.H.. McDowell, T.L. Bott, J.W. Elwood, C. Ishizaki, J.M. Melack, B.L. Peckarsky, B.J. Peterson, and R.A. Rublee. 1988. Elemental dynamics in streams. Journal of the North American Benthological Society 7: 410–432.

Meyer, O. 1994. Functional groups of microorganisms. pp. 67–96. In E.-D. Schulze and H.A. Mooney (eds.), *Biodiversity and Ecosystem Function.* Springer-Verlag, New York.

Michaels, A.F., N.R. Bates, K.O. Buesseler, C.A. Carlson, and A.H. Knap. 1994. Carbon-cycle imbalances in the Sargasso sea. Nature 372: 537–540.

Michalopoulos, P. and R.C. Aller. 1995. Rapid clay mineral formation in Amazon Delta sediments: Reverse weathering and oceanic elemental cycles. Science 270: 614–617.

Michel, R.L. and H.E. Suess. 1975. Bomb tritium in the Pacific Ocean. Journal of Geophysical Research 80: 4139–4152.

Middleton, K.R. and G.S. Smith. 1979. A comparison of ammoniacal and nitrate nutrition of perennial ryegrass through a thermodynamic model. Plant and Soil 53: 487–504.

Migdisov, A.A., A.B. Ronov, and V.A. Grinenko. 1983. The sulphur cycle in the lithosphere. pp. 25–127. In M.V. Ivanov and J.R. Freney (eds.), *The Global Biogeochemical Sulphur Cycle.* Wiley, New York.

Miller, E.K., J.A. Panek, A.J. Friedland, J. Kadlecek, and V.A. Mohnen. 1993a. Atmospheric deposition to a high-elevation forest at Whiteface Mountain, New York, USA. Tellus 45B: 209–227.

Miller, E.K., J.D. Blum, and A.J. Friedland. 1993b. Determination of soil exchangeable-cation loss and weathering rates using Sr isotopes. Nature 362: 438–441.

Miller, H.G., J.M. Cooper, and J.D. Miller. 1976. Effects of nitrogen supply on nutrients in litter fall and crown leaching in a stand of corsican pine. Journal of Applied Ecology 13: 233–248.

Miller, J.R. and G.L. Russell. 1992. The impact of global warming on river runoff. Journal of Geophysical Research 97: 2757–2764.

Miller, S.L. 1953. A production of amino acids under possible primitive Earth conditions. Science 117: 528–529.

Miller, S.L. 1957. The formation of organic compounds on the primitive Earth. Annals of the New York Academy of Sciences 69: 260–275.

Miller, W.R. and J.I. Drever. 1977. Chemical weathering and related controls on surface water chemistry in the Absaroka Mountains, Wyoming. Geochimica et Cosmochimica Acta 41: 1693–1702.

Milliman, J.D. 1993. Production and accumulation of calcium carbonate in the ocean: Budget of a nonsteady state. Global Biogeochemical Cycles 7: 927–957.

Milliman, J.D. and R.H. Meade. 1983. World-wide delivery of river sediment to the oceans. Journal of Geology 91: 1–21.

Milliman, J.D. and P.M. Syvitski. 1992. Geomorphic/tectonic control of sediment discharge to the ocean: The importance of small mountainous rivers. Journal of Geology 100: 525–544.

Minarik, W.G., F.J. Ryerson, and E.B. Watson. 1996. Textural entrapment of core-forming melts. Science 272: 530–533.

Minnis, P., E.F. Harrison, L.L. Stowe, G.G. Gibson, F.M. Denn, D.R. Doelling, and W.L. Smith. 1993. Radiative climate forcing by the Mount Pinatubo eruption. Science 259: 1411–1415.

Minoura, H. and Y. Iwasaka. 1996. Rapid change in nitrate and sulfate concentrations observed in early stage of precipitation and their deposition processes. Journal of Atmospheric Chemistry 24: 39–55.

Minschwander, K., R.J. Salawitch, and M.B. McElroy. 1993. Absorption of solar radiation by O_2: Implications for O_3 and lifetimes of N_2O, $CFCl_3$ and CF_2Cl_2. Journal of Geophysical Research 98: 10543–10561.

Miskimmin, B.M., J.W.M. Rudd, and C.A. Kelly. 1992. Influence of dissolved organic carbon, pH, and microbial respiration rates on mercury methylation and demethylation in lake water. Canadian Journal of Fisheries and Aquatic Sciences 49: 17–22.

Mispagel, M.E. 1978. The ecology and bioenergetics of the acridid grasshopper, *Bootettix punctatus,* on creosotebush, *Larrea tridentata,* in the northern Mojave Desert. Ecology 59: 779–788.

Mitchell, J.F.B. 1983. The hydrological cycle as simulated by an atmospheric general circulation model. pp. 429–446. In A. Street-Perrott and M. Beran (eds.), *Variations in the Global Water Budget.* Reidel, Hingham, Massachusetts.

Mitchell, J.F.B., T.C. Johns, J.M. Gregory, and S.F.B. Tett. 1995. Climate response to increasing levels of greenhouse gases and sulphate aerosols. Nature 376: 501–504.

Mitchell, M.J., M.B. David, D.G. Maynard, and S.A. Telang. 1986. Sulfur constituents in soils and streams of a watershed in the Rocky Mountains of Alberta. Canadian Journal of Forest Research 16: 315–320.

Mitchell, M.J., C.T. Driscoll, R.D. Fuller, M.B. David, and G.E. Likens. 1989. Effect of whole-tree harvesting on the sulfur dynamics of a forest soil. Soil Science Society of America Journal 53: 933–940.

Mitchell, M.J., N.W. Foster, J.P. Shepard, and I.K. Morrison. 1992. Nutrient cycling in Huntington Forest and Turkey Lakes deciduous stands: Nitrogen and sulfur. Canadian Journal of Forest Research 22: 457–464.

Mitsch, W.J. and J.G. Gosselink. 1986. *Wetlands.* Van Nostrand Reinhold, New York.

Mitsch, W.J., C.L. Dorge, and J.R. Wiemhoff. 1979. Ecosystem dynamics and a phosphorus budget of an alluvial cypress swamp in southern Illinois. Ecology 60: 1116–1124.

Miyoshi, A., S. Hatakeyama, and N. Washida. 1994. OH radical-initiated photooxidation of isoprene: An estimate of global CO production. Journal of Geophysical Research 99: 18779–18787.

Mizutani, H. and E. Wada. 1988. Nitrogen and carbon isotope ratios in seabird rookeries and their ecological implications. Ecology 69: 340–349.

Mizutani, H., H. Hasegawa, and E. Wada. 1986. High nitrogen isotope ratio for soils of seabird rookeries. Biogeochemistry 2: 221–247.

Moeller, J.R., G.W. Minshall, K.W. Cummins, R.C. Peterson, C.E. Cushing, J.R. Sedell, R.A. Larson, and R.L. Vannote. 1979. Transport of dissolved organic carbon in streams of differing physiographic characteristics. Organic Geochemistry 1: 139–150.

Moeslund, L., B. Thamdrup, and B.B. Jørgensen. 1994. Sulfur and iron cycling in a coastal sediment: Radiotracer studies and seasonal dynamics. Biogeochemistry 27: 129–152.

Moffett, J.W. and L.E. Brand. 1996. Production of strong, extracellular Cu chelators by marine cyanobacteria in response to Cu stress. Limnology and Oceanography 41: 388–395.

Mohren, G.M.J., J. Van den Burg, and F.W. Burger. 1986. Phosphorus deficiency induced by nitrogen input in Douglas fir in the Netherlands. Plant and Soil 95: 191–200.

Mojzsis, S.J., G. Arrhenius, K.D. McKeegan, T.M. Harrison, A.P. Nutman, and C.R.L. Friend. 1996. Evidence for life on Earth before 3,800 million years ago. Nature 384: 55–59.

Molina, M.J. and R.S. Rowland. 1974. Stratospheric sink for chlorofluoromethanes: Chlorine atom-catalysed destruction of ozone. Nature 249: 810–812.

Molina, M.J., T.-L. Tso, L.T. Molina, and F.C.-Y. Wang. 1987. Antarctic stratospheric chemistry of chlorine nitrate, hydrogen chloride, and ice: Release of active chlorine. Science 238: 1253–1257.

Möller, D. 1984. On the global natural sulphur emission. Atmospheric Environment 18: 29–39.

Möller, D. 1990. The Na/Cl ratio in rainwater and the seasalt chloride cycle. Tellus 42B: 254–262.

Molles, M.C. and C.N. Dahm. 1990. A perspective on El Niño and La Niña: Global implications for stream ecology. Journal of the North American Benthological Society 9: 68–76.

Molot, L.A. and P.J. Dillon. 1993. Nitrogen mass balances and denitrification rates in central Ontario lakes. Biogeochemistry 20: 195–212.

Monger, H.C., L.A. Daugherty, W.C. Lindemann, and C.M. Liddell. 1991. Microbial precipitation of pedogenic calcite. Geology 19: 997–1000.

Monk, C.D. 1966. An ecological significance of evergreenness. Ecology 47: 504–505.

Mooney, H.A. and W.D. Billings. 1961. Comparative physiological ecology of arctic and alpine populations of *Oxyria digyna*. Ecological Monographs 31: 1–29.

Mooney, H.A., P.M. Vitousek, and P.A. Matson. 1987. Exchange of materials between terrestrial ecosystems and the atmosphere. Science 238: 926–932.

Moore, T.R. and M. Dalva. 1993. The influence of temperature and water table position on carbon dioxide and methane emissions from laboratory columns of peatland soils. Journal of Soil Science 44: 651–664.

Moore, T.R. and R. Knowles. 1989. The influence of water table levels on methane and carbon dioxide emissions from peatland soils. Canadian Journal of Soil Science 69: 33–38.

Moorhead, D.L., J.F. Reynolds, and P.J. Fonteyn. 1989. Patterns of stratified soil water loss in a Chihuahuan desert community. Soil Science 148: 244–249.

Moorhead, K.K. and M.M. Brinson. 1995. Response of wetlands to rising sea level in the lower coastal plain of North Carolina. Ecological Applications 5: 261–271.

Mora, C.I., S.G. Driese, and L.A. Colarusso. 1996. Middle to late Paleozoic atmospheric CO_2 levels from soil carbonate and organic matter. Science 271: 1105–1107.

Moran, M.A. and R.E. Hodson. 1994. Dissolved humic substances of vascular plant origin in a coastal marine environment. Limnology and Oceanography 39: 762–771.

Moran, S.B. and K.O. Buesseler. 1992. Short residence time of colloids in the upper ocean estimated from ^{238}U-^{234}Th disequilibria. Nature 359: 221–223.

Moran, S.B. and R.M. Moore. 1988. Evidence from mesocosm studies for biological removal of dissolved aluminum from sea water. Nature 335: 706–708.

Morel, F.M.M., J.R. Reinfelder, S.B. Roberts, C.P. Chamberlain, J.G. Lee, and D. Yee. 1994. Zinc and carbon co-limitation of marine phytoplankton. Nature 369: 740–742.

Morowitz, H.J. 1968. *Energy Flow in Biology: Biological Organization as a Problem in Thermal Physics*. Academic Press, New York.

Morris, A.W., J.I. Allen, R.J.M. Howland, and R.G. Wood. 1995. The estuary plume zone: Source or sink for land-derived nutrient discharges? Estuarine, Coastal and Shelf Science 40: 387–402.

Morrow, P.A. and V.C. LaMarche. 1978. Tree ring evidence for chronic insect suppression of productivity in subalpine *Eucalyptus*. Science 201: 1244–1246.

Mortimer, C.H. 1941. The exchange of dissolved substances between mud and water in lakes. Journal of Ecology 29: 280–329.

Mortimer, C.H. 1942. The exchange of dissolved substances between mud and water in lakes. Journal of Ecology 30: 147–201.

Mosier, A., D.S. Schimel, D.W. Valentine, K.F. Bronson, and W.J. Parton. 1991. Methane and nitrous oxide fluxes in native, fertilized and cultivated grasslands. Nature 350: 330–332.

Mosier, A.R., W.D. Guenzi, and E.E. Schweizer. 1986. Field denitrification estimation by nitrogen-15 and acetylene inhibition techniques. Soil Science Society of America Journal 50: 831–833.

Mount, G.H. 1992. The measurement of tropospheric OH by long path absorption. I. Instrumentation. Journal of Geophysical Research 97: 2427–2444.

Mowbray, T. and W.H. Schlesinger. 1988. The buffer capacity of organic soils of the Bluff Mountain fen, North Carolina. Soil Science 146: 73–79.

Moyers, J.L., L.E. Ranweiler, S.B. Hopf, and N.E. Korte. 1977. Evaluation of particulate trace species in southwest desert atmosphere. Environmental Science and Technology 11: 789–795.

Mulder, J. and A. Stein. 1994. The solubility of aluminum in acidic forest soils: Long-term changes due to acid deposition. Geochimica et Cosmochimica Acta 58: 85–94.

Mulholland, P.J. 1981. Deposition of riverborne organic carbon in floodplain wetlands and deltas. pp. 142–172. In G.E. Likens, F.T. MacKenzie, J.E. Richey, J.R. Sedell, and K.K. Turekian (eds.), *Flux of Organic Carbon by Rivers to the Oceans.* CONF-8009140. U.S. Department of Energy, Washington, D.C.

Mulholland, P.J. 1992. Regulation of nutrient concentrations in a temperate forest stream: Roles of upland, riparian, and instream processes. Limnology and Oceanography 37: 1512–1526.

Mulholland, P.J. and E.J. Kuenzler. 1979. Organic carbon export from upland and forested wetland watersheds. Limnology and Oceanography 24: 960–966.

Mulholland, P.J. and J.A. Watts. 1982. Transport of organic carbon to the oceans by rivers of North America: A synthesis of existing data. Tellus 34: 176–186.

Müller, J.-F. 1992. Geographic distribution and seasonal variation of surface emissions and deposition velocities of atmospheric trace gases. Journal of Geophysical Research 97: 3787–3804.

Mummey, D.L., J.L. Smith, and H. Bolton. 1994. Nitrous oxide flux from a shrub-steppe ecosystem: Sources and regulation. Soil Biology and Biochemistry 26: 279–286.

Munger, J.W. and S.J. Eisenreich. 1983. Continental-scale variations in precipitation chemistry. Environmental Science and Technology 17: 32–42.

Murphy, E.M., S.N. Davis, A. Long, D. Donahue, and A.J.T. Jull. 1989. ^{14}C in fractions of dissolved organic carbon in ground water. Nature 337: 153–155.

Murphy, T.P. and B.G. Brownlee. 1981. Ammonia volatilization in a hypertrophic prairie lake. Canadian Journal of Fisheries and Aquatic Sciences 38: 1035–1039.

Murray, J.W. and V. Grundmanis. 1980. Oxygen consumption in pelagic marine sediments. Science 209: 1527–1530.

Murray, J.W. and K.M. Kuivila. 1990. Organic matter diagenesis in the northeast Pacific: Transition from aerobic red clay to suboxic hemipelagic sediments. Deep Sea Research 37: 59–80.

Murray, J.W., R.T. Barber, M.R. Roman, M.P. Bacon, and R.A. Feely. 1994. Physical and biological controls on carbon cycling in the Equatorial Pacific. Science 266: 58–65.

Murray, J.W., E. Johnson, and C. Garside. 1995. A U.S. JGOFS process study in the equatorial Pacific (EqPac): Introduction. Deep Sea Research 42: 275–293.

Murray, T.E. 1995. The correlation between iron sulfide precipitation and hypolimnetic phosphorus accumulation during one summer in a softwater lake. Canadian Journal of Fisheries and Aquatic Sciences 52: 1190–1194.

Muzio, L.J. and J.C. Kramlich. 1988. An artifact in the measurement of N_2O from combustion sources. Geophysical Research Letters 15: 1369–1372.

Myrold, D.D., P.A. Matson, and D.L. Peterson. 1989. Relationships between soil microbial properties and aboveground stand characteristics of conifer forests in Oregon. Biogeochemistry 8: 265–281.

Nadelhoffer, K.J. and B. Fry. 1988. Controls on natural nitrogen-15 and carbon-13 abundances in forest soil organic matter. Soil Science Society of America Journal 52: 1633–1640.

Nadelhoffer, K.J., and J.W. Raich. 1992. Fine root production estimates and below-ground carbon allocation in forest ecosystems. Ecology 73: 1139–1147.

Nadelhoffer, K.J., J.D. Aber, and J.M. Melillo. 1984. Seasonal patterns of ammonium and nitrate uptake in ine temperate forest ecosystems. Plant and Soil 80: 321–335.

Nadelhoffer, K.J., M.R. Downs, B. Fry, J.D. Aber, A.H. Magill, and J.M. Melillo. 1995. The fate of ^{15}N-labelled nitrate additions to a northern hardwood forest in eastern Maine, USA. Oecologia 103: 292–301.

Naegeli, M.W., U. Hartmann, E.I. Meyer, and U. Uehlinger. 1995. POM-dynamics and community respiration in the sediments of a floodprone prealpine river (Necker, Switzerland). Archiv für Hydrobiologie 133: 339–347.

Naiman, R.J. 1982. Characteristics of sediment and organic carbon export from pristine boreal forest watersheds. Canadian Journal of Fisheries and Aquatic Sciences 39: 1699–1718.

Naiman, R.J. and J.R. Sedell. 1981. Stream ecosystem research in a watershed perspective. VerhandlungenInternationale Vereinigung fuer Theoretische und Angewandte Limnologie 21: 804–811.

Najjar, R.G., D.J. Erickson and S. Madronich. 1995. Modeling the air-sea fluxes of gases formed from the decomposition of dissolved organic matter: Carbonyl sulfide and carbon monoxide. pp. 107–132. In R.G. Zepp and C. Sonntag (eds.), *The Role of Nonliving Organic Matter in the Earth's Carbon Cycle.* Wiley, New York.

Nance, J.D., P.V. Hobbs, L.F. Radke, and D.E. Ward. 1993. Airborne measurements of gases and particles from an Alaskan wildfire. Journal of Geophysical Research 98: 14873–14882.

Nazaret, S., W.H. Jeffrey, E. Saouter, R. von Haven, and T. Barkay. 1994. *mer*A gene expression in aquatic environments measured by mRNA production and Hg(II) volatilization. Applied and Environmental Microbiology 60: 4059–4065.

Nealson, K.H. and C.R. Myers. 1992. Microbial reduction of manganese and iron: New approaches to carbon cycling. Applied and Environmental Microbiology 58: 439–443.

Neary, A.J. and W.I. Gizyn. 1994. Throughfall and stemflow chemistry under deciduous and coniferous forest canopies in south-central Ontario. Canadian Journal of Forest Research 24: 1089–1100.

Nedwell, D.B. and A. Watson. 1995. CH_4 production, oxidation and emission in a U.K. ombrotrophic peat bog: Influence of SO_4^{2-} from acid rain. Soil Biology and Biochemistry 27: 893–903.

Neilsen, W.A., W. Pataczek, T. Lynch, and R. Pyrke. 1992. Growth response of *Pinus radiata* to multiple applications of nitrogen fertilizer and evaluation of the quantity of added nitrogen remaining in the forest system. Plant and Soil 144: 207–217.

Neilson, R.P. 1995. A model for predicting continental-scale vegetation distribution and water balance. Ecological Applications 5: 362–385.

Neilson, R.P. and D. Marks. 1994. A global perspective of regional vegetation and hydrologic sensitivities from climatic change. Journal of Vegetation Science 5: 715–730.

Nelson, D.M., P. Tréguer, M.A. Brzezinski, A. Leynaert, and B. Quéguiner. 1995. Production and dissolution of biogenic silica in the ocean: Revised global estimates, comparison with regional data and relationship to biogenic sedimentation. Global Biogeochemical Cycles 9: 359–372.

Nelson, D.W. 1982. Gaseous losses of nitrogen other than through denitrification. pp. 327–363. In F.J. Stevenson (ed.), *Nitrogen in Agricultural Soils*. American Society of Agronomy, Madison, Wisconsin.

Nelson, P.N., J.A. Baldock, and J.M. Oades. 1993. Concentration and composition of dissolved organic carbon in streams in relation to catchment soil properties. Biogeochemistry 19: 27–50.

Nepstad, D.C., C.R. de Carvalho, E.A. Davidson, P.H. Jipp, P.A. Lefebvre, G.H. Negreiros, E.D. da Silva, T.A. Stone, S.E. Trumbore, and S. Vieira. 1994. The role of deep roots in the hydrological and carbon cycles of Amazonian forests and pastures. Nature 372: 666–669.

Nerem, R.S. 1995. Global mean sea level variations from TOPEX/POSEIDON altimeter data. Science 268: 708–710.

Nettleton, W.D., J.E. Witty, R.E. Nelson, and J.W. Hawley. 1975. Genesis of argillic horizons in soils of desert areas of the southwestern United States. Soil Science Society of America Proceedings 39: 919–926.

Neukum, G. 1977. Lunar cratering. Philosophical Transactions of the Royal Society of London 285A: 267–272.

Nevison, C.D., R.F. Weiss, and D.J. Erickson. 1995. Global oceanic emissions of nitrous oxide. Journal of Geophysical Research 100: 15809–15820.

Nevison, C.D., G. Esser, and E.A. Holland. 1996. A global model of changing N_2O emissions from natural and perturbed soils. Climatic Change 32: 327–378.

Newbold, J.D., R.V. O'Neill, J.W. Elwood, and W. Van Winkle. 1982. Nutrient spiralling in streams: Implications for nutrient limitation and invertebrate activity. American Naturalist 120: 628–652.

Newbold, J.D., J.W. Elwood, R.V. O'Neill, and A.L. Sheldon. 1983. Phosphorus dynamics in a woodland stream ecosystem: A study of nutrient spiralling. Ecology 64: 1249–1265.

Newman, E.I. 1995. Phosphorus inputs to terrestrial ecosystems. Journal of Ecology 83: 713–726.

Newman, E.I. and R.E. Andrews. 1973. Uptake of phosphorus and potassium in relation to root growth and root density. Plant and Soil 38: 49–69.

Newsom, N.E. and K.W.W. Sims. 1991. Core formation during early accretion of the Earth. Science 252: 926–933.

Nguyen, B.C., N. Mihalopoulos, J.P. Putaud, and B. Bonsang. 1995. Carbonyl sulfide emissions from biomass burning in the tropics. Journal of Atmospheric Chemistry 22: 55–65.

Nielsen, D.R., M.T. Van Genuchten, and J.W. Biggar. 1986. Water flow and solute transport processes in the unsaturated zone. Water Resources Research 22: 89S–108S.

Nielsen, H. 1974. Isotopic composition of the major contributors to atmospheric sulfur. Tellus 26: 213–221.

Niemann, H.B., S.K. Atreya, G.R. Carignan, T.M. Donahue, J.A. Haberman, D.N. Harpold, R.E. Hartle, D.M. Hunten, W.T. Kasprzak, P.R. Mahaffy, T.C. Owen,

N.W. Spencer, and S.H. Way. 1996. The Galileo probe mass spectrometer: Composition of Jupiter's atmosphere. Science 272: 846–849.
Nisbet, E.G. and B. Ingham. 1995. Methane output from natural and quasinatural sources: A review of the potential for change and for biotic and abiotic feedbacks. pp. 189–218. In G.M. Woodwell and F.T. MacKenzie (eds.), *Biotic Feedbacks in the Global Climatic System.* Oxford University Press, New York.
Nixon, S.W. 1980. Between coastal marshes and coastal waters—a review of twenty years of speculation and research on the role of salt marshes in estuarine productivity and water chemistry. pp. 437–525. In P. Hamilton and K.B. MacDonald (eds.), *Estuarine and Wetland Processes.* Plenum, New York.
Nixon, S.W. 1987. Chesapeake Bay nutrient budgets—a reassessment. Biogeochemistry 4: 77–90.
Nixon, S.W., S.L. Granger, and B.L. Nowicki. 1995. An assessment of the annual mass balance of carbon, nitrogen, and phosphorus in Narragansett Bay. Biogeochemistry 31: 15–61.
Nodvin, S.C., C.T. Driscoll, and G.E. Likens. 1988. Soil processes and sulfate loss at the Hubbard Brook Experimental Forest. Biogeochemistry 5: 185–199.
Nohrstedt, H.-O., U. Sikström, E. Ring, T. Näsholm, P. Högberg, and T. Persson. 1996. Nitrate in soil water in three Norway spruce stands in southwest Sweden as related to N-deposition and soil, stand, and foliage properties. Canadian Journal of Forest Research 26: 836–848.
Norby, R.J. 1996. Forest canopy productivity index. Nature 381: 564.
Norby, R.J., C.A. Gunderson, S.D. Wullschleger, E.G. O'Neill, and M.K. McCracken. 1992. Productivity and compensatory responses of yellow-poplar trees in elevated CO_2. Nature 357: 322–324.
Northup, R.R., Z. Yu, R.A. Dahlgren, and K.A. Vogt. 1995. Polyphenol control of nitrogen release from pine litter. Nature 377: 227–229.
Novelli, P.C., L.P. Steele, and P.P. Tans. 1992. Mixing ratios of carbon monoxide in the troposphere. Journal of Geophysical Research 97: 20731–20750.
Novelli, P.C., K.A. Masarie, P.P. Tans, and P.M. Lang. 1994. Recent changes in atmospheric carbon monoxide. Science 263: 1587–1590.
Nozette, S. and J.S. Lewis. 1982. Venus: Chemical weathering of igneous rocks and buffering of atmospheric composition. Science 216: 181–183.
Nriagu, J.O. 1989. A global assessment of natural sources of atmospheric trace metals. Nature 338: 47–49.
Nriagu, J.O. and R.D. Coker. 1978. Isotopic composition of sulfur in precipitation within the Great Lakes Basin. Tellus 30: 365–375.
Nriagu, J.O. and D.A. Holdway. 1989. Production and release of dimethyl sulfide from the Great Lakes. Tellus 41B: 161–169.
Nriagu, J.O., D.A. Holdway, and R.D. Coker. 1987. Biogenic sulfur and the acidity of rainfall in remote areas of Canada. Science 237: 1189–1192.
Nriagu, J.O., R.D. Coker, and L.A. Barrie. 1991. Origin of sulphur in Canadian Arctic haze from isotope measurements. Nature 349: 142–145.
Nürnberg, G.K. 1995. Quantifying anoxia in lakes. Limnology and Oceanography 40: 1100–1111.
Nye, P.H. 1977. The rate-limiting step in plant nutrient absorption from soil. Soil Science 123: 292–297.

Nye, P.H. 1981. Changes of pH across the rhizosphere induced by roots. Plant and Soil 61: 7–26.

Oades, J.M. 1988. The retention of organic matter in soils. Biogeochemistry 5: 35–70.

Oaks, A. 1992. A re-evaluation of nitrogen assimilation in roots. BioScience 42: 103–111.

Oaks, A. 1994. Primary nitrogen assimilation in higher plants and its regulation. Canadian Journal of Botany 72: 739–750.

O'Brien, B.J. and J.D. Stout. 1978. Movement and turnover of soil organic matter as indicated by carbon isotope measurements. Soil Biology and Biochemistry 10: 309–317.

O'Connell, A.M. 1988. Nutrient dynamics in decomposing litter in karri (*Eucalyptus diversicolor* F. Muell.) forests of south-western Australia. Journal of Ecology 76: 1186–1203.

Odada, E.O. 1992. Growth rates of ferromanganese encrustations on rocks from the Romanche Fracture zone, equatorial Atlantic. Deep Sea Research 39: 235–244.

Odum, E.P. 1969. The strategy of ecosystem development. Science 164: 262–270.

Odum, W.E. 1988. Comparative ecology of tidal freshwater and salt marshes. Annual Review of Ecology and Systematics 19: 147–176.

Oechel, W.C., S.J. Hastings, G. Vourlitis, M. Jenkins, G. Riechers, and N. Grulke. 1993. Recent change of arctic tundra ecosystems from a net carbon dioxide sink to a source. Nature 361: 520–523.

Oechel, W.C., S. Cowles, N. Grulke, S.J. Hastings, B. Lawrence, T. Prudhomme, G. Riechers, B. Strain, D. Tissue, and G. Vourlitis. 1994. Transient nature of CO_2 fertilization in arctic tundra. Nature 371: 500–503.

Oechel, W.C., G.L. Vourlitis, S.J. Hastings, and S.A. Bochkarev. 1995. Change in arctic CO_2 flux over two decades: Effects of climate change at Barrow, Alaska. Ecological Applications 5: 846–855.

Oerlemans, J. 1994. Quantifying global warming from the retreat of glaciers. Science 264: 243–245.

Officer, C.B., R.B. Biggs, J.L. Talf, L.E. Cronin, M.A. Tyler, and W.R. Boynton. 1984. Chesapeake Bay anoxia: Origin, development, and significance. Science 223: 22–27.

Oglesby, R.T. 1977. Phytoplankton summer standing crop and annual productivity as functions of phosphorus loading and various physical factors. Journal of the Fisheries Research Board of Canada 34: 2255–2270.

O'Leary, M.H. 1988. Carbon isotopes in photosynthesis. BioScience 38: 328–336.

Olive, K.A. and D.N. Schramm. 1992. Astrophysical ^{7}Li as a product of Big Bang nucleosynthesis and galactic cosmic-ray spallation. Nature 360: 439–442.

Ollinger, S.V., J.D. Aber, G.M. Lovett, S.E. Millham, R.G. Lathrop, and J.M. Ellis. 1993. A spatial model of atmospheric deposition for the northeastern U.S. Ecological Applications 3: 459–472.

Olsen, S.R., C.V. Cole, F.S. Watanabe, and L.A. Dean. 1954. Estimation of available phosphorus in soils by extraction with sodium bicarbonate. U.S. Department of Agriculture, Circular 939, Washington, D.C.

Olson, J.S. 1963. Energy storage and the balance of producers and decomposers in ecological systems. Ecology 44: 322–331.

Olson, J.S., J.A. Watts, and L.J. Allison. 1983. *Carbon in Live Vegetation of Major World Ecosystems.* DOE/NBB-0037. National Technical Information Service, Springfield, Virginia.

Olson, J.S., R.M. Garrels, R.A. Berner, T.V. Armentano, M.I. Dyer, and D.H. Yaalon. 1985. The natural carbon cycle. pp. 175–213. In J.R. Trabalka (ed.), *Atmospheric Carbon Dioxide and the Global Carbon Cycle.* U.S. Department of Energy, Washington, D.C.

Olson, R.K. and W.A. Reiners. 1983. Nitrification in subalpine balsam fir soils: Tests for inhibitory factors. Soil Biology and Biochemistry 15: 413–418.

Olson, R.K., W.A. Reiners, C.S. Cronan, and G.E. Lang. 1981. The chemistry and flux of throughfall and stemflow in subalpine balsam fir forests. Holarctic Ecology 4: 291–300.

Olsson, M. and P.-A. Melkerud. 1989. Chemical and mineralogical changes during genesis of a podzol from till in southern Sweden. Geoderma 45: 267–287.

Oltmans, S.J. and D.J. Hofmann. 1995. Increase in lower-stratospheric water vapour at a mid-latitude northern hemisphere site from 1981 to 1994. Nature 374: 146–149.

Oltmans, S.J. and H. Levy. 1992. Seasonal cycle of surface ozone over the western North Atlantic. Nature 358: 392–394.

O'Neill, R.V. and D.L. De Angelis. 1981. Comparative productivity and biomass relations of forest ecosystems. pp. 411–449. In D.E. Reichle (ed.), *Dynamic Properties of Forest Ecosystems.* Cambridge University Press, Cambridge.

Oort, A.H. 1970. The energy cycle of the Earth. Scientific American 223(3): 54–63.

Oort, A.H., L.A. Anderson, and J.P. Peixoto. 1994. Estimates of the energy cycle of the oceans. Journal of Geophysical Research 99: 7665–7688.

Oppenheimer, M., C.B. Epstein, and R.E. Yuhnke. 1985. Acid deposition, smelter emissions, and the linearity issue in the western United States. Science 229: 859–862.

Oremland, R.S. and C.W. Culbertson. 1992. Evaluation of methyl fluoride and dimethyl ether as inhibitors of aerobic methane oxidation. Applied and Environmental Microbiology 58: 2983–2992.

Oremland, R.S. and B.F. Taylor. 1978. Sulfate reduction and methanogenesis in marine sediments. Geochimica et Cosmochimica Acta 42: 209–214.

Oremland, R.S., J.T. Hollibaugh, A.S. Maest, T.S. Presser, L.G. Miller, and C.W. Culbertson. 1989. Selenate reduction to elemental selenium by anaerobic bacteria in sediments and culture: Biogeochemical significance of a novel sulfate-independent respiration. Applied and Environmental Microbiology 55: 2333–2343.

Oren, R., K.S. Werk, E.-D. Schulze, J. Meyer, B.U. Schneider, and P. Schramel. 1988. Performance of two *Picea abies* (L.) Karst. stands at different stages of decline. VI. Nutrient concentration. Oecologia 77: 151–162.

Orgel, L.E. 1992. Molecular replication. Nature 358: 203–209.

Orgel, L.E. 1994. The origin of life on the Earth. Scientific American 271(4): 76–83.

Orians, K.J. and K.W. Bruland. 1985. Dissolved aluminum in the central North Pacific. Nature 316: 427–429.

Orians, K.J. and K.W. Bruland. 1986. The biogeochemistry of aluminum in the Pacific Ocean. Earth and Planetary Science Letters 78: 397–410.

Orians, K.J., E.A. Boyle, and K.W. Bruland. 1990. Dissolved titanium in the open ocean. Nature 348: 322–325.

Osmond, C.B., K. Winter, and H. Ziegler. 1982. Functional significance of different pathways of CO_2 fixation in photosynthesis. pp. 479–547. In A. Person and M.H.

Zimmerman (eds.), *Encyclopedia of Plant Physiology.* Vol. 12B. Springer-Verlag, New York.

Ostendorf, B. and J.F. Reynolds. 1993. Relationships between a terrain-based hydrologic model and patch-scale vegetation patterns in an arctic tundra landscape. Landscape Ecology 8: 229–237.

Ostlund, H.G. 1983. *Tritium and Radiocarbon, TTO Western North Atlantic Section, GEOSECS Reoccupation.* Rosentiel School of Marine and Atmospheric Sciences, University of Miami, Coral Gables, Florida.

Ostman, N.L. and G.T. Weaver. 1982. Autumnal nutrient transfers by retranslocation, leaching, and litter fall in a chestnut oak forest in southern Illinois. Canadian Journal of Forest Research 12: 40–51.

Oswalt, T.D., J.A. Smith, M.A. Wood, and P. Hintzen. 1996. A lower limit of 9.5 Gyr on the age of the Galactic disk from the oldest white dwarf stars. Nature 382: 692–694.

Ottow, J.C.G. 1971. Iron reduction and gley formation by nitrogen-fixing *Clostridia.* Oecologia 6: 164–175.

Oudot, C., C. Andrie, and Y. Montel. 1990. Nitrous oxide production in the tropical Atlantic Ocean. Deep Sea Research 37: 183–202.

Overpeck, J.T., P.J. Bartlein, and T. Webb. 1991. Potential magnitude of future vegetation change in eastern North America: Comparisons with the past. Science 254: 692–694.

Owen, D.F. and R.G. Wiegert. 1976. Do consumers maximize plant fitness? Oikos 27: 488–492.

Owen, R.M. and D.K. Rea. 1985. Sea-floor hydrothermal activity links climate to tectonics: The Eocene carbon dioxide greenhouse. Science 227: 166–169.

Owen, T. and K. Biemann. 1976. Composition of the atmosphere at the surface of Mars. Detection of Argon-36 and preliminary analysis. Science 193: 801–803.

Owen, T., K. Biemann, D.R. Rushneck, J.E. Biller, D.W. Howarth, and A.L. LaFleur. 1977. The composition of the atmosphere at the surface of Mars. Journal of Geophysical Research 82: 4635–4639.

Owen, T., J.P. Maillard, C. de Bergh, and B.L. Lutz. 1988. Deuterium on Mars: The abundance of HDO and the value of D/H. Science 240: 1767–1770.

Owen, T., A. Bar-Nun, and I. Kleinfeld. 1992. Possible cometary origin of heavy noble gases in the atmospheres of Venus, Earth and Mars. Nature 358: 43–46.

Owens, N.J.P., J.N. Galloway, and R.A. Duce. 1992. Episodic atmospheric nitrogen deposition to oligotrophic oceans. Nature 357: 397–399.

Oyama, V.I., G.C. Carle, F. Woeller, and J.B. Pollack. 1979. Venus lower atmospheric composition: Analysis by gas chromotography. Science 203: 802–805.

Ozima, M. and F.A. Podosek. 1983. *Noble Gas Geochemistry.* Cambridge University Press, Cambridge.

Paerl, H. 1985a. Microzone formation: Its role in the enhancement of aquatic N_2 fixation. Limnology and Oceanography 30: 1246–1252.

Paerl, H. 1985b. Enhancement of marine primary production by nitrogen-enriched acid rain. Nature 315: 747–749.

Paerl, H.W. and B.M. Bebout. 1988. Direct measurement of O_2-depleted microzones in marine *Oscillatoria:* Relation to N_2 fixation. Science 241: 442–445.

Paerl, H.W. and R.G. Carlton. 1988. Control of nitrogen fixation by oxygen depletion in surface-associated microzones. Nature 332: 260–262.

Paerl, H.W., K.M. Crocker, and L.E. Prufert. 1987. Limitation of N_2 fixation in coastal marine waters: Relative importance of molybdenum, iron, phosphorus, and organic matter availability. Limnology and Oceanography 32: 525–536.

Pagel, B.E.J. 1993. Abundances of light elements. Proceedings of the National Academy of Sciences, U.S.A. 90: 4789–4792.

Paine, R.T. 1971. The measurement and application of the calorie to ecological problems. Annual Review of Ecology and Systematics 2: 145–164.

Pakulski, J.D. and R. Benner. 1994. Abundance and distribution of carbohydrates in the ocean. Limnology and Oceanography 39: 930–940.

Palit, S., A. Sharma, and G. Talukder. 1994. Effects of cobalt on plants. Botanical Review 60: 149–181.

Paolini, J. 1995. Particulate organic carbon and nitrogen in the Orinoco River (Venezuela). Biogeochemistry 29: 59–70.

Parfitt, R.L. and R. St.C. Smart. 1978. The mechanism of sulfate adsorption on iron oxides. Soil Science Society of America Journal 42: 48–50.

Parker, G.G. 1983. Throughfall and stemflow in the forest nutrient cycle. Advances in Ecological Research 13: 57–133.

Parker, R.S. and B.M. Troutman. 1989. Frequency distribution for suspended sediment loads. Water Resources Research 25: 1567–1574.

Parkes, R.J., B.A. Cragg, S.J. Bale, J.M. Getliff, K. Goodman, P.A. Rochelle, J.C. Fry, A.J. Weightman, and S.M. Harvey. 1994. Deep bacterial biosphere in Pacific Ocean sediments. Nature 371: 410–413.

Parkin, T.B. 1987. Soil microsites as a source of denitrification variability. Soil Science Society of America Journal 51: 1194–1199.

Parkin, T.B., A.J. Sexstone, and J.M. Tiedje. 1985. Comparison of field denitrification rates determined by acetylene-based soil core and nitrogen-15 methods. Soil Science Society of America Journal 49: 94–99.

Parkin, T.B., J.L. Starr, and J.J. Meisinger. 1987. Influence of sample size on measurement of soil denitrification. Soil Science Society of America Journal 51: 1492–1501.

Parrilla, G., A. Lavín, H. Bryden, M. García, and R. Millard. 1994. Rising temperatures in the subtropical North Atlantic Ocean over the past 35 years. Nature 369: 48–51.

Parrington, J.R., W.H. Zoller, and N.K. Aras. 1983. Asian dust: Seasonal transport to the Hawaiian islands. Science 220: 195–197.

Parrish, D.D., J.S. Holloway, M. Trainer, P.C. Murphy, G.L. Forbes, and F.C. Fehsenfeld. 1993. Export of North American ozone pollution to the North Atlantic ocean. Science 259: 1436–1439.

Parton, W.J., D.S. Schimel, C.V. Cole, and D.S. Ojima. 1987. Analysis of factors controlling soil organic matter levels in Great Plains grasslands. Soil Science Society of America Journal 51: 1173–1179.

Parton, W.J., J.W.B. Stewart, and C.V. Cole. 1988. Dynamics of C, N, P and S in grassland soils: A model. Biogeochemistry 5: 109–131.

Parton, W.J., J.M.O. Scurlock, D.S. Ojima, D.S. Schimel, and D.O. Hall. 1995. Impact of climate change on grassland production and soil carbon worldwide. Global Change Biology 1: 13–22.

Paruelo, J.M. and O.E. Sala. 1995. Water losses in the Patagonian steppe: A modelling approach. Ecology 76: 510–520.

Pastor, J. and W.M. Post. 1986. Influence of climate, soil moisture, and succession on forest carbon and nitrogen cycles. Biogeochemistry 2: 3–27.

Pastor, J. and W.M. Post. 1993. Linear regressions do not predict the transient response of eastern North American forests to CO_2-induced climate change. Climatic Change 23: 111–119.

Pastor, J., J.D. Aber, C.A. McClaugherty, and J.M. Mellilo. 1984. Aboveground production and N and P cycling along a nitrogen mineralization gradient on Blackhawk Island, Wisconsin. Ecology 65: 256–268.

Patric, J.H., J.E. Douglass, and J.D. Hewlett. 1965. Soil water absorption by mountain and piedmont forests. Soil Science Society of America Proceedings 29: 303–308.

Patrick, W.H. and A. Jugsujinda. 1992. Sequential reduction and oxidation of inorganic nitrogen, manganese, and iron in flooded soil. Soil Science Society of America Journal 56: 1071–1073.

Patrick, W.H. and M.E. Tusneem. 1972. Nitrogen loss from flooded soil. Ecology 53: 735–737.

Paulsen, D.M., H.W. Paerl, and P.E. Bishop. 1991. Evidence that molybdenum-dependent nitrogen fixation is not limited by sulfate concentrations in marine environments. Limnology and Oceanography 36: 1325–1334.

Pauly, D. and V. Christensen. 1995. Primary production required to sustain global fisheries. Nature 374: 255–257.

Pearson, J.A., D.H. Knight, and T.J. Fahey. 1987. Biomass and nutrient accumulation during stand development in Wyoming lodgepole pine forests. Ecology 68: 1966–1973.

Pecoraro, V.L. 1988. Structural proposals for the manganese centers of the oxygen evolving complex: An inorganic chemist's perspective. Photochemistry and Photobiology 48: 249–264.

Pedro, G., M. Jamagne, and J.C. Begon. 1978. Two routes in genesis of strongly differentiated acid soils under humid, cool-temperate conditions. Geoderma 20: 173–189.

Peierls, B.L., N.F. Caraco, M.L. Pace, and J.J. Cole. 1991. Human influence on river nitrogen. Nature 350: 386–387.

Peltier, L.C. 1950. The geographic cycle in periglacial regions as it is related to climatic geomorphology. Annals of the Association of American Geographers 40: 214–236.

Peltier, W.R. and A.M. Tushingham. 1989. Global sea level rise and the greenhouse effect: Might they be connected? Science 244: 806–810.

Pennell, K.D., H.L. Allen, and W.A. Jackson. 1990. Phosphorus uptake capacity of 14-year-old loblolly pine as indicated by a ^{32}P root bioassay. Forest Science 36: 358–366.

Penner, J.E., C.S. Atherton, J. Dignon, S.J. Ghan, and J.J. Walton. 1991. Tropospheric nitrogen: A three-dimensional study of sources, distributions, and deposition. Journal of Geophysical Research 96: 959–990.

Penner, J.E., R.E. Dickinson, and C.A. O'Neill. 1992. Effects of aerosol from biomass burning on the global radiation budget. Science 256: 1432–1434.

Pennington, W. 1981. Records of a lake's life in time: The sediments. Hydrobiologia 79: 197–219.

Peñuelas, J. and J. Azcón-Bieto. 1992. Changes in leaf $\Delta^{13}C$ of herbarium plant species during the last 3 centuries of CO_2 increase. Plant, Cell and Environment 15: 485–489.

Peñuelas, J. and R. Matamala. 1990. Changes in N and S leaf content, stomatal density and specific leaf area of 14 plant species during the last three centuries of CO_2 increase. Journal of Experimental Botany 41: 1119–1124.

Penzias, A.A. 1979. The origin of the elements. Science 205: 549–554.

Perdue, E.M., K.C. Beck, and J.H. Reuter. 1976. Organic complexes of iron and aluminium in natural waters. Nature 260: 418–420.

Peterjohn, W.T. and D.L. Correll. 1984. Nutrient dynamics in an agricultural watershed: Observations on the role of a riparian forest. Ecology 65: 1466–1475.

Peterjohn, W.T. and W.H. Schlesinger. 1991. Factors controlling denitrification in a Chihuahuan desert ecosystem. Soil Science Society of America Journal 55: 1694–1701.

Peterjohn, W.T., J.M. Melillo, P.A. Steudler, K.M. Newkirk, F.P. Bowles, and J.D. Aber. 1994. Responses of trace gas fluxes and N availability to experimentally elevated soil temperatures. Ecological Applications 4: 617–625.

Peterjohn, W.T., M.B. Adams, and F.S. Gilliam. 1996. Symptoms of nitrogen saturation in two central Appalachian hardwood forest ecosystems. Biogeochemistry 35: 507–522.

Peters, V. and R. Conrad. 1996. Sequential reduction processes and initiation of CH_4 production upon flooding of oxic upland soils. Soil Biology and Biochemistry 28: 371–382.

Peterson, B.J. 1980. Aquatic primary productivity and the ^{14}C-CO_2 method: A history of the productivity problem. Annual Review of Ecology and Systematics 11: 359–385.

Peterson, B.J. 1981. Perspectives on the importance of the oceanic particulate flux in the global carbon cycle. Ocean Science and Engineering 6: 71–108.

Peterson, B.J. and R.W. Howarth. 1987. Sulfur, carbon, and nitrogen isotopes used to trace organic matter flow in the salt-marsh estuaries of Sapelo Island, Georgia. Limnology and Oceanography 32: 1195–1213.

Peterson, B.J. and J.M. Melillo. 1985. The potential storage of carbon caused by eutrophication of the biosphere. Tellus 37B: 117–127.

Peterson, B.J., J.E. Hobbie, A.E. Hershey, M.A. Lock, T.E. Ford, J.R. Vestal, V.L. McKinley, M.A.J. Hullar, M.C. Miller, R.M. Ventullo, and G.S. Volk. 1985. Transformation of a tundra river from heterotrophy to autotrophy by addition of phosphorus. Science 229: 1383–1386.

Peterson, B.J., R.W. Howarth, and R.H. Garritt. 1986. Sulfur and carbon isotopes as tracers of salt-march organic matter flow. Ecology 67: 865–874.

Peterson, D.L., M.A. Spanner, S.W. Running, and K.T. Teuber. 1987. Relationship of thematic mapper simulator data to leaf area index of temperate coniferous forests. Remote Sensing of Environment 22: 323–341.

Peterson, T.C., V.S. Golubev, and P.Y. Groisman. 1995. Evaporation losing its strength. Nature 377: 687–688.

Petit, J.R., L. Mounier, J. Jouzel, Y.S. Korotkevich, V.I. Kotlyakov, and C. Lorius. 1990. Palaeoclimatological and chronological implications of the Vostok core dust record. Nature 343: 56–58.

Petty, G.W. 1995. The status of satellite-based rainfall estimation over land. Remote Sensing of Environment 51: 125–137.

Phelps, T.J., E.M. Murphy, S.M. Pfiffner, and D.C. White. 1994. Comparison between geochemical and biological estimates of subsurface microbial activities. Microbial Ecology 28: 335–349.

Philander, G. 1989. El Niño and La Niña. American Scientist 77: 451–459.

Phillips, F.M., J.L. Mattick, T.A. Duval, D. Elmore, and P.W. Kubik. 1988. Chlorine 36 and tritium from nuclear weapons fallout as tracers for long-term liquid and vapor movement in desert soils. Water Resources Research 24: 1877–1891.

Piccot, S.D., J.J. Watson, and J.W. Jones. 1992. A global inventory of volatile organic compound emissions from anthropogenic sources. Journal of Geophysical Research 97: 9897–9912.

Pierce, R.S., J.W. Hornbeck, G.E. Likens, and F.H. Bormann. 1970. Effects of elimination of vegetation on stream water quantity and quality. pp. 311–328. In *Proceedings of the International Association of Scientific Hydrology*. Wellington, New Zealand.

Pigott, C.D. and K. Taylor. 1964. The distribution of some woodland herbs in relation to the supply of nitrogen and phosphorus in the soil. Journal of Ecology 52(Suppl.): 175–185.

Pimentel, D., C. Harvey, P. Resosudarmo, K. Sinclair, D. Kurz, M. McNair, S. Crist, L. Shpritz, L. Fitton, R. Saffouri, and R. Blair. 1995. Environmental and economic costs of soil erosion and conservation benefits. Science 267: 1117–1123.

Pinay, G., L. Roques, and A. Febre. 1993. Spatial and temporal patterns of denitrification in a riparian forest. Journal of Applied Ecology 30: 581–591.

Pingitore, N.E. and M.P. Eastman. 1986. The coprecipitation of Sr^{2+} with calcite at 25°C and 1 atm. Geochimica et Cosmochimica Acta 50: 2195–2203.

Piñol, J., J.M. Alcañiz, and F. Rodà. 1995. Carbon dioxide efflux and pCO_2 in soils of three *Quercus ilex* montane forests. Biogeochemistry 30: 191–215.

Pinto, J.P., G.R. Gladstone, and Y.L. Yung. 1980. Photochemical production of formaldehyde in Earth's primitive atmosphere. Science 210: 183–185.

Pirozynski, K.A. and D.W. Malloch. 1975. The origin of land plants: A matter of mycotrophism. BioSystems 6: 153–164.

Pitsch, S., A. Eschenmoser, B. Gedulin, S. Hui, and G. Arrhenius. 1995. Mineral induced formation of sugar phosphates. Origins of Life and Evolution of the Biosphere 25: 297–334.

Platt, T. and M.R. Lewis. 1987. Estimation of phytoplankton production by remote sensing. Advances in Space Research 7: 131–135.

Platt, T. and S. Sathyendranath. 1988. Oceanic primary production: Estimation by remote sensing at local and regional scales. Science 241: 1613–1620.

Platt, U., M. Rateike, W. Junkermann, J. Rudolph, and D.H. Ehhalt. 1988. New tropospheric OH measurements. Journal of Geophysical Research 93: 5159–5166.

Pletscher, D.H., F.H. Bormann, and R.S. Miller. 1989. Importance of deer compared to other vertebrates in nutrient cycling and energy flow in a northern hardwood ecosystem. American Midland Naturalist 121: 302–311.

Pocklington, R. and F.C. Tan. 1987. Seasonal and annual variations in the organic matter contributed by the St. Lawrence River to the Gulf of St. Lawrence. Geochimica et Cosmochimica Acta 51: 2579–2586.

Poiani, K.A. and W.C. Johnson. 1991. Global warming and prairie wetlands. BioScience 41: 611–618.

Polglase, P.J., P.M. Attiwill, and M.A. Adams. 1992. Nitrogen and phosphorus cycling in relation to stand age of *Eucalyptus regnans* F. Muell. Plant and Soil 142: 177–185.

Pollack, J.B. and D.C. Black. 1982. Noble gases in planetary atmospheres: Implications for the origin and evolution of atmospheres. Icarus 51: 169–198.

Pollack, J.B., J.F. Kasting, S.M. Richardson, and K. Poliakoff. 1987. The case for a wet, warm climate on early Mars. Icarus 71: 203–224.

Pollack, J.B., T. Roush, F. Witteborn, J. Bregman, D. Wooden, C. Stoker, O.B. Toon, D. Rank, B. Dalton, and R. Freedman. 1990. Thermal emission spectra of Mars (5.4–10.5 μm.): Evidence for sulfates, carbonates, and hydrates. Journal of Geophysical Research 95: 14595–14627.

Pollman, C.D., T.M. Lee, W.J. Andrews, L.A. Sacks, S.A. Gherini, and R.K. Munson. 1991. Preliminary analysis of the hydrologic and geochemical controls on acid-neutralizing capacity in two acidic seepage lakes in Florida. Water Resources Research 27: 2321–2335.

Polovina, J.J., G.T. Mitchum, and G.T. Evans. 1995. Decadal and basin-scale variation in mixed layer depth and the impact on biological production in the Central and North Pacific, 1960–88. Deep Sea Research 42: 1701–1716.

Polubesova, T.A., J. Chorover, and G. Sposito. 1995. Surface charge characteristics of podzolized soil. Soil Science Society of America Journal 59: 772–777.

Pomeroy, L.R. and D. Deibel. 1986. Temperature regulation of bacterial activity during the spring bloom in Newfoundland coastal waters. Science 233: 359–361.

Pomeroy, L.R., W.M. Darley, E.L. Dunn, J.L. Gallagher, E.B. Haines, and D.M. Whitney. 1981. Primary production. pp. 39–67. In L.R. Pomeroy and R.G. Wiegert (eds.), *The Ecology of a Salt Marsh.* Springer-Verlag, New York.

Ponnamperuma, F.N. 1972. The chemistry of submerged soils. Advances in Agronomy 24: 29–96.

Ponnamperuma, F.N., E.M. Tianco, and T. Loy. 1967. Redox equilibria in flooded soils. I. The iron hydroxide systems. Soil Science 103: 374–382.

Poorter, H. 1993. Interspecific variation in the growth response of plants to an elevated ambient CO_2 concentration. Vegetatio 104: 77–97.

Porter, K.G. 1976. Enhancement of algal growth and productivity by grazing zooplankton. Science 192: 1332–1334.

Post, W.M., W.R. Emanuel, P.J. Zinke, and A.G. Stangenberger. 1982. Soil carbon pools and world life zones. Nature 298: 156–159.

Post, W.M., J. Pastor, P.J. Zinke, and A.G. Stangenberger. 1985. Global patterns of soil nitrogen storage. Nature 317: 613–616.

Postel, S.L., G.C. Daily, and P.R. Ehrlich. 1996. Human appropriation of renewable fresh water. Science 271: 785–788.

Postgate, J.R., H.M. Kent, and R.L. Robson. 1988. Nitrogen fixation by *Desulfovibrio.* pp. 457–471. In J.A. Cole and S.J. Ferguson (eds.), *The Nitrogen and Sulphur Cycles.* Cambridge University Press, Cambridge.

Potter, C.S., J.T. Randerson, C.B. Field, P.A. Matson, P.M. Vitousek, H.A. Mooney, and S.A. Klooster. 1993. Terrestrial ecosystem production: A process model based on global satellite and surface data. Global Biogeochemical Cycles 7: 811–841.

Potter, C.S., P.A. Matson, P.M. Vitousek, and E.A. Davidson. 1996. Process modeling of controls on nitrogen trace gas emissions from soils worldwide. Journal of Geophysical Research 101: 1361–1377.

Potts, M.J. 1978. Deposition of air-borne salt on *Pinus radiata* and the underlying soil. Journal of Applied Ecology 15: 543–550.

Powell, T.M., J.E. Cloern, and L.M. Huzzey. 1989. Spatial and temporal variability in South San Francisco Bay (USA). I. Horizontal distribution of salinity, suspended sediments, and phytoplankton biomass and productivity. Estuarine, Coastal and Shelf Science 28: 583–597.

Prahl, F.G., J.R. Ertel, M.A. Goni, M.A. Sparrow, and B. Eversmeyer. 1994. Terrestrial organic carbon contributions to sediments on the Washington margin. Geochimica et Cosmochimica Acta 58: 3035–3048.

Prather, M.J. 1985. Continental sources of halocarbons and nitrous oxide. Nature 317: 221–225.

Prather, M.J., R. Derwent, D. Ehhalt, P. Fraser, E. Sanhueza, and X. Zhou. 1995. Other trace gases and atmospheric chemistry. pp. 73–126. In J.T. Houghton, L.G. Meira Filho, J. Bruce, H. Lee, B.A. Callender, E. Haites, N. Harris, and K. Maskell (eds.), *Climate Change 1994*. Cambridge University Press, Cambridge.

Pražák, J., M. Šír, and M. Tesař. 1994. Estimation of plant transpiration from meteorological data under conditions of sufficient soil moisture. Journal of Hydrology 162: 409–427.

Pregitzer, K.S., R.L. Hendrick, and R. Fogel. 1993. The demography of fine roots in response to patches of water and nitrogen. New Phytologist 125: 575–580.

Premuzic, E.T., C.M. Benkovitz, J.S. Gaffney, and J.J. Walsh. 1982. The nature and distribution of organic matter in the surface sediments of the world's oceans and seas. Organic Geochemistry 4: 63–77.

Prentice, I.C. and M.T. Sykes. 1995. Vegetation geography and global carbon storage changes. pp. 304–312. In G.M. Woodwell and F.T. MacKenzie (eds.), *Biotic Feedbacks in the Global Climatic System.* Oxford University Press, New York.

Prentice, I.C., W. Cramer, S.P. Harrison, R. Leemans, R.A. Monserud, and A.M. Solomon. 1992. A global biome model based on plant physiology and dominance, soil properties and climate. Journal of Biogeography 19: 117–134.

Prentice, K.C. 1990. Bioclimatic distribution of vegetation for general circulation model studies. Journal of Geophysical Research 95: 11811–11830.

Press, F. and R. Siever. 1986. *Earth.* 4th ed. Freeman, New York.

Preston, C.M. and D.J. Mead. 1994. Growth response and recovery of ^{15}N-fertilizer one and eight growing seasons after application to lodgepole pine in British Columbia. Forest Ecology and Management 65: 219–229.

Prézelin, B.B. and B.A. Boczar. 1986. Molecular bases of cell absorption and fluorescence in phytoplankton: Potential applications to studies in optical oceanography. Progress in Phycological Research 4:349–464.

Price, N.M. and F.M.M. Morel. 1990. Cadmium and cobalt substitution for zinc in a marine diatom. Nature 334: 658–660.

Price, N.M. and F.M.M. Morel. 1991. Colimitation of phytoplankton growth by nickel and nitrogen. Limnology and Oceanography 36: 1071–1077.

Prinn, R., D. Cunnold, R. Rasmussen, P. Simmonds, F. Alyea, A. Crawford, P. Fraser, and R. Rosen. 1990. Atmospheric emissions and trends of nitrous oxide deduced from 10 years of ALE-GAGE data. Journal of Geophysical Research 95: 18369–18385.

Prinn, R., D. Cunnold, P. Simmonds, F. Alyea, R. Boldi, A. Crawford, P. Fraser, D. Gutzler, D. Hartley, R. Rosen, and R. Rasmussen. 1992. Global average concentration and trend for hydroxyl radicals deduced from ALE/GAGE trichloroethane (methyl chloroform) data for 1978–1990. Journal of Geophysical Research 97: 2445–2461.

Prinn, R.G., R.F. Weiss, B.R. Miller, J. Huang, F.N. Alyea, D.M. Cunnold, P.J. Fraser, D.E. Hartley, and P.G. Simmonds. 1995. Atmospheric trends and lifetime of CH_3CCl_3 and global OH concentrations. Science 269: 187–192.

Prins, E.M. and W.P. Menzel. 1994. Trends in South American biomass burning detected with the GOES visible infrared spin scan radiometer atmospheric sounder from 1983 to 1991. Journal of Geophysical Research 99: 16719–16735.

Probst, J.L. and Y. Tardy. 1987. Long range streamflow and world continental runoff fluctuations since the beginning of this century. Journal of Hydrology 94: 289–311.

Proe, M.F., J. Dutch, H.G. Miller, and J. Sutherland. 1992. Long-term partitioning of biomass and nitrogen following application of nitrogen fertilizer to Corsican pine. Canadian Journal of Forest Research 22: 82–87.

Prospero, J.M. and D.L. Savoie. 1989. Effect of continental sources on nitrate concentrations over the Pacific Ocean. Nature 339: 687–689.

Prospero, J.M., D.L. Savoie, E.S. Saltzman, and R. Larsen. 1991. Impact of oceanic sources of biogenic sulphur on sulphate aerosol concentrations at Mawson, Antarctica. Nature 350: 221–223.

Protz, R., G.J. Ross, I.P. Martini, and J. Terasmae. 1984. Rate of podzolic soil formation near Hudson Bay, Ontario. Canadian Journal of Soil Science 64: 31–49.

Protz, R., G.J. Ross, M.J. Shipitalo, and J. Terasmae. 1988. Podzolic soil development in the southern James Bay lowlands, Ontario. Canadian Journal of Soil Science 68: 287–305.

Pugnaire, F.I. and F.S. Chapin. 1993. Controls over nutrient resorption from leaves of evergreen Mediterranean species. Ecology 74: 124–129.

Pulliam, W.M. 1993. Carbon dioxide and methane exports from a southeastern floodplain swamp. Ecological Monographs 63: 29–53.

Putaud, J.-P., S. Belviso, B.C. Nguyen, and N. Mihalopoulos. 1993. Dimethylsulfide, aerosols, and condensation nuclei over the tropical northeastern Atlantic Ocean. Journal of Geophysical Research 98: 14863–14871.

Pye, K. 1987. *Aeolian Dust and Dust Deposits*. Academic Press, London.

Quade, J., T.E. Cerling, and J.R. Bowman. 1989. Development of Asian monsoon revealed by marked ecological shift during the latest Miocene in northern Pakistan. Nature 342: 163–166.

Qualls, R.G. 1984. The role of leaf litter nitrogen immobilization in the nitrogen budget of a swamp stream. Journal of Environmental Quality 13: 640–644.

Qualls, R.G., B.L. Haines, and W.T. Swank. 1991. Fluxes of dissolved organic nutrients and humic substances in a deciduous forest. Ecology 72: 254–266.

Quay, P.D., S.L. King, J.M. Lansdown, and D.O. Wilbur. 1988. Isotopic composition of methane released from wetlands: Implications for the increase in atmospheric methane. Global Biogeochemical Cycles 2: 385–397.

Quay, P.D., S.L. King, J. Stutsman, D.O. Wilbur, L.P. Steele, I. Fung, R.H. Gammon, T.A. Brown, F.W. Farrell, P.M. Gootes, and F.H. Schmidt. 1991. Carbon isotopic composition of atmospheric CH_4: Fossil and biomass burning source strengths. Global Biogeochemical Cycles 5: 25–47.

Quay, P.D., B. Tilbrook, and C.S. Wong. 1992. Oceanic uptake of fossil fuel CO_2: Carbon-13 evidence. Science 256: 74–79.

Quay, P.D., D.O. Wilbur, J.E. Richey, A.H. Devol, R. Benner, and B.R. Forsberg. 1995. The ^{18}O:^{16}O of dissolved oxygen in rivers and lakes in the Amazon basin:

Determining the ratio of respiration to photosynthesis rates in freshwaters. Limnology and Oceanography 40: 718–729.

Quinn, P.K., R.J. Charlson, and W.H. Zoller. 1987. Ammonia, the dominant base in the remote marine atmosphere: A review. Tellus 39B: 413–425.

Quinn, P.K., R.J. Charlson, and T.S. Bates. 1988. Simultaneous observations of ammonia in the atmosphere and ocean. Nature 335: 336–338.

Raaimakers, D., R.G.A. Boot, P. Dijkstra, S. Pot, and T. Pons. 1995. Photosynthetic rates in relation to leaf phosphorus content in pioneer versus climax tropical rainforest trees. Oecologia 102: 120–125.

Rabenhorst, M.C. and K.C. Haering. 1989. Soil micromorphology of a Chesapeake Bay tidal marsh: Implications for sulfur accumulation. Soil Science 147: 339–347.

Rabinowitz, J., J. Flores, R. Krebsbach, and G. Rogers. 1969. Peptide formation in the presence of linear or cyclic polyphosphates. Nature 224: 795–796.

Raghubanshi, A.S. 1992. Effect of topography on selected soil properties and nitrogen mineralization in a dry tropical forest. Soil Biology and Biochemistry 24: 145–150.

Rahn, K.A. and D.H. Lowenthal. 1984. Elemental tracers of distant regional pollution aerosols. Science 223: 132–139.

Raich, J.W. and K.J. Nadelhoffer. 1989. Belowground carbon allocation in forest ecosystems: Global trends. Ecology 70: 1346–1354.

Raich, J.W. and C.S. Potter. 1995. Global patterns of carbon dioxide emissions from soils. Global Biogeochemical Cycles 9: 23–36.

Raich, J.W. and W.H. Schlesinger. 1992. The global carbon dioxide flux in soil respiration and its relationship to vegetation and climate. Tellus 44B: 81–99.

Raison, R.J. 1979. Modification of the soil environment by vegetation fires, with particular reference to nitrogen transformations: A review. Plant and Soil 51: 73–108.

Raison, R.J., P.K. Khanna, and P.V. Woods. 1985. Mechanisms of element transfer to the atmosphere during vegetation fires. Canadian Journal of Forest Research 15: 132–140.

Raison, R.J., M.J. Connell, and P.K. Khanna. 1987. Methodology for studying fluxes of soil mineral-N *in situ*. Soil Biology and Biochemistry 19: 521–530.

Raiswell, R. and R.A. Berner. 1986. Pyrite and organic matter in Phanerozoic normal marine shales. Geochimica et Cosmochimica Acta 50: 1967–1976.

Ralph, B.J. 1979. Oxidative reactions in the sulfur cycle. pp. 369–400. In P.A. Trudinger and D.J. Swaine (eds.), *Biogeochemical Cycling of Mineral-Forming Elements*. Elsevier Scientific Publishers, Amsterdam.

Ramanathan, V. 1988. The greenhouse theory of climate change: A test by an inadvertent global experiment. Science 240: 293–299.

Ramanathan, V., R.D. Cess, E.F. Harrison, P. Minnis, B.R. Barkstrom, E. Ahmad, and D. Hartmann. 1989. Cloud-radiative forcing and climate: Results from the Earth radiation budget experiment. Science 243: 57–63.

Ramirez, A.J. and A.W. Rose. 1992. Analytical geochemistry of organic phosphorus and its correlation with organic carbon in marine and fluvial sediments and soils. American Journal of Science 292: 421–454.

Randlett, D.L., D.R. Zak, and N.W. MacDonald. 1992. Sulfate adsorption and microbial immobilization in northern hardwood forests along an atmospheric deposition gradient. Canadian Journal of Forest Research 22: 1843–1850.

Rank, D.M., P.A. Pinto, S.E. Woosley, J.D. Bregman, F.C. Witteborn, T.S. Axelrod, and M. Cohen. 1988. Nickel, argon, and cobalt in the infrared spectrum of SN1987A: The core becomes visible. Nature 331: 505–506.

Raper, C.D., D.L. Osmond, M. Wann, and W.W. Weeks. 1978. Interdependence of root and shoot activities in determining nitrogen uptake rate of roots. Botanical Gazette 139: 289–294.

Rapport, D.J., H.A. Regier, and T.C. Hutchinson. 1985. Ecosystem behavior under stress. American Naturalist 125: 617–640.

Rasmussen, B. 1996. Early-diagenetic REE-phosphate minerals (florencite, gorceixite, crandallite, and xenotime) in marine sandstones: A major sink for oceanic phosphorus. American Journal of Science 296: 601–632.

Rasmussen, R.A. 1974. Emission of biogenic hydrogen sulfide. Tellus 26: 254–260.

Rasmussen, R.A. and M.A.K. Khalil. 1986. Atmospheric trace gases: Trends and distributions over the last decade. Science 232: 1623–1624.

Rastetter, E.B., R.B. McKane, G.R. Shaver, and J.M. Melillo. 1992. Changes in C storage by terrestrial ecosystems: How C-N interactions restrict responses to CO_2 and temperature. Water, Air, and Soil Pollution 64: 327–344.

Raval, A. and V. Ramanathan. 1989. Observational determination of the greenhouse effect. Nature 342: 758–761.

Raven, J.A., A.A. Franco, E. Lino de Jesus, and J. Jacob-Neto. 1990. H^+ extrusion and organic-acid synthesis in N_2-fixing symbioses involving vascular plants. New Phytologist 114: 369–389.

Raven, J.A., B. Wollenweber, and L.L. Handley. 1992. A comparison of ammonium and nitrate as nitrogen sources for photolithotrophs. New Phytologist 121: 19–32.

Raynaud, D., J. Jouzel, J.M. Barnola, J. Chappellaz, D.J. Delmas, and C. Lorius. 1993. The ice record of greenhouse gases. Science 259: 926–934.

Reader, R.J. and J.M. Stewart. 1972. The relationship between net primary production and accumulation for a peatland in southeastern Manitoba. Ecology 53: 1024–1037.

Reddy, K.J., W.L. Lindsay, S.M. Workman, and J.I. Drever. 1990. Measurment of calcite ion activity products in soils. Soil Science Society of America Journal 54: 67–71.

Reddy, K.R. and W.H. Patrick. 1975. Effect of alternate aerobic and anaerobic conditions on redox potential, organic matter decomposition and nitrogen loss in a flooded soil. Soil Biology and Biochemistry 7: 87–94.

Reddy, K.R. and W.H. Patrick. 1976. Effect of frequent changes in aerobic and anaerobic conditions on redox potential and nitrogen loss in a flooded soil. Soil Biology and Biochemistry 8: 491–495.

Redfield, A.C. 1958. The biological control of chemical factors in the environment. American Scientist 46: 205–221.

Redfield, A.C., B.H. Ketchum, and F.A. Richards. 1963. The influence of organisms on the composition of sea-water. pp. 26–77. In M.N. Hill (ed.), *The Sea*. Vol. 2. Wiley, New York.

Redmond, K.T. and R.W. Koch. 1991. Surface climate and streamflow variability in the western United States and their relationship to large-scale circulation indices. Water Resources Research 27: 2381–2399.

Reeburgh, W.S. 1983. Rates of biogeochemical processes in anoxic sediments. Annual Review of Earth and Planetary Sciences 11: 269–298.

Reeburgh, W.S., S.C. Whalen, and M.J. Alperin. 1993. The role of microbially-mediated oxidation in the global CH_4 budget. pp. 1–14. In J.C. Murrell and D.P. Kelley (eds.), *Microbiology of C_1 Compounds*. Intercept, Andover, U.K.

Rees, C.E., W.J. Jenkins, and J. Monster. 1978. The sulphur isotopic composition of ocean water sulphate. Geochimica et Cosmochimica Acta 42: 377–381.

Reeves, H. 1994. On the origin of the light elements ($Z < 6$). Reviews of Modern Physics 66: 193–216.

Regnell, O. 1994. The effect of pH and dissolved oxygen levels on methylation and partitioning of mercury in freshwater model systems. Environmental Pollution 84: 7–13.

Reheis, M.C. and R. Kihl. 1995. Dust deposition in southern Nevada and California, 1984–1989: Relations to climate, source area, and source lithology. Journal of Geophysical Research 100: 8893–8918.

Reich, P.B. and A.W. Schoettle. 1988. Role of phosphorus and nitrogen in photosynthetic and whole plant carbon gain and nutrient use efficiency in eastern white pine. Oecologia 77: 25–33.

Reich, P.B., M.B. Walters, and D.S. Ellsworth. 1992. Leaf life-span in relation to leaf, plant, and stand characteristics among diverse ecosystems. Ecological Monographs 62: 365–392.

Reich, P.B., M.B. Walters, D.S. Ellsworth, and C. Uhl. 1994. Photosynthesis-nitrogen relations in Amazonian tree species. I. Patterns among species and communities. Oecologia 97: 62–72.

Reich, P.B., B.D. Kloeppel, D.S. Ellsworth, and M.B. Walters. 1995. Different photosynthesis-nitrogen relations in deciduous hardwood and evergreen coniferous tree species. Oecologia 104: 24–30.

Reichle, D.E., B.E. Dinger, N.T. Edwards, W.F. Harris, and P. Sollins. 1973a. Carbon flow and storage in a forest ecosystem. pp. 345–365. In G.M. Woodwell and E.V. Pecan (eds.), *Carbon and the Biosphere.* CONF 720510. National Technical Information Service, Washington, D.C.

Reichle, D.E., R.A. Goldstein, R.I. Van Hook, and C.J. Dodson. 1973b. Analysis of insect consumption in a forest canopy. Ecology 54: 1076–1084.

Reiners, W.A. 1972. Structure and energetics of three Minnesota forests. Ecological Monographs 42: 71–94.

Reiners, W.A. 1986. Complementary models for ecosystems. American Naturalist 127: 59–73.

Reiners, W.A. 1992. Twenty years of ecosystem reorganization following experimental deforestation and regrowth suppression. Ecological Monographs 62: 505–523.

Reiners, W.A. and D.O. Anderson. 1968. CO_2 concentrations in forests along a topographic gradient. American Midland Naturalist 80: 111–117.

Reiners, W.A., L.L. Strong, P.A. Matson, I.C. Burke, and D.S. Ojima. 1989. Estimating biogeochemical fluxes across sagebrush-steppe landscapes with thematic mapper imagery. Remote Sensing of Environment 28: 121–129.

Remde, A. and R. Conrad. 1991. Role of nitrification and denitrification for NO metabolism in soil. Biogeochemistry 12: 189–205.

Renberg, I. and M. Wik. 1984. Dating recent lake sediments by soot particle counting. VerhandlungenInternationale Vereinigung fuer Theoretische und Angewandte Limnologie 22: 712–718.

Renberg, I., T. Korsman, and H.J.B. Birks. 1993. Prehistoric increases in the pH of acid-sensitive Swedish lakes caused by land-use changes. Nature 362: 824–827.

Rennenberg, H. 1991. The significance of higher plants in the emission of sulfur compounds from terrestrial ecosystems. pp. 217–260. In T.D. Sharkey, E.A. Holland, and H.A. Mooney (eds.), *Trace Gas Emissions by Plants.* Academic Press, Orlando, Florida.

Reuss, J.O. 1980. Simulation of soil nutrient losses resulting from rainfall acidity. Ecological Modelling 11: 15–38.

Reuss, J.O. and D.W. Johnson. 1986. *Acid Deposition and the Acidification of Soils and Waters.* Springer-Verlag, New York.

Reuss, J.O., B.J. Cosby, and R.F. Wright. 1987. Chemical processes governing soil and water acidification. Nature 329: 27–32.

Reuss, R.W. and S.W. Seagle. 1994. Landscape patterns in soil microbial processes in the Serengeti National Park, Tanzania. Ecology 75: 892–904.

Revelle, R.R. 1983. Methane hydrates in continental slope sediments and increasing atmospheric carbon dioxide. pp. 252–261. In *Changing Climate.* U.S. National Academy Press, Washington, D.C.

Reynolds, R.C. 1978. Polyphenol inhibition of calcite precipitation in Lake Powell. Limnology and Oceanography 23: 585–597.

Rice, E.L. and S.K. Pancholy. 1972. Inhibition of nitrification by climax ecosystems. American Journal of Botany 59: 1033–1040.

Rich, P.H. and R.G. Wetzel. 1978. Detritus in the lake ecosystem. American Naturalist 112: 57–71.

Richards, S.R., C.A. Kelly, and J.W.M. Rudd. 1991. Organic volatile sulfur in lakes of the Canadian Shield and its loss to the atmosphere. Limnology and Oceanography 36: 468–482.

Richardson, C.J. 1985. Mechanisms controlling phosphorus retention capacity in freshwater wetlands. Science 228: 1424–1427.

Richey, J.E. and R.L. Victoria. 1993. C, N, and P export dynamics in the Amazon river. pp. 123–139. In R. Wollast, F.T. MacKenzie, and L. Chou (eds.), *Interactions of C, N, P and S in Biogeochemical Cycles.* Springer-Verlag, Berlin.

Richey, J.E., A.H. Devol, S.C. Wofsy, R. Victoria, and M.N.G. Riberio. 1988. Biogenic gases and the oxidation and reduction of carbon in Amazon River and floodplain waters. Limnology and Oceanography 33: 551–561.

Richter, D.D. and L.I. Babbar. 1991. Soil diversity in the tropics. Advances in Ecological Research 21: 315–389.

Richter, D.D. and D. Markewitz. 1995. How deep is soil? BioScience 45: 600–609.

Richter, D.D., C.W. Ralston, and W.R. Harms. 1982. Prescribed fire: Effects on water quality and forest nutrient cycling. Science 215: 661–663.

Richter, D.D., D. Markewitz, C.G. Wells, H.L. Allen, R. April, P.R. Heine, and B. Urrego. 1994. Soil chemical change during three decades in an old-field loblolly pine (*Pinus taeda* L.) ecosystem. Ecology 75: 1463–1473.

Richter, F.M., D.B. Rowley, and D.J. DePaolo. 1992. Sr isotope evolution of seawater: The role of tectonics. Earth and Planetary Science Letters 109: 11–23.

Ridley, B.A., J.E. Dye, J.G. Walega, J. Zheng, F.E. Grahek, and W. Rison. 1996. On the production of active nitrogen by thunderstorms over New Mexico. Journal of Geophysical Research 101: 20985–21005.

Ridley, W.P., L.J. Dizikes, and J.M. Wood. 1977. Biomethylation of toxic elements in the environment. Science 197: 329–332.

Riemann, B. and R.T. Bell. 1990. Advances in estimating bacterial biomass and growth in aquatic systems. Archiv für Hydrobiologie 118: 385–402.

Riggan, P.J., R.N. Lockwood, and E.N. Lopez. 1985. Deposition and processing of airborne nitrogen pollutants in Mediterranean-type ecosystems of southern California. Environmental Science and Technology 19: 781–789.

Rind, D., R. Goldberg, J. Hansen, C. Rosenzweig, and R. Ruedy. 1990. Potential evapotranspiration and the likelihood of future drought. Journal of Geophysical Research 95: 9983–10004.

Rind, D., E.-W. Chiou, W. Chu, J. Larsen, S. Oltmans, J. Lerner, M.P. McCormick, and L. McMaster. 1991. Positive water vapour feedback in climate models confirmed by satellite data. Nature 349: 500–503.

Rinsland, C.P., R. Zander, E. Mahieu, P. Demoulin, A. Goldman, D.H. Ehhalt, and J. Rudolph. 1992. Ground-based infrared measurements of carbonyl sulfide total column abundances: Long-term trends and variability. Journal of Geophysical Research 97: 5995–6002.

Risley, L.S. and D.A. Crossley. 1988. Herbivore-caused greenfall in the southern Appalachians. Ecology 69: 1118–1127.

Roberts, T.L., J.W.B. Stewart, and J.R. Bettany. 1985. The influence of topography on the distribution of organic and inorganic soil phosphorus across a narrow environmental gradient. Canadian Journal of Soil Science 65: 651–665.

Robertson, D.M., R.A. Ragotzkie, and J.J. Magnuson. 1992. Lake ice records used to detect historical and future climatic changes. Climatic Change 21: 407–427.

Robertson, G.P. 1982a. Nitrification in forested ecosystems. Philosophical Transactions of the Royal Society of London 296B: 445–457.

Robertson, G.P. 1982b. Factors regulating nitrification in primary and secondary succession. Ecology 63: 1561–1573.

Robertson, G.P. 1984. Nitrification and nitrogen mineralization in a lowland rainforest succession in Costa Rica, central America. Oecologia 61: 99–104.

Robertson, G.P. and T. Rosswall. 1986. Nitrogen in West Africa: the regional cycle. Ecological Monographs 56: 43–72.

Robertson, G.P. and J.M. Tiedje. 1984. Denitrification and nitrous oxide production in successional and old-growth Michigan forests. Soil Science Society of America Journal 48: 383–389.

Robertson, G.P. and J.M. Tiedje. 1988. Deforestation alters denitrification in a lowland tropical rain forest. Nature 336: 756–759.

Robertson, G.P. and P.M. Vitousek. 1981. Nitrification potentials in primary and secondary succession. Ecology 62: 376–386.

Robertson, G.P., M.A. Huston, F.C. Evans, and J.M Tiedje. 1988. Spatial variability in a successional plant community: Patterns of nitrogen availability. Ecology 69: 1517–1524.

Robertson, J.E. and A.J. Watson. 1992. Thermal skin effect of the surface ocean and its implications for CO_2 uptake. Nature 358: 738–740.

Robertson, J.E., C. Robinson, D.R. Turner, P. Holligan, A.J. Watson, P. Boyd, E. Fernandez, and M. Finch. 1994. The impact of coccolithophore bloom on oceanic carbon uptake in the northeast Atlantic during summer 1991. Deep Sea Research 41: 297–314.

Robertson, M.P. and S.L. Miller. 1995. An efficient prebiotic synthesis of cytosine and uracil. Nature 375: 772–774.

Robertson, W.D. and S.L. Schiff. 1994. Fractionation of sulphur isotopes during biogenic sulphate reduction below a sandy forested recharge area in south-central Canada. Journal of Hydrology 158: 123–134.

Robinson, D. 1986. Limits to nutrient inflow rates in roots and root systems. Physiologia Plantarum 68: 551–559.

Robinson, D. 1994. The responses of plants to non-uniform supplies of nutrients. New Phytologist 127: 635–674.

Rodhe, H., P. Crutzen, and A. Vanderpol. 1981. Formation of sulfuric and nitric acid in the atmosphere during long-range transport. Tellus 33: 132–141.

Rogers, H.H., G.B. Runion, and S.V. Krupa. 1994. Plant responses to atmospheric CO_2 enrichment with emphasis on roots and the rhizosphere. Environmental Pollution 83: 155–189.

Romanek, C.S., M.M. Grady, I.P. Wright, D.W. Mittlefehldt, R.A. Socki, C.T. Pillinger, and E.K. Gibson. 1994. Record of fluid-rock interactions on Mars from the meteorite ALH84001. Nature 372: 655–657.

Romell, L.G. 1935. Ecological problems of the humus layer in the forest. Cornell University, Agricultural Experiment Station Memoir 170, Ithaca, New York.

Rondón, A. and L. Granat. 1994. Studies on the dry deposition of NO_2 to coniferous species at low NO_2 concentrations. Tellus 46B: 339–352.

Ronen, D., M. Magaritz, and E. Almon. 1988. Contaminated aquifers are a forgotten component of the global N_2O budget. Nature 335: 57–59.

Rose, S.L. and C.T. Youngberg. 1981. Tripartite associations in snowbrush (*Ceanothus velutinus*): Effect of vesicular-arbuscular mycorrhizae on growth, nodulation, and nitrogen fixation. Canadian Journal of Botany 59: 34–39.

Rosen, M.R., J.V. Turner, L. Coshell, and V. Gailitis. 1995. The effects of water temperature, stratification, and biological activity on the stable isotope composition and timing of carbonate precipitation in a hypersaline lake. Geochimica et Cosmochimica Acta 59: 979–990.

Rosenzweig, C. and M.L. Parry. 1994. Potential impact of climate change on world food supply. Nature 367: 133–138.

Rosenzweig, M.L. 1968. Net primary productivity of terrestrial communities: Prediction from climatological data. American Naturalist 102: 67–74.

Roskoski, J.P. 1980. Nitrogen fixation in hardwood forests of the northeastern United States. Plant and Soil 54: 33–44.

Ross, D.S. and R.J. Bartlett. 1996. Field-extracted Spodosol solutions and soils: Aluminum, organic carbon, and pH interrelationships. Soil Science Society of America Journal 60: 589–595.

Ross, J.E. and L.H. Aller. 1976. The chemical composition of the Sun. Science 191: 1223–1229.

Rosswall, T. 1982. Microbiological regulation of the biogeochemical nitrogen cycle. Plant and Soil 67: 15–34.

Rotty, R.M. and C.D. Masters. 1985. Carbon dioxide from fossil fuel combustion: Trends, resources, and technological implications. pp. 63–80. In J.R. Trabalka (ed.), *Atmospheric Carbon Dioxide and the Global Carbon Cycle.* DOE/Er-0239. U.S. Department of Energy, Washington, D.C.

Rouhier, H., G. Billès, A. El Kohen, M. Mousseau, and P. Bottner. 1994. Effect of elevated CO_2 on carbon and nitrogen distribution within a tree (*Castanea sativa* Mill.)-soil system. Plant and Soil 162: 281–292.

Roulet, N.T., R. Ash, and T.R. Moore. 1992. Low boreal wetlands as a source of atmospheric methane. Journal of Geophysical Research 97: 3739–3749.

Routson, R.C., R.E. Wildung, and T.R. Garland. 1977. Mineral weathering in an arid watershed containing soil developed from mixed basaltic-felsic parent materials. Soil Science 124: 303–308.

Rowland, F.S. 1989. Chlorofluorocarbons and the depletion of stratospheric ozone. American Scientist 77: 36–45.

Rowland, F.S. 1991. Stratospheric ozone depletion. Annual Review of Physical Chemistry 42: 731–768.

Rudaz, A.O., E.A. Davidson, and M.K. Firestone. 1991. Sources of nitrous oxide production following wetting of dry soil. FEMS Microbiology Ecology 85: 117–124.

Rudd, J.W.M. and C.D. Taylor. 1980. Methane cycling in aquatic environments. Advances in Aquatic Microbiology 2: 77–150.

Rudd, J.W.M., C.A. Kelly, and A. Furutani. 1986a. The role of sulfate reduction in long term accumulation of organic and inorganic sulfur in lake sediments. Limnology and Oceanography 31: 1281–1291.

Rudd, J.W.M., C.A. Kelly, V. St. Louis, R.H. Hesslein, A. Furutani, and M.H. Holoka. 1986b. Microbial consumption of nitric and sulfuric acids in acidified north temperate lakes. Limnology and Oceanography 31: 1267–1280.

Rudd, J.W.M., C.A. Kelly, D.W. Schindler, and M.A. Turner. 1988. Disruption of the nitrogen cycle in acidified lakes. Science 240: 1515–1517.

Rudolph, J., A. Khedim, R. Koppmann, and B. Bonsang. 1995. Field study of the emissions of methyl chloride and other halocarbons from biomass burning in western Africa. Journal of Atmospheric Chemistry 22: 67–80.

Ruhe, R.V. 1984. Soil-climate system across the prairies in midwestern U.S.A. Geoderma 34: 201–219.

Running, S.W., R.R. Nemani, D.L. Peterson, L.E. Band, D.F. Potts, L.L. Pierce, and M.A. Spanner. 1989. Mapping regional forest evapotranspiration and photosynthesis by coupling satellite data with ecosystem simulation. Ecology 70: 1090–1101.

Runyon, J., R.H. Waring, S.N. Goward, and J.M. Welles. 1994. Environmental limits on net primary production and light-use efficiency across the Oregon transect. Ecological Applications 4: 226–237.

Russel, J.C. 1929. Organic matter problems under dry farming conditions. Journal of the American Society of Agronomy 21: 960–969.

Russell, G.L. and J.R. Miller. 1990. Global river runoff calculated from a global atmospheric general circulation model. Journal of Hydrology 117: 241–254.

Russell, J.M., M. Luo, R.J. Cicerone, and L.E. Deaver. 1996. Satellite confirmation of the dominance of chlorofluorocarbons in the global stratospheric chlorine budget. Nature 379: 526–529.

Rustad, L.E. 1994. Element dynamics along a decay continuum in a red spruce ecosystem in Maine, USA. Ecology 75: 867–879.

Ruttenberg, K.C. 1993. Reassessment of the oceanic residence time of phosphorus. Chemical Geology 107: 405–409.

Ruttenberg, K.C. and R.A. Berner. 1993. Authigenic apatite formation and burial in sediments from non-upwelling, continental margin environments. Geochimica et Cosmochimica Acta 57: 991–1007.

Ryan, D.F. and F.H. Bormann. 1982. Nutrient resorption in northern hardwood forests. BioScience 32: 29–32.

Ryan, M.G. 1991. Effects of climate change on plant respiration. Ecological Applications 1: 157–167.

Ryan, M.G. 1995. Foliar maintenance respiration of subalpine and boreal trees and shrubs in relation to nitrogen content. Plant, Cell and Environment 18: 765–772.

Ryan, M.G., R.M. Hubbard, D.A. Clark, and R.L. Sanford. 1994. Woody-tissue respiration for *Simarouba amara* and *Minquartia guianensis,* two tropical wet forest trees with different growth habits. Oecologia 100: 213–220.

Ryan, M.G., S.T. Gower, R.M. Hubbard, R.H. Waring, H.L. Gholz, W.P. Cropper, and S.W. Running. 1995. Woody tissue maintenance respiration of four conifers in contrasting climates. Oecologia 101: 133–140.

Ryan, P.F., G.M. Hornberger, B.J. Cosby, J.N. Galloway, J.R. Webb, and E.B. Rastetter. 1989. Changes in the chemical composition of stream water in two catchments in the Shenandoah National Park, Virginia, in response to atmospheric deposition of sulfur. Water Resources Research 25: 2091–2099.

Rychert, R., J. Skujiņs, D. Sorensen, and D. Porcella. 1978. Nitrogen fixation by lichens and free-living microorganisms in deserts. pp. 20–30. In N.E. West and J. Skujins (eds.), *Nitrogen in Desert Ecosystems.* Dowden, Hutchinson and Ross, Stroudsburg, Pennsylvania.

Rycroft, D.W., D.J.A. Williams, and H.A.P. Ingram. 1975. The transmission of water through peat. I. Review. Journal of Ecology 63: 535–556.

Ryden, J.C. 1981. N_2O exchange between a grassland soil and the atmosphere. Nature 292: 235–237.

Rye, R., P.H. Kuo, and H.D. Holland. 1995. Atmospheric carbon dioxide concentrations before 2.2 billion years ago. Nature 378: 603–605.

Rygiewicz, P.T. and C.P. Andersen. 1994. Mycorrhizae alter quality and quantity of carbon allocated below ground. Nature 369: 58–60.

Rysgaard, S., N. Risgaard-Peterson, N.P. Sloth, K. Jensen, and L.P. Nielsen. 1994. Oxygen regulation of nitrification and denitrification in sediments. Limnology and Oceanography 39: 1643–1652.

Saá, A., M.C. Trasar-Cepeda, F. Gil-Sotres, and T. Carballas. 1993. Changes in soil phosphorus and acid phosphatase activity immediately following fires. Soil Biology and Biochemistry 25: 1223–1230.

Saá, A., M.C. Trasar-Cepeda, B. Soto, F. Gil-Sotres, and F. Díaz-Fierros. 1994. Forms of phosphorus in sediments eroded from burnt soils. Journal of Environmental Quality 23: 739–746.

Sagan, C., W.R. Thompson, R. Carlson, D. Gurnett, and C. Hord. 1993. A search for life on Earth from the Galileo spacecraft. Nature 365: 715–721.

Saggar, S., J.R. Bettany, and J.W.B. Stewart. 1981. Sulfur transformations in relation to carbon and nitrogen in incubated soils. Soil Biology and Biochemistry 13: 499–511.

Sahagian, D.L., F.W. Schwartz, and D.K. Jacobs. 1994. Direct anthropogenic contributions to sea level rise in the twentieth century. Nature 367: 54–57.

Saini, H.S., J.M. Attieh, and A.D. Hanson. 1995. Biosynthesis of halomethanes and methanethiol by higher plants via a novel methyltransferase reaction. Plant, Cell and Environment 18: 1027–1033.

Sala, O.E., W.J. Parton, L.A. Joyce, and W.K. Lauenroth. 1988. Primary production of the central grassland region of the United States. Ecology 69: 40–45.

Salati, E. and P.B. Vose. 1984. Amazon Basin: A system in equilibrium. Science 225: 129–138.

Sambrotto, R.N., G. Savidge, C. Robinson, P. Boyd, T. Takahashi, D.M. Karl, C. Langdon, D. Chipman, J. Marrs, and L. Codispoti. 1993. Elevated consumption of carbon relative to nitrogen in the surface ocean. Nature 363: 248–250.

Sanborn, P.T. and T.M. Ballard. 1991. Combustion losses of sulphur from conifer foliage; Implications of chemical form and soil nitrogen status. Biogeochemistry 12: 129–134.

Sanchez, P.A., D.E. Bandy, J.H. Villachica, and J.J. Nicholaides. 1982a. Amazon Basin soils: Management for continuous crop production. Science 216: 821–827.

Sanchez, P.A., M.P. Gichuru, and L.B. Katz. 1982b. Organic matter in major soils of the tropical and temperate regions. International Congress of Soil Science 12: 99–114.

Sanford, R.L., J. Saldarriaga, K.E. Clark, C. Uhl, and R. Herrera. 1985. Amazon rain-forest fires. Science 227: 53–55.

Sanhueza, E., L. Cárdenas, L. Donoso, and M. Santana. 1994. Effects of plowing on CO_2, CO, CH_4, N_2O, and NO fluxes from tropical savannah soils. Journal of Geophysical Research 99: 16429–16434.

Santer, B.D., K.E. Taylor, T.M.L. Wigley, T.C. Johns, P.D. Jones, D.J. Karoly, J.F.B. Mitchell, A.H. Oort, J.E. Penner, V. Ramaswamy, M.D. Schwarzkopf, R.J. Stouffer, and S. Tett. 1996. A search for human influences on the thermal structure of the atmosphere. Nature 382: 39–46.

Santos, P.F., N.Z. Elkins, Y. Steinberger, and W.G. Whitford. 1984. A comparison of surface and buried *Larrea tridentata* leaf litter decomposition in North American hot deserts. Ecology 65: 278–284.

Sanyal, A., N.G. Hemming, G.N. Hanson, and W.S. Broecker. 1995. Evidence for a higher pH in the glacial ocean from boron isotopes in foraminifera. Nature 373: 234–236.

Sarmiento, J.L. and E.T. Sundquist. 1992. Revised budget for the oceanic uptake of anthropogenic carbon dioxide. Nature 356: 589–593.

Sass, R.L., F.M. Fisher, S.T. Lewis, N.F. Jund, and F.T. Turner. 1994. Methane emissions from rice fields: Effect of soil properties. Global Biogeochemical Cycles 8: 135–140.

Sathyendranath, S., T. Platt, E.P.W. Horne, W.G. Harrison, O. Ulloa, R. Outerbridge, and N. Hoepffner. 1991. Estimation of new production in the ocean by compound remote sensing. Nature 353: 129–133.

Saunders, J.F. and W.M. Lewis. 1988. Transport of phosphorus, nitrogen, and carbon by the Apure River, Venezuela. Biogeochemistry 5: 323–342.

Savoie, D.L. and J.M. Prospero. 1989. Comparison of oceanic and continental sources of non-sea-salt sulphate over the Pacific ocean. Nature 339: 685–687.

Savoie, D.L., J.M. Prospero, and R.T. Nees. 1987. Nitrate, non-sea-salt sulfate, and mineral aerosol over the northwestern Indian ocean. Journal of Geophysical Research 92: 933–942.

Savoie, D.L., J.M. Prospero, R.J. Larsen, F. Huang, M.A. Izaguirre, T. Huang, T.H. Snowdon, L. Custals, and C.G. Sanderson. 1993. Nitrogen and sulfur species in

Antarctic aerosols at Mawson, Palmer Station, and Marsh (King George Island). Journal of Atmospheric Chemistry 17: 95–122.

Saxton, K.E., W.J. Rawls, J.S. Romberger, and R.I. Papendick. 1986. Estimating generalized soil-water characteristics from texture. Soil Science Society of America Journal 50: 1031–1036.

Sayles, F.L. 1981. The composition and diagenesis of interstitial solutions. II: Fluxes and diagenesis at the water-sediment interface in the high latitude North and South Atlantic. Geochimica et Cosmochimica Acta 45: 1061–1086.

Sayles, F.L. and P.C. Mangelsdorf. 1977. The equilibration of clay minerals with seawater: Exchange reactions. Geochimica et Cosmochimica Acta 41: 951–960.

Sayles, F.L., W.R. Martin, and W.G. Deuser. 1994. Response of benthic oxygen demand to particulate organic carbon supply in the deep sea near Bermuda. Nature 371: 686–689.

Schaefer, D.A. and W.A. Reiners. 1989. Throughfall chemistry and canopy processing mechanisms. pp. 241–284. In S.E. Lindberg, A.L. Page, and S.A. Norton (eds.), *Acid Precipitation.* Vol. 3. *Sources, Deposition and Canopy Interactions.* Springer-Verlag, New York.

Schaefer, D.A. and W.G. Whitford. 1981. Nutrient cycling by the subterranean termite *Gnathamitermes tubiformans* in a Chihuahuan desert ecosystem. Oecologia 48: 277–283.

Schaefer, D.A., Y. Steinberger, and W.G. Whitford. 1985. The failure of nitrogen and lignin control of decomposition in a North American desert. Oecologia 65: 382–386.

Schaefer, M. 1990. The soil fauna of a beech forest on limestone: Trophic structure and energy budget. Oecologia 82: 128–136.

Schelske, C.L. 1985. Biogeochemical silica mass balances in Lake Michigan and Lake Superior. Biogeochemistry 1: 197–218.

Schelske, C.L. 1988. Historic trends in Lake Michigan silica concentrations. Internationale Revue der Gesamten Hydrobiologie 73: 559–591.

Schelske, C.L., E.F. Stoermer, D.J. Conley, J.A. Robbins, and R.M. Glover. 1983. Early eutrophication in the lower Great Lakes: New evidence from biogenic silica in sediments. Science 222: 320–322.

Schelske, C.L., J.A. Robbins, W.S. Gardner, D.J. Conley, and R.A. Bourbonniere. 1988. Sediment record of biogeochemical responses to anthropogenic perturbations of nutrient cycles in Lake Ontario. Canadian Journal of Fisheries and Aquatic Sciences 45: 1291–1303.

Schidlowski, M. 1980. The atmosphere. pp. 1–16. In O. Hutzinger (ed.). *The Handbook of Environmental Chemistry.* Vol. 1, Part A. *The Natural Environment and the Biogeochemical Cycles.* Springer-Verlag, New York.

Schidlowski, M. 1983. Evolution of photoautotrophy and early atmospheric oxygen levels. Precambrian Research 20: 319–335.

Schidlowski, M. 1988. A 3,800–million-year isotopic record of life from carbon in sedimentary rocks. Nature 333: 313–318.

Schidlowski, M., J.M. Hayes, and I.R. Kaplan. 1983. Isotopic inferences of ancient biochemistries: Carbon, sulfur, hydrogen, and nitrogen. pp. 149–186. In J.W. Schopf (ed.), *Earth's Earliest Biosphere.* Princeton University Press, Princeton, New Jersey.

Schiffman, P.M. and W.C. Johnson. 1989. Phytomass and detrital carbon storage during forest regrowth in the southeastern United States piedmont. Canadian Journal of Forest Research 19: 69–78.

Schimel, D.S. 1995. Terrestrial ecosystems and the carbon cycle. Global Change Biology 1: 77–91.

Schimel, D.S., M.A. Stillwell, and R.G. Woodmansee. 1985. Biogeochemistry of C, N, and P in a soil catena of the shortgrass steppe. Ecology 66: 276–282.

Schimel, D.S., B.H. Braswell, E.A. Holland, R. McKeown, D.S. Ojima, T.H. Painter, W.J. Parton, and A.R. Townsend. 1994. Climatic, edaphic, and biotic controls over storage and turnover of carbon in soils. Global Biogeochemical Cycles 8: 279–293.

Schimel, D.S., I.G. Enting, M. Heimann, T.M.L. Wigley, D. Raynaud, D. Alves, and U. Siegenthaler. 1995. CO_2 and the carbon cycle. pp. 35–71. In J.T. Houghton, L.G. Meira Filho, J. Bruce, H. Lee, B.A. Callander, E. Haites, N. Harris, and K. Maskell (eds.), *Climate Change 1994*. Cambridge University Press, Cambridge.

Schimel, J.P. and M.K. Firestone. 1989. Nitrogen incorporation and flow through a coniferous forest soil profile. Soil Science Society of America Journal 53: 779–784.

Schimel, J.P., M.K. Firestone, and K.S. Killham. 1984. Identification of heterotrophic nitrification in a Sierran forest soil. Applied and Environmental Microbiology 48: 802–806.

Schindler, D.W. 1974. Eutrophication and recovery in experimental lakes: Implications for lake management. Science 184: 897–899.

Schindler, D.W. 1977. Evolution of phosphorus limitation in lakes. Science 195: 260–262.

Schindler, D.W. 1978. Factors regulating phytoplankton production and standing crop in the world's freshwaters. Limnology and Oceanography 23: 478–486.

Schindler, D.W., G.J. Brunskill, S. Emerson, W.S. Broecker, and T.-H. Peng. 1972. Atmospheric carbon dioxide: Its role in maintaining phytoplankton standing crops. Science 177: 1192–1194.

Schindler, D.W., M.A. Turner, M.P. Stainton, and G.A. Linsey. 1986. Natural sources of acid neutralizing capacity in low alkalinity lakes of the Precambrian shield. Science 232: 844–847.

Schindler, D.W., K.G. Beaty, E.J. Fee, D.R. Cruikshank, E.R. DeBruyn, D.L. Findlay, G.A. Linsey, J.A. Shearer, M.P. Stainton, and M.A. Turner. 1990. Effects of climatic warming on lakes of the central boreal forest. Science 250: 967–970.

Schindler, S.C., M.J. Mitchell, T.J. Scott, R.D. Fuller, and C.T. Driscoll. 1986. Incorporation of ^{35}S-sulfate into inorganic and organic constituents of two forest soils. Soil Science Society of America Journal 50: 457–462.

Schipper, L.A., A.B. Cooper, C.G. Harfoot, and W.J. Dyck. 1993. Regulators of denitrification in an organic riparian soil. Soil Biology and Biochemistry 25: 925–933.

Schleser, G.H. 1982. The response of CO_2 evolution from soils to global temperature changes. Zeitschrift fuer Naturforschung 37*a*: 287–291.

Schlesinger, W.H. 1977. Carbon balance in terrestrial detritus. Annual Review of Ecology and Systematics 8: 51–81.

Schlesinger, W.H. 1978. Community structure, dynamics and nutrient cycling in the Okefenokee cypress swamp-forest. Ecological Monographs 48: 43–65.

Schlesinger, W.H. 1982. Carbon storage in the caliche of arid soils: A case study from Arizona. Soil Science 133: 247–255.

Schlesinger, W.H. 1984. Soil organic matter: A source of atmospheric CO_2. pp. 111–127. In G.M. Woodwell (ed.), *The Role of Terrestrial Vegetation in the Global Carbon Cycle.* Wiley, New York.

Schlesinger, W.H. 1985a. Decomposition of chaparral shrub foliage. Ecology 66: 1353–1359.

Schlesinger, W.H. 1985b. The formation of caliche in soils of the Mojave Desert, California. Geochimica et Cosmochimica Acta 49: 57–66.

Schlesinger, W.H. 1986. Changes in soil carbon storage and associated properties with disturbance and recovery. pp. 194–220. In J.R. Trabalka and D.E. Reichle (eds.), *The Changing Carbon Cycle: A Global Analysis.* Springer-Verlag, New York

Schlesinger, W.H. 1989. Discussion: Ecosystem structure and function. pp. 268–274. In J. Roughgarden, R.M. May, and S.A. Levin (eds.), *Perspectives in Ecological Theory.* Princeton University Press, Princeton, New Jersey.

Schlesinger, W.H. 1990. Evidence from chronosequence studies for a low carbon-storage potential of soils. Nature 348: 232–234.

Schlesinger, W.H. 1994. The vulnerability of biotic diversity. pp. 245–260. In R. Socolow, C. Andrews, F. Berkhout, and V. Thomas (eds.), *Industrial Ecology and Global Change.* Cambridge University Press, Cambridge.

Schlesinger, W.H. and N. Gramenopoulos. 1996. Archival photographs show no climate-induced changes in woody vegetation in the Sudan, 1943–1994. Global Change Biology 2: 137–141.

Schlesinger, W.H. and A.E. Hartley. 1992. A global budget for atmospheric NH_3. Biogeochemistry 15: 191–211.

Schlesinger, W.H. and P.L. Marks. 1977. Mineral cycling and the niche of Spanish moss, *Tillandsia usneoides* L. American Journal of Botany 64: 1254–1262.

Schlesinger, W.H. and J.M. Melack. 1981. Transport of organic carbon in the world's rivers. Tellus 33: 172–187.

Schlesinger, W.H. and W.T. Peterjohn. 1988. Ion and sulfate-isotope ratios in arid soils subject to wind erosion in the southwestern USA. Soil Science Society of America Journal 52: 54–58.

Schlesinger, W.H. and W.T. Peterjohn. 1991. Processes controlling ammonia volatilization from Chihuahuan desert soils. Soil Biology and Biochemistry 23: 637–642.

Schlesinger, W.H., J.T. Gray, and F.S. Gilliam. 1982. Atmospheric deposition processes and their importance as sources of nutrients in a chaparral ecosystem of southern California. Water Resources Research 18: 623–629.

Schlesinger, W.H., P.J. Fonteyn, and G.M. Marion. 1987. Soil moisture content and plant transpiration in the Chihuahuan desert of New Mexico. Journal of Arid Environments 12: 119–126.

Schlesinger, W.H., E.H. DeLucia, and W.D. Billings. 1989. Nutrient-use efficiency of woody plants on contrasting soils in the western Great Basin, Nevada. Ecology 70: 105–113.

Schlesinger, W.H., J.F. Reynolds, G.L. Cunningham, L.F. Huenneke, W.M. Jarrell, R.A. Virginia, and W.G. Whitford. 1990. Biological feedbacks in global desertification. Science 247: 1043–1048.

Schlesinger, W.H., J.A. Raikes, A.E. Hartley, and A.F. Cross. 1996. On the spatial pattern of soil nutrients in desert ecosystems. Ecology 77: 364–374.

Schmitz, W.J. 1995. On the interbasin-scale thermohaline circulation. Reviews of Geophysics 33: 151–173.

Schneider, S.H. 1994. Detecting climatic change signals: Are there any "fingerprints"? Science 263: 341–347.

Schneider, S.H. and P.J. Boston (eds.). 1991. *Scientists on Gaia.* MIT Press, Cambridge, Massachusetts.

Schnell, S. and G.M. King. 1994. Mechanistic analysis of ammonium inhibition of atmospheric methane consumption in forest soils. Applied and Environmental Microbiology 60: 3514–3521.

Schnitzer, M. and H.R. Schulten. 1992. The analysis of soil organic matter by pyrolysis-field ionization mass spectrometry. Soil Science Society of America Journal 56: 1811–1817.

Schoenau, J.J. and J.R. Bettany. 1987. Organic matter leaching as a component of carbon, nitrogen, phosphorus, and sulfur cycles in a forest, grassland, and gleyed soil. Soil Science Society of America Journal 51: 646–651.

Scholes, R.L. and M.C. Scholes. 1995. The effect of land use on nonliving organic matter in the soil. pp. 209–225. In R.G. Zepp and C. Sonntag (eds.), *The Role of Nonliving Organic Matter in the Earth's Carbon Cycle.* Wiley, New York.

Schönheit, P., J.P. Krisjansson, and R.K. Thauer. 1982. Kinetic mechanism for the ability of sulfate reducers to out-compete methanogens for acetate. Archives of Microbiology 132: 285–288.

Schopf, J.W. 1993. Microfossils of the early Archean Apex chert: New evidence of the antiquity of life. Science 260: 640–646.

Schuffert, J.D., R.A. Jahnke, M. Kastner, J. Leather, A. Sturz, and M.R. Wing. 1994. Rates of formation of modern phosphorite off western Mexico. Geochimica et Cosmochimica Acta 58: 5001–5010.

Schulten, H.-R. and M. Schnitzer. 1993. A state of the art structural concept for humic substances. Naturwissenschaften 80: 29–30.

Schulthess, C.P. and C.P. Huang. 1991. Humic and fulvic acid adsorption by silicon and aluminum oxide surfaces on clay minerals. Soil Science Society of America Journal 55: 34–42.

Schultz, R.C., P.P. Kormanik, W.C. Bryan, and G.H. Brister. 1979. Vesicular-arbuscular mycorrhiza influence growth but not mineral concentrations in seedlings of eight sweetgum familes. Canadian Journal of Forest Research 9: 218–223.

Schulze, E.-D. 1989. Air pollution and forest decline in a spruce (*Picea abies*) forest. Science 244: 776–783.

Schütz, H., R. Conrad, S. Goodwin, and W. Seiler. 1988. Emission of hydrogen from deep and shallow freshwater environments. Biogeochemistry 5: 295–311.

Schütz, H., W. Seiler, and R. Conrad. 1989. Processes involved in formation and emission of methane in rice paddies. Biogeochemsitry 7: 33–53.

Schütz, H., P. Schröder, and H. Rennenberg. 1991. Role of plants in regulating the methane flux to the atmosphere. pp. 29–63. In T.D. Sharkey, E.A. Holland, and H.A. Mooney (eds.), *Trace Gas Emissions by Plants.* Academic Press, San Diego.

Schütz, L. 1980. Long range transport of desert dust with special emphasis on the Sahara. Annals of the New York Academy of Sciences 338: 515–532.

Schwartz, S.E. 1988. Are global cloud albedo and climate controlled by marine phytoplankton? Nature 336: 441–445.

Schwartz, S.E. 1989. Acid deposition: Unraveling a regional phenomenon. Science 243: 753–763.

Schwartzman, D.W. and T. Volk. 1989. Biotic enhancement of weathering and the habitability of Earth. Nature 340: 457–460.

Schwertmann, U. 1966. Inhibitory effect of soil organic matter on the crystallization of amorphous ferric hydroxide. Nature 212: 645–646.

Schwintzer, C.R. 1983. Nonsymbiotic and symbiotic nitrogen fixation in a weakly minerotrophic peatland. American Journal of Botany 70: 1071–1078.

Sclater, F.R., E. Boyle, and J.M. Edmond. 1976. On the marine geochemistry of nickel. Earth and Planetary Science Letters 31: 119–128.

Scott, R.M. 1962. Exchangeable bases of mature, well-drained soils in relation to rainfall in East Africa. Journal of Soil Science 13: 1–9.

Scranton, M.I. and P.G. Brewer. 1977. Occurrence of methane in the near-surface waters of the western subtropical North-Atlantic. Deep Sea Research 24: 127–138.

Scriber, J.M. 1977. Limiting effects of low leaf-water content on the nitrogen utilization, energy budget, and larval growth of *Hyalophora cecropia* (Lepidoptera: Saturniidae). Oecologia 28: 269–287.

Sears, S.O. and D. Langmuir. 1982. Sorption and mineral equilibria controls on moisture chemistry in a c-horizon soil. Journal of Hydrology 56: 287–308.

Seastedt, T.R. 1985. Maximization of primary and secondary productivity by grazers. American Naturalist 126: 559–564.

Seastedt, T.R. 1988. Mass, nitrogen, and phosphorus dynamics in foliage and root detritus of tallgrass prairie. Ecology 69: 59–65.

Seastedt, T.R. and D.A. Crossley. 1980. Effects of microarthropods on the seasonal dynamics of nutrients in forest litter. Soil Biology and Biochemistry 12: 337–342.

Seastedt, T.R. and D.C. Hayes. 1988. Factors influencing nitrogen concentrations in soil water in a North American tallgrass prairie. Soil Biology and Biochemistry 20: 725–729.

Seastedt, T.R. and C.M. Tate. 1981. Decomposition rates and nutrient contents of arthropod remains in forest litter. Ecology 62: 13–19.

Seastedt, T.R., C.C. Coxwell, D.S. Ojima, and W.J. Parton. 1994. Controls of plant and soil carbon in a semihumid temperate grassland. Ecological Applications 4: 344–353.

Sebacher, D.I., R.C. Harriss, and K.B. Bartlett. 1985. Methane emissions to the atmosphere through aquatic plants. Journal of Environmental Quality 14: 40–46.

Sebacher, D.I., R.C. Harriss, K.B. Bartlett, S.M. Sebacher, and S.S. Grice. 1986. Atmospheric methane sources: Alaskan tundra bogs, an alpine fen, and a subarctic boreal marsh. Tellus 38B: 1–10.

Seemann, J.R., T.D. Sharkey, J. Wang, and C.B. Osmond. 1987. Environmental effects on photosynthesis, nitrogen-use efficiency, and metabolic pools in leaves of sun and shade plants. Plant Physiology 84: 796–802.

Sehmel, G.A. 1980. Particle and gas dry deposition: A review. Atmospheric Environment 14: 983–1011.

Seiler, W. and R. Conrad. 1987. Contribution of tropical ecosystems to the global budgets of trace gases, especially CH_4, H_2, CO, and N_2O. pp. 133–162. In R.E. Dickinson (ed.), *The Geophysiology of Amazonia.* Wiley, New York.

Seinfeld, J.H. 1989. Urban air pollution: State of the science. Science 243: 745–752.

Seitzinger, S.P. 1988. Denitrification in freshwater and coastal marine ecosystems: Ecological and geochemical significance. Limnology and Oceanography 33: 702–724.

Seitzinger, S.P. 1991. The effect of pH on the release of phosphorus from Potomac estuary sediments: Implications for blue-green algal blooms. Estuarine, Coastal and Shelf Science 33: 409–418.

Seitzinger, S.P. 1994. Linkages between organic matter mineralization and denitrification in eight riparian wetlands. Biogeochemistry 25: 19–39.

Seitzinger, S., S. Nixon, M.E.Q. Pilson, and S. Burke. 1980. Denitrification and N_2O production in near-shore marine sediments. Geochimica et Cosmochimica Acta 44: 1853–1860.

Seitzinger, S., S.W. Nixon, and M.E.Q. Pilson. 1984. Denitrification and nitrous oxide production in a coastal marine ecosystem. Limnology and Oceanography 29: 73–83.

Seitzinger, S.P., L.P. Nielsen, J. Caffrey, and P.B. Christensen. 1993. Denitrification measurements in aquatic sediments: A comparison of three methods. Biogeochemistry 23: 147–167.

Semtner, A.J. 1995. Modeling ocean circulation. Science 269: 1379–1385.

Sequeira, R. 1993. On the large-scale impact of arid dust on precipitation chemistry of the continental northern hemisphere. Atmospheric Environment 27A: 1553–1565.

Serrasolsas, I. and P.K. Khanna. 1995. Changes in heated and autoclaved forest soils of S.E. Australia. II. Phosphorus and phosphatase activity. Biogeochemistry 29: 25–41.

Servais, P. 1995. Measurement of the incorporation rates of four amino acids into proteins for estimating bacterial production. Microbial Ecology 29: 115–128.

Servant, J. 1986. The burden of the sulphate layer of the stratosphere during volcanic "quiescent" periods. Tellus 38B: 74–79.

Servant, J. 1989. Les sources et les puits d'oxysulfure de carbone (COS) à l'échelle mondiale. Atmospheric Research 23: 105–116.

Servant, J. 1991. Les sources naturelles et artificielles de méthane dans l'atmosphére a l'échelle mondiale. Atmospheric Research 26: 525–542.

Servant, J. 1994. The continental carbon cycle during the last glacial maximum. Atmospheric Research 31: 253–268.

Sexstone, A.J., N.P. Revsbech, T.B. Parkin, and J.M. Tiedje. 1985a. Direct measurement of oxygen profiles and denitrification rates in soil aggregates. Soil Science Society of America Journal 49: 645–651.

Sexstone, A.J., T.B. Parkin, and J.M. Tiedje. 1985b. Temporal response of soil denitrification rates to rainfall and irrigation. Soil Science Society of America Journal 49: 99–105.

Shackleton, N.J. 1977. Carbon-13 in Uvigerina: Tropical rainforest history and the equatorial Pacific carbonate dissolution cycles. pp. 201–277. In N.R. Andersen and A. Malahoff (eds.), *The Fate of Fossil Fuel CO_2 in the Oceans*. Plenum, New York.

Shaffer, G. 1993. Effects of the marine biota on global carbon cycling. pp. 431–455. In M. Heimann (ed.), *The Global Carbon Cycle*. Springer-Verlag, New York.

Shaffer, G. and J.L. Sarmiento. 1995. Biogeochemical cycling in the global ocean. I. A new, analytical model with continuous vertical resolution and high-latitude dynamics. Journal of Geophysical Research 100: 2659–2672.

Shaffer, P.W. and M.R. Church. 1989. Terrestrial and in-lake contributions to alkalinity budgets of drainage lakes: An assessment of regional differences. Canadian Journal of Fisheries and Aquatic Sciences 46: 509–515.

Shanks, A.L. and J.D. Trent. 1979. Marine snow: Microscale nutrient patches. Limnology and Oceanography 24: 850–854.

Shannon, J.D. and D.L. Sisterson. 1992. Estimation of S and NO_x-N deposition budgets for the United States and Canada. Water, Air, and Soil Pollution 63: 211–235.

Shannon, R.D. and J.R. White. 1994. A three-year study of controls on methane emissions from two Michigan peatlands. Biogeochemistry 27: 35–60.

Sharkey, T.D. 1985. Photosynthesis in intact leaves of C_3 plants: Physics, physiology and rate limitations. Botanical Review 51: 53–105.

Sharkey, T.D. 1988. Estimating the rate of photorespiration in leaves. Physiologia Plantarum 73: 147–152.

Sharkey, T.D. and E.L. Singsaas. 1995. Why plants emit isoprene. Nature 374: 769.

Sharpley, A.N. 1985. The selective erosion of plant nutrients in runoff. Soil Science Society of America Journal 49: 1527–1534.

Sharpley, A.N., H. Tiessen, and C.V. Cole. 1987. Soil phosphorus forms extracted by soil tests as a function of pedogenesis. Soil Science Society of America Journal 51: 362–365.

Shaver, G.R. and J.M. Melillo. 1984. Nutrient budgets of marsh plants: Efficiency concepts and relation to availability. Ecology 65: 1491–1510.

Shaver, G.R., F. Chapin, and B.L. Gartner. 1986. Factors limiting seasonal growth and peak biomass accumulation in *Eriophorum vaginatum* in Alaskan tussock tundra. Journal of Ecology 74: 257–278.

Shaw, R.W. 1987. Air pollution by particles. Scientific American 255(8): 96–103.

Shear, C.B., H.L. Crane, and A.T. Myers. 1946. Nutrient-element balance: A fundamental concept in plant nutrition. Proceedings of the American Society for Horticultural Science 47: 239–248.

Shearer, G. and D.H. Kohl. 1988. Natural ^{15}N abundance as a method of estimating the contribution of biologically fixed nitrogen to N_2-fixing systems: Potential for non-legumes. Plant and Soil 110: 317–327.

Shearer, G. and D.H. Kohl. 1989. Estimates of N_2 fixation in ecosystems: The need for and basis of the ^{15}N natural abundance method. pp. 342–374. In P.W. Rundel, J.R. Ehleringer, and K.A. Nagy (eds.), *Stable Isotopes in Ecological Research*. Springer-Verlag, New York.

Shearer, G., D.H. Kohl, R.A. Virginia, B.A. Bryan, J.L. Skeeters, E.T. Nilsen, M.R. Sharifi, and P.W. Rundel. 1983. Estimates of N_2-fixation from variation in the natural abundance of ^{15}N in Sonoran desert ecosystems. Oecologia 56: 365–373.

Shen, S.M., G. Pruden, and D.S. Jenkinson. 1984. Mineralization and immobilization of nitrogen in fumigated soil and the measurement of microbial biomass nitrogen. Soil Biology and Biochemistry 16: 437–444.

Shepherd, M.F., S. Barzetti, and D.R. Hastie. 1991. The production of atmospheric NO_x and N_2O from a fertilized agricultural soil. Atmospheric Environment 25A: 1961–1969.

Shimshock, J.P. and R.G. de Pena. 1989. Below-cloud scavenging of tropospheric ammonia. Tellus 41B: 296–304.

Shneour, E.A. 1966. Oxidation of graphitic carbon in certain soils. Science 151: 991–992.

Shoji, S., M. Nanzyo, Y. Shirato, and T. Ito. 1993. Chemical kinetics of weathering in young Andisols from northeastern Japan using soil age normalized to 10°C. Soil Science 155: 53–60.

Sholkovitz, E.R. 1976. Flocculation of dissolved organic and inorganic matter during the mixing of river water and seawater. Geochimica et Cosmochimica Acta 40: 831–845.

Shorter, J.H., C.E. Kolb, P.M. Crill, R.A. Kerwin, R.W. Talbot, M.E. Hines, and R.C. Harriss. 1995. Rapid degradation of atmospheric methyl bromide in soils. Nature 377: 717–719.

Shortle, W.C. and K.T. Smith. 1988. Aluminium-induced calcium deficiency syndrome in declining red spruce. Science 240: 1017–1018.

Shukla, J. and Y. Mintz. 1982. Influence of land-surface evapotranspiration on the Earth's climate. Science 215: 1498–1501.

Shukla, J., C. Nobre, and P. Sellers. 1990. Amazon deforestation and climate change. Science 247: 1322–1325.

Shuttleworth, W.J. 1988. Evaporation from Amazonian rainforest. Proceedings of the Royal Society of London 233B: 321–346.

Siegenthaler, U. and H. Oeschger. 1987. Biospheric CO_2 emissions during the past 200 years reconstructed by deconvolution of ice core data. Tellus 39B: 140–154.

Siegenthaler, U. and S.L. Sarmiento. 1993. Atmospheric carbon dioxide and the ocean. Nature 365: 119–125.

Siever, R. 1974. The steady state of the Earth's crust, atmosphere, and oceans. Scientific American 230(6): 72–79.

Sigg, A. and A. Neftel. 1991. Evidence for a 50% increase in H_2O_2 over the past 200 years from a Greenland ice core. Nature 351: 557–559.

Sillén, L.G. 1966. Regulation of O_2, N_2 and CO_2 in the atmosphere: Thoughts of a laboratory chemist. Tellus 18: 198–206.

Silver, W.L. 1994. Is nutrient availability related to plant nutrient use in humid tropical forests? Oecologia 98: 336–343.

Silvester, W.B. 1989. Molybdenum limitation of asymbiotic nitrogen fixation in forests of Pacific Northwest America. Soil Biology and Biochemistry 21: 283–289.

Silvester, W.B., P. Sollins, T. Verhoeven, and S.P. Cline. 1982. Nitrogen fixation and acetylene reduction in decaying conifer boles: Effects of incubation time, aeration, and moisture content. Canadian Journal of Forest Research 12: 646–652.

Silvola, J., J. Alm, U. Ahlholm, H. Nykänen, and P.J. Martikainen. 1996. CO_2 fluxes from peat in boreal mires under varying temperature and moisture conditions. Journal of Ecology 84: 219–228.

Simkiss, K. and K.M. Wilbur. 1989. *Biomineralization: Cell Biology and Mineral Deposition.* Academic Press, San Diego.

Simon, L., J. Bousquet, R.C. Lévesque, and M. Lalonde. 1993. Origin and diversification of endomycorrhizal fungi and coincidence with vascular land plants. Nature 363: 67–69.

Simon, N.S. 1988. Nitrogen cycling between sediment and the shallow-water column in the transition zone of the Potomac River and estuary. I. Nitrate and ammonium fluxes. Estuarine, Coastal and Shelf Science 26: 483–497.

Simon, N.S. 1989. Nitrogen cycling between sediment and the shallow-water column in the transition zone of the Potomac River and estuary. II. The role of wind-driven resuspension and adsorbed ammonium. Estuarine, Coastal and Shelf Science 28: 531–547.

Simonson, R.W. 1995. Airborne dust and its significance to soils. Geoderma 65: 1–43.

Simpkins, W.W. and T.B. Parkin. 1993. Hydrogeology and redox geochemistry of CH_4 in a late Wisconsinan till and loess sequence of central Iowa. Water Resources Research 29: 3643–3657.

Sinclair, A. R. E., D. S. Hik, O. J. Schmitz, G. G. G. Scudder, D. H. Turpin, and N. C. Larter. 1995. Biodiversity and the need for habitat renewal. Ecological Applications 5: 579–587.

Singer, A. 1989. Illite in the hot-aridic soil environment. Soil Science 147: 126–133.

Singh, J.S. and S.R. Gupta. 1977. Plant decomposition and soil respiration in terrestrial ecosystems. Botanical Review 43: 449–528.

Singh, J.S., W.K. Lauenroth, and R.K. Steinhorst. 1975. Review and assessment of various techniques for estimating net aerial primary production in grasslands from harvest data. Botanical Review 41: 181–232.

Singleton, G.A. and L.M. Lavkulich. 1987. A soil chronosequence on beach sands, Vancouver Island, British Columbia. Canadian Journal of Soil Science 67: 795–810.

Sinke, A.J.C., A.A. Cornelese, T.E. Cappenberg, and A.J.B. Zehnder. 1992. Seasonal variation in sulfate reduction and methanogenesis in peaty sediments of eutrophic Lake Loosdrecht, The Netherlands. Biogeochemistry 16: 43–61.

Sinsabaugh, R.L., R.K. Antibus, A.E. Linkins, C.A. McClaugherty, L. Rayburn, D. Repert, and T. Weiland. 1993. Wood decomposition: Nitrogen and phosphorus dynamics in relation to extracellular enzyme activity. Ecology 74: 1586–1593.

Sirois, A. and L.A. Barrie. 1988. An estimate of the importance of dry deposition as a pathway of acidic substances from the atmosphere to the biosphere in eastern Canada. Tellus 40B: 59–80.

Sjöberg, A. 1976. Phosphate analysis of anthropic soils. Journal of Field Archaeology 3: 447–454.

Sjöström, J. and U. Qvarfort. 1992. Long-term changes of soil chemistry in central Sweden. Soil Science 154: 450–457.

Skiba, U., K.A. Smith, and D. Fowler. 1993. Nitrification and denitrification as sources of nitric oxide and nitrous oxide in a sandy loam soil. Soil Biology and Biochemistry 25: 1527–1536.

Skole, D., and C. Tucker. 1993. Tropical deforestation and habitat fragmentation in the Amazon: Satellite data from 1978 to 1988. Science 260: 1905–1910.

Skyring, G.W. 1987. Sulfate reduction in coastal ecosystems. Geomicrobiology Journal 5: 295–374.

Sleep, N.H., K.J. Zahnle, J.F. Kasting, and H.J. Morowitz. 1989. Annihilation of ecosystems by large asteroid impacts on the early Earth. Nature 342: 139–142.

Slemr, F. and E. Langer. 1992. Increase in global atmospheric concentrations of mercury inferred from measurements over the Atlantic ocean. Nature 355: 434–437.

Slemr, F. and W. Seiler. 1991. Field study of environmental variables controlling the NO emissions from soil and the NO compensation point. Journal of Geophysical Research 96: 13017–13031.

Slemr, F., G. Schuster, and W. Seiler. 1985. Distribution, speciation, and budget of atmospheric mercury. Journal of Atmospheric Chemistry 3: 407–434.

Slinn, W.G.N. 1988. A simple model for Junge's relationship between concentration fluctuations and residence times for tropospheric trace gases. Tellus 40B: 229–232.

Smeck, N.E. 1985. Phosphorus dynamics in soils and landscapes. Geoderma 36: 185–199.

Smedley, S.R. and T. Eisner. 1995. Sodium uptake by puddling in a moth. Science 270: 1816–1818.

Smil, V. 1991. Nitrogen and phosphorus. pp. 423–436. In B.L. Turner, W.C. Clark, R.W. Kates, J.F. Richards, J.T. Mathews, and W.B. Meyer (eds.), *The Earth as Transformed by Human Action.* Cambridge University Press, Cambridge.

Smirnoff, N., P. Todd, and G.R. Stewart. 1984. The occurrence of nitrate reduction in the leaves of woody plants. Annals of Botany 54: 363–374.

Smith, C.J., R.D. DeLaune, and W.H. Patrick. 1983. Nitrous oxide emission from Gulf coast wetlands. Geochimica et Cosmochimica Acta 47: 1805–1814.

Smith, K.L. 1992. Benthic boundary layer communities and carbon cycling at abyssal depths in the central North Pacific. Limnology and Oceanography 37: 1034–1056.

Smith, M.S. and J.M. Tiedje. 1979. Phases of denitrification following oxygen depletion in soil. Soil Biology and Biochemistry 11: 261–267.

Smith, R.A., R.B. Alexander, and M.G. Wolman. 1987. Water-quality trends in the nation's rivers. Science 235: 1607–1615.

Smith R.B., R.H. Waring, and D.A. Perry. 1981. Interpreting foliar analyses from Douglas-fir as weight per unit of leaf area. Canadian Journal of Forest Research 11: 593–598.

Smith, R.C., B.B. Prézelin, K.S. Baker, R.R. Bidigare, N.P. Boucher, T. Coley, D. Karentz, S. MacIntyre, H.A. Matlick, D. Menzies, M. Ondrusek, Z. Wan, and K.J. Waters. 1992. Ozone depletion: Ultraviolet radiation, and phytoplankton biology in Antarctic waters. Science 255: 952–959.

Smith, R.L. and M.J. Klug. 1981. Reduction of sulfur compounds in the sediments of a eutrophic lake basin. Applied and Environmental Microbiology 41: 1230–1237.

Smith, S.D., C.A. Herr, K.L. Leary, and J.M. Piorkowski. 1995. Soil-plant water relations in a Mojave Desert mixed shrub community: A comparison of three geomorphic surfaces. Journal of Arid Environments 29: 339–351.

Smith, S.J., J.F. Power, and W.D. Kemper. 1994. Fixed ammonium and nitrogen availability indexes. Soil Science 158: 132–140.

Smith, S.V. 1981. Marine macrophytes as a global carbon sink. Science 211: 838–840.

Smith, S.V. and J.T. Hollibaugh. 1989. Carbon-controlled nitrogen cycling in a marine 'macrocosm': An ecosystem-scale model for managing cultural eutrophication. Marine Ecology: Progress Series 52: 103–109.

Smith, S.V. and F.T. MacKenzie. 1987. The ocean as a net heterotrophic system: Implications from the carbon biogeochemical cycle. Global Biogeochemical Cycles 1: 187–198.

Smith, T.M. and H.H. Shugart. 1993. The transient response of terrestrial carbon storage to a perturbed climate. Nature 361: 523–526.

Smith, T.M., R. Leemans, and H.H. Shugart. 1992. Sensitivity of terrestrial carbon storage to CO_2-induced climate change: Comparison of four scenarios based on general circulation models. Climatic Change 21: 367–384.

Smith, V.H. 1982. The nitrogen and phosphorus dependence of algal biomass in lakes: An empirical and theoretical analysis. Limnology and Oceanography 27: 1101–1112.

Smith, V.H. 1983. Low nitrogen to phosphorus ratios favor dominance by blue-green algae in lake phytoplankton. Science 221: 669–671.

Smith, V.H. 1992. Effects of nitrogen:phosphorus supply ratios on nitrogen fixation in agricultural and pastoral ecosystems. Biogeochemistry 18: 19–35.

Smith, W.H. 1976. Character and significance of forest tree root exudates. Ecology 57: 324–331.

Smith, W.H. and T.G. Siccama. 1981. The Hubbard Brook Ecosystem Study: Biogeochemistry of lead in the northern hardwood forest. Journal of Environmental Quality 10: 323–333.

Snaydon, R.W. 1962. Micro-distribution of *Trifolium repens* L. and its relation to soil factors. Journal of Ecology 50: 133–143.

Snow, T.P. and A.N. Witt. 1995. The interstellar carbon budget and the role of carbon in dust and large molecules. Science 270: 1455–1460.

Søballe, D.M. and B.L. Kimmel. 1987. A large-scale comparison of factors influencing phytoplankton abundance in rivers, lakes, and impoundments. Ecology 68: 1943–1954.

Soil Science Society of America. 1984. *Glossary of Soil Science Terms.* Madison, Wisconsin.

Sokolik, I.N. and O.B. Toon. 1996. Direct radiative forcing by anthropogenic airborne mineral aerosols. Nature 381: 681–683.

Sollins, P., G. Spycher, and C.A. Glassman. 1984. Net nitrogen mineralization from light- and heavy-fraction forest soil organic matter. Soil Biology and Biochemistry 16: 31–37.

Sollins, P., G.P. Robertson, and G. Uehara. 1988. Nutrient mobility in variable- and permanent-charge soils. Biogeochemistry 6: 181–199.

Solomon, D.K. and T.E. Cerling. 1987. The annual carbon dioxide cycle in a montane soil: Observations, modeling, and implications for weathering. Water Resources Research 23: 2257–2265.

Solomon, S. 1990. Progress towards a quantitative understanding of Antarctic ozone depletion. Nature 347: 347–354.

Solomon, S., R.R. Garcia, F.S. Rowland, and D.J. Wuebbles. 1986. On the depletion of Antarctic ozone. Nature 321: 755–758.

Sonzogni, W.C., S.C. Chapra, D.E. Armstrong, and T.J. Logan. 1982. Bioavailability of phosphorus inputs to lakes. Journal of Environmental Quality 11: 555–563.

Sowers, T. and M. Bender. 1995. Climate records covering the last deglaciation. Science 269: 210–214.

Spalding, R.F. and M.E. Exner. 1993. Occurrence of nitrate in groundwatera review. Journal of Environmental Quality 22: 393–402.

Spalding, R.F. and J.D. Parrott. 1994. Shallow groundwater denitrification. Science of the Total Environment 141: 17–25.

Speidel, D.H. and A.F. Agnew. 1982. *The Natural Geochemistry of Our Environment.* Westview Press, Boulder, Colorado.

Speir, T.W., H.A. Kettles, and R.D. More. 1995. Aerobic emissions of N_2O and N_2 from soil cores: Measurement procedures using ^{13}N-labelled NO_3^- and NH_4^+. Soil Biology and Biochemistry 27: 1289–1298.

Spencer, R.W. and J.R. Christy. 1990. Precise monitoring of global temperature trends from satellites. Science 247: 1558–1562.

Spiro, P.A., D.J. Jacob, and J.A. Logan. 1992. Global inventory of sulfur emissions with 1° × 1° resolution. Journal of Geophysical Research 97: 6023–6036.

Spivack, A.J., C.-F. You, and H.J. Smith. 1993. Foraminiferal boron isotope ratios as a proxy for surface ocean pH over the past 21 Myr. Nature 363: 149–151.

Sposito, G. 1984. *The Surface Chemistry of Soils*. Oxford University Press, Oxford.

Spycher, G., P. Sollins, and S. Rose. 1983. Carbon and nitrogen in the light fraction of a forest soil: Vertical distribution and seasonal patterns. Soil Science 135: 79–87.

Srivastava, S.C. and J.S. Singh. 1988. Carbon and phosphorus in the soil biomass of some tropical soils of India. Soil Biology and Biochemistry 20: 743–747.

Staaf, H. and B. Berg. 1982. Accumulation and release of plant nutrients in decomposing Scots pine needle litter: Long-term decomposition in a Scots pine forest. II. Canadian Journal of Botany 60: 1561–1568.

Staffelbach, T., A. Neftel, B. Stauffer, and D. Jacob. 1991. A record of the atmospheric methane sink from formaldehyde in polar ice cores. Nature 349: 603–605.

Stallard, R.F. and J.M. Edmond. 1981. Geochemistry of the Amazon. 1. Precipitation chemistry and the marine contribution to the dissolved load at the time of peak discharge. Journal of Geophysical Research 86: 9844–9858.

Stallard, R.F. and J.M. Edmond. 1983. Geochemistry of the Amazon. 2. The influence of geology and weathering environment on the dissolved load. Journal of Geophysical Research 88: 9671–9688.

Stanley, D.W. and J.E. Hobbie. 1981. Nitrogen recycling in a North Carolina coastal river. Limnology and Oceanography 26: 30–42.

Stanley, S.R. and E.J. Ciolkosz. 1981. Classification and genesis of Spodosols in the central Appalachians. Soil Science Society of America Journal 45: 912–917.

Starkel, L. 1989. Global paleohydrology. Quaternary International 2: 25–33.

Staubes, R., H.-W. Georgii, and G. Ockelmann. 1989. Flux of COS, DMS and CS_2 from various soils in Germany. Tellus 41B: 305–313.

Stauffer, R.E. 1987. Effects of oxygen transport on the areal hypolimnetic oxygen deficit. Water Resources Research 23: 1887–1892.

Steele, L.P., P.J. Fraser, R.A. Rasmussen, M.A.K. Khalil, T.J. Conway, A.J. Crawford, R.H. Gammon, K.A. Masarie, and K.W. Thoning. 1987. The global distribution of methane in the troposphere. Journal of Atmospheric Chemistry 5: 125–171.

Steele, L.P., E.J. Dlugokencky, P.M. Lang, P.P. Tans, R.C. Martin, and K.A. Masarie. 1992. Slowing down of the global accumulation of atmospheric methane during the 1980s. Nature 358: 313–316.

Sterner, R.W. and D.O. Hessen. 1994. Algal nutrient limitation and the nutrition of aquatic herbivores. Annual Review of Ecology and Systematics 25: 1–29.

Stetter, K.O., G. Lauerer, M. Thomm, and A. Neuner. 1987. Isolation of extremely thermophilic sulfate reducers: Evidence for a novel branch of Archaebacteria. Science 236: 822–824.

Steudler, P.A. and B.J. Peterson. 1985. Annual cycle of gaseous sulfur emissions from a New England *Spartina alterniflora* marsh. Atmospheric Environment 19: 1411–1416.

Steudler, P.A., R.D. Bowden, J.M. Melillo, and J.D. Aber. 1989. Influence of nitrogen fertilization on methane uptake in temperate forest soils. Nature 341: 314–316.

Stevens, T.O. and J.P. McKinley. 1995. Lithoautotrophic microbial ecosystems in deep basalt aquifers. Science 270: 450–454.

Stevenson, D.J. 1983. The nature of the Earth prior to the oldest known rock record: The Hadean Earth. pp. 32–40. In J.W. Schopf (ed.), *Earth's Earliest Biosphere.* Princeton University Press, Princeton, New Jersey.

Stevenson, F.J. 1982. Origin and distribution of nitrogen in soil. pp. 1–42. In F.J. Stevenson (ed.), *Nitrogen in Agricultural Soils.* American Society of Agronomy, Madison, Wisconsin.

Stevenson, F.J. 1986. *Cycles of Soil.* Wiley, New York.

Stewart, A.J. and R.G. Wetzel. 1982. Phytoplankton contribution to alkaline phosphatase activity. Archiv für Hydrobiologie 93: 265–271.

Stiefel, E.I. 1996. Molybdenum bolsters the bioinorganic brigade. Science 272: 1599–1600.

St. Louis, V.L., J.W.M. Rudd, C.A. Kelly, K.G. Beaty, N.S. Bloom, and R.J. Flett. 1994. Importance of wetlands as sources of methyl mercury to boreal forest ecosystems. Canadian Journal of Fisheries and Aquatic Sciences 51: 1065–1076.

Stock, J.B., A.M. Stock, and J.M. Mottonen. 1990. Signal transduction in bacteria. Nature 344: 395–400.

Stockner, J.G. and N.J. Antia. 1986. Algal picoplankton from marine and freshwater ecosystems: A multidisciplinary perspective. Canadian Journal of Fisheries and Aquatic Sciences 43: 2472–2503.

Stoiber, R.E., S.N. Williams, and B. Huebert. 1987. Annual contribution of sulfur dioxide to the atmosphere by volcanoes. Journal of Volcanology and Geothermal Research 33: 1–8.

Stolarski, R., R. Bojkov, L. Bishop, C. Zerefos, J. Staehelin, and J. Zawodny. 1992. Measured trends in stratospheric ozone. Science 256: 342–349.

Stolt, M.H., J.C. Baker, and T.W. Simpson. 1992. Characterization and genesis of saprolite derived from gneissic rocks of Virginia. Soil Science Society of America Journal 56: 531–539.

Stolzy, L.H., D.D. Focht, and H. Flühler. 1981. Indicators of soil aeration status. Flora 171: 236–265.

Stone, E.L. and R. Kszystyniak. 1977. Conservation of potassium in the *Pinus resinosa* ecosystem. Science 198: 192–194.

Stouffer, R.J., S. Manabe, and K.Y. Vinnikov. 1994. Model assessment of the role of natural variability in recent global warming. Nature 367: 634–636.

Straub, K.L., M. Beng, B. Schink, and F. Widdel. 1996. Anaerobic, nitrate-dependent microbial oxidation of ferrous iron. Applied and Environmental Microbiology 62: 1458–1460.

Strain, B.R. and J.D. Cure (eds.). 1985. *Direct Effects of Increasing Carbon Dioxide on Vegetation.* U.S. Department of Energy, Publication DOE/Er-0238, Washington, D.C.

Striegl, R.G., T.A. McConnaughey, D.C. Thorstenson, E.P. Weeks, and J.C. Woodward. 1992. Consumption of atmospheric methane by desert soils. Nature 357: 145–147.

Strong, A.E. 1989. Greater global warming revealed by satellite-derived sea-surface-temperature trends. Nature 338: 642–645.

Stuart, D.M. and R.M. Dixon. 1973. Water movement and caliche formation in layered arid and semiarid soils. Soil Science Society of America Proceedings 37: 323–324.

Stuiver, M. 1980. ^{14}C distribution in the Atlantic ocean. Journal of Geophysical Research 85: 2711–2718

Stuiver, M., P.D. Quay, and H.G. Ostlund. 1983. Abyssal water carbon-14 distribution and the age of the world oceans. Science 219: 849–851.

Stumm, W. and J.J. Morgan. 1981. *Aquatic Chemistry* 2nd ed. Wiley, New York.

Sturges, W.T., G.F. Cota, and P.T. Buckley. 1992. Bromoform emission for Arctic ice algae. Nature 358: 660–662.

Suarez, D.L., J.D. Wood, and I. Ibrahim. 1992. Reevaluation of calcite supersaturation in soils. Soil Science Society of America Journal 56: 1776–1784.

Subak, S., P. Raskin, and D. von Hippel. 1993. National greenhouse gas accounts: Current anthropogenic sources and sinks. Climatic Change 25: 15–58.

Subba Rao, D.V. 1981. Effect of boron on primary production of nannoplankton. Canadian Journal of Fisheries and Aquatic Sciences 38: 52–58.

Suberkropp, K., G.L. Godshalk, and M.J. Klug. 1976. Changes in the chemical composition of leaves during processing in a woodland stream. Ecology 57: 720–727.

Suchet, P.A. and J.L. Probst. 1995. A global model for present-day atmospheric/soil CO_2 consumption by chemical erosion of continental rocks (GEM-CO_2). Tellus 47B: 273–380.

Suess, E. 1980. Particulate organic carbon flux in the oceans–Surface productivity and oxygen utilization. Nature 288: 260–263.

Sullivan, C.W., K.R. Arrigo, C.R. McClain, J.C. Comiso, and J. Firestone. 1993. Distributions of phytoplankton blooms in the Southern Ocean. Science 262: 1832–1837.

Suman, D.O., T.A.J. Kuhlbusch, and B. Lim. 1997. Marine Sediments: A reservoir for black carbon and their use as spatial and temporal records of combustion. In J.S. Clark (ed.), *Biomass Burning Emissions and Global Change.* Springer-Verlag, Berlin.

Sunda, W.G. and S.A. Huntsman. 1992. Feedback interactions between zinc and phytoplankton in seawater. Limnology and Oceanography 37: 25–40.

Sundby, B., C. Gobeil, N. Silverberg, and A. Mucci. 1992. The phosphorus cycle in coastal marine sediments. Limnology and Oceanography 37: 1129–1145.

Sundquist, E.T. 1993. The global carbon dioxide budget. Science 259: 934–941.

Sundquist, E.T., L.N. Plummer, and T.M.L. Wigley. 1979. Carbon dioxide in the ocean surface: The homogeneous buffer factor. Science 204: 1203–1205.

Sutherland, R.A., C. van Kessel, R.E. Farrell, and D.J. Pennock. 1993. Landscape-scale variations in plant and soil nitrogen-15 natural abundance. Soil Science Society of America Journal 57: 169–178.

Sutton, M.A., C.E.R. Pitcairn, and D. Fowler. 1993. The exchange of ammonia between the atmosphere and plant communities. Advances in Ecological Research 24: 301–393.

Svedrup, H.U., M.W. Johnson, and R.H. Fleming. 1942. *The Oceans.* Prentice-Hall, New York.

Swain, E.B., D.R. Engstrom, M.E. Brigham, T.A. Henning, and P.L. Brezonik. 1992. Increasing rates of atmospheric mercury deposition in midcontinental North America. Science 257: 784–787.

Swank, W.T. and J.E. Douglass. 1977. Nutrient budgets for undisturbed and manipulated hardwood forest ecosystems in the mountains of North Carolina. pp. 343–

362. In D.L. Correll (ed.), *Watershed Research in Eastern North America.* Vol. 1. Smithsonian Institution, Edgewater, Maryland.

Swank, W.T. and G.S. Henderson. 1976. Atmospheric input of some cations and anions to forest ecosystems in North Carolina and Tennessee. Water Resources Research 12: 541–546.

Swank, W.T., J.B. Waide, D.A. Crossley, and R.L. Todd. 1981. Insect defoliation enhances nitrate export from forest ecosystems. Oecologia 51: 297–299.

Swank, W.T., J.W. Fitzgerald, and J.T. Ash. 1984. Microbial transformation of sulfate in forest soils. Science 223: 182–184.

Swanson, F.J., R.L. Fredriksen, and F.M. McCorison. 1982. Material transfer in a western Oregon forested watershed. pp. 233–266. In R.L. Edmonds (ed.), *Analysis of Coniferous Forest Ecosystems in the Western United States.* Dowden, Hutchinson and Ross, Stroudsburg, Pennsylvania.

Swap, R., M. Garstang, S. Greco, R. Talbot, and P. Kållberg. 1992. Saharan dust in the Amazon basin. Tellus 44B: 133–149.

Sweerts, J.-P.R.A., M.-J. Bär-Gilissen, A.A. Cornelese, and T.E. Cappenberg. 1991. Oxygen-consuming processes at the profundal and littoral sediment-water interface of a small meso-eutrophic lake (Lake Vechten, The Netherlands). Limnology and Oceanography 36: 1124–1133.

Swetnam, T.W. and J.L. Betancourt. 1990. Fire-Southern oscillation relations in the southwestern United States. Science 249: 1017–1020.

Swift, M.J., O.W. Heal, and J.M. Anderson. 1979. *Decomposition in Terrestrial Ecosystems.* University of California Press, Berkeley.

Szikszay, M., A.A. Kimmelmann, R. Hypolito, R.M. Figueira, and R.H. Sameshima. 1990. Evolution of the chemical composition of water passing through the unsaturated zone to ground water at an experimental site at the University of São Paulo, Brazil. Journal of Hydrology 118: 175–190.

Tabatabai, M.A. and W.A. Dick. 1979. Distribution and stability of pyrophosphatase in soils. Soil Biology and Biochemistry 11: 655–659.

Tabazadeh, A. and R.P. Turco. 1993. Stratospheric chlorine injection by volcanic eruptions: HCl scavenging and implications for ozone. Science 260: 1082–1086.

Takahashi, T., D. Chipman, and T. Volk. 1983. Geographical, seasonal, and secular variations of the partial pressure of CO_2 in the surface waters of the North Atlantic Ocean. The results of the Atlantic TTO Program. pp. 125–143. In *Proceedings: Carbon Dioxide, Science, and Consensus.* CONF 820970. U.S. Department of Energy, Washington, D.C.

Takahashi, T., W.S. Broecker, and S. Langer. 1985. Redfield ratio based on chemical data from isopycnal surfaces. Journal of Geophysical Research 90: 6907–6924.

Takamura, N., A. Otsuki, M. Aizaki and Y. Nojiri. 1992. Phytoplankton species shift accompanied by transition from nitrogen dependence to phosphorus dependence of primary production in Lake Kasumigaura, Japan. Archiv für Hydrobiologie 124: 129–148.

Talbot, R.W., R.C. Harriss, E.V. Browell, G.L. Gregory, D.I. Sebacher, and S.M. Beck. 1986. Distribution and geochemistry of aerosols in the tropical North Atlantic troposphere: Relationship to Saharan dust. Journal of Geophysical Research 91: 5173–5182.

Talukdar, R.K., A. Mellouki, A.-M. Schmoltner, T. Watson, S. Montzka, and A.R. Ravishankara. 1992. Kinetics of the OH reaction with methyl chloroform and its atmospheric implications. Science 257: 227–230.

Tamm, C.O. and L. Hällbäcken. 1986. Change in soil pH over a 50-year period under different forest canopies in SW Sweden. Water, Air, and Soil Pollution 31: 337–341.

Tamrazyan, G.P. 1989. Global peculiarities and tendencies in river discharge and wash-down of the suspended sediments—the Earth as a whole. Journal of Hydrology 107: 113–131.

Tan, K.H. 1980. The release of silicon, aluminum, and potassium during decomposition of soil minerals by humic acid. Soil Science 129: 5–11.

Tan, K.H. and P.S. Troth. 1982. Silica-sesquioxide ratios as aids in characterization of some temperate region and tropical soil clays. Soil Science Society of America Journal 46: 1109–1114.

Tanaka, N. and K.K. Turekian. 1991. Use of cosmogenic ^{35}S to determine the rates of removal of atmospheric SO_2. Nature 352: 226–228.

Tanaka, N. and K.K. Turekian. 1995. Determination of the dry deposition flux of SO_2 using cosmogenic ^{35}S and ^{7}Be measurements. Journal of Geophysical Research 100: 2841–2848.

Tans, P.P., I.Y. Fung, and T. Takahashi. 1990. Observational constraints on the global atmospheric CO_2 budget. Science 247: 1431–1438.

Tarafdar, J.C. and N. Claassen. 1988. Organic phosphorus compounds as a phosphorus source for higher plants through the activity of phosphatases produced by plant roots and microorganisms. Biology and Fertility of Soils 5: 308–312.

Taran, Y.A., J.W. Hedenquist, M.A. Korzhinsky, S.I. Tkachenko, and K.I. Shmulovich. 1995. Geochemistry of magmatic gases from Kudryavy volcano, Iturup, Kuril Islands. Geochimica et Cosmochimica Acta 59: 1749–1761.

Tarrasón, L., S. Turner, and I. Fløisand. 1995. Estimation of seasonal dimethyl sulphide fluxes over the North Atlantic Ocean and their contribution to European pollution levels. Journal of Geophysical Research 100: 11623–11639.

Tate, R.L. 1980. Microbial oxidation of organic matter of Histosols. Advances in Microbial Ecology 4: 169–201.

Taylor, A.B. and M.A. Velbel. 1991. Geochemical mass balances and weathering rates in forested watersheds of the southern Blue Ridge. II. Effects of botanical uptake terms. Geoderma 51: 29–50.

Taylor, A.H., A.J. Watson, and J.E. Robertson. 1992. The influence of the spring phytoplankton bloom on carbon dioxide and oxygen concentrations in the surface waters of the northeast Atlantic during 1989. Deep Sea Research 39: 137–152.

Taylor, B.R., D. Parkinson, and W.F.J. Parsons. 1989. Nitrogen and lignin content as predictors of litter decay rates: A microcosm test. Ecology 70: 97–104.

Tegen, I. and I. Fung. 1995. Contribution to the atmospheric mineral aerosol load from land surface modification. Journal of Geophysical Research 100: 18707–18726.

Tegen, I., A.A. Lacis, and I. Fung. 1996. The influence on climate forcing of mineral aerosols from disturbed soils. Nature 380: 419–422.

Temple, K.L. and A.R. Colmer. 1951. The autotrophic oxidation of iron by a new bacterium: *Thiobacillus ferrooxidans*. Journal of Bacteriology 62: 605–611.

Terman, G.L. 1977. Quantitative relationships among nutrients leached from soils. Soil Science Society of America Journal 41: 935–940.

Terman, G.L. 1979. Volatilization losses of nitrogen as ammonia from surface-applied fertilizers, organic amendments, and crop residues. Advances in Agronomy 31: 189–223.

Tezuka, Y. 1990. Bacterial regeneration of ammonium and phosphate as affected by the carbon : nitrogen : phosphorus ratio of organic substrates. Microbial Ecology 19: 227–238.

Thamdrup, B., H. Fossing, and B.B. Jørgensen. 1994. Manganese, iron, and sulfur cycling in a coastal marine sediment, Aarhus Bay, Denmark. Geochimica et Cosmochimica Acta 58: 5115–5129

Thiemens, M.H. and W.C. Trogler. 1991. Nylon production: An unknown source of atmospheric nitrous oxide. Science 251: 932–934.

Thode-Andersen, S. and B.B. Jørgensen. 1989. Sulfate reduction and the formation of ^{35}S-labeled FeS, FeS_2, and S° in coastal marine sediments. Limnology and Oceanography 34: 793–806.

Thomas, G.E., J.J. Olivero, E.J. Jensen, W. Schroeder, and O.B. Toon. 1989. Relation between increasing methane and the presence of ice clouds at the mesopause. Nature 338: 490–492.

Thomas, R.B., J.D. Lewis, and B.R. Strain. 1994. Effects of leaf nutrient status on photosynthetic capacity in loblolly pine (*Pinus taeda* L.) seedlings grown in elevated atmospheric CO_2. Tree Physiology 14: 947–960.

Thompson, A.M. 1992. The oxidizing capacity of the Earth's atmosphere: Probable past and future changes. Science 256: 1157–1165.

Thomson, J., N.C. Higgs, I.W. Croudance, S. Colley, and D.J. Hydes. 1993. Redox zonation of elements at an oxic/post-oxic boundary in deep-sea sediments. Geochimica et Cosmochimica Acta 57: 579–595.

Thorneloe, S.A., M.A. Barlaz, R. Peer, L.C. Huff, L. Davis, and J. Mangino. 1993. Waste management. pp. 362–398. In M.A.K. Khalil (ed.), *Atmospheric Methane: Sources, Sinks and Role in Global Change.* Springer-Verlag, New York.

Thorpe, S.A. 1985. Small-scale processes in the upper ocean boundary layer. Nature 318: 519–522.

Tiedje, J.M., A.J. Sexstone, T.B. Parkin, N.P. Revsbech, and D.R. Shelton. 1984. Anaerobic processes in soil. Plant and Soil 76: 197–212.

Tiedje, J.M., S. Simkins, and P.M. Groffman. 1989. Perspectives on measurement of denitrification in the field including recommended protocols for acetylene based methods. Plant and Soil 115: 261–284.

Tiessen, H. and J.W.B. Stewart. 1983. Particle-size fractions and their use in studies of soil organic matter. II. Cultivation effects on organic matter composition in size fractions. Soil Science Society of America Journal 47: 509–514.

Tiessen, H. and J.W.B. Stewart. 1988. Light and electron microscopy of stained microaggregates: The role of organic matter and microbes in soil aggregation. Biogeochemistry 5: 312–322.

Tiessen, H., J.W.B. Stewart, and J.R. Bettany. 1982. Cultivation effects on the amounts and concentration of carbon, nitrogen, and phosphorus in grassland soils. Agronomy Journal 74: 831–835.

Tiessen, H., J.W.B. Stewart, and C.V. Cole. 1984. Pathways of phosphorus transformations in soils of differing pedogenesis. Soil Science Society of America Journal 48: 853–858.

Tietze, K., M. Geyh, H. Muller, W. Stahl, and H. Wehner. 1980. The genesis of the methane in Lake Kivu (Central Africa). Geologische Rundschau 69: 452–472.

Tilman, D. 1982. *Resource Competition and Community Structure.* Princeton University Press, Princeton, New Jersey.

Tilman, D. 1985. The resource-ratio hypothesis of plant succession. American Naturalist 125: 827–852.

Tilman, D. 1987. Secondary succession and the pattern of plant dominance along experimental nitrogen gradients. Ecological Monographs 57: 189–214.

Tilman, D., D. Wedin, and J.Knops. 1996. Productivity and sustainability influenced by biodiversity in grassland ecosystems. Nature 379: 718–720.

Tilton, D.L. 1978. Comparative growth and foliar element concentrations of *Larix laricina* over a range of wetland types in Minnesota. Journal of Ecology 66: 499–512.

Timmer, V.R. and E.L. Stone. 1978. Comparative foliar analysis of young balsam fir fertilized with nitrogen, phosphorus, potassium, and lime. Soil Science Society of America Journal 42: 125–130.

Ting, J.C. and M. Chang. 1985. Soil-moisture depletion under three southern pine plantations in east Texas. Forest Ecology and Management 12: 179–193.

Tisdall, J.M. and J.M. Oades. 1982. Organic matter and water-stable aggregates in soils. Journal of Soil Science 33: 141–163.

Titman, D. 1976. Ecological competition between algae: Experimental confirmation of resource-based competition theory. Science 192: 463–465.

Toggweiler, J.R. and B. Samuels. 1993. New radiocarbon constraints on the upwelling of abyssal water to the ocean's surface. pp. 333–366. In M. Heimann (ed.), *The Global Carbon Cycle.* Springer-Verlag, New York.

Toon, O.B., J.F. Kasting, R.P. Turco, and M.S. Liu. 1987. The sulfur cycle in the marine atmosphere. Journal of Geophysical Research 92: 943–963.

Topp, G.C., W.D. Zebchuk, and J. Dumanski. 1980. The variation of in situ measured soil water properties within soil map units. Canadian Journal of Soil Science 60: 497–509.

Torres, A.L. and H. Buchan. 1988. Tropospheric nitric oxide measurements over the Amazon basin. Journal of Geophysical Research 93: 1396–1406.

Towe, K.M. 1990. Aerobic respiration in the Archaean? Nature 348: 54–56.

Townsend, A.R., B.H. Braswell, E.A. Holland, and J.E. Penner. 1996. Spatial and temporal patterns in terrestrial carbon storage due to deposition of fossil fuel nitrogen. Ecological Applications 6:806–814.

Travis, C.C. and E.L. Etnier. 1981. A survey of sorption relationships for reactive solutes in soil. Journal of Environmental Quality 10: 8–17.

Trefry, J.H. and S. Metz. 1989. Role of hydrothermal precipitates in the geochemical cycling of vanadium. Nature 342: 531–533.

Trefry, J.H., S. Metz, R.P. Trocine, and T.A. Nelsen. 1985. A decline in lead transport by the Mississippi River. Science 230: 439–441.

Tréguer, P., D.M. Nelson, A.J. Van Bennekom, D.J. DeMaster, A. Leynaert, and B. Quéguiner. 1995. The silica balance in the world ocean: A reestimate. Science 268: 375–379.

Trenberth, K.E. and C.J. Guillemot. 1994. The total mass of the atmosphere. Journal of Geophysical Research 99: 23079–23088.

Trimble, S.W. 1977. The fallacy of stream equilibrium in contemporary denudation studies. American Journal of Science 277: 876–887.

Triska, F.J., J.R. Sedell, K. Cromack, S.V. Gregory, and F.M. McCorison. 1984. Nitrogen budget for a small coniferous forest stream. Ecological Monographs 54: 119–140.

Trlica, M.J. and M.E. Biondini. 1990. Soil water dynamics, transpiration, and water losses in a crested wheatgrass and native shortgrass ecosystem. Plant and Soil 126: 187–201.

Trumbore, S.E. 1993. Comparison of carbon dynamics in tropical and temperate soils using radiocarbon measurements. Global Biogeochemical Cycles 7: 275–290.

Trumbore, S.E., O.A. Chadwick, and R. Amundson. 1996. Rapid exchange between soil carbon and atmospheric carbon dioxide driven by temperature change. Science 272: 393–396.

Tsonis, A.A. 1996. Widespread increases in low-frequency variability of precipitation over the past century. Nature 382: 700–702.

Tucker, C.J., H.E. Dregne, and W.W. Newcomb. 1991. Expansion and contraction of the Sahara desert from 1980 to 1990. Science 253: 299–301.

Tukey, H.B. 1970. The leaching of substances from plants. Annual Review of Plant Physiology 21: 305–324.

Turco, R.P., R.C. Whitten, O.B. Toon, J.B. Pollack, and P. Hamill. 1980. OCS, stratospheric aerosols and climate. Nature 283: 283–286.

Turekian, K.K. 1977. The fate of metals in the oceans. Geochimica et Cosmochimica Acta 41: 1139–1144.

Turner, D.P., G.J. Koerper, M.E. Harmon, and J.J. Lee. 1995. A carbon budget for forests of the conterminous United States. Ecological Applications 5: 421–436.

Turner, F.T. and W.H. Patrick. 1968. Chemical changes in waterlogged soils as a result of oxygen depletion. Transactions of the 9th International Congress of Soil Science 4: 53–65.

Turner, J. 1982. The mass flow component of nutrient supply in three western Washington forest types. Acta Oecologica [Series]: Oecologia Plantarum 3: 323–329.

Turner, J. and P.R. Olson. 1976. Nitrogen relations in a douglas-fir plantation. Annals of Botany 40: 1185–1193.

Turner, J., D.W. Johnson, and M.J. Lambert. 1980. Sulphur cycling in a douglas-fir forest and its modification by nitrogen applications. Acta Oecologica [Series]: Oecologia Plantarum 1: 27–35.

Turner, S.M., G. Malin, P.S. Liss, D.S. Harbour, and P.M. Holligan. 1988. The seasonal variation of dimethyl sulfide and dimethylsulfoniopropionate concentrations in nearshore waters. Limnology and Oceanography 33: 364–375.

Turner, S. M., P. D. Nightingale, L. J. Spokes, M. I. Liddicoat, and P. S. Liss. 1996. Increased dimethyl sulphide concentrations in sea water from *in situ* iron enrichment. Nature 383: 513–517.

Tyler, G. 1994. A new approach to understanding the calcifuge habit of plants. Annals of Botany 73: 327–330.

Tyler, G. and L. Ström. 1995. Differing organic acid exudation pattern explains calcifuge and acidfuge behaviour of plants. Annals of Botany 75: 75–78.

Tyler, S.C., D.R. Blake, and F.S. Rowland. 1987. $^{13}C/^{12}C$ ratio in methane from the flooded Amazon forest. Journal of Geophysical Research 92: 1044–1048.

Uehara, G. and G. Gillman. 1981. *The Mineralogy, Chemistry and Physics of Tropical Soils with Variable Charge Clays.* Westview Press, Boulder, Colorado

Ugolini, F.C. and R. Dahlgren. 1987. The mechanism of podzolization as revealed by soil solution studies. pp. 195–203. In D. Righi and A. Chauvel (eds.), *Podzols and Podzolisation.* Institut National de la Recherche Agronomique, Paris.

Ugolini, F.C. and R.S. Sletten. 1991. The role of proton donors in pedogenesis as revealed by soil solution studies. Soil Science 151: 59–75.

Ugolini, F.C., R. Minden, H. Dawson, and J. Zachara. 1977. An example of soil processes in the *Abies amabilis* zone of central Cascades, Washington. Soil Science 124: 291–302.

Ugolini, F.C., M.G. Stoner, and D.J. Marrett. 1987. Arctic pedogenesis: I. Evidence for contemporary podzolization. Soil Science 144: 90–100.

Uhl, C. and C.F. Jordan. 1984. Succession and nutrient dynamics following forest cutting and burning in Amazonia. Ecology 65: 1476–1490.

Urban, N.R. and S.J. Eisenreich. 1988. Nitrogen cycling in a forested Minnesota bog. Canadian Journal of Botany 66: 435–449.

Urban, N.R., S.J. Eisenreich, and D.F. Grigal. 1989. Sulfur cycling in a forested *Sphagnum* bog in northern Minnesota. Biogeochemistry 7: 81–109.

Urban, N.R., P.L. Brezonik, L.A. Baker, and L.A. Sherman. 1994. Sulfate reduction and diffusion in sediments of Little Rock Lake, Wisconsin. Limnology and Oceanography 39: 797–815.

U.S. Department of Agriculture. 1994. *Keys to Soil Taxonomy*. Washington, D.C.

Vadstein, O., Y. Olsen, H. Reinertsen, and A. Jensen. 1993. The role of planktonic bacteria in phosphorus cycling in lakessink and link. Limnology and Oceanography 38: 1539–1544.

Valentine, D.W., E.A. Holland, and D.S. Schimel. 1994. Ecosystem and physiological controls over methane production in northern wetlands. Journal of Geophysical Research 99: 1563–1571.

Valentini, R., P. De Angelis, G. Matteucci, R. Monaco, S. Dore, and G.E. Scarascia Mungnozza. 1996. Seasonal net carbon dioxide exchange of a beech forest with the atmosphere. Global Change Biology 2: 199–207.

Valiela, I. and J.M. Teal. 1979. The nitrogen budget of a salt marsh ecosystem. Nature 280: 652–656.

Vallentyne, J.R. 1974. *The Algal Bowl*. Department of Environment, Miscellaneous Special Publication 22, Ottawa, Canada.

van Breemen, N., P.A. Burrough, E.J. Velthorst, H.F. van Dobben, T. de Wit, T.B. Ridder, and H.F.R. Reijnders. 1982. Soil acidification from atmospheric ammonium sulphate in forest canopy throughfall. Nature 299: 548–550.

van Cappellen, P. and E.D. Ingall. 1996. Redox stabilization of the atmosphere and oceans by phosphorus-limited marine productivity. Science 271: 493–496.

Vance, G.F. and M.B. David. 1991. Chemical characteristics and acidity of soluble organic substances from a northern hardwood forest floor, central Maine, USA. Geochimica et Cosmochimica Acta 55: 3611–3625.

Van Cleve, K. and R. White. 1980. Forest-floor nitrogen dynamics in a 60–year-old paper birch ecosystem in interior Alaska. Plant and Soil 54: 359–381.

Van Cleve, K., R. Barney, and R. Schlentner. 1981. Evidence of temperature control of production and nutrient cycling in two interior Alaska black spruce ecosystems. Canadian Journal of Forest Research 11: 258–273.

Van Cleve, K., L. Oliver, R. Schlentner, L.A. Viereck, and C.T. Dyrness. 1983. Productivity and nutrient cycling in taiga forest ecosystems. Canadian Journal of Forest Research 13: 747–766.

Van Cleve, K., W.C. Oechel, and J.L. Hom. 1990. Response of black spruce (*Picea mariana*) ecosystems to soil temperature modification in interior Alaska. Canadian Journal of Forest Research 20: 1530–1535.

Vandal, G.M., W.F. Fitzgerald, C.F. Boutron, and J.-P. Candelone. 1993. Variations in mercury deposition to Antarctica over the past 34,000 years. Nature 362: 621–623.

Vandenberg, J.J. and K.R. Knoerr. 1985. Comparison of surrogate surface techniques for estimation of sulfate dry deposition. Atmospheric Environment 19: 627–635.

van den Bergh, S. 1992. The age and size of the Universe. Science 258: 421–424.

van den Driessche, R. 1974. Prediction of mineral nutrient status of trees by foliar analysis. Botanical Review 40: 347–394.

Van Devender, T.R. and W.G. Spaulding. 1979. Development of vegetation and climate in the southwestern United States. Science 204: 701–710.

Van de Water, P.K., S.W. Leavitt, and J.L. Betancourt. 1994. Trends in stomatal density and $^{13}C/^{12}C$ ratios of *Pinus flexilis* needles during last glacial-interglacial cycle. Science 264: 239–243.

van Dijk, H.F.G., M.H.J. de Louw, J.G.M. Roelofs, and J.J. Verburgh. 1990. Impact of artificial, ammonium-enriched rainwater on soils and young coniferous trees in a greenhouse. Part II-Effects on the trees. Environmental Pollution 63: 41–59.

van Kessel, C., R.E. Farrell, J.P. Roskoski, and K.M. Keane. 1994. Recycling of the naturally-occurring 15N in an established stand of *Leucaena leucocephala.* Soil Biology and Biochemistry 26: 757–762.

Van Sickle, J. 1981. Long-term distributions of annual sediment yields from small watersheds. Water Resources Research 17: 659–663.

Van Trump, J.E. and S.L. Miller. 1972. Prebiotic synthesis of methionine. Science 178: 859–860.

van Veen, J.A., J.N. Ladd, J.K. Martin, and M. Amato. 1987. Turnover of carbon, nitrogen and phosphorus through the microbial biomass in soils incubated with ^{14}C-, ^{15}N- and ^{32}P-labelled bacterial cells. Soil Biology and Biochemistry 19: 559–565.

Van Vuuren, M.M.I., R. Aerts, F. Berendse, and W. de Visser. 1992. Nitrogen mineralization in heathland ecosystems dominated by different plant species. Biogeochemistry 16: 151–166.

Vasconcelos, C., J.A. McKenzie, S. Bernasconi, D. Grujic, and A.J. Tien. 1995. Microbial mediation as a possible mechanism for natural dolomite formation at low temperatures. Nature 377: 220–222.

Vaughan, D.G. and C.S.M. Doake. 1996. Recent atmospheric warming and retreat of ice shelves on the Antarctic peninsula. Nature 379: 328–331.

Velbel, M.A. 1990. Mechanisms of saprolitization, isovolumetric weathering, and pseudomorphous replacement during rock weathering—a review. Chemical Geology 84: 17–18.

Velbel, M.A. 1992. Geochemical mass balances and weathering rates in forested watersheds of the southern Blue Ridge. III. Cation budgets and the weathering rate of amphibole. American Journal of Science 292: 58–78.

VEMAP Members. 1995. Vegetation/ecosystem modeling and analysis project: Comparing biogeography and biogeochemistry models in a continental-scale study of terrestrial ecosystem responses to climate change and CO_2 doubling. Global Biogeochemical Cycles 9: 407–437.

Vengosh, A., Y. Kolodny, A. Starinsky, A.R. Chivas, and M.T. McCulloch. 1991. Coprecipitation and isotopic fractionation of boron in modern biogenic carbonates. Geochimica et Cosmochimica Acta 55: 2901–2910.

Venrick, E.L., J.A. McGowan, D.R. Cayan, and T.L. Hayward. 1987. Climate and chlorophyll a: Long-term trends in the central North Pacific Ocean. Science 238: 70–72.

Vermes, J.-F. and D.D. Myrold. 1992. Denitrification in forest soils of Oregon. Canadian Journal of Forest Research 22: 504–512.

Verry, E.S. and D.R. Timmons. 1982. Waterborne nutrient flow through an upland-peatland watershed in Minnesota. Ecology 63: 1456–1467.

Verstraten, J.M., J.C.R. Dopheide, J.J.H.M. Duysings, A. Tietema, and W. Bouten. 1990. The proton cycle of a deciduous forest ecosystem in the Netherlands and its implications for soil acidification. Plant and Soil 127: 61–69.

Viereck, L.A. 1966. Plant succession and soil development on gravel outwash of the Muldrow Glacier, Alaska. Ecological Monographs 36: 181–199.

Virginia, R.A. and C.C. Delwiche. 1982. Natural ^{15}N abundance of presumed N_2-fixing and non-N_2-fixing plants from selected ecosystems. Oecologia 54: 317–325.

Virginia, R.A. and W.M. Jarrell. 1983. Soil properties in a mesquite-dominated Sonoran desert ecosystem. Soil Science Society of America Journal 47: 138–144.

Virginia, R.A., W.M. Jarrell, and E. Franco-Vizcaino. 1982. Direct measurement of denitrification in a *Prosopis* (mesquite) dominated Sonoran desert ecosystem. Oecologia 53: 120–122.

Vitousek, P.M. 1977. The regulation of element concentrations in mountain streams in the northeastern United States. Ecological Monographs 47: 65–87.

Vitousek, P.M. 1982. Nutrient cycling and nutrient-use efficiency. American Naturalist 119: 553–572.

Vitousek, P.M. 1984. Litterfall, nutrient cycling, and nutrient limitation in tropical forests. Ecology 65: 285–298.

Vitousek, P.M. 1994. Beyond global warming: Ecology and global change. Ecology 75: 1861–1876.

Vitousek, P.M. and R.W. Howarth. 1991. Nitrogen limitation on land and in the sea: How can it occur? Biogeochemistry 13: 87–115.

Vitousek, P.M. and P.A. Matson. 1984. Mechanisms of nitrogen retention in forest ecosystems: A field experiment. Science 225: 51–52.

Vitousek, P.M. and P.A. Matson. 1988. Nitrogen transformations in a range of tropical forest soils. Soil Biology and Biochemistry 20: 361–367.

Vitousek, P.M. and J.M. Melillo. 1979. Nitrate losses from disturbed forests: Patterns and mechanisms. Forest Science 25: 605–619.

Vitousek, P.M. and W.A. Reiners. 1975. Ecosystem succession and nutrient retention: A hypothesis. BioScience 25: 376–381.

Vitousek, P.M. and R.L. Sanford. 1986. Nutrient cycling in moist tropical forest. Annual Review of Ecology and Systematics 17: 137–167.

Vitousek, P.M., J.R. Gosz, C.C. Grier, J.M. Melillo, and W.A. Reiners. 1982. A comparative analysis of potential nitrification and nitrate mobility in forest ecosystems. Ecological Monographs 52: 155–177.

Vitousek, P.M., P.R. Ehrlich, A.H. Ehrlich, and P.A. Matson. 1986. Human appropriation of the products of photosynthesis. BioScience 36: 368–373.

Vitousek, P.M., L.R. Walker, L.D. Whiteaker, D. Mueller-Dombois, and P.A. Matson. 1987. Biological invasion of *Myrica faya* alters ecosystem development in Hawaii. Science 238: 802–804.

Vitousek, P.M., T. Fahey, D.W. Johnson, and M.J. Swift. 1988. Element interactions in forest ecosystems: Succession, allometry and input-output budgets. Biogeochemistry 5: 7–34.

Vogt, K.A., C.C. Grier, C.E. Meier, and R.L. Edmonds. 1982. Mycorrhizal role in net primary production and nutrient cycling in *Abies amabilis* ecosystems in western Washington. Ecology 63: 370–380.

Vogt, K.A., C.C. Grier, C.E. Meier, and M.R. Keyes. 1983. Organic matter and nutrient dynamics in forest floors of young and mature *Abies amabilis* stands in western Washington, as affected by fine-root input. Ecological Monographs 53: 139–157.

Vogt, K.A., C.C. Grier, and D.J. Vogt. 1986. Production, turnover, and nutrient dynamics of above- and belowground detritus of world forests. Advances in Ecological Research 15: 303–377.

Vollenweider, R.A. 1968. Scientific fundamentals of the eutrophication of lakes and flowing waters, with particular reference to nitrogen and phosphorus as factors in eutrophication. Organization for Economic Cooperative Development, Technical Report DAS/CSI/68.27, Paris.

Vollenweider, R.A., M. Munawar, and P. Stadelmann. 1974. A comparative review of the phytoplankton and primary production in the Laurentian Great Lakes. Journal of the Fisheries Research Board of Canada 31: 739–763.

Volz, A. and D. Kley. 1988. Evaluation of the Montsouris series of ozone measurements made in the nineteenth century. Nature 332: 240–242.

Vonder Haar, T.H. and A.H. Oort. 1973. New estimate of annual poleward energy transport by northern hemisphere oceans. Journal of Physical Oceanography 3: 169–172.

von Gunten, H.R., and C. Lienert. 1993. Decreased metal concentrations in ground water caused by controls of phosphate emissions. Nature 364: 220–222.

Voroney, R.P. and E.A. Paul. 1984. Determination of k_c and k_n *in situ* for calibration of the chloroform fumigation-incubation method. Soil Biology and Biochemistry 16: 9–14.

Vörösmarty, C.J., B. Moore, A.L. Grace, M.P. Gildea, J.M. Melillo, B.J. Peterson, E.B. Rastetter, and P.A. Steudler. 1989. Continental scale models of water balance and fluvial transport: An application to South America. Global Biogeochemical Cycles 3: 241–265.

Vossbrinck, C.R., D.C. Coleman, and T.A. Woolley. 1979. Abiotic and biotic factors in litter decomposition in a semiarid grassland. Ecology 60: 265–271.

Wagner, C., A. Grieẞhammer, and H.L. Drake. 1996. Acetogenic capacities and the anaerobic turnover of carbon in a Kansas prairie soil. Applied and Environmental Microbiology 62: 494–500.

Wahlen, M., N. Tanaka, R. Henry, B. Deck, J. Zeglen, J.S. Vogel, J. Southon, A. Shemesh, R. Fairbanks, and W. Broecker. 1989. Carbon-14 in methane sources and in atmospheric methane: The contribution from fossil carbon. Science 245: 286–290.

Wakatsuki, T. and A. Rasyidin. 1992. Rates of weathering and soil formation. Geoderma 52: 251–263.

Walbridge, M.R. and B.G. Lockaby. 1994. Effects of forest management on biogeochemical functions in southern forested wetlands. Wetlands 14: 10–17.

Walbridge, M.R. and P.M. Vitousek. 1987. Phosphorus mineralization potentials in acid organic soils: Processes affecting $^{32}PO_4^{3-}$ isotope dilution measurements. Soil Biology and Biochemistry 19: 709–717.

Walbridge, M.R., C.J. Richardson, and W.T. Swank. 1991. Vertical distribution of biological and geochemical phosphorus subcycles in two southern Appalachian forest soils. Biogeochemistry 13: 61–85.

Waldman, J.M., J.W. Munger, D.J. Jacob, R.C. Flagan, J.J. Morgan, and M.R. Hoffmann. 1982. Chemical composition of acid fog. Science 218: 677–680.

Waldman, J.M., J.W. Munger, D.J. Jacob, and M.R. Hoffmann. 1985. Chemical characterization of stratus cloudwater and its role as a vector for pollutant deposition in a Los Angeles pine forest. Tellus 37B: 91–108.

Walker, J.C.G. 1977. *Evolution of the Atmosphere.* Macmillan, New York.

Walker, J.C.G. 1980. The oxygen cycle. pp. 87–104. In O. Hutzinger (ed.), *Handbook of Environmental Chemistry.* Vol. 1, Part A. *The Natural Environment and the Biogeochemical Cycles.* Springer-Verlag, New York.

Walker, J.C.G. 1983. Possible limits on the composition of the Archaean ocean. Nature 302: 518–520.

Walker, J.C.G. 1984. How life affects the atmosphere. BioScience 34: 486–491.

Walker, J.C.G. 1985. Carbon dioxide on the early Earth. Origins of Life 16: 117–127.

Walker, J.C.G. and P. Brimblecombe. 1985. Iron and sulfur in the pre-biologic ocean. Precambrian Research 28: 205–222.

Walker, J.C.G., C. Klein, M. Schidlowski, J.W. Schopf, D.J. Stevenson and M.R. Walter. 1983. Environmental evolution of the Archean-early Proterozoic Earth. pp. 260–290. In J.W. Schopf (ed.), *Earth's Earliest Biosphere.* Princeton University Press, Princeton, New Jersey.

Walker, L.R. 1989. Soil nitrogen changes during primary succession on a floodplain in Alaska, U.S.A. Arctic and Alpine Research 21: 341–349.

Walker, T.W. and A.F.R. Adams. 1958. Studies on soil organic matter: I. Influence of phosphorus content of parent materials on accumulations of carbon, nitrogen, sulfur, and organic phosphorus in grassland soils. Soil Science 85: 307–318.

Walker, T.W. and J.K. Syers. 1976. The fate of phosphorus during pedogenesis. Geoderma 15: 1–19.

Wallace, A., S.A. Bamburg, and J.W. Cha. 1974. Quantitative studies of roots of perennial plants in the Mojave Desert. Ecology 55: 1160–1162.

Wallerstein, G. 1988. Mixing in stars. Science 240: 1743–1750.

Walsh, J.J. 1984. The role of ocean biota in accelerated ecological cycles: A temporal view. BioScience 34: 499–507.

Walsh, J.J. 1991. Importance of continental margins in the marine biogeochemical cycling of carbon and nitrogen. Nature 350: 53–55.

Walsh, J.J. and D.A. Dieterle. 1988. Use of satellite ocean colour observations to refine understanding of global geochemical cycles. pp. 287–318. In T. Rosswall, R.G. Woodmansee, and P.G. Risser (eds.), *Scales and Global Change.* Wiley, New York.

Wanninkhof, R. 1992. Relationship between wind speed and gas exchange over the ocean. Journal of Geophysical Research 97: 7373–7382.

Ward, B.B., K.A. Kilpatrick, P.C. Novelli, and M.I. Scranton. 1987. Methane oxidation and methane fluxes in the ocean surface layer and deep anoxic waters. Nature 327: 226–229.

Ward, R.C. 1967. *Principles of Hydrology*. McGraw-Hill, London.

Wardle, D.A. 1992. A comparative assessment of factors which influence microbial biomass carbon and nitrogen levels in soil. Biological Reviews 67: 321–358.

Warembourg, F.R. and E.A. Paul. 1977. Seasonal transfers of assimilated ^{14}C in grassland: Plant production and turnover, soil and plant respiration. Soil Biology and Biochemistry 9: 295–301.

Waring, R.H. and W.H. Schlesinger. 1985. *Forest Ecosystems*. Academic Press, Orlando, Florida.

Waring, R.H., J.J. Rogers, and W.T. Swank. 1981. Water relations and hydrologic cycles. pp. 205–264. In D.E. Reichle (ed.), *Dynamic Properties of Forest Ecosystems*. Cambridge University Press, Cambridge.

Warneck, P. 1988. *Chemistry of the Natural Atmosphere*. Academic Press, London.

Watanabe, S., H. Yamamoto, and S. Tsunogai. 1995. Relation between the concentrations of DMS in surface seawater and air in the temperate North Pacific region. Journal of Atmospheric Chemistry 22: 271–283.

Waters, J.W., L. Froidevaux, W.G. Read, G.L. Manney, L.S. Elson, D.A. Flower, R.F. Jarnot, and R.S. Harwood. 1993. Stratospheric ClO and ozone from the Microwave Limb Sounder on the Upper Atmosphere Research Satellite. Nature 362: 597–602.

Watkins, N.D., R.S.J. Sparks, H. Sigurdsson, T.C. Huang, A. Federman, S. Carey, and D. Ninkovich. 1978. Volume and extent of the Minoan tephra from Santorini Volcano: New evidence from deep-sea sediment cores. Nature 271: 122–126.

Watson, A.J. and J.E. Lovelock. 1983. Biological homeostasis of the global environment: The parable of Daisyworld. Tellus 35B: 284–289.

Watson, A.J., J.E. Lovelock, and L. Margulis. 1978. Methanogenesis, fires and the regulation of atmospheric oxygen. BioSystems 10: 293–298.

Watson, A.J., C. Robinson, J.E. Robinson, P.J. le B. Williams, and M.J.R. Fasham. 1991. Spatial variability in the sink for atmospheric carbon dioxide in the North Atlantic. Nature 350: 50–53.

Watson, R.T., H. Rodhe, H. Oeschger, and U. Siegenthaler. 1990. Greenhouse gases and aerosols. pp. 1–40. In J.T. Houghton, G.J. Jenkins, and J.J. Ephraums (eds.), *Climate Change*. Cambridge University Press, Cambridge.

Watwood, M.E. and J.W. Fitzgerald. 1988. Sulfur transformations in forest litter and soil: Results of laboratory and field incubations. Soil Science Society of America Journal 52: 1478–1483.

Watwood, M.E., J.W. Fitzgerald, W.T. Swank, and E.R. Blood. 1988. Factors involved in potential sulfur accumulation in litter and soil from a coastal pine forest. Biogeochemistry 6: 3–19.

Waughman, G.J. 1980. Chemical aspects of the ecology of some South German peatlands. Journal of Ecology 68: 1025–1046.

Waughman, G.J. and D.J. Bellamy. 1980. Nitrogen fixation and the nitrogen balance in peatland ecosystems. Ecology 61: 1185–1198.

Weathers, K.C., G.E. Likens, F.H. Bormann, J.S. Eaton, W.B. Bowden, J.L. Andersen, D.A. Cass, J.N. Galloway, W.C. Keene, K.D. Kimball, P. Huth, and D. Smiley. 1986. A regional acidic cloud/fog water event in the eastern United States. Nature 319: 657–658.

Webb, R.S., C.E. Rosenzweig, and E.R. Levine. 1993. Specifying land surface characteristics in general circulation models: Soil profile data set and derived water-holding capacities. Global Biogeochemical Cycles 7: 97–108

Webb, W.L., S. Szarek, W. Lauenroth, R. Kinerson, and M. Smith. 1978. Primary productivity and water use in native forest, grassland, and desert ecosystems. Ecology 59: 1239–1247.

Webb, W.L., W.K. Lauenroth, S.R. Szarek, and R.S. Kinerson. 1983. Primary production and abiotic controls in forests, grasslands, and desert ecosystems in the United States. Ecology 64: 134–151.

Wedepohl, K.H. 1995. The composition of the continental crust. Geochimica et Cosmochimica Acta 59: 1217–1232.

Weier, K.L. and J.W. Gilliam. 1986. Effect of acidity on denitrification and nitrous oxide evolution from Atlantic coastal plain soils. Soil Science Society of America Journal 50: 1202–1206.

Weier, K.L., J.W. Doran, J.F. Power, and D.T. Walters. 1993. Denitrification and dinitrogen/nitrous oxide ratio as affected by soil water, available carbon, and nitrate. Soil Science Society of America Journal 57: 66–73.

Weir, J.S. 1972. Spatial distribution of elephants in an African National Park in relation to environmental sodium. Oikos 23: 1–13.

Weiss, H., M.-A. Courty, W. Wetterstrom, F. Guichard, L. Senior, R. Meadow, and A. Curnow. 1993. The genesis and collapse of third millennium north Mesopotamian civilization. Science 261: 995–1004.

Weiss, H.V., M. Koide, and E.D. Goldberg. 1971. Mercury in a Greenland ice sheet: Evidence of recent input by man. Science 174: 692–694.

Weiss, P.S., J.E. Johnson, R.H. Gammon, and T.S. Bates. 1995. Reevaluation of the open ocean source of carbonyl sulfide to the atmosphere. Journal of Geophysical Research 100: 23083–23092.

Welch, S.A. and W.J. Ullman. 1993. The effect of organic acids on plagioclase dissolution rates and stoichiometry. Geochimica et Cosmochimica Acta 57: 2725–2736.

Wells, P.V. 1983. Paleobiogeography of montane islands in the Great Basin since the last glaciopluvial. Ecological Monographs 53: 341–382.

Weng, F., R.R. Ferraro, and N.C. Grody. 1994. Global precipitation estimations using Defense Meteorological Satellite Program F10 and F11 special sensor microwave imager data. Journal of Geophysical Research 99: 14493–14502.

Wennberg, P.O., R.C. Cohen, R.M. Stimpfle, J.P. Koplow, J.G. Anderson, R.J. Salawitch, D.W. Fahey, E.L. Woodbridge, E.R. Klein, R.S. Gao, C.R. Webster, R.D. May, D.W. Toohey, L.M. Avallone, M.H. Proffitt, M. Loewenstein, J.R. Podolske, K.R. Chan, and S.C. Wofsy. 1994. Removal of stratospheric O_3 by radicals: In situ measurements of OH, HO_2, NO, NO_2, ClO, and BrO. Science 266: 398–404.

Wentz, D.A.,, W.J. Rose, and K.E. Webster. 1995. Long-term hydrologic and biogeochemical responses of a soft water seepage lake in north central Wisconsin. Water Resources Research 31: 199–212.

Wessman, C.A., J.D. Aber, D.L. Peterson, and J.M. Melillo. 1988a. Remote sensing of canopy chemistry and nitrogen cycling in temperate forest ecosystems. Nature 335: 154–156.

Wessman, C.A., J.D. Aber, D.L. Peterson, and J.M. Melillo. 1988b. Foliar analysis using near infrared reflectance spectroscopy. Canadian Journal of Forest Research 18: 6–11.

Westheimer, F.H. 1987. Why nature chose phosphates. Science 235: 1173–1178.

Wetherill, G.W. 1985. Occurrence of giant impacts during the growth of the terrestrial planets. Science 228: 877–879.

Wetherill, G.W. 1994. Provenance of the terrestrial planets. Geochimica et Cosmochimica Acta 58: 4513–4520.

Wetselaar, R. 1968. Soil organic nitrogen mineralization as affected by low soil water potentials. Plant and Soil 29: 9–17.

Weyl, P.K. 1978. Micropaleontology and ocean surface climate. Science 202: 475–481.

Whalen, S.C. and J.R. Cornwell. 1985. Nitrogen, phosphorus, and organic carbon cycling in an Arctic lake. Canadian Journal of Fisheries and Aquatic Sciences 42: 797–808.

Whalen, S.C. and W.S. Reeburgh. 1990. Consumption of atmospheric methane by tundra soils. Nature 346: 160–162.

Whalen, S.C., W.S. Reeburgh, and K.S. Kizer. 1991. Methane consumption and emission by taiga. Global Biogeochemical Cycles 5: 261–273.

Whelpdale, D.M. and J.N. Galloway. 1994. Sulfur and reactive nitrogen oxide fluxes in the North Atlantic atmosphere. Global Biogeochemical Cycles 8: 481–493.

Whelpdale, D.M. and R.W. Shaw. 1974. Sulphur dioxide removal by turbulent transfer over grass, snow, and water surfaces. Tellus 26: 196–205.

White, A.F. and A.E. Blum. 1995. Effects of climate on chemical weathering in watersheds. Geochimica et Cosmochimica Acta 59: 1729–1747.

White, A.F., A.E. Blum, M.S. Schulz, T.D. Bullen, J.W. Harden, and M.L. Peterson. 1996. Chemical weathering rates of a soil chronosequence on granitic alluvium: I. Quantification of mineralogical and surface area changes and calculation of primary silicate reaction rates. Geochimica et Cosmochimica Acta 60: 2533–2550.

White, C.S. 1986. Volatile and water-soluble inhibitors of nitrogen mineralization and nitrification in a ponderosa pine ecosystem. Biology and Fertility of Soils 2: 97–104.

White, C.S. 1988. Nitrification inhibition by monoterpenoids: Theoretical mode of action based on molecular structures. Ecology 69: 1631–1633.

White, D.S. and B.L. Howes. 1994. Long-term ^{15}N-nitrogen retention in the vegetated sediments of a New England salt marsh. Limnology and Oceanography 39: 1878–1892.

White, E.J. and F. Turner. 1970. A method of estimating income of nutrients in catch of airborne particles by a woodland canopy. Journal of Applied Ecology 7: 441–461.

White, T.C.R. 1984. The abundance of invertebrate herbivores in relation to the availability of nitrogen in stressed food plants. Oecologia 63: 90–105.

Whitehead, D.C., D.R. Lockyer, and N. Raistrick. 1988. The volatilization of ammonia from perennial ryegrass during decomposition, drying and induced senescence. Annals of Botany 61: 567–571.

Whitehead, D.R. 1981. Late-Pleistocene vegetational changes in northeastern North Carolina. Ecological Monographs 51: 451–471.

Whitehead, D.R., H. Rochester, S.W. Rissing, C.B. Douglass, and M.C. Sheehan. 1973. Late glacial and postglacial productivity changes in a New England pond. Science 181: 744–747.

Whitfield, M. and D.R. Turner. 1979. Water-rock partition coefficients and the composition of seawater and river water. Nature 278: 132–137.

Whitfield, P.H. and H. Schreier. 1981. Hysteresis in relationships between discharge and water chemistry in the Fraser River basin, British Columbia. Limnology and Oceanography 26: 1179–1182.

Whiticar, M.J. 1993. Stable isotopes and global budgets. pp. 138–167. In M.A.K. Khalil (ed.), *Atmospheric Methane: Sources, Sinks, and Role in Global Change.* Springer-Verlag, New York.

Whiticar, M.J., E. Faber, and M. Schoell. 1986. Biogenic methane formation in marine and freshwater environments: CO_2 reduction vs. acetate fermentation—Isotope evidence. Geochimica et Cosmochimica Acta 50: 693–709.

Whiting, G.J. and J.P. Chanton. 1993. Primary production control of methane emission from wetlands. Nature 364: 794–795.

Whitlow, S., P. Mayewski, J. Dibb, G. Holdsworth, and M. Twickler. 1994. An ice-core-based record of biomass burning in the arctic and subarctic, 1750–1980. Tellus 46B: 234–242.

Whittaker, R.H. 1970. *Communities and Ecosystems.* Macmillan, New York.

Whittaker, R.H. 1975. *Communities and Ecosystems.* 2nd ed. Macmillan, New York.

Whittaker, R.H. and G.E. Likens. 1973. Carbon in the biota. pp. 281–302. In G.M. Woodwell and E.V. Pecan (eds.), *Carbon and the Biosphere.* CONF 720510. National Technical Information Service, Washington, D.C.

Whittaker, R.H. and P.L. Marks. 1975. Methods of assessing terrestrial productivity. pp. 55–118. In H. Lieth and R.H. Whittaker (eds.), *Primary Productivity of the Biosphere.* Springer-Verlag, New York.

Whittaker, R.H. and W.A. Niering. 1975. Vegetation of the Santa Catalina Mountains, Arizona. V. Biomass, production, and diversity along the elevation gradient. Ecology 56: 771–790.

Whittaker, R.H. and G.M. Woodwell. 1968. Dimension and production relations of trees and shrubs in the Brookhaven forest, New York. Journal of Ecology 56: 1–25.

Whittaker, R.H., F.H. Bormann, G.E. Likens, and T.G. Siccama. 1974. The Hubbard Brook Ecosystem Study: Forest biomass and production. Ecological Monographs 44: 233–254.

Widdel, F., S. Schnell, S. Heising, A. Ehrenreich, B. Assmus, and B. Schink. 1993. Ferrous iron oxidation by anoxygenic phototrophic bacteria. Nature 362: 834–836.

Wieder, R.K. and G.E. Lang. 1988. Cycling of inorganic and organic sulfur in peat from Big Run Bog, West Virginia. Biogeochemistry 5: 221–242.

Wieder, R.K., J.B. Yavitt, and G.E. Lang. 1990. Methane production and sulfate reduction in two Appalachian peatlands. Biogeochemistry 10: 81–104.

Wiegert, R.G. and F.C. Evans. 1964. Primary production and the disappearance of dead vegetation on an old field in southeastern Michigan. Ecology 45: 49–63.

Wiegert, R.G., L.R. Pomeroy, and W.J. Wiebe. 1981. Ecology of salt marshes: An introduction. pp. 3–19. In L.R. Pomeroy and R.G. Wiegert (eds.), *The Ecology of a Salt Marsh.* Springer-Verlag, New York.

Wigley, T.M.L. 1989. Possible climate change due to SO_2-derived cloud condensation nuclei. Nature 339: 365–367.

Wildung, R.E., T.R. Garland, and R.L. Buschbom. 1975. The interdependent effects of soil temperature and water content on soil respiration rate and plant root decomposition in arid grassland soils. Soil Biology and Biochemistry 7: 373–378.

Wilhelm, S.W., D.P. Maxwell, and C.G. Trick. 1996. Growth, iron requirements and siderophore production in iron-limited *Synechococcus* PCC 7002. Limnology and Oceanography 41: 89–97.

Williams, E.J., G.L. Hutchinson, and F.C. Fehsenfeld. 1992. No_x and N_2O emissions from soil. Global Biogeochemical Cycles 6: 351–388.

Williams, P.M. and E.R.M. Druffel. 1987. Radiocarbon in dissolved organic matter in the central North Pacific Ocean. Nature 330: 246–248.

Williams, S.N., R.E. Stoiber, N. Garcia, P.A. Londoño, C.J.B. Gemmell, D.R. Lowe, and C.B. Connor. 1986. Eruption of the Nevado del Ruiz Volcano, Columbia, on 13 November 1985: Gas flux and fluid geochemistry. Science 233: 964–967.

Williams, S.N., S.J. Schaefer, M.L. Calvache V., and D. Lopez. 1992. Global carbon dioxide emission to the atmosphere by volcanoes. Geochimica et Cosmochimica Acta 56: 1765–1770.

Willison, T.W., K.W.T. Goulding, and D.S. Powlson. 1995. Effect of land-use change and methane mixing ratio on methane uptake from United Kingdom soil. Global Change Biology 1: 209–212.

Wilson, A.T. 1978. Pioneer agriculture explosion and CO_2 levels in the atmosphere. Nature 273: 40–41.

Wilson, G.V., J.M. Alfonsi, and P.M. Jardine. 1989. Spatial variability of saturated hydraulic conductivity of the subsoil of two forested watersheds. Soil Science Society of America Journal 53: 679–685.

Wilson, J.O., P.M. Crill, K.B. Bartlett, D.I. Sebacher, R.C. Harriss, and R.L. Sass. 1989. Seasonal variation of methane emissions from a temperate swamp. Biogeochemistry 8: 55–71.

Windolf, J., E. Jeppesen, J.P. Jensen, and P. Kristensen. 1996. Modelling of seasonal variation in nitrogen retention and in-lake concentration: A four-year mass balance study in 16 shallow Danish lakes. Biogeochemistry 33: 25–44.

Winner, W.E., C.L. Smith, G.W. Koch, H.A. Mooney, J.D. Bewley, and H.R. Krouse. 1981. Rates of emission of H_2S from plants and patterns of stable sulphur isotope fractionation. Nature 289: 672–673.

Wissmar, R.C., J.E. Richey, R.F. Stallard, and J.M. Edmond. 1981. Plankton metabolism and carbon processes in the Amazon River, its tributaries, and floodplain waters, Peru-Brazil, May-June 1977. Ecology 62: 1622–1633.

Wofsy, S.C., R.C. Harriss, and W.A. Kaplan. 1988. Carbon dioxide in the atmosphere over the Amazon Basin. Journal of Geophysical Research 93: 1377–1387.

Wogelius, R.A. and J.V. Walther. 1991. Olivine dissolution at 25°C: Effects of pH, CO_2, and organic acids. Geochimica et Cosmochimica Acta 55: 943–954.

Wolin, M.J. and T.L. Miller. 1987. Bioconversion of organic carbon to CH_4 and CO_2. Geomicrobiology Journal 5: 239–259.

Wollast, R. 1981. Interactions between major biogeochemical cycles in marine ecosystems. pp. 125–142. In G.E. Likens (ed.), *Some Perspectives of the Major Biogeochemical Cycles.* Wiley, New York.

Wollast, R. 1991. The coastal organic carbon cycle: Fluxes, sources, and sinks. pp. 365–381. In R.F.C. Mantoura, J.-M. Martin, and R. Wollast (eds.), *Ocean Margin Processes in Global Change.* Wiley, New York.

Wollast, R. 1993. Interactions of carbon and nitrogen cycles in the coastal zone. pp. 195–210. In R. Wollast, F.T. MacKenzie, and L. Chou (eds.), *Interactions of C, N, P and S Biogeochemical Cycles and Global Change.* Springer-Verlag, New York.

Woltemate, I., M.J. Whiticar, and M. Schoell. 1984. Carbon and hydrogen isotopic composition of bacterial methane in a shallow freshwater lake. Limnology and Oceanography 29: 985–992.

Wong, C.S., Y.-H. Chan, J.S. Page, G.E. Smith, and R.D. Bellegay. 1993. Changes in equatorial CO_2 flux and new production estimated from CO_2 and nutrient levels in Pacific surface waters during the 1986/87 El Niño. Tellus 45B: 64–79.

Wong, M.T.F., R. Hughes, and D.L. Rowell. 1990. Retarded leaching of nitrate in acid soils from the tropics: Measurement of the effective anion exchange capacity. Journal of Soil Science 41: 655–663.

Wood, C.W., H.A. Torbert, H.H. Rogers, G.B. Runion, and S.A. Prior. 1994. Free-air CO_2 enrichment effects on soil carbon and nitrogen. Agricultural and Forest Meteorology 70: 103–116.

Wood, T., F.H. Bormann, and G.K. Voigt. 1984. Phosphorus cycling in a northern hardwood forest: Biological and chemical control. Science 223: 391–393.

Wood, W.W. and M.J. Petraitis. 1984. Origin and distribution of carbon dioxide in the unsaturated zone of the southern High Plains of Texas. Water Resources Research 20: 1193–1208.

Woodmansee, R.G. 1978. Additions and losses of nitrogen in grassland ecosystems. BioScience 28: 448–453.

Woodmansee, R.G. and L.S. Wallach. 1981. Effects of fire regimes on biogeochemical cycles. pp. 649–669. In F.E. Clark and T. Rosswall (eds.), *Terrestrial Nitrogen Cycles.* Swedish Natural Science Research Council, Stockholm.

Woodmansee, R.G., J.L. Dodd, R.A. Bowman, F.E. Clark, and C.E. Dickinson. 1978. Nitrogen budget of a shortgrass prairie ecosystem. Oecologia 34: 363–376.

Woodward, F.I. 1987. Stomatal numbers are sensitive to increases in CO_2 from pre-industrial levels. Nature 327: 617–618.

Woodward, F.I. 1993. Plant responses to past concentrations of CO_2. Vegetatio 104: 145–155.

Woodward, F.I., G.B. Thompson, and I.F. McKee. 1991. The effects of elevated concentrations of carbon dioxide on individual plants, populations, communities, and ecosystems. Annals of Botany 67(Suppl.): 23–38.

Woodwell, G.M. 1970. Effects of pollution on the structure and physiology of ecosystems. Science 168: 429–433.

Woodwell, G.M. 1974. Variation in the nutrient content of leaves of *Quercus alba, Quercus coccinea,* and *Pinus rigida* in the Brookhaven forest from bud-break to abscission. American Journal of Botany 61: 749–753.

Woodwell, G.M. and W.R. Dykeman. 1966. Respiration of a forest measured by carbon dioxide accumulation during temperature inversions. Science 154: 1031–1034.

Woosley, S.E. 1986. Nucleosynthesis and stellar evolution. pp. 1–195. In J. Audouze, C. Chiosi, and S.E. Woosley (eds.), *Nucleosynthesis and Chemical Evolution.* The Geneva Observatory, Switzerland.

Woosley, S.E. and M.M. Phillips. 1988. Supernova 1987A! Science 240: 750–759.

Worsley, T.R. and T.A. Davies. 1979. Sea-level fluctuations and deep-sea sedimentation rates. Science 203: 455–456.

Wright, J.R., A. Leahey, and H.M. Rice. 1959. Chemical, morphological and mineralogical characteristics of a chronosequence of soils on alluvial deposits in the Northwest Territories. Canadian Journal of Soil Science 39: 32–43.

Wright, R.F. 1976. The impact of forest fire on the nutrient influxes to small lakes in northeastern Minnesota. Ecology 57: 649–663.

Wright, R.F., E. Lotse, and A. Semb. 1994. Experimental acidification of alpine catchments at Sogndal, Norway: Results after 8 years. Water, Air, and Soil Pollution 72: 297–315.

Wu, J. 1981. Evidence of sea spray produced by bursting bubbles. Science 212: 324–326.

Wu, J., A.G. O'Donnell, and J.K. Syers. 1995. Influences of glucose, nitrogen and plant residues on the immobilization of sulphate-S in soil. Soil Biology and Biochemistry 27: 1363–1370.

Wuebbles, D.J. and J.S. Tamaresis. 1993. The role of methane in the global environment. pp. 469–513. In M.A.K. Khalil (ed.), *Atmospheric Methane: Sources, Sinks, and Role in Global Change.* Springer-Verlag, Berlin.

Wullschleger, S.D., W.M. Post, and A.W. King. 1995. On the potential for a CO_2 fertilization effect in forests: Estimates of the biotic growth factor based on 58 controlled-exposure studies. pp. 85–107. In G.M. Woodwell and F.T. MacKenzie (eds.), *Biotic Feedbacks in the Global Climatic System.* Oxford University Press, Oxford.

Wullstein, L.H. and S.A. Pratt. 1981. Scanning electron microscopy of rhizosheaths of *Oryzopsis hymenoides.* American Journal of Botany 68: 408–419.

Wuosmaa, A.M. and L.P. Hager. 1990. Methyl chloride transferase: A carbocation route for biosynthesis of halometabolites. Science 249: 160–162.

Yaalon, D.H. 1965. Downward movement and distribution of anions in soil profiles with limited wetting. pp. 157–164. In E.D. Hallsworth and D.V. Crawford (eds.), *Experimental Pedology.* Butterworth, London.

Yachandra, V.K., V.J. DeRose, M.J. Latimer, I. Mukerji, K. Sauer, and M.P. Klein. 1993. Where plants make oxygen: A structural model for the photosynthetic oxygen-evolving manganese cluster. Science 260: 675–679.

Yagi, K., J. Williams, N.-Y. Wang, and R.J. Cicerone. 1995. Atmospheric methyl bromide (CH_3Br) from agricultural soil fumigations. Science 267: 1979–1981.

Yanagawa, H., Y. Ogawa, K. Kojima, and M.-H. Ito. 1988. Construction of protocellular structures under simulated primitive Earth conditions. Origins of Life and Evolution of the Biosphere 18: 179–207.

Yanai, R.D. 1992. Phosphorus budget of a 70–year-old northern hardwood forest. Biogeochemistry 17: 1–22.

Yavitt, J.B. and T.J. Fahey. 1993. Production of methane and nitrous oxide by organic soils within a northern hardwood forest ecosystem. pp. 261–277. In R.S. Oremland (ed.), *Biogeochemistry and Global Change.* Chapman and Hall, New York.

Yavitt, J.B. and A.K. Knapp. 1995. Methane emission to the atmosphere through emergent cattail (*Typha latifolia* L.) plants. Tellus 47B: 521–534.

Ye, R.W., B.A. Averill, and J.M. Tiedje. 1994. Denitrification: Production and consumption of nitric oxide. Applied and Environmental Microbiology 60: 1053–1058.

Yienger, J.J. and H. Levy. 1995. Empirical model of global soil-biogenic NO_x emissions. Journal of Geophysical Research 100: 11447–11464.

Yin, X. 1993. Variation in foliar nitrogen concentration by forest type and climatic gradients in North America. Canadian Journal of Forest Research 23: 1587–1602.

Yoh, M., H. Terai, and Y. Saijo. 1988. Nitrous oxide in freshwater lakes. Archiv für Hydrobiologie 113: 273–294.

Yoneyama, T., T. Muraoka, T. Murakami, and N. Boonkerd. 1993. Natural abundance of ^{15}N in tropical plants with emphasis on tree legumes. Plant and Soil 153: 295–304.

Yoshinari, T. and R. Knowles. 1976. Acetylene inhibition of nitrous oxide reduction by denitrifying bacteria. Biochemical and Biophysical Research Communications 69: 705–710.

Young, J.R., E.C. Ellis, and G.M. Hidy. 1988. Deposition of air-borne acidifiers in the western environment. Journal of Environmental Quality 17: 1–26.

Youngberg, C.T. and A.G. Wollum. 1976. Nitrogen accretion in developing *Ceanothus velutinus* stands. Soil Science Society of America Proceedings 40: 109–112.

Youngs, E.G. 1987. Estimating hydraulic conductivity values from ring infiltrometer measurements. Journal of Soil Science 38: 623–632.

Yuan, G. and L.M. Lavkulich. 1994. Phosphate sorption in relation to extractable iron and aluminum in Spodosols. Soil Science Society of America Journal 58: 343–346.

Yung, Y.L. and M.B. McElroy. 1979. Fixation of nitrogen in the prebiotic atmosphere. Science 203: 1002–1004.

Yung, Y.L., T. Lee, C.-H. Wang, and Y.-T. Shieh. 1996. Dust: A diagnostic of the hydrologic cycle during the last glacial maximum. Science 271: 962–963.

Zagal, E. and J. Persson. 1994. Immobilization and remineralization of nitrate during glucose decomposition at four rates of nitrogen addition. Soil Biology and Biochemistry 26: 1313–1321.

Zak, D.R., G.E. Host, and K.S. Pregitzer. 1989. Regional variability in nitrogen mineralization, nitrification, and overstory biomass in northern Lower Michigan. Canadian Journal of Forest Research 19: 1521–1526.

Zak, D.R., K.S. Pregitzer, P.S. Curtis, J.A. Teeri, F. Fogel, and D.L. Randlett. 1993. Elevated atmospheric CO_2 and feedback between carbon and nitrogen cycles. Plant and Soil 151: 105–117.

Zak, D.R., D. Tilman, R.R. Parmenter, C.W. Rice, F.M. Fisher, J. Vose, D. Milchunas, and C.W. Martin. 1994. Plant production and soil microorganisms in late-successional ecosystems: A continental-scale study. Ecology 75: 2333–2347.

Zander, R., C.P. Rinsland, E. Mahieu, M.R. Gunson, C.B. Farmer, M.C. Abrams, and M.K.W. Ko. 1994. Increase of carbonyl flouride (COF_2) in the stratosphere and its contribution to the 1992 budget of inorganic fluorine in the upper stratosphere. Journal of Geophysical Research 99: 16737–16743.

Zardini, D., R. Raynaud, S. Scharffe, and W. Seiler. 1989. N_2O measurements of air extracted from Antarctic ice cores: Implication on atmospheric N_2O back to the last glacial-interglacial transition. Journal of Atmospheric Chemistry 8: 189–201.

Zechmeister-Boltenstern, S. and H. Kinzel. 1990. Non-symbiotic nitrogen fixation associated with temperate soils in relation to soil properties and vegetation. Soil Biology and Biochemistry 22: 1075–1084.

Zehnder, A.J.B. and S.H. Zinder. 1980. The sulfur cycle. pp. 105–145. In O. Hutzinger (ed.), *The Handbook of Environmental Chemistry*. Vol. 1, Part A. *The Natural Environment and the Biogeochemical Cycles*. Springer-Verlag, New York.

Zektser, I.S. and H.A. Loaiciga. 1993. Groundwater fluxes in the global hydrologic cycle: Past, present and future. Journal of Hydrology 144: 405–427.

Zent, A.P. 1996. The evolution of the Martian climate. American Scientist 84: 442–451.

Zimmerman, P.R., J.P. Greenberg, and C.E. Westberg. 1988. Measurements of atmospheric hydrocarbons and biogenic emission fluxes in the Amazon boundary layer. Journal of Geophysical Research 93: 1407–1416.

Zinke, P.J. 1980. Influence of chronic air pollution on mineral cycling in forests. U.S. Forest Service, Pacific Southwest Forest and Range Experiment Station, General Technical Report 43: 88–99.

Zobel, D.B. and J.A. Antos. 1991. 1980 tephra from Mount St. Helens: spatial and temporal variation beneath forest canopies. Biology and Fertility of Soils 12: 60–66.

Zwally, H.J., A.C. Brenner, J.A. Major, R.A. Bindschadler, and J.G. Marsh. 1989. Growth of the Greenland ice sheet: Measurement. Science 246: 1587–1589.

Index

A

B

C

D

E

F

G

H

I

M

N

O

S

W

Z